Useful Combinations of Constants

$\hbar = h/2\pi = 1.0546 \times 10^{-34}\,\text{J}\cdot\text{s} = 6.5821 \times 10^{-16}\,\text{eV}\cdot\text{s}$

$hc = 1.9864 \times 10^{-25}\,\text{J}\cdot\text{m} = 1239.8\,\text{eV}\cdot\text{nm}$

$\hbar c = 3.1615 \times 10^{-26}\,\text{J}\cdot\text{m} = 197.33\,\text{eV}\cdot\text{nm}$

$\dfrac{1}{4\pi\epsilon_0} = 8.9876 \times 10^{9}\,\text{N}\cdot\text{m}^2\cdot\text{C}^{-2}$

Compton wavelength $\lambda_c = \dfrac{h}{m_e c} = 2.4263 \times 10^{-12}\,\text{m}$

$\dfrac{e^2}{4\pi\epsilon_0} = 2.3071 \times 10^{-28}\,\text{J}\cdot\text{m} = 1.4400 \times 10^{-9}\,\text{eV}\cdot\text{m}$

Fine structure constant $\alpha = \dfrac{e^2}{4\pi\epsilon_0 \hbar c} = 0.0072974 \approx \dfrac{1}{137}$

Bohr magneton $\mu_{\text{B}} = \dfrac{e\hbar}{2m_e} = 9.2740 \times 10^{-24}\,\text{J/T} = 5.7884 \times 10^{-5}\,\text{eV/T}$

Nuclear magneton $\mu_{\text{N}} = \dfrac{e\hbar}{2m_p} = 5.0508 \times 10^{-27}\,\text{J/T}$
$$= 3.1525 \times 10^{-8}\,\text{eV/T}$$

Bohr radius $a_0 = \dfrac{4\pi\epsilon_0 \hbar^2}{m_e e^2} = 5.2918 \times 10^{-11}\,\text{m}$

Hydrogen ground state $E_0 = \dfrac{e^2}{8\pi\epsilon_0 a_0} = 13.606\,\text{eV} = 2.1799 \times 10^{-18}\,\text{J}$

Rydberg constant $R_\infty = \dfrac{\alpha^2 m_e c}{2h} = 1.09737 \times 10^{7}\,\text{m}^{-1}$

Hydrogen Rydberg $R_{\text{H}} = \dfrac{\mu}{m_e} R_\infty = 1.09678 \times 10^{7}\,\text{m}^{-1}$

Gas constant $R = N_A k = 8.3145\,\text{J}\cdot\text{mol}^{-1}\cdot\text{K}^{-1}$

Magnetic flux quantum $\Phi_0 = \dfrac{h}{2e} = 2.0678 \times 10^{-15}\,\text{T}\cdot\text{m}^2$

Classical electron radius $r_e = \alpha^2 a_0 = 2.8179 \times 10^{-15}\,\text{m}$

$kT = 2.5249 \times 10^{-2}\,\text{eV} \approx \dfrac{1}{40}\,\text{eV}$ at $T = 293\,\text{K}$

Note: The latest values of the fundamental constants can be found at the National Institute of Standards and Technology website at http://physics.nist.gov/cuu/Constants/index.html

MODERN PHYSICS

For Scientists and Engineers

Third Edition

STEPHEN T. THORNTON

University of Virginia

ANDREW REX

University of Puget Sound

BROOKS/COLE
CENGAGE Learning™

Australia • Brazil • Japan • Korea • Mexico • Singapore • Spain • United Kingdom • United States

BROOKS/COLE
CENGAGE Learning™

Modern Physics: For Scientists and Engineers, Third Edition

Stephen T. Thornton, Andrew Rex

Physics Acquisition Editor: Chris Hall

Publisher: David Harris

Vice President, Editor-in-Chief: Michelle Julet

Development Editor: Rebecca Heider

Editorial Assistants: Brandi Kirksey,
   Kyra Engelberg

Technology Project Manager: Sam Subity

Executive Marketing Manager: Julie Conover

Marketing Assistant: Leyla Jowza

Advertising Project Manager:
   Stacey Purviance

Project Manager, Editorial Production:
   Teri Hyde

Print/Media Buyer: Karen Hunt

Permissions Editor: Sarah Harkrader

Production Service: G&S Book Services

Text Designer: Stuart Paterson

Photo Researcher: Kathleen Olson

Copy Editor: Anita Wagner

Illustrator: G&S Book Services

Cover Designer: Bill Reuter

Cover Image: Courtesy D. Hornbaker and
   A. Yazdani

Compositor: G&S Book Services

For product information and technology assistance, contact us at
Cengage Learning Customer & Sales Support, 1-800-354-9706

For permission to use material from this text or product, submit all requests online at **cengage.com/permissions**
Further permissions questions can be emailed to
permission request@cengage.com

Library of Congress Catalog Card Number: 2004112116

ISBN-10: 0-534-41781-7

ISBN-13: 978-0-534-41781-9

Brooks/Cole
10 Davis Drive
Belmont, CA 94002
USA

Cengage Learning is a leading provider of customized learning solutions with office locations around the globe, including Singapore, the United Kingdom, Australia, Mexico, Brazil, and Japan. Locate your local office at: **international**.cengage.com/region

Cengage Learning products are represented in Canada by Nelson Education, Ltd.

For your course and learning solutions, visit **www.cengage.com**

Purchase any of our products at your local college store or at ourpreferred online store **www.ichapters.com**

Printed in the Canada
4 5 6 7 8 9 10 11 10 09

Contents Overview

Contents

Preface

Our objective in writing this book was to produce a textbook for a modern physics course of either one or two semesters for physics and engineering students. Such a course normally follows a full-year, introductory, calculus-based physics course for freshmen or sophomores. Before proceeding with this third edition, we had the publisher send a questionnaire to users of modern physics books. We received many replies, and our users encouraged us not to make significant changes in the book's overall structure. More comments concerned how material is presented for a one- or two-semester course than anything else, and we have endeavored to make sure the book is easily covered in either version. We will present later our recommendations for a one- versus a two-semester modern physics course. We have made no major changes in the order of subjects in the third edition.

Coverage

The first edition of our text established a trend for a contemporary approach to the exciting, thriving, and changing field of modern science. After briefly visiting the status of physics at the turn of the last century, we cover relativity and quantum theory, the basis of any study of modern physics. Almost all areas of science depend on quantum theory and the methods of experimental physics. We have included the name Quantum Mechanics in two of our chapter titles (Chapters 5 and 6) to emphasize the quantum connection. The latter part of the book is devoted to the subfields of physics (atomic, condensed matter, nuclear, and particle) and the exciting field of cosmology. Our experience is that science and engineering majors particularly enjoy the study of modern physics after the sometimes-laborious study of classical mechanics, thermodynamics, electricity, magnetism, and optics. The level of mathematics is not difficult for the most part, and students feel they are finally getting to the frontiers of physics. We have brought the study of modern physics alive by presenting many current applications and challenges in physics, for example, nanoscience, high-temperature superconductors, quantum teleportation, neutrino mass and oscillations, age of the universe, gamma-ray bursts, holography, and nuclear fusion. Modern physics texts need to be updated periodically to include recent advances. Although we have emphasized modern applications, we also provide the sound theoretical basis for quantum theory that will be needed by physics majors in their upper division and graduate courses.

Changes for the Third Edition

We have made a number of significant changes for the third edition to aid the student in learning modern physics. We have added 16% more end-of-chapter questions and problems to those already in the second edition, for a larger number than in any competing textbook.

Examples

We continue to add to the already large number of examples in the text. We have made pedagogical changes in this edition based on the results of physics education research. First, we have added conceptual examples to aid students in their conceptual understanding. Second, we have added structure to the example solutions to aid students in their thought processes when solving problems. In each worked example, we discuss the strategy for solving the problem. This includes a discussion of what needs to be done in the example, the procedure we will go through to find the desired answer, and relevant equations that will be needed. Then we go through the solution in some detail.

Famous Scientists

We have added a large number of photos and biographies of scientists who have made significant contributions to modern physics. We have done this to give students a perspective of the background, education, trials, and efforts of these scientists.

Chapter-by-Chapter Changes

We have rewritten sections in which we hoped to make the explanations clearer to the student. Some material has been deleted, and new material has been added. Complete sections have been added where appropriate. In particular we added new results that have been reported since the second edition. We have covered the most important subjects of modern physics, but we realize that in order to cover everything, the book would have to be much longer, which is not what our users want. Our intention was to keep the level of the textbook at the sophomore/junior undergraduate level. We think it is important for instructors to be able to supplement the book whenever they choose—especially to cover those topics in which they themselves are expert. Particular changes by chapter include the following:

Chapter 2 includes a new subsection on testing Lorentz symmetry, because of the recent increase in experiments to test special relativity. We have also made a change in the symbol used for frequency from ν to f, to avoid confusion between the italic v for velocity and the Greek nu ν.

Chapter 3 now includes the Rayleigh-Jeans formula for the classical theory of blackbody radiation and its comparison with the Planck law. We have included Millikan's original data for the photoelectric effect to show how well Einstein's theory worked and that the Planck constant h can be obtained from Millikan's data.

Chapter 4 includes additional discussion and examples about Rutherford finding evidence for the nuclear radius through his alpha-particle scattering experiments.

Chapter 5 has an expanded section that includes the Copenhagen interpretation of quantum mechanics. We have moved the uncertainty principle discussion to precede the expanded section on probability, wave functions, and the beginning of quantum theory. Material has been added about the group and phase velocity for a particle moving as a wave.

Chapter 6 includes new material on tunneling, with two new examples and a more understandable estimate for the tunneling probability.

Chapter 7 includes a new Special Topic box on the 21-cm line transition of the hydrogen atom. New examples help explain spherical harmonics and the normalization of the three-dimensional hydrogen wave functions.

Chapter 8 includes new figures to help students understand total angular momentum and precession, as well as an expanded description of Hund's rules.

Chapter 9 contains new examples with a greater variety in level of difficulty. The section on Bose-Einstein condensation is updated and expanded.

Chapter 10 includes updated material on laser applications and an improved semiclassical model of diamagnetism. Superconductivity records and applications are updated.

Chapter 11 contains numerous updates from the world of semiconductor technology, and a new section on nanotechnology.

Chapter 12 includes an updated Special Topic box on neutrino detection with new results. We have added a new Special Topic box on the Earth with an emphasis on radio-activity. New problems were added about the shell model.

Chapter 13 includes additional discussion of problems associated with nuclear power and comparison with other sources of energy. We have updated the discussions of current uranium enrichment processes and the use of radioisotopes in medicine. The exciting search for new elements has been made current.

Chapter 14 includes discussion of current elementary particle physics topics including the search for the Higgs boson, neutrino oscillations and mass, matter/antimatter, superstrings and supersymmetry, M-theory, and the status of the Large Hadron Collider at CERN. A new section called Beyond the Standard Model summarizes several current questions in particle physics.

Chapter 15 includes new discussion and a figure on spacetime curvature. Additional discussion on gravitational waves includes experimental searches at LIGO, Gravity Probe B (a new Special Topic box), and the future LISA spacecraft experiment. We have included new discussion about GPS and Hawking radiation and updated our description of black holes. Discussion about the Friedmann equation has been added.

Chapter 16 has been extensively rewritten and updated to include new results in cosmology. This includes new results on the age of the universe and the accelerating rate of the expansion of the universe. Discussion of quasars, gamma-ray bursts, supernovae, active galactic nuclei, extrasolar planets, and other astronomical objects has been updated. The results from the Wilkinson Microscope Anisotropy Probe are included. The current problems of cosmology are summarized, including dark matter, dark energy, Einstein's cosmological constant, and quintessence. The current and planned space telescopes are summarized.

Teaching Suggestions

The text has been used extensively in its first two editions in courses at our home institutions. These include a one-semester course for engineering students at the University of Virginia and a two-semester course for physics and pre-engineering students at the University of Puget Sound. These are representative of the one- and two-semester modern physics courses taught elsewhere. Both one- and two-semester courses should cover the material through the establishment of the periodic table in Chapter 8 with few exceptions. We have eliminated the denoting of optional sections, because we believe that depends on the wishes of the instructor, but we feel Sections 2.4, 3.9, 4.2, 6.4, 6.6, 7.2, 7.6, 8.2, and 8.3 from the first nine chapters might be optional. Our suggestions for the one- and two-semester courses (3 or 4 credit hours per semester) are then

One-semester: Chapters 1–9 and selected other material as chosen by the instructor

Two-semester: Chapters 1–16 with supplementary material as desired, with possible student projects

An Internet-based, distance-learning version of the course is offered by one of the authors every summer (Physics 252, 4 credit hours) at the University of Virginia that covers all chapters of the textbook, with emphasis on Chapters 1–8. Homework problems and exams are given on WebAssign. The course can be taken by a student located anywhere there is an Internet connection. See http://modern.physics.virginia.edu/course/ for details.

Features

End-of-Chapter Problems

The 1061 questions and problems (248 questions and 813 problems) are significantly more than in any of the competing textbooks. Such a large number of questions and problems allows the instructor to make different homework assignments year after year without having to repeat problems. Many of our problems have been coded in WebAssign, the Internet-based grading system. We have tried to provide thought-provoking questions that have actual answers. The end-of-chapter problems have been separated by section, and general problems are included at the end to allow assimilation of the material. The easier problems are generally listed first within a section, and the more difficult ones are noted by a shaded blue square behind the problem number. A few computer-based problems are given in the text, but no computer disk supplement is provided, because many computer software programs are commercially available.

Solutions Manuals

An **Instructor's Solutions Manual** is available to the instructor (by contacting your local Brooks/Cole—Cengage sales representative). This manual contains the *solution to every end-of-chapter problem* and has been checked by at least two physics professors. The answers to selected odd-numbered problems are given at the end of the textbook itself. A **Student Solutions Manual** that contains the solutions to about 25% of the end-of-chapter problems is also available for sale to the students.

Text Format

The two-color format helps to present clear illustrations and to highlight material in the text; for example, important and useful equations are highlighted in blue, and the most important part of each illustration is rendered in thick blue lines. Blue margin notes help guide the student to the important points, and the margins allow students to make their own notes. The first time key words or topics are introduced they are set in **boldface**, and *italics* are used for emphasis.

Examples

Although we had a large number of worked examples in the second edition, we have added new ones in this edition. The examples have been rewritten in a more pedagogical way and presented in the manner in which students are expected to work the end-of-chapter problems: that is, to develop a conceptual understanding and strategy before attempting a numerical solution. Problem solving does not come easily for most students, especially the problems requiring several steps (that is, not simply plugging numbers into one equation). We expect that the many text examples with varying degrees of difficulty will help students.

Special Topic Boxes

Upon the advice of our users, we have deleted the guest essays, but have retained the Special Topic boxes. We hope both students and professors find them interesting, and we have added new ones to enliven interest in current and historical subjects, which range from the discovery of new elements to the 21-cm line transition in the hydrogen atom.

History

We include historical aspects of modern physics that some students will find interesting and that others can simply ignore. A new feature of this edition is the addition of considerably more photos and biographies of scientists who have made significant contributions to modern physics. We believe this helps to enliven and humanize the material.

The Multimedia Manager Instructor's Resource CD

This easy-to-use multimedia lecture tool allows instructors to quickly assemble art and multimedia with notes to create fluid lectures. It includes PowerPoint lectures with art from the text as well as electronic files of the Instructor's Solutions Manual.

Website

Students can access the book's companion website at http://physics.brookscole .com/mpse. This site features student study aids such as outlines and summaries of each chapter. Instructors will also find downloadable PowerPoint lectures and images for use in classroom lecture presentation. Students may also access the authors' websites at http:// www.modern.physics.virginia.edu/ and http://www.ups.edu/physics/faculty/rex/ where the authors will post errata, present new exciting results, and give links to sites that have particularly interesting features like simulations and photos, among other things.

Acknowledgments

We acknowledge the assistance of many persons who have helped with this text. There are too many that helped us with the first and second editions to list here, but the book would not have been possible without them. We acknowledge the professional staff at Brooks/ Cole Publishing who helped make this third edition a useful, popular, and attractive text. They include Developmental Editor Rebecca Heider and Production Manager Teri Hyde, who kept the production process on track, and Physics Editor Chris Hall for her support, guidance, and encouragement. Anita Wagner did a superb job with the copyediting. We also want to thank the many individuals who gave us critical reviews and suggestions since the first edition. In preparing this third edition, we owe a special debt of gratitude to the following reviewers:

Bernard Bates, University of Puget Sound
Tom Carter, College of DuPage
Allen Flora, Hood College
Nancy M. Haegel, Fairfield University
David Olsgaard, Simpson College
Chad Orzel, Union College
Gabriel Spalding, Illinois Wesleyan University
Scott Wissink, Indiana University—Bloomington
Rand S. Worland, University of Puget Sound

We especially want to acknowledge the valuable help of Richard R. Bukrey of Loyola University of Chicago who helped us in many ways through his enlightening reviews, careful manuscript proofing, and checking of the end-of-chapter problem solutions in the first two editions. We are grateful to Gabriel Spalding for his accuracy review of the third edition. We also thank Allen Flora of Hood College for assuming the task of preparing and checking problem solutions for the third edition.

Stephen T. Thornton
University of Virginia
Charlottesville, Virginia
stt@virginia.edu

Andrew Rex
University of Puget Sound
Tacoma, Washington
rex@ups.edu

The Birth of Modern Physics

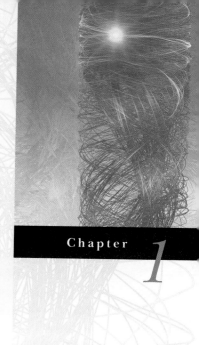

The more important fundamental laws and facts of physical science have all been discovered, and these are now so firmly established that the possibility of their ever being supplanted in consequence of new discoveries is exceedingly remote. . . . Our future discoveries must be looked for in the sixth place of decimals.

Albert A. Michelson, 1894

There is nothing new to be discovered in physics now. All that remains is more and more precise measurement.

William Thomson (Lord Kelvin), 1900

Physics has existed as a science for hundreds of years. Sometimes it has been indistinguishable from other sciences such as chemistry and astronomy—a similarity that continues in some areas even today. In physics we can easily distinguish *classical physics*, which was mostly developed before the year 1895, from *modern physics*, which is based on discoveries made after 1895. The precise year is unimportant, but monumental changes occurred in physics around 1900.

The long reign of Queen Victoria of England, from 1837 to 1901, saw considerable changes in social, political, and intellectual realms, but perhaps none so important as the remarkable achievements that occurred in physics. For example, the description and predictions of electromagnetism by Maxwell are partly responsible for the instantaneous telecommunications of today. It was also during this period that thermodynamics rose to become an exact science. But none of these have had the ramifications of the discoveries and applications of modern physics in the century to follow. The world would never be the same.

In this chapter we briefly review the status of physics around 1895, including Newton's laws, Maxwell's equations, and the laws of thermodynamics. These results are just as important today as they were a little over a hundred years ago. Arguments by scientists concerning the interpretation of experimental data using wave and particle descriptions that seemed to have been resolved 200 years ago were reopened in the twentieth century. Today we look back on the evidence of the late nineteenth century and wonder how anyone could have doubted the validity of the atomic view of matter. Fundamental interactions like gravity,

1

electricity, and magnetism were thought to be well understood in 1895. Physicists continued to be driven by the goal of understanding fundamental laws throughout the twentieth century. This is demonstrated by the fact that other fundamental forces (specifically the nuclear and weak interactions) have been added and in some cases—curious as it may seem—various forces have even been combined. The search for the holy grail of fundamental interactions continues unabated today.

We finish this chapter with a status report on physics just before 1900. The few problems not then understood would be the basis for decades of fruitful investigations and discoveries continuing into the twenty-first century. We hope you find this chapter interesting both for the physics learned and for the historical account of some of the most exciting scientific discoveries of the modern era.

1.1   Classical Physics of the 1890s

Scientists and engineers of the late nineteenth century were indeed rather smug. They thought they had just about everything under control (see the quotes from Michelson and Kelvin on page 1). The best scientists of the day were highly recognized and rewarded. Public lectures were frequent. Some scientists had easy access to their political leaders, partly because science and engineering had benefited their war machines, but also because of the many useful technical advances. Basic research was recognized as important because of the commercial and military applications of scientific discoveries. Although there were only primitive automobiles and no airplanes in 1895, advances in these modes of transportation were soon to follow. A few people already had telephones, and plans for widespread distribution of electricity were under way.

Based on their success with what we now call macroscopic classical results, scientists felt that given enough time and resources, they could explain just about anything. They did recognize some difficult questions they still couldn't answer; for example, they didn't clearly understand the structure of matter—that was under intensive investigation. Nevertheless, on a macroscopic scale, they knew how

Early successes of science

to build efficient engines. Ships plied the lakes, seas, and oceans of the world. Travel between the countries of Europe was frequent and easy by train. Many scientists were born in one country, educated in one or two others, and eventually worked in still other countries. The most recent ideas traveled quickly among the centers of research. Except for some isolated scientists, of whom Einstein is the most notable example, discoveries were quickly and easily shared. Scientific journals were becoming accessible.

The ideas of classical physics are just as important and useful today as they were at the end of the nineteenth century. For example, they allow us to build automobiles and produce electricity. The conservation laws of energy, linear momentum, angular momentum, and charge can be stated as follows:

Classical conservation laws

Conservation of energy: The total sum of energy (in all its forms) is conserved in all interactions.

Conservation of linear momentum: In the absence of external forces, linear momentum is conserved in all interactions (vector relation).

Conservation of angular momentum: In the absence of external torque, angular momentum is conserved in all interactions (vector relation).

Conservation of charge: Electric charge is conserved in all interactions.

A nineteenth-century scientist might even add the **conservation of mass** to the ones above, but we know it not to be valid today (you will find out why in Chapter 2). These conservation laws are reflected in the laws of mechanics, electromagnetism, and thermodynamics. Electricity and magnetism, separate subjects for hundreds of years, had been combined by James Clerk Maxwell (1831–1879) in his four equations. Maxwell showed optics to be a special case of electromagnetism. Waves, which permeated mechanics and optics, were believed to be an important component of nature. Many natural phenomena could be explained by wave motion using the laws of physics.

Mechanics

The laws of mechanics were developed over hundreds of years by many researchers. Important contributions were made by astronomers because of the great interest in the heavenly bodies. Galileo (1564–1642) may rightfully be called the first great experimenter. His experiments and observations laid the groundwork for the important discoveries to follow during the next 200 years.

Galileo, the first great experimenter

Isaac Newton (1642–1727) was certainly the greatest scientist of his time and one of the best the world has ever seen. His discoveries were in the fields of mathematics, astronomy, and physics and include gravitation, optics, motion, and forces.

Newton, the greatest scientist of his time

We owe to Newton our present understanding of motion. He understood clearly the relationships among position, displacement, velocity, and acceleration. He understood how motion was possible and that a body at rest was just a special case of a body having constant velocity. It may not be so apparent to us today, but we should not forget the tremendous unification that Newton made when he pointed out that the motions of the planets about our sun can be understood by the same laws that explain motion on Earth, like apples falling from trees or a soccer ball being shot toward a goal. Newton was able to elucidate carefully the relationship between force and acceleration, and his concepts were stated in three laws that bear his name today:

Galileo Galilei (1564–1642) was born, educated, and worked in Italy. Often said to be the "father of physics" because of his careful experimentation, he is shown here performing experiments of rolling balls down an inclined plane. He is perhaps best known for his experiments on motion, the development of the telescope, and his many astronomical discoveries. He came into disfavor with the Catholic Church for his belief in the Copernican theory. He was finally cleared of heresy by Pope John Paul II in 1992, 350 years after his death. *Scala/Art Resource, New York.*

Newton's laws

Isaac Newton (1642–1727), the great English physicist and mathematician, did most of his work at Cambridge where he was educated and became the Lucasian Professor of Mathematics. He was known not only for his work on the laws of motion, but also as a founder of optics. His useful works are too numerous to list here, but it should be mentioned that he spent a considerable amount of his time on alchemy, theology, and the spiritual universe. His writings on these subjects, which were dear to him, were quite unorthodox. This painting shows him performing experiments with light. *Courtesy of Bausch & Lomb Optical Co. and the AIP Niels Bohr Library.*

Newton's first law: *An object in motion with a constant velocity will continue in motion unless acted upon by some net external force.* A body at rest is just a special case of Newton's first law with zero velocity. Newton's first law is often called the *law of inertia* and is also used to describe inertial reference frames.

Newton's second law: *The acceleration $\vec{a}$ of a body is proportional to the net external force $\vec{F}$ and inversely proportional to the mass m of the body.* It is stated mathematically as

$$\vec{F} = m\vec{a} \tag{1.1a}$$

A more general statement* relates force to the time rate of change of the linear momentum $\vec{p}$.

$$\vec{F} = \frac{d\vec{p}}{dt} \tag{1.1b}$$

Newton's third law: *The force exerted by body 1 on body 2 is equal in magnitude and opposite in direction to the force that body 2 exerts on body 1.* If the force on body 2 by body 1 is denoted by $\vec{F}_{21}$, then Newton's third law is written as

$$\vec{F}_{21} = -\vec{F}_{12} \tag{1.2}$$

It is often called the *law of action and reaction.*

These three laws can be used to define the concept of force. Using that concept together with the concepts of velocity $\vec{v}$, acceleration $\vec{a}$, linear momentum $\vec{p}$, rotation (angular velocity $\vec{\omega}$ and angular acceleration $\vec{\alpha}$), and angular momentum $\vec{L}$, we can describe the complex motion of bodies.

Electromagnetism

Electromagnetism developed over a period of several hundred years. Important contributions were made by Coulomb (1736–1806), Oersted (1777–1851), Young (1773–1829), Ampère (1775–1836), Faraday (1791–1867), Henry (1797–1878), Maxwell (1831–1879), and Hertz (1857–1894). Maxwell showed that electricity and magnetism were intimately connected and were related by a change in the inertial frame of reference. His work also led to the understanding of electromagnetic radiation, of which light and optics are special cases. Maxwell's four equations, together with the Lorentz force law, explain much of electromagnetism.

Maxwell's equations

Gauss's law for electricity	$\oint \vec{E} \cdot d\vec{A} = \dfrac{q}{\epsilon_0}$	(1.3)
Gauss's law for magnetism	$\oint \vec{B} \cdot d\vec{A} = 0$	(1.4)
Faraday's law	$\oint \vec{E} \cdot d\vec{s} = -\dfrac{d\Phi_B}{dt}$	(1.5)
Generalized Ampère's law	$\oint \vec{B} \cdot d\vec{s} = \mu_0\epsilon_0 \dfrac{d\Phi_E}{dt} + \mu_0 I$	(1.6)
Lorentz force law	$\vec{F} = q\vec{E} + q\vec{v} \times \vec{B}$	(1.7)

*It is a remarkable fact that Newton wrote his second law not as $\vec{F} = m\vec{a}$, but as $\vec{F} = d(m\vec{v})/dt$, thus taking into account mass flow and change in velocity. This has applications in both fluid mechanics and rocket propulsion.

Maxwell's equations indicate that charges and currents create fields, and in turn, these fields can create other fields, both electric and magnetic.

Thermodynamics

Thermodynamics deals with temperature T, heat Q, work W, and the internal energy of systems U. The understanding of the concepts used in thermodynamics—such as pressure P, volume V, temperature, thermal equilibrium, heat, entropy, and especially energy—was slow in coming. We can understand the concepts of pressure and volume as mechanical properties, but the concept of temperature has to be carefully considered. We have learned that the internal energy of a system of noninteracting point masses depends only on the temperature.

Important contributions to thermodynamics were made by Benjamin Thompson (Count Rumford, 1753–1814), Carnot (1796–1832), Joule (1818–1889), Clausius (1822–1888), and Lord Kelvin (1824–1907). The primary results of thermodynamics can be described in two laws:

First law of thermodynamics: *The change in the internal energy ΔU of a system is equal to the heat Q added to the system plus the work W done on the system.*   **Laws of thermodynamics**

$$\Delta U = Q + W \tag{1.8}$$

The first law of thermodynamics generalizes the conservation of energy by including heat.

Second law of thermodynamics: *It is not possible to convert heat completely into work without some other change taking place.* Various forms of the second law state similar, but slightly different, results. For example, it is not possible to build a perfect engine or a perfect refrigerator. It is not possible to build a perpetual motion machine. Heat does not spontaneously flow from a colder body to a hotter body without some other change taking place. The second law forbids all these from happening. The first law states the conservation of energy, but the second law says what kinds of energy processes cannot take place. For example, it is possible to completely convert work into heat, but not vice versa, without some other change taking place.

Two other "laws" of thermodynamics are sometimes expressed. One is called the "zeroth" law, and it is useful in understanding temperature. It states that *if two thermal systems are in thermodynamic equilibrium with a third system, they are in equilibrium with each other.* We can state it more simply by saying that *two systems at the same temperature as a third system have the same temperature as each other.* This concept was not explicitly stated until the twentieth century, which is why it is called *zeroth.* The "third" law of thermodynamics expresses that *it is not possible to achieve an absolute zero temperature.*

1.2   The Kinetic Theory of Gases

We understand now that gases are composed of atoms and molecules in rapid motion, bouncing off each other and the walls, but in the 1890s this had just gained acceptance. The kinetic theory of gases is related to thermodynamics and to the atomic theory of matter, which we will discuss in Section 1.5. Experiments were relatively easy to perform on gases, and the Irish chemist Robert Boyle

(1627–1691) showed around 1662 that the pressure times the volume of a gas was constant for a constant temperature. The relation PV = constant (for constant T) is now referred to as *Boyle's law*. The French physicist Jacques Charles (1746–1823) found that V/T = constant (at constant pressure), referred to as *Charles's law*. Joseph Louis Gay-Lussac (1778–1850) later produced the same result, and the law is sometimes associated with his name. If we combine these two laws, we obtain the ideal gas equation

Ideal gas equation

$$PV = nRT \tag{1.9}$$

where n is the number of moles and R is the ideal gas constant, 8.31 J/mol · K.

In 1811 the Italian physicist Amedeo Avogadro (1776–1856) proposed that equal volumes of gases at the same temperature and pressure contained equal numbers of molecules. This hypothesis was so far ahead of its time that it was not accepted for many years. The famous English chemist John Dalton opposed this idea because he apparently misunderstood the difference between atoms and molecules. Considering the rudimentary nature of the atomic theory of matter at the time, this was not surprising.

Daniel Bernoulli (1700–1782) apparently originated the kinetic theory of gases in 1738, but his results were generally ignored. Many scientists, including Newton, Laplace, Davy, Herapath, and Waterston, had contributed to the development of kinetic theory by 1850. Theoretical calculations were being compared to experiment, and by 1895 the kinetic theory of gases was widely accepted. The statistical interpretation of thermodynamics was made in the latter half of the nineteenth century by the great Scottish mathematical physicist Maxwell, the Austrian physicist Ludwig Boltzmann (1844–1906), and the American physicist J. Willard Gibbs (1839–1903).

In introductory physics classes, the kinetic theory of gases is usually taught by applying Newton's laws to the collisions that a molecule makes with other molecules and with the walls. A representation of a few molecules colliding is shown in Figure 1.1. In the simple model of an ideal gas, only elastic collisions are considered. By taking averages over the collisions of many molecules, the derivation of the ideal gas law, Equation (1.9), is revealed. The average kinetic energy of the molecules is interpreted as being linearly proportional to the temperature, and the internal energy U is

Statistical thermodynamics

$$U = nN_A\langle K \rangle = \frac{3}{2}nRT \tag{1.10}$$

where n is the number of moles of gas, N_A is Avogadro's number, $\langle K \rangle$ is the average kinetic energy of a molecule, and R is the ideal gas constant. This relation ignores any nontranslational contributions to the molecular energy, such as rotations and vibrations.

However, energy is not represented only by translational motion. It became clear that all *degrees of freedom*, including rotational and vibrational, were also capable of carrying energy. The *equipartition theorem* states that each degree of freedom of a molecule has an average energy of $kT/2$, where k is the Boltzmann constant ($k = R/N_A$). Translational motion has three degrees of freedom, and rotational and vibrational modes can also be excited at higher temperatures. If there are f degrees of freedom, then Equation (1.10) becomes

Equipartition theorem

Internal energy

$$U = \frac{f}{2}nRT \tag{1.11}$$

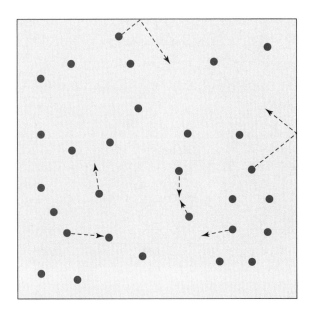

Figure 1.1 Molecules inside a closed container are shown colliding with the walls and with each other. The motions of a few molecules are indicated by the arrows. The number of molecules inside the container is huge.

The molar ($n = 1$) heat capacity c_V at constant volume for an ideal gas is the rate of change in internal energy with respect to change in temperature and is given by

$$c_V = \frac{3}{2} R \tag{1.12}$$

Heat capacity

The experimental quantity c_V/R is plotted versus temperature for hydrogen in Figure 1.2. The ratio c_V/R is equal to 3/2 for low temperatures, where only translational kinetic energy is important, but it rises to 5/2 at 300 K, where rotations occur for H_2, and finally reaches 7/2, due to vibrations at still higher temperatures, before the molecule dissociates. Although the kinetic theory of gases fails to predict specific heats for real gases, it leads to models that can be used on a gas-by-gas basis. Kinetic theory is also able to provide useful information on other properties such as diffusion, speed of sound, mean free path, and collision frequency.

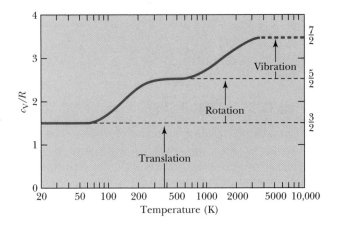

Figure 1.2 The molar heat capacity at constant volume (c_V) divided by R (c_V/R is dimensionless) is displayed as a function of temperature for hydrogen gas. Note that as the temperature increases, the rotational and vibrational modes become important. This experimental result is consistent with the equipartition theorem, which adds $kT/2$ of energy per molecule ($RT/2$ per mole) for each degree of freedom.

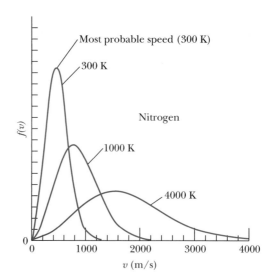

Figure 1.3 The Maxwell distribution of molecular speeds (for nitrogen), $f(v)$, is shown as a function of speed for three temperatures.

In the 1850s Maxwell derived a relation for the distribution of speeds of the molecules in gases. The distribution of speeds $f(v)$ is given as a function of the speed and the temperature by the equation

Maxwell's speed distribution

$$f(v) = 4\pi N\left(\frac{m}{2\pi kT}\right)^{3/2} v^2 e^{-mv^2/2kT} \qquad (1.13)$$

where m is the mass of a molecule and T is the temperature. This result is plotted for nitrogen in Figure 1.3 for temperatures of 300 K, 1000 K, and 4000 K. The peak of each distribution is the most probable speed of a gas molecule for the given temperature. In 1895 measurement was not precise enough to completely confirm Maxwell's distribution, and it was not confirmed experimentally until 1921.

By 1895 Boltzmann had made Maxwell's calculation more rigorous, and the general relation is called the *Maxwell-Boltzmann distribution*. The distribution can be used to find the *root-mean-square* speed v_{rms},

$$v_{rms} = \sqrt{\langle v^2 \rangle} = \sqrt{\frac{3kT}{m}} \qquad (1.14)$$

which shows the relationship of the energy to the temperature for an ideal gas:

$$U = nN_A\langle K\rangle = nN_A\frac{m\langle v^2 \rangle}{2} = nN_A\frac{m3kT}{2m} = \frac{3}{2}nRT \qquad (1.15)$$

This was the result of Equation (1.10).

1.3   Waves and Particles

We first learned the concepts of velocity, acceleration, force, momentum, and energy in introductory physics by using a single particle with its mass concentrated in one small point. In order to adequately describe nature, we add two- and three-dimensional bodies and rotations and vibrations. However, many aspects of physics can still be treated as if the bodies are simple particles. In particular, the kinetic energy of a moving particle is one way in which energy can be transported from one place to another.

But we have found that many natural phenomena can be explained only in terms of *waves*, which are traveling disturbances that carry energy. This description includes standing waves, which are superpositions of traveling waves. Most waves, like water waves and sound waves, need an elastic medium in which to move. Curiously enough, matter is not transported in waves—but energy is. Mass may oscillate, but it doesn't actually propagate along with the wave. Two examples are a cork and a boat on water. As a water wave passes, the cork gains energy as it moves up and down, and after the wave passes the cork remains. The boat also reacts to the wave, but it primarily rocks back and forth, throwing around things that are not fixed on the boat. The boat obtains considerable kinetic energy from the wave. After the wave passes, the boat eventually returns to rest.

Energy transport

Waves and particles were the subject of disagreement as early as the seventeenth century, when there were two competing theories of the nature of light. Newton supported the idea that light consisted of corpuscles (or particles). He performed extensive experiments on light for many years, and finally published his book *Opticks* in 1704. *Geometrical optics* uses straight-line, particle-like trajectories called *rays* to explain familiar phenomena like reflection and refraction. Geometrical optics was also able to explain the apparent observation of sharp shadows. The other competing theory considered light as a wave phenomenon. Its strongest proponent was the Dutch physicist Christian Huygens (1629–1695), who presented his theory in 1678. The wave theory could also explain reflection and refraction, but it could not explain the sharp shadows observed. Experimental physics of the 1600s and the 1700s was not able to discern between the two competing theories. Huygens's poor health and other duties kept him from working on optics much after 1678. Although Newton did not feel strongly about his corpuscular theory, the magnitude of his reputation caused it to be almost universally accepted for over a hundred years and throughout most of the eighteenth century.

Nature of light: waves or particles?

Finally, in 1802, the English physician Thomas Young (1773–1829) announced the results of his two-slit interference experiment, indicating that light behaved as a wave. Even after this singular event, the corpuscular theory had its supporters. During the next few years Young and, independently, Augustin Fresnel (1788–1827) performed several experiments that clearly showed that light behaved as a wave. By 1830 most physicists believed in the wave theory—some 150 years after Newton performed his first experiments on light.

One final experiment indicated that the corpuscular theory was difficult to accept. Let c be the speed of light in vacuum and v be the speed of light in another medium. If light behaves as a particle, then in order to explain refraction, light must speed up when going through denser material ($v > c$). The wave theory of Huygens predicts just the opposite ($v < c$). The measurements of the speed of light in various media were slowly improving, and finally, in 1850, Foucault showed that *light traveled more slowly in water than in air*. The corpuscular theory seemed incorrect. Newton would probably have been surprised that his weakly held beliefs lasted as long as they did. Now we realize that geometrical optics is correct only if the wavelength of light is much smaller than the size of the obstacles and apertures that the light encounters.

Figure 1.4 shows the "shadows" or *diffraction patterns* from light falling on sharp edges. In Figure 1.4a the alternating black and white lines can be seen all around the razor blade's edges. Figure 1.4b is a highly magnified photo of the diffraction from a sharp edge. The bright and dark regions can be understood only if light is a wave and not a particle. The physicists of 200–300 years ago apparently did not observe such phenomena. They believed that shadows were sharp, and only the particle nature of light could explain their observations.

(a)   (b)

Figure 1.4 In contradiction to what scientists thought in the seventeenth century, shadows are not sharp, but show dramatic diffraction patterns—as seen here (a) for a razor blade and (b) for a highly magnified sharp edge. *(a) Courtesy Ken Kay/Fundamental Photographs, New York; (b)* The Atlas of Optical Phenomena, *by Michael Cagnet, Mauric Francon, and Jean Claude Thrierr, p. 32, © 1962 Springer-Verlag, New York.*

In the 1860s Maxwell showed that electromagnetic waves consist of oscillating electric and magnetic fields. Visible light covers just a narrow range of the total electromagnetic spectrum, and all electromagnetic radiation travels at the speed of light c in free space, given by

$$c = \frac{1}{\sqrt{\mu_0 \epsilon_0}} = \lambda f \tag{1.16}$$

where λ is the wavelength and f is the frequency. The fundamental constants μ_0 and ϵ_0 are defined in electricity and magnetism and reveal the connection to the speed of light. In 1887 the German physicist Heinrich Hertz (1857–1894) succeeded in generating and detecting electromagnetic waves having wavelengths far outside the visible range ($\lambda \approx 5$ m). The properties of these waves were just as Maxwell had predicted. His results continue to have far-reaching effects in modern telecommunications: cable TV, cell phones, lasers, fiber optics, wireless Internet, and so on.

Some unresolved issues about electromagnetic waves in the 1890s eventually led to one of the two great modern theories, the *theory of relativity* (see Section 1.6 and Chapter 2). Waves play a central and essential role in the other great modern physics theory, *quantum mechanics*, which is sometimes called *wave mechanics*. Because waves play such a central role in modern physics, we will review their properties in Chapter 5.

1.4   Conservation Laws and Fundamental Forces

Conservation laws are the guiding principles of physics. The application of a few laws explains a vast quantity of physical phenomena. We listed the conservation laws of classical physics in Section 1.1. They include energy, linear momentum, angular momentum, and charge. Each of these is extremely useful in introductory physics. We use linear momentum when studying collisions, and the conservation laws when examining dynamics. We have seen the concept of the conservation of energy change. At first we had only the conservation of kinetic energy

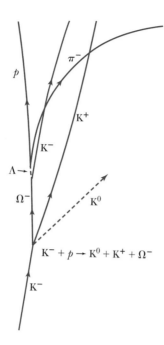

$$K^- + p \rightarrow K^0 + K^+ + \Omega^-$$

Figure 1.5 The conservation laws of momentum and energy are invaluable in untangling complex particle reactions like the one shown here, where a 5-GeV K^- meson interacts with a proton at rest to produce an Ω^- in a bubble chamber. The uncharged K^0 is not observed. Notice the curved paths of the charged particles in the magnetic field. Such reactions will be studied in Chapter 14.

in a force-free region. Then we added potential energy and formed the conservation of mechanical energy. In our study of thermodynamics, we added internal energy, and so on. The study of electrical circuits was made easier by the conservation of charge flow at each junction and the conservation of energy throughout all the circuit elements.

Much of what we know about conservation laws and fundamental forces has been learned within the last hundred years. In our study of modern physics we will find that mass is added to the conservation of energy, and the result is sometimes called the *conservation of mass-energy*, although the term *conservation of energy* is still sufficient and generally used. When we study elementary particles we will add the conservation of baryons and the conservation of leptons. Closely related to conservation laws are invariance principles. Some parameters are invariant in some interactions or in specific systems but not in others. Examples include time reversal, parity, and distance. We will study the Newtonian or Galilean invariance and find it lacking in our study of relativity; a new invariance principle will be needed. In our study of nuclear and elementary particles, conservation laws and invariance principles will often be used (see Figure 1.5).

Fundamental Forces

In introductory physics, we often begin our study of forces by examining the reaction of a mass at the end of a spring, because the spring force can be easily calibrated. We subsequently learn about tension, friction, gravity, surface, electrical, and magnetic forces. Despite the seemingly complex array of forces, we presently believe there are only three fundamental forces. All the other forces can be derived from them. These three forces are the **gravitational**, **electroweak**, and **strong** forces. Some physicists refer to the electroweak interaction as separate electromagnetic and weak forces because the unification occurs only at very high energies. The strengths and ranges of the three fundamental forces are listed in Table 1.1. Physicists sometimes use the term *interaction* when referring to the

Table 1.1   Fundamental Forces

Interaction		Relative Strength*	Range
Strong		1	Short, $\sim 10^{-15}$ m
Electroweak {	Electromagnetic	10^{-2}	Long, $1/r^2$
	Weak	10^{-9}	Short, $\sim 10^{-15}$ m
Gravitational		10^{-39}	Long, $1/r^2$

*These strengths are quoted for neutrons and/or protons in close proximity.

fundamental forces because it is the overall interaction among the constituents of a system that is of interest.

The gravitational force is the weakest. It is the force of mutual attraction between masses and, according to Newton, is given by

Gravitational interaction

$$\vec{F}_g = -G\frac{m_1 m_2}{r^2}\hat{r} \tag{1.17}$$

where m_1 and m_2 are two point masses, G is the gravitational constant, r is the distance between the masses, and $\hat{r}$ is a unit vector directed along the line between the two point masses (attractive force). The gravitational force is effective only on a macroscopic scale, but it has tremendous importance: it is the force that keeps Earth rotating about our source of life energy—the sun—and that keeps us and our atmosphere anchored to the ground. Gravity is a long-range force that diminishes as $1/r^2$.

The primary component of the electroweak force is *electromagnetic*. The other component is the *weak* interaction, which is responsible for beta decay in nuclei, among other processes. In the 1970s Glashow, Weinberg, and Salam predicted that the electromagnetic and weak forces were in fact facets of the same force. Their theory predicted the existence of new particles, called W and Z bosons, which were discovered in 1983. We will discuss bosons and the experiment in Chapter 14. For all practical purposes, the weak interaction is effective in the nucleus only over distances the size of 10^{-15} m. Except when dealing with very high energies, physicists mostly treat nature as if the electromagnetic and weak forces were separate. Therefore, you will sometimes see references to the *four* fundamental forces (gravity, strong, electromagnetic, and weak).

Weak interaction

Electromagnetic interaction

The electromagnetic force is probably the force that most affects us. It is responsible for holding atoms together, for friction, for contact forces, for tension, and for electrical and optical signals. It is responsible for all chemical and biological processes, including cellular structure and nerve processes. The list is long because the electromagnetic force is responsible for practically all nongravitational forces that we experience. The electrostatic, or Coulomb, force between two point charges q_1 and q_2, separated by a distance r, is given by

Coulomb force

$$\vec{F}_C = \frac{1}{4\pi\epsilon_0}\frac{q_1 q_2}{r^2}\hat{r} \tag{1.18}$$

The easiest way to remember the vector direction is that like charges repel and unlike charges attract. Moving charges also create and react to magnetic fields [see Equation (1.7)].

The third fundamental force, the strong force, is the one holding the nucleus together. It is the strongest of all the forces, but it is effective only over short distances—up to about 10^{-15} m. The strong force is so strong that it easily binds two protons together inside a nucleus even though the electrical force of repulsion over the tiny confined space is huge. The strong force is able to contain dozens of protons inside the nucleus before the electrical force of repulsion becomes strong enough to cause nuclear decay. We will study the strong force extensively in this book, learning that neutrons and protons are composed of *quarks*, and that the part of the strong force acting between quarks has the unusual name of *color force*.

Strong interaction

Physicists strive to combine forces into more fundamental ones. Centuries ago the forces responsible for friction, contact, and tension were all believed to be different. Today we know they are all part of the electroweak force. Two hundred years ago scientists thought the electrical and magnetic forces were independent, but after a series of experiments, physicists slowly began to see their connection. This culminated in the 1860s in Maxwell's work, which clearly showed they were but part of one force and at the same time explained light and other radiation. Figure 1.6 is a diagram of the unification of forces over time. Newton certainly had an inspiration when he was able to unify the planetary motions with the apple falling from the tree. We will see in Chapter 15 that Einstein was even able to link gravity with space and time.

Unification of forces

The further unification of forces currently remains one of the most active research fields. Considerable efforts have been made to unify the electroweak and strong forces through the *grand unified theories* or GUTs. A leading GUT is the mathematically complex *string theory*. Several predictions of these theories have not yet been verified experimentally (for example, the instability of the proton and the existence of magnetic monopoles). We present some of the exciting research areas in present-day physics throughout this book, because these topics

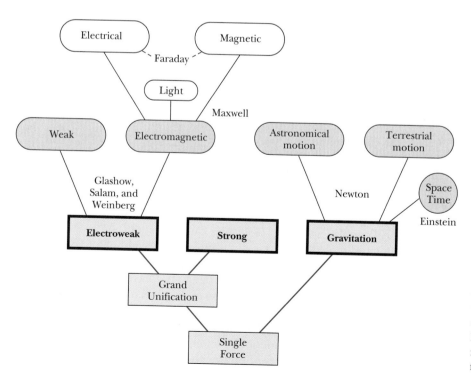

Figure 1.6 The three fundamental forces (shown in the heavy boxes) are themselves unifications of forces that were once believed to be fundamental. Present research is under way (see blue lines) to further unify the fundamental forces into one single force.

are the ones you will someday read about on the front pages of newspapers and in the weekly news magazines and perhaps contribute to in your own careers.

1.5   The Atomic Theory of Matter

It is not possible to reach college, or even high school, today without hearing about the atomic theory of matter. The idea that matter is composed of tiny particles called *atoms* is taught in elementary school and expounded throughout later schooling. We are told that the Greek philosophers Democritus and Leucippus proposed the concept of atoms as early as 450 B.C. The smallest piece of matter, which could not be subdivided further, was called an *atom*, after the Greek word *atomos*, meaning "indivisible." Physicists do not discredit the early Greek philosophers for thinking that the basic entity of life consisted of atoms. For centuries, scientists were called "natural philosophers," and in this tradition the highest university degree American scientists receive is a Ph.D., which stands for doctor of philosophy.

Not many new ideas were proposed about atoms until the seventeenth century, when scientists started trying to understand the properties and laws of gases. The work of Boyle, Charles, and Gay-Lussac presupposed the interactions of tiny particles in gases. Chemists and physical chemists made many important advances. In 1799 the French chemist Proust (1754–1826) proposed the *law of definite proportions*, which states that when two or more elements combine to form a compound, the proportions by weight (or mass) of the elements are always the same. Water (H_2O) is always formed of one part hydrogen and eight parts oxygen by mass.

Dalton, the father of the atomic theory

The English chemist John Dalton (1766–1844) is given most of the credit for originating the modern atomic theory of matter. In 1803 he proposed that the atomic theory of matter could explain the law of definite proportions if the elements are composed of atoms. Each element has atoms that are physically and chemically characteristic. The concept of atomic weights (or masses) was the key to the atomic theory.

In 1811 the Italian physicist Avogadro proposed the existence of molecules, consisting of individual or combined atoms. He stated without proof that *all gases contain the same number of molecules in equal volumes at the same temperature and pressure*. Avogadro's ideas were ridiculed by Dalton and others who could not imagine that atoms of the same element could combine together. If this could happen, they argued, then all the atoms of a gas would combine to form a liquid. The concept of molecules and atoms was indeed difficult to imagine, but finally, in 1858, the Italian chemist Cannizzaro (1826–1910) solved the problem and showed how Avogadro's ideas could be used to find atomic masses. Today we think of an atom as the smallest unit of matter that can be identified with a particular element. A molecule can be a single atom or a combination of two or more atoms of either like or dissimilar elements. Molecules can consist of thousands of atoms.

Avogadro's number

The number of molecules in one gram-molecular weight of a particular element (6.023×10^{23} molecules/mol) is called Avogadro's number (N_A) in honor of Avogadro. For example, one mole of hydrogen (H_2) has a mass of 2 g and one mole of carbon has a mass of about 12 g; one mole of each substance consists of 6.023×10^{23} atoms. Avogadro's number was not even estimated until 1865, and it was finally accurately measured by Perrin, as we will discuss later.

During the mid-1800s the kinetic theory of gases was being developed, and because it was based on the concept of atoms, its successes gave validity to the atomic theory. The experimental results of specific heats, Maxwell speed distribution, and transport phenomena (see the discussion in Section 1.2) all supported the concept of the atomic theory.

In 1827 the English botanist Robert Brown (1773–1858) observed with a microscope the motion of tiny pollen grains suspended in water. The pollen appeared to dance around in random motion, while the water was still. At first the motion (now called *Brownian motion*) was ascribed to convection or organic matter, but eventually it was observed to occur for any tiny particle suspended in liquid. The explanation according to the atomic theory is that the molecules in the liquid are constantly bombarding the tiny grains. A satisfactory explanation was not given until the twentieth century (by Einstein).

Although it may appear, according to the preceding discussion, that the atomic theory of matter was universally accepted by the end of the nineteenth century, that was not the case. Certainly most physicists believed in it, but there was still opposition. A principal leader in the anti-atomic movement was the renowned Austrian physicist Ernst Mach. Mach was an absolute positivist, believing in the reality of nothing but our own sensations. A simplified version of his line of reasoning would be that because we have never *seen* an atom, we cannot say anything about its reality. The Nobel Prize–winning German physical chemist Wilhelm Ostwald supported Mach philosophically but also had more practical arguments on his side. In 1900 there were difficulties in understanding radioactivity, x rays, discrete spectral lines, and how atoms formed molecules and solids. Ostwald contended that we should therefore think of atoms as hypothetical constructs, useful for bookkeeping in chemical reactions.

Opposition to atomic theory

On the other hand, there were many believers in the atomic theory. Max Planck, the originator of the quantum theory, grudgingly accepted the atomic theory of matter because his radiation law supported the existence of submicroscopic quanta. Boltzmann was convinced that atoms must exist, mainly because they were necessary in his statistical mechanics. It is said that Boltzmann committed suicide in 1905 partly because he was despondent that so many people rejected his theory. Today we have pictures of the atom (see Figure 1.7) that would

Figure 1.7 This scanning tunneling microscope photo, called the "stadium corral," shows 76 individually placed iron atoms on a copper surface. The IBM researchers were trying to contain and modify electron density, observed by the wave patterns, by surrounding the electrons inside the quantum "corral." Researchers are thus able to study the quantum behavior of electrons. See also the Special Topic on Scanning Probe Microscopes in Chapter 6. *Courtesy of International Business Machines.*

undoubtedly have convinced even Mach, who died in 1916 still unconvinced of the validity of the atomic theory.

Overwhelming evidence of atomic theory

Overwhelming evidence for the existence of atoms was finally presented in the first decade of the twentieth century. First, Einstein in one of his three famous papers published in 1905 (the others were about special relativity and the photoelectric effect) provided an explanation of the Brownian motion observed almost 80 years earlier by Robert Brown. Einstein explained the motion in terms of molecular motion and presented theoretical calculations for the *random walk* problem. Random walk (often called the *drunkard's walk*) is a statistical process that determines how far from its initial position a tiny grain may be after many random molecular collisions. Einstein was able to determine the approximate mass and sizes of atoms and molecules from experimental data.

Finally, in 1908, the French physicist Jean Perrin (1870–1942) presented data from an experiment designed using kinetic theory that agreed with Einstein's predictions. Perrin's experimental method of observing many particles of different sizes is a classic work, for which he received the Nobel Prize for Physics in 1926. His experiment utilized four different types of measurements. Each was consistent with the atomic theory and each gave a quantitative determination of Avogadro's number—the first accurate measurements that had been made. Since 1908 the atomic theory of matter has been accepted by practically everyone.

1.6 Unresolved Questions of 1895 and New Horizons

We choose the year 1895 as a convenient time to separate the periods of classical and modern physics, although we admit this is an arbitrary choice based on discoveries made in 1895–1897. The thousand or so physicists living in 1895 were rightfully proud of the status of their profession. The precise experimental method was firmly established. Theories were available that could explain many observed phenomena. In large part, scientists were busy measuring and understanding such physical parameters as specific heats, densities, compressibility, resistivity, indices of refraction, and permeabilities. The pervasive feeling was that, given enough time, everything in nature could be understood by applying the careful thinking and experimental techniques of physics. The field of mechanics was in particularly good shape, and its application had led to the stunning successes of the kinetic theory of gases and statistical thermodynamics.

Experiment and reasoning

In hindsight we can see now that this euphoria of success applied only to the macroscopic world. Objects of human dimensions like automobiles, steam engines, airplanes, telephones, and electric lights either existed or were soon to appear and were triumphs of science and technology. However, the atomic theory of matter was not universally accepted, and what made up an atom was purely conjecture. The structure of matter was unknown.

Clouds on the horizon

There were certainly problems that physicists could not resolve. Only a few of the deepest thinkers seemed to be concerned with them. Lord Kelvin, in a speech in 1900 to the Royal Institution, referred to "two clouds on the horizon." These two clouds were the electromagnetic medium and the failure of classical physics to explain blackbody radiation. We mention these and other problems here. Their solutions were soon to lead to two of the greatest breakthroughs in human thought ever recorded—the theories of quantum physics and relativity.

Electromagnetic Medium. The waves that were well known and understood by physicists all had media in which the waves propagated. Water waves traveled in water, and sound waves traveled in any material. It was natural for nineteenth-century physicists to assume that electromagnetic waves also traveled in a medium, and this medium was called the *ether*. Several experiments, the most notable of which were done by Michelson, had sought to detect the ether without success. An extremely careful experiment by Michelson and Morley in 1887 was so sensitive it should have revealed the effects of the ether. Subsequent experiments to check other possibilities were also negative. In 1895 some physicists were concerned that the elusive ether could not be detected. Was there an alternative explanation?

Electrodynamics. The other difficulty with Maxwell's electromagnetic theory had to do with the electric and magnetic fields as seen and felt by moving bodies. What appears as an electric field in one reference system may appear as a magnetic field in another system moving with respect to the first. Although the relationship between electric and magnetic fields seemed to be understood by using Maxwell's equations, the equations do not keep the same form under a Galilean transformation [see Equations (2.1) and (2.2)], a situation that concerned both Hertz and Lorentz. Hertz unfortunately died in 1894 at the young age of 36 and never experienced the modern physics revolution. The Dutch physicist Lorentz (1853–1928), on the other hand, proposed a radical idea that solved the electrodynamics problem: space was contracted along the direction of motion of the body. FitzGerald in Ireland independently proposed the same concept. The Lorentz-FitzGerald hypothesis, proposed in 1892, was a precursor to Einstein's theory advanced in 1905 (see Chapter 2).

Blackbody Radiation. In 1895 thermodynamics was on a strong footing; it had achieved much success. One of the interesting experiments in thermodynamics concerns an object, called a *blackbody*, that absorbs the entire electromagnetic radiation incident on it. An enclosure with a small hole serves as a blackbody, because all the radiation entering the hole is absorbed. A blackbody also emits radiation, and the *emissivity* is the electromagnetic power emitted per unit area. The radiation emitted covers all frequencies, each with its own intensity. Precise measurements were carried out to determine the spectrum of blackbody radiation, such as that shown in Figure 1.8. Blackbody radiation was a fundamental issue,

Lord Kelvin (William Thomson, 1824–1907) was born in Belfast, Ireland, and at age 10 entered the University of Glasgow in Scotland where his father was a professor of mathematics. He graduated from the University of Cambridge and at age 22 accepted the chair of natural philosophy (later called physics) at the University of Glasgow where he finished his illustrious 53-year career, finally resigning in 1899 at age 75. Lord Kelvin's contributions to nineteenth-century science were far reaching, and he made contributions in electricity, magnetism, thermodynamics, hydrodynamics, and geophysics. He was involved in the successful laying of the transatlantic cable. He was arguably the preeminent scientist of the latter part of the nineteenth century. He was particularly well known for his prediction of the Earth's age, which would later turn out to be inaccurate (see Chapter 12). *AIP Emilio Segrè Visual Archives, Brittle Books Collection.*

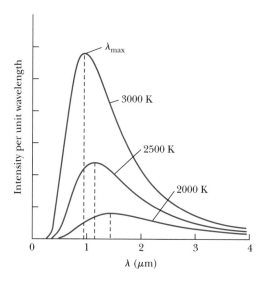

Figure 1.8 The blackbody spectrum, showing the emissivity of radiation emitted from a blackbody as a function of the radiation wavelength. Different curves are produced for different temperatures, but they are independent of the type of blackbody cavity. The intensity peaks at λ_{max}.

because the emissivity is independent of the body itself—it is characteristic of all blackbodies.

Many physicists of the period—including Kirchhoff, Stefan, Boltzmann, Rubens, Pringsheim, Lummer, Wien, Lord Rayleigh, Jeans, and Planck—had worked on the problem. It was possible to understand the spectrum both at the low-frequency end and at the high-frequency end, but no single theory could account for the entire spectrum. When the most modern theory of the day (the equipartition of energy applied to standing waves in a cavity) was applied to the problem, the result led to an *infinite* emissivity (or energy density) for high frequencies. The failure of the theory was known as the "ultraviolet catastrophe." The solution of the problem by Max Planck in 1900 was to shake the very foundations of physics.

Ultraviolet catastrophe: infinite emissivity

On the Horizon

Discovery of x rays

During the years 1895–1897 there were four discoveries that were all going to require a deep understanding of the atom. The first was the discovery of x rays by the German physicist Röntgen (1845–1923) in November 1895. Next came the accidental discovery of radioactivity by the French physicist Henri Becquerel (1852–1908), who in February 1896 placed uranium salt next to a carefully wrapped photographic plate. When the plate was developed, a silhouette of the uranium salt was evident—indicating the presence of a very penetrating ray.

Discovery of radioactivity

The third discovery, that of the electron, was actually due to several physicists over a period of years. Michael Faraday, as early as 1833, observed a gas discharge glow—evidence of electrons. Over the next few years, several scientists detected evidence of particles, called *cathode rays*, being emitted from charged cathodes. In 1896 Perrin proved that cathode rays were negatively charged. The discovery of the electron, however, is generally credited to the British physicist J. J. Thomson (1856–1940), who in 1897 isolated the electron (cathode ray) and measured its velocity and its ratio of charge to mass.

Discovery of the electron

The final important discovery of the period was made by the Dutch physicist Pieter Zeeman (1865–1943), who in 1896 found that a single spectral line was sometimes separated into two or three lines when the sample was placed in a magnetic field. The (normal) *Zeeman effect* was quickly explained by Lorentz as the result of light being emitted by the motion of electrons inside the atom. Zeeman and Lorentz showed that the frequency of the light was affected by the magnetic field according to the classical laws of electromagnetism.

Discovery of the Zeeman effect

The unresolved issues of 1895 and the important discoveries of 1895–1897 bring us to the subject of this book, *Modern Physics*. In 1900 Max Planck completed his radiation law, which solved the blackbody problem but required that energy be quantized. In 1905 Einstein presented his three important papers on Brownian motion, the photoelectric effect, and special relativity. While the work of Planck and Einstein may have solved the problems of the nineteenth-century physicists, they broadened the horizons of physics and have kept physicists active ever since.

Summary

Physicists of the 1890s felt that almost anything in nature could be explained by the application of careful experimental methods and intellectual thought. The application of mechanics to the kinetic theory of gases and statistical thermodynamics, for example, was a great success.

The nature of light was apparently resolved in the early 1800s in favor of waves. For over 100 years the particle viewpoint had prevailed, mostly due to the weakly held belief of the great Newton. In the 1860s Maxwell showed that his electromagnetic theory predicted a much wider frequency range of electromagnetic radiation than the visible optical phenomena. In the twentieth century, the question of waves versus particles was to reappear.

The conservation laws of energy, momentum, angular momentum, and charge are well established. The three fundamental forces are gravitational, electroweak, and strong. Over the years many forces have been unified into these three. Physicists are actively pursuing attempts to unify these three forces into only two or even just one single fundamental force.

The atomic theory of matter assumes atoms are the smallest unit of matter that is identified with a characteristic element. Molecules are composed of atoms, and these atoms can be from different elements. The kinetic theory of gases assumes the atomic theory is correct, and the development of the two theories proceeded together. The atomic theory of matter was not fully accepted until around 1910, by which time Einstein had explained Brownian motion and Perrin had published overwhelming experimental evidence.

The year 1895 saw several outstanding problems that seemed to worry only a few physicists. These problems included the inability to detect an electromagnetic medium, the difficulty in understanding the electrodynamics of moving bodies, and blackbody radiation. Four important discoveries during the period 1895–1897 were to signal the atomic age: x rays, radioactivity, the electron, and the splitting of spectral lines (Zeeman effect). The understanding of these problems and discoveries (among others) is the object of this book on modern physics.

Special Theory of Relativity

It was found that there was no displacement of the interference fringes, so that the result of the experiment was negative and would, therefore, show that there is still a difficulty in the theory itself. . . .

Albert Michelson, Light Waves and Their Uses, 1907

One of the great theories of physics appeared early in the twentieth century when Albert Einstein presented his special theory of relativity in 1905. We learned in introductory physics that Newton's laws of motion must be measured relative to some reference frame. A reference frame is called an **inertial frame** if Newton's laws are valid in that frame. For example, if a body subject to no net external force moves in a straight line with constant velocity, then the coordinate system establishing this fact is an inertial frame. If Newton's laws are valid in one reference frame, then they are also valid in a reference frame moving at a uniform velocity relative to the first system. This is known as the **Newtonian principle of relativity** or **Galilean invariance**.

Newton showed that it was not possible to determine absolute motion in space by any experiment so he decided to use relative motion. In addition, the Newtonian concepts of time and space are completely separable. Consider two inertial reference frames K and K′ that move along their x and x' axes, respectively, with uniform relative velocity $\vec{v}$ as shown in Figure 2.1. We show system K′ moving to the right with velocity $\vec{v}$ with respect to system K, which is fixed or

Inertial frame (margin)

Galilean invariance (margin)

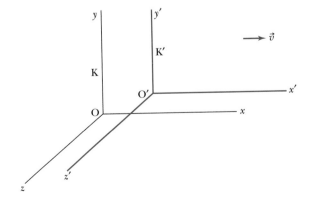

Figure 2.1 Two inertial systems are moving with relative speed v along their x axes. We show the system K at rest and the system K′ moving with speed v relative to the system K.

stationary somewhere, such as on Earth. One result of the relativity theory is that there are no fixed, absolute frames of reference. We use the term *fixed* to refer to a system that is fixed on a particular object, such as a planet, star, or spaceship that itself is moving in space. The transformation of the coordinates of a point in one system to the other system is given by

$$x' = x - vt$$
$$y' = y \qquad\qquad (2.1)$$
$$z' = z$$

Similarly, the inverse transformation is given by

$$x = x' + vt$$
$$y = y' \qquad\qquad (2.2)$$
$$z = z'$$

where we have set $t = t'$ because Newton considered time to be absolute. Equations (2.1) and (2.2) are known as the **Galilean transformation**. Newton's laws of motion are invariant under a Galilean transformation; that is, they have the same form in both systems K and K′.

Galilean transformation

In the late nineteenth century, Albert Einstein was concerned that although Newton's laws of motion had the same form under a Galilean transformation, Maxwell's equations did not. Einstein believed so strongly in Maxwell's equations that he showed there was a significant problem in our understanding of the Newtonian principle of relativity. In 1905 he published ideas that rocked the very foundations of physics and science. He proposed that space and time are not separate and that Newton's laws are only an approximation. This special theory of relativity and its ramifications are the subject of this chapter. We begin by presenting the experimental situation historically—showing why a problem existed and what was done to try to rectify the situation. Then we discuss Einstein's two postulates on which the special theory is based. The interrelation of space and time is discussed, and several amazing and remarkable predictions based on the new theory are shown.

As the concepts of relativity became more used in everyday research and development, it became essential to understand the transformation of momentum, force, and energy. Here we study relativistic dynamics and the relationship between mass and energy, which leads to one of the most famous equations in physics and a new conservation law of mass-energy. Finally, we return to electromagnetism to investigate the effects of relativity. We learn that Maxwell's equations don't require change, and electric and magnetic effects are relative, depending on the observer. We leave until Chapter 15 our discussion of Einstein's general theory of relativity.

2.1   The Need for Ether

Thomas Young, an English physicist and physician, performed his famous experiments on the interference of light in 1802. A decade later, the French physicist and engineer Augustin Fresnel published his calculations showing the detailed understanding of interference, diffraction, and polarization. Because all known waves (other than light) require a medium in which to propagate (water waves have water, sound waves have, for example, air, and so on), it was naturally

The concept of ether

assumed that light also required a medium, even though light was apparently able to travel in vacuum through outer space. This medium was called the *luminiferous ether* or just **ether** for short, and it must have some amazing properties. The ether had to have such a low density that planets could pass through it, seemingly for eternity, with no apparent loss of orbit position. Its elasticity must be strong enough to pass waves of incredibly high speeds!

The electromagnetic theory of light (1860s) of the Scottish mathematical physicist James Clerk Maxwell shows that the speed of light in different media depends only on the electric and magnetic properties of matter. In vacuum, the speed of light is given by $v = c = 1/\sqrt{\mu_0 \epsilon_0}$, where μ_0 and ϵ_0 are the permeability and permittivity of free space, respectively. The properties of the ether, as proposed by Maxwell in 1873, must be consistent with electromagnetic theory, and the feeling was that to be able to discern the ether's various properties required only an experiment that was sensitive enough. The concept of ether was well accepted by 1880.

When Maxwell presented his electromagnetic theory, scientists were so confident in the laws of classical physics that they immediately pursued the aspects of Maxwell's theory that were in contradiction with those laws. As it turned out, this investigation led to a new, deeper understanding of nature. Maxwell's equations predict the velocity of light in a vacuum to be c. If we have a flashbulb go off in the moving system K′, an observer in system K′ measures the speed of the light pulse to be c. However, if we make use of Equation (2.1) to find the relation between speeds, we find the speed measured in system K to be $c + v$, where v is the relative speed of the two systems. However, Maxwell's equations don't differentiate between these two systems. Physicists of the late nineteenth century proposed that there must be one preferred inertial reference frame in which the ether was stationary and that in this system the speed of light was c. In the other systems, the speed of light would indeed be affected by the relative speed of the reference system. Because the speed of light was known to be so enormous, 3×10^8 m/s, no experiment had as yet been able to discern an effect due to the relative speed v. The ether frame would in fact be an absolute standard, from which other measurements could be made. Scientists set out to find the effects of the ether.

Albert A. Michelson (1852–1931) shown at his desk at the University of Chicago in 1927. He was born in Prussia, but came to the United States when he was two years old. He was educated at the U.S. Naval Academy and later returned on the faculty. Michelson had appointments at several American universities including the Case School of Applied Science, Cleveland, in 1883; Clark University, Worcester, Massachusetts, in 1890; and the University of Chicago in 1892 until his retirement in 1929. During World War I he returned to the U.S. Navy where he developed a rangefinder for ships. He spent his retirement years in Pasadena, California, where he continued to measure the speed of light at Mount Wilson. *AIP Emilio Segrè Visual Archives.*

2.2   The Michelson-Morley Experiment

The Earth orbits around the sun at a high orbital speed, about $10^{-4}c$, so an obvious experiment is to try to find the effects of the Earth's motion through the ether. Even though we don't know how fast the sun might be moving through the ether, certainly at various times during the solar year the Earth must be passing rapidly through the ether.

Albert Michelson (1852–1931) performed perhaps the most significant American physics experiment of the 1800s. Michelson, who was the first U.S. citizen to receive the Nobel Prize for Physics (1907), was an ingenious scientist who built an extremely precise device called an *interferometer* that measures the phase difference between two light waves. Michelson used his interferometer to detect the difference in the speed of light passing through the ether in different directions. The basic technique is shown in Figure 2.2. Initially, it is assumed that one of the interferometer arms (AC) is parallel to the motion of the Earth through the ether. Light leaves the source S and passes through the glass plate at A. Because the back of A is partially silvered, part of the light is reflected, eventually going to the mirror at D, and part of the light travels through A on to the mirror

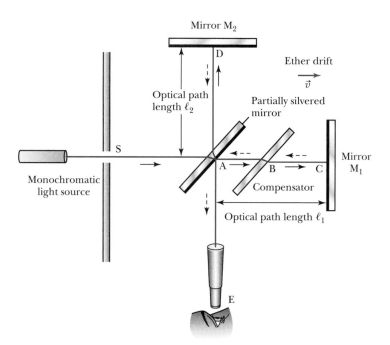

Figure 2.2 A schematic diagram of Michelson's interferometer experiment. Light of a single wavelength is partially reflected and partially transmitted by the glass at A. The light is subsequently reflected by mirrors at C and D, and, after reflection or transmission again at A, enters the telescope at E. Interference fringes will appear at E.

at C. The light is reflected at the mirrors C and D and comes back to the partially silvered mirror A, where part of the light from each path passes on to the telescope and eye at E. The compensator is added at B to make sure both light paths pass through equal thicknesses of glass. Interference fringes can be found by using a bright light source like sodium, and the apparatus is adjusted for maximum intensity of the light at E. The fringe pattern should shift if the apparatus is rotated through 90° such that arm AD becomes parallel to the motion of the Earth through the ether and arm AC is perpendicular to the motion.

We let the optical path lengths of AC and AD be denoted by ℓ_1 and ℓ_2, respectively. The observed interference pattern consists of alternating bright and dark bands, corresponding to constructive and destructive interference, respectively (Figure 2.3). For constructive interference, the difference between the two

Figure 2.3 Interference fringes as they would appear in the eyepiece of the Michelson-Morley experiment. *From L. S. Swenson, Jr., Invention and Discovery* **43** *(Fall 1987).*

path lengths (to and from the mirrors) is given by some number of wavelengths, $2(\ell_1 - \ell_2) = n\lambda$, where λ is the wavelength of the light and n is an integer.

The expected shift in the interference pattern can be calculated by determining the time difference between the two paths. When the light travels from A to C, the velocity of light according to the Galilean transformation is $c + v$, because the ether carries the light along with it. On the return journey from C to A the velocity is $c - v$, because the light travels opposite to the path of the ether. The total time for the round-trip journey to mirror M_1 is t_1:

$$t_1 = \frac{\ell_1}{c + v} + \frac{\ell_1}{c - v} = \frac{2c\ell_1}{c^2 - v^2} = \frac{2\ell_1}{c}\left(\frac{1}{1 - v^2/c^2}\right)$$

Now imagine what happens to the light that is reflected from mirror M_2. If the light is pointed directly at point D, the ether will carry the light with it and the light misses the mirror, much as the wind can affect the flight of an arrow. If a swimmer (who can swim with speed v_2 in still water) wants to swim across a swiftly moving river (speed v_1), the swimmer must start heading upriver, so that when the current carries her downstream, she will move directly across the river. Careful reasoning shows that the swimmer's velocity is $\sqrt{v_2^2 - v_1^2}$ throughout her journey (Problem 4). Thus the time t_2 for the light to pass to mirror M_2 at D and back is given by

$$t_2 = \frac{2\ell_2}{\sqrt{c^2 - v^2}} = \frac{2\ell_2}{c}\frac{1}{\sqrt{1 - v^2/c^2}}$$

The time difference between the two journeys Δt is

$$\Delta t = t_2 - t_1 = \frac{2}{c}\left(\frac{\ell_2}{\sqrt{1 - v^2/c^2}} - \frac{\ell_1}{1 - v^2/c^2}\right) \tag{2.3}$$

We now rotate the apparatus by $90°$ so that the ether passes along the length ℓ_2 toward the mirror M_2. We denote the new quantities by primes and carry out an analysis similar to that just done. The time difference $\Delta t'$ is now

$$\Delta t' = t_2' - t_1' = \frac{2}{c}\left(\frac{\ell_2}{1 - v^2/c^2} - \frac{\ell_1}{\sqrt{1 - v^2/c^2}}\right) \tag{2.4}$$

Michelson looked for a shift in the interference pattern when his apparatus was rotated by $90°$. The time difference is

$$\Delta t' - \Delta t = \frac{2}{c}\left(\frac{\ell_1 + \ell_2}{1 - v^2/c^2} - \frac{\ell_1 + \ell_2}{\sqrt{1 - v^2/c^2}}\right)$$

Because we know $c \gg v$, we can use the binomial expansion* to expand the terms involving v^2/c^2, keeping only the lowest terms.

$$\Delta t' - \Delta t = \frac{2}{c}(\ell_1 + \ell_2)\left[\left(1 + \frac{v^2}{c^2} + \cdots\right) - \left(1 + \frac{v^2}{2c^2} + \cdots\right)\right]$$

$$\approx \frac{v^2(\ell_1 + \ell_2)}{c^3} \tag{2.5}$$

Michelson left his position at the U.S. Naval Academy in 1880 and took his interferometer to Europe for postgraduate studies with some of Europe's best physicists, particularly Hermann Helmholtz in Berlin. After a few false starts he finally

---

*See Appendix 3 for the binomial expansion.

Figure 2.4 An adaptation of the Michelson and Morley 1887 experiment taken from their publication [A. A. Michelson and E. M. Morley, *Philosophical Magazine* **190**, 449 (1887)]. (a) A perspective view of the apparatus. To reduce vibration, the experiment was done on a massive soapstone, 1.5 m square and 0.3 m thick. This stone was placed on a wooden float that rested on mercury inside the annular piece shown underneath the stone. The entire apparatus rested on a brick pier. (b) The incoming light is focused by the lens and is both transmitted and reflected by the partly silvered mirror. The adjustable mirror allows fine adjustments in the interference fringes. The stone was slowly and uniformly rotated on the mercury to look for the interference effects of the ether.

was able to perform a measurement in Potsdam (near Berlin) in 1881. In order to use Equation (2.5) for an estimate of the expected time difference, the value of the Earth's orbital speed around the sun, 3×10^4 m/s, was used. Michelson's apparatus had $\ell_1 \approx \ell_2 \approx \ell = 1.2$ m. Thus Equation (2.5) predicts a time difference of 8×10^{-17} s. This is an exceedingly small time, but for a visible wavelength of 6×10^{-7} m, the period of one wavelength amounts to $T = 1/f = \lambda/c = 2 \times 10^{-15}$ s. Thus the time period of 8×10^{-17} s represents 0.04 fringes in the interference pattern. Michelson reasoned that he should be able to detect a shift of at least half this value, but found none. Although disappointed, Michelson concluded that the hypothesis of the stationary ether must be incorrect. **Michelson in Europe**

The result of Michelson's experiment was so surprising that he was asked by several well-known physicists to repeat it. In 1882 Michelson accepted a position at the then-new Case School of Applied Science in Cleveland. Together with Edward Morley (1838–1923), a professor of chemistry at nearby Western Reserve College who had become interested in Michelson's work, he put together the more sophisticated experiment shown in Figure 2.4. The new experiment had an optical path length of 11 m, created by reflecting the light for eight round trips. The new apparatus was mounted on soapstone that floated on mercury to eliminate vibrations and was so effective that Michelson and Morley believed they could detect a fraction of a fringe shift as small as 0.005. With their new apparatus they expected the ether to produce a shift as large as 0.4 of a fringe. They reported in 1887 a *null result*—no effect whatsoever! The ether does not seem to exist. It is this famous experiment that has become known as the *Michelson-Morley experiment*. **Null result of Michelson-Morley experiment**

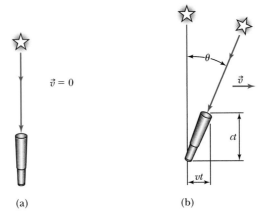

(a)   (b)

Figure 2.5 The effect of stellar aberration. (a) If a telescope is at rest, light from a distant star will pass directly into the telescope. (b) However, if the telescope is traveling at speed v (because it is fixed on the Earth, which has a motion about the sun), it must be slanted slightly to allow the starlight to enter the telescope. This leads to an apparent circular motion of the star as seen by the telescope, as the motion of the Earth about the sun changes throughout the solar year.

The measurement so shattered a widely held belief that many suggestions were made to explain it. What if the Earth just happened to have a zero motion through the ether at the time of the experiment? Michelson and Morley repeated their experiment during night and day and for different seasons throughout the year. It is unlikely that at least sometime during these many experiments, the Earth would not be moving through the ether. Michelson and Morley even took their experiment to a mountaintop to see if the effects of the ether might be different. There was no change.

Of the many possible explanations of the null ether measurement, the one **Ether drag** taken most seriously was the *ether drag* hypothesis. Some scientists proposed that the Earth somehow dragged the ether with it as the Earth rotates on its own axis and revolves around the sun. However, the ether drag hypothesis contradicts re- **Stellar aberration** sults from several experiments, including that of *stellar aberration* noted by the British astronomer James Bradley in 1728. Bradley noticed that the apparent position of the stars seems to rotate in a circular motion with a period of one year. The angular diameter of this circular motion with respect to the Earth is 41 seconds of arc. This effect can be understood by an analogy. From the viewpoint of a person sitting in a car during a rainstorm, the raindrops appear to fall vertically when the car is at rest but appear to be slanted toward the windshield when the car is moving forward. The same effect occurs for light coming from stars directly above the Earth's orbital plane. If the telescope and star are at rest with respect to the ether, the light enters the telescope as shown in Figure 2.5a. However, because the Earth is moving in its orbital motion, the apparent position of the star is at an angle θ as shown in Figure 2.5b. The telescope must actually be slanted at an angle θ in order to observe the light from the overhead star. During a time period t the starlight moves a vertical distance ct while the telescope moves a horizontal distance vt, so that the tangent of the angle θ is

$$\tan \theta = \frac{vt}{ct} = \frac{v}{c}$$

The orbital speed of the Earth is about 3×10^4 m/s; therefore, the angle θ is 10^{-4} rad or 20.6 seconds of arc, with a total opening of $2\theta = 41$ s as the Earth rotates—in agreement with Bradley's observation. The aberration reverses itself every six months as the Earth orbits about the sun, in effect giving a circular motion to the star's position. This observation is in disagreement with the hypothesis of the Earth dragging the ether. If the ether were dragged with the Earth, there would be no need to tilt the telescope! The experimental observation of stellar aberration together with the null result of the Michelson and Morley experiment is enough evidence to refute the suggestions that the ether exists. Many other experimental observations have now been made that also confirm this conclusion.

The inability to detect the ether was a serious blow to reconciling the invariant form of the electromagnetic equations of Maxwell. There seems to be no single reference inertial system in which the speed of light is actually c. H. A. Lorentz and G. F. FitzGerald suggested, apparently independently, that the results of the Michelson-Morley experiment could be understood if length is contracted by the factor $\sqrt{1 - v^2/c^2}$ in the direction of motion, where v is the speed in the direction of travel. For this situation, the length ℓ_1, in the direction of motion, will be contracted by the factor $\sqrt{1 - v^2/c^2}$, whereas the length ℓ_2, perpendicular to v, will not. The result in Equation (2.3) is that t_1 will have the extra factor $\sqrt{1 - v^2/c^2}$, making Δt precisely zero as determined experimentally by Michelson. This contraction postulate, which became known as the *Lorentz-FitzGerald contraction*, was not proven from first principles using Maxwell's equations, and its true significance was not understood for several years until Einstein presented his explanation. An obvious problem with the Lorentz-FitzGerald contraction is that it is an ad hoc assumption that cannot be directly tested. Any measuring device would presumably be shortened by the same factor.

2.3   Einstein's Postulates

At the turn of the twentieth century, the Michelson-Morley experiment had laid to rest the idea of finding a preferred inertial system for Maxwell's equations, yet the Galilean transformation, which worked for the laws of mechanics, was invalid for Maxwell's equations. This quandary represented a turning point for physics.

Albert Einstein (1879–1955) was only two years old when Michelson reported his first null measurement for the existence of the ether. Einstein said that he began thinking at the age of 16 about the form of Maxwell's equations in moving inertial systems, and in 1905, when he was 26 years old, he published his startling proposal* about the principle of relativity, which he believed to be fundamental. Working without the benefit of discussions with colleagues outside his small circle of friends, Einstein was apparently unaware of the interest concerning the null result of Michelson and Morley.† Einstein instead looked at the problem in a more

Albert Einstein (1879–1955), shown here sailing on Long Island Sound, was born in Germany and studied in Munich and Zurich. After having difficulty finding a position, he served seven years in the Swiss Patent Office in Bern (1902–1909), where he did some of his best work. He obtained his doctorate at the University of Zurich in 1905. His fame quickly led to appointments in Zurich, Prague, back to Zurich, and then to Berlin in 1914. In 1933, after Hitler came to power, Einstein left for the Institute for Advanced Study at Princeton University where he became a U.S. citizen in 1940 and remained until his death in 1955. Einstein's total contributions to physics are rivaled only by those of Isaac Newton. *AIP Emilio Segrè Visual Archives.*

---

*In one issue of the German journal *Annalen der Physik* **17**, No. 4 (1905), Einstein published three remarkable papers. The first, on the quantum properties of light, explained the photoelectric effect; the second, on the statistical properties of molecules, included an explanation of Brownian motion; and the third was on special relativity. All three papers contained predictions that were subsequently confirmed experimentally.

†The question of whether Einstein knew of Michelson and Morley's null result before he produced his special theory of relativity is somewhat uncertain. For example, see J. Stachel, "Einstein and Ether Drift Experiments," *Physics Today* (May 1987), p. 45.

Figure 2.6 The problem of simultaneity. Flashbulbs positioned in system K at one meter on either side of Frank go off simultaneously in (a). Frank indeed sees both flashes simultaneously in (b). However, Mary, at rest in system K′ moving to the right with speed v, does not see the flashes simultaneously despite the fact that she was alongside Frank when the flashbulbs went off. During the finite time it took light to travel the one meter, Mary has moved slightly, as shown in exaggerated form in (b).

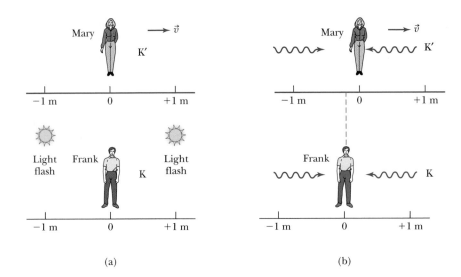

formal manner and believed that Maxwell's equations must be valid in all inertial frames. With piercing insight and genius, Einstein was able to bring together seemingly inconsistent results concerning the laws of mechanics and electromagnetism with two postulates (as he himself called them; today we would call them laws). These postulates are

Einstein's two postulates

1. **The principle of relativity:** The laws of physics are the same in all inertial systems. There is no way to detect absolute motion, and no preferred inertial system exists.
2. **The constancy of the speed of light:** Observers in all inertial systems measure the same value for the speed of light in a vacuum.

The first postulate indicates that the laws of physics will be the same in all coordinate systems moving with uniform relative motion to each other. Einstein showed that postulate 2 actually follows from the first one. He returned to the principle of relativity as espoused by Newton. Although Newton's principle referred only to the laws of mechanics, Einstein expanded it to include all laws of physics—including those of electromagnetism. We can now modify our previous

Inertial frames of reference revisited

definition of *inertial frames of reference* to be those frames of reference in which *all the laws of physics* are valid.

Einstein's solution requires us to take a careful look at time. Return to the two systems of Figure 2.1 and remember that we had previously assumed that $t = t'$. We assumed that events occurring in system K′ and in system K could easily be synchronized. Einstein realized that each system must have its own observers with their own clocks and metersticks. *An event in a given system must be specified by stating both its space and time coordinates.* Consider the flashing of two bulbs fixed in system K as shown in Figure 2.6a. **M**ary, in system K′ (the **M**oving system) is beside **F**rank, who is in system K (the **F**ixed system), when the bulbs flash. As seen in Figure 2.6b the light pulses travel the same distance in system K and arrive at Frank

Simultaneity

simultaneously. Frank sees the two flashes at the same time. However, the two light pulses do not reach Mary simultaneously, because system K′ is moving to the right, and she has moved closer to the bulb on the right by the time the flash reaches her. The light flash coming from the left will reach her at some later time. We conclude that

Two events that are simultaneous in one reference frame (K) are not necessarily simultaneous in another reference frame (K') moving with respect to the first frame.

We must be careful when comparing the same event in two systems moving with respect to one another. Time comparison can be accomplished by sending light signals from one observer to another, but this information can travel only as fast as the finite speed of light. It is best if each system has its own observers with clocks that are synchronized. How can we do this? We place observers with clocks throughout a given system. If, when we bring all the clocks together at one spot at rest, all the clocks agree, then the clocks are said to be **synchronized**. However, we have to move the clocks relative to each other in order to reposition them, and that might affect the synchronization. A better way would be to flash a bulb half-way between each pair of clocks at rest and make sure the pulses arrive simultaneously at each clock. This will require many measurements, but it is a safe way of synchronizing the clocks. We can determine the time of an event occurring far away from us by having a colleague at the event, with a clock fixed at rest, measure the time of the particular event and send us the results, for example, by telephone or even by mail. If we need to check our clocks we can always send light signals to each other over known distances at some predetermined time.

Synchronization of clocks

In the next section we will derive the correct transformation, called the **Lorentz transformation**, that makes the laws of physics invariant between inertial frames of reference. We use the coordinate systems described by Figure 2.1. At $t = t' = 0$, the origins of the two coordinate systems are coincident, and the system K' is traveling along the x and x' axes. For this special case, the Lorentz transformation equations are

$$x' = \frac{x - vt}{\sqrt{1 - v^2/c^2}}$$

$$y' = y$$

$$z' = z \qquad\qquad (2.6)$$

$$t' = \frac{t - (vx/c^2)}{\sqrt{1 - v^2/c^2}}$$

Lorentz transformation equations

We commonly use the symbols β and the *relativistic factor* γ to represent two longer terms:

Relativistic factor

$$\beta = \frac{v}{c} \qquad\qquad (2.7)$$

$$\gamma = \frac{1}{\sqrt{1 - v^2/c^2}} \qquad\qquad (2.8)$$

which allows the Lorentz transformation equations to be rewritten in compact form as

$$x' = \gamma(x - \beta ct)$$

$$y' = y$$

$$z' = z \qquad\qquad (2.6)$$

$$t' = \gamma(t - \beta x/c)$$

Note that $\gamma \geq 1$ ($\gamma = 1$ when $v = 0$).

2.4   The Lorentz Transformation

In this section we will use Einstein's two postulates to find a transformation between inertial frames of reference such that all the physical laws, including Newton's laws of mechanics and Maxwell's electrodynamics equations, will have the same form. We use the fixed system K and moving system K' of Figure 2.1. At $t = t' = 0$ the origins and axes of both systems are coincident and system K' is moving to the right along the x axis. A flashbulb goes off at the origins when $t = t' = 0$. According to postulate 2, the speed of light will be c in both systems, and the wavefronts observed in both systems must be spherical and described by

$$x^2 + y^2 + z^2 = c^2 t^2 \tag{2.9a}$$

$$x'^2 + y'^2 + z'^2 = c^2 t'^2 \tag{2.9b}$$

These two equations are inconsistent with a Galilean transformation because a wavefront can be spherical in only one system when the second is moving at speed v with respect to the first. The Lorentz transformation *requires* both systems to have a spherical wavefront centered on each system's origin.

Another clear break with Galilean and Newtonian physics is that we do not assume that $t = t'$. Each system must have its own clocks and metersticks as indicated in a two-dimensional system in Figure 2.7. Because the systems move only along their x axes, observers in both systems agree by direct observation that

$$y' = y$$

$$z' = z$$

We know that the Galilean transformation $x' = x - vt$ is incorrect, but what is the correct transformation? We require a linear transformation so that each event in system K corresponds to one, and only one, event in system K'. The simplest *linear* transformation is of the form

$$x' = \gamma(x - vt) \tag{2.10}$$

We will see if such a transformation suffices. The parameter γ cannot depend on x or t because the transformation must be linear. The parameter γ must be close to 1 for $v \ll c$ in order for Newton's laws of mechanics to be valid for most of our measurements. We can use similar arguments from the standpoint of an observer stationed in system K' to obtain an equation similar to Equation (2.10).

$$x = \gamma'(x' + vt') \tag{2.11}$$

Because postulate 1 requires that the laws of physics be the same in both reference systems, we demand that $\gamma' = \gamma$. Notice that the only difference between Equations (2.10) and (2.11) other than the primed and unprimed quantities being switched is that $v \rightarrow -v$, which is reasonable because according to the observer in each system, the other observer is moving either forward or backward.

According to postulate 2 the speed of light must be c in both systems. Therefore in each system the wavefront of the flashbulb light pulse along the respective x axes must be described by $x = ct$ and $x' = ct'$, which we substitute into Equations (2.10) and (2.11) to obtain

$$ct' = \gamma(ct - vt) \tag{2.12a}$$

and

$$ct = \gamma(ct' + vt') \tag{2.12b}$$

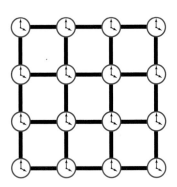

Figure 2.7 In order to make sure accurate event measurements can be obtained, synchronized clocks and measuring sticks are placed throughout a system.

We divide each of these equations by c and obtain

$$t' = \gamma t\left(1 - \frac{v}{c}\right) \qquad (2.13)$$

and

$$t = \gamma t'\left(1 + \frac{v}{c}\right) \qquad (2.14)$$

We substitute the value of t from Equation (2.14) into Equation (2.13).

$$t' = \gamma^2 t'\left(1 - \frac{v}{c}\right)\left(1 + \frac{v}{c}\right) \qquad (2.15)$$

We solve this equation for γ^2 and obtain

$$\gamma^2 = \frac{1}{1 - v^2/c^2}$$

or

$$\gamma = \frac{1}{\sqrt{1 - v^2/c^2}} \qquad (2.16)$$

In order to find a transformation for time t', we substitute x' from Equation (2.10) into Equation (2.11) and obtain

$$x = \gamma^2(x - vt) + \gamma vt'$$

We solve this equation for t'

$$t' = \frac{x}{\gamma v}(1 - \gamma^2) + \gamma t$$

If we use the value of γ in Equation (2.16) and β from Equation (2.7), a little algebra shows that $(1 - \gamma^2)/\gamma v = -\gamma\beta^2/v$, which we substitute in the previous equation to finally obtain

$$t' = \frac{t - (vx/c^2)}{\sqrt{1 - \beta^2}}$$

We are now able to write the complete Lorentz transformation equations as

$$x' = \frac{x - vt}{\sqrt{1 - \beta^2}}$$

$$y' = y$$

$$z' = z \qquad (2.17)$$

$$t' = \frac{t - (vx/c^2)}{\sqrt{1 - \beta^2}}$$

The inverse transformation equations are obtained by replacing v by $-v$ as discussed previously and by exchanging the primed and unprimed quantities.

Inverse Lorentz transformation equations

$$x = \frac{x' + vt'}{\sqrt{1 - \beta^2}}$$

$$y = y'$$

$$z = z'$$

$$t = \frac{t' + (vx'/c^2)}{\sqrt{1 - \beta^2}}$$

(2.18)

Notice that Equations (2.17) and (2.18) both reduce to the Galilean transformation when $v \ll c$. It is only for speeds that approach the speed of light that the Lorentz transformation equations become significantly different from the Galilean equations. In our studies of mechanics we normally do not consider such high speeds, and our previous results probably require no corrections. The laws of mechanics credited to Newton are still valid over the region of their applicability. Even for a speed as high as the Earth orbiting about the sun, 30 km/s, the value of the relativistic factor γ is 1.000000005. We show a plot of the relativistic parameter γ versus speed in Figure 2.8. As a rule of thumb, we should consider using the relativistic equations when $v/c > 0.1$ ($\gamma \approx 1.005$).

Finally, consider the implications of the Lorentz transformation. The linear transformation equations ensure that a single event in one system is described by a single event in another inertial system. However, space and time are not separate. In order to express the position of x in system K', we must use both x' and t'. We have also found that the Lorentz transformation does not allow a speed greater than c; the transformation equations become imaginary in this case. We show later in this chapter that no object of nonzero mass can have a speed greater than c.

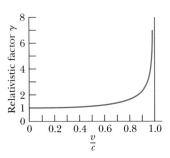

Figure 2.8 A plot of the relativistic factor γ as a function of speed v/c, showing that γ becomes large quickly as v approaches c.

2.5   Time Dilation and Length Contraction

The Lorentz transformations have immediate consequences with respect to time and length measurements made by observers in different inertial frames. We shall consider time and length measurements separately and then see how they are related to one another.

Time Dilation

Consider again our two systems K and K' with system K fixed and system K' moving along the x axis with velocity $\vec{v}$ as shown in Figure 2.9a. Frank lights a sparkler at position x_1 in system K. A clock placed beside the sparkler indicates the time to be t_1 when the sparkler is lit and t_2 when the sparkler goes out (Figure 2.9b). The sparkler burns for time T_0, where $T_0 = t_2 - t_1$. The time difference between two events occurring at the same position in a system as measured by a clock at rest in the system is called the **proper time**. We use the subscript zero on the time difference T_0 to denote the proper time.

Proper time

Now what is the time as determined by Mary who is passing by (but at rest in her own system K')? All the clocks in both systems have been synchronized when the systems are at rest with respect to one another. The two events (sparkler lit

(a)   (b)

Figure 2.9 Frank measures the proper time for the time interval that a sparkler stays lit. His clock is at the same position in system K when the sparkler is lit in (a) and when it goes out in (b). Mary, in the moving system K′, is beside the sparkler at position x_1' when it is lit in (a), but by the time it goes out in (b), she has moved away. Melinda, at position x_2', measures the time in system K′ when the sparkler goes out in (b).

and then going out) do not occur at the same place according to Mary. She is beside the sparkler when it is lit, but she has moved far away from the sparkler when it goes out (Figure 2.9b). Her friend Melinda, also at rest in system K′, is beside the sparkler when it goes out. Mary and Melinda measure the two times for the sparkler to be lit and to go out in system K′ as times t_1' and t_2'. The Lorentz transformation relates these times to those measured in system K as

$$t_2' - t_1' = \frac{(t_2 - t_1) - (v/c^2)(x_2 - x_1)}{\sqrt{1 - v^2/c^2}}$$

In system K the clock is fixed at x_1, so $x_2 - x_1 = 0$; that is, the two events occur at the same position. The time $t_2 - t_1$ is the proper time T_0, and we denote the time difference $t_2' - t_1' = T'$ as measured in the moving system K′.

$$T' = \frac{T_0}{\sqrt{1 - v^2/c^2}} = \gamma T_0 \qquad (2.19) \quad \textbf{Time dilation}$$

Thus the time interval measured in the moving system K′ is greater than the time interval measured in system K where the sparkler is at rest. This effect is known as **time dilation** and is a direct result of Einstein's two postulates. The time measured by Mary and Melinda in their system K′ for the time difference was greater than T_0 by the relativistic factor γ ($\gamma > 1$). The two events, sparkler being lit and then going out, did not occur at the same position ($x_2' \neq x_1'$) in system K′ (see Figure 2.9b). This result occurs because of the absence of simultaneity. The events do not occur at the same space and time coordinates in the two systems. It requires three clocks to perform the measurement: one in system K and two in system K'.

The time dilation result is often interpreted by saying that *moving clocks run slow* by the factor γ^{-1}, and sometimes this is a useful way to remember the effect. The moving clock in this case can be any kind of clock. It can be the time that sand takes to pass through an hourglass, the time a sparkler stays lit, the time between heartbeats, the time between ticks of a clock, or the time spent in a class lecture. In all cases, the actual time interval on a moving clock is greater than the proper time as measured on a clock at rest. The proper time is always the smallest possible time interval between two events.

Each person will claim the clock in the other (moving) system is running slow. If Mary had a sparkler in her system K' at rest, Frank (fixed in system K) would also measure a longer time interval on his clock in system K because the sparkler would be moving with respect to his system.

Moving clocks run slow

Example 2.1

Show that Frank in the fixed system will also determine the time dilation result by having the sparkler be at rest in the system K'.

Strategy We should be able to proceed similarly to the derivation we did before when the sparkler was at rest in system K. In this case the sparkler is lit by Mary in the moving system K'. The time interval over which the sparkler is lit is given by $T'_0 = t'_2 - t'_1$, and the sparkler is placed at the position $x'_1 = x'_2$ so that $x'_2 - x'_1 = 0$. In this case T'_0 is the proper time. We use the Lorentz transformation from Equation (2.18) to de-

termine the time difference $T = t_2 - t_1$ as measured by the clocks of Frank and his colleagues.

Solution We use Equation (2.18) to find $t_2 - t_1$:

$$T = t_2 - t_1 = \frac{(t'_2 - t'_1) + (v/c^2)(x'_2 - x'_1)}{\sqrt{1 - v^2/c^2}}$$

$$= \frac{T'_0}{\sqrt{1 - v^2/c^2}} = \gamma T'_0$$

The time interval is still smaller in the system where the sparkler is at rest.

Gedanken experiments

The preceding results naturally seem a little strange to us. In relativity we often carry out thought (or *gedanken* from the German word) experiments, because the actual experiments would be somewhat impractical. Consider the following *gedanken* experiment. Mary, in the moving system K', flashes a light at her origin along her y' axis (Figure 2.10). The light travels a distance L, reflects off a mirror, and returns. Mary says that the total time for the journey is $T'_0 = t'_2 - t'_1 = 2L/c$, and this is indeed the proper time, because the clock in K' beside Mary is at rest.

What do Frank and other observers in system K measure? Let T be the round-trip time interval measured in system K for the light to return to the x axis. The light is flashed when the origins are coincident, as Mary passes by Frank with relative velocity v. When the light reaches the mirror in the system K' at time $T/2$, the system K' will have moved a distance $vT/2$ down the x axis. When the light is reflected back to the x axis, Frank will not even see the light return, because it will return a distance vT away, where another observer, Fred, is positioned. Because observers Frank and Fred have previously synchronized their clocks, they can still measure the total elapsed time for the light to be reflected from the mirror and return. According to observers in the K system, the total distance the light travels

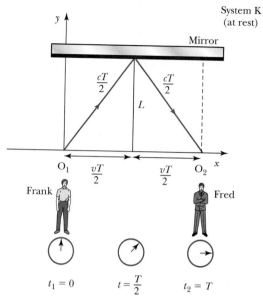

Figure 2.10 Mary, in system K′, flashes a light along her y' axis and measures the proper time $T_0' = 2L/c$ for the light to return. In system K Frank will see the light travel partially down his x axis, because system K′ is moving. Fred times the arrival of the light in system K. The time interval T that Frank and Fred measure is related to the proper time by $T = \gamma T_0'$.

(as shown in Figure 2.10) is $2\sqrt{(vT/2)^2 + L^2}$. And according to postulate 2 the light must travel at the speed of light, so the total time interval T measured in system K is

$$T = \frac{\text{distance}}{\text{speed}} = \frac{2\sqrt{(vT/2)^2 + L^2}}{c}$$

As can be determined from above, $L = cT_0'/2$, so we have

$$T = \frac{2\sqrt{(vT/2)^2 + (cT_0'/2)^2}}{c}$$

which reduces to

$$T = \frac{T_0'}{\sqrt{1 - v^2/c^2}} = \gamma T_0'$$

This is consistent with the earlier result. In this case $T > T_0'$. The proper time is always the shortest time interval, and we find that the clock in Mary's system K′ is "running slow."

Example 2.2

It is the year 2050 and the United Nations Space Federation has finally perfected the storage of antiprotons for use as fuel in a spaceship. (Antiprotons are the antiparticles of protons.

We will discuss antiprotons in Chapter 3.) Preparations are under way for a manned spacecraft visit to the planetary system of the star Alpha Centauri, some 4 lightyears away.

Provisions are placed on board to allow a trip of 16 years' total duration. How fast must the spacecraft travel if the provisions are to last? Neglect the period of acceleration, turn-around, and visiting times, because they are negligible compared to the actual travel time.

Strategy The time interval as measured by the astronauts on the spacecraft can be no longer than 16 years, because that is how long the provisions will last. However, from Earth we realize that the spacecraft will be moving at a high relative speed v to us, and that according to our clock in the stationary system K, the trip will last $T = 2L/v$, where L is the distance to the star.

Because provisions on board the spaceship will last for only 16 years, we let the proper time T'_0 in system K′ be 16 years. Using the time dilation result, we determine the relationship between T, the time measured on Earth, and the proper time T'_0 to be

$$T = \frac{2L}{v} = \frac{T'_0}{\sqrt{1 - v^2/c^2}} \qquad (2.20)$$

We then solve this equation for the required speed v.

Solution A lightyear is a convenient way to measure large distances. It is the distance light travels in one year and is denoted by ly:

$$1\ \text{ly} = \left(3 \times 10^8\ \frac{\text{m}}{\text{s}}\right)(1\ \text{year})\left(365\ \frac{\text{days}}{\text{year}}\right)\left(24\ \frac{\text{h}}{\text{day}}\right)\left(3600\ \frac{\text{s}}{\text{h}}\right)$$

$$= 9.5 \times 10^{15}\ \text{m}$$

Note that the distance of one lightyear is the speed of light, c, multiplied by the time of one year. The dimension of a lightyear works out to be length. For example, $4\ \text{ly} = c(4\ \text{y}) = 38 \times 10^{15}\ \text{m}$.

We insert the appropriate numbers into Equation (2.20) and obtain

$$\frac{2(4\ \text{ly})(9.5 \times 10^{15}\ \text{m/ly})}{v} = \frac{16\ \text{y}}{\sqrt{1 - v^2/c^2}}$$

The solution to this equation is $v = 0.447c = 1.34 \times 10^8\ \text{m/s}$. The time interval as measured on Earth will be $\gamma T'_0 = 17.9\ \text{y}$. Notice that the astronauts will age only 16 years (their clocks run slow), while their friends remaining on Earth will age 17.9 years. Can this really be true? We shall discuss this question again in Section 2.8.

Length Contraction

Now let's consider what might happen to the length of objects in relativity. Let an observer in each system K and K′ have a meterstick at rest in his or her own respective system. Each observer lays the stick down along his or her respective x axis, putting the left end at x_ℓ (or x'_ℓ) and the right end at x_r (or x'_r). Thus, Frank in system K measures his stick to be $L_0 = x_r - x_\ell$. Similarly, in system K′, Mary measures her stick at rest to be $L'_0 = x'_r - x'_\ell = L_0$. Every observer measures a meterstick at rest in his or her own system to have the same length, namely one meter. The length as measured at rest is called the **proper length**.

Proper length

Let system K be at rest, and system K′ move along the x axis with speed v. Frank, who is at rest in system K, measures the length of the stick moving in K′. The difficulty is to measure the ends of the stick simultaneously. We insist that Frank measure the ends of the stick at the same time so that $t = t_r = t_\ell$. The events denoted by (x, t) are (x_ℓ, t) and (x_r, t). We use Equation (2.17) and find

$$x'_r - x'_\ell = \frac{(x_r - x_\ell) - v(t_r - t_\ell)}{\sqrt{1 - v^2/c^2}}$$

The meterstick is at rest in system K′, so the length $x'_r - x'_\ell$ must be the proper length L'_0. Denote the length measured by Frank as $L = x_r - x_\ell$. The times t_r and t_ℓ are identical, as we insisted, so $t_r - t_\ell = 0$. Notice that the times of measurement by Mary in her system, t'_ℓ and t'_r, are *not* identical. It makes no difference when Mary makes the measurements in her own system, because the stick is at rest. It does, however, make a big difference when Frank makes his measurements, because the stick is moving with speed v with respect to him. The measurements must be done simultaneously! With these results, the previous equation becomes

$$L_0' = \frac{L}{\sqrt{1 - v^2/c^2}} = \gamma L$$

or, because $L_0' = L_0$,

$$L = L_0\sqrt{1 - v^2/c^2} = \frac{L_0}{\gamma} \qquad (2.21) \qquad \textbf{Length contraction}$$

Notice that $L_0 > L$, so the moving meterstick shrinks according to Frank. This effect is known as **length** or **space contraction** and is characteristic of relative motion. This effect is also sometimes called the *Lorentz-FitzGerald contraction* because Lorentz and FitzGerald independently suggested the contraction as a way to solve the electrodynamics problem. This effect, like time dilation, is also reciprocal. Each observer will say that the other moving stick is shorter. There is no length contraction perpendicular to the relative motion, however, because $y' = y$ and $z' = z$. Observers in both systems can check the length of the other meterstick placed perpendicular to the direction of motion as the metersticks pass each other. They will agree that both metersticks are one meter long.

We can also perform another *gedanken* experiment to arrive at the same result. This time we lay the meterstick along the x' axis in the moving system K' (Figure 2.11a). The two systems K and K' are aligned at $t = t' = 0$. A mirror is placed at the end of the meterstick, and a flashbulb goes off at the origin at

(a)   (b)

Figure 2.11 (a) Mary, in system K', flashes a light down her x' axis along a stick at rest in her system of length L_0, which is the proper length. The time interval for the light to travel down the stick and back is $2L_0/c$. (b) Frank, in system K, sees the stick moving, and the mirror has moved a distance vt_1 by the time the light is reflected. By the time the light returns to the beginning of the stick, the stick has moved a total distance of vt_2. The times can be compared to show that the moving stick has been length contracted by $L = L_0\sqrt{1 - v^2/c^2}$.

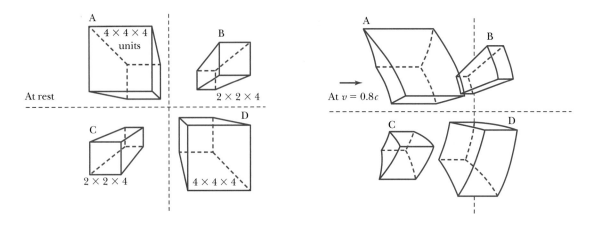

At rest

A

4 × 4 × 4
units

B

2 × 2 × 4

C

2 × 2 × 4

D

4 × 4 × 4

At v = 0.8c

A

B

C

D

Figure 2.12 In this computer simulation, the rectangular boxes are drawn as if the observer were 5 units in front of the near plane of the boxes and directly in front of the origin. The boxes are shown at rest on the left. On the right side, the boxes are moving to the right at a speed of $v = 0.8c$. The horizontal lines are only length contracted, but notice that the vertical lines become hyperbolas. The objects appear to be slightly rotated in space. The objects that are further away from the origin appear earlier because they are photographed at an earlier time, and because the light takes longer to reach the camera (or our eyes). *Reprinted with permission from G. D. Scott and M. R. Viner, American Journal of Physics* **33**, *534 (1965).* © *1965 American Association of Physics Teachers.*

$t = t' = 0$, sending a light pulse down the x' axis, where it is reflected and returned. Mary sees the stick at rest in system K′ and measures the proper length L_0 (which should of course be one meter). Mary uses the same clock fixed at $x' = 0$ for the time measurements. The stick is moving at speed v with respect to Frank in the fixed system K. The clocks at $x = x' = 0$ both read zero when the origins are aligned just when the flashbulb goes off. Notice the situation shown in system K (Figure 2.11b), where by the time the light reaches the mirror, the entire stick has moved a distance vt_1. By the time the light has been reflected back to the front of the stick again, the stick has moved a total distance vt_2. We leave the solution in terms of length contraction to Problem 18.

The effect of length contraction along the direction of travel may strongly affect the appearances of two- and three-dimensional objects. We see such objects when the light reaches our eyes, not when the light actually leaves the object. Thus, if the objects are moving rapidly, we will not see them as they appear at rest. Figure 2.12 shows the appearance of several such objects as they move. Note that not only do the horizontal lines become contracted, but the vertical lines also become hyperbolas. We show in Figure 2.13 a row of bars moving to the right with speed $v = 0.9c$. The result is quite surprising.

Figure 2.13 (a) An array of rectangular bars is seen from above at rest. (b) The bars are moving to the right at $v = 0.9c$. The bars appear to contract and rotate. *Quoted from P.-K. Hsuing and R. H. P. Dunn,* Science News **137**, *232 (1990).*

(a)

(b)

Example 2.3

Consider the solution of Example 2.2 from the standpoint of length contraction.

Strategy The astronauts have only enough provisions for a trip lasting 16 years. Thus they expect to travel for 8 years each way. If the star Alpha Centauri is 4 lightyears away, it appears that they may need to travel at a velocity of $0.5c$ in order to make the trip. We want to consider this example as if the astronauts are at rest. Alpha Centauri will appear to be moving toward them, and the distance to the star is length contracted. The distance measured by the astronauts will be less than 4 ly.

Solution The contracted distance according to the astronauts in motion is $(4 \text{ ly}) \sqrt{1 - v^2/c^2}$. The velocity they need to make this journey is the contracted distance divided by 8 years.

$$v = \frac{\text{distance}}{\text{time}} = \frac{(4 \text{ ly}) \sqrt{1 - v^2/c^2}}{8 \text{ y}}$$

If we divide by c, we obtain

$$\beta = \frac{v}{c} = \frac{(4 \text{ ly}) \sqrt{1 - v^2/c^2}}{c(8 \text{ y})} = \frac{(4 \text{ ly}) \sqrt{1 - v^2/c^2}}{(8 \text{ ly})}$$

$$\beta = \frac{\sqrt{1 - \beta^2}}{2}$$

which gives

$$\beta = 0.447$$

$$v = 0.447c$$

which is just what we found in the previous example. The effects of time dilation and length contraction give identical results.

2.6 Addition of Velocities

A spaceship launched from a space station (see Figure 2.14) quickly reaches its cruising speed of $0.60c$ with respect to the space station when a band of asteroids is observed straight ahead of the ship. Mary, the commander, reacts quickly and

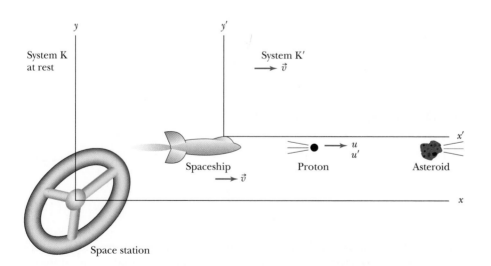

Figure 2.14 The space station is at rest at the origin of system K. The spaceship is moving to the right with speed v with respect to the space station and is in system K′. An asteroid is moving to the left toward both the spaceship and space station, so Mary, the commander of the spaceship, orders that the proton gun shoot protons to break up the asteroid. The speed of the protons is u and $u′$ with respect to systems K and K′, respectively.

orders her crew to blast away the asteroids with the ship's proton gun to avoid a catastrophic collision. Frank, the admiral on the space station, listens with apprehension to the communications because he fears the asteroids may eventually destroy his space station as well. Will the high-energy protons of speed $0.99c$ be able to successfully blast away the asteroids and save both the spaceship and space station? If $0.99c$ is the speed of the protons with respect to the spaceship, what speed will Frank measure for the protons?

We will use the letter u to denote speeds of objects as measured in various coordinate systems. In this case, Frank (in the fixed, stationary system K on the space station) will measure the speed of the protons to be u, whereas Mary, the commander of the spaceship (the moving system K'), will measure $u' = 0.99c$. We reserve the letter v to express the speed of the coordinate systems with respect to each other. The speed of the spaceship with respect to the space station is $v = 0.60c$.

Newtonian mechanics teaches us that to find the speed of the protons with respect to the space station, we simply add the speed of the spaceship with respect to the space station ($0.60c$) to the speed of the protons with respect to the spaceship ($0.99c$) to determine the simple result $u = v + u' = 0.60c + 0.99c = 1.59c$. However, this result is not in agreement with the results of the Lorentz transformation. We use Equation (2.18), letting x be along the direction of motion of the spaceship (and high-speed protons), and take the differentials, with the results

$$dx = \gamma(dx' + v\,dt')$$

$$dy = dy'$$

$$dz = dz' \tag{2.22}$$

$$dt = \gamma[dt' + (v/c^2)\,dx']$$

Velocities are defined by $u_x = dx/dt$, $u_y = dy/dt$, $u'_x = dx'/dt'$, and so on. Therefore we determine u_x by

Relativistic velocity addition

$$u_x = \frac{dx}{dt} = \frac{\gamma(dx' + v\,dt')}{\gamma[dt' + (v/c^2)\,dx']} = \frac{u'_x + v}{1 + (v/c^2)u'_x} \tag{2.23a}$$

Similarly, u_y and u_z are determined to be

$$u_y = \frac{u'_y}{\gamma[1 + (v/c^2)u'_x]} \tag{2.23b}$$

$$u_z = \frac{u'_z}{\gamma[1 + (v/c^2)u'_x]} \tag{2.23c}$$

Equations (2.23) are referred to as the **Lorentz velocity transformations**. Notice that although the relative motion of the systems K and K' is only along the x direction, the velocities along y and z are affected as well. This contrasts with the Lorentz transformation equations, where $y = y'$ and $z = z'$. However, the difference in velocities is simply ascribed to the transformation of time, which depends on v and x'. Thus, the transformations for u_y and u_z depend on v and u'_x. The inverse transformations for u'_x, u'_y, and u'_z can be determined by simply switching primed and unprimed variables and changing v to $-v$. The results are

$$u'_x = \frac{u_x - v}{1 - (v/c^2)u_x}$$

$$u'_y = \frac{u_y}{\gamma[1 - (v/c^2)u_x]}$$

$$u'_z = \frac{u_z}{\gamma[1 - (v/c^2)u_x]}$$

(2.24)

Note that we found the velocity transformation equations for the situation corresponding to the inverse Lorentz transformation, Equations (2.18), before finding the velocity transformation for Equations (2.17).

What is the correct result for the speed of the protons with respect to the space station? We have $u'_x = 0.99c$ and $v = 0.60c$, so Equation (2.23a) gives us the result

$$u_x = \frac{0.990c + 0.600c}{1 + \dfrac{(0.600c)(0.990c)}{c^2}} = 0.997c$$

where we have assumed we know the speeds to three significant figures. Therefore, the result is a speed only slightly less than c. The Lorentz transformation does not allow a material object to have a speed greater than c. Only massless particles, such as light, can have speed c. If the crew of the spaceship spots the asteroids far enough in advance, their reaction times should allow them to shoot down the uncharacteristically swiftly moving asteroids and save both the spaceship and the space station.

Although no particle with mass can carry energy faster than c, we can imagine a signal being processed faster than c. Consider the following *gedanken* experiment. A giant floodlight placed on a space station above the Earth revolves at 100 Hz, as shown in Figure 2.15. Light spreads out in the radial direction from the floodlight at speeds of c. On the surface of the moon, the light beam sweeps across at speeds far exceeding c (Problem 36). However, the light itself does not reach the moon at speeds faster than c. No energy is associated with the beam of light sweeping across the moon's surface. The energy (and linear momentum) is only along the radial direction from the space station to the moon.

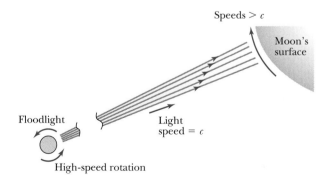

Speeds > c

Moon's surface

Floodlight

Light speed = c

High-speed rotation

Figure 2.15 A floodlight revolving at high speeds can sweep a light beam across the surface of the moon at speeds exceeding c, but the speed of the light still does not exceed c.

Example 2.4

Mary, the commander of the spaceship just discussed, is holding target practice for junior officers by shooting protons at small asteroids and space debris off to the side (perpendicular to the direction of spaceship motion) as the spaceship passes by. What speed will an observer in the space station measure for these protons?

Strategy We use the coordinate systems and speeds of the spaceship and proton gun as described previously. Let the direction of the protons now be perpendicular to the direction of the spaceship—along the y' direction. We already know in the spaceship's K′ system that $u'_y = 0.99c$ and $u'_x = u'_z = 0$, and that the speed of the K′ system (spaceship) with respect to the space station is $v = 0.60c$. We use Equations (2.23) to determine u_x, u_y, and u_z and finally the speed u.

Solution In order to find the speeds in the system K, we first need to find γ.

$$\gamma = \frac{1}{\sqrt{1 - v^2/c^2}} = \frac{1}{\sqrt{1 - 0.600^2}} = 1.25$$

Next we are able to determine the components of $\vec{u}$.

$$u_x(\text{protons}) = \frac{0 + 0.600c}{[1 + (0.600c)(0c)/c^2]} = 0.600c$$

$$u_y(\text{protons}) = \frac{0.990c}{1.25[1 + (0.600c)(0c)/c^2]} = 0.792c$$

$$u_z(\text{protons}) = \frac{0}{1.25[1 + (0.600c)(0c)/c^2]} = 0$$

$$u(\text{protons}) = \sqrt{u_x^2 + u_y^2 + u_z^2} = \sqrt{(0.600c)^2 + (0.792c)^2}$$

$$= 0.994c$$

We have again assumed we know the velocity components to three significant figures. Mary and her junior officers only observe the protons moving perpendicular to their motion. But because there are both u_x and u_y components, Frank (on the space station) sees the protons moving at an angle with respect to both his x and his y directions.

Example 2.5

By the early 1800s, experiments had shown that light slows down when passing through liquids. A. J. Fresnel suggested in 1818 that there would be a partial drag on light by the medium through which the light was passing. Fresnel's suggestion explained the problem of stellar aberration if the Earth was at rest in the ether. In a famous experiment in 1851, H. L. Fizeau measured the "ether" drag coefficient for light passing in opposite directions through flowing water. Let a moving system K′ be at rest in the flowing water and let v be the speed of the flowing water with respect to a fixed observer in K (see Figure 2.16). The speed of light in the water at rest (that is, in system K′) is u', and the speed of light as measured in K is u. If the index of refraction of the water is n, Fizeau found experimentally that

$$u = u' + \left(1 - \frac{1}{n^2}\right)v \qquad (2.25)$$

which was in agreement with Fresnel's prediction. This result was considered an affirmation of the ether concept. The factor $1 - 1/n^2$ became known as *Fresnel's drag coefficient*. Show that this result can be explained using relativistic velocity addition *without the ether concept*.

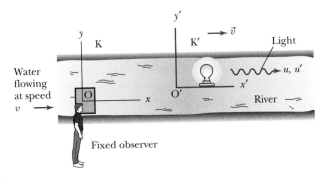

Figure 2.16 A stationary system K is fixed on shore, and a moving system K′ floats down the river at speed v. Light emanating from a source under water in system K′ has speed u, u' in systems K, K′, respectively.

Strategy We note from introductory physics that the velocity of light in a medium of index of refraction n gives $u' = c/n$. We use Equation (2.23a) to solve for u.

Solution We have to calculate the speed only in the x-direction, so we dispense with the subscripts. We utilize Equation (2.23a) to determine

$$u = \frac{u' + v}{1 + u'v/c^2} = \frac{c/n + v}{1 + v/nc} = \frac{c}{n}\frac{\left(1 + \dfrac{nv}{c}\right)}{\left(1 + \dfrac{v}{nc}\right)}$$

Because $v \ll c$ in this case, we can expand the denominator $(1 + x)^{-1} = 1 - x + \cdots$ keeping only the lowest term in $x = v/c$. The above equation becomes

$$u = \frac{c}{n}\left(1 + \frac{nv}{c}\right)\left(1 - \frac{v}{nc} + \cdots\right)$$

$$= \frac{c}{n}\left(1 + \frac{nv}{c} - \frac{v}{nc} + \cdots\right)$$

$$= \frac{c}{n} + v - \frac{v}{n^2} = u' + \left(1 - \frac{1}{n^2}\right)v$$

which is in agreement with Fizeau's experimental result and Fresnel's prediction, Equation (2.25). This relativistic calculation is another stunning success of the special theory of relativity. There is no need to consider the existence of the ether.

2.7 Experimental Verification

We have used the special theory of relativity to describe some unusual phenomena. The special theory has also made some startling predictions concerning length contraction, time dilation, and velocity addition. In this section we will discuss only a few of the many experiments that have been done to confirm the special theory of relativity.

Muon Decay

When high-energy particles called *cosmic rays* enter the Earth's atmosphere from outer space, they interact with particles in the upper atmosphere (see Figure 2.17), creating additional particles in a *cosmic shower*. Many of the particles in the shower are π-*mesons* (pions), which decay into other unstable particles called *muons*. The properties of muons are described later when we discuss nuclear and particle physics. Because muons are unstable, they decay according to the radioactive decay law

$$N = N_0 \exp\left(-\frac{\ln(2)t}{t_{1/2}}\right) = N_0 \exp\left(-\frac{0.693t}{t_{1/2}}\right)$$

Radioactive decay law

where N_0 and N are the number of muons at times $t = 0$ and $t = t$, respectively, and $t_{1/2}$ is the half-life of the muons. This means that in the time period $t_{1/2}$ half

Figure 2.17 Much of what we know about muons in cosmic rays was learned from balloon flights carrying sophisticated detectors. This balloon is being prepared for launch in NASA's Ultra Long Duration Balloon program for a mission that may last up to 100 days. The payload will hang many feet below the balloon. Victor Hess began the first such balloon flights in 1912 (when he discovered cosmic rays), and much improved versions are still launched today from all over the world to study cosmic rays, the atmosphere, the sun, and the universe. *Photo courtesy of NASA.*

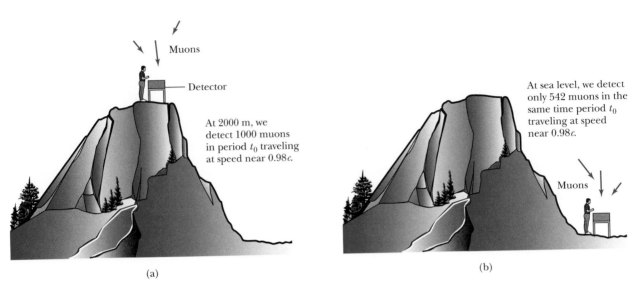

Muons

Detector

At 2000 m, we detect 1000 muons in period t_0 traveling at speed near $0.98c$.

At sea level, we detect only 542 muons in the same time period t_0 traveling at speed near $0.98c$.

Muons

(a)

(b)

Figure 2.18 The number of muons detected with speeds near $0.98c$ is much different (a) on top of a mountain than (b) at sea level, because of the muon's decay. The experimental result agrees with our time dilation equation.

of the muons will decay to other particles. The half-life of muons (1.52×10^{-6} s) is long enough that many of them survive the trip through the atmosphere to the Earth's surface.

We perform an experiment by placing a muon detector on top of a mountain 2000 m high and counting the number of muons traveling at a speed near $v = 0.98c$ (see Figure 2.18a). Suppose we count 10^3 muons during a given time period t_0. We then move our muon detector to sea level (see Figure 2.18b), and we determine experimentally that approximately 540 muons survive the trip without decaying. We ignore any other interactions that may remove muons.

Classically, muons traveling at a speed of $0.98c$ cover the 2000-m path in 6.8×10^{-6} s, and according to the radioactive decay law, only 45 muons should survive the trip. There is obviously something wrong with the classical calculation, because we counted a factor of 12 more muons surviving than the classical calculation predicts.

Because the classical calculation does not agree with the experimental result, we should consider a relativistic calculation. The muons are moving at a speed of $0.98c$ with respect to us on Earth, so the effects of time dilation will be dramatic. In the muon rest frame, the time period for the muons to travel 2000 m (on a clock fixed with respect to the mountain) is calculated from Equation (2.19) to be $(6.8/5.0) \times 10^{-6}$ s, because $\gamma = 5.0$ for $v = 0.98c$. For the time $t = 1.36 \times 10^{-6}$ s, the radioactive decay law predicts that 538 muons will survive the trip, in agreement with the observations. An experiment similar to this was performed by B. Rossi and D. B. Hall* in 1941 on the top of Mount Washington in New Hampshire.

*B. Rossi and D. B. Hall, *Physical Review* **50**, 223 (1941). An excellent, though now dated, film recreating this experiment (*Time Dilation—An Experiment with μ-mesons* by D. H. Frisch and J. H. Smith) is available from the Education Development Center, Newton, Mass. See also D. H. Frisch and J. H. Smith, *American Journal of Physics* **31**, 342 (1963).

It is useful to examine the muon decay problem from the perspective of an observer traveling with the muon. This observer would not measure the distance from the top of the 2000-m mountain to sea level to be 2000 m. Rather, this observer would say that the distance is contracted and is only (2000 m)/5.0 = 400 m. The time to travel the 400-m distance would be $(400 \text{ m})/0.98c = 1.36 \times 10^{-6}$ s according to a clock at rest with a muon. Using the radioactive decay law, an observer traveling with the muons would still predict 538 muons to survive. Therefore, we obtain the identical result whether we consider time dilation or space contraction, and both are in agreement with experiment, thus confirming the special theory of relativity.

Atomic Clock Measurement

An extremely accurate measurement of time can be made using a well-defined transition in the ^{133}Cs atom ($f = 9{,}192{,}631{,}770$ Hz). In 1971 two American physicists, J. C. Hafele and Richard E. Keating (Figure 2.19), used four cesium beam atomic clocks to test the time dilation effect. They flew the four portable cesium clocks eastward and westward on regularly scheduled commercial jet airplanes around the world and compared the time with a reference atomic time scale at rest at the U.S. Naval Observatory in Washington, D.C. (Figure 2.20).

The trip eastward took 65.4 hours with 41.2 flight hours, whereas the westward trip, taken a week later, took 80.3 hours with 48.6 flight hours. The comparison with the special theory of relativity is complicated by the rotation of the Earth and by a gravitational effect arising from the general theory of relativity.

Figure 2.19 Joseph Hafele and Richard Keating are shown unloading one of their atomic clocks and the associated electronics from an airplane in Tel Aviv, Israel, during a stopover in November 1971 on their round-the-world trip to test special relativity. *AP/Wide World Photos.*

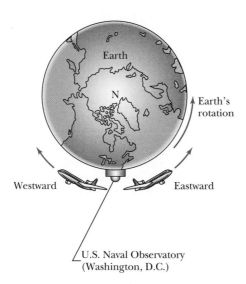

Figure 2.20 Two airplanes took off (at different times) from Washington, D.C., where the U.S. Naval Observatory is located. The airplanes traveled east and west around Earth as it rotated. Atomic clocks on the airplanes were compared with similar clocks kept at the observatory to show that the moving clocks in the airplanes ran slower.

The actual relativistic predictions and experimental observations for the time differences* are

Travel	Predicted	Observed
Eastward	-40 ± 23 ns	-59 ± 10 ns
Westward	275 ± 21 ns	273 ± 7 ns

A negative time indicates that the time on the moving clock is less than the reference clock. The moving clocks lost time (ran slower) during the eastward trip, but gained time (ran faster) during the westward trip. This occurs because of the rotation of the Earth, indicating that the flying clocks ticked faster or slower than the reference clocks on Earth. The special theory of relativity is verified within the experimental uncertainties.

*See J. C. Hafele and R. E. Keating, *Science* **177**, 166–170 (1972).

Example 2.6

In 1985 the space shuttle *Challenger* flew a cesium clock and compared its time with a fixed clock left on Earth. The shuttle orbited at approximately 330 km above Earth with a speed of 7712 m/s (~17,250 mph). (a) Calculate the expected time lost per second for the moving clock and compare with the measured result of -295.02 ± 0.29 ps/s, which includes a predicted effect due to general relativity of 35.0 ± 0.06 ps/s. (b) How much time would the clock lose due to special relativity alone during the entire shuttle flight that lasted for 7 days?

Strategy This should be a straightforward application of the time dilation effect, but we have the complicating fact that the space shuttle is moving in a non-inertial system (orbiting around Earth). We don't want to consider this now, so we make the simplifying assumption that the space shuttle travels in a straight line with respect to Earth and the two events in the calculations are the shuttle passing the starting point (launch) and the ending point (landing). We are not considering the effects of general relativity.

We know the orbital speed of the shuttle with respect to Earth, which allows us to determine β and the relativistic factor γ. We let T be the time measured by the clock fixed on Earth. Then we can use the time dilation effect given by Equation (2.19) to determine the proper time T'_0 measured by the clock in the space shuttle. The time difference is $\Delta T = T - T'_0$. We have $T'_0 = T\sqrt{1 - \beta^2}$ and $\Delta T = T - T'_0 = T(1 - \sqrt{1 - \beta^2})$. For part (b) we need to find the total time lost for the moving clock for 7 days.

Solution (a) We have $\beta = v/c = (7712 \text{ m/s})/(2.998 \times 10^8 \text{ m/s}) = 2.572 \times 10^{-5}$. Because β is such a small quantity, we can use a power series expansion of the square root $\sqrt{1 - \beta^2}$, keeping only the lowest term in β^2 for ΔT.

$$\Delta T = T\left[1 - \left(1 - \frac{\beta^2}{2} + \cdots\right)\right] = \frac{\beta^2 T}{2}$$

Now we have

$$\frac{\Delta T}{T} = \frac{\beta^2}{2} = \frac{1}{2}(2.572 \times 10^{-5})^2 = 330.76 \times 10^{-12}$$

In this case ΔT is positive, which indicates that the space shuttle clock lost this fraction of time, so the moving clock lost 330.76 ps for each second of motion. How does this compare with the measured time? The total measured result was a loss of 295.02 ± 0.29 ps/s, but we must add the general relativity prediction of 35.0 ± 0.06 ps/s to the measured value to obtain the result due only to special relativity. So the measured special relativity result is close to 330.02 ps/s, which differs from our calculated result by only 0.2%!

(b) The total time of the seven-day mission was 6.05×10^5 s, so the total time difference between clocks is $(330.76 \times 10^{-12})(6.05 \times 10^5 \text{ s}) = 0.2$ ms, which is easily detected by cesium clocks.

Velocity Addition

An interesting test of the velocity addition relations was made by T. Alväger and colleagues* at the CERN nuclear and particle physics research facility on the border of Switzerland and France. They used a beam of almost 20-GeV (20×10^9 eV) protons to strike a target to produce neutral pions (π^0) having energies of more than 6 GeV. The π^0 ($\beta \approx 0.99975$) have a very short half-life and soon decay into two γ rays. In the rest frame of the π^0 the two γ rays go off in opposite directions. The experimenters measured the velocity of the γ rays going in the forward direction in the laboratory (actually 6°, but we will assume 0° for purposes of calculation because there is little difference). The Galilean addition of velocities would require the velocity of the γ rays to be $u = 0.99975c + c = 1.99975c$, because the velocity of γ rays is already c. However, the relativistic velocity addition, in which $v = 0.99975c$ is the velocity of the π^0 rest frame with respect to the laboratory and $u' = c$ is the velocity of the γ rays in the rest frame of the π^0, predicts the velocity u of the γ rays measured in the laboratory to be, according to Equation (2.23a),

Pion decay experiment

$$u = \frac{c + 0.99975c}{1 + \dfrac{(0.99975c)(c)}{c^2}} = c$$

The experimental measurement was accomplished by measuring the time taken for the γ rays to travel between two detectors placed about 30 m apart and was in excellent agreement with the relativistic prediction, but not the Galilean one. We again have conclusive evidence of the need for the special theory of relativity.

Testing Lorentz Symmetry

Although we have mentioned only three rather interesting experiments, physicists performing experiments with nuclear and particle accelerators have examined thousands of cases that verify the correctness of the concepts discussed here. It may seem that a violation of the Lorentz transformation is far fetched, but we should keep an open mind. Quantum electrodynamics (QED) includes special relativity in its framework, and QED has been tested to one part in 10^{12}.

Lorentz symmetry requires the laws of physics to be the same for all observers, and Lorentz symmetry is important at the very foundation of our description of fundamental particles and forces. Lorentz symmetry, together with the principles of quantum mechanics that are discussed in much of the remainder of this book, form the framework of relativistic quantum field theory. Many interactions that could be added to our best theories of physics (see the Standard Model in Chapter 14) are excluded, because they would violate Lorentz symmetry. In the last decade, physicists have conceived and performed many experiments that test Lorentz symmetry. These tests have been done for several particles, but no violations have been discovered to date. For example, tests done with electrons have shown no violations to one part in 10^{32}, with neutrons one part in 10^{31}, and with protons one part in 10^{27}. These are phenomenal numbers, but many more experiments are presently underway, and more are planned. There is considerable interest in Lorentz symmetry.[†]

*See T. Alväger, F. J. M. Farley, J. Kjellman, and I. Wallin, *Physics Letters* **12**, 260 (1964). See also article by J. M. Bailey, *Arkiv Fysik* **31**, 145 (1966).

[†]See "Lorentz Invariance on Trial," Maxim Pospelov and Michael Romalis, *Physics Today* (July 2004) p. 40. See also *Scientific American* (September 2004) Special Issue on "Beyond Einstein."

2.8   Twin Paradox

One of the most interesting topics in relativity is the twin (or clock) paradox. Almost from the time of publication of Einstein's famous paper in 1905, this subject has received considerable attention, and many variations exist. Let's summarize the paradox. Suppose twins, Mary and Frank, choose different career paths. Mary (the **M**oving twin) becomes an astronaut and Frank (the **F**ixed twin) a stockbroker. At age 30, Mary sets out on a spaceship to study a star system 8 ly from Earth. Mary travels at very high speeds to reach the star and returns during her life span. According to Frank's understanding of special relativity, Mary's biological clock ticks more slowly than his own, so he claims that Mary will return from her trip younger than he. The paradox is that Mary similarly claims that it is Frank who is moving rapidly with respect to her, so that when she returns, Frank will be the younger. To further complicate the paradox, one could argue that because nature cannot allow both possibilities, it must be true that symmetry prevails and that the twins will still be the same age. Which is the correct solution?

Who is the younger twin?

The correct answer is that Mary returns from her space journey as the younger twin. According to Frank, Mary's spaceship takes off from Earth and quickly reaches its travel speed of $0.8c$. She travels the distance of 8 ly to the star system, slows down and turns around quickly, and returns to Earth at the same speed. The accelerations (positive and negative) take negligible times compared to the travel times between Earth and the star system. According to Frank, Mary's travel time to the star is 10 years $[(8\ \mathrm{ly})/0.8c = 10\ \mathrm{y}]$ and the return is also 10 years, for a total travel time of 20 years, so that Frank will be $30 + 10 + 10\ \mathrm{y} = 50$ years old when Mary returns. However, since Mary's clock is ticking more slowly, her travel time to the star is only $10\sqrt{1 - 0.8^2}\ \mathrm{y} = 6$ years. Frank calculates that Mary will only be $30 + 6 + 6\ \mathrm{y} = 42$ years old when she returns with respect to his own clock at rest.

The important fact here is that Frank's clock is in an inertial system* during the entire trip; however, Mary's clock is not. As long as Mary is traveling at constant speed away from Frank, both of them can argue that the other twin is aging less rapidly. However, when Mary slows down to turn around, she leaves her original inertial system and eventually returns in a completely different inertial system. Mary's claim is no longer valid, because she does not remain in the same inertial system. There is also no doubt as to who is in the inertial system. Frank feels no acceleration during Mary's entire trip, but Mary will definitely feel acceleration during her reversal time, just as we do when we step hard on the brakes of a car. The acceleration at the beginning and the deceleration at the end of her trip present little problem, because the fixed and moving clocks could be compared if Mary were just passing by Frank each way. It is Mary's acceleration at the star system that is the key. If we invoke the two postulates of special relativity, there is no paradox. The instantaneous rate of Mary's clock is determined by her instantaneous speed, but she must account for the acceleration effect when she turns around. A careful analysis of Mary's entire trip using special relativity, including acceleration, will be in agreement with Frank's assessment that Mary is younger.

Mary is both younger and rich

Mary returns to Earth rich as well as famous, because her stockbroker brother has invested her salary wisely during the 20-year period (for which she only worked 12 years!).

We follow A. P. French's excellent book, *Special Relativity*, to present Table 2.1, which analyzes the twin paradox. Both Mary and Frank send out signals at a

---

*The rotating and orbiting Earth is only an approximate inertial system.

Item	Measured by Frank (remains on Earth)	Measured by Mary (traveling astronaut)
Table 2.1 Twin Paradox Analysis		
Time of total trip	$T = 2L/v$	$T' = 2L/\gamma v$
Total number of signals sent	$fT = 2fL/v$	$fT' = 2fL/\gamma v$
Frequency of signals received at beginning of trip f'	$f\sqrt{\dfrac{1-\beta}{1+\beta}}$	$f\sqrt{\dfrac{1-\beta}{1+\beta}}$
Time of detecting Mary's turnaround	$t_1 = L/v + L/c$	$t_1' = L/\gamma v$
Number of signals received at the rate f'	$f't_1 = \dfrac{fL}{v}\sqrt{1-\beta^2}$	$f't_1' = \dfrac{fL}{v}(1-\beta)$
Time for remainder of trip	$t_2 = L/v - L/c$	$t_2' = L/\gamma v$
Frequency of signals received at end of trip f''	$f\sqrt{\dfrac{1+\beta}{1-\beta}}$	$f\sqrt{\dfrac{1+\beta}{1-\beta}}$
Number of signals received at rate f''	$f''t_2 = \dfrac{fL}{v}\sqrt{1-\beta^2}$	$f''t_2' = \dfrac{fL}{v}(1+\beta)$
Total number of signals received	$2fL/\gamma v$	$2fL/v$
Conclusion as to other twin's measure of time taken	$T' = 2L/\gamma v$	$T = 2L/v$

After A. French, *Special Relativity*, New York: Norton (1968), p. 158.

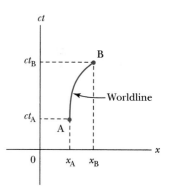

Figure 2.21 A spacetime diagram is used to specify events. The worldline denoting the path from event A to event B is shown.

frequency f (as measured by their own clock). We include in the table the various journey timemarks and signals received during the trip, with one column for the twin Frank who stayed at home and one for the astronaut twin Mary who went on the trip. Let the total time of the trip as measured on Earth be T. The speed of Mary's spaceship is v (as measured on Earth), which gives a relativistic factor γ. The distance Mary's spaceship goes before turning around (as measured on Earth) is L. Much of this table is best analyzed by using spacetime (see the next section) and the Doppler effect (see Section 2.10).

2.9   Spacetime

When describing events in relativity, it is convenient sometimes to represent events on a **spacetime** diagram like that shown in Figure 2.21. For convenience we use only one spatial coordinate x and specify position in this one dimension. We use ct instead of time so that both coordinates will have dimensions of length. Spacetime diagrams were first used by H. Minkowski in 1908 and are often called **Minkowski diagrams**. We have learned in relativity that we must denote both space and time in order to specify an event. This is the origin of the term *fourth dimension* for time. The events for A and B in Figure 2.21 are denoted by the respective coordinates (x_A, ct_A) and (x_B, ct_B), respectively. The line connecting events A and B is the path from A to B and is called a **worldline**. A spaceship launched from $x = 0$, $ct = 0$ with velocity v will have a particular worldline as shown in

Spacetime (Minkowski) diagrams

Worldline

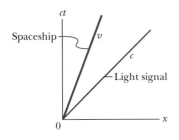

Figure 2.22 A light signal has the slope of 45° on a spacetime diagram. A spaceship moving along the x axis with speed v is a straight line on the spacetime diagram with a slope c/v.

Figure 2.22. It will be a straight line in the spacetime diagram with a slope c/v. For example, a light signal sent out from the origin with speed c is represented on a spacetime graph with a worldline that has a slope $c/c = 1$, so that line makes an angle of 45° with both the x and ct axes. Any real motion in the spacetime diagram cannot have a slope of less than 1 (angle $< 45°$), because that motion would have a speed greater than c. The Lorentz transformation does not allow such a speed.

Let us consider two events that occur at the same time ($ct = 0$), but at different positions, x_1 and x_2. We denote the events (x, ct) as $(x_1, 0)$ and $(x_2, 0)$, and we show them in Figure 2.23 in an inertial system with an origin fixed at $x = 0$ and $ct = 0$. How can we be certain that the two events happen simultaneously if they occur at different positions? We must devise a method that will allow us to determine experimentally that the events occurred simultaneously. Let us place clocks at positions x_1 and x_2 and place a flashbulb at position x_3 halfway between x_1 and x_2. The two clocks have been previously synchronized and keep identical time. At time $t = 0$, the flashbulb explodes and sends out light signals from position x_3. The light signals proceed along their worldlines as shown in Figure 2.23. The two light signals arrive at positions x_1 and x_2 at identical times t as shown on the spacetime diagram. By using such techniques we can be sure that events occur simultaneously in our inertial reference system.

But what about other inertial reference systems? We realize that the two events will not be simultaneous in a reference system K′ moving at speed v with respect to our (x, ct) system. Because the two events have different spatial coordinates, x_1 and x_2, the Lorentz transformation will preclude them from occurring at the same time t' simultaneously in the moving coordinate systems. We can see this by supposing that events 1, 2, and 3 take place on a spaceship moving with velocity v. The worldlines for x_1 and x_2 are the two slanted lines beginning at x_1 and x_2 in Figure 2.24. However, when the flashbulb goes off, the light signals from x_3 still proceed at 45° in the (x, ct) reference system. The light signals intersect the worldlines from positions x_1 and x_2 at different times, so we do not see the events as being simultaneous in the moving system. Spacetime diagrams can be useful in showing such phenomena.

Anything that happened earlier in time than $t = 0$ is called the *past* and anything that occurs after $t = 0$ is called the *future*. The spacetime diagram in Figure 2.25a shows both the past and the future. Notice that only the events within the shaded area below $t = 0$ can affect the present. Events outside this area

Figure 2.23 Clocks positioned at x_1 and x_2 can be synchronized by sending a light signal from a position x_3 halfway between. The light signals intercept the worldlines of x_1 and x_2 at the same time t.

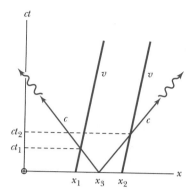

Figure 2.24 If the positions x_1 ($= x'_1$) and x_2 ($= x'_2$) of the previous figure are on a moving system K′ when the flashbulb goes off, the times will not appear simultaneously in system K, because the worldlines for x'_1 and x'_2 are slanted.

cannot affect the present because of the limitation $v \leq c$; this region is called *elsewhere*. Similarly, the present cannot affect any events occurring outside the shaded area above $t = 0$, again because of the limitation of the speed of light. If we add another spatial coordinate y to our spacetime coordinates, we will have a cone as shown in Figure 2.25b, which we refer to as the **light cone**. All causal events related to the present ($x = 0$, $ct = 0$) must be within the light cone. In Figure 2.25b, anything occurring at present ($x = 0$, $ct = 0$) cannot possibly affect an event at position A; however, the event B can easily affect event A because A would be within the range of light signals emanating from B.

Light cone

Invariant quantities have the same value in all inertial frames. They serve a special role in physics because their values do not change from one system to another. For example, the speed of light c is invariant. We are used to defining distances by $d^2 = x^2 + y^2 + z^2$, and in Euclidean geometry, we obtain the same result for d^2 in any inertial frame of reference. Is there a quantity, similar to d^2, that is also invariant in the special theory? If we refer to Equations (2.9), we have similar equations in both systems K and K′. Let us look more carefully at the quantity s^2 defined as

$$s^2 = x^2 - (ct)^2 \tag{2.26}$$

and also

$$s'^2 = x'^2 - (ct')^2$$

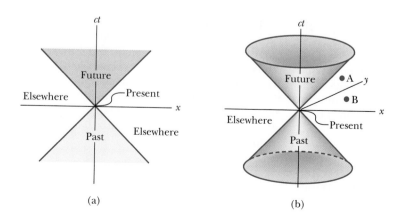

(a)

(b)

Figure 2.25 (a) The spacetime diagram can be used to show the past, present, and future. Only causal events are placed inside the shaded area. Events outside the shaded area below $t = 0$ cannot affect the present. (b) If we add an additional spatial coordinate y, a space cone can be drawn. The present cannot affect event A, but event B can.

Another invariant quantity

If we use the Lorentz transformation for x and t, we find that $s^2 = s'^2$, so the quantity s^2 is an invariant quantity. This relationship can be extended to include the two other spatial coordinates, y and z, so that

$$s^2 = x^2 + y^2 + z^2 - (ct)^2 \qquad (2.27)$$

Because this is similar to the length squared of a position vector in Cartesian coordinates, it is useful to think of time as the fourth dimension of spacetime. We have no need here to use the four-dimensional approach, but it is useful to link momentum and energy together as well as space and time. For simplicity, we will continue to use only the single spatial coordinate x.

If we consider two events, we can determine the quantity Δs^2 where

$$\Delta s^2 = \Delta x^2 - c^2\,\Delta t^2 \qquad (2.28)$$

Spacetime interval

between the two events, and we find that it is invariant in any inertial frame. The quantity Δs is known as the **spacetime interval** between two events. There are three possibilities for the invariant quantity Δs^2.

1. **$\Delta s^2 = 0$:** In this case $\Delta x^2 = c^2\,\Delta t^2$, and the two events can be connected only by a light signal. The events are said to have a **lightlike** separation.

Lightlike

2. **$\Delta s^2 > 0$:** Here we must have $\Delta x^2 > c^2\,\Delta t^2$, and no signal can travel fast enough to connect the two events. The events are not causally connected and are said to have a **spacelike** separation. In this case we can always find an inertial frame traveling at a velocity less than c in which the two events can occur simultaneously in time but at different places in space.

Spacelike

3. **$\Delta s^2 < 0$:** Here we have $\Delta x^2 < c^2\,\Delta t^2$, and the two events can be causally connected. The interval is said to be **timelike**. In this case we can find an inertial frame traveling at a velocity less than c in which the two events occur at the same position in space but at different times. The two events can never occur simultaneously.

Timelike

Example 2.7

Draw the spacetime diagram for the motion of the twins discussed in Section 2.8. Draw light signals being emitted from each twin at annual intervals and count the number of light signals received by each twin from the other.

Strategy We shall let Mary leave Earth at the origin $(x, ct) = (0, 0)$. She will return to Earth at $x = 0$, but at a later time $ct = 20$ ly. Her worldlines will be described by two lines of slope $+c/v$ and $-c/v$, whereas Frank's worldline remains fixed at $x = 0$. Frank's and Mary's signals have slopes of ± 1 on the spacetime diagram. We pay close attention to when the light signals sent out by Frank and Mary reach their twin's worldlines.

Solution We show in Figure 2.26 the spacetime diagram. The line representing Mary's trip has a slope $c/0.8c = 1.25$ on the outbound trip and -1.25 on the return trip. During

the trip to the star system, Mary does not receive the second annual light signal from Frank until she reaches the star system. This occurs because the light signal takes considerable time to catch up with Mary. However, during the return trip Mary receives Frank's light signals at a rapid rate, receiving the last one (number 20) just as she returns. Because Mary's clock is running slow, we see the light signals being sent less often on the spacetime diagram in the fixed system. Mary sends out her sixth annual light signal when she arrives at the star system. However, this signal does not reach Frank until the 18th year! During the remaining two years, however, Frank receives Mary's signals at a rapid rate, finally receiving all 12 of them. Frank receives the last 6 signals during a time period of only 2 years.

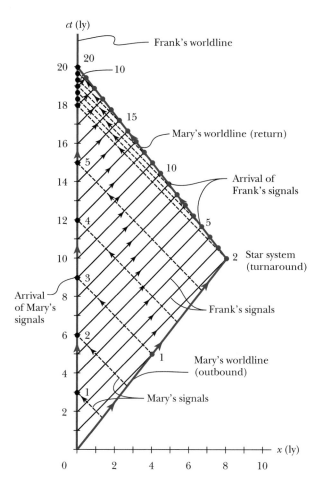

Figure 2.26 The spacetime diagram for Mary's trip to the star system and back. Notice that Frank's worldline is a vertical line at $x = 0$, and Mary's two worldlines do indeed have the correct slope given by the magnitude c/v. The black dashed lines represent light signals sent at annual intervals from Mary to Frank. Frank's annual signals to Mary are solid black. The solid dots denote the time when the light signals arrive.

2.10 Doppler Effect

You probably have already studied the Doppler effect of sound in introductory physics. It is represented by an increased frequency of sound as a source such as a train (with whistle blowing) approaches a receiver (our eardrum), and a decrease in frequency as the source recedes. A change in sound frequency also occurs when the source is fixed and the receiver is moving. The change in frequency of the sound wave depends on whether the source or receiver is moving. On first thought it seems that the Doppler effect in sound violates the principle of relativity, until we realize that there is in fact a special frame for sound waves. Sound waves depend on media such as air, water, or a steel plate in order to propagate. For light, however, there is no such medium. It is only relative motion of the source and receiver that is relevant, and we expect some differences between the relativistic Doppler effect for light waves and the normal Doppler effect for sound. It is not possible for a source of light to travel faster than light in a vacuum, but it is possible for a source of sound to travel faster than the speed of sound. Similarly, in a medium such as water in which light travels slower than c, a light source can travel faster than the speed of light.

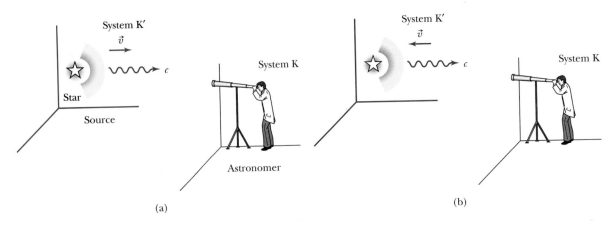

(a)

(b)

Figure 2.27 (a) The source (star) is approaching the receiver (astronomer) with velocity v while it emits starlight signals with speed c. (b) Here the source and receiver are receding with velocity v. The Doppler effect for light is different than that for sound, because of relativity and no medium to carry the light waves.

Consider a source of light (for example, a star) and a receiver (an astronomer) approaching one another with a relative velocity v. First we consider the receiver fixed (Figure 2.27a) in system K and the light source in system K′ moving toward the receiver with velocity v. The source emits n waves during the time interval T. Because the speed of light is always c and the source is moving with velocity v, the total distance between the front and rear of the wave train emitted during the time interval T is

$$\text{Length of wave train} = cT - vT$$

Because there are n waves emitted during this time period, the wavelength must be

$$\lambda = \frac{cT - vT}{n}$$

and the frequency, $f = c/\lambda$, is

$$f = \frac{cn}{cT - vT} \tag{2.29}$$

In its rest frame, the source emits n waves of frequency f_0 during the proper time T_0'.

$$n = f_0 T_0' \tag{2.30}$$

The proper time interval T_0' measured on the clock at rest in the moving system is related to the time interval T measured on a clock fixed by the receiver in system K by

$$T_0' = \frac{T}{\gamma} \tag{2.31}$$

where γ is the relativistic factor of Equation (2.16). The clock moving with the source measures the proper time because it is present with both the beginning and end of the wave.

We substitute the proper time T_0' from Equation (2.31) into Equation (2.30) to determine the number of waves n. Then n is substituted into Equation (2.29) to determine the frequency.

$$f = \frac{cf_0 T/\gamma}{cT - vT}$$

$$= \frac{1}{1 - v/c} \frac{f_0}{\gamma} = \frac{\sqrt{1 - v^2/c^2}}{1 - v/c} f_0$$

where we have inserted the equation for γ. If we use $\beta = v/c$, we can write the previous equation as

$$f = \frac{\sqrt{1 + \beta}}{\sqrt{1 - \beta}} f_0 \qquad \text{Source and receiver approaching} \qquad (2.32)$$

It is straightforward to show that Equation (2.32) is also valid when the source is fixed and the receiver approaches it with velocity v. It is the relative velocity v, of course, that is important (Problem 49).

But what happens if the source and receiver are receding from each other with velocity v (see Figure 2.27b)? The derivation is similar to the one just done, except that the distance between the beginning and end of the wave train becomes

$$\text{Length of wave train} = cT + vT$$

because the source and receiver are receding rather than approaching. This change in sign is propagated throughout the derivation (Problem 50), with the final result

$$f = \frac{\sqrt{1 - \beta}}{\sqrt{1 + \beta}} f_0 \qquad \text{Source and receiver receding} \qquad (2.33)$$

Equations (2.32) and (2.33) can be combined into one equation if we agree to use a $+$ sign for β ($+v/c$) when the source and receiver are approaching each other and a $-$ sign for β ($-v/c$) when they are receding. The final equation becomes

$$f = \frac{\sqrt{1 + \beta}}{\sqrt{1 - \beta}} f_0 \qquad \text{Relativistic Doppler effect} \qquad (2.34)$$

The Doppler effect is useful in many areas of science including astronomy, atomic physics, and nuclear physics. One of its many applications includes an effective radar system for locating airplane position and speed (see Special Topic, "Applications of the Doppler Effect").

Elements absorb and emit characteristic frequencies of light due to the existence of particular atomic levels. We will learn more about this later. Scientists have observed these characteristic frequencies in starlight and have observed shifts in the frequencies. One reason for these shifts is the Doppler effect, and the frequency changes are used to determine the speed of the emitting object with respect to us. This is the source of the **redshifts** of starlight caused by objects   **Redshifts**

Special Topic

APPLICATIONS OF THE DOPPLER EFFECT

The Doppler effect is not just a curious result of relativity. It has many practical applications, three of which are discussed here, and others are mentioned in various places in this text.

Astronomy

Perhaps the best-known application is in astronomy, where the Doppler shifts of known atomic transition frequencies determine the relative velocities of astronomical objects with respect to us. Such measurements continue to be used today to find the distances of such unusual objects as quasars (objects having incredibly large masses that produce tremendous amounts of radiation; see Chapter 16). The Doppler effect has been used to discover other mundane effects in astronomy, for example the rate of rotation of Venus and the fact that Venus rotates in the opposite direction of Earth—the sun rises in the west on Venus. This was determined by observing light reflected from both sides of Venus—on one side it is blueshifted and on the other side it is redshifted, as shown in Figure A. The

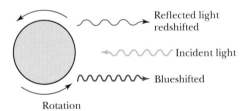

Reflected light redshifted

Incident light

Blueshifted

Rotation

Figure A

same technique has been used to determine the rate of rotation of stars.

Radar

The Doppler effect is nowhere more important than it is in radar. When an electromagnetic radar signal reflects off of a moving target, the so-called *echo* signal will be shifted in frequency by the Doppler effect. Very small frequency shifts can be determined by examining the beat frequency of the echo signal with a reference signal. The frequency shift is proportional to the radial component of the target's velocity. Navigation radar is quite complex, and ingenious techniques have been devised to determine the target position and velocity using multiple radar beams. By using pulsed Doppler radar it is possible to separate moving targets from stationary targets, called clutter.

Doppler radar is also extensively used in meteorology. Vertical motion of airdrafts, sizes and motion of raindrops, motion of thunderstorms, and detailed patterns of wind distribution have all been studied with Doppler radar.

X rays and gamma rays emitted from moving atoms and nuclei will have their frequencies shifted by the Doppler effect. Such phenomena tend to broaden radiation frequencies emitted by stationary atoms and nuclei and add to the natural spectral widths observed.

Laser Cooling

In order to perform fundamental measurements in atomic physics, it is useful to limit the effects of thermal motion and to isolate single atoms. A method taking advantage of the Doppler effect can slow down even neutral atoms and eventually isolate them. Atoms

moving away from us. These data have been used to ascertain that the universe is expanding. The farther away the star, the higher the redshift. This observation is what led Harlow Shapley and Edwin Hubble to the idea that the universe started with a Big Bang.*

*Excellent references are "The Cosmic Distance Scale" by Paul Hodge, *American Scientist* **72**, 474 (1984) and "Origins" by S. Weinberg, *Science* **230**, 15 (1985). This subject is discussed in Chapter 16.

Figure B

emitted from a hot oven will have a spread of velocities around some average value. If these atoms form a beam as shown in Figure B, a laser beam impinging on the atoms from the right can slow them down by transferring momentum.

Atoms have characteristic energy levels that allow them to absorb and emit radiation of specific frequencies. Atoms moving with respect to the laser beam will "see" a shift in the laser frequency because of the Doppler effect. For example, atoms moving toward the laser beam will encounter a laser with high frequency, and atoms moving away from the laser beam will encounter a laser with low frequency. Even atoms moving in the same direction within the beam of atoms will see slightly different frequencies depending on the velocities of the various atoms. Now, if the frequency of the laser beam is tuned to the precise frequency seen by the faster atoms so that those atoms can be excited by absorbing the radiation, then those faster atoms will be slowed down by absorbing the momentum of the laser radiation. The slower atoms will "see" a laser beam that has been Doppler shifted to a lower frequency than is needed to absorb the radiation, and these atoms are not as likely to absorb the laser radiation. The net ef-

fect is that the atoms as a whole are *slowed down* and their *velocity spread is reduced*.

As the atoms slow down, they see that the Doppler-shifted frequencies of the laser change, and the atoms no longer absorb the laser radiation. They continue with the same lower velocity and velocity spread. There are two methods to make the deceleration process continue. In one, the frequency of the laser beam is increased to keep the radiation consistent with the Doppler-shifted frequency needed to excite the atoms. In the other method, the laser frequency is kept constant, but carefully shaped magnetic fields are used to vary the frequencies needed to excite the atoms by changing the excited atomic energy levels. By using six intersecting laser beams coming in at different angles, an "optical molasses" has been created in which the atoms are essentially trapped and the average velocity is zero. In a remarkable series of experiments by various researchers improving on the basic technique, atoms have been cooled to temperatures of 0.5 nK (5×10^{-10} K). The 1997 Nobel Prize in Physics was awarded to Steven Chu, Claude Cohen-Tannoudji, and William Phillips for this technique. Important, fundamental physics measurements have been performed utilizing these methods including making more accurate clocks, manipulating DNA strands, and making ultraprecise atom interferometers. See Steven Chu, "Laser Trapping of Neutral Particles," *Scientific American* **266**, 70 (February 1992). See also in Chapter 9 a discussion of how laser cooling is used to produce an ultracold state of matter known as a Bose-Einstein condensate.

So far in this section we have only considered the source and receiver to be directly approaching or receding. Of course, it is also possible for the two to be moving at an angle with respect to one another, as shown in Figure 2.28. We omit the derivation here* but present the results. The angles θ and θ' are the

*See Robert Resnick, *Introduction to Special Relativity*, New York: Wiley (1968).

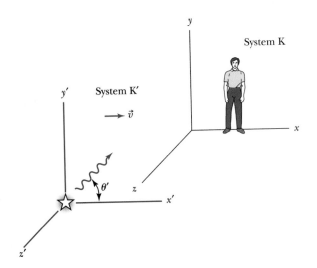

Figure 2.28 The light signals in system K′ are emitted at an angle θ' from the x' axis and remain in the $x'y'$ plane.

angles the light signals make with the x axes in the K and K′ systems. They are related by

$$f\cos\theta = \frac{f_0(\cos\theta' + \beta)}{\sqrt{1 - \beta^2}}$$ (2.35)

and

$$f\sin\theta = f_0 \sin\theta'$$ (2.36)

The generalized Doppler shift equation becomes

$$f = \frac{1 + \beta\cos\theta'}{\sqrt{1 - \beta^2}} f_0$$ (2.37)

Note that Equation (2.37) gives Equation (2.32) when $\theta' = 0$ (source and receiver approaching) and gives Equation (2.33) when $\theta' = 180°$ (source and receiver receding). This situation is known as the *longitudinal Doppler effect*.

When $\theta = 90°$ the emission is purely transverse to the direction of motion, and we have the *transverse Doppler effect*, which is purely a relativistic effect that does not occur classically. The transverse Doppler effect is directly due to time dilation and has been verified experimentally. Equations (2.35) through (2.37) can also be used to understand stellar aberration.

Example 2.8

In Section 2.8 we discussed what happened when Mary traveled on a spaceship away from her twin brother Frank, who remained on Earth. Analyze the light signals sent out by Frank and Mary by using the relativistic Doppler effect.

Strategy We will use Equation (2.34) for both the outbound and return trip to analyze the frequency of the light signals sent and received. During the outbound trip the source (Frank) and receiver (Mary) are receding so that

$\beta = -0.8$. For the return trip, we have $\beta = +0.8$. The frequency f_0 will be the signals that Frank sends; the frequency f will be those that Mary receives.

Solution First, we analyze the frequency of the light signals that Mary receives from Frank. Equation (2.34) gives

$$f = \frac{\sqrt{1 + (-0.8)}}{\sqrt{1 - (-0.8)}} f_0 = \frac{f_0}{3}$$

Because Frank sends out signals annually, Mary will receive the signals only every 3 years. Therefore during the 6-year trip in Mary's system to the star system, she will receive only 2 signals.

During the return trip, $\beta = 0.8$ and Equation (2.34) gives

$$f = \frac{\sqrt{1 + 0.8}}{\sqrt{1 - 0.8}} f_0 = 3f_0$$

so that Mary receives 3 signals each year for a total of 18 signals during the return trip. Mary receives a total of 20 annual light signals from Frank, and she concludes that Frank has aged 20 years during her trip.

Now let's analyze the light signals that Mary sends Frank. During the outbound trip the frequency at which Frank receives signals from Mary will also be $f_0/3$. During the 10 years that it takes Mary to reach the star system on his clock, he will receive 10/3 signals—3 signals plus 1/3 of the time to the next one. Frank continues to receive signals from Mary for another 8 years, because that is how long it takes the sixth signal she sent him to reach Earth. Therefore, for the first 18 years of her journey, according to his own clock he receives 18/3 = 6 signals. Frank has no way of knowing that Mary has turned around and is coming back until he starts receiving signals at frequency $3f_0$. During Mary's return trip Frank will receive signals at the frequency $3f_0$ or 3 per year. However, in his system, Mary returns 2 years after he has received her sixth signal and turned around to come back. During this 2-year period he will receive 6 more signals, so he concludes she has aged a total of only 12 years.

Notice that this analysis is in total agreement with the spacetime diagram of Figure 2.26 and is somewhat easier to obtain. Although geometrical constructions like spacetime diagrams are sometimes useful, an analytical calculation is usually easier.

2.11 Relativistic Momentum

Newton's second law, $\vec{F} = d\vec{p}/dt$, keeps its same form under a Galilean transformation, but we might not expect it to do so under a Lorentz transformation. There may be similar transformation difficulties with the conservation laws of linear momentum and energy. We need to take a careful look at our previous definition of linear momentum to see if it is still valid at high speeds. According to Newton's second law, for example, an acceleration of a particle already moving at very high speeds could lead to a velocity greater than the speed of light. That would be in conflict with the Lorentz transformation, so we expect that Newton's second law might somehow be modified at high speeds.

Because physicists believe the conservation of linear momentum is fundamental, we begin by considering a collision that has no external forces. Frank (**Fixed** or stationary system) is at rest in system K holding a ball of mass m. Mary (**Moving** system) holds a similar ball in system K′ that is moving in the x direction with velocity v with respect to system K as shown in Figure 2.29a. Frank throws his ball along his y axis, and Mary throws her ball with exactly the same speed along her negative y' axis. The two balls collide in a perfectly elastic collision, and each of them catches their own ball as it rebounds. Each twin measures the speed of his or her own ball to be u_0 both before and after the collision.

We show the collision according to both observers in Figure 2.29. Consider the conservation of momentum according to Frank as seen in system K. The velocity of the ball thrown by Frank has components in his own system K of

$$u_{Fx} = 0$$

$$u_{Fy} = u_0 \tag{2.38}$$

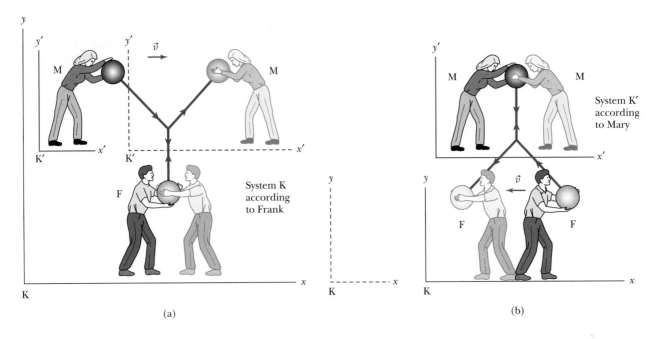

Figure 2.29 Frank is in the fixed K system, and Mary is in the moving K' system. Frank throws his ball along his +y axis, and Mary throws her ball along her −y' axis. The balls collide. The event is shown in Frank's system in (a) and in Mary's system in (b). (Because it is awkward to show the twins as they catch the ball, we have drawn them faintly and in a reversed position.)

If we use the definition of momentum, $\vec{p} = m\vec{v}$, the momentum of the ball thrown by Frank is entirely in the y direction:

$$p_{Fy} = mu_0 \qquad (2.39)$$

Because the collision is perfectly elastic, the ball returns to Frank with speed u_0 along the −y axis. The change of momentum of his ball as observed by Frank in system K is

$$\Delta p_F = \Delta p_{Fy} = -2mu_0 \qquad (2.40)$$

In order to confirm the conservation of linear momentum, we need to determine the change in the momentum of Mary's ball *as measured by Frank*. We will let the primed speeds be measured by Mary and the unprimed speeds be measured by Frank (except that u_0 is always the speed of the ball as measured by the twin in his or her own system). Mary measures the initial velocity of her own ball to be $u'_{Mx} = 0$ and $u'_{My} = -u_0$, because she throws it along her own −y' axis. In order to determine the velocity of Mary's ball as measured by Frank we need to use the velocity transformation equations of Equation (2.23). If we insert the appropriate values for the speeds just discussed, we obtain

$$u_{Mx} = v$$
$$\qquad\qquad\qquad\qquad\qquad (2.41)$$
$$u_{My} = -u_0\sqrt{1 - v^2/c^2}$$

Before the collision, the momentum of Mary's ball as measured by Frank becomes

$$\text{Before} \qquad p_{Mx} = mv$$
$$\text{Before} \qquad p_{My} = -mu_0\sqrt{1 - v^2/c^2} \tag{2.42}$$

For a perfectly elastic collision, the momentum after the collision is

$$\text{After} \qquad p_{Mx} = mv$$
$$\text{After} \qquad p_{My} = +mu_0\sqrt{1 - v^2/c^2} \tag{2.43}$$

The change in momentum of Mary's ball according to Frank is

$$\Delta p_M = \Delta p_{My} = 2mu_0\sqrt{1 - v^2/c^2} \tag{2.44}$$

The conservation of linear momentum requires the total change in momentum of the collision, $\Delta p_F + \Delta p_M$, to be zero. The addition of Equations (2.40) and (2.44) clearly does not give zero. *Linear momentum is not conserved if we use the conventions for momentum from classical physics even if we use the velocity transformation equations from the special theory of relativity.* There is no problem with the x direction, but there is a problem with the y direction along the direction the ball is thrown in each system.

Difficulty with classical linear momentum

Rather than abandon the conservation of linear momentum, let us look for a modification of the definition of linear momentum that preserves both it and Newton's second law. We follow a procedure similar to the one we used in deriving the Lorentz transformation; we assume the simplest, most reasonable change that may preserve the conservation of momentum. We assume that the classical form of momentum $m\vec{u}$ is multiplied by a factor that may depend on velocity. Let the factor be $\Gamma(u)$. Our trial definition for linear momentum now becomes

$$\vec{p} = \Gamma(u)m\vec{u} \tag{2.45}$$

In Example 2.9 we show that momentum is conserved in the collision just described for the value of $\Gamma(u)$ given by

$$\Gamma(u) = \frac{1}{\sqrt{1 - u^2/c^2}} \tag{2.46}$$

Notice that the *form* of Equation (2.46) is the same as that found earlier for the Lorentz transformation. We even give $\Gamma(u)$ the same symbol: $\Gamma(u) = \gamma$. However, this γ is different; it contains the speed of the particle u, whereas the Lorentz transformation contains the relative speed v between the two inertial reference frames. This distinction should be kept in mind because it can cause confusion. Because the usage is so common among physicists, we will use γ for both purposes. However, when there is any chance of confusion, we will write out $1/\sqrt{1 - u^2/c^2}$ and use $\gamma = 1/\sqrt{1 - v^2/c^2}$ for the Lorentz transformation. We will write out $1/\sqrt{1 - u^2/c^2}$ often in order to avoid confusion.

We can make a plausible determination for the correct form of the momentum if we use the proper time discussed previously to determine the velocity. The momentum becomes

$$\vec{p} = m\frac{d\vec{r}}{d\tau} = m\frac{d\vec{r}}{dt}\frac{dt}{d\tau} \tag{2.47}$$

We retain the velocity $\vec{u} = d\vec{r}/dt$ as used classically, where $\vec{r}$ is the position vector. All observers do not agree as to the value of $d\vec{r}/dt$, but they do agree as to the value of $d\vec{r}/d\tau$, where $d\tau$ is the proper time measured in the moving system K'. The value of $dt/d\tau\ (= \gamma)$ is obtained from Equation (2.31), where the speed u is used in the relation for γ to represent the relative speed of the moving (Mary's) frame and the fixed (Frank's) frame.

The definition of the **relativistic momentum** becomes, from Equation (2.47),

Relativistic momentum

$$\vec{p} = m\frac{d\vec{r}}{dt}\gamma$$

$$\vec{p} = \gamma m\vec{u} \qquad \text{Relativistic momentum} \qquad (2.48)$$

where

$$\gamma = \frac{1}{\sqrt{1 - u^2/c^2}} \qquad (2.49)$$

Figure 2.30 The linear momentum of a particle of mass m is plotted versus its velocity (v/c) for both the classical and relativistic momentum results. As $v \to c$ the relativistic momentum becomes quite large, but the classical momentum continues its linear rise. The relativistic result is the correct one.

This result for the relativistic momentum reduces to the classical result for small values of u/c. The classical momentum expression is good to an accuracy of 1% as long as $u < 0.14c$. We show both the relativistic and classical momentum in Figure 2.30.

Some physicists like to refer to the mass in Equation (2.48) as the *rest mass* m_0 and call the term $m = \gamma m_0$ the *relativistic mass*. In this manner the classical form of momentum, $m\vec{u}$, is retained. The mass is then imagined to increase at high speeds. Most physicists prefer to keep the concept of mass as an invariant, intrinsic property of an object. We adopt this latter approach and will use the term *mass* exclusively to mean *rest mass*. Although we may use the terms *mass* and *rest mass* synonymously, we will not use the term *relativistic mass*. The use of relativistic mass too often leads the student into mistakenly inserting the term into classical expressions where it does not apply.

Rest and relativistic mass

Example 2.9

Show that linear momentum is conserved for the collision just discussed and shown in Figure 2.29.

Strategy We use the relativistic momentum to modify the expressions obtained for the momentum of the balls thrown by Frank and Mary. We will then check to see if momentum is conserved according to Frank. We leave to Problem 62 the question of whether momentum is conserved according to Mary's system.

Solution From Equation (2.39), the momentum of the ball thrown by Frank becomes

$$p_{Fy} = \gamma m u_0 = \frac{m u_0}{\sqrt{1 - u_0^2/c^2}}$$

For an elastic collision, the magnitude of the momentum for this ball is the same before and after the collision. After the collision, the momentum will be the negative of this value, so the change in momentum becomes, from Equation (2.40),

$$\Delta p_F = \Delta p_{Fy} = -2\gamma m u_0 = -\frac{2 m u_0}{\sqrt{1 - u_0^2/c^2}} \qquad (2.50)$$

Now we consider the momentum of Mary's ball as measured by Frank. Even with the addition of the γ factor for the momentum in the x direction, we still have $\Delta p_{Mx} = 0$. We must look more carefully at Δp_{My}. First, we find the speed of the ball thrown by Mary as measured by Frank. We use Equations (2.41) to determine

$$u_M = \sqrt{u_{Mx}^2 + u_{My}^2} = \sqrt{v^2 + u_0^2(1 - v^2/c^2)} \qquad (2.51)$$

The relativistic factor γ for the momentum for this situation is

$$\gamma = \frac{1}{\sqrt{1 - u_M^2/c^2}}$$

The value of p_{My} is now found by modifying Equation (2.42) with this value of γ.

$$p_{My} = -\gamma m u_0 \sqrt{1 - v^2/c^2} = \frac{-mu_0\sqrt{1 - v^2/c^2}}{\sqrt{1 - u_M^2/c^2}}$$

We insert the value of u_M from Equation (2.51) into this equation to give

$$p_{My} = \frac{-mu_0\sqrt{1 - v^2/c^2}}{\sqrt{(1 - u_0^2/c^2)(1 - v^2/c^2)}} = \frac{-mu_0}{\sqrt{1 - u_0^2/c^2}} \qquad (2.52)$$

The momentum after the collision will still be the negative of this value, so the change in momentum becomes

$$\Delta p_M = \Delta p_{My} = \frac{2mu_0}{\sqrt{1 - u_0^2/c^2}} \qquad (2.53)$$

The change in the momentum of the two balls as measured by Frank is given by the sum of Equations (2.50) and (2.53):

$$\Delta p = \Delta p_F + \Delta p_M = 0$$

Thus Frank indeed finds that momentum is conserved. Mary should also determine that linear momentum is conserved (see Problem 62).

2.12   Relativistic Energy

We now turn to the concepts of energy and force. When forming the new theories of relativity and quantum physics, physicists resisted changing the well-accepted ideas of classical physics unless absolutely necessary. In this same spirit we also choose to keep intact as many definitions from classical physics as possible and let experiment dictate when we are incorrect. In practice, the concept of force is best defined by its use in Newton's laws of motion, and we retain here the classical definition of force as used in Newton's second law. In the previous section we studied the concept of momentum and found a relativistic expression in Equation (2.48). Therefore, we modify Newton's second law to include our new definition of linear momentum, and force becomes

$$\vec{F} = \frac{d\vec{p}}{dt} = \frac{d}{dt}(\gamma m\vec{u}) = \frac{d}{dt}\left(\frac{m\vec{u}}{\sqrt{1 - u^2/c^2}}\right) \qquad (2.54) \qquad \textbf{Relativistic force}$$

Aspects of this force will be examined in the problems (see Problems 55–58).

Introductory physics presents kinetic energy as the work done on a particle by a net force. We retain here the same definitions of kinetic energy and work. The work W_{12} done by a force $\vec{F}$ to move a particle from position 1 to position 2 along a path $\vec{s}$ is defined to be

$$W_{12} = \int_1^2 \vec{F} \cdot d\vec{s} = K_2 - K_1 \qquad (2.55)$$

where K_1 is defined to be the kinetic energy of the particle at position 1.

For simplicity, let the particle start from rest under the influence of the force $\vec{F}$ and calculate the final kinetic energy K after the work is done. The force is related to the dynamic quantities by Equation (2.54). The work W and kinetic energy K are

$$W = K = \int \frac{d}{dt}(\gamma m\vec{u}) \cdot \vec{u}\, dt \qquad (2.56)$$

where the integral is performed over the differential path $d\vec{s} = \vec{u}\, dt$. Because the mass is invariant, it can be brought outside the integral. The relativistic factor γ depends on u and cannot be brought outside the integral. Equation (2.56) becomes

$$K = m \int dt \, \frac{d}{dt}(\gamma \vec{u}) \cdot \vec{u} = m \int u \, d(\gamma u)$$

The limits of integration are from an initial value of 0 to a final value of γu.

$$K = m \int_0^{\gamma u} u \, d(\gamma u) \tag{2.57}$$

The integral in Equation (2.57) is straightforward if done by the method of integration by parts. The result, called the *relativistic kinetic energy*, is

Relativistic kinetic energy

$$K = \gamma mc^2 - mc^2 = mc^2 \left(\frac{1}{\sqrt{1 - u^2/c^2}} - 1 \right) = mc^2(\gamma - 1) \tag{2.58}$$

Equation (2.58) does not seem to resemble the classical result for kinetic energy, $K = \frac{1}{2}mu^2$. However, if it is correct, we expect it to reduce to the classical result for low speeds. Let's see if it does. For speeds $u \ll c$, we expand γ in a binomial series as follows:

$$K = mc^2 \left(1 - \frac{u^2}{c^2} \right)^{-1/2} - mc^2$$

$$= mc^2 \left(1 + \frac{1}{2}\frac{u^2}{c^2} + \cdots \right) - mc^2$$

where we have neglected all terms of power $(u/c)^4$ and greater, because $u \ll c$. This gives the following equation for the relativistic kinetic energy at low speeds:

$$K = mc^2 + \frac{1}{2}mu^2 - mc^2 = \frac{1}{2}mu^2 \tag{2.59}$$

Figure 2.31 The kinetic energy as a fraction of rest energy (K/mc^2) of a particle of mass m is shown versus its velocity (v/c) for both the classical and relativistic calculations. Only the relativistic result is correct. Like the momentum, the kinetic energy rises rapidly as $v \to c$.

which is the expected classical result. We show both the relativistic and classical kinetic energies in Figure 2.31. They diverge considerably above a velocity of $0.6c$.

A common mistake students make when first studying relativity is to use either $\frac{1}{2}mu^2$ or $\frac{1}{2}\gamma mu^2$ for the relativistic kinetic energy. It is important to *use only Equation (2.58) for the relativistic kinetic energy*. Although Equation (2.58) looks much different than the classical result, it is the only correct one, and *neither $\frac{1}{2}mu^2$ nor $\frac{1}{2}\gamma mu^2$ is a correct relativistic result*.

Equation (2.58) is particularly useful when dealing with particles accelerated to high speeds. For example, the fastest speeds produced in the United States have been in the 3-kilometer-long electron accelerator at the Stanford Linear Accelerator Laboratory. This accelerator produces electrons with a kinetic energy of 8×10^{-9} J (50 GeV) or 50×10^9 eV. The electrons have velocities so close to the speed of light that the tiny difference from c is difficult to measure directly. The speed of the electrons is inferred from the relativistic kinetic energy of Equation (2.58) and is given by $0.99999999995c$. Such calculations are difficult to do with calculators because of significant-figure limitations. As a result, we use kinetic energy or momentum to express the motion of a particle moving near the speed of light and rarely use its speed.

Conceptual Example 2.10

Determine whether an object with mass can ever have the speed of light.

Solution If we examine Equation (2.58), we see that when $u \rightarrow c$, the kinetic energy $K \rightarrow \infty$. Because there is not an infinite amount of energy available, we agree that no object with mass can have the speed of light. The classical and relativistic speeds for electrons are shown in Figure 2.32 as a function of their kinetic energy. Physicists have found that experimentally it does not matter how much energy we give an object having mass. Its speed can never quite reach c.

Figure 2.32 The velocity (v/c) of electrons is shown versus kinetic energy for both classical (incorrect) and relativistic calculations. The experimentally measured data points agree with the relativistic results. *Adapted with permission from data by W. Bertozzi,* American Journal of Physics *32, 551 (1964). © 1964 American Association of Physics Teachers.*

Example 2.11

Electrons in a television set are accelerated by a potential difference of 25,000 volts before striking the screen. Calculate the velocity of the electrons and determine the error in using the classical kinetic energy result.

Strategy We calculate the velocity from the kinetic energy, which we determine both classically and relativistically and then compare the results. In order to determine the correct speed of the electrons, we must use the relativistically correct kinetic energy given by Equation (2.58). The work done to accelerate an electron across a potential difference V is given by qV, where q is the charge of the particle. The work done to accelerate the electron from rest is the final kinetic energy K of the electron.

Solution The kinetic energy is given by

$$K = W = qV = (1.6 \times 10^{-19}\,\text{C})(25 \times 10^3\,\text{V})$$

$$= 4.0 \times 10^{-15}\,\text{J}$$

We first determine γ from Equation (2.58) and from that, the speed. We have

$$K = (\gamma - 1)mc^2 \qquad (2.60)$$

From this equation, γ is found to be

$$\gamma = 1 + \frac{K}{mc^2} \qquad (2.61)$$

The quantity mc^2 for the electron is determined to be

$$mc^2(\text{electron}) = (9.11 \times 10^{-31}\,\text{kg})(3 \times 10^8\,\text{m/s})^2$$

$$= 8.19 \times 10^{-14}\,\text{J}$$

The relativistic factor is then $\gamma = 1 + [(4.0 \times 10^{-15}\,\text{J})/(8.19 \times 10^{-14}\,\text{J})] = 1.049$. Equation (2.8) can be rearranged to determine β^2 as a function of γ^2, where $\beta = u/c$.

$$\beta^2 = \frac{\gamma^2 - 1}{\gamma^2} = \frac{(1.049)^2 - 1}{(1.049)^2} = 0.091 \qquad (2.62)$$

The value of β is 0.30, and the correct velocity, $u = \beta c$, is 0.90×10^8 m/s.

We determine the error in using the classical result by calculating the velocity using the nonrelativistic expression. The nonrelativistic expression is $K = \frac{1}{2}mu^2$, and the velocity is given by

$$u = \sqrt{\frac{2(4.0 \times 10^{-15}\,\mathrm{J})}{9.11 \times 10^{-31}\,\mathrm{kg}}}$$

$$= 0.94 \times 10^8 \text{ m/s} \qquad \text{(nonrelativistic)}$$

The (incorrect) classical speed is about 4% greater than the (correct) relativistic speed. Such an error is significant enough to be important in designing electronic equipment and in making test measurements. Relativistic calculations are particularly important for electrons, because they have such a small mass and are easily accelerated to speeds very close to c.

Total Energy and Rest Energy

We rewrite Equation (2.58) in the form

$$\gamma mc^2 = \frac{mc^2}{\sqrt{1 - u^2/c^2}} = K + mc^2 \tag{2.63}$$

The term mc^2 is called the **rest energy** and is denoted by E_0.

Rest energy
$$E_0 = mc^2 \tag{2.64}$$

This leaves the sum of the kinetic energy and rest energy to be interpreted as the **total energy** of the particle. The total energy is denoted by E and is given by

Total energy
$$E = \gamma mc^2 = \frac{mc^2}{\sqrt{1 - u^2/c^2}} = \frac{E_0}{\sqrt{1 - u^2/c^2}} = K + E_0 \tag{2.65}$$

Equivalence of Mass and Energy

These last few equations suggest the equivalence of mass and energy, a concept attributed to Einstein. The result that energy $= mc^2$ is one of the most famous equations in physics. Even when a particle has no velocity, and thus no kinetic energy, we still believe that the particle has energy through its mass, $E_0 = mc^2$. Nuclear reactions are certain proof that mass and energy are equivalent. The concept of motion as being described by *kinetic energy* is preserved in relativistic dynamics, but a particle with no motion still has energy through its mass.

In order to establish the equivalence of mass and energy, we must modify two of the conservation laws that we learned in classical physics. Mass and energy are no longer two separate conserved quantities. We must combine them into one law of the **conservation of mass-energy**. We will see ample proof during the remain-

Conservation of mass-energy

der of this book of the validity of this basic conservation law.

Even though we often say "energy is turned into mass" or "mass is converted into energy" or "mass and energy are interchangeable," what we mean is that mass and energy are *equivalent*; this is important to understand. Mass is another form of energy, and we use the terms *mass-energy* and *energy* interchangeably. This is not the first time we have had to change our understanding of energy. In the late eighteenth century it became clear that heat was another form of energy, and the

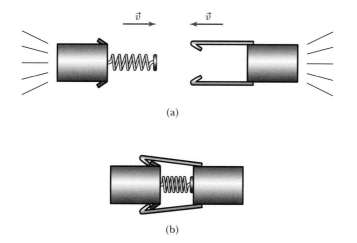

(a)

(b)

Figure 2.33 (a) Two blocks of wood, one with a spring attached and both having mass m, move with equal speeds v and kinetic energies K toward a head-on collision. (b) The two blocks collide, compressing the spring, which locks in place. The system now has increased mass, $M = 2m + 2K/c^2$, with the kinetic energy being converted into the potential energy of the spring.

nineteenth-century experiments of James Joule showed that heat loss or gain was related to work.

Consider two blocks of wood, each of mass m and having kinetic energy K, moving toward each other as shown in Figure 2.33. A spring placed between them is compressed and locks in place as they collide. Let's examine the conservation of mass-energy. The energy before the collision is

$$\text{Mass-energy before:} \quad E = 2mc^2 + 2K \tag{2.66}$$

and the energy after the collision is

$$\text{Mass-energy after:} \quad E = Mc^2 \tag{2.67}$$

where M is the (rest) mass of the system. Because energy is conserved, we have $E = 2mc^2 + 2K = Mc^2$, and the new mass M is greater than the individual masses $2m$. The kinetic energy went into compressing the spring, so the spring has increased potential energy. Kinetic energy has been converted into mass, the result being that the potential energy of the spring has caused the system to have more mass. We find the difference in mass ΔM by setting the previous two equations for energy equal and solving for $\Delta M = M - 2m$.

$$\Delta M = M - 2m = \frac{2K}{c^2} \tag{2.68}$$

Linear momentum is conserved in this head-on collision.

The fractional mass increase in this case is quite small and is given by $f_r = \Delta M/2m$. If we use Equation (2.68), we have

$$f_r = \frac{M - 2m}{2m} = \frac{2K/c^2}{2m} = \frac{K}{mc^2} \tag{2.69}$$

For typical masses and kinetic energies of blocks of wood, this fractional increase in mass is too small to measure. For example, if we have blocks of wood of mass 0.1 kg moving at 10 m/s, Equation (2.69) gives

$$f_r = \frac{\frac{1}{2}mv^2}{mc^2} = \frac{1}{2}\frac{v^2}{c^2} = \frac{1}{2}\frac{(10 \text{ m/s})^2}{(3 \times 10^8 \text{ m/s})^2} = 6 \times 10^{-16}$$

where we have used the nonrelativistic expression for kinetic energy because the speed is so low. This very small numerical result indicates that questions of mass increase are inappropriate for macroscopic objects like blocks of wood and automobiles crashing into one another. Such small increases cannot even be measured, but in the next section, we will look at the collision of two high-energy protons, in which considerable energy is available to create additional mass. Mass-energy relations are essential in such reactions.

Relationship of Energy and Momentum

Physicists believe that linear momentum is a more fundamental concept than kinetic energy. There is no conservation of kinetic energy, whereas the conservation of linear momentum in isolated systems is inviolate as far as we know. A more fundamental result for the total energy in Equation (2.65) might include momentum rather than kinetic energy. Let's proceed to find a useful result. We begin with Equation (2.48) for the relativistic momentum written in magnitude form only.

$$p = \gamma mu = \frac{mu}{\sqrt{1 - u^2/c^2}}$$

We square this result, multiply by c^2, and rearrange the result.

$$p^2 c^2 = \gamma^2 m^2 u^2 c^2$$

$$= \gamma^2 m^2 c^4 \left(\frac{u^2}{c^2}\right) = \gamma^2 m^2 c^4 \beta^2$$

We use Equation (2.62) for β^2 and find

$$p^2 c^2 = \gamma^2 m^2 c^4 \left(1 - \frac{1}{\gamma^2}\right)$$

$$= \gamma^2 m^2 c^4 - m^2 c^4$$

The first term on the right-hand side is just E^2, and the second term is E_0^2. The last equation becomes

$$p^2 c^2 = E^2 - E_0^2$$

We rearrange this last equation to find the result we are seeking, a relation between energy and momentum.

Momentum-energy relation

$$E^2 = p^2 c^2 + E_0^2 \tag{2.70}$$

or

$$E^2 = p^2 c^2 + m^2 c^4 \tag{2.71}$$

Equation (2.70) is a useful result to relate the total energy of a particle with its momentum. The quantities $(E^2 - p^2 c^2)$ and m are invariant quantities. Note that when a particle's velocity is zero and it has no momentum, Equation (2.70) correctly gives E_0 as the particle's total energy.

Massless Particles

Equation (2.70) can also be used to determine the total energy for particles having no mass. For example, Equation (2.70) predicts that the total energy of a photon is

$$E = pc \qquad \text{Photon} \qquad (2.72)$$

The energy of a photon is completely due to its motion and not at all to its rest energy.

We can show that the previous relativistic equations correctly predict that the speed of a photon must be the speed of light c. We use Equations (2.65) and (2.72) for the total energy of a photon and set the two equations equal.

$$E = \gamma mc^2 = pc$$

If we insert the value of the relativistic momentum from Equation (2.48), we have

$$\gamma mc^2 = \gamma muc$$

The fact that $u = c$ follows directly from this equation after careful consideration of letting $m \to 0$ and realizing that $\gamma \to \infty$.

$$u = c \qquad \text{Massless particle} \qquad (2.73)$$

Massless particles must travel at the speed of light

Conceptual Example 2.12

Tachyons are postulated particles that travel faster than the speed of light. (The word tachyon is derived from the Greek word *tachus*, which means "speedy.") They were first seriously proposed and investigated in the 1960s. Use what we have learned thus far in this chapter and discuss several properties that tachyons might have.

Solution Let's first examine Equation (2.65) for energy:

$$E = \gamma mc^2 = \frac{mc^2}{\sqrt{1 - u^2/c^2}} \qquad (2.65)$$

Because $u > c$, the energy must be imaginary if the mass is real, or conversely, if we insist that energy be real, we must have an imaginary mass! For purposes of discussion, we will henceforth assume that energy is real and tachyon mass is imaginary. Remember that ordinary matter must always travel at speed less than c, light must travel at the speed of light, and tachyons must always have speed greater than c. In order to slow down a tachyon, we must give it *more* energy, according to Equation (2.65). Note that the energy must be-

come infinite if we want to slow down a tachyon to speed c. If the tachyon's energy is reduced, it speeds up!

Because tachyons travel faster than c, we have a problem with causality. Consider a tachyon leaving Earth at time $t = 0$ that arrives at a distant galaxy at time T. A spaceship traveling at speed less than c from Earth to the galaxy could conceivably find that the tachyon arrived at the galaxy before it left Earth!

It has been proposed that tachyons might be created in high-energy particle collisions or in cosmic rays. No confirming evidence has been found. Tachyons, if charged, could also be detected from *Cerenkov radiation*. When we refer to speed c, we always mean in a vacuum. When traveling in a medium, the speed must be less than c. When particles have speed greater than light travels in a medium, characteristic electromagnetic radiation is emitted. The effect of the blue glow in swimming pool nuclear reactors is due to this Cerenkov radiation. One of this book's authors has designed, constructed, and used a Cerenkov detector in an experiment at the Stanford Linear Accelerator Center (SLAC). A charged tachyon moving through vacuum should emit such radiation, but none has ever been observed.

2.13   Computations in Modern Physics

We were taught in introductory physics that the international system of units is preferable when doing calculations in science and engineering. This is generally true, but in modern physics a somewhat different, more convenient set of units is often used. The smallness of quantities often used in modern physics suggests some practical changes. We introduce those new units in this section and indicate their practicality through several examples. Recall that the work done in accelerating a charge through a potential difference is given by $W = qV$. For a proton, with the charge $e = 1.602 \times 10^{-19}$ C, being accelerated across a potential difference of 1 V, the work done is

$$W = (1.602 \times 10^{-19})(1 \text{ V}) = 1.602 \times 10^{-19} \text{ J}$$

In modern physics calculations, the amount of charge being considered is almost always some multiple of the electron charge. Atoms and nuclei all have an exact multiple of the electron charge (or neutral). For example, some charges are proton $(+e)$, electron $(-e)$, neutron (0), pion $(0, \pm e)$, and a singly ionized carbon atom $(+e)$. The work done to accelerate the proton across a potential difference of 1 V could also be written as

$$W = (1 \text{ } e)(1 \text{ V}) = 1 \text{ eV}$$

Use eV for energy   where e stands for the electron charge. Thus eV, pronounced "electron volt," is also a unit of energy. It is related to the SI (*Système International*) unit joule by the two previous equations.

$$1 \text{ eV} = 1.602 \times 10^{-19} \text{ J} \qquad (2.74)$$

The eV unit is used more often in modern physics than the SI unit J. The term eV is often used with the SI prefixes where applicable. For example, in atomic and solid state physics, eV itself is mostly used, whereas in nuclear physics MeV (10^6 eV, *mega*-electron-volt) and GeV (10^9 eV, *giga*-electron-volt) are predominant, and in particle physics GeV and TeV (10^{12} eV, *tera*-electron-volt) are used. When we speak of a particle having a certain amount of energy, the common usage is to refer to the kinetic energy. A 6-GeV proton has a *kinetic* energy of 6 GeV, not a *total* energy of 6 GeV. Because the rest energy of a proton is about 1 GeV, this proton would have a total energy of about 7 GeV.

Like the SI unit for energy, the SI unit for mass, kilogram, is a very large unit of mass in modern physics calculations. For example, the mass of a proton is only 1.6726×10^{-27} kg. Two other mass units are commonly used in modern physics. First, the rest energy E_0 is given by Equation (2.64) as mc^2. The rest energy of the proton is given by

$$E_0(\text{proton}) = (1.67 \times 10^{-27} \text{ kg})(3 \times 10^8 \text{ m/s})^2 = 1.50 \times 10^{-10} \text{ J}$$

$$= 1.50 \times 10^{-10} \text{ J} \frac{1 \text{ eV}}{1.602 \times 10^{-19} \text{ J}} = 9.38 \times 10^8 \text{ eV}$$

The rest energies of the elementary particles are usually quoted in MeV or GeV. (To five significant figures, the rest energy of the proton is 938.27 MeV.) Because
Use MeV/c^2 for mass $E_0 = mc^2$, the mass is often quoted in units of MeV/c^2; for example, the mass of the proton is given by 938.27 MeV/c^2. We will find that the mass unit of MeV/c^2 is quite useful. The masses of several elementary particles are given on the inside of the front book cover. Although we will not do so, research physicists often quote the mass in units of just eV (or MeV, etc.).

The other commonly used mass unit is the (unified) **atomic mass unit**. It is   **Atomic mass unit** based on the definition that the mass of the neutral carbon-12 (^{12}C) atom is exactly 12 u, where u is one atomic mass unit.* We obtain the conversion between kilogram and atomic mass units u by comparing the mass of one ^{12}C atom.

$$\text{Mass}(^{12}\text{C atom}) = \frac{12 \text{ g/mol}}{6.02 \times 10^{23} \text{ atoms/mol}}$$

$$= 1.99 \times 10^{-23} \text{ g/atom} \tag{2.75}$$

$$\text{Mass}(^{12}\text{C atom}) = 1.99 \times 10^{-26} \text{ kg} = 12 \text{ u/atom}$$

Therefore, the conversion is (when properly done to 6 significant figures)

$$1 \text{ u} = 1.66054 \times 10^{-27} \text{ kg} \tag{2.76}$$

$$1 \text{ u} = 931.494 \text{ MeV}/c^2 \tag{2.77}$$

We have added the conversion from atomic mass units to MeV/c^2 for completeness.

From Equations (2.70) and (2.72) we see that a convenient unit of momentum is energy divided by the speed of light, or eV/c. We will use the unit eV/c for momentum when appropriate. Remember also that we often quote β ($= v/c$) for velocity, so that c itself is an appropriate unit of velocity.

*To avoid confusion between velocity and atomic mass unit, we will henceforth use v for velocity when the possibility exists for confusing the mass unit u with the velocity variable u.

Example 2.13

A 2-GeV proton hits another 2-GeV proton in a head-on collision. (a) Calculate v, β, p, K, and E for each of the initial protons. (b) What happens to the kinetic energy?

Strategy (a) By the convention just discussed, a 2-GeV proton has a kinetic energy of 2 GeV. We use Equation (2.65) to determine the total energy and Equation (2.70) to determine momentum if we know the total energy. In order to determine β and v, it helps to first determine the relativistic factor γ, which we can use Equation (2.65) to find. Then we use Equation (2.62) to find β and v. These are all typical calculations that are performed when doing relativistic computations.

Solution (a) We use $K = 2$ GeV and the proton rest energy, 938 MeV, to find the total energy from Equation (2.65),

$$E = K + E_0 = 2 \text{ GeV} + 938 \text{ MeV} = 2.938 \text{ GeV}$$

The momentum is determined from Equation (2.70).

$$p^2c^2 = E^2 - E_0^2 = (2.938 \text{ GeV})^2 - (0.938 \text{ GeV})^2$$

$$= 7.75 \text{ GeV}^2$$

The momentum is calculated to be

$$p = \sqrt{7.75(\text{GeV}/c)^2} = 2.78 \text{ GeV}/c$$

Notice how naturally the unit of GeV/c arises in our calculation.

In order to find β we first find the relativistic factor γ. There are several ways to determine γ; one is to compare the rest energy with the total energy. From Equation (2.65) we have

$$E = \gamma E_0 = \frac{E_0}{\sqrt{1 - u^2/c^2}}$$

$$\gamma = \frac{E}{E_0} = \frac{2.938 \text{ GeV}}{0.938 \text{ GeV}} = 3.13$$

We use Equation (2.62) to determine β.

$$\beta = \sqrt{\frac{\gamma^2 - 1}{\gamma^2}} = \sqrt{\frac{3.13^2 - 1}{3.13^2}} = 0.948$$

The velocity of a 2-GeV proton is 0.95c or 2.8×10^8 m/s.

(b) When the two protons collide head-on, the situation is similar to the case when the two blocks of wood collided head-on with one important exception. The time for the two protons to interact is less than 10^{-20} s. If the two protons did momentarily stop at rest, then the two-proton system would have its rest mass increased by an amount given by Equation (2.68), $2K/c^2$ or 4 GeV/c^2. The result would be a highly excited system. In fact, the collision between the protons happens very quickly, and there are several possible outcomes. The two protons may either remain or disappear, and new additional particles may be created. Two of the possibilities are

$$p + p \rightarrow p + p + p + \bar{p} \tag{2.78}$$

$$p + p \rightarrow \pi^+ + d \tag{2.79}$$

where the symbols are p (proton), $\bar{p}$ (antiproton), π (pion), and d (deuteron). We will learn more about the possibilities later when we study nuclear and particle physics. Whatever happens must be consistent with the conservation laws of charge, energy, and momentum, as well as with other conservation laws to be learned. Such experiments are routinely done in particle physics. In the analysis of these experiments, the equivalence of mass and energy is taken for granted.

Binding Energy

The equivalence of mass and energy becomes apparent when we study the binding energy of systems like atoms and nuclei that are formed from individual particles. For example, the hydrogen atom is formed from a proton and electron bound together by the electrical (Coulomb) force. A deuteron is a proton and neutron bound together by the nuclear force. The potential energy associated with the force keeping the system together is called the **binding energy** E_B. The binding energy is the work required to pull the particles out of the bound system into separate, free particles at rest. The conservation of energy is written as

$$M_{\text{bound system}} c^2 + E_B = \sum_i m_i c^2 \tag{2.80}$$

where the m_i values are the masses of the free particles. The binding energy is *the difference between the rest energy of the individual particles and the rest energy of the combined, bound system.*

$$E_B = \sum_i m_i c^2 - M_{\text{bound system}} c^2 \tag{2.81}$$

For the case of two final particles having masses m_1 and m_2, we have

$$E_B = (m_1 + m_2 - M_{\text{bound system}})c^2 = \Delta M c^2 \tag{2.82}$$

where ΔM is the difference between the final and initial masses.

When two particles (for example, a proton and neutron) are bound together to form a composite (like a deuteron), part of the rest energy of the individual particles is lost, resulting in the binding energy of the system. The rest energy of the combined system must be reduced by this amount. The deuteron is a good example. The rest energies of the particles are

Proton $\qquad E_0 = 1.007276c^2$ u $= 938.27$ MeV

Neutron $\qquad E_0 = 1.008665c^2$ u $= 939.57$ MeV

Deuteron $\qquad E_0 = 2.01355c^2$ u $= 1875.61$ MeV

The binding energy E_B is determined from Equation (2.81) to be

$$E_B(\text{deuteron}) = 938.27 \text{ MeV} + 939.57 \text{ MeV} - 1875.61 \text{ MeV} = 2.23 \text{ MeV}$$

Conceptual Example 2.14

Why can we ignore the binding energy of the proton and electron when making mass determinations for nuclei, but not the binding energy of a proton and neutron?

Solution The binding energy of the proton and electron in the hydrogen atom is only 13.6 eV, which is so much smaller than the 1-GeV rest energy of a neutron and proton that it can be neglected when making mass determinations. The

deuteron binding energy of 2.23 MeV, however, represents a much larger fraction of the rest energies and is extremely important. The binding energies of heavy nuclei like uranium can be more than 1000 MeV, and even that much energy is not large enough to keep uranium from decaying to lighter nuclei. The Coulomb repulsion between the many protons in heavy nuclei is mostly responsible for their instability. Nuclear stability is addressed in Chapter 12.

Example 2.15

What is the minimum kinetic energy the protons must have in the head-on collision of Equation (2.79), $p + p \rightarrow \pi^+ + d$, in order to produce the positively charged pion and deuteron? The mass of π^+ is 139.6 MeV/c^2.

Strategy For the minimum kinetic energy K required, we need just enough energy to produce the rest energies of the final particles. We let the final kinetic energies of the pion and deuteron be zero. Because the collision is head-on, the momentum will be zero before and after the collision, so the pion and deuteron will truly be at rest with no kinetic energy. We use the conservation of energy to determine the kinetic energy.

Solution Conservation of energy requires

$$m_p c^2 + K + m_p c^2 + K = m_d c^2 + m_{\pi^+} c^2$$

The rest energies of the proton and deuteron were given in this section, so we solve the previous equation for the kinetic energy.

$$K = \tfrac{1}{2}(m_d c^2 + m_{\pi^+} c^2 - 2 m_p c^2)$$

$$= \tfrac{1}{2}[1875.6 \text{ MeV} + 139.6 \text{ MeV} - 2(938.3 \text{ MeV})]$$

$$= 69 \text{ MeV}$$

Nuclear experiments like this are normally done with fixed targets, not head-on collisions, and much more energy than 69 MeV is required, because linear momentum must also be conserved.

Example 2.16

The atomic mass of the ^{4}He atom is 4.002603 u. Find the binding energy of the ^{4}He nucleus.

Strategy This is a straightforward application of Equation (2.81), and we will need to determine the atomic masses.

Solution Equation (2.81) gives

$$E_B(^4\text{He}) = 2 m_p c^2 + 2 m_n c^2 - M_{^4\text{He}} c^2$$

Later we will learn to deal with atomic masses in cases like this, but for now we will subtract the two electron masses from the atomic mass of ^{4}He to obtain the mass of the ^{4}He nucleus.

The mass of the electron is given on the inside of the front cover, along with the masses of the proton and neutron.

$$M_{^4\text{He}}(\text{nucleus}) = 4.002603 \text{ u} - 2(0.000549 \text{ u})$$

$$= 4.001505 \text{ u}$$

We determine the binding energy of the ^{4}He nucleus to be

$$E_B(^4\text{He}) = [2(1.007276 \text{ u}) + 2(1.008665 \text{ u}) - 4.001505 \text{ u}]c^2$$

$$= 0.0304 c^2 \text{ u}$$

$$E_B(^4\text{He}) = (0.0304 c^2 \text{ u})\frac{931.5 \text{ MeV}}{c^2 \text{ u}} = 28.3 \text{ MeV}$$

The binding energy of the ^{4}He nucleus is large, almost 1% of its rest energy.

Example 2.17

The molecular binding energy is called the *dissociation energy*. It is the energy required to separate the atoms in a molecule. The dissociation energy of the NaCl molecule is 4.24 eV. Determine the fractional mass increase of the Na and Cl atoms when they are not bound together in NaCl. What is the mass increase for a mole of NaCl?

Strategy Binding energy is a concept that applies to various kinds of bound objects, including a nucleus, an atom, a molecule, and others. We can use Equation (2.82) in the present case to find ΔM, the change in mass, in terms of the binding energy E_B/c^2. We then divide ΔM by M to find the fractional mass increase.

Solution From Equation (2.82) we have $\Delta M = E_B/c^2$ (the binding energy divided by c^2) as the mass difference between the molecule and separate atoms. The mass of NaCl is 58.44 u. The fractional mass increase is

$$f_r = \frac{\Delta M}{M} = \frac{E_B/c^2}{M} = \frac{4.24 \text{ eV}/c^2}{58.44 \text{ u}} \frac{c^2 \text{ u}}{931 \text{ MeV}} \frac{1 \text{ MeV}}{10^6 \text{ eV}}$$

$$= 7.8 \times 10^{-11}$$

One mole of NaCl has a mass of 58.44 g, so the mass decrease for a mole of NaCl is $f_r \times 58.44$ g or only 4.6×10^{-9} g. Such small masses cannot be directly measured, which is why nonconservation of mass was not observed for chemical reactions—the changes are too small.

Example 2.18

A positively charged sigma particle (symbol Σ^+) produced in a particle physics experiment decays very quickly into a neutron and positively charged pion before either its energy or momentum can be measured. The neutron and pion are observed to move in the same direction as the Σ^+ was originally moving, with momenta of 4702 MeV/c and 169 MeV/c, respectively. What was the kinetic energy of the Σ^+ and its mass?

Strategy The decay reaction is

$$\Sigma^+ \rightarrow n + \pi^+$$

where n is a neutron. Obviously the Σ^+ has more mass than the sum of the masses of n and π^+, or the decay would not occur. We have to conserve both momentum and energy for this reaction. We use Equation (2.70) to find the total energy of the neutron and positively charged pion, but in order to determine the rest energy of Σ^+, we need to know the momentum. We can determine the Σ^+ momentum from the conservation of momentum.

Solution The rest energies of n and π^+ are 940 MeV and 140 MeV, respectively. The total energies of E_n and E_{π^+} are, from $E = \sqrt{p^2 c^2 + E_0^2}$,

$$E_n = \sqrt{(4702 \text{ MeV})^2 + (940 \text{ MeV})^2} = 4795 \text{ MeV}$$

$$E_{\pi^+} = \sqrt{(169 \text{ MeV})^2 + (140 \text{ MeV})^2} = 219 \text{ MeV}$$

The sum of these energies gives the total energy of the reaction, 4795 MeV + 219 MeV = 5014 MeV, both before and after the decay of Σ^+. Because all the momenta are along the same direction, we must have

$$p_{\Sigma^+} = p_n + p_{\pi^+} = 4702 \text{ MeV}/c + 169 \text{ MeV}/c$$

$$= 4871 \text{ MeV}/c$$

This must be the momentum of the Σ^+ before decaying, so now we can find the rest energy of Σ^+ from Equation (2.70).

$$E_0^2(\Sigma^+) = E^- - \pi^- \psi^- = (="+ \div \text{ MeV})^2 - (4871 \text{ MeV})^2$$

$$= (1189 \text{ MeV})^2$$

The rest energy of the Σ^+ is 1189 MeV, and its mass is 1189 MeV/c^2.

We find the kinetic energy of Σ^+ from Equation (2.65).

$$K = E - E_0 = 5014 \text{ MeV} - 1189 \text{ MeV} = 3825 \text{ MeV}$$

2.14　Electromagnetism and Relativity

We have been concerned mostly with the kinematical and dynamical aspects of the special theory of relativity strictly from the mechanics aspects. However, recall that Einstein first approached relativity through electricity and magnetism. He was convinced that Maxwell's equations were invariant (have the same form) in all inertial frames. Einstein wrote in 1952,

> *What led me more or less directly to the special theory of relativity was the conviction that the electromagnetic force acting on a body in motion in a magnetic field was nothing else but an electric field.*

Einstein's conviction about electromagnetism

Einstein was convinced that magnetic fields appeared as electric fields observed in another inertial frame. That conclusion is the key to electromagnetism and relativity.

Maxwell's equations and the Lorentz force law are invariant in different inertial frames. In fact, with the proper Lorentz transformations of the electric and magnetic fields (from relativity theory) together with Coulomb's law (force between stationary charges), Maxwell's equations can be obtained. We certainly do not intend to accomplish that fairly difficult mathematical task here, nor do we intend to obtain the Lorentz transformation of the electric and magnetic fields. These subjects are studied in more advanced physics classes. However, we will show qualitatively that the magnetic force that one observer sees is simply an electric force according to an observer in another inertial frame. The electric field arises from charges, whereas the magnetic field arises from *moving* charges.

Magnetism and electricity are relative

Electricity and magnetism were well understood in the late 1800s. Maxwell predicted that all electromagnetic waves travel at the speed of light, and he combined electricity, magnetism, and optics all into one successful theory. This classical theory has withstood the onslaught of time and experimental tests.* There were, however, some troubling aspects of the theory when it was observed from different Galilean frames of reference. In 1895 H. A. Lorentz "patched up" the difficulties with the Galilean transformation by developing a new transformation that now bears his name, the Lorentz transformation. However, Lorentz did not understand the full implication of what he had done. It was left to Einstein, who in 1905 published a paper titled "On the Electrodynamics of Moving Bodies," to fully merge relativity and electromagnetism. Einstein did not even mention the famous Michelson-Morley experiment in this classic 1905 paper, which we take as the origin of the special theory of relativity, and the Michelson-Morley experiment apparently played little role in his thinking. Einstein's belief that *Maxwell's equations describe electromagnetism in any inertial frame* was the key that led Einstein to the Lorentz transformations. Maxwell's assertion that all electromagnetic waves travel at the speed of light and Einstein's postulate that the speed of light is invariant in all inertial frames seem intimately connected.

We now proceed to discuss qualitatively the relative aspects of electric and magnetic fields and their forces. Consider a positive test charge q_0 moving to the right with speed v outside a neutral, conducting wire as in Figure 2.34a in the frame of the inertial system K, where the positive charges are at rest and the negative electrons in the wire have speed v to the right. The conducting wire is long and has the same number of positive ions and conducting electrons. For

---

*The meshing of electricity and magnetism together with quantum mechanics, called the *theory of quantum electrodynamics* (QED), is one of the most successful theories in physics.

Figure 2.34 (a) A positive charge is placed outside a neutral, conducting wire. The figure is shown in the system where the positive charges in the wire are at rest. Note that the charge q_0 has the same velocity as the electrons. (b) The moving electrons produce a magnetic field, which causes a force $\vec{F}_B$ on q_0. (c) This is similar to (a), but in this system the electrons are at rest. (d) Now there is an abundance of positive charges due to length contraction, and the resulting electric field repels q_0. There is also a magnetic field, but this causes no force on q_0, which is at rest in this system.

simplicity, we have taken the electrons and the positive charges to have the same speed, but the argument can be generalized.

What is the force on the positive test charge q_0 outside the wire? The total force is given by the Lorentz force

$$\vec{F} = q_0(\vec{E} + \vec{v} \times \vec{B}) \qquad (2.83)$$

and can be due to an electric field, a magnetic field, or both. Because the total charge inside the wire is zero, the electric force on the test charge q_0 in Figure 2.34a is also zero. But we learned in introductory physics that the moving electrons in the wire (current) produce a magnetic field $\vec{B}$ at the position of q_0 that is into the page (Figure 2.34b). The moving charge q_0 *will be repelled upward by the magnetic force* ($q_0\vec{v} \times \vec{B}$) due to the magnetic field of the wire.

Let's now see what happens in a different inertial frame K′ moving at speed v to the right with the test charge (see Figure 2.34c). Both the test charge q_0 and the negative charges in the conducting wire are at rest in system K′. In this system an observer at the test charge q_0 observes the same density of negative ions in the wire as before. However, in system K′ the positive ions are now moving to the left with speed v. Due to length contraction, the positive ions will appear to be closer together to a stationary observer in K′. Because the positive charges appear to be closer together, there is a higher density of positive charges than of negative charges in the conducting wire. The result is an electric field as shown in Figure 2.34d. The test charge q_0 will now be *repelled* in the presence of the electric field. What about the magnetic field now? The moving charges in Figure 2.34c also produce a magnetic field that is into the page, but this time the charge q_0 is at rest with respect to the magnetic field, so charge q_0 feels no magnetic force.

What appears as a magnetic force in one inertial frame (Figure 2.34b) appears as an electric force in another (Figure 2.34d). Electric and magnetic fields are *relative* to the coordinate system in which they are observed. The Lorentz contraction of the moving charges accounts for the difference. This example can be extended to two conducting wires with electrons moving, and a similar result will be obtained (see Problem 86). It is this experiment, on the force between two parallel, conducting wires, in which current is defined. Because charge is defined using current, the experiment is also the basis of the definition of the electric charge.

We have come full circle in our discussion of the special theory of relativity. The laws of electromagnetism represented by Maxwell's equations have a special place in physics. The equations themselves are invariant in different inertial systems; only the interpretations as electric and magnetic fields are relative.

Summary

Efforts by Michelson and Morley proved in 1887 that either the elusive ether does not exist or there must be significant problems with our understanding of nature.

Albert Einstein solved the problem in 1905 by applying two postulates:

1. The principle of relativity: The laws of physics are the same in all inertial systems.
2. The constancy of the speed of light: Observers in all inertial systems measure the same value for the speed of light in vacuum.

Einstein's two postulates are used to derive the Lorentz transformation relating the space and time coordinates of events viewed from different inertial systems. If system K′ is moving at speed v along the $+x$ axis with respect to system K, the two sets of coordinates are related by

$$x' = \frac{x - vt}{\sqrt{1 - \beta^2}}$$
$$y' = y$$
$$z' = z \tag{2.17}$$
$$t' = \frac{t - (vx/c^2)}{\sqrt{1 - \beta^2}}$$

The inverse transformation is obtained by switching the primed and unprimed quantities and changing v to $-v$.

The time interval between two events occurring at the same position in a system as measured by a clock at rest is called the proper time T_0. The time interval T' between the same two events measured by a moving observer is related to the proper time T_0 by the time dilation effect.

$$T' = \frac{T_0}{\sqrt{1 - v^2/c^2}} \tag{2.19}$$

We say that moving clocks run slow, because the shortest time is always measured on clocks at rest.

The length of an object measured by an observer at rest relative to the object is called the proper length L_0. The length of the same object measured by an observer who sees the object moving at speed v is L, where

$$L = L_0\sqrt{1 - v^2/c^2} \tag{2.21}$$

This effect is known as length or space contraction, because moving objects are contracted in the direction of their motion.

If u and u' are the velocities of an object measured in systems K and K′, respectively, and v is the relative velocity between K and K′; the relativistic addition of velocities (Lorentz velocity transformation) is

$$u_x = \frac{dx}{dt} = \frac{u'_x + v}{1 + (v/c^2)u'_x}$$
$$u_y = \frac{u'_y}{\gamma[1 + (v/c^2)u'_x]} \tag{2.23}$$
$$u_z = \frac{u'_z}{\gamma[1 + (v/c^2)u'_x]}$$

where

$$\gamma = \frac{1}{\sqrt{1 - v^2/c^2}} \tag{2.8}$$

The Lorentz transformation has been tested for a hundred years, and no violation has yet been detected. Nevertheless, physicists continue to test its validity, because it is one of the most important results in science.

Spacetime diagrams are useful to represent events geometrically. Time may be considered to be a fourth dimension for some purposes. The spacetime interval for two events defined by $\Delta s^2 = \Delta x^2 + \Delta y^2 + \Delta z^2 - c^2\,\Delta t^2$ is an invariant between inertial systems.

The relativistic Doppler effect for light frequency f is given by

$$f = \frac{\sqrt{1+\beta}}{\sqrt{1-\beta}}\,f_0 \tag{2.34}$$

where β is positive when source and receiver are approaching one another and negative when they are receding.

The classical form for linear momentum is replaced by the special relativity form:

$$\vec{p} = \gamma m\vec{u} = \frac{m\vec{u}}{\sqrt{1-u^2/c^2}} \tag{2.48}$$

The relativistic kinetic energy is given by

$$K = \gamma mc^2 - mc^2 = mc^2\left(\frac{1}{\sqrt{1-u^2/c^2}} - 1\right) \tag{2.58}$$

The *total energy* E is given by

$$E = \gamma mc^2 = \frac{mc^2}{\sqrt{1-u^2/c^2}} = \frac{E_0}{\sqrt{1-u^2/c^2}} = K + E_0 \tag{2.65}$$

where $E_0 = mc^2$. This equation denotes the equivalence of mass and energy. The laws of the conservation of mass and of energy are combined into one conservation law: the conservation of mass-energy.

Energy and momentum are related by

$$E^2 = p^2c^2 + E_0^2 \tag{2.70}$$

In the case of massless particles (for example, the photon), $E_0 = 0$, so $E = pc$. Massless particles must travel at the speed of light.

The electron volt, denoted by eV, is equal to 1.602×10^{-19} J. The unified atomic mass unit u is based on the mass of the ^{12}C atom.

$$1\,\mathrm{u} = 1.66054 \times 10^{-27}\,\mathrm{kg} = 931.494\,\mathrm{MeV}/c^2 \tag{2.76, 2.77}$$

Momentum is often quoted in units of eV/c, and the velocity is often given in terms of $\beta\ (= v/c)$.

The difference between the rest energy of individual particles and the rest energy of the combined, bound system is called the binding energy.

Maxwell's equations are invariant under transformations between any inertial reference frames. What appears as electric and magnetic fields is relative to the reference frame of the observer.

Questions

1. Michelson used the motion of the Earth around the sun to try to determine the effects of the ether. Can you think of a more convenient experiment with a higher speed that Michelson might have used in the 1880s? What about today?

2. If you wanted to set out today to find the effects of the ether, what experimental apparatus would you want to use? Would a laser be included? Why?

3. For what reasons would Michelson and Morley repeat their experiment on top of a mountain? Why would they perform the experiment in summer and winter?

4. Does the fact that Maxwell's equations do not need to be modified due to the special theory of relativity, whereas Newton's laws of motion do, mean that Maxwell's work is somehow greater or more significant than Newton's? Explain.

5. The special theory of relativity has what effect on measurements done today? (a) None whatsoever, because any correction would be negligible. (b) We need to consider the effects of relativity when objects move close to the speed of light. (c) We should always make a correction due to relativity, because Newton's laws are basically wrong. (d) It doesn't matter, because we can't make measurements where relativity would matter.

6. Why did it take so long to discover the theory of relativity? Why didn't Newton figure it out?

7. Can you think of a way by which you can make yourself older than those born on your same birthday?

8. Will metersticks manufactured on Earth work correctly on spaceships moving at high speed? Explain.

9. Devise a system for you and three other colleagues, at rest with you, to synchronize your clocks if your clocks are too large to move and are separated by hundreds of miles.

10. In the experiment to verify time dilation by flying the cesium clocks around the Earth, what is the order of the speed of the four clocks in a system fixed at the center of the Earth, but not rotating?

11. Can you think of an experiment to directly verify length contraction? Explain.

12. Would it be easier to perform the muon decay experiment in the space shuttle orbiting above Earth and

then compare with the number of muons on Earth? Explain.

13. On a spacetime diagram, can events above $t = 0$ but not in the shaded area in Figure 2.25 affect the future? Explain.

14. Why don't we also include the spatial coordinate z when drawing the light cone?

15. What would be a suitable name for events connected by $\Delta s^2 = 0$?

16. Is the relativistic Doppler effect valid only for light waves? Can you think of another situation where it might be valid?

17. In Figure 2.22, why cannot a real worldline have a slope less than one?

18. Explain how in the twin paradox, we might arrange to compare clocks at the beginning and end of Mary's journey and not have to worry about acceleration effects.

19. In each of the following pairs, which is the more massive: a relaxed or compressed spring, a charged or uncharged capacitor, a piston-cylinder when closed or open?

20. In the fission of ^{235}U, the masses of the final products are less than the mass of ^{235}U. Does this make sense? What happens to the mass?

21. In the fusion of deuterium and tritium nuclei to produce a thermonuclear reaction, where does the kinetic energy that is produced come from?

22. Mary, the astronaut, wants to travel to Alpha Centauri, our nearest star, which is 4.3 lightyears away. She wants to leave on her 30th birthday, travel to Alpha Centauri but not stop, and return in time for her wedding to Vladimir on her 35th birthday. What is most likely to happen? (a) Vladimir is a lucky man, because he will marry Mary after she completes her journey. (b) Mary will have to hustle to get in her wedding gown, and the wedding is likely to be watched by billions of people. (c) It is a certainty that Mary will not reach Alpha Centauri if she wants to marry Vladimir as scheduled. (d) Mary does reach Alpha Centauri before her 35th birthday and sends a radio message to Vladimir from Alpha Centauri that she will be back on time. Vladimir is relieved to receive the message before the wedding date.

23. A salesman driving a very fast car was arrested for driving through a traffic light while it was red according to a policeman parked near the traffic light. The salesman said that the light was actually green to him, because it was Doppler shifted. Is he likely to be found innocent? Explain.

Problems

Note: The more challenging problems are indicated by a shaded box.

2.1 The Need for Ether

1. Show that the form of Newton's second law is invariant under the Galilean transformation.

2. Show that the definition of linear momentum, $p = mv$, has the same form $p' = mv'$ under a Galilean transformation.

2.2 The Michelson-Morley Experiment

3. Show that the equation for t_2 in Section 2.2 expresses the time required for the light to travel to the mirror D and back in Figure 2.2. In this case the light is traveling perpendicular to the supposed direction of the ether. In what direction must the light travel in order to be reflected by the mirror if the light must pass through the ether?

4. A swimmer wants to swim straight across a river with current flowing at a speed of $v_1 = 0.32$ km/h. If the swimmer swims in still water with speed $v_2 = 0.94$ km/h, at what angle should the swimmer point

upstream from the shore and at what speed will the swimmer swim across the river?

5. Show that the time difference $\Delta t'$ given by Equation (2.4) is correct when the Michelson interferometer is rotated by 90°.

6. In the 1887 experiment by Michelson and Morley, the length of each arm was 11 m. The experimental limit for the fringe shift was 0.005 fringes. If sodium light was used with the interferometer ($\lambda = 589$ nm), what lower limit did the null experiment place on the speed of the Earth through the expected ether?

7. Show that if length is contracted by the factor $\sqrt{1 - v^2/c^2}$ in the direction of motion, then the result in Equation (2.3) will have the factor needed to make $\Delta t = 0$ as needed by Michelson and Morley.

2.3 Einstein's Postulates

8. Explain why Einstein argued that the constancy of the speed of light (postulate 2) actually follows from the principle of relativity (postulate 1).

9. Prove that the constancy of the speed of light (postulate 2) is inconsistent with the Galilean transformation.

2.4   The Lorentz Transformation

10. Use the spherical wavefronts of Equations (2.9) to derive the Lorentz transformation given in Equations (2.17). Supply all the steps.

11. Show that both Equations (2.17) and (2.18) reduce to the Galilean transformation when $v \ll c$.

12. Determine the ratio $\beta = v/c$ for the following: (a) A car traveling 100 km/h. (b) A commercial jet airliner traveling 290 m/s. (c) A supersonic airplane traveling at Mach 2.3 (Mach number $= v/v_{\text{sound}}$). (d) The space shuttle, traveling 27,000 km/h. (e) An electron traveling 25 cm in 2 ns. (f) A proton traveling across a nucleus (10^{-14} m) in 0.35×10^{-22} s.

13. Two events occur in an inertial system K as follows:

Event 1: $x_1 = a$, $t_1 = 2a/c$, $y_1 = 0$, $z_1 = 0$

Event 2: $x_2 = 2a$, $t_2 = 3a/2c$, $y_2 = 0$, $z_2 = 0$

In what frame K′ will these events appear to occur at the same time? Describe the motion of system K′.

14. An event occurs in system K′ at $x' = 2$ m, $y' = 3.5$ m, $z' = 3.5$ m, and $t' = 0$. System K′ and K have their axes coincident at $t = t' = 0$, and system K′ travels along the x axis of system K with a speed $0.8c$. What are the coordinates of the event in system K?

15. A light signal is sent from the origin of a system K at $t = 0$ to the point $x = 3$ m, $y = 5$ m, $z = 10$ m. (a) At what time t is the signal received? (b) Find (x', y', z', t') for the receipt of the signal in a frame K′ that is moving along the x axis of K at a speed of $0.8c$. (c) From your results in (b) verify that the light traveled with a speed c as measured in the K′ frame.

16. Is there a frame K′ in which the two events described in Problem 13 occur at the same place? Explain.

17. Find the relativistic factor γ for each of the parts of Problem 12.

2.5   Time Dilation and Length Contraction

18. Show that the experiment depicted in Figure 2.11 and discussed in the text leads directly to the derivation of length contraction.

19. A rocket ship carrying passengers blasts off to go from New York to Los Angeles, a distance of about 5000 km. How fast must the rocket ship go to have its own length shortened by 1%?

20. Astronomers discover a planet orbiting around a star similar to our sun that is 20 lightyears away. How fast must a rocket ship go if the round trip is to take no longer than 40 years in time for the astronauts aboard? How long will the trip take as measured on Earth?

21. Particle physicists use particle track detectors to determine the lifetime of short-lived particles. A muon has a mean lifetime of 2.2 μs and makes a track 9.5 cm long before decaying into an electron and two neutrinos. What was the speed of the muon?

22. The Apollo astronauts returned from the moon under the Earth's gravitational force and reached speeds of almost 25,000 mi/h with respect to Earth. Assuming (incorrectly) they had this speed for the entire trip from the moon to Earth, what was the time difference for the trip between their clocks and clocks on Earth?

23. A clock in a spaceship is observed to run at a speed of only 3/5 that of a similar clock at rest on Earth. How fast is the spaceship moving?

24. A spaceship of length 40 m at rest is observed to be 20 m long when in motion. How fast is it moving?

25. The Concorde traveled 8000 km between two places in North America and Europe at an average speed of 375 m/s. What is the total difference in time between two similar atomic clocks, one on the airplane and one at rest on Earth during a one-way trip? Consider only time dilation and ignore other effects like the rotation of the Earth.

26. A mechanism on Earth used to shoot down geosynchronous satellites that house laser-based weapons is finally perfected and propels golf balls at $0.94c$. (a) How far will a detector placed inside the golf ball initially measure the distance to the satellite? (Geosynchronous satellites are placed 3.58×10^4 km above the surface of the Earth.) (b) How much time will it take the golf ball to make the journey to the satellite in the Earth's frame? How much time will it take in the golf ball's frame?

27. Two events occur in an inertial system K at the same time but 4 km apart. What is the time difference measured in a system K′ moving between these two events when the distance separation of the events is measured to be 5 km?

28. Imagine that in another universe the speed of light is only 100 m/s. (a) A person traveling along an interstate highway at 120 km/h ages how much slower than a person at rest? (b) This person passes by a meterstick at rest on the highway. How long does the meterstick appear?

29. In another universe where the speed of light is only 100 m/s, an airplane that is 40 m long at rest and flies at 300 km/h will appear to be how long to an observer at rest?

30. Two systems K and K′ synchronize their clocks at $t = t' = 0$ when their origins are aligned as system K′ passes by system K along the x axis at relative speed $0.8c$. At time $t = 3$ ns, Frank (in system K) shoots a proton gun having proton speeds of $0.98c$ along his x axis. The protons leave the gun at $x = 1$ m and arrive at a target 120 m away. Determine the event coordinates (x, t) of the gun firing and of the protons arriving as measured by observers in both systems K and K′.

2.6   Addition of Velocities

31. A spaceship is moving at a speed of $0.8c$ away from an observer at rest. A boy in the spaceship shoots a proton gun with protons having a speed of $0.7c$. What is the speed of the protons measured by the observer at rest when the gun is shot (a) away from the observer and (b) toward the observer?

32. A proton and an antiproton are moving toward each other in a head-on collision. If each has a speed of $0.8c$ with respect to the collision point, how fast are they moving with respect to each other?

33. Imagine the speed of light in another universe to be only 100 m/s. Two cars are traveling along an interstate highway in opposite directions. Person 1 is traveling 110 km/h, and person 2 is traveling 140 km/h. How fast does person 1 measure person 2 to be traveling? How fast does person 2 measure person 1 to be traveling?

34. In the Fizeau experiment described in Example 2.5, suppose that the water is flowing at a speed of 5 m/s. Find the difference in the speeds of two beams of light, one traveling in the same direction as the water and the other in the opposite direction. Use $n = 1.33$ for water.

35. Three galaxies are aligned along an axis in the order A, B, C. An observer in galaxy B is in the middle and observes that galaxies A and C are moving in opposite directions away from him, both with speeds $0.60c$. What is the speed of galaxies B and C as observed by someone in galaxy A?

36. Consider the *gedanken* experiment discussed in Section 2.6 in which a giant floodlight stationed 400 km above the Earth's surface shines its light across the moon's surface. How fast does the light flash across the moon?

2.7 Experimental Verification

37. A group of scientists decide to repeat the muon decay experiment at the Mauna Kea telescope site in Hawaii, which is 4205 m (13,796 feet) above sea level. They count 10^4 muons during a certain time period. Repeat the calculation of Section 2.7 and find the classical and relativistic number of muons expected at sea level. Why did they decide to count as many as 10^4 muons instead of only 10^3?

38. Consider a reference system placed at the U.S. Naval Observatory in Washington, D.C. Two planes take off from Washington Dulles airport, one going eastward and one going westward, both carrying a cesium atomic clock. The distance around the Earth at 39° latitude (Washington, D.C.) is 31,000 km, and Washington rotates about the Earth's axis at a speed of 360 m/s. Calculate the predicted differences between the clock left at the observatory and the two clocks in the airplanes (each traveling at 300 m/s) when the airplanes return to Washington. Include the rotation of the Earth but no general relativistic effects. Compare with the predictions given in the text.

2.8 Twin Paradox

39. Derive the results in Table 2.1 for the frequencies f' and f''. During what time period do Frank and Mary receive these frequencies?

40. Derive the results in Table 2.1 for the time of the total trip and the total number of signals sent in the frame of both twins.

2.9 Spacetime

41. Using the Lorentz transformation, prove that $s^2 = s'^2$.

42. Prove that for a timelike interval, two events can never be considered to occur simultaneously.

43. Prove that for a spacelike interval, two events cannot occur at the same place in space.

44. Given two events, (x_1, t_1) and (x_2, t_2), use a spacetime diagram to find the speed of a frame of reference in which the two events occur simultaneously. What values may Δs^2 have in this case?

45. (a) Draw on a spacetime diagram in the fixed system a line expressing all the events in the moving system that occur at $t' = 0$ if the clocks are synchronized at $t = t' = 0$. (b) What is the slope of this line? (c) Draw lines expressing events occurring for the four times $t'_4, t'_3, t'_2,$ and t'_1 where $t'_4 < t'_3 < 0 < t'_2 < t'_1$. (d) How are these four lines related geometrically?

46. Consider a fixed and a moving system with their clocks synchronized and their origins aligned at $t = t' = 0$. (a) Draw on a spacetime diagram in the fixed system a line expressing all the events occurring at $t' = 0$. (b) Draw on this diagram a line expressing all the events occurring at $x' = 0$. (c) Draw all the worldlines for light that pass through $t = t' = 0$. (d) Are the x' and ct' axes perpendicular? Explain.

47. Use the results of the two previous problems to show that events simultaneous in one system are not simultaneous in another system moving with respect to the first. Consider a spacetime diagram with x, ct and x', ct' axes drawn such that the origins coincide and the clocks were synchronized at $t = t' = 0$. Then consider events 1 and 2 that occur simultaneously in the fixed system. Are they simultaneous in the moving system?

2.10 Doppler Effect

48. An astronaut is said to have tried to get out of a traffic violation for running a red light ($\lambda = 670$ nm) by telling the judge that the light appeared green ($\lambda = 540$ nm) to her as she passed by in her high-powered transport. If this is true, how fast was the astronaut going?

49. Derive Equation (2.32) for the case where the source is fixed, but the receiver approaches it with velocity v.

50. Do the complete derivation for Equation (2.33) when the source and receiver are receding with relative velocity v.

51. A spacecraft traveling out of the solar system at a speed of $0.92c$ sends back information at a rate of 400 Hz. At what rate do we receive the information?

52. Three radio-equipped plumbing vans are broadcasting on the same frequency f_0. Van 1 is moving east of van 2 with speed v, van 2 is fixed, and van 3 is moving west of van 2 with speed v. What is the frequency of each van as received by the others?

53. Three radio-equipped plumbing vans are broadcasting on the same frequency f_0. Van 1 is moving north of van 2 with speed v, van 2 is fixed, and van 3 is moving west of van 2 with speed v. (a) What change in frequency $(f - f_0)$ does van 3 hear from van 2; (b) from van 1?

54. A spaceship moves radially away from Earth with acceleration 25 m/s². How long does it take for the sodium streetlamps ($\lambda = 589$ nm) on Earth to be invisible (with a powerful telescope) to the human eye of the astronauts? The range of visible wavelengths is about 400–700 nm.

2.11   Relativistic Momentum

55. Newton's second law is given by $\vec{F} = d\vec{p}/dt$. If the force is always perpendicular to the velocity, show that $\vec{F} = m\gamma\vec{a}$, where $\vec{a}$ is the acceleration.

56. Using the result of the previous problem, show that the radius of a particle's circular path having charge q traveling with speed v in a magnetic field perpendicular to the particle's path is $r = p/qB$. What happens to the radius as the speed increases as in a cyclotron?

57. Newton's second law is given by $\vec{F} = d\vec{p}/dt$. If the force is always parallel to the velocity, show that $\vec{F} = \gamma^3 m\vec{a}$.

58. Find the force necessary to give a proton an acceleration of 10^{19} m/s² when the proton has a velocity (along the same direction as the force) of (a) $0.01c$, (b) $0.1c$, (c) $0.9c$, and (d) $0.99c$.

59. A particle having a speed of $0.92c$ has a momentum of 10^{-16} kg·m/s. What is its mass?

60. A particle initially has a speed of $0.5c$. At what speed does its momentum increase by (a) 1%, (b) 10%, (c) 100%?

61. What magnetic field would be necessary to require 230-MeV protons to revolve in a circle of 15 m? (See Problem 56.)

62. Show that linear momentum is conserved in Example 2.9 as measured by Mary.

2.12   Relativistic Energy

63. Show that $\frac{1}{2}m\gamma v^2$ does not give the correct kinetic energy.

64. How much ice must melt at 0°C in order to gain 2 g of mass? Where does this mass come from? The heat of fusion for water is 334 J/g.

65. Physicists at the Stanford Linear Accelerator Center (SLAC) want to bombard 9-GeV electrons head-on with 3.1-GeV positrons to create B mesons and anti-B mesons. What speeds will the electron and positron have when they collide?

66. The Tevatron accelerator at the Fermi National Accelerator Laboratory (Fermilab) outside Chicago boosts protons to 1 TeV (1000 GeV) in five stages (the numbers given in parentheses represent the total kinetic

energy at the end of each stage): Cockcroft-Walton (750 keV), Linac (400 MeV), Booster (8 GeV), Main ring or injector (150 GeV), and finally the Tevatron itself (1 TeV). What is the speed of the proton at the end of each stage?

67. Calculate the momentum, kinetic energy, and total energy of an electron traveling at a speed of (a) $0.01c$, (b) $0.1c$, and (c) $0.9c$.

68. The total energy of a body is found to be twice its rest energy. How fast is it moving with respect to the observer?

69. A system is devised to exert a constant force of 8 N on an 80-kg body of mass initially at rest. The force pushes the mass horizontally on a frictionless table. How far does the body have to be pushed to increase its mass-energy by 25%?

70. What is the speed of a proton when its kinetic energy is equal to twice its rest energy?

71. What is the speed of an electron when its kinetic energy is (a) 10% of its rest energy, (b) equal to the rest energy, and (c) 10 times the rest energy?

72. Derive the following equation:

$$\beta = \frac{v}{c} = \sqrt{1 - \left(\frac{E_0}{E_0 + K}\right)^2}$$

73. Prove that $\beta = pc/E$. This is a useful relation to find the velocity of a highly energetic particle.

74. A good rule of thumb is to use relativistic equations whenever the kinetic energies determined classically and relativistically differ by more than 1%. Find the speeds when this occurs for (a) electrons, and (b) protons.

75. How much mass-energy (in joules) is contained in a peanut weighing 0.1 ounce? What happens if you eat 10 ounces of peanuts?

76. Calculate the energy needed to accelerate a spaceship of mass 10,000 kg to a speed of $0.3c$ for intergalactic space exploration. Compare this with a projected annual energy usage on Earth of 10^{21} J.

77. Derive Equation (2.58) for the relativistic kinetic energy and show all the steps, especially the integration by parts.

78. A test automobile of mass 1000 kg moving at high speed crashes into a wall. The average temperature of the car is measured to rise by 0.5°C after the wreck. What is the change in mass of the car? Where does this change in mass come from? (Assume the average specific heat of the automobile is close to that of steel, 0.11 cal·g⁻¹·°C⁻¹.)

2.13   Computations in Modern Physics

79. A helium nucleus has a mass of 4.001505 u. What is its binding energy?

80. A free neutron is an unstable particle and beta decays into a proton with the emission of an electron. How much kinetic energy is available in the decay?

81. Protons of 10^{12} eV (TeV) energy are produced at the Fermilab. Calculate the speed, momentum, and total energy of the protons.

82. What is the kinetic energy of (a) an electron having a momentum of 30 GeV/c? (b) a proton having a momentum of 30 GeV/c?

83. A muon has a mass of 106 MeV/c^2. Calculate the speed, momentum, and total energy of a 200-MeV muon.

84. The reaction ^{2}H + ^{2}H → n + ^{3}He (where n is a neutron) is one of the reactions useful for producing energy through nuclear fusion. (a) Assume the deuterium nuclei (^{2}H) are at rest and use the atomic mass units of the masses in Appendix 8 to calculate the mass-energy imbalance in this reaction. (Note: you can use atomic masses for this calculation, because the electron masses cancel out.) This amount of energy is given up when this nuclear reaction occurs. (b) What percentage of the initial rest energy is given up?

85. The reaction ^{2}H + ^{3}H → n + ^{4}He is one of the reactions useful for producing energy through nuclear fusion. (a) Assume the deuterium (^{2}H) and tritium (^{3}H) nuclei are at rest and use the atomic mass units of the masses in Appendix 8 to calculate the mass-energy imbalance in this reaction. This amount of energy is given up when this nuclear reaction occurs. (b) What percentage of the initial rest energy is given up?

2.14 Electromagnetism and Relativity

86. Instead of one positive charge outside a conducting wire, as was discussed in Section 2.14 and shown in Figure 2.34, consider a second conducting wire parallel to the first one. Both wires have positive and negative charges, and the wires are electrically neutral. Assume that in both wires the positive charges travel to the right and negative charges to the left. (a) Consider an inertial frame moving with the negative charges of wire 1. Show that the second wire is attracted to the first wire in this frame. (b) Now consider an inertial frame moving with the positive charges of the second wire. Show that the first wire is attracted to the second. (c) Use this argument to show that electrical and magnetic forces are relative.

General Problems

87. An Ω^- particle has rest energy 1672 MeV and mean lifetime 8.2×10^{-11} s. It is created and decays in a particle track detector and leaves a track 24 mm long. What is the total energy of the Ω^- particle?

88. Show that the following form of Newton's second law satisfies the Lorentz transformation. Assume the force is parallel to the velocity.

$$F = m\frac{dv}{dt}\frac{1}{[1-(v^2/c^2)]^{3/2}}$$

89. Using the results listed in Table 2.1, find the time of detecting Mary's turnaround and the number of signals received at the rate f'.

90. Using the previous results listed in Table 2.1, find the time for the remainder of the trip, the number of signals received at the rate f'', and the total number of signals received. Determine the conclusion as to the other twin's measure of the time taken for the trip.

91. Frank and Mary are twins. Mary jumps on a spaceship and goes to Alpha Centauri (4 lightyears away) and returns. She travels at a speed of $0.8c$ with respect to Earth and emits a radio signal every week. Frank also sends out a radio signal to Mary once a week. (a) How many signals does Mary receive from Frank before she turns around? (b) At what time does the frequency of signals Frank receives suddenly change? How many signals has he received at this time? (c) How many signals do Frank and Mary receive for the entire trip? (d) How much time does the trip take according to Frank and to Mary? (e) How much time does each twin say the other twin will measure for the trip? Do the answers agree with those for (d)?

92. An electron has a total energy that is 200 times its rest energy. Determine its (a) kinetic energy, (b) speed, and (c) momentum.

93. A proton moves with a speed of $0.9c$. Find the speed of an electron that has (a) the same momentum as the proton, and (b) the same kinetic energy.

94. A high-speed K^0 meson is traveling at a speed of $0.9c$ when it decays into a π^+ and a π^- meson. What are the greatest and least speed that the mesons may have?

95. Frank and Mary are twins, and Mary wants to travel to our nearest star, the binary star Alpha Centauri (4.3 lightyears away). Mary leaves on her 30th birthday and intends to return to Earth on her 52nd birthday. (a) Assuming her spaceship returns from Alpha Centauri without stopping, how fast must her spaceship travel? (b) How old will Frank be when she returns?

96. The International Space Federation constructs a new spaceship that can travel at a speed of $0.995c$. Mary, the astronaut, boards the spaceship to travel to Barnard's star, which is the second nearest star to our solar system after Alpha Centauri and is 5.98 ly away. After reaching Barnard's star, the spaceship travels slowly around the star system for 3 years doing research before returning back to Earth. (a) How much time does her journey take? (b) How much older is her twin Frank when she returns?

97. A powerful laser on Earth rotates its laser beam in a circle at a frequency of 0.03 Hz. (a) How fast does the spot that the laser makes on the moon move across the moon's landscape? (b) With what rotation frequency should the laser rotate if the laser spot moves across the moon's landscape at speed c?

98. The Lockheed SR-71 Blackbird may be the fastest non-research airplane ever built; it traveled at 2200 miles/hour (983 m/s) and was in operation from 1966 to 1990. Its length is 32.74 m. (a) By what percentage would it appear to be length contracted while in flight? (b) How much time difference would occur on an atomic clock in the plane compared to a similar clock on Earth during a flight of the Blackbird over its range of 3200 km?

99. A spaceship is coming directly toward you while you are in the International Space Station. You are told that the spaceship is shining sodium light (with an intense yellow doublet of wavelengths 588.9950 and 589.5924 nm). You have an apparatus that can resolve two closely spaced wavelengths if the difference is $\Delta\lambda < 0.55$ nm. If you find that you can just resolve the doublet, how fast is the spaceship traveling with respect to you?

100. Quasars are among the most distant objects in the universe and are moving away from us at the highest known speeds, as discussed in Chapter 16. Astrophysicists use the redshift parameter z to determine the redshift of such rapidly moving objects. The parameter z is determined by observing a wavelength λ' of a known spectral line of wavelength λ_{source} on Earth; $z = \Delta\lambda/\lambda_{source} = (\lambda' - \lambda_{source})/\lambda_{source}$. Find the speed of two quasars having z values of 1.9 and 4.9.

101. One of the possible decay modes of the neutral kaon is $K^0 \rightarrow \pi^0 + \pi^0$. The rest energies of the K^0 and π^0 are 498 MeV and 135 MeV, respectively. The kaon is initially at rest when it decays. (a) How much energy is released in the decay? (b) What are the momentum and relative directions of the two neutral pions (π^0)?

The Experimental Basis of Quantum Theory

As far as I can see, our ideas are not in contradiction to the properties of the photoelectric effect observed by Mr. Lenard.

Max Planck, 1905

As discussed in Chapter 1, during the final decades of the 1800s scientists discovered phenomena that could not always be explained by what we now call classical physics. Despite the prevalent confidence in the laws of classical physics, the few exceptions to these laws discovered during the latter part of the nineteenth century led to the fabulous 30-year period of 1900–1930, when our understanding of the laws of physics was dramatically changed. One of these exceptions led to the special theory of relativity, which was introduced by Einstein in 1905 and successfully explained the null result of the Michelson-Morley experiment. The other great conceptual advance of twentieth-century physics, the quantum theory that is the subject of this chapter, began in 1900 when Max Planck introduced his explanation of blackbody radiation.

We begin this chapter with Wilhelm Röntgen's discovery of the x ray and J. J. Thomson's discovery of the electron. Robert Millikan later determined the electron's charge. We shall see that, although it was necessary to assume that certain physical quantities may be quantized, scientists found this idea hard to accept. We will discuss the difficulties of explaining blackbody radiation with classical physics and how Planck's proposal solved the problem. Finally, we will see that Einstein's explanation of the photoelectric effect and Arthur Compton's understanding of data on x-ray scattering made the quantum hypothesis difficult to refute. After many difficult and painstaking experiments, it became clear that quantization was not only necessary; it was also the correct description of nature.

3.1 Discovery of the X Ray and the Electron

In the 1890s scientists and engineers were familiar with the "cathode rays" that were generated from one of the metal plates in an evacuated tube across which a large electric potential had been established. The origin and constitution of these cathode rays were not known. The concept of an atomic substructure of matter was widely accepted because of its use in explaining the results of chemical experiments. Therefore, it was surmised that cathode rays had something to do with atoms. It was known, for example, that cathode rays could penetrate matter, and their properties were of great interest and under intense investigation in the 1890s.

New penetrating ray: x ray

In 1895 Wilhelm Röntgen, who had received early training as a mechanical engineer but was at the time a professor of physics at the University of Würzburg in Germany, was studying the effects of cathode rays passing through various materials. During one such experiment he noticed that a nearby phosphorescent screen was glowing vividly in the darkened room. Röntgen soon realized he was observing a new kind of ray, one that, unlike cathode rays, was unaffected by magnetic fields and was far more penetrating than cathode rays. These **x rays**, as he called them, were apparently produced by the cathode rays bombarding the glass walls of his vacuum tube. Röntgen studied their transmission through many materials and even showed that he could obtain an image of the bones in a hand when the x rays were allowed to pass through as shown in Figure 3.1. This experiment created tremendous excitement, and medical applications of x rays were quickly developed. For this discovery, Röntgen received the first Nobel Prize for Physics in 1901.

For several years before the discovery of x rays, J. J. Thomson (1856–1940), professor of experimental physics at Cambridge University, had been studying the properties of electrical discharges in gases. Thomson's apparatus was similar to that used by Röntgen and many other scientists because of its simplicity (Figure 3.2). Thomson believed that cathode rays were particles, whereas several respected German scientists (such as Heinrich Hertz) believed they were wave phenomena.

Wilhelm Röntgen (1845–1923), born in Germany but raised in the Netherlands, studied mechanical engineering at the University of Zurich. After holding several university appointments, he went to the University of Munich as Chair of Physics in 1900, where he remained for the rest of his life. As a professor at the University of Würzburg in 1895, he discovered x rays while investigating the passage of electric current through low-pressure gases. He preferred working alone and built most of his own apparatus. He refused to benefit from his many discoveries and died nearly bankrupt after World War I. *Bettmann/Corbis.*

Thomson was able to prove in 1897 that the charged particles emitted from a heated electrical cathode were in fact the same as cathode rays. The main features of Thomson's experiment are shown in the schematic apparatus of Figure 3.2. The rays from the cathode are attracted to the positive potential on aperture A (anode) and are further collimated by aperture B to travel in a straight line to impinge on the fluorescent screen in the rear of the tube, where they can be visually detected by a flash of light. A voltage across the deflection plates sets up an electric field that can deflect charged particles. Previously, in a similar experiment, Hertz had observed no effect on the cathode rays due to the deflecting voltage. Thomson at first found the same result, but upon further evacuating the glass tube, he observed the deflection and proved that cathode rays had a negative charge. The previous experiment, in a poorer vacuum, had failed because the cathode rays had interacted with and ionized the residual gas. Thomson also studied the effects of a magnetic field upon the cathode rays and proved convincingly that the cathode rays acted as charged particles (electrons) in both electric and magnetic fields, for which he received the Nobel Prize for Physics in 1906.

Thomson's method of measuring the ratio of the electron's charge to mass, e/m, is now a standard technique and generally studied as an example of charged particles passing through perpendicular electric and magnetic fields as shown schematically in Figure 3.3. With the magnetic field turned off, the electron entering the region between the plates is accelerated upward by the electric field

$$F_y = ma_y = qE \qquad (3.1)$$

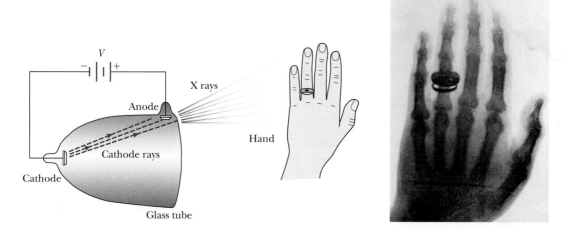

Figure 3.1 In Röntgen's experiment, "x rays" were produced by cathode rays (electrons) hitting the glass near the anode. He studied the penetration of the x rays through several substances and even noted that if the hand was held between the glass tube and a screen, the darker shadow of the bones could be discriminated from the less dark shadow of the hand. *Photo courtesy of Bettmann/ Corbis.*

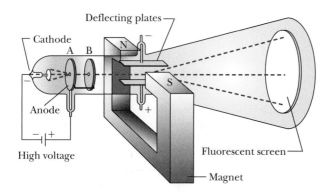

Figure 3.2 Apparatus of Thomson's cathode-ray experiment. Thomson proved that the rays emitted from the cathode were negatively charged particles (electrons) by deflecting them in electric and magnetic fields. The key to the experiment was to evacuate the glass tube.

Figure 3.3 Thomson's method of measuring the ratio of the electron's charge to mass was to send electrons through a region containing a magnetic field ($\vec{B}$ into paper) perpendicular to an electric field ($\vec{E}$ down). The electrons having $v = E/B$ go through undeflected. Then, using electrons of the same energy, the magnetic field is turned off and the electric field deflects the electrons, which exit at angle θ. The ratio of e/m can be determined from $\vec{B}, \vec{E}, \theta$, and ℓ, where ℓ is the length of the field distance and θ is the emerging angle. See Equation (3.5).

Sir Joseph John Thomson (1856–1940), universally known as "J.J.," went to Cambridge University at age 20 and remained there for the rest of his life. Thomson's career with the Cavendish Laboratory spanned a period of over 50 years, and he served as director from 1884 until 1918 when he stepped down in favor of Ernest Rutherford. Thomson was exceptional in designing apparatus and diagnosing problems, although he was not a particularly gifted experimentalist with his hands. His guidance at the Cavendish Laboratory was partly instrumental in the award of seven Nobel Prizes in Physics to him and his peers during his 50 years at the lab. *Science Museum.*

where m and q are, respectively, the mass and charge of the electron, and a_y is its resulting acceleration. The time for the electron to traverse the deflecting plates (length $= \ell$) is $t \approx \ell / v_0$. The exit angle θ of the electron is then given by

$$\tan \theta = \frac{v_y}{v_x} = \frac{a_y t}{v_0} = \frac{qE}{m} \frac{\ell}{v_0^2} \tag{3.2}$$

The ratio q/m can be determined if the velocity is known. By turning on the magnetic field and adjusting the strength of $\vec{B}$ so that no deflection of the electron occurs, the velocity can be determined. The condition for zero deflection is that the net force on the electron is exactly zero.

$$\vec{F} = q\vec{E} + q\vec{v} \times \vec{B} = 0 \tag{3.3}$$

Hence,

$$\vec{E} = -\vec{v} \times \vec{B}$$

Because $\vec{v}$ and $\vec{B}$ are perpendicular, the electric and magnetic field strengths are related by

$$|\vec{E}| = |v_x| |\vec{B}|$$

so that

$$v_x = \frac{E}{B} = v_0 \tag{3.4}$$

where we have used $E = |\vec{E}|$ for the magnitude of the electric field and similarly B for the magnitude of $\vec{B}$. If we insert this value for v_0 into Equation (3.2), we extract the ratio q/m.

$$\frac{q}{m} = \frac{v_0^2 \tan \theta}{E\ell} = \frac{E \tan \theta}{B^2 \ell} \tag{3.5}$$

Example 3.1

In an experiment similar to Thomson's, we use deflecting plates 5.0 cm in length with an electric field of 1.00×10^4 V/m. Without the magnetic field we find an angular deflection of 30°, and with a magnetic field of 8.0×10^{-4} T (8.0 gauss) we find no deflection. What is the initial velocity of the electron and its q/m?

Strategy Because we know the values of E and B for which there is no deflection, we use Equation (3.4) to determine the electron's velocity v_0. Then we can use Equation (3.5) to determine q/m for the situation with no magnetic field.

Solution We insert the values of E and B into Equation (3.4) to find

$$v_0 = \frac{E}{B} = \frac{1.00 \times 10^4 \text{ V/m}}{8.0 \times 10^{-4} \text{ T}} = 1.25 \times 10^7 \text{ m/s}$$

Because all our units for E and B are in the international system (SI), the value for v_0 is in meters/second. Equation (3.5) gives the following result for q/m:

$$\frac{q}{m} = \frac{E \tan \theta}{B^2 \ell} = \frac{(1.00 \times 10^4 \text{ V/m})(\tan 30°)}{(8.0 \times 10^{-4} \text{ T})^2 (0.050 \text{ m})}$$

$$= 1.80 \times 10^{11} \text{ C/kg}$$

Thomson's actual experiment, done in the manner of the previous example, obtained a result about 35% lower than the presently accepted value of 1.76×10^{11} C/kg for e/m. Thomson realized that the value of e/m (e = absolute value of electron charge) for an electron was much larger than had been anticipated and a factor of 1000 larger than any value of q/m that had been previously measured (for the hydrogen atom). He concluded that either m was small or e was large (or both), and the "carriers of the electricity" were quite penetrating as compared to atoms or molecules, which must be much larger in size.

3.2   Determination of Electron Charge

After Thomson's measurement of e/m and the confirmation of the cathode ray as the charge carrier (called *electron*), several investigators attempted to determine the electron's charge, which was poorly known in 1897. In 1911 the American physicist Robert A. Millikan (1868–1953) reported convincing evidence for an accurate determination of the electron's charge. Millikan's classic experiment began in 1907 at the University of Chicago. The experiment consisted of visual observation of the motion of uncharged and both positively and negatively charged oil drops moving under the influence of electrical and gravitational forces. The essential parts of the apparatus are shown in Figure 3.4. As the drops emerge from the nozzle, frictional forces sometimes cause them to be charged. Millikan's method consisted of balancing the upward force of the electric field between the plates against the downward force of the gravitational field.

When an oil drop falls downward through the air, it experiences a frictional force $\vec{F}_f$ proportional to its velocity due to the air's viscosity:

$$\vec{F}_f = -b\vec{v} \tag{3.6}$$

This force has a minus sign because a drag force always opposes the velocity. The constant b is determined by Stokes's law and is proportional to the oil drop's

Three great physicists (foreground), 1931: Michelson, Einstein, and Robert A. Millikan (1868–1953). Millikan received his degrees from Oberlin College and Columbia University and was at the University of Chicago from 1896 to 1921 before leaving to join the California Institute of Technology, where he was chair of the Executive Council from 1921 to 1945 (de facto president) and helped Caltech become a leading research institution. His important work included the famous oil-drop experiment to determine the electron charge, a confirmation of Einstein's photoelectric theory in which Millikan measured Planck's constant h, and Brownian motion. He received the Nobel Prize in Physics in 1923 for the first two experiments. He also did important work in cosmic ray physics and is given credit for the name *cosmic rays*. In later life he became particularly interested in teaching and was a prolific textbook author. *Courtesy of California Institute of Technology.*

(a)

(b)

Figure 3.4 (a) Diagram of the Millikan oil-drop experiment to measure the charge of the electron. Some of the oil drops from the atomizer emerge charged, and the electric field (voltage) is varied to slow down or reverse the direction of the oil drops, which can have positive or negative charges. (b) A student looking through the microscope is adjusting the voltage between the plates to slow down a tiny plastic ball that serves as the oil drop. *Stephen T. Thornton.*

radius. Millikan showed that Stokes's law for the motion of a small sphere through a resisting medium was incorrect for small-diameter spheres because of the atomic nature of the medium, and he found the appropriate correction. The buoyancy of the air produces an upward force on the drop, but we can neglect this effect for a first-order calculation.

To suspend the oil drop at rest between the plates, the upward electric force must equal the downward gravitational force. The frictional force is then zero because the velocity of the oil drop is zero.

$$\vec{F}_E = q\vec{E} = -m\vec{g} \qquad \text{(when } v = 0\text{)} \tag{3.7}$$

The magnitude of the electric field is $E = V/d$, and V is the voltage across large, flat plates separated by a small distance d. The magnitude of the electron charge q may then be extracted as

$$q = \frac{mgd}{V} \tag{3.8}$$

To calculate q we have to know the mass m of the oil drops. Millikan found he could determine m by turning off the electric field and measuring the terminal velocity of the oil drop. The radius of the oil drop is related to the terminal velocity by Stokes's law (see Problem 7). The mass of the drop can then be determined by knowing the radius r and density ρ of the type of oil used in the experiment:

$$m = \tfrac{4}{3}\pi r^3 \rho \tag{3.9}$$

If the power supply has a switch to reverse the polarity of the voltage and an adjustment for the voltage magnitude, the oil drop can be moved up and down in the apparatus at will. Millikan reported that in some cases he was able to observe a given oil drop for up to six hours and that the drop changed its charge several times.

Millikan made thousands of measurements using different oils and showed that there is a basic quantized electron charge. Millikan's value of e was very close to our presently accepted value of 1.602×10^{-19} C. Notice that we always quote a positive number for the charge e. The charge on an electron is then $-e$.

Measurement of electron charge

Example 3.2

For an undergraduate physics laboratory experiment we often make two changes in Millikan's procedure. First, we use plastic balls of about 1 micrometer (μm or micron) in diameter, for which we can measure the mass easily and accurately. This avoids the measurement of the oil drop's terminal velocity and the dependence on Stokes's law. The small plastic balls are sprayed through an atomizer in liquid solution, but the liquid soon evaporates in air. The plastic balls are observed by looking through a microscope. One other improvement is to occasionally bombard the region between the plates with ionizing radiation, such as an electron (beta particle) from a radioactive source. This radiation ionizes the air and makes it easier for the charge on a ball to change. By making many measurements we can determine whether the charges determined from Equation (3.8) are multiples of some basic charge unit.

In an actual undergraduate laboratory experiment the mass of the balls was $m = 5.7 \times 10^{-16}$ kg and the spacing be-

tween the plates was $d = 4.0$ mm. Therefore q can be found from Equation (3.8).

$$q = \frac{mgd}{V} = \frac{(5.7 \times 10^{-16} \text{ kg})(9.8 \text{ m/s}^2)(4.0 \times 10^{-3} \text{ m})}{V}$$

$$q = \frac{(2.23 \times 10^{-17} \text{ V})}{V} \text{ C}$$

where V is the voltage between plates when the observed ball is stationary. Two students observed 30 balls and found the values of V shown in Table 3.1 for a stationary ball. In this experiment the voltage polarity can be easily changed, and a positive voltage represents a ball with a positive charge. Notice that charges of both signs are observed.

The values of $|q|$ are plotted on a histogram in units of $\Delta q = 0.2 \times 10^{-19}$ C in Figure 3.5 (solid area). When 70 additional measurements from other students are added, a clear pattern of quantization develops with a charge $q = nq_0$,

Table 3.1 Student Measurements in Millikan Experiment

Particle	Voltage (V)	q ($\times 10^{-19}$ C)	Particle	Voltage (V)	q	Particle	Voltage (V)	q
1	−30.0	−7.43	11	−126.3	−1.77	21	−31.5	−7.08
2	+28.8	+7.74	12	−83.9	−2.66	22	−66.8	−3.34
3	−28.4	−7.85	13	−44.6	−5.00	23	+41.5	+5.37
4	+30.6	+7.29	14	−65.5	−3.40	24	−34.8	−6.41
5	−136.2	−1.64	15	−139.1	−1.60	25	−44.3	−5.03
6	−134.3	−1.66	16	−64.5	−3.46	26	−143.6	−1.55
7	+82.2	+2.71	17	−28.7	−7.77	27	+77.2	+2.89
8	+28.7	+7.77	18	−30.7	−7.26	28	−39.9	−5.59
9	−39.9	−5.59	19	+32.8	+6.80	29	−57.9	−3.85
10	+54.3	+4.11	20	−140.8	+1.58	30	+42.3	+5.27

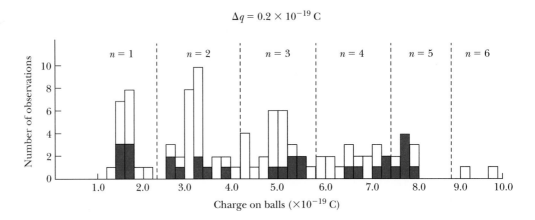

Figure 3.5 A histogram of the number of observations for the charge on a ball in a student Millikan experiment. The histogram is plotted for $\Delta q = 0.2 \times 10^{-19}$ C. The solid area refers to the first group's 30 measurements, and the open area to another 70 measurements. Notice the peaks, especially for the first three ($n = 1, 2, 3$) groups, indicating the electron charge quantization. When the basic charge q_0 is found from $q = nq_0$ (n = integer), $q_0 = 1.6 \times 10^{-19}$ C was determined in this experiment from all 100 observations.

especially for the first three groups. The areas of the histogram can be separated for the various n values, and the value of q_0 found for each measurement is then averaged. For the histogram shown we find $q_0 = 1.7 \times 10^{-19}$ C for the first 30 measurements and $q_0 = 1.6 \times 10^{-19}$ C for all 100 observations.

3.3   Line Spectra

In contrast to the smooth, continuous radiation spectrum obtained from thermal bodies, chemical elements produce unique wavelengths (colors) when burned in a flame or when excited in an electrical discharge, a fact already known in the early 1800s. Prisms had been used to investigate these early sources of spectra, and optical spectroscopy became an important area of experimental physics, primarily because of the modern development of diffraction gratings by Henry Rowland* (1848–1901) of Johns Hopkins University in the 1880s.

An example of a spectrometer used to observe optical spectra is shown in Figure 3.6. An electrical discharge excites atoms of a low-pressure gas contained in the tube. The collimated light passes through a diffraction grating with thousands of ruling lines per centimeter, and the diffracted light is separated at angle θ according to its wavelength. The equation expressing diffraction maxima is

$$d \sin \theta = n\lambda \tag{3.10}$$

where d is the distance between rulings, and n (an integer) is called the order number ($n = 1$ has the strongest scattered intensity). The resulting pattern of light bands and dark areas on the screen is called a *line spectrum*. By 1860 Bunsen and Kirchhoff realized that the wavelengths of these line spectra would

*Rowland was one of the first six professors chosen in 1875 for the founding of Johns Hopkins University and, together with Albert Michelson, was one of the foremost American physicists of the last part of the nineteenth century. He was a founder and was elected the first president of the American Physical Society in 1899. Albert Michelson was the vice president; neither had formally earned a Ph.D. degree.

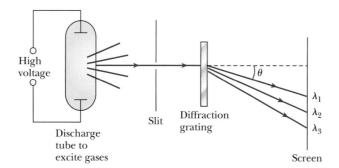

Figure 3.6 Schematic of an optical spectrometer. Light produced by a high-voltage discharge in the glass tube is collimated and passed through a diffraction grating, where it is deflected according to its wavelength. See Equation (3.10).

Figure 3.7 The Balmer series of line spectra of the hydrogen atom with wavelengths indicated in nanometers. The four visible lines are noted as well as the lower limit of the series.

allow identification of the chemical elements and the composition of materials. It was discovered that each element had its own characteristic wavelengths (see examples shown on the inside back cover). The field of spectroscopy flourished because finer and more evenly ruled gratings became available, and improved experimental techniques allowed more spectral lines to be observed and catalogued. Particular attention was paid to the sun's spectrum in hopes of understanding the origin of sunlight. The helium atom was actually "discovered" by its line spectra from the sun before it was recognized on Earth (see Special Topic, "The Discovery of Helium").

Characteristic line spectra of elements

Many scientists believed that the spectral lines somehow reflected the complicated internal structure of the atom, and that by carefully investigating the wavelengths for many elements, the structure of atoms and matter could be understood. That belief was eventually partially realized.

For much of the nineteenth century scientists attempted to find some simple underlying order for the characteristic wavelengths of line spectra. Hydrogen appeared to have an especially simple-looking spectrum, and because some chemists thought hydrogen atoms might be the constituents of heavier atoms, hydrogen was singled out for intensive study. Finally, in 1885, Johann Balmer, a Swiss schoolteacher, succeeded in obtaining a simple empirical formula that fit the wavelengths of the four lines then known in the hydrogen spectrum and several ultraviolet lines that had been identified in the spectra of white stars. This series of lines, called the *Balmer series*, is shown in Figure 3.7. Balmer found that the expression

Balmer's empirical result

$$\lambda = 364.56 \frac{k^2}{k^2 - 4} \text{ nm} \qquad (3.11)$$

(where $k = 3, 4, 5, \ldots; k > 2$) fit all the visible hydrogen lines. Wavelengths are normally given in units of nanometers* (nm). It is more convenient to take the

---

*Wavelengths were formerly listed in units of angstroms [one angstrom (Å) $= 10^{-10}$ m], named after Anders Ångström (1817–1874), one of the first persons to observe and measure the wavelengths of the four visible lines of hydrogen.

Special Topic

THE DISCOVERY OF HELIUM

*I*t might seem that the discovery of helium, the second simplest of all elements, would have occurred centuries ago. As we shall see, this is not the case, and in fact the discovery occurred over a period of several years in the latter part of the nineteenth century as scientists were scrambling to understand unexpected results. The account here is taken from *Helium*, by William H. Keesom.*

Spectroscopes are optical devices used to measure wavelengths of light and normally consist of a slit, a collimating lens, and a prism to refract the light. Their first use in a solar eclipse was on August 18, 1868, to investigate the sun's atmosphere. Several people (including P. J. C. Janssen, G. Rayet, C. T. Haig, and J. Herschel) traveling to the total eclipse region in India and Malaysia reported observing, either directly or indirectly, an unusual yellow line in the spectra that would later prove to be due to helium. It occurred to Janssen the day of the eclipse that it should be possible to see the sun's spectrum directly without the benefit of the eclipse, and he did so with a spectroscope on the days following the eclipse. The same idea had occurred to J. N. Lockyer earlier, but he did not succeed in measuring the sun's spectrum until October 1868, a month or so after Janssen. This method of observing the sun's atmosphere at any time was considered to be an important discovery, and Janssen and Lockyer are prominently recognized not only for their role in the evolution of helium's discovery, but also for their method of studying the sun's atmosphere.

The actual discovery of helium was delayed by the fact that the new yellow line seen in the sun's atmosphere was very close in wavelength to two well-known

*W. H. Keesom, *Helium*, Amsterdam, London, and New York: Elsevier (1942).

yellow lines of sodium. This is apparent in the atomic line spectra of both helium and sodium seen on the inside back cover of this text. By December 1868, Lockyer, A. Secchi, and Janssen each independently recognized that the yellow line was different than that of sodium.

Another difficulty was to prove that the new yellow line, called D3, was not due to some other known element, especially hydrogen. For many years Lockyer thought that D3 was related to hydrogen, and he and E. Frankland performed several experiments in an unsuccessful attempt to prove his thesis. Lockyer wrote as late as 1887 that D3 was a form of hydrogen. Despite Lockyer's convictions, Lord Kelvin reported in 1871 during his presidential address to the British Association that Frankland and Lockyer could not find the D3 line to be related to any terrestrial (from Earth) flame. Kelvin reported that it seemed to represent a new substance, which Frankland and Lockyer proposed to call helium (from the Greek word for "sun").

It was not until 1895 that helium was finally clearly observed on Earth by Sir William Ramsay, who had received a letter reporting that W. F. Hillebrand had produced nitrogen gas by boiling uranium ores (*pitchblende*) in dilute sulfuric acid. Ramsay was skeptical of the report and proceeded to reproduce it. He was astounded, after finding a small amount of nitrogen and the expected argon gas, to see a brilliant yellow line that he compared with those from sodium, finding the wavelengths to be slightly different. Sir William Crookes measured the wavelength and reported the following day that it was the D3 line, proving the terrestrial existence of helium. Later in 1895 H. Kayser found the helium line in spectra taken from a gas that had evolved from a spring in Germany's Black Forest. Eventually, in 1898, helium was confirmed in the Earth's atmosphere by E. C. Baly. No one person can be credited with the discovery of helium.

The remarkable properties of *liquid* helium are discussed in Section 9.7.

Table 3.2 Hydrogen Series of Spectral Lines			
Discoverer (year)	**Wavelength**	**n**	**k**
Lyman (1916)	Ultraviolet	1	>1
Balmer (1885)	Visible, ultraviolet	2	>2
Paschen (1908)	Infrared	3	>3
Brackett (1922)	Infrared	4	>4
Pfund (1924)	Infrared	5	>5

reciprocal of Equation (3.11) and write Balmer's formula in the form

$$\frac{1}{\lambda} = \frac{1}{364.56 \text{ nm}} \frac{k^2 - 4}{k^2} = \frac{4}{364.56 \text{ nm}} \left(\frac{1}{2^2} - \frac{1}{k^2} \right) = R_H \left(\frac{1}{2^2} - \frac{1}{k^2} \right) \quad (3.12)$$

where R_H is called the *Rydberg constant* (for hydrogen) and has the more accurate value 1.096776×10^7 m^{-1}, and k is an integer greater than two ($k > 2$).

Efforts by Johannes Rydberg and particularly Walther Ritz resulted by 1890 in a more general empirical equation for calculating the wavelengths, called the *Rydberg equation*.

$$\frac{1}{\lambda} = R_H \left(\frac{1}{n^2} - \frac{1}{k^2} \right) \quad (3.13) \quad \textbf{Rydberg equation}$$

where $n = 2$ corresponds to the Balmer series and $k > n$ always. In the next 20 years after Balmer's contribution, other series of the hydrogen atom's spectral lines were discovered, and by 1925 five series had been discovered, each having a different integer n (Table 3.2). The understanding of the Rydberg equation (3.13) and the discrete spectrum of hydrogen were important research topics early in the twentieth century.

Example 3.3

The visible lines of the Balmer series were observed first because they are most easily seen. Show that the wavelengths of spectral lines in the *Lyman* ($n = 1$) and *Paschen* ($n = 3$) series are not in the visible region. Find the wavelengths of the four visible atomic hydrogen lines. Assume the visible wavelength region is $\lambda = 400–700$ nm.

Strategy We use Equation (3.13) to determine the various wavelengths for $n = 1$, 2, and 3. If the wavelengths are between 400 and 700 nm, we conclude they are in the visible region. Otherwise, they are not visible.

Solution We use Equation (3.13) first to examine the Lyman series ($n = 1$):

$$\frac{1}{\lambda} = R_H \left(1 - \frac{1}{k^2} \right)$$

$$= 1.0968 \times 10^7 \left(1 - \frac{1}{k^2} \right) \text{ m}^{-1}$$

$k = 2$ $\dfrac{1}{\lambda} = 1.0968 \times 10^7 \left(1 - \dfrac{1}{4} \right) \text{ m}^{-1}$

$\lambda = 1.216 \times 10^{-7}$ m $= 121.6$ nm   (Ultraviolet)

$k = 3$ $\dfrac{1}{\lambda} = 1.0968 \times 10^7 \left(1 - \dfrac{1}{9} \right) \text{ m}^{-1}$

$\lambda = 1.026 \times 10^{-7}$ m $= 102.6$ nm   (Ultraviolet)

Because the wavelengths are decreasing for higher k values, all the wavelengths in the Lyman series are in the ultraviolet region and not visible by eye.

For the Balmer series ($n = 2$) we find

$$k = 3 \quad \frac{1}{\lambda} = 1.0968 \times 10^7 \left(\frac{1}{4} - \frac{1}{9} \right) \text{ m}^{-1}$$

$$\lambda = 6.565 \times 10^{-7} \text{ m} = 656.5 \text{ nm} \quad \text{(Red)}$$

$$k = 4 \quad \frac{1}{\lambda} = 1.0968 \times 10^7 \left(\frac{1}{4} - \frac{1}{16} \right) \text{ m}^{-1}$$

$$\lambda = 4.863 \times 10^{-7} \text{ m} = 486.3 \text{ nm} \quad \text{(Blue-green)}$$

$$k = 5 \quad \frac{1}{\lambda} = 1.0968 \times 10^7 \left(\frac{1}{4} - \frac{1}{25} \right) \text{ m}^{-1}$$

$$\lambda = 4.342 \times 10^{-7} \text{ m} = 434.2 \text{ nm} \quad \text{(Violet)}$$

$$k = 6 \quad \frac{1}{\lambda} = 1.0968 \times 10^7 \left(\frac{1}{4} - \frac{1}{36} \right) \text{ m}^{-1}$$

$$\lambda = 4.103 \times 10^{-7} \text{ m} = 410.3 \text{ nm} \quad \text{(Violet)}$$

$$k = 7 \quad \frac{1}{\lambda} = 1.0968 \times 10^7 \left(\frac{1}{4} - \frac{1}{49} \right) \text{ m}^{-1}$$

$$\lambda = 3.971 \times 10^{-7} \text{ m} = 397.1 \text{ nm} \quad \text{(Ultraviolet)}$$

Therefore $k = 7$ and higher k values will be in the ultraviolet region. The four lines $k = 3$, 4, 5, and 6 of the Balmer series are visible, although the 410-nm ($k = 6$) line is difficult to see because it is barely in the visible region and is weak in intensity.

The next series, $n = 3$, named after Paschen, has wavelengths of

$$k = 4 \quad \frac{1}{\lambda} = 1.0968 \times 10^7 \left(\frac{1}{9} - \frac{1}{16} \right) \text{ m}^{-1}$$

$$\lambda = 1.876 \times 10^{-6} \text{ m} = 1876 \text{ nm} \quad \text{(Infrared)}$$

$$k = 5 \quad \frac{1}{\lambda} = 1.0968 \times 10^7 \left(\frac{1}{9} - \frac{1}{25} \right) \text{ m}^{-1}$$

$$\lambda = 1.282 \times 10^{-6} \text{ m} = 1282 \text{ nm} \quad \text{(Infrared)}$$

$$k = \infty \quad \frac{1}{\lambda} = 1.0968 \times 10^7 \left(\frac{1}{9} - \frac{1}{\infty} \right) \text{ m}^{-1}$$

$$\lambda = 8.206 \times 10^{-7} \text{ m} = 820.6 \text{ nm} \quad \text{(Infrared)}$$

Thus the Paschen series has wavelengths entirely in the infrared region. The *series limit* is the smallest wavelength that can occur for each series (see Problem 9). Notice that the series limit is found for $k = \infty$ and is equal to 820.6 nm for the Paschen series. The higher series, $n \geq 4$, will all have wavelengths longer than the visible region.

3.4   Quantization

As we discussed in Chapter 1, some early Greek philosophers believed that matter must be composed of fundamental units that could not be divided further. The word *atom* means "not further divisible." Today some scientists believe, as these ancient philosophers did, that matter must eventually be indivisible. However, as we have encountered new experimental facts, our ideas about the fundamental, indivisible "building blocks" of matter have changed. "Elementary" particles will be discussed further in Chapter 14.

Is matter indivisible?

Whatever the elementary units of matter may turn out to be, we suppose there are some basic units of mass-energy of which matter is composed. This idea is hardly foreign to us: we have already seen that Millikan's oil-drop experiment showed the quantization of electric charge. Current theories predict that charges are quantized in units (called **quarks**) of $\pm e/3$ and $\pm 2e/3$, but quarks are not directly observed experimentally. The charges of particles that have been directly observed are quantized in units of $\pm e$.

Electric charge is quantized

In nature we see other examples of quantization. The measured atomic weights are not continuous—they have only discrete values, which are close to integral multiples of a unit mass. Molecules are formed from an integral number of atoms. The water molecule is made up of exactly two atoms of hydrogen and one of oxygen. The fact that an organ pipe produces one fundamental musical note with overtones is a form of quantization arising from fitting a precise number (or fractions) of sound waves into the pipe.

Quantization occurs often in nature

Line spectra provide a prime example of quantization. We have learned that the hydrogen line spectra have precise wavelengths that can be described empirically by simple equations. We will see in the next chapter that Niels Bohr used some simple assumptions based on the new quantum theory to model the atom and successfully predict these wavelengths. By the end of the nineteenth century, radiation spectra had been well studied. There certainly didn't appear to be any quantization effects observed in blackbody radiation spectra emitted by hot bodies. However, the explanation of radiation spectra was to have a tremendous influence on the discovery of quantum physics.

3.5 Blackbody Radiation

It has been known for many centuries that when matter is heated, it emits radiation. We can feel the heat radiation emitted by the heating element of an electric stove as it warms up. As the heating element reaches 550°C, its color becomes dark red, turning to bright red around 700°C. If the temperature were increased still further, the color would progress through orange, yellow, and finally white. We can determine experimentally that a broad spectrum of wavelengths is emitted when matter is heated. This process was of great interest to physicists of the nineteenth century. They measured the intensity of radiation being emitted as a function of material, temperature, and wavelength.

All bodies simultaneously emit and absorb radiation. When a body's temperature is constant in time, the body is said to be in *thermal equilibrium* with its surroundings. In order for the temperature to be constant, the body must absorb thermal energy at the same rate as it emits it. This implies that a good thermal emitter is also a good absorber.

Radiation emission and absorption

Physicists generally try to study first the simplest or most idealized case of a problem in order to gain the insight that is needed to analyze more complex situations. For thermal radiation the simplest case is a **blackbody**, which has the ideal property that it absorbs all the radiation falling on it and reflects none. The simplest way to construct a blackbody is to drill a small hole in the wall of a hollow container as shown in Figure 3.8. Radiation entering the hole will be reflected around inside the container and then finally absorbed. Only a small fraction of the entering rays will be re-emitted through the hole. If the blackbody is in thermal equilibrium, then it must also be an excellent emitter of radiation.

Blackbody radiation is theoretically interesting because of its universal character: the radiation properties of the blackbody (that is, the cavity) are independent of the particular material of which the container is made. Physicists can study the previously mentioned properties of intensity versus wavelength (called *spectral distribution*) at fixed temperatures without having to understand the details of emission or absorption by a particular kind of atom. The question of precisely what the thermal radiation actually consisted of was also of interest, although it was assumed, for lack of evidence to the contrary (and correctly, it turned out!), to be electromagnetic radiation.

Blackbody radiation is unique

The intensity $\mathcal{I}(\lambda, T)$ is the total power radiated per unit area per unit wavelength at a given temperature. Measurements of $\mathcal{I}(\lambda, T)$ for a blackbody are displayed in Figure 3.9. Two important observations should be noted:

1. The maximum of the distribution shifts to smaller wavelengths as the temperature is increased.
2. The total power radiated increases with the temperature.

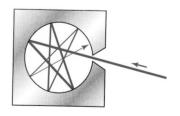

Figure 3.8 Blackbody radiation. Electromagnetic radiation (for example, light) entering a small hole reflects around inside the container before eventually being absorbed.

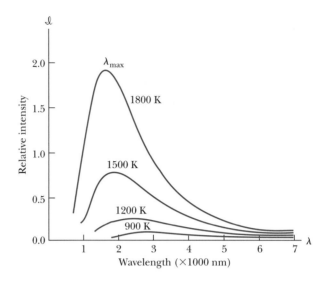

Figure 3.9 Spectral distribution of radiation emitted from a blackbody for different blackbody temperatures.

The first observation is commonly referred to as **Wien's displacement law**:

Wien's displacement law

$$\lambda_{\max} T = 2.898 \times 10^{-3}\ \text{m} \cdot \text{K} \tag{3.14}$$

where $\lambda_{\max}$ is the wavelength of the peak of the spectral distribution at a given temperature. We can see in Figure 3.9 that the position of $\lambda_{\max}$ varies with temperature as prescribed by Equation (3.14). Wilhelm Wien received the Nobel Prize in 1911 for his discoveries concerning radiation. We can quantify the second observation by integrating the quantity $\mathcal{I}(\lambda,\ T)$ over all wavelengths to find the power per unit area at T.

$$R(T) = \int_0^\infty \mathcal{I}(\lambda,\ T)\ d\lambda \tag{3.15}$$

Josef Stefan found empirically in 1879, and Boltzmann demonstrated theoretically several years later, that $R(T)$ is related to the temperature by

Stefan-Boltzmann law

$$R(T) = \epsilon \sigma T^4 \tag{3.16}$$

This is known as the **Stefan-Boltzmann law**, with the constant σ experimentally measured to be $5.6705 \times 10^{-8}\ \text{W}/(\text{m}^2 \cdot \text{K}^4)$. The Stefan-Boltzmann law equation can be applied to any material for which the emissivity is known. The **emissivity ϵ** ($\epsilon = 1$ for an idealized blackbody) is simply the ratio of the emissive power of an object to that of an ideal blackbody and is always less than 1. Thus, Equation (3.16) is a useful and valuable relation for practical scientific and engineering work.

Example 3.4

A furnace has walls of temperature 1600°C. What is the wavelength of maximum intensity emitted when a small door is opened?

Strategy We assume the furnace with a small door open is a blackbody so that we can determine λ_{max} from Equation (3.14).

Solution We first convert the temperature to kelvins.

$$T = (1600 + 273) \text{ K} = 1873 \text{ K}$$

Equation (3.14) gives

$$\lambda_{max}(1873 \text{ K}) = 2.898 \times 10^{-3} \text{ m} \cdot \text{K}$$

$$\lambda_{max} = 1.55 \times 10^{-6} \text{ m} = 1550 \text{ nm}$$

The peak wavelength is in the infrared region.

Example 3.5

The wavelength of maximum intensity of the sun's radiation is observed to be near 500 nm. Assume the sun to be a blackbody and calculate (a) the sun's surface temperature, (b) the power per unit area $R(T)$ emitted from the sun's surface, and (c) the energy received by the Earth each day from the sun's radiation.

Strategy (a) We use Equation (3.14) with λ_{max} to determine the sun's surface temperature. (b) We assume the sun is a blackbody. We use the temperature T with Equation (3.16) to determine the power per unit area $R(T)$. (c) Because we know $R(T)$, we can determine the amount of the sun's energy intercepted by the Earth each day.

Solution (a) From Equation (3.14) we calculate the sun's surface temperature with $\lambda_{max} = 500$ nm.

$$(500 \text{ nm}) T_{sun} = 2.898 \times 10^{-3} \text{ m} \cdot \text{K} \frac{10^9 \text{ nm}}{\text{m}}$$

$$T_{sun} = \frac{2.898 \times 10^6}{500} \text{ K} = 5800 \text{ K} \quad (3.17)$$

(b) The power per unit area $R(T)$ radiated by the sun is found to be

$$R(T) = \sigma T^4 = 5.67 \times 10^{-8} \frac{\text{W}}{\text{m}^2 \cdot \text{K}^4} (5800 \text{ K})^4$$

$$= 6.42 \times 10^7 \text{ W/m}^2 \quad (3.18)$$

(c) Because this is the power per unit surface area, we need to multiply it by $4\pi r^2$, the surface area of the sun. The radius of the sun is 6.96×10^5 km.

$$\text{Surface area (sun)} = 4\pi (6.96 \times 10^8 \text{ m})^2 = 6.09 \times 10^{18} \text{ m}^2$$

Thus the total power, P_{sun}, radiated from the sun's surface is

$$P_{sun} = 6.42 \times 10^7 \frac{\text{W}}{\text{m}^2} (6.09 \times 10^{18} \text{ m}^2)$$

$$= 3.91 \times 10^{26} \text{ W} \quad (3.19)$$

The fraction F of the sun's radiation received by Earth is given by the fraction of the total area over which the radiation is spread.

$$F = \frac{\pi r_E^2}{4\pi R_{Es}^2}$$

where r_E = radius of Earth = 6.37×10^6 m, and R_{Es} = mean Earth-sun distance = 1.49×10^{11} m. Then

$$F = \frac{\pi r_E^2}{4\pi R_{Es}^2} = \frac{(6.37 \times 10^6 \text{ m})^2}{4(1.49 \times 10^{11} \text{ m})^2} = 4.57 \times 10^{-10}$$

Thus the radiation received by the Earth from the sun is

$$P_{Earth}(\text{received}) = (4.57 \times 10^{-10})(3.91 \times 10^{26} \text{ W})$$

$$= 1.79 \times 10^{17} \text{ W}$$

and in one day the Earth receives

$$U_{Earth} = 1.79 \times 10^{17} \frac{\text{J}}{\text{s}} \frac{60 \text{ s}}{\text{min}} \frac{60 \text{ min}}{\text{h}} \frac{24 \text{ h}}{\text{day}}$$

$$= 1.55 \times 10^{22} \text{ J} \quad (3.20)$$

The power received by the Earth per unit of exposed area is

$$R_{Earth} = \frac{1.79 \times 10^{17} \text{ W}}{\pi (6.37 \times 10^6 \text{ m})^2} = 1400 \text{ W/m}^2 \quad (3.21)$$

This is the source of most of our energy on Earth. Measurements of the sun's radiation outside the Earth's atmosphere give a value near 1400 W/m², so our calculation is fairly accurate. Apparently the sun does act as a blackbody, and the energy received by the Earth comes primarily from the surface of the sun.

Attempts to understand and derive from basic principles the shape of the blackbody spectral distribution (Figure 3.9) were unsuccessful throughout the 1890s and presented a serious dilemma to the best scientists of the day. The nature of the dilemma can be understood from classical electromagnetic theory, together with statistical thermodynamics. The radiation emitted from a blackbody can be expressed as a superposition of electromagnetic waves of different frequencies within the cavity. That is, radiation of a given frequency is represented by a standing wave inside the cavity. The equipartition theorem of thermodynamics assigns equal average energy kT to each possible wave configuration.

Lord Rayleigh used the classical theories of electromagnetism and thermodynamics to show that the blackbody spectral distribution should be

Rayleigh-Jeans formula

$$\mathcal{I}(\lambda, T) = \frac{2\pi c k T}{\lambda^4} \tag{3.22}$$

This result is known as the **Rayleigh-Jeans formula**, and it is the best description that classical theory can provide to describe blackbody radiation. For long wavelengths there are few configurations whereby a standing wave can form inside the cavity. However, as the wavelength becomes shorter the number of standing wave possibilities increases, and as $\lambda \to 0$ the number of possible configurations increases without limit. This means the total energy of all configurations is infinite, because each standing wave configuration has the nonzero energy kT. We show a graph of the Rayleigh-Jeans result compared with experimental data in Figure 3.10, and although it approaches the data at longer wavelengths, it deviates badly at short wavelengths. This problem for small wavelengths became known as "the ultraviolet catastrophe" and was one of the outstanding exceptions that classical physics could not explain.

In the late 1890s the German theoretical physicist Max Planck became interested in this problem. By this time, different empirical expressions for the blackbody spectrum had been separately fit to the data for both short wavelengths and long wavelengths, but no one had explained the whole spectrum. Planck tried

Max Planck (1858–1947) spent most of his productive years as a professor at the University of Berlin (1889–1928). Planck was one of the early theoretical physicists and did work in optics, thermodynamics, and statistical mechanics. His theory of the *quantum of action* was slow to be accepted. Finally, after Einstein's photoelectric effect explanation and Rutherford and Bohr's atomic model, Planck's contribution became widely acclaimed. He received many honors, among them the Nobel Prize in Physics in 1918. *AIP Emilio Segrè Visual Archives.*

Figure 3.10 The spectral distribution calculated by the Rayleigh-Jeans formula is compared with blackbody radiation experimental data at 1200 K. The theory approaches the data at large wavelengths, but disagrees badly at low wavelengths.

various functions of wavelength and temperature until he found a single formula that fit the measurements of $\mathscr{I}(\lambda, T)$ over the entire wavelength range. When Planck announced this result in October of 1900, his theoretical formula was in better agreement over the entire blackbody spectrum than the previous empirical ones, which were only valid for either short or long wavelengths.

Planck became quite intrigued and started working to find a sound theoretical basis for his empirical equation. He was an expert in thermodynamics and statistical mechanics. Following Hertz's work using oscillators to confirm the existence of Maxwell's electromagnetic waves, and lacking detailed information about the atomic composition of the cavity walls, Planck assumed that the radiation in the cavity was emitted (and absorbed) by some sort of "oscillators" that were contained in the walls. Whereas we would now refer to the radiation of the electromagnetic field in the cavity, Planck referred to the radiation produced by the "oscillators," a term we will briefly continue to use. When adding up the energies of the oscillators, he assumed (for convenience) that each one had an energy that was an integral multiple of hf, where f is the frequency of the oscillating wave. He was applying a technique invented by Boltzmann, and Planck ultimately expected to take the limit $h \rightarrow 0$, in order to include all the possibilities. However, he noticed that by keeping h nonzero he arrived at the equation needed for $\mathscr{I}(\lambda, T)$:

$$\mathscr{I}(\lambda, T) = \frac{2\pi c^2 h}{\lambda^5} \frac{1}{e^{hc/\lambda kT} - 1} \qquad (3.23)$$

Planck's radiation law

Equation (3.23) is **Planck's radiation law**. The derivation of Equation (3.23) is sufficiently complicated that we have omitted it here, but we will revisit it in Chapter 9. No matter what Planck tried, he could arrive at agreement with the experimental data only by making two important modifications of classical theory:

1. The oscillators (of electromagnetic origin) can only have certain discrete energies determined by $E_n = nhf$, where n is an integer, f is the frequency, and h is called **Planck's constant** and has the value

Planck's constant h

$$h = 6.6261 \times 10^{-34} \, \text{J} \cdot \text{s} \qquad (3.24)$$

2. The oscillators can absorb or emit energy in discrete multiples of the fundamental quantum of energy given by

$$\Delta E = hf \qquad (3.25)$$

Planck found these results quite disturbing and spent several years trying to find a way to keep the agreement with experiment while letting $h \rightarrow 0$. Each attempt failed, and Planck's quantum result became one of the cornerstones of modern science.

Example 3.6

Show that Wien's displacement law follows from Planck's radiation law.

Strategy Wien's law, Equation (3.14), refers to the wavelength for which $\mathscr{I}(\lambda, T)$ is a maximum for a given temperature. From calculus we know we can find the maximum value of a function for a certain parameter by taking the derivative of the function with respect to the parameter, set the derivative to zero, and solve for the parameter.

Solution Therefore, to find the value of the Planck radiation law for a given wavelength we find $d\mathscr{I}/d\lambda = 0$ and solve for λ.

$$\frac{d\mathcal{I}(\lambda, T)}{d\lambda} = 0 \qquad \text{for } \lambda = \lambda_{\max}$$

$$2\pi c^2 h \frac{d}{d\lambda} \left[\lambda^{-5} (e^{hc/\lambda kT} - 1)^{-1} \right] \Big|_{\lambda_{\max}} = 0$$

$$-5\lambda_{\max}^{-6} (e^{hc/\lambda_{\max} kT} - 1)^{-1} - \lambda_{\max}^{-5} (e^{hc/\lambda_{\max} kT} - 1)^{-2}$$

$$\times \left(\frac{-hc}{kT\lambda_{\max}^2} \right) e^{hc/\lambda_{\max} kT} = 0$$

Multiplying by $\lambda_{\max}^6 (e^{hc/\lambda_{\max} kT} - 1)$ results in

$$-5 + \frac{hc}{\lambda_{\max} kT} \left(\frac{e^{hc/\lambda_{\max} kT}}{e^{hc/\lambda_{\max} kT} - 1} \right) = 0$$

Let

$$x = \frac{hc}{\lambda_{\max} kT}$$

Then

$$-5 + \frac{xe^x}{e^x - 1} = 0$$

and

$$xe^x = 5(e^x - 1)$$

This is a transcendental equation and can be solved numerically (try it!) with the result $x \approx 4.966$, and therefore

$$\frac{hc}{\lambda_{\max} kT} = 4.966$$

$$\lambda_{\max} T = \frac{hc}{4.966\, k} = \frac{1240 \text{ eV} \cdot \text{nm}}{4.966 \left(8.617 \times 10^{-5} \dfrac{\text{eV}}{\text{K}} \right)} \frac{10^{-9} \text{ m}}{\text{nm}}$$

and finally,

$$\lambda_{\max} T = 2.898 \times 10^{-3} \text{ m} \cdot \text{K}$$

which is Wien's displacement law.

Example 3.7

Use Planck's radiation law to derive the Stefan-Boltzmann law.

Strategy We determine $R(T)$ by integrating $\mathcal{I}(\lambda, T)$ over all wavelengths.

Solution

$$R(T) = \int_0^\infty \mathcal{I}(\lambda, T)\, d\lambda$$

$$= 2\pi c^2 h \int_0^\infty \frac{1}{\lambda^5} \frac{1}{e^{hc/\lambda kT} - 1}\, d\lambda$$

Let

$$x = \frac{hc}{\lambda kT}$$

Then

$$dx = -\frac{hc}{kT} \frac{d\lambda}{\lambda^2}$$

Now we have

$$R(T) = -2\pi c^2 h \int_\infty^0 \left(\frac{kT}{hc} \right)^6 x^5 \frac{1}{e^x - 1} \frac{1}{x^2} \left(\frac{hc}{kT} \right)^2 dx$$

$$= +2\pi c^2 h \left(\frac{kT}{hc} \right)^4 \int_0^\infty \frac{x^3}{e^x - 1}\, dx$$

We look up this integral in Appendix 7 and find it to be $\pi^4/15$.

$$R(T) = 2\pi c^2 h \left(\frac{kT}{hc} \right)^4 \frac{\pi^4}{15}$$

$$R(T) = \frac{2\pi^5 k^4}{15 h^3 c^2} T^4$$

Putting in the values for the constants k, h, and c results in

$$R(T) = 5.67 \times 10^{-8}\, T^4 \frac{\text{W}}{\text{m}^2 \cdot \text{K}^4}$$

Example 3.8

Show that the Planck radiation law agrees with the Rayleigh-Jeans formula for large wavelengths.

$$\frac{1}{e^{hc/\lambda kT} - 1} = \frac{1}{\left[1 + \dfrac{hc}{\lambda kT} + \left(\dfrac{hc}{\lambda kT} \right)^2 \dfrac{1}{2} + \cdots \right] - 1} \rightarrow \frac{\lambda kT}{hc}$$

for large λ

Strategy We use Equation (3.23) for the Planck radiation law, letting $\lambda \rightarrow \infty$ for the term involving the exponential, and see if the result agrees with Equation (3.22).

Equation (3.23) now becomes

$$\mathfrak{L}(\lambda, T) = \frac{2\pi c^2 h}{\lambda^5} \frac{\lambda kT}{hc} = \frac{2\pi ckT}{\lambda^4}$$

Solution We follow the strategy and find the result for the term involving the exponential.

which is the same as the Rayleigh-Jeans result in Equation (3.22).

Example 3.9

Show that Planck's radiation law resolves the *ultraviolet catastrophe* .

Strategy The ultraviolet catastrophe occurs because the number of configurations whereby a standing wave can form inside the cavity in the classical calculation becomes infinite as $\lambda \rightarrow 0$. We want to find out what happens to $\mathfrak{L}(\lambda, T)$ if we let $\lambda \rightarrow 0$. We also need to investigate the total energy of the system, especially for the large number of small-wavelength oscillators.

Solution If we let $\lambda \rightarrow 0$ in Equation (3.23), the value of $e^{hc/\lambda kT} \rightarrow \infty$. The exponential term dominates the λ^5 term as $\lambda \rightarrow 0$, so the denominator in Equation (3.23) is infinite, and the value of $\mathfrak{L}(\lambda, T) \rightarrow 0$. Note that as the wavelength gets smaller, the frequency becomes larger ($f = c/\lambda$), and $hf \gg kT$. Few oscillators will be able to obtain such large energies, partly because of the large energy necessary to take the energy step from 0 to hf. The probability of occupying the states with small wavelengths (large frequency and high energy) is vanishingly small, so the total energy of the system remains finite. The ultraviolet catastrophe is avoided.

3.6   Photoelectric Effect

Perhaps the most compelling, and certainly the simplest, evidence for the quantization of radiation energy comes from the only acceptable explanation of the **photoelectric effect**. While Heinrich Hertz was performing his famous experiment in 1887 that confirmed Maxwell's electromagnetic wave theory of light, he noticed that when ultraviolet light fell on a metal electrode a charge was produced that separated the leaves of his electroscope. Although Hertz recognized this discovery of what would become the photoelectric effect, it was of little use to him at the time and he left the exploitation of the effect to others, particularly Philipp Lenard. The photoelectric effect is one of several ways in which electrons can be emitted by materials. By the early 1900s it was known that electrons are bound to matter. The valence electrons in metals are "free"—they are able to move easily from atom to atom, but are not able to leave the surface of the material. The methods known now by which electrons can be made to completely leave the material include

Table 3.3 Work Functions					
Element	ϕ (eV)	Element	ϕ (eV)	Element	ϕ (eV)
Ag	4.64	K	2.29	Pd	5.22
Al	4.20	Li	2.93	Pt	5.64
C	5.0	Na	2.36	W	4.63
Cs	1.95	Nd	3.2	Zr	4.05
Cu	4.48	Ni	5.22		
Fe	4.67	Pb	4.25		

From *Handbook of Chemistry and Physics*, 83rd ed. Boca Raton, Fla.: CRC Press (2002–03), p. 12-130.

Methods of electron emission

1. Thermionic emission: Application of heat allows electrons to gain enough energy to escape.
2. Secondary emission: The electron gains enough energy by transfer from another high-speed particle that strikes the material from outside.
3. Field emission: A strong external electric field pulls the electron out of the material.
4. Photoelectric effect: Incident light (electromagnetic radiation) shining on the material transfers energy to the electrons, allowing them to escape.

It is not surprising that electromagnetic radiation interacts with electrons within metals and gives the electrons increased kinetic energy. Because electrons in metals are weakly bound, we expect that light can give electrons enough **Photoelectrons** extra kinetic energy to allow them to escape. We call the ejected electrons **photoelectrons**. The minimum extra kinetic energy that allows electrons to escape the **Work function** material is called the **work function** ϕ. The work function is the minimum binding energy of the electron to the material (see Table 3.3 for work function values for several elements).

Experimental Results of Photoelectric Effect

Experiments carried out around 1900 showed that photoelectrons are produced when visible and/or ultraviolet light falls on clean metal surfaces. Photoelectricity was studied using an experimental apparatus shown schematically in Figure 3.11. Incident light falling on the **emitter** (also called the **photocathode** or **cathode**)

Figure 3.11 Photoelectric effect. Electrons emitted when light shines on a surface are collected, and the photocurrent I is measured. A negative voltage, relative to that of the emitter, can be applied to the collector. When this retarding voltage is sufficiently large, the emitted electrons are repelled, and the current to the collector drops to zero.

Incident light
Collector
e^-
Emitter
Vacuum tube
I
A) Ammeter
Power supply
(Voltage V)

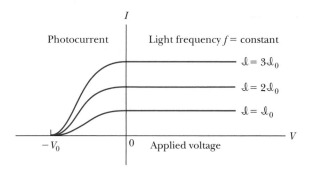

Figure 3.12 The photoelectric current I is shown as a function of the voltage V applied between the emitter and collector for a given frequency f of light for three different light intensities. Notice that no current flows for a retarding potential more negative than $-V_0$ and that the photocurrent is constant for potentials near or above zero (this assumes that the emitter and collector are closely spaced or in spherical geometry to avoid loss of photoelectrons).

ejects electrons. Some of the electrons travel toward the **collector** (also called the **anode**), where either a negative (retarding) or positive (accelerating) applied voltage V is imposed by the power supply. The current I measured in the ammeter (photocurrent) arises from the flow of photoelectrons from emitter to collector.

The pertinent experimental facts about the photoelectric effect are these:

1. The kinetic energies of the photoelectrons are independent of the light intensity. In other words, a stopping potential (applied voltage) of $-V_0$ is sufficient to stop all photoelectrons, *no matter what the light intensity*, as shown in Figure 3.12. For a given light intensity there is a maximum photocurrent, which is reached as the applied voltage increases from negative to positive values.

2. The maximum kinetic energy of the photoelectrons, for a given emitting material, depends only on the frequency of the light. In other words, for light of different frequency (Figure 3.13) a different retarding potential $-V_0$ is required to stop the most energetic photoelectrons. The value of V_0 depends on the frequency f but not on the intensity (see Figure 3.12).

3. The smaller the work function ϕ of the emitter material, the smaller is the threshold frequency of the light that can eject photoelectrons. No photoelectrons are produced for frequencies below this threshold frequency, no matter what the intensity. Data similar to Millikan's results (discussed later)

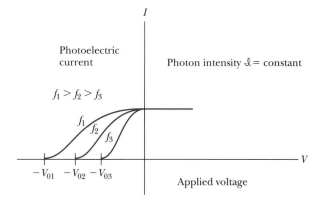

Figure 3.13 The photoelectric current I is shown as a function of applied voltage for three different light frequencies. The retarding potential $-V_0$ is different for each f and is more negative for larger f.

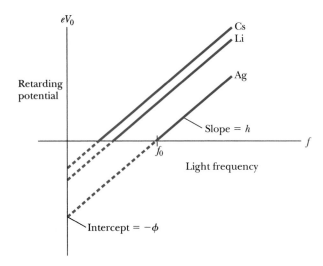

Figure 3.14 The retarding potential eV_0 (maximum electron kinetic energy) is plotted versus light frequency for three different emitter materials.

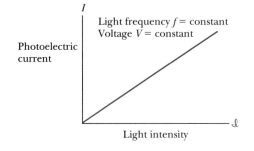

Figure 3.15 The photoelectric current I is a linear function of the light intensity for a constant f and V.

are shown in Figure 3.14, where the threshold frequencies f_0 are measured for three different metals.

4. When the photoelectrons are produced, however, their number is proportional to the intensity of light as shown in Figure 3.15. That is, the maximum photocurrent is proportional to the light intensity.

5. The photoelectrons are emitted almost instantly ($\leq 3 \times 10^{-9}$ s) following illumination of the photocathode, independent of the intensity of the light.

Except for result 5, these experimental facts were known in rudimentary form by 1902, primarily due to the work of Philipp Lenard, who had been an assistant to Hertz in 1892 after Hertz had moved from Karlsruhe to Bonn. Lenard, who extensively studied the photoelectric effect, received the Nobel Prize in Physics in 1905 for this and other research on the identification and behavior of electrons.

Classical Interpretation

As stated previously, classical theory allows electromagnetic radiation to eject photoelectrons from matter. However, classical theory predicts that the total amount of energy in a light wave increases as the light intensity increases. Therefore, according to classical theory, the electrons should have more kinetic energy if the light intensity is increased. However, according to experimental result 1 above and Figure 3.12, a characteristic retarding potential $-V_0$ is sufficient to stop all photoelectrons for a given light frequency f, no matter what the intensity. Classical electromagnetic theory is unable to explain this result. Similarly, classical theory cannot explain result 2, because the maximum kinetic energy of the photoelectrons depends on the value of the light frequency f and not on the intensity.

The existence of a threshold frequency, shown in experimental result 3, is completely inexplicable in classical theory. Classical theory cannot predict the results shown in Figure 3.14. Classical theory does predict that the number of photoelectrons ejected will increase with intensity in agreement with experimental result 4.

Finally, classical theory would predict that for extremely low light intensities, a long time would elapse before any one electron could obtain sufficient energy to escape. We observe, however, that the photoelectrons are ejected almost immediately. For example, experiments have shown that a light intensity equivalent to the illumination produced over a 1-cm^2 area by a 100-watt incandescent bulb at a distance of 1000 km is sufficient to produce photoelectrons within a second.

Example 3.10

Photoelectrons may be emitted from sodium ($\phi = 2.36$ eV) even for light intensities as low as 10^{-8} W/m^2. Calculate classically how much time the light must shine in order to produce a photoelectron of kinetic energy 1.00 eV.

Strategy We will assume that all of the light is absorbed in the first layer of atoms in the surface. Then we calculate the number of sodium atoms per unit area in a layer one atom thick. We assume that each atom in a single atomic layer absorbs equal energy, but a single electron in each of these atoms receives all the energy. We then calculate how long it takes these electrons to attain the energy (2.36 eV + 1.00 eV = 3.36 eV) needed for the electron to escape.

Solution We first find the number of Na atoms/volume:

$$\frac{\text{Avogadro's number}}{\text{Na gram molecular weight}} \times \text{density}$$

$$= \frac{\text{number of Na atoms}}{\text{volume}}$$

$$\frac{6.02 \times 10^{23} \text{ atoms/mol}}{23 \text{ g/mol}} \times 0.97 \frac{\text{g}}{\text{cm}^3}$$

$$= 2.54 \times 10^{22} \frac{\text{atoms}}{\text{cm}^3} = 2.54 \times 10^{28} \frac{\text{atoms}}{\text{m}^3} \quad (3.26)$$

To estimate the thickness of one layer of atoms, we assume a cubic structure.

$$\frac{1 \text{ atom}}{d^3} = 2.54 \times 10^{28} \frac{\text{atoms}}{\text{m}^3}$$

$$d = 3.40 \times 10^{-10} \text{ m}$$

$$= \text{thickness of one layer of sodium atoms}$$

If all the light is absorbed in the first layer of atoms, the number of exposed atoms per m^2 is

$$2.54 \times 10^{28} \frac{\text{atoms}}{\text{m}^3} \times 3.40 \times 10^{-10} \text{ m} = 8.64 \times 10^{18} \frac{\text{atoms}}{\text{m}^2}$$

With the intensity of 10^{-8} W/m^2, each atom will receive energy at the rate of

$$1.00 \times 10^{-8} \frac{\text{W}}{\text{m}^2} \times \frac{1}{8.64 \times 10^{18} \text{ atoms/m}^2}$$

$$= 1.16 \times 10^{-27} \text{ W}$$

$$= 1.16 \times 10^{-27} \frac{\text{J}}{\text{s}} \times \frac{1}{1.6 \times 10^{-19} \text{ J/eV}}$$

$$= 7.25 \times 10^{-9} \text{ eV/s}$$

If energy is absorbed at the rate of 7.25×10^{-9} eV/s for a single electron, we can calculate the time T needed to absorb 3.36 eV:

$$T = \frac{3.36 \text{ eV}}{7.25 \times 10^{-9} \text{ eV/s}} = 4.63 \times 10^8 \text{ s} = 14.7 \text{ years}$$

Based on classical calculations, the time required to eject a photoelectron should be 15 years!

Einstein's Theory

Albert Einstein was intrigued by Planck's hypothesis that the electromagnetic radiation field had to be absorbed and emitted in quantized amounts. Einstein took Planck's idea one step further and suggested that the *electromagnetic radiation field itself is quantized*, and that "the energy of a light ray spreading out from a point source is not continuously distributed over an increasing space but consists of a finite number of energy quanta which are localized at points in space, which move without dividing, and which can only be produced and absorbed as com-

Photons   plete units."* We now call these energy quanta of light **photons**. According to Einstein each photon has the energy quantum

Energy quantum
$$E = hf \tag{3.27}$$

where f is the frequency of the electromagnetic wave associated with the light, and h is Planck's constant. Notice that Equation (3.27) is consistent with Planck's relation for quantum of energy presented in Equation (3.25). The photon travels at the speed of light in a vacuum, and its wavelength is given by

$$\lambda f = c \tag{3.28}$$

In other words, Einstein proposed that in addition to its well-known wavelike aspect, amply exhibited in interference phenomena, light should also be considered to have a particle-like aspect. Einstein suggested that the photon (quantum of light) delivers its entire energy hf to a single electron in the material. In order to leave the material, the struck electron must give up an amount of energy ϕ to overcome its binding in the material. The electron may lose some additional energy by interacting with other electrons on its way to the surface. Whatever energy remains will then appear as kinetic energy of the electron as it leaves the emitter. The conservation of energy requires that

$$\text{Energy before (photon)} = \text{energy after (electron)}$$

$$hf = \phi + \text{K.E. (electron)} \tag{3.29}$$

Because the energies involved here are on the order of electron volts, we are safe in using the nonrelativistic form of the electron's kinetic energy, $\frac{1}{2}mv^2$. The electron's energy will be degraded as it passes through the emitter material, so, strictly speaking, we want to experimentally detect the maximum value of the kinetic energy.

$$hf = \phi + \frac{1}{2}mv_{max}^2 \tag{3.30}$$

The retarding potentials measured in the photoelectric effect are thus the opposing potentials needed to stop the most energetic electrons.

$$eV_0 = \frac{1}{2}mv_{max}^2 \tag{3.31}$$

*For an English translation of A. Einstein, *Annalen der Physik* **17**, 132 (1905), see A. B. Arons and M. B. Peppard, *American Journal of Physics* **33**, 367 (1965).

Quantum Interpretation

We should now reexamine the experimental results of the photoelectric effect to see whether Einstein's quantum interpretation can explain all the data. The first and second experimental results (which indicate that the kinetic energies of the photoelectrons depend on the light frequency, but not the light intensity) can be explained. The kinetic energy of the electrons, K.E. (electron) $= hf - \phi$ [see Equation (3.29)], does not depend on the light intensity at all, but only on the light frequency and the work function of the material.

$$\tfrac{1}{2} m v_{max}^2 = eV_0 = hf - \phi \qquad (3.32)$$

A potential slightly more positive than $-V_0$ will not be able to repel all the electrons, and, for a close geometry of the emitter and collector, practically all the electrons will be collected when the retarding voltage is near zero. For very large positive potentials all the electrons will be collected, and the photocurrent will level off as shown in Figure 3.12. If the light intensity increases, there will be more photons per unit area, more electrons ejected, and therefore a higher photocurrent, as displayed in Figure 3.12.

If a different light frequency is used, say f_2, then a different stopping potential is required to stop the most energetic electrons [see Equation (3.32)], $eV_{02} = hf_2 - \phi$. For a constant light intensity (more precisely, a constant number of photons/area/time), a different stopping potential V_0 is required for each f, but the maximum photocurrent will not change, because the number of photoelectrons ejected is constant (see Figure 3.13). The quantum theory easily explains Figure 3.15, because the number of photons increases linearly with the light intensity, producing more photoelectrons and hence more photocurrent.

Equation (3.32), proposed by Einstein in 1905, predicts that the stopping potential will be linearly proportional to the light frequency, with a slope h, *the same constant found by Planck*. The slope is independent of the metal used to construct the photocathode. Equation (3.32) can be rewritten

$$eV_0 = \tfrac{1}{2} m v_{max}^2 = hf - hf_0 = h(f - f_0) \qquad (3.33)$$

where $\phi = hf_0$ represents the negative of the y intercept. The frequency f_0 represents the threshold frequency for the photoelectric effect (when the kinetic energy of the electron is precisely zero). The data available in 1905 were not sufficiently accurate either to prove or to disprove Einstein's theory, and even Planck himself, among others, viewed the theory with skepticism. R. A. Millikan, then at the University of Chicago, tried to show Einstein was wrong by undertaking a series of elegant experiments that required almost 10 years to complete. In 1916 Millikan reported data shown in Figure 3.16 that confirmed Einstein's prediction. Millikan found the value of h from the slope of the line in Figure 3.16 to be 4.1×10^{-15} eV·s, in good agreement with the value of h determined for blackbody radiation by Planck. Einstein's theory of the photoelectric effect was gradually accepted after 1916; finally in 1922 he received the Nobel Prize for the year 1921, primarily for his explanation of the photoelectric effect.*

---

*R. A. Millikan also received the Nobel Prize in Physics in 1923, partly for his precise study of the photoelectric effect and partly for measuring the charge of the electron. Millikan's award was the last in a series of Nobel Prizes spanning 18 years that honored the fundamental efforts to measure and understand the photoelectric effect: Lenard, Einstein, and Millikan.

Figure 3.16 Millikan published data in 1916 for the photoelectric effect in which he shone light of varying frequency on a sodium electrode and measured the maximum kinetic energies of the photoelectrons. He found that no photoelectrons were emitted below a frequency of 4.39×10^{14} Hz (or longer than a wavelength of 683 nm). The results were independent of light intensity, and the slope of a straight line drawn through the data produced a value of Planck's constant in excellent agreement with Planck's theory. Even though Millikan admitted his own data were sufficient proof of Einstein's photoelectric effect equation, Millikan was not convinced of the photon concept for light and its role in quantum theory.

Quantization of electromagnetic radiation field

We should summarize what we have learned about the quantization of the electromagnetic radiation field. First, electromagnetic radiation consists of photons, which are particle-like (or corpuscular), each consisting of energy

$$E = hf = \frac{hc}{\lambda} \tag{3.34}$$

where f and λ are the frequency and wavelength of the light, respectively. The total energy of a beam of light is the sum total of the energy of all the photons and for monochromatic light is an integral multiple of hf (generally the integer is very large).

This representation of the photon picture must be true over the entire electromagnetic spectrum from radio waves to visible light, x rays, and even high-energy gamma rays. This must be true because, as we saw in Chapter 2, a photon of given frequency, observed from a moving system, can be redshifted or blue-shifted by an arbitrarily large amount, depending on the system's speed and direction of motion. We will examine these possibilities later. During emission or absorption of any form of electromagnetic radiation (light, x rays, gamma rays, and so on), photons must be created or absorbed. The photons have only one speed: the speed of light ($= c$ in vacuum).

Example 3.11

Light of wavelength 400 nm is incident upon lithium ($\phi = 2.93$ eV). Calculate (a) the photon energy and (b) the stopping potential V_0.

Strategy (a) Light is normally described by wavelengths in nm, so it is useful to have an equation to calculate the energy in terms of λ.

$$E = hf = \frac{hc}{\lambda}$$

$$= \frac{(6.626 \times 10^{-34}\,\text{J} \cdot \text{s})(2.998 \times 10^8\,\text{m/s})}{\lambda(1.602 \times 10^{-19}\,\text{J/eV})(10^{-9}\,\text{m/nm})}$$

$$E = \frac{1.240 \times 10^3\,\text{eV} \cdot \text{nm}}{\lambda} \tag{3.35}$$

(b) We use Equation (3.32) to determine the stopping potential once we know the frequency f and work function ϕ.

Solution (a) For a wavelength of $\lambda = 400$ nm we have for the photon's energy

$$E = \frac{1.240 \times 10^3\,\text{eV} \cdot \text{nm}}{400\,\text{nm}} = 3.10\,\text{eV}$$

(b) For the stopping potential, Equation (3.32) gives

$$eV_0 = hf - \phi = E - \phi = 3.10\,\text{eV} - 2.93\,\text{eV} = 0.17\,\text{eV}$$

$$V_0 = 0.17\,\text{V}$$

A retarding potential of 0.17 V will stop all photoelectrons.

Example 3.12

What frequency of light is needed to produce electrons of kinetic energy 3.0 eV from illumination of lithium?

Strategy We have enough information to determine the photon energy needed from Equation (3.30), and we can determine the frequency from $E = hf$.

Solution From Equation (3.30), we have

$$hf = \phi + \tfrac{1}{2}mv_{\text{max}}^2$$

$$= 2.93\,\text{eV} + 3.0\,\text{eV} = 5.9\,\text{eV}$$

The photon frequency is now found to be

$$f = \frac{E}{h} = \frac{(5.9\,\text{eV})(1.6 \times 10^{-19}\,\text{J/eV})}{(6.626 \times 10^{-34}\,\text{J} \cdot \text{s})}$$

$$= 1.42 \times 10^{15}\,\text{s}^{-1} = 1.42 \times 10^{15}\,\text{Hz}$$

Example 3.13

For the light intensity of Example 3.10, $\mathcal{I} = 10^{-8}$ W/m², a wavelength of 350 nm is used. What is the number of photons/(m²·s) in the light beam?

Strategy We first find the photon energy, and because we know the intensity, we will be able to determine the photon flux.

Solution From Equation (3.35) we find the photon energy E_γ:

$$E_\gamma = \frac{1.240 \times 10^3 \text{ eV} \cdot \text{nm}}{350 \text{ nm}} = 3.5 \text{ eV}$$

The intensity $\mathcal{I}$ is the product of the photon flux N and photon energy E_γ:

$$\text{Intensity } \mathcal{I} = \left[N\left(\frac{\text{photons}}{\text{m}^2 \cdot \text{s}}\right) \right]\left[E_\gamma\left(\frac{\text{energy}}{\text{photon}}\right) \right]$$

$$= N E_\gamma\left(\frac{\text{energy}}{\text{m}^2 \cdot \text{s}}\right)$$

where we have put the units of N and E_γ in parentheses. We solve this for N:

$$N = \frac{\mathcal{I}}{E_\gamma} = \frac{1.0 \times 10^{-8} \text{ J} \cdot \text{s}^{-1} \text{ m}^{-2}}{(1.6 \times 10^{-19} \text{ J/eV})(3.5 \text{ eV/photon})}$$

$$= 1.8 \times 10^{10} \frac{\text{photons}}{\text{m}^2 \cdot \text{s}}$$

Thus even a low-intensity light beam has a large flux of photons, and even a few photons can produce a photocurrent (albeit a very small one!).

The photoelectric effect is responsible for many applications in the detection of light. These include the photomultiplier tube for counting individual light pulses, photoelectric cells for light-activated devices (such as door openers and intrusion alarms), and solar panels.

3.7 X-Ray Production

In the photoelectric effect, a photon gives up all of its energy to an electron, which may then escape from the material to which it was bound. Can the inverse process occur? Can an electron (or any charged particle) give up its energy and create a photon? The answer is yes, but the process must be consistent with other laws of physics. Recall that photons must be created or absorbed as whole units. A photon cannot give up half its energy; it must give up all its energy. If in some physical process only part of the photon's energy were required, then a *new* photon would be created to carry away the remaining energy.

However, electrons do not act as photons. An electron may give up part or all of its kinetic energy and still be the same electron. When an electron interacts with the strong electric field of the atomic nucleus and is consequently accelerated, the electron radiates electromagnetic energy. According to classical electromagnetic theory, it should do so continuously. In the quantum picture we must think of the electron as emitting a series of photons with varying energies; this is the only way that the inverse photoelectric effect can occur. An energetic electron passing through matter will radiate photons and lose kinetic energy. The process by which photons are emitted by an electron slowing down is called **bremsstrahlung**, from the German word for "braking radiation." The process is shown schematically in Figure 3.17 where an electron of energy E_i passing

Bremsstrahlung process

Figure 3.17 Bremsstrahlung is a process by which an electron is accelerated while under the influence of the nucleus. The accelerated electron emits a photon.

through the electric field of a nucleus is accelerated and produces a photon of energy $E = hf$. The final energy of the electron is determined from the conservation of energy to be

$$E_f = E_i - hf \qquad (3.36)$$

Because linear momentum must be conserved, the nucleus absorbs very little energy, and it is ignored. One or more photons may be created in this way as electrons pass through matter.

In Section 3.1 we mentioned Röntgen's discovery of x rays. The x rays are produced by the bremsstrahlung effect in an apparatus shown schematically in Figure 3.18. Current passing through a filament produces copious numbers of electrons by thermionic emission. These electrons are focused by the cathode structure into a beam and are accelerated by potential differences of thousands of volts until they impinge on a metal anode surface, producing x rays by bremsstrahlung (and other processes) as they stop in the anode material. Much of the electron's kinetic energy is lost by heating the anode material and not by bremsstrahlung. The x-ray tube is evacuated so that the air between the filament and anode will not scatter the electrons. The x rays produced pass through the sides of the tube and can be used for a large number of applications, including medical diagnosis and therapy, fundamental research in crystal and liquid structure, and engineering diagnoses of flaws in large welds and castings.

X rays from a standard tube include photons of many wavelengths. By scattering x rays from crystals we can produce strongly collimated monochromatic (single-wavelength) x-ray beams. Early x-ray spectra produced by x-ray tubes of accelerating potential 35 kV are shown in Figure 3.19. These particular tubes had targets of tungsten, molybdenum, and chromium. The smooth, continuous x-ray spectra are those produced by bremsstrahlung, and the sharp "characteristic x rays" are produced by atomic excitations and will be explained in Section 4.6. X-ray wavelengths typically range from 0.01 to 1 nm. However, high-energy accelerators can produce x rays with wavelengths as short as 10^{-6} nm.

Notice that in Figure 3.19 the minimum wavelength λ_{min} for all three targets is the same. The minimum wavelength λ_{min} corresponds to the maximum frequency f_{max}. If the electrons are accelerated through a voltage V_0, then their kinetic energy is eV_0. The maximum photon energy therefore occurs when the

Figure 3.18 Schematic of x-ray tube where x rays are produced by the bremsstrahlung process of energetic electrons.

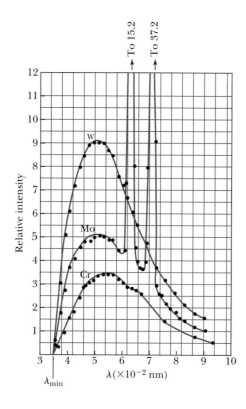

Figure 3.19 The relative intensity of x rays produced in an x-ray tube is shown for an accelerating voltage of 35 kV. Notice that λ_{min} is the same for all three targets. *From C. T. Ulrey*, Physical Review *11, 405 (1918).*

electron gives up all of its kinetic energy and creates one photon (this is relatively unlikely, however). This process is the **inverse photoelectric effect**. The conservation of energy requires that the electron kinetic energy equal the maximum photon energy (where we neglect the work function ϕ because it is normally so small compared to eV_0).

$$eV_0 = hf_{max} = \frac{hc}{\lambda_{min}}$$

or

$$\lambda_{min} = \frac{hc}{e}\frac{1}{V_0} = \frac{1.240 \times 10^{-6}\,\text{V} \cdot \text{m}}{V_0} \qquad (3.37) \qquad \textbf{Duane-Hunt rule}$$

The relation Equation (3.37) was first found experimentally and is known as the **Duane-Hunt limit** (or **rule**). Its explanation in 1915 by the quantum theory is now considered further evidence of Einstein's photon concept. The value λ_{min} depends only on the accelerating voltage and is the same for all targets.

Only the quantum hypothesis explains all of these data. Because the heavier elements have stronger nuclear electric fields, they are more effective in accelerating electrons and making them radiate. The intensity of the x rays increases with the square of the atomic number of the target. The intensity is also approximately proportional to the square of the voltage used to accelerate the electrons. This is why high voltages and tungsten anodes are so often used in x-ray machines. Tungsten also has a very high melting temperature and can withstand high electron-beam currents.

Conceptual Example 3.14

Explain how the Duane-Hunt limit can be used to determine the electron bombarding energy in a device like a scanning electron microscope.

Solution If we look closely at Equation (3.37), we can see that any reduction in the acceleration voltage V_0 will lead to an increase in the value of λ_{min}. A careful analysis of the minimum value of the wavelength should be in agreement with the expected voltage V_0. If the value of λ_{min} varies over time, for example depending on the electron beam current, it may be due to anomalous charging effects in the beam acceleration/transport system. Solutions for problems like this require painstaking efforts and may dictate the experimental conditions, such as using lower beam currents to avoid problems.

Example 3.15

If we have a tungsten anode (work function $\phi = 4.63$ eV) and electron acceleration voltage of 35 kV, why do we ignore in Equation (3.36) the initial kinetic energy of the electrons from the filament and the work functions of the filaments and anodes? What is the minimum wavelength of the x rays?

Strategy We can ignore the initial electron kinetic energies and the work functions, because they are on the order of a few electron volts (eV), whereas the kinetic energy of the electrons due to the accelerating voltage is 35,000 eV. The error in neglecting everything but eV_0 is small. We will use Equation (3.37) to determine the minimum wavelength.

Solution We use the Duane-Hunt rule of Equation (3.37) to determine

$$\lambda_{min} = \frac{1.240 \times 10^{-6}\,\text{V}\cdot\text{m}}{35.0 \times 10^{3}\,\text{V}} = 3.54 \times 10^{-11}\,\text{m}$$

which is in good agreement with the data of Figure 3.19.

3.8   Compton Effect

When a photon enters matter, it is likely to interact with one of the atomic electrons. According to classical theory, the electrons will oscillate at the photon frequency because of the interaction of the electron with the electric and magnetic field of the photon and will reradiate electromagnetic radiation (photons) at this same frequency. This is called **Thomson scattering**. However, in the early 1920s Arthur Compton experimentally confirmed an earlier observation by J. A. Gray that, especially at backward-scattering angles, there appeared to be a component of the emitted radiation (called a **modified** wave) that had a longer wavelength than the original primary (**unmodified**) wave. Classical electromagnetic theory cannot explain this modified wave. Compton then attempted to understand theoretically such a process and could find only one explanation: Einstein's photon particle concept must be correct. The scattering process is shown in Figure 3.20.

Thomson scattering

Compton proposed in 1923 that the photon is scattered from only one electron, rather than from all the electrons in the material, and that the laws of the conservation of energy and momentum apply as in any elastic collision between two particles. We recall from Chapter 2 that the momentum of a particle moving at the speed of light (photon) is given by

$$p = \frac{E}{c} = \frac{hf}{c} = \frac{h}{\lambda} \tag{3.38}$$

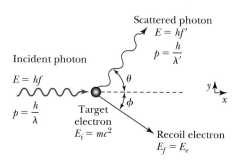

Figure 3.20 Compton scattering of a photon by an electron essentially at rest.

We treat the photon as a particle with a definite energy and momentum. Scattering takes place in a plane, which we take to be the xy plane in Figure 3.20. Both x and y components must be conserved, because of the vector nature of the linear momentum. The energy and momentum before and after the collision (treated relativistically) are given in Table 3.4.

In the final system the electron's total energy is related to its momentum by

$$E_e^2 = (mc^2)^2 + p_e^2 c^2 \qquad (3.39)$$

We can write the conservation laws now as

$$\text{Energy} \qquad hf + mc^2 = hf' + E_e \qquad (3.40a)$$

$$p_x \qquad \frac{h}{\lambda} = \frac{h}{\lambda'}\cos\theta + p_e\cos\phi \qquad (3.40b)$$

$$p_y \qquad \frac{h}{\lambda'}\sin\theta = p_e\sin\phi \qquad (3.40c)$$

We will relate the change in wavelength $\Delta\lambda = \lambda' - \lambda$ to the scattering angle θ of the photon. We first eliminate the recoil angle ϕ by squaring Equations (3.40b) and (3.40c) and adding them together, resulting in

$$p_e^2 = \left(\frac{h}{\lambda}\right)^2 + \left(\frac{h}{\lambda'}\right)^2 - 2\left(\frac{h}{\lambda}\right)\left(\frac{h}{\lambda'}\right)\cos\theta \qquad (3.41)$$

Professor Arthur Compton (1892–1962) is shown here in 1931 looking into an ionization chamber that he designed to study cosmic rays in the atmosphere. These complex detectors had to be quite sturdy and were carefully tested and calibrated on Earth before being sent up with balloons. Compton received his degrees from the College of Wooster and Princeton University. He spent most of his career at the University of Chicago and the Washington University, St. Louis. After his early work with x rays for which he received the Nobel Prize in 1927, he was a pioneer in high-energy physics through his cosmic ray studies.
UPI/Bettmann/Corbis.

Table 3.4 Results of Compton Scattering		
Energy or Momentum	**Initial System**	**Final System**
Photon energy	hf	hf'
Photon momentum in x direction (p_x)	$\dfrac{h}{\lambda}$	$\dfrac{h}{\lambda'}\cos\theta$
Photon momentum in y direction (p_y)	0	$\dfrac{h}{\lambda'}\sin\theta$
Electron energy	mc^2	$E_e = mc^2 + \text{K.E.}$
Electron momentum in x direction (p_x)	0	$p_e\cos\phi$
Electron momentum in y direction (p_y)	0	$-p_e\sin\phi$

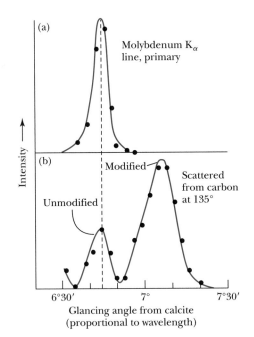

Figure 3.21 Compton's original data showing (a) the primary x-ray beam from Mo unscattered , and (b) the scattered spectrum from carbon at 135° showing both the modified and unmodified wave. *Adapted from Arthur H. Compton, Physical Review* **22**, *409 (1923).*

Then we substitute E_e from Equation (3.40a) and p_e from Equation (3.41) into Equation (3.39) (setting $\lambda = c/f$).

$$[h(f - f') + mc^2]^2 = m^2c^4 + (hf)^2 + (hf')^2 - 2(hf)(hf')\cos\theta$$

Squaring the left-hand side and canceling terms leaves

$$mc^2(f - f') = hff'(1 - \cos\theta)$$

Rearranging terms gives

$$\frac{h}{mc^2}(1 - \cos\theta) = \frac{f - f'}{ff'} = \frac{\dfrac{c}{\lambda} - \dfrac{c}{\lambda'}}{\dfrac{c^2}{\lambda\lambda'}} = \frac{1}{c}(\lambda' - \lambda)$$

or

Compton effect

$$\Delta\lambda = \lambda' - \lambda = \frac{h}{mc}(1 - \cos\theta) \qquad (3.42)$$

which is the result Compton found in 1923 for the increase in wavelength of the scattered photon.

Compton then proceeded to check the validity of his theoretical result by performing a very careful experiment in which he scattered x rays of wavelength 0.071 nm from carbon at several angles. He showed that the modified wavelength was in good agreement with his prediction.* A part of his data is shown in Figure 3.21, where both the modified (λ') and unmodified (λ) scattered waves are identified.

*An interesting personal account of Compton's discovery can be found in A. H. Compton, *American Journal of Physics* **29**, 817–820 (1961).

The kinetic energy and scattering angle of the recoiling electron can also be predicted. Experiments in which the recoiling electrons were detected were soon carried out, thus completely confirming Compton's theory. The process of elastic photon scattering from electrons is now called the **Compton effect**. Note that the difference in wavelength, $\Delta\lambda = \lambda' - \lambda$, depends only on the constants h, c, and m_e in addition to the scattering angle θ. The quantity $\lambda_c = h/m_e c = 2.426 \times 10^{-3}$ nm is called the **Compton wavelength** of the electron. Only for wavelengths on the same order as λ_c (or shorter) will the fractional shift $\Delta\lambda/\lambda$ be large. For visible light, for example with $\lambda = 500$ nm, the maximum $\Delta\lambda/\lambda$ is on the order of 10^{-5}, and $\Delta\lambda$ would be difficult to detect. The probability of the occurrence of the Compton effect for visible light is also quite small. However, for the x rays of wavelength 0.071 nm used by Compton, the ratio of $\Delta\lambda/\lambda$ is ~0.03 and could easily be observed. Thus, the Compton effect is important only for x rays or γ-ray photons and is small for visible light.

Compton wavelength

The physical process of the Compton effect can be described as follows. The photon elastically scatters from an essentially free electron in the material. (The photon's energy is so much larger than the binding energy of the almost free electron that the atomic binding energy can be neglected.) The newly created scattered photon then has a modified, longer wavelength. What happens if the photon scatters from one of the tightly bound inner electrons? Then the binding energy is not negligible, and the electron might not be dislodged. The scattering is then effectively from the entire atom (nucleus + electrons). Then the mass in Equation (3.42) would be several thousand times larger than m_e, and $\Delta\lambda$ would be correspondingly smaller. Scattering from tightly bound electrons results in the unmodified photon scattering ($\lambda \approx \lambda'$), which is also observed in Figure 3.21. Thus, the quantum picture also explains the existence of the unmodified wavelength predicted by the classical theory (Thomson scattering) alluded to earlier.

The success of the Compton theory convincingly demonstrated the correctness of both the quantum concept and the particle nature of the photon. The use of the laws of the conservation of energy and momentum applied relativistically to pointlike scattering of the photon from the electron finally convinced the great majority of scientists of the validity of the new modern physics. Compton received the Nobel Prize in Physics for this discovery in 1927.

Example 3.16

An x ray of wavelength 0.050 nm scatters from a gold target. (a) Can the x ray be Compton-scattered from an electron bound by as much as 62,000 eV? (b) What is the largest wavelength of scattered photon that can be observed? (c) What is the kinetic energy of the most energetic recoil electron and at what angle does it occur?

Strategy We first determine the x-ray energy to see if it has enough energy to dislodge the electron. We use Equation (3.42) with both the atomic and electron mass to determine the scattered photon wavelength. We then use the conservation of energy to determine the recoil electron kinetic energy.

Solution From Equation (3.35) the x-ray energy is

$$E_{x\,ray} = \frac{1.240 \times 10^3 \text{ eV} \cdot \text{nm}}{0.050 \text{ nm}} = 24,800 \text{ eV} = 24.8 \text{ keV}$$

Therefore, the x ray does not have enough energy to dislodge the inner electron, which is bound by 62 keV. In this case we have to use the atomic mass in Equation (3.42), which results in little change in the wavelength (Thomson scattering).

Scattering may still occur from outer electrons, so we examine Equation (3.42) with the electron mass. The longest wavelength $\lambda' = \lambda + \Delta\lambda$ occurs when $\Delta\lambda$ is a maximum or when $\theta = 180°$.

$$\lambda' = \lambda + \frac{h}{m_e c}(1 - \cos 180°) = \lambda + \frac{2h}{m_e c}$$

$$= 0.050 \text{ nm} + 2(0.00243 \text{ nm}) = 0.055 \text{ nm}$$

The energy of the scattered photon is then a minimum and has the value

$$E'_{x\text{ ray}} = \frac{1.240 \times 10^3 \text{ eV} \cdot \text{nm}}{0.055 \text{ nm}} = 2.25 \times 10^4 \text{ eV} = 22.5 \text{ keV}$$

The difference in energy of the initial and final photon must equal the kinetic energy of the electron (neglecting binding energies). The recoil electron must scatter in the forward di-

rection at $\phi = 0°$ when the final photon is in the backward direction ($\theta = 180°$) in order to conserve momentum. The kinetic energy of the electron is then a maximum.

$$E_{x\text{ ray}} = E'_{x\text{ ray}} + \text{K.E. (electron)}$$

$$\text{K.E. (electron)} = E_{x\text{ ray}} - E'_{x\text{ ray}}$$

$$= 24.8 \text{ keV} - 22.5 \text{ keV} = 2.3 \text{ keV}$$

Because $\Delta\lambda$ does not depend on λ or λ', we can determine the wavelength (and energy) of the incident photon by merely observing the kinetic energy of the electron at forward angles (see Problem 58).

3.9 Pair Production and Annihilation

A guiding principle of scientific investigation, if not a general rule of nature, is that if some process is not absolutely forbidden (by some law like conservation of energy, momentum, or charge), then we might expect that it will eventually occur. In the photoelectric effect, bremsstrahlung, and the Compton effect, we have studied exchanges of energy between photons and electrons. Have we covered all possible mechanisms? For example, can the kinetic energy of a photon be converted into particle mass and vice versa? It would appear that if none of the conservation laws are violated, then such a process should be possible.

First, let us consider the conversion of photon energy into mass. The electron, which has a mass ($m = 0.511 \text{ MeV}/c^2$), is the lightest particle within an atom. If a photon can create an electron, it must also create a positive charge to balance charge conservation. In 1932, C. D. Anderson (Nobel Prize in Physics, 1936) observed a positively charged electron (e^+) in cosmic radiation. This particle, called **Positron** a **positron**, had been predicted to exist several years earlier by P. A. M. Dirac (Nobel Prize in Physics, 1933). It has the same mass as the electron but an opposite charge. Positrons are also observed when high-energy gamma rays (photons) pass through matter. Experiments show that a photon's energy can be converted entirely into an electron and a positron in a process called **pair production**. The reaction is

Pair production
$$\gamma \to e^+ + e^- \tag{3.43}$$

However, this process occurs only when the photon passes through matter, because energy and momentum would not be conserved if the reaction took place in isolation. The missing momentum must be supplied by interaction with a nearby massive object such as a nucleus.

Example 3.17

Show that a photon cannot produce an electron-positron pair in free space as shown in Figure 3.22a.

Strategy We need to look carefully at the conservation of momentum and energy to see if pair production can occur in free space.

Solution Let the total energy and momentum of the electron and the positron be E_-, p_- and E_+, p_+, respectively. The conservation laws are then

Energy $\qquad hf = E_+ + E_-$ (3.44a)

Momentum, $p_x \qquad \dfrac{hf}{c} = p_- \cos\theta_- + p_+ \cos\theta_+$ (3.44b)

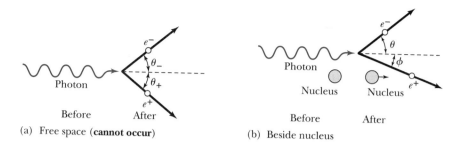

Figure 3.22 (a) A photon cannot decay into an electron-positron pair in free space, but (b) if a nucleus is nearby, the nucleus can absorb sufficient linear momentum to allow the process to proceed.

Momentum, p_y $0 = p_- \sin \theta_- - p_+ \sin \theta_+$ (3.44c)

Equation (3.44b) can be written as

$$hf = p_-c \cos \theta_- + p_+c \cos \theta_+ \qquad (3.45)$$

If we insert $E_\pm^2 = p_\pm^2 c^2 + m^2 c^4$ into Equation (3.44a), we have

$$hf = \sqrt{p_+^2 c^2 + m^2 c^4} + \sqrt{p_-^2 c^2 + m^2 c^4} \qquad (3.46)$$

The maximum value of hf is, from Equation (3.45),

$$hf_{max} = p_-c + p_+c$$

But from Equation (3.46), we also have

$$hf > p_-c + p_+c$$

Equations (3.45) and (3.46) are inconsistent and cannot simultaneously be valid. Equations (3.44), therefore, do not describe a possible reaction. The reaction displayed in Figure 3.22a is not possible, because energy and momentum are not simultaneously conserved.

Consider the conversion of a photon into an electron and a positron that takes place inside an atom where the electric field of a nucleus is large. The nucleus recoils and takes away a negligible amount of energy but a considerable amount of momentum. The conservation of energy will now be

$$hf = E_+ + E_- + \text{K.E. (nucleus)} \qquad (3.47)$$

A diagram of the process is shown in Figure 3.22b. The photon energy must be at least equal to $2m_e c^2$ in order to create the rest masses.

$$hf > 2m_e c^2 = 1.022 \text{ MeV} \qquad \text{(for pair production)} \qquad (3.48)$$

The probability of pair production increases dramatically both with higher photon energy and with higher atomic number Z of the nearby nucleus because of the correspondingly higher electric field that mediates the process.

The next question concerns the new particle, the positron. Why is it not commonly found in nature? We also need to answer the question posed earlier: can mass be converted to pure kinetic energy?

Positrons are found in nature. They are detected in cosmic radiation and as products of radioactivity from several radioactive elements. However, their lives are doomed because of their interaction with electrons. When positrons and electrons are in proximity for even a short time they annihilate each other, producing photons. A positron passing through matter will quickly lose its kinetic energy through atomic collisions and will likely **annihilate** with an electron. After a positron slows down, it is drawn to an electron by their mutual electric attraction, and the electron and positron may then form an atomlike configuration called **positronium**, in which they orbit around their common center of mass. Eventually the electron and positron come together and annihilate each other (typically in 10^{-10} s), producing electromagnetic radiation (photons). The process $e^+ + e^- \rightarrow \gamma + \gamma$ is called **pair annihilation**.

Pair annihilation

(a)   Positronium,
before decay
(schematic only)

(b)   After annihilation

Figure 3.23 Annihilation of positronium atom (consisting of an electron and positron), producing two photons.

Consider a positronium "atom" in free space. It must emit at least two photons in order to conserve energy and momentum. If the positronium annihilation takes place near a nucleus, it is possible that only one photon will be created, because the missing momentum can be supplied by nucleus recoil as in pair production. Under certain conditions even three photons may be produced. Because the emission of two photons is by far the most likely annihilation mode, let us consider this mode, displayed in Figure 3.23. The conservation laws for the process $(e^+e^-)_{\text{atom}} \rightarrow \gamma + \gamma$ will be (we neglect the atomic binding energy of about 6.8 eV)

$$\text{Energy} \qquad 2m_ec^2 \approx hf_1 + hf_2 \qquad (3.49a)$$

$$\text{Momentum} \qquad 0 = \frac{hf_1}{c} - \frac{hf_2}{c} \qquad (3.49b)$$

Conceptual Example 3.18

Fluorine-18 is a radioactive nuclide that is a e^+ emitter and is used with blood flow within the brain to study brain disorders. Positron emission tomography (PET) scans rely on gamma rays (photons) being emitted in opposite directions for detection in scans. How is it possible for gamma rays to be simultaneously emitted in opposite directions?

Solution When fluorine-18 emits a e^+, the e^+ soon binds with an electron to form a positronium atom, which has a relatively low energy and linear momentum. When the annihilation occurs, 1.022 MeV is available, and both the conservation of energy and linear momentum must be obeyed. Conservation of momentum requires the photons (gamma rays) to emerge in precisely opposite directions with equal energies if the initial linear momentum is zero.

The frequency of the two photons will be almost identical, so that $f_1 = f_2 = f$. Thus Equation (3.49a) becomes

$$2m_ec^2 = 2hf$$

or

$$hf = m_ec^2 = 0.511 \text{ MeV} \qquad (3.50)$$

In other words, the two photons from positronium annihilation will move in opposite directions, each with energy 0.511 MeV. This is exactly what is observed experimentally.

The production of two photons in opposite directions with energies just over 0.5 MeV is so characteristic a signal of the presence of a positron that it has **PET scan** useful applications. Positron emission tomography (PET) scanning has become

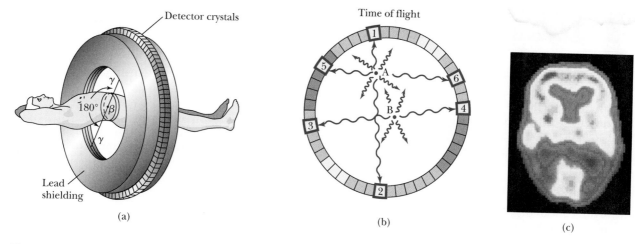

Figure 3.24 Positron emission tomography is a useful medical diagnostic tool to study the path and location of a positron-emitting radiopharmaceutical in the human body. (a) Appropriate radiopharmaceuticals are chosen to concentrate by physiological processes in the region to be examined. (b) The positron travels only a few millimeters before annihilation, which produces two photons that can be detected to give the positron position. (c) PET scan of a normal brain. *(a) and (b) are after G. L. Brownell et al.,* Science **215**, *619 (1982); (c) National Institutes of Health/SPL/Photo Researchers, Inc.*

a standard diagnostic technique in medicine. A positron-emitting radioactive chemical (containing a nucleus such as ^{15}O, ^{11}C, ^{13}N, or ^{18}F) injected into the body causes two characteristic annihilation photons to be emitted from the points where the chemical has been concentrated by physiological processes. The location in the body where the photons originate is identified by measuring the directions of two gamma-ray photons of the correct energy that are detected in coincidence, as shown in Figure 3.24. Measurement of blood flow in the brain is an example of a diagnostic tool used in the evaluation of strokes, brain tumors, and other brain lesions.

Before leaving the subject of positrons we should pursue briefly the idea of **antiparticles**. The positron is the antiparticle of the electron, having the opposite charge but the same mass.* In 1955 the antiproton was discovered by E. G. Segrè and O. Chamberlain (Nobel Prize, 1959), and today, many antiparticles are known. We now believe that every particle has an antiparticle. In some cases, as for photons or neutral pi-mesons, the particle and antiparticle are the same, but for most other particles (for example, the neutron and proton), particle and antiparticle are distinct.

Antiparticles

We know that matter and antimatter cannot exist together in our world, because their ultimate fate will be annihilation. However, we may let our speculation run rampant! If we believe in symmetry, might there not be another world, perhaps in a distant galaxy that is made of antimatter? Because galaxies are so far apart in space, annihilation would be infrequent. Modern cosmology predicts that the universe should be made up almost entirely of real particle matter and explains the obvious asymmetry this involves. However, if a large chunk of antimatter ever struck the Earth, it would tend to restore the picture of a symmetric universe. As we see from Problem 57, however, in such an event there would be no one left to receive the appropriate Nobel Prize.

*Other particle properties (for example, spin) are described later (particularly in Chapters 7 and 14) and also need to be considered.

Summary

In 1895 Röntgen discovered x rays, and in 1897 Thomson proved the existence of electrons and measured their charge-to-mass ratio. Finally, in 1911 Millikan reported an accurate determination of the electron's charge. Experimental studies resulted in the empirical Rydberg equation to calculate the wavelengths of the hydrogen atom's spectrum:

$$\frac{1}{\lambda} = R_H \left(\frac{1}{n^2} - \frac{1}{k^2} \right) \qquad k > n \qquad (3.13)$$

where $R_H = 1.096776 \times 10^7 \text{ m}^{-1}$.

In order to explain blackbody radiation Planck proposed his quantum theory of radiation in 1900, which signaled the era of modern physics. From Planck's theory we can derive Wien's displacement law:

$$\lambda_{max} T = 2.898 \times 10^{-3} \text{ m} \cdot \text{K} \qquad (3.14)$$

and the Stefan-Boltzmann law:

$$R(T) = \epsilon \sigma T^4 \qquad (3.16)$$

Planck's radiation law gives the power radiated per unit area per unit wavelength from a blackbody.

$$\mathcal{I}(\lambda, T) = \frac{2\pi c^2 h}{\lambda^5} \frac{1}{e^{hc/\lambda kT} - 1} \qquad (3.23)$$

The oscillators of the electromagnetic radiation field can change energy only by quantized amounts given by $\Delta E = hf$, where $h = 6.6261 \times 10^{-34} \text{ J} \cdot \text{s}$ is called *Planck's constant*.

Classical theory could not explain the photoelectric effect, but in 1905 Einstein proposed that the electromagnetic radiation field itself is quantized. We call these particle-like quanta of light *photons*, and they each have energy $E = hf$ and momentum $p = h/\lambda$. The photoelectric effect is easily explained by the photons each interacting with only one electron. The conservation of energy gives

$$hf = \phi + \tfrac{1}{2} m v^2_{max} \qquad (3.30)$$

where ϕ is the work function of the emitter. The retarding potential required to stop all electrons depends only on the photon's frequency

$$eV_0 = \tfrac{1}{2} m v^2_{max} = hf - hf_0 \qquad (3.33)$$

where $\phi = hf_0$. Millikan proved experimentally in 1916 that Einstein's theory was correct.

Bremsstrahlung radiation (x rays) is emitted when charged particles (for example, electrons) pass through matter and are accelerated by the nuclear field. These x rays have a minimum wavelength

$$\lambda_{min} = \frac{hc}{eV_0} \qquad (3.37)$$

where electrons accelerated by a voltage of V_0 impinge on a target.

In the Compton effect a photon scatters from an electron with a new photon created, and the electron recoils. For an incident and an exit photon of wavelength λ and λ', respectively, the change in wavelength is

$$\Delta\lambda = \lambda' - \lambda = \frac{h}{mc}(1 - \cos\theta) \qquad (3.42)$$

when the exit photon emerges at angle θ to the original photon direction. The Compton wavelength of the electron is $\lambda_c = h/m_e c = 2.426 \times 10^{-3}$ nm. The success of the Compton theory in 1923 convincingly demonstrated the particle-like nature of the photon.

Finally, photon energy can be converted into mass in pair production:

$$\gamma \rightarrow e^+ + e^- \qquad (3.43)$$

where e^+ is the positron, the antiparticle of the electron. Similarly, a particle and antiparticle annihilate catastrophically in the process

$$e^+ + e^- \rightarrow \gamma + \gamma$$

called pair annihilation.

Questions

1. How did the ionization of gas by cathode rays prevent H. Hertz from discovering the true character of electrons?

2. Why do television tubes generally deflect electrons with magnetic fields rather than with electric fields, as is done in cathode-ray-tube oscilloscopes?

3. In Thomson's e/m experiment, does it matter whether the electron passing through interacts first with the electric field or with the magnetic field or both? Explain.

4. Women in the late 1890s were terrified about the possible misuse of the new Röntgen x rays. What use do you think they envisioned?

5. In the late 1890s many people had x rays taken of their body. X-ray machines were common in shoe stores in the late 1940s and early 1950s for people to examine how their shoes fit; customers enjoyed seeing pictures of their bones. Discuss the safety of these undertakings.

6. Parents tell their children not to sit close to the television screen. Can x rays be produced in TVs? Explain.

7. In Example 3.2, why would you be concerned about observing a cluster of several balls in the Millikan electron charge experiment?

8. In Figure 3.5 why are the histogram peaks harder to identify as the charge increases?

9. How is it possible for the plastic balls in Example 3.2 to have both positive and negative charges? What is happening?

10. Why do you suppose Millikan tried several different kinds of oil, as well as H_2O and Hg, for his oil-drop experiment?

11. In the experiment of Example 3.2, how could you explain an experimental value of $q = 0.8 \times 10^{-19}$ C?

12. Why do you suppose scientists worked so hard to develop better diffraction gratings?

13. Why was helium discovered in the sun's spectrum before being observed on Earth? Why was hydrogen observed on Earth first?

14. Do you believe there is any relation between the wavelengths of the Paschen (1908) and Pfund (1924) series and the respective dates they were discovered? Explain.

15. It is said that no two snowflakes look exactly alike, but we know that snowflakes have a quite regular, although complex, crystal structure. Discuss how this could be due to quantized behavior.

16. Why do we say that the elementary units of matter or "building blocks" must be some basic unit of mass-energy rather than only of mass?

17. Why is a red-hot object cooler than a white-hot one of the same material?

18. Why did scientists choose to study blackbody radiation from something as complicated as a hollow container rather than the radiation from something simple like a thin solid cylinder (such as a dime)?

19. Why does the sun apparently act as a blackbody?

20. Astronomers determine the surface temperature of a star by measuring its brightness at different frequencies. Explain how they can then use the Planck radiation law to obtain the surface temperature.

21. In a typical photoelectric effect experiment, consider replacing the metal photocathode by a gas. What difference would you expect?

22. What do the work functions of Table 3.3 tell us about the properties of particular metals? Which have the most tightly and least tightly bound electrons?

23. Why is it important to produce x-ray tubes with high accelerating voltages that are also able to withstand electron currents?

24. For a given beam current and target thickness, why would you expect a tungsten target to produce a higher x-ray intensity than targets of molybdenum or chromium?

25. List all possible known interactions between photons and electrons discussed in this chapter. Can you think of any more?

26. Discuss why it is difficult to see the Compton effect using visible light.

27. What do you believe to be an optimum lifetime for a positron-emitting radioactive nuclide used in brain tumor diagnostics? Explain.

Problems

Note: The more challenging problems are indicated by a shaded box.

3.1 Discovery of the X Ray and the Electron

1. Design an apparatus that will produce the correct magnetic field needed in Figure 3.2.

2. For an electric field of 2×10^5 V/m, what is the strength of the magnetic field needed to pass an electron of speed 2×10^6 m/s with no deflection? Draw $\vec{v}$, $\vec{E}$, and $\vec{B}$ directions for this to occur.

3. Across what potential difference does an electron have to be accelerated in order to reach the speed $v = 2 \times 10^7$ m/s? Should you use relativistic calculations?

4. An electron entering Thomson's e/m apparatus (Figures 3.2 and 3.3) has an initial velocity (in horizontal direction only) of 0.50×10^7 m/s. Lying around the lab is a permanent horseshoe magnet of strength 1.3×10^{-2} T, which you would like to use. (a) What electric field will you need in order to produce zero deflection of the electrons as they travel through the

apparatus? (b) The length of nonzero $\vec{E}$ and $\vec{B}$ fields is 2.0 cm. When the magnetic field is turned off, but the same electric field remains, how far in the vertical direction will the electron beam be deflected over this length?

3.2 Determination of Electron Charge

5. Consider the following possible forces on an oil drop in Millikan's experiment: gravitational, electrical, frictional, and buoyant. Draw a diagram indicating the forces on the oil drop (a) when the electric field is turned off and the droplet is falling freely, and (b) when the electric field causes the droplet to rise.

6. Neglecting the buoyancy force on an oil droplet, show that the terminal speed of the droplet is $v_t = mg/b$, where b is the coefficient of friction when the droplet is in free fall. (Remember that the frictional force $\vec{F}_f$ is given by $\vec{F}_f = -b\vec{v}$ where velocity is a vector.)

7. Stokes's law relates the coefficient of friction b to the radius r of the oil drop and the viscosity η of the medium the droplet is passing through: $b = 6\pi\eta r$. Show that the radius of the oil drop is given in terms of the terminal velocity v_t (see previous problem), η, g, and the density of the oil ρ by $r = 3\sqrt{\eta v_t/2g\rho}$.

8. In a Millikan oil-drop experiment the terminal velocity of the droplet is observed to be 1.3 mm/s. The density of the oil is $\rho = 900$ kg/m^3 and the viscosity of air is $\eta = 1.82 \times 10^{-5}$ kg/m·s. Using the results of the two previous problems, calculate (a) the droplet radius, (b) the mass of the droplet, and (c) the coefficient of friction.

3.3 Line Spectra

9. What is the series limit (that is, the smallest wavelength) for (a) the Lyman series and (b) the Balmer series?

10. Light from a slit passes through a transmission diffraction grating of 400 lines/mm, which is located 2.5 m from a screen. What are the distances on the screen (from the unscattered slit image) of the three brightest visible (first-order) hydrogen lines?

11. A transmission diffraction grating of 420 lines/mm is used to study the light intensity of different orders (n). A screen is located 2.5 m from the grating. What is the separation on the screen between the three brightest red lines for a hydrogen source?

12. Calculate the four largest wavelengths for the Brackett and Pfund series for hydrogen.

13. Josef von Fraunhofer made the first diffraction grating in 1821 and used it to measure the wavelengths of specific colors as well as the dark lines in the solar spectrum. His first diffraction grating consisted of 262 parallel wires. Assume that the wires were 0.20 mm apart and that Fraunhofer could resolve two spectral lines that were deflected at angles 0.50 min of arc apart. Using this grating, what is the minimum separation (in wavelength) of two first-order spectral lines near a wavelength of 400 nm that can be resolved?

14. Suppose that a detector in the Hubble Space Telescope was capable of detecting visible light in the wavelength range of 400–700 nm. (a) List all the wavelengths for the hydrogen atom that are in this range and their series name. (b) The detector measures visible wavelengths of 537.5 nm, 480.1 nm, 453.4 nm that researchers believe are due to the hydrogen atom. Why are all the known visible hydrogen lines not detected? (c) Use these data to calculate the speed of the stellar object that emitted the spectra. Assume that the object is not rotating. Why might rotation be an issue?

15. The Spitzer Space Telescope was launched in 2003 to detect infrared radiation. Suppose that a particular detector on the telescope is sensitive over part of the near infrared region of wavelengths 980–1920 nm. Astronomers want to detect the radiation being emitted from a red giant star and decide to concentrate on wavelengths from the Paschen series of the hydrogen atom. (a) What are the known wavelengths in this wavelength region? (b) The detector measures wavelengths of 1334.5, 1138.9, 1046.1 nm believed to be from the Paschen series. Why are these wavelengths different than those found in part (a)? (c) How fast is the star moving with respect to us?

3.4 Quantization

16. Quarks have charges $\pm e/3$ and $\pm 2e/3$. What combination of three quarks could yield (a) a proton, (b) a neutron?

3.5 Blackbody Radiation

17. Calculate λ_{max} for blackbody radiation for (a) liquid helium (4.2 K), (b) room temperature (293 K), and (c) a steel furnace (2500 K).

18. Calculate the temperature of a blackbody if the spectral distribution peaks at (a) gamma rays, $\lambda = 1.00 \times 10^{-14}$ m, (b) x rays, 1.00 nm, (c) red light, 670 nm, (d) broadcast television waves, 1.00 m, and (e) AM radio waves, 204 m.

19. A blackbody's temperature is increased from 900 K to 1900 K. By what factor does the total power radiated per unit area increase?

20. (a) At what wavelength will the human body radiate the maximum radiation? (b) Estimate the total power radiated by a person of medium build (assume an area given by a cylinder of 165-cm height and 13-cm radius).

21. White dwarf stars have been observed with a surface temperature as hot as 200,000°C. What is the wavelength of the maximum intensity produced by this star?

22. For a temperature of 5800 K (the sun's surface temperature) find the wavelength λ for which the spectral distribution calculated by the Planck and Rayleigh-Jeans results differ by 5%.

23. A tungsten filament of a typical incandescent light-bulb operates at a temperature near 3000 K. At what wavelength is the intensity a maximum?

24. Use a computer to calculate Planck's radiation law for a temperature of 3000 K, which is the temperature of a typical tungsten filament in an incandescent light-bulb. Plot the intensity versus wavelength. (a) How much of the power is in the visible region (400–700 nm) compared with the ultraviolet and infrared? (b) What is the ratio of the intensity at 400 nm and 700 nm to the wavelength with maximum intensity?

25. Show that the ultraviolet catastrophe is avoided for short wavelengths ($\lambda \to 0$) with Planck's radiation law by calculating the limiting intensity $\mathscr{I}(\lambda, T)$ as $\lambda \to 0$.

26. Estimate the power radiated by (a) a basketball at $20°C$, (b) the human body (assume a temperature of $37°C$).

27. At what wavelength is the radiation emitted by the human body a maximum? Assume a temperature of $37°C$.

28. If we have waves in a one-dimensional box, such that the wave displacement $\Psi(x, t) = 0$ for $x = 0$ and $x = L$, where L is the length of the box, and

$$\frac{1}{c^2}\frac{\partial^2 \Psi}{\partial t^2} - \frac{\partial^2 \Psi}{\partial x^2} = 0 \qquad \text{(wave equation)}$$

show that the solutions are of the form

$$\Psi(x, t) = a(t)\sin\left(\frac{n\pi x}{L}\right) \qquad (n = 1, 2, 3, \dots)$$

and $a(t)$ satisfies the (harmonic-oscillator) equation

$$\frac{d^2 a(t)}{dt^2} + \Omega_n^2 a(t) = 0$$

where $\Omega_n = n\pi c/L$ is the angular frequency, $2\pi f$.

29. If the angular frequencies of waves in a three-dimensional box of sides L generalize to

$$\Omega = \frac{\pi c}{L}(n_x^2 + n_y^2 + n_z^2)^{1/2}$$

where all n are integers, show that the number of distinct states in the frequency interval f to $f + \Delta f$ ($f = \Omega/2\pi$) is given by (where f is large)

$$dN = 4\pi \frac{L^3}{c^3} f^2\, df$$

30. Let the energy density in the frequency interval f to $f + df$ within a blackbody at temperature T be $dU(f, T)$. Show that the power emitted through a small hole of area ΔA in the container is

$$\frac{c}{4} dU(f, T)\, \Delta A$$

31. Derive the Planck radiation law emitted by a black-body. Remember that light has two directions of polarization and treat the waves as an ensemble of harmonic oscillators.

3.6 Photoelectric Effect

32. An FM radio station of frequency 107.7 MHz puts out a signal of 50,000 W. How many photons/s are emitted?

33. How many photons/s are contained in a beam of electromagnetic radiation of total power 150 W if the source is (a) an AM radio station of 1100 kHz, (b) 8.0-nm x rays, and (c) 4.0-MeV gamma rays?

34. What is the threshold frequency for the photoelectric effect on lithium ($\phi = 2.93$ eV)? What is the stopping potential if the wavelength of the incident light is 400 nm?

35. What is the maximum wavelength of incident light that can produce photoelectrons from silver ($\phi = 4.64$ eV)? What will be the maximum kinetic energy of the photoelectrons if the wavelength is halved?

36. A 2.0-mW laser ($\lambda = 530$ nm) shines on a cesium photocathode ($\phi = 1.95$ eV). Assume an efficiency of 10^{-5} for producing photoelectrons (that is, one photoelectron produced for every 10^5 incident photons) and determine the photoelectric current.

37. An experimenter finds that no photoelectrons are emitted from tungsten unless the wavelength of light is less than 270 nm. Her experiment will require photoelectrons of maximum kinetic energy 2.0 eV. What frequency of light should be used to illuminate the tungsten?

38. The human eye is sensitive to a pulse of light containing as few as 100 photons. For yellow light of wavelength 580 nm how much energy is contained in the pulse?

39. In a photoelectric experiment it is found that a stopping potential of 1.00 V is needed to stop all the electrons when incident light of wavelength 260 nm is used and 2.30 V is needed for light of wavelength 207 nm. From these data determine Planck's constant and the work function of the metal.

40. What is the limit of energies and frequencies for visible light of wavelengths 400–700 nm?

3.7 X-Ray Production

41. What is the minimum x-ray wavelength produced for an x-ray machine operated at 30 kV?

42. The Stanford Linear Accelerator can accelerate electrons to 50 GeV (50×10^9 eV). What is the minimum wavelength of photon it can produce by bremsstrahlung? Is this photon still called an x ray?

43. A television tube operates at 20,000 V. What is λ_{min} for the continuous x-ray spectrum produced when the electrons hit the phosphor?

44. Calculate λ_{min} for all three elements shown in Figure 3.19. Use the value of the work function for tungsten in Table 3.3 and calculate the percentage error in neglecting the work function for the Duane-Hunt rule using the data of Figure 3.19.

45. The two peaks for the molybdenum spectra of Figure 3.19 are *characteristic* spectral lines for the molyb-

denum element. What would be the minimum voltage needed to accelerate electrons in an x-ray tube to produce both of these lines?

3.8 Compton Effect

46. Calculate the maximum $\Delta\lambda/\lambda$ of Compton scattering for green light ($\lambda = 530$ nm). Could this be easily observed?

47. A photon having 40 keV scatters from a free electron at rest. What is the maximum energy that the electron can obtain?

48. If a 6.0-keV photon scatters from a free proton at rest, what is the change in the photon's wavelength if the photon recoils at 90°?

49. Is it possible to have a scattering similar to Compton scattering from a proton in H_2 gas? What would be the Compton wavelength for a proton? What energy photon would have this wavelength?

50. An instrument has resolution $\Delta\lambda/\lambda = 0.40\%$. What wavelength of incident photons should be used in order to resolve the modified and unmodified scattered photons for scattering angles of (a) 30°, (b) 90°, and (c) 170°?

51. Derive the relation for the recoil kinetic energy of the electron and its recoil angle ϕ in Compton scattering. Show that

$$\text{K.E. (electron)} = \frac{\Delta\lambda/\lambda}{1 + (\Delta\lambda/\lambda)} hf$$

$$\cot\phi = \left(1 + \frac{hf}{mc^2}\right)\tan\frac{\theta}{2}$$

52. A gamma ray of 700-keV energy Compton-scatters from an electron. Find the energy of the photon scattered at 110°, the kinetic energy of the scattered electron, and the recoil angle of the electron.

53. A photon of wavelength 2.0 nm Compton-scatters from an electron at an angle of 90°. What is the modified wavelength and the percentage change, $\Delta\lambda/\lambda$?

3.9 Pair Production and Annihilation

54. How much photon energy would be required to produce a proton-antiproton pair? Where could such a high-energy photon come from?

55. What is the minimum photon energy needed to create an e^--e^+ pair when a photon collides (a) with a free electron at rest and (b) with a free proton at rest?

General Problems

56. What wavelength of photons are needed to produce 30.0-keV electrons in a Compton scattering?

57. The gravitational energy of the Earth is approximately $\frac{1}{2}(GM_E^2/R_E)$ where M_E is the mass of the Earth. This is approximately the energy needed to blow the Earth into small fragments (the size of asteroids). How large

would an antimatter meteorite the density of nickel-iron ($\rho \approx 5 \times 10^3$ kg/m^3) have to be in order to blow up the Earth when it strikes? Compute the energy involved in the particle-antiparticle annihilation and compare it with the total energy in all the nuclear arsenals of the world [~2000 megatons (MT), where 1 MT = 4.2×10^{15} J].

58. Show that the maximum kinetic energy of the recoil electron in Compton scattering is given by

$$\text{K.E.}_{max}\text{ (electron)} = hf\frac{\frac{2hf}{mc^2}}{1 + \frac{2hf}{mc^2}}$$

At what angles θ and ϕ does this occur? If we detect a scattered electron at $\phi = 0°$ of 100 keV, what energy photon was scattered?

59. Using the Wien displacement law, make a log-log plot of λ_{max} (from 10^{-8} m to 10^{-2} m) versus temperature (from 10^0 K to 10^5 K). Mark on the plot the regions of visible, ultraviolet, infrared, and microwave wavelengths. Put the following points on the line: sun (5800 K), furnace (1900 K), room temperature (300 K), and the background radiation of the universe (2.7 K). Discuss the electromagnetic radiation that is emitted from each of these sources. Does it make sense?

60. (a) What is the maximum possible energy for a Compton-backscattered x ray ($\theta = 180°$)? Express your answer in terms of λ, the wavelength of the incoming photon. (b) Evaluate numerically if the incoming photon's energy is 100 keV.

61. The naked eye can detect a stellar object of sixth magnitude in the night sky. With binoculars, we can see an object of the ninth magnitude. The sun's brightness at Earth is 1400 W/m^2. The Hubble Space Telescope can detect an object of the 30th magnitude, which amounts to a brightness of about 2×10^{-20} W/m^2. (a) Consider a detector in the Hubble Space Telescope with a collection area of 0.30 m^2. If you assume hydrogen light of frequency 486 nm (blue-green), how many photons/s enter the telescope from a 30th-magnitude star? (b) An increase of magnitude one represents a decrease in brightness by a factor of 2.5119. Estimate how many photons/s from a sixth-magnitude star would enter your eye if the diameter of your pupil is 6.5 mm.

62. Original data from Millikan's pivotal photoelectric experiment that confirmed Einstein's quantum explanation is shown in Figure 3.16 [from R. A. Millikan, *Physical Review* 7, 362 (1916)]. Sodium was the photocathode. Use the data to find the work function for sodium and Planck's constant.

Structure of the Atom

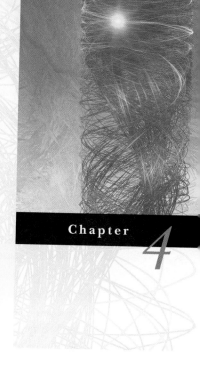

In the present first part of the paper the mechanism of the binding of electrons by a positive nucleus is discussed in relation to Planck's theory. It will be shown that it is possible from the point of view taken to account in a simple way for the law of the line spectrum of hydrogen.

Niels Bohr, 1913

$\mathbf{B}$y the end of the nineteenth century most physicists and chemists (with a few notable exceptions) believed in an atomic theory of matter, even though no one had ever directly observed an atom. Origins of atomic theory date back to the Greek philosophers who imagined atoms to be featureless hard spheres. Even though scientists of the late nineteenth century did not have technology to see things as small as atoms, they believed that atoms were composite structures having an internal structure. There are considerable similarities between how physicists addressed their atomic theories in the late nineteenth century and how elementary particle physicists still search for the underlying structure of the building blocks of matter.

We mention three pieces of evidence that physicists and chemists had in 1900 to indicate that the atom was not a fundamental unit. First, there seemed to be too many kinds of atoms, each belonging to a distinct chemical element. The original Greek idea was that there were four types of atoms—earth, air, water, and fire—which combined together to make the various kinds of matter we observe. But the development of chemistry made it clear that there were at least 70 kinds of atoms, far too many for them all to be the ultimate elementary constituents of matter.

Second, it was found experimentally that atoms and electromagnetic phenomena were intimately related. For example, molecules can be dissociated into their component atoms by electrolysis. Some kinds of atoms form magnetic materials, and others form electrical conductors and insulators. All kinds of atoms emit light (which was known to be electromagnetic in nature) when they are heated, as well as when an electrical discharge passes through them. The visible light emitted by free or nearly free atoms of the chemical elements is not a continuum of frequencies, but rather a discrete set of characteristic colors, so substances can be analyzed according to their chemical composition using their flame spectra. The existence of **characteristic spectra** pointed to an internal structure distinguishing the elements.

Third, there was the problem of **valence**—why certain elements combine with some elements but not with others, and when they do combine, why they do so in varying proportions determined by the valences of the atoms. The laws of valence suggested that the forces between atoms are specific in nature, a characteristic that hinted at an internal atomic structure.

Finally, there were the discoveries of radioactivity, of x rays, and of the electron, all of which were at variance with earlier ideas of indivisible and elementary atoms. Because of these tantalizing indirect hints that the atom had a structure, the most exciting frontier of science in the early part of the twentieth century developed into an investigation of the atom and its internal composition.

The subject of this chapter is the beginning of quantum physics and its relation to the first cohesive theories of atomic structure. Although we now have a more complete theoretical framework with which to understand the early experiments than was available to scientists at the time, it is worth repeating some of the scientists' reasoning, both for historical interest and to illustrate how science progresses by trying to extend well-established ideas into unknown terrain.

In this chapter we discuss the atomic models of Thomson and Rutherford and learn how Rutherford discerned the correct structure of the atom by performing alpha-particle scattering experiments. We see that Bohr presented a model of the hydrogen atom based on the new quantum concept that correctly produced the Rydberg equation, and we study the successes and failures of Bohr's theory. We also learn the origin of characteristic x-ray spectra and the concept of atomic number. Finally, we show that electron scattering (the Franck-Hertz experiment) also confirmed the quantized structure of the atom.

4.1   The Atomic Models of Thomson and Rutherford

In the years immediately following J. J. Thomson's discovery of the electron in 1897, Thomson and others tried to unravel the mystery of the atomic structure. In just a few years it was learned that electrons were much less massive than atoms and that the number of electrons was equal to about half the number representing the atomic mass. The central question was, "How are the electrons arranged and where are the positive charges that make the atom electrically neutral?" (Remember that protons had not been yet discovered.) Thomson proposed a model

Thomson's "plum-pudding" model of the atom

wherein the positive charges were spread uniformly throughout a sphere the size of the atom, with electrons embedded in the uniform background. His model, which was likened to raisins in plum pudding, is shown schematically in Figure 4.1. The arrangement of charges had to be in stable equilibrium. In Thomson's view, when the atom was heated, the electrons could vibrate about their equilibrium positions, thus producing electromagnetic radiation. The emission frequencies of this radiation would fall in the range of visible light if the sphere of positive charges were of diameter $\sim 10^{-10}$ m, which was known to be the approximate size of an atom. Nevertheless, even though he tried for several years, Thomson was unable to calculate the light spectrum of hydrogen using his model.

Figure 4.1 Schematic of J. J. Thomson's model of the atom (later proven to be incorrect). The electrons are embedded in a homogeneous positively charged mass much like raisins in plum pudding. The electric force on the electrons is zero, so the electrons do not move around rapidly. The oscillations of the electrons give rise to electromagnetic radiation.

The small size of the atoms made it impossible to directly see their internal structure. In order to make further progress in deciphering atomic structure, a new approach was needed. The new direction was supplied by Ernest Rutherford, who was already famous for his Nobel Prize–winning work on radioactivity. Rutherford projected very small particles onto atoms in a process called *scattering* and detected particles that exit the collision at different angles. Rutherford, assisted by Hans Geiger, conceived a new technique for investigating the structure of matter by scattering energetic alpha (α) particles* (emitted by radioactive sources) from atoms. Together with a young student, Ernest Marsden, and working in Rutherford's lab, Geiger showed in 1909 that surprisingly many α particles were scattered from thin gold-leaf targets at backward angles greater than 90° (see Figure 4.2).

For several years Rutherford had pondered the structure of the atom. He was well aware of Thomson's model because he had worked for Thomson at the Cavendish Laboratory as a research student from 1895 to 1898, after receiving his undergraduate education in his native New Zealand. Although he greatly respected Thomson, Rutherford could see that Thomson's model agreed neither with spectroscopy nor with Geiger's latest experiment with α particles.

The experiments of Geiger and Marsden were instrumental in the development of Rutherford's model. A simple thought experiment with a .22-caliber rifle that fires a bullet into a thin black box is a model for understanding the problem. If the box contains a homogeneous material such as wood or water (as in Thomson's plum-pudding model), the bullet will pass through the box with little or no deviation in its path. However, if the box contains a few massive steel ball bearings, then occasionally a bullet will be deflected backward, similar to what Geiger and Marsden observed with α scattering.

*Rutherford had already demonstrated that the α particle is an ionized helium atom.

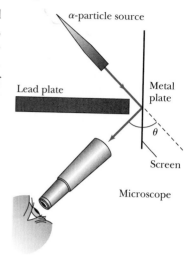

Figure 4.2 Schematic diagram of apparatus used by Geiger and Marsden to observe scattering of α particles past 90°. "A small fraction of the α particles falling upon a metal plate have their directions changed to such an extent that they emerge again at the side of incidence." *From H. Geiger and E. Marsden,* Proceedings of Royal Society (London) *82, 495 (1909).*

Example 4.1

Geiger and Marsden (1909) observed backward-scattered ($\theta \geq 90°$) α particles when a beam of energetic α particles was directed at a piece of gold foil as thin as 6.0×10^{-7} m. Assuming an α particle scatters from an electron in the foil, what is the maximum scattering angle?

Strategy We consider elastic scattering between the incident α particle and an electron in the gold foil. The collision must obey the laws of conservation of momentum and energy. We find the maximum scattering angle corresponding to the maximum momentum change for the α particle. Assume the incident α particle has mass M_α and velocity v_α, while the mass of the electron is m_e. The maximum momentum transfer occurs when the α particle hits the electron (at rest) head-on, as shown in Figure 4.3.

Solution Conservation of momentum (nonrelativistically) gives

$$M_\alpha \vec{v}_\alpha = M_\alpha \vec{v}'_\alpha + m_e \vec{v}'_e$$

Figure 4.3 Schematic diagram (before and after) of an α particle of speed $v = v_\alpha$ and mass M_α making a head-on collision with an electron initially at rest. Because the α particle is so much more massive than the electron, the α particle's velocity is hardly reduced.

Because the α particle is so much more massive than the electron ($M_\alpha/m_e \approx 4 \times 1837 \approx 7000$), the α particle's velocity is hardly affected and $v'_\alpha \approx v_\alpha$. In an elastic collision with such unequal masses, $v'_e \approx 2v_\alpha$ in order to conserve both energy and linear momentum (see Problem 3). Thus the maximum momentum change of the α particle is simply

$$\Delta \vec{p}_\alpha = M_\alpha \vec{v}_\alpha - M_\alpha \vec{v}'_\alpha = m_e \vec{v}'_e$$

or

$$\Delta p_{\max} = 2m_e v_\alpha$$

Although this maximum momentum change is along the direction of motion, let's determine an upper limit for the angular deviation θ by letting $\Delta p_{\max}$ be perpendicular to the direction of motion as shown in Figure 4.4. (This value of θ is larger than can actually be observed because we know that the Δp_α we calculated was for a head-on collision, and the Δp_α for a glancing collision would be smaller.) Thus

$$\theta_{\max} = \frac{\Delta p_\alpha}{p_\alpha} = \frac{2m_e v_\alpha}{M_\alpha v_\alpha} = \frac{2m_e}{M_\alpha} = 2.7 \times 10^{-4}\ \text{rad} = 0.016°$$

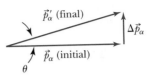

Figure 4.4 Vector diagram illustrating the change in momentum $\Delta \vec{p}_\alpha$ of the α particle after scattering from the electron.

Thus it is impossible for an α particle to be deflected through a large angle by a single encounter with an electron.

Multiple scattering from electrons

What would happen if an α particle were scattered by *many* electrons in the target? Multiple scattering is possible, and a calculation for random multiple scattering from N electrons results in an average scattering angle $\langle\theta\rangle_{\text{total}} \approx \sqrt{N}\theta$. The α particle is as likely to scatter on one side of its direction as the other side for each collision. We can estimate the number of atoms across the thin gold layer of 6×10^{-7} m used by Geiger and Marsden.

$$\frac{\text{Number of molecules}}{\text{cm}^3} = [\text{Avogadro's no. (molecules/mol)}]$$

$$\times \left[\frac{1}{\text{gram} - \text{molecular weight}}\left(\frac{\text{mol}}{\text{g}}\right)\right]\left[\text{density}\left(\frac{\text{g}}{\text{cm}^3}\right)\right]$$

$$= \left(6.02 \times 10^{23}\ \frac{\text{molecules}}{\text{mol}}\right)\left(\frac{1\ \text{mol}}{197\ \text{g}}\right)\left(19.3\ \frac{\text{g}}{\text{cm}^3}\right)$$

$$= 5.9 \times 10^{22}\ \frac{\text{molecules}}{\text{cm}^3} = 5.9 \times 10^{28}\ \frac{\text{atoms}}{\text{m}^3}$$

If there are 5.9×10^{28} atoms/m^3, then each atom occupies $(5.9 \times 10^{28})^{-1}$ m^3 of space. Assuming the atoms are equidistant, the distance d between centers is $d = (5.9 \times 10^{28})^{-1/3}$ m $= 2.6 \times 10^{-10}$ m. In the foil, then, there are

$$N = \frac{6 \times 10^{-7}\ \text{m}}{2.6 \times 10^{-10}\ \text{m}} = 2300\ \text{atoms}$$

along the α particle's path. If we assume the α particle interacts with one electron from each atom, then

$$\langle\theta\rangle_{\text{total}} = \sqrt{2300}\,(0.016°) = 0.8°$$

where we have used the result for $\theta_{\max}$ from Example 4.1. Even if the α particle scattered from all 79 electrons in each atom of gold, $\langle\theta\rangle_{\text{total}} = 6.8°$.

Rutherford's atomic model

Rutherford reported[*] in 1911 that the experimental results were not consistent with α-particle scattering from the atomic structure proposed by Thomson and that "it seems reasonable to suppose that the deflection through a large angle is due to a single atomic encounter." Rutherford proposed that an atom consisted mostly of empty space with a central charge, either positive or negative. Rutherford

[*]E. Rutherford, *Philosophical Magazine* **21**, 669 (1911).

wrote in 1911, "Considering the evidence as a whole, it seems simplest to suppose that the atom contains a central charge distributed through a very small volume, and that the large single deflections are due to the central charge as a whole, and not to its constituents." Rutherford worked out the scattering expected for the α particles as a function of angle, thickness of material, velocity, and charge. Geiger and Marsden immediately began an experimental investigation of Rutherford's ideas and reported* in 1913, "we have completely verified the theory given by Prof. Rutherford." In that same year, Rutherford was the first to use the word **nucleus** for the central charged core and definitely decided that the core (containing most of the mass) was *positively* charged, surrounded by the negative electrons.

The popular conception of an atom today, often depicted as in Figure 4.5, is due to Rutherford. An extremely small positively charged core provides a Coulomb attraction for the negatively charged electrons flying at high speeds around the nucleus; this is the so-called "solar system" or "planetary" model. We now know that the nucleus is composed of positively charged protons and neutral neutrons, each having approximately the same mass, and the electrons do not execute prescribed orbital paths.

The nucleus

Figure 4.5 Solar or planetary model of the atom. Rutherford proposed that there is a massive, central core with a highly electric positive charge. According to Bohr, the electrons orbit around this nucleus. Although this is a useful graphic, we now know this schematic is too simplistic.

4.2   Rutherford Scattering

Rutherford's "discovery of the nucleus" laid the foundation for many of today's atomic and nuclear scattering experiments. By means of scattering experiments similar in concept to those of Rutherford and his assistants, scientists have elucidated the electron structure of the atom, the internal structure of the nucleus, and even the internal structures of the nuclear constituents, protons and neutrons. Rutherford's calculations and procedures are well worth studying in some detail because of their applicability to many areas of physical and biological science.

Scattering experiments help us study matter on an atomic scale, which is too small to be observed directly. The material to be studied is bombarded with rapidly moving particles (like the 5- to 8-MeV α particles used by Geiger and Marsden) in a well-defined and collimated beam. Although the present discussion is limited to charged-particle beams, the general procedure applies also to neutral particles such as neutrons; only the interaction between the beam particles and the target material is different.

Basic scattering experiments

The scattering of charged particles by matter is called **Coulomb** or **Rutherford scattering** when it takes place at low energies, where only the Coulomb force is important. At higher beam energies other forces (for example, nuclear interactions) may become important. A typical scattering experiment is diagrammed in Figure 4.6. A charged particle of mass m, charge $Z_1 e$, and speed v_0 is incident on the target material or scatterer of charge $Z_2 e$. The distance b is called the classical *impact parameter*; it is the closest distance of approach between the beam particle and scatterer if the projectile had continued in a straight line. The angle θ between the incident beam direction and the direction of the deflected particle is called the *scattering angle*. Normally detectors are positioned at one or more scattering angles to count the particles scattered into the small cones of solid angle subtended by the detectors (see Figure 4.7).

Rutherford or Coulomb scattering

Scattering angle

Depending on the functional form of the interaction between the particle and the scatterer, there will be a particular relationship between the impact parameter b and the scattering angle θ. In the case of Coulomb scattering between

---

*Hans Geiger and Ernest Marsden, *Philosophical Magazine* **25**, 604 (1913).

Figure 4.6 Representation of
Coulomb or Rutherford scattering. The projectile of mass m and
charge Z_1e scatters from a particle
of charge Z_2e at rest. The parameters r and ϕ, which describe the
projectile's orbit, are defined as
shown. The angle $\phi = 0$ represents
the position of closest approach.
The impact parameter b and scattering angle θ are also displayed.

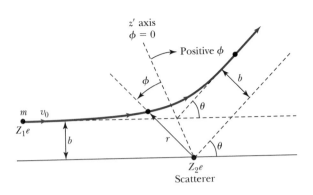

a positively charged α particle and a positively charged nucleus, the trajectories
will resemble those in Figure 4.7. When the impact parameter is small, the distance of closest approach r_{min} is small, and the Coulomb force is large. Therefore,
the scattering angle can be large and the particle can be repelled backward. Conversely, for large impact parameters the particles never get close together, so the
Coulomb force is small and the scattering angle is also small.

An important relationship for any interaction is that between b and θ. We
wish to find this dependence for the Coulomb force. We will make the same assumptions as Rutherford:

Scattering assumptions

1. The scatterer is so massive that it does not significantly recoil; therefore
   the initial and final kinetic energies of the α particle are practically equal.
2. The target is so thin that only a single scattering occurs.
3. The bombarding particle and target scatterer are so small that they may
   be treated as point masses and charges.
4. Only the Coulomb force is effective.

Assumption 1 means that $K \equiv \text{K.E.}_{\text{initial}} \approx \text{K.E.}_{\text{final}}$ for the α particle. For central forces like the Coulomb force, the angular momentum, mv_0b, where v_0 is the
initial velocity of the particle, is also conserved (see Problem 52). This means that
the trajectory of the scattered particle lies in a plane.

We define the instantaneous position of the particle by the angle ϕ and the
distance r from the force center, where $\phi = 0$ (which defines the z' axis) when the
distance r is a minimum, as shown in Figure 4.6. The change in momentum is
equal to the impulse.

$$\Delta \vec{p} = \int \vec{F}_{\Delta p} \, dt \qquad (4.1)$$

where $\vec{F}_{\Delta p}$ is the force along the direction of $\Delta \vec{p}$. The massive scatterer absorbs
this (small) momentum change without gaining any appreciable kinetic energy

Figure 4.7 The relationship
between the impact parameter b
and scattering angle θ. Particles
with small impact parameters approach the nucleus most closely
(r_{min}) and scatter to the largest
angles. Particles within the range
of impact parameters Δb will be
scattered within $\Delta\theta$.

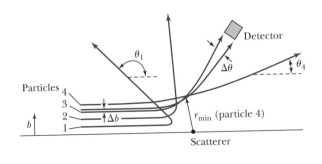

(no recoil). We use the diagram of Figure 4.8 to show

$$\Delta \vec{p} = \vec{p}_f - \vec{p}_i \tag{4.2}$$

where the subscripts i and f indicate the initial and final values of the projectile's momentum, respectively. Because $p_f \approx p_i = mv_0$, the triangle between $\vec{p}_f$, $\vec{p}_i$, and $\Delta \vec{p}$ is isosceles. We redraw the triangle in Figure 4.8b, indicating the bisector of angle θ. The magnitude Δp of the vector $\Delta \vec{p}$ is now

$$\Delta p = 2 m v_0 \sin \frac{\theta}{2} \tag{4.3}$$

The direction of $\Delta \vec{p}$ is the z' axis (where $\phi = 0$), so we need the component of $\vec{F}$ along z' in Equation (4.1). The Coulomb force $\vec{F}$ is along the instantaneous direction of the position vector $\vec{r}$ (unit vector $\hat{e}_r$, where $\hat{\ }$ indicates a unit vector).

$$\vec{F} = \frac{1}{4\pi\epsilon_0} \frac{Z_1 Z_2 e^2}{r^2} \hat{e}_r \tag{4.4}$$

and

$$F_{\Delta p} = F \cos \phi \tag{4.5}$$

where $F_{\Delta p}$ is the component of the force $\vec{F}$ along the direction of $\Delta \vec{p}$ that we need.

Substituting the magnitudes from Equations (4.3) and (4.5) into the components of Equation (4.1) along the z' axis ($\phi = 0$) gives

$$\Delta p = 2 m v_0 \sin \frac{\theta}{2} = \int F \cos \phi \, dt = \frac{Z_1 Z_2 e^2}{4\pi\epsilon_0} \int \frac{\cos \phi}{r^2} \, dt$$

The instantaneous angular momentum must be conserved, so

$$m r^2 \frac{d\phi}{dt} = m v_0 b$$

and

$$r^2 = \frac{v_0 b}{d\phi / dt}$$

Therefore,

$$2 m v_0 \sin \frac{\theta}{2} = \frac{Z_1 Z_2 e^2}{4\pi\epsilon_0} \int \frac{\cos \phi}{v_0 b} \frac{d\phi}{dt} \, dt$$

$$= \frac{Z_1 Z_2 e^2}{4\pi\epsilon_0 v_0 b} \int_{\phi_i}^{\phi_f} \cos \phi \, d\phi$$

We let the initial angle ϕ_i be on the negative side and the final angle ϕ_f be on the positive side of the z' axis ($\phi = 0$, see Figure 4.6). Then $\phi_i = -\phi_f$, $-\phi_i + \phi_f + \theta = \pi$, so $\phi_i = -(\pi - \theta)/2$ and $\phi_f = +(\pi - \theta)/2$.

$$\frac{8\pi\epsilon_0 m v_0^2 b}{Z_1 Z_2 e^2} \sin \frac{\theta}{2} = \int_{-(\pi-\theta)/2}^{+(\pi-\theta)/2} \cos \phi \, d\phi = 2 \cos \frac{\theta}{2}$$

We now solve this equation for the impact parameter b.

$$b = \frac{Z_1 Z_2 e^2}{4\pi\epsilon_0 m v_0^2} \cot \frac{\theta}{2}$$

(a)

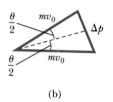

(b)

Figure 4.8 (a) The scattering angle θ and momentum change $\Delta \vec{p}$ are determined from the initial and final values of the α particle's momentum. (b) Because $\vec{p}_f$, $\vec{p}_i$, and $\vec{p}$ almost exactly form an isosceles triangle, we determine the magnitude of $\Delta \vec{p}$ by bisecting the angle θ and finding the length of the triangle leg opposite the angle $\theta/2$.

Special Topic

LORD RUTHERFORD OF NELSON

Ernest Rutherford* was born of Scottish parents on August 30, 1871, near the town of Nelson, New Zealand. Rutherford, one of eleven children, obtained both a bachelor's and master's degree in 1894 from the University of New Zealand in Christchurch. He constructed a magnetic detector that was able to receive electromagnetic waves over a distance of 60 feet through walls, quite a feat at the time.

In 1895 Rutherford won a competition to bring able men to British universities and went to work for the famous Professor J. J. Thomson of the Cavendish Laboratory at Cambridge University. Thomson was quite impressed by Rutherford's continuing research on the transmission and detection of "wireless waves" and encouraged him to publish his results and make presentations at scientific meetings. Rutherford also began investigations with Thomson on the effects of x-ray radiation from uranium in various gases. In 1898 Rutherford applied for and received a chaired appointment as Professor of Physics at McGill University in Montreal, Canada.

*This account is taken from A. S. Eve, *Rutherford*, New York: Macmillan (1939); and H. A. Boorse and L. Motz, eds., *The World of the Atom*, New York: Basic Books (1966).

Barely 27 years of age, he arrived in Montreal in 1898 to take up his new duties in the Macdonald Physics Laboratories. Rutherford's studies on radioactivity continued, and he had frequent correspondence and occasional visits with scientists from abroad. Early in 1900 he published a paper in the *Philosophical Magazine* in which he named *alpha*, *beta*, and *gamma* as the three types of radiation from thorium and uranium. Rutherford attracted the aid of a young research chemist, Frederick Soddy of Oxford, who had arrived at McGill in 1900. Rutherford and Soddy discovered in 1902 that the elements, heretofore considered immute, actually decayed to other elements. During the next few years Rutherford investigated α particles and the radioactive decay chains of radium, thorium, and uranium.

In 1907 Rutherford returned to England as professor of physics at the University of Manchester where he did his greatest work. His first success was the proof that α particles were indeed helium ions. In 1908 Rutherford received word that he had won the Nobel Prize for the work he and Soddy had done, but Rutherford was startled and amused to learn it was in chemistry, not physics.

It was during the next few years that Rutherford carried out his research into the nature of the atom that culminated with his discovery of the nucleus. In 1912 Rutherford wrote to a colleague, "Bohr, a Dane, has pulled out of Cambridge and turned up here to get some experience in radioactive work." That

---

This equation becomes

Relation between b and θ

$$b = \frac{Z_1 Z_2 e^2}{8\pi\epsilon_0 K}\cot\frac{\theta}{2} \tag{4.6}$$

where $K = mv_0^2/2$ is the kinetic energy of the bombarding particle. This is the fundamental relationship between the impact parameter b and scattering angle θ that we have been seeking for the Coulomb force.

We are not able to select individual impact parameters b in a given experiment. When we put a detector at a particular angle θ, we cover a finite $\Delta\theta$, which corresponds to a range of impact parameters Δb. The bombarding particles are incident at varied impact parameters all around the scatterer as shown in Figure 4.9. All the particles with impact parameters less than b_0 will be scattered at angles greater than θ_0. Any particle with an impact parameter inside the area of momentous trip resulted in the "Rutherford-Bohr atom." The work of Rutherford at Manchester together with Hans Geiger, Ernest Marsden, and Henry Moseley was to have dramatic consequences. Rutherford used the word *proton* to describe the fast hydrogen nuclei produced when he bombarded hydrogen and nitrogen with fast α particles.

World War I broke up the family of research students working at Manchester, and in 1919 Rutherford accepted the Cavendish Professorship at Cambridge, the post just vacated by J. J. Thomson, who remained as Master of Trinity College. Being the successor at Cambridge to Maxwell, Rayleigh, and Thomson was no small feat, and Rutherford continued to do important research until the time of his death in 1937 at age 66. Rutherford was knighted in 1914, and in 1931 he was made a baron, choosing the town of Nelson near his boyhood home to become "Lord Rutherford of Nelson." He was the greatest experimental physicist of his day, yet he was said to have "never made an enemy and never lost a friend."

Figure A Ernest Rutherford, on the right, talking with J. A. Ratcliffe in the Cavendish Laboratory in 1936. The sign above Rutherford reads "TALK SOFTLY PLEASE" because the detectors being used were very sensitive to vibrations and noise. Rutherford, whose deep booming voice disturbed the detectors more than anyone else's, didn't seem to think the warning applied to him and was in a loud conversation when this photo was taken. *AIP Emilio Segrè Visual Archives.*

the circle of area πb_0^2 (radius b_0) will be similarly scattered. For the case of Coulomb scattering, we denote the cross section by the symbol σ, where

$$\sigma = \pi b^2 \tag{4.7}$$

is the cross section for scattering through an angle θ or more. The cross section σ is related to the *probability* for a particle being scattered by a nucleus. If we have

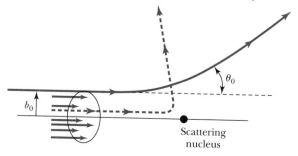

Scattering nucleus

Figure 4.9 All particles with impact parameters *less* than b_0 will scatter at angles *greater* than θ_0.

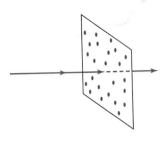

Figure 4.10 The target is assumed to be so thin that all nuclei are exposed to the bombarding particles. No nucleus is hidden behind another.

a target foil of thickness t with n atoms/volume, the number of target nuclei per unit area is nt. Because we assumed a thin target of area A and all nuclei are exposed as shown in Figure 4.10, the number of target nuclei is simply ntA. The value of n is the density ρ (g/cm^3) times Avogadro's number N_A (molecules/mol) times the number of atoms/molecule N_M divided by the gram-molecular weight M_g (g/mol).

$$n = \frac{\rho\left(\dfrac{g}{cm^3}\right)N_A\left(\dfrac{molecules}{mol}\right)N_M\left(\dfrac{atoms}{molecule}\right)}{M_g\left(\dfrac{g}{mol}\right)} = \frac{\rho N_A N_M}{M_g}\frac{atoms}{cm^3} \tag{4.8}$$

The number of scattering nuclei per unit area is nt.

$$nt = \frac{\rho N_A N_M t}{M_g}\frac{atoms}{cm^2} \tag{4.9}$$

If we have a foil of area A, the number of target nuclei N_s is

$$N_s = ntA = \frac{\rho N_A N_M t A}{M_g}\ atoms \tag{4.10}$$

The probability of the particle being scattered is equal to the total target area exposed for all the nuclei divided by the total target area A. If σ is the cross section for each nucleus, then $ntA\sigma$ is the total area exposed by the target nuclei, and the fraction of incident particles scattered by an angle of θ or greater is

$$f = \frac{\text{target area exposed by scatterers}}{\text{total target area}} = \frac{ntA\sigma}{A}$$

$$= nt\sigma = nt\pi b^2 \tag{4.11}$$

$$f = \pi nt\left(\frac{Z_1 Z_2 e^2}{8\pi\epsilon_0 K}\right)^2 \cot^2\frac{\theta}{2} \tag{4.12}$$

Example 4.2

Find the fraction of 7.7-MeV α particles that is deflected at an angle of 90° or more from a gold foil of 10^{-6} m thickness.

Strategy We can use Equation (4.12) to calculate the fraction, but first we need to calculate n, the number of atoms/cm^3. We do that using Equation (4.8).

Solution The density of gold is 19.3 g/cm^3 and the atomic weight is 197 u. Equation (4.8) determines n.

$$n = \frac{\left(19.3\dfrac{g}{cm^3}\right)\left(6.02\times10^{23}\dfrac{molecules}{mol}\right)\left(1\dfrac{atom}{molecule}\right)}{197\ g/mol}$$

$$= 5.90\times10^{22}\frac{atoms}{cm^3} = 5.90\times10^{28}\frac{atoms}{m^3}$$

We insert this value of n into Equation (4.12) and find

$$f = \pi\left(5.90\times10^{28}\frac{atoms}{m^3}\right)(10^{-6}\ m)$$

$$\times\left[\frac{(79)(2)(1.6\times10^{-19}\ C)^2(9\times10^9\ N\cdot m^2/C^2)}{2(7.7\ MeV)(1.60\times10^{-13}\ J/MeV)}\right]^2$$

$$\times(\cot 45°)^2$$

$$= 4\times10^{-5}$$

One α particle in 25,000 is deflected by 90° or greater.

In an actual experiment, however, a detector is positioned over a range of angles from θ to $\theta + \Delta\theta$, as shown in Figure 4.11. Thus we need to find the number of particles scattered between θ and $\theta + d\theta$ that corresponds to incident particles with impact parameters between b and $b + db$ as displayed in Figure 4.12. The fraction of the incident particles scattered between θ and $\theta + d\theta$ is df. The derivative of Equation (4.12) is

$$df = -\pi nt \left(\frac{Z_1 Z_2 e^2}{8\pi\epsilon_0 K} \right)^2 \cot\frac{\theta}{2} \csc^2\frac{\theta}{2} \, d\theta$$

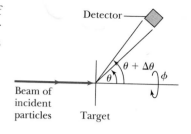

If the total number of incident particles is N_i, the number of particles scattered into the ring of angular width $d\theta$ is $N_i |df|$. The area dA into which the particles scatter is $(r \, d\theta)(2\pi r \sin\theta) = 2\pi r^2 \sin\theta \, d\theta$. Therefore, the number of particles scattered per unit area, $N(\theta)$, into the ring at scattering angle θ is

$$N(\theta) = \frac{N_i |df|}{dA} = \frac{N_i \pi nt \left(\dfrac{Z_1 Z_2 e^2}{8\pi\epsilon_0 K} \right)^2}{2\pi r^2 \sin\theta \, d\theta} \cot\frac{\theta}{2} \csc^2\frac{\theta}{2} \, d\theta$$

$$N(\theta) = \frac{N_i nt}{16} \left(\frac{e^2}{4\pi\epsilon_0} \right)^2 \frac{Z_1^2 Z_2^2}{r^2 K^2 \sin^4(\theta/2)} \tag{4.13}$$

This is the famous **Rutherford scattering equation**. The important points are

Rutherford scattering equation

1. The scattering is proportional to the square of the atomic number of both the incident particle (Z_1) and the target scatterer (Z_2).
2. The number of scattered particles is inversely proportional to the square of the kinetic energy K of the incident particle.
3. The scattering is inversely proportional to the fourth power of $\sin(\theta/2)$, where θ is the scattering angle.
4. The scattering is proportional to the target thickness for thin targets.

These very specific predictions by Rutherford in 1911 were confirmed experimentally by Geiger and Marsden in 1913. The angular dependence is particularly characteristic and can be verified in a well-equipped undergraduate physics laboratory, as we see from some actual data shown in Figure 4.13.

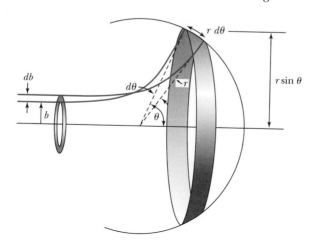

Figure 4.12 Particles over the range of impact parameters from b to $b + db$ will scatter into the angular range θ to $\theta + d\theta$ (with db positive, $d\theta$ will be negative).

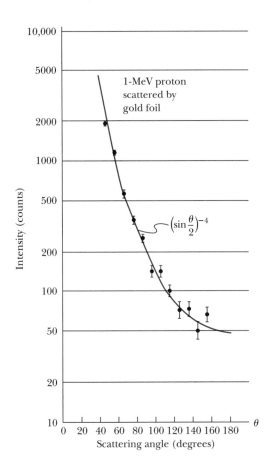

Figure 4.13 Results of under-graduate laboratory experiment of scattering 1-MeV protons from a gold target. The solid line shows the $1/\sin^4(\theta/2)$ angular depen-dence of the data, verifying Ruth-erford's calculation.

Example **4.3**

Calculate the fraction per mm^2 area of 7.7-MeV α particles scattered at 45° from a gold foil of thickness 2.1×10^{-7} m at a distance of 1.0 cm from the target.

Strategy We use Equation (4.13) to determine the frac-tion per unit area $N(\theta)/N_i$. We calculated $n = 5.90 \times 10^{28}$ atoms/m^3 in Example 4.2.

Solution We insert the values into Equation (4.13).

$$\frac{N(\theta)}{N_i} = \frac{\left(5.90 \times 10^{28}\,\frac{\text{atoms}}{\text{m}^3}\right)(2.1 \times 10^{-7}\,\text{m})\left[(1.6 \times 10^{-19}\,\text{C})^2\left(9 \times 10^9\,\frac{\text{N} \cdot \text{m}^2}{\text{C}^2}\right)\right]^2}{16}$$

$$\times \frac{(2)^2(79)^2}{(1.0 \times 10^{-2}\,\text{m})^2\left(7.7\,\text{MeV} \times \frac{10^6\,\text{eV}}{\text{MeV}} \times \frac{1.6 \times 10^{-19}\,\text{J}}{\text{eV}}\right)^2}\frac{1}{\sin^4(45°/2)}$$

$$\frac{N(\theta)}{N_i} = 3.2 \times 10^{-1}\,\text{m}^{-2} = 3.2 \times 10^{-7}\,\text{mm}^{-2}$$

This is the theoretical basis for the experiment per-formed by Geiger and Marsden in 1913 to check the validity of Rutherford's calculation. Our calculated result agrees with their experimental result of 3.7×10^{-7} mm^{-2} when the experimental uncertainty is taken into account.

Conceptual Example 4.4

How can we find the distance of closest approach between a bombarding particle and a target scatterer of like charge?

Solution We can find this distance of closest approach for a given kinetic energy K and impact parameter b. The minimum separation occurs for a head-on collision. The bombarding particle turns around and scatters backward at 180°. At the instant the particle turns around, the entire kinetic energy has been converted into Coulomb potential energy. By setting the original (maximum) kinetic energy equal to the Coulomb potential energy when $r = r_{min}$, we can then solve the resulting equation for r_{min}. Let K be the original kinetic energy of the bombarding particle.

$$K = \frac{(Z_1 e)(Z_2 e)}{4\pi\epsilon_0 r_{min}} \qquad (4.14)$$

We solve this equation to determine r_{min}.

$$r_{min} = \frac{Z_1 Z_2 e^2}{4\pi\epsilon_0 K} \qquad (4.15)$$

Example 4.5

Rutherford found deviations from his Equation (4.13) at backward angles when he scattered 7.7-MeV α particles ($Z_1 = 2$) on aluminum ($Z_2 = 13$). He suspected this was because the α particle might be affected by approaching the nucleus so closely. Estimate the size of the nucleus based on these data.

Strategy We have just determined the distance of closest approach for the α particle, which occurs for a head-on collision or scattering angle of 180°. We propose that the r_{min} in this case is close to the sum of the α particle (^{4}He nucleus) radius and the aluminum nuclear radius.

Solution We insert the values for the α particle incident on aluminum into Equation (4.15) to find r_{min}.

$$r_{min} = \frac{Z_1 Z_2 e^2}{4\pi\epsilon_0 K}$$

$$= \frac{(2)(13)(1.60 \times 10^{-19}\,\text{C})^2 (8.99 \times 10^9\,\text{N}\cdot\text{m}^2\cdot\text{C}^{-2})}{(7.7\,\text{MeV})(1.60 \times 10^{-13}\,\text{J/MeV})}$$

$$= 4.9 \times 10^{-15}\,\text{m}$$

We find the sum of the ^{4}He and aluminum nuclear radii to be about 5×10^{-15} m.

We will see in Chapter 12 that the aluminum nuclear radius is about twice as large as that of ^{4}He, and our approximate result here is in fair agreement with modern data. We now know that nuclear radii vary from 1×10^{-15} to 8×10^{-15} m. Thus when α particles scatter from aluminum, an α particle may approach the nucleus close enough to be affected by the nuclear force (see Chapter 12).

4.3   The Classical Atomic Model

After Rutherford presented his calculations of charged-particle scattering in 1911 and the experimental verification by his group in 1913, it was generally conceded that the atom consisted of a small, massive, positively charged "nucleus" surrounded by moving electrons. Thomson's plum-pudding model was definitively excluded by the data. Actually, Thomson had previously considered a planetary model resembling the solar system (where the planets move in elliptical orbits about the sun) but had rejected it because, although both gravitational and Coulomb forces vary inversely with the square of the distance, the planets *attract* one another while orbiting around the sun, whereas the electrons would *repel* one another. Thomson considered this to be a fatal flaw from his knowledge of planetary theory.

In order to examine the failure of the planetary model, let us examine the simplest atom, hydrogen. We will assume circular electron orbits for simplicity rather than the more general elliptical ones. The force of attraction on the electron due to the nucleus (charge $= +e$) is

$$\vec{F}_e = \frac{-1}{4\pi\epsilon_0} \frac{e^2}{r^2} \hat{e}_r \qquad (4.16)$$

where the negative sign indicates the force is attractive and $\hat{e}_r$ is a unit vector in the direction from the nucleus to the electron. This electrostatic force provides the centripetal force needed for the electron to move in a circular orbit at constant speed. Its radial acceleration is

$$a_r = \frac{v^2}{r} \qquad (4.17)$$

where v is the tangential velocity of the electron. Newton's second law now gives

$$\frac{1}{4\pi\epsilon_0} \frac{e^2}{r^2} = \frac{mv^2}{r} \qquad (4.18)$$

and

$$v = \frac{e}{\sqrt{4\pi\epsilon_0 mr}} \qquad (4.19)$$

where we are using m without a subscript to be the electron's mass. When it is not clear what particle m refers to, we write the electron mass as m_e.

Example 4.6

Are we justified in using a nonrelativistic treatment for the speed of an electron in the hydrogen atom?

Strategy We use Equation (4.19) to calculate the electron's speed. If it is less than 1% of the speed of light, we are justified in using a nonrelativistic treatment. One difficulty is knowing the radius of the hydrogen atom. The size of an atom was thought to be about 10^{-10} m, so we let $r = 0.5 \times 10^{-10}$ m in order to estimate the electron's velocity.

Solution Equation (4.19) gives

$$v \approx \frac{(1.6 \times 10^{-19} \text{ C})(9 \times 10^9 \text{ N} \cdot \text{m}^2/\text{C}^2)^{1/2}}{(9.11 \times 10^{-31} \text{ kg})^{1/2}(0.5 \times 10^{-10} \text{ m})^{1/2}}$$

$$\approx 2.2 \times 10^6 \text{ m/s} < 0.01c$$

This justifies a nonrelativistic treatment.

The kinetic energy of the system is due to the electron, $K = mv^2/2$. The nucleus is so massive compared to the electron ($m_{\text{proton}} = 1836m$) that the nucleus may be considered to be at rest. The potential energy V is simply $-e^2/4\pi\epsilon_0 r$, so the total mechanical energy is

$$E = K + V = \frac{1}{2}mv^2 - \frac{e^2}{4\pi\epsilon_0 r} \qquad (4.20)$$

If we substitute for v from Equation (4.19), we have

$$E = \frac{e^2}{8\pi\epsilon_0 r} - \frac{e^2}{4\pi\epsilon_0 r} = \frac{-e^2}{8\pi\epsilon_0 r} \qquad (4.21)$$

The total energy is negative, indicating a bound system.

Thus far, the classical atomic model seems plausible. The problem arises when we consider that the electron is accelerating due to its circular motion about the nucleus. We know from classical electromagnetic theory that an accelerated electric charge continuously radiates energy in the form of electromagnetic radiation. If the electron is radiating energy, then the total energy E of the system, Equation (4.21), must decrease continuously. In order for this to happen, the radius r must decrease. The electron will continuously radiate energy as the electron orbit becomes smaller and smaller until the electron crashes into the nucleus! This process, displayed in Figure 4.14, would occur in about 10^{-9} s (see Problem 18).

Thus the classical theories of Newton and Maxwell, which had served Rutherford so well in his analysis of α-particle scattering and had thereby enabled him to discover the nucleus, also led to the failure of the planetary model of the atom. Physics had reached a decisive turning point like that encountered in 1900 with Planck's revolutionary hypothesis of the quantum behavior of radiation. In the early 1910s, however, the answer would not be long in coming, as we shall see in the next section.

Planetary model is doomed.

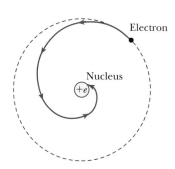

Figure 4.14 The electromagnetic radiation of an orbiting electron in the planetary model of the atom will cause the electron to spiral inward until it crashes into the nucleus.

4.4 The Bohr Model of the Hydrogen Atom

Shortly after receiving his Ph.D. from the University of Copenhagen in 1911, the 26-year-old Danish physicist Niels Bohr traveled to Cambridge University to work with J. J. Thomson and then with Rutherford at Manchester for a few months in 1912 where he became particularly involved in the mysteries of the new Rutherford model of the atom. Bohr returned to the University of Copenhagen the summer of 1912 with many questions about atomic structure. Like several others, he believed that a fundamental length about the size of an atom (10^{-10} m) was needed for an atomic model. This fundamental length might somehow be connected to Planck's new constant h. The pieces finally came together during the fall and winter of 1912 when Bohr learned of new precise measurements of the hydrogen spectrum and of the empirical formulas describing them. He set out to find a fundamental basis from which to derive the Balmer formula [Equation (3.12)], the Rydberg equation [Equation (3.13)], and Ritz's combination principles (see Problem 19).

Bohr was well acquainted with Planck's work on the quantum nature of radiation. Like Einstein, Bohr believed that quantum principles should govern more phenomena than just the blackbody spectrum. He was impressed by Einstein's application of the quantum theory to the photoelectric effect and to the specific heat of solids (see Chapter 9 for the latter) and wondered how the quantum theory might affect atomic structure.

In 1913, following several discussions with Rutherford during 1912 and 1913, Bohr published the paper* "On the Constitution of Atoms and Molecules." He

*Niels Bohr, *Philosophical Magazine* **26**, 1 (1913) and **30**, 394 (1915).

subsequently published several other papers refining and restating his "assumptions" and their predicted results. We will generally follow Bohr's papers in our discussion.

Bohr assumed that electrons moved around a massive, positively charged nucleus. We will assume for simplicity (as did Bohr at first) that the electron orbits are circular rather than elliptical and that the nuclear mass is so much greater than the electron's mass that it may be taken to be infinite. The electron has charge $-e$ and mass m and revolves around a nucleus of charge $+e$ in a circle of radius a. The size of the nucleus is small compared to the atomic radius a.

Bohr's model may best be summarized by the following "*general assumptions*" of his 1915 paper:

Bohr's general assumptions

A. Certain "stationary states" exist in atoms, which differ from the classical stable states in that the orbiting electrons do not continuously radiate electromagnetic energy. The stationary states are states of definite total energy.

B. The emission or absorption of electromagnetic radiation can occur only in conjunction with a transition between two stationary states. The frequency of the emitted or absorbed radiation is proportional to the difference in energy of the two stationary states (1 and 2):

$$E = E_1 - E_2 = hf$$

where h is Planck's constant.

C. The dynamical equilibrium of the system in the stationary states is governed by classical laws of physics, but these laws do not apply to transitions between stationary states.

D. The mean value K of the kinetic energy of the electron-nucleus system is given by $K = nhf_{orb}/2$, where f_{orb} is the frequency of rotation. For a circular orbit, Bohr pointed out that this assumption is equivalent to the angular momentum of the system in a stationary state being an integral multiple of $h/2\pi$. (This combination of constants occurs so often that we give it a separate symbol, $\hbar \equiv h/2\pi$, pronounced "h bar.")

These four assumptions were all that Bohr needed to derive the Rydberg equation. Bohr believed that Assumptions A and C were self-evident because atoms were stable: atoms exist and do not continuously radiate energy (therefore Assumption A). It also seemed that the classical laws of physics could not explain the observed behavior of the atom (therefore Assumption C).

Bohr later stated (1915) that Assumption B "appears to be necessary in order to account for experimental facts." Assumption D was the hardest for Bohr's critics to accept. It is central to the derivation of the binding energy of the hydrogen atom in terms of fundamental constants; hence Bohr restated and defended it in several ways in his papers. We have emphasized here the quantization of angular momentum aspect of Assumption D. This leads to a particularly simple derivation of the Rydberg equation.

Bohr chose his four assumptions to keep as much as possible of classical physics by introducing just those new ideas that were needed to explain experimental data. Bohr's recognition that something new was needed and his attempt to tie this to Planck's quantum hypothesis represented an advance in understanding perhaps even greater than Einstein's theory of the photoelectric effect.

Let us now proceed to derive the Rydberg equation using Bohr's assumptions. The total energy (potential plus kinetic) of a hydrogen atom was derived

Niels Bohr (1885–1962) was more than just a discoverer of modern physics theories. Born in Denmark, he was the son of a university professor and began high school at about the time Planck announced his results. After his education in Denmark, Bohr traveled to England in 1911 where he worked first with J. J. Thomson and later with Ernest Rutherford. Bohr nurtured many young theoretical physicists in his Institute of Theoretical Physics (now called the Niels Bohr Institute) formed in Copenhagen in 1921, the year before Bohr won the Nobel Prize. *AIP Niels Bohr Library.*

previously in Equation (4.21). For circular motion, the magnitude of the angular momentum L of the electron is

$$L = |\vec{r} \times \vec{p}| = mvr$$

Assumption D states this should equal $n\hbar$:

$$L = mvr = n\hbar \tag{4.22a}$$

where n is an integer called the **principal quantum number**. We solve the previous equation for the velocity and obtain

Principal quantum number

$$v = \frac{n\hbar}{mr} \tag{4.22b}$$

Equation (4.19) yields an independent relation between v and r. If we determine v^2 from Equations (4.19) and (4.22b) and set them equal, we find

$$v^2 = \frac{e^2}{4\pi\epsilon_0 mr} = \frac{n^2\hbar^2}{m^2 r^2} \tag{4.23}$$

From Equation (4.23) we see that only certain values of r are allowed.

$$r_n = \frac{4\pi\epsilon_0 n^2\hbar^2}{me^2} \equiv n^2 a_0 \tag{4.24}$$

where the **Bohr radius** a_0 is given by

$$a_0 = \frac{4\pi\epsilon_0\hbar^2}{me^2}$$

$$= \frac{(1.055 \times 10^{-34}\,\text{J}\cdot\text{s})^2}{\left(8.99 \times 10^9\,\dfrac{\text{N}\cdot\text{m}^2}{\text{C}^2}\right)(9.11 \times 10^{-31}\,\text{kg})(1.6 \times 10^{-16}\,\text{C})^2}$$

$$= 0.53 \times 10^{-10}\,\text{m}$$

Bohr radius

Notice that the smallest diameter of the hydrogen atom is $2r_1 = 2a_0 \approx 10^{-10}$ m, the suspected (now known) size of the hydrogen atom! Bohr had found the fundamental length a_0 that he sought in terms of the fundamental constants ϵ_0, h, e, and m. This fundamental length is determined for the value $n = 1$. Note from Equation (4.24) that the atomic radius is now quantized. The quantization of various physical values arises because of the principal quantum number n. The value $n = 1$ gives the radius of the hydrogen atom in its lowest energy state (called the "ground" state). The values of $n > 1$ determine other possible radii where the hydrogen atom is in an "excited" state.

The energies of the stationary states can now be determined from Equations (4.21) and (4.24).

$$E_n = -\frac{e^2}{8\pi\epsilon_0 r_n} = -\frac{e^2}{8\pi\epsilon_0 a_0 n^2} \equiv -\frac{E_0}{n^2} \tag{4.25}$$

Quantized energy states

The lowest energy state ($n = 1$) is $E_1 = -E_0$ where

$$E_0 = \frac{e^2}{8\pi\epsilon_0 a_0} = \frac{e^2}{(8\pi\epsilon_0)}\frac{me^2}{4\pi\epsilon_0\hbar^2} = \frac{me^4}{2\hbar^2(4\pi\epsilon_0)^2} = 13.6\,\text{eV} \tag{4.26}$$

n E (eV)

∞   0.00
4   −0.85
3   n_u −1.51
2   n_ℓ −3.40

Energy

1   −13.6

Figure 4.15 The energy-level diagram of the hydrogen atom. The principal quantum numbers n are shown on the left, with the energy of each level indicated on the right. The ground-state energy is −13.6 eV; negative total energy indicates a bound, attractive system. When an atom is in an excited state (for example, $n_u = 3$) and decays to a lower stationary state (for example, $n_\ell = 2$), the hydrogen atom must emit the energy difference in the form of electromagnetic radiation; that is, a photon emerges.

This is the experimentally measured ionization energy of the hydrogen atom. Bohr's Assumptions C and D imply that the atom can exist only in "stationary states" with definite, quantized energies E_n, displayed in the **energy-level diagram** of Figure 4.15. Emission of a quantum of light occurs when the atom is in an excited state (quantum number $n = n_u$) and decays to a lower energy state ($n = n_\ell$). A transition between two energy levels is schematically illustrated in Figure 4.15. According to Assumption B we have

$$hf = E_u - E_\ell \tag{4.27}$$

where f is the frequency of the emitted light quantum (photon). Because $\lambda f = c$, we have

$$\frac{1}{\lambda} = \frac{f}{c} = \frac{E_u - E_\ell}{hc}$$

$$= \frac{-E_0}{hc}\left(\frac{1}{n_u^2} - \frac{1}{n_\ell^2}\right) = \frac{E_0}{hc}\left(\frac{1}{n_\ell^2} - \frac{1}{n_u^2}\right) \tag{4.28}$$

where

$$\frac{E_0}{hc} = \frac{me^4}{4\pi c\hbar^3(4\pi\epsilon_0)^2} \equiv R_\infty \tag{4.29}$$

This constant R_∞ is called the **Rydberg constant** (for an infinite nuclear mass). Equation (4.28) becomes

$$\frac{1}{\lambda} = R_\infty\left(\frac{1}{n_\ell^2} - \frac{1}{n_u^2}\right) \tag{4.30}$$

which is similar to the Rydberg equation (3.13). The value of $R_\infty = 1.097373 \times 10^7$ m^{-1} calculated from Equation (4.29) agrees well with the experimental values given in Chapter 3, and we will obtain an even more accurate result in the next section.

Bohr's model predicts the frequencies (and wavelengths) of all possible transitions in atomic hydrogen. Several of the series are shown in Figure 4.16. The Lyman series represents transitions to the lowest state with $n_\ell = 1$; the Balmer series results from downward transitions to the stationary state $n_\ell = 2$; and the Paschen series represents transitions to $n_\ell = 3$. As mentioned in Section 3.3, not all of these series were known experimentally in 1913, but it was clear that Bohr had successfully accounted for the known spectral lines of hydrogen.

Bohr predicted new hydrogen wavelengths

The frequencies of the photons in the emission spectrum of an element are directly proportional to the differences in energy of the stationary states. When we pass white light (composed of all visible photon frequencies) through atomic hydrogen gas, we find that certain frequencies are absent. This pattern of dark lines is called an **absorption spectrum**. The missing frequencies are *precisely* the ones observed in the corresponding **emission spectrum**. In absorption, certain photons of light are absorbed, giving up energy to the atom and enabling the electron to move from a lower (ℓ) to a higher (u) stationary state. Equations (4.27) and (4.30) describe the frequencies and wavelengths of the absorbed photons. The atom will remain in the excited state for only a very short time (on the order of 10^{-10} s) before emitting a photon and returning to a lower stationary state. Thus, at ordinary temperatures practically all hydrogen atoms exist in the lowest possible energy state, $n = 1$, and only the absorption spectral lines of the Lyman series are

Absorption and emission spectrum

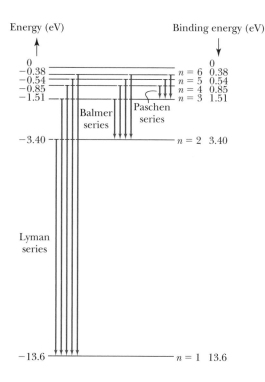

Figure 4.16 Transitions between many of the stationary states in the hydrogen atom are indicated. Transitions (ultraviolet) to the $n = 1$ state from the higher-lying states are called the Lyman series. The transitions shown to the $n = 2$ state (Balmer series) were discovered first because they are in the visible wavelength range. The Paschen series (transitions to $n = 3$) are in the infrared. The energies of each state as well as the binding energies are noted.

normally observed. However, these lines are not in the visible region. The sun produces electromagnetic radiation over a wide range of wavelengths, including the visible region. When sunlight passes through our atmosphere, hydrogen atoms in water vapor absorb the wavelengths of the Balmer series (visible region), and the absorption spectrum has dark lines at the known wavelengths of the Balmer series.

We can determine the electron's velocity in the Bohr model from Equations (4.22b) and (4.23).

$$v_n = \frac{n\hbar}{mr_n} = \frac{n\hbar}{mn^2 a_0} = \frac{1}{n}\frac{\hbar}{ma_0} \tag{4.31}$$

or

$$v_n = \frac{1}{n}\frac{e^2}{4\pi\epsilon_0 \hbar}$$

The value of v_1 is $\hbar/ma_0 = 2.2 \times 10^6$ m/s, which is less than 1% of the speed of light. We define the dimensionless quantity ratio of v_1 to c as

$$\alpha \equiv \frac{v_1}{c} = \frac{\hbar}{ma_0 c} = \frac{e^2}{4\pi\epsilon_0 \hbar c} \approx \frac{1}{137} \tag{4.32}$$ **Fine structure constant**

This ratio is called the **fine structure constant**. It appears often in atomic structure calculations.

We insert a word of caution at this point. Bohr's atomic model of quantized energy levels represented a significant step forward in understanding the structure of the atom. Although it had many successes, we know now that, in principle, it is wrong. We will discuss some of its successes and failures in the next section and will discuss the correct quantum theory in Chapter 6. Nevertheless, Bohr's atomic model is useful in our first attempt in understanding the structure of the atom.

Example 4.7

Atomic hydrogen in its lowest energy state absorbs a photon, raising the electron to an $n = 3$ state. If we assume the lifetime of the excited state is 10^{-10} s, and if we make the rudimentary assumption that the electron orbits around the proton, how many revolutions does the electron make in the excited state before returning to a lower energy state?

Strategy We know both the electron's speed v_3 and the radius r_3 of the orbit. From these we can find the time T it takes the electron to go one orbit. To find the number of rotations, we divide the lifetime by the time T.

Solution The velocity of the electron will be

$$v = \frac{\text{circumference}}{\text{period}} = \frac{2\pi r}{T}$$

Therefore

$$T = \frac{2\pi r_3}{v_3} = \frac{(2\pi)(9a_0)(3)(4\pi\epsilon_0\hbar)}{e^2}$$

$$T = \frac{(54\pi)(0.53 \times 10^{-10}\,\text{m})(1.055 \times 10^{-34}\,\text{J}\cdot\text{s})}{(1.6 \times 10^{-19}\,\text{C})^2\left(9 \times 10^9\,\dfrac{\text{N}\cdot\text{m}^2}{\text{C}^2}\right)}$$

$$= 4.1 \times 10^{-15}\,\text{s}$$

$$\text{Number of revolutions} = \frac{10^{-10}\,\text{s}}{4.1 \times 10^{-15}\,\text{s}} = 2.4 \times 10^4$$

The electron revolves many times in the excited state before decaying to a lower energy state.

The Correspondence Principle

Early in the 1900s physicists had trouble reconciling well-known and well-understood classical physics results with the new quantum ones. Sometimes completely different results were valid in their own domains. For example, there were two radiation laws: one used classical electrodynamics to determine the properties of radiation from an accelerated charge, but another explanation was expressed in Bohr's atomic model. Physicists proposed various kinds of correspondence principles to relate the new modern results with the old classical ones that had worked so well in their own domain. In his 1913 paper Bohr proposed perhaps the best *correspondence principle* to guide physicists in developing new theories. This principle was refined several times over the next few years.

Bohr's correspondence principle   *Bohr's correspondence principle: In the limits where classical and quantum theories should agree, the quantum theory must reduce to the classical result.*

To illustrate this principle, let us examine the predictions of the two radiation laws. The frequency of the radiation produced by the atomic electrons in the Bohr model of the hydrogen atom should agree with that predicted by classical electrodynamics in a region where the finite size of Planck's constant is unimportant—for large quantum numbers n where quantization effects are minimized. To see how this works we recall that classically the frequency of the radiation emitted is equal to the orbital frequency f_{orb} of the electron around the nucleus:

$$f_{\text{classical}} = f_{\text{orb}} = \frac{\omega}{2\pi} = \frac{1}{2\pi}\frac{v}{r} \tag{4.33a}$$

where for circular motion the angular velocity is $\omega = v/r$. If we substitute for v from Equation (4.19), we find

$$f_{\text{classical}} = \frac{1}{2\pi}\left(\frac{e^2}{4\pi\epsilon_0 mr^3}\right)^{1/2} \tag{4.33b}$$

We make the connection to the Bohr model by inserting the orbital radius r from Equation (4.24) into Equation (4.33b). We then know the classical frequency in terms of fundamental constants and the principal quantum number n.

$$f_{\text{classical}} = \frac{me^4}{4\epsilon_0^2 h^3} \frac{1}{n^3} \tag{4.34}$$

In the Bohr model, the nearest we can come to continuous radiation is a cascade of transitions from a level with principal quantum number $n + 1$ to the next lowest and so on:

$$n + 1 \rightarrow n \rightarrow n - 1 \rightarrow \cdots$$

The frequency of the transition from $n + 1 \rightarrow n$ is

$$f_{\text{Bohr}} = \frac{E_0}{h}\left[\frac{1}{n^2} - \frac{1}{(n+1)^2}\right]$$

$$= \frac{E_0}{h}\left[\frac{n^2 + 2n + 1 - n^2}{n^2(n+1)^2}\right] = \frac{E_0}{h}\left[\frac{2n+1}{n^2(n+1)^2}\right]$$

which for large n becomes

$$f_{\text{Bohr}} \approx \frac{2nE_0}{hn^4} = \frac{2E_0}{hn^3}$$

If we substitute E_0 from Equation (4.26), the result is

$$f_{\text{Bohr}} = \frac{me^4}{4\epsilon_0^2 h^3} \frac{1}{n^3} = f_{\text{classical}} \tag{4.35}$$

Equivalence of Bohr and classical frequencies

so the frequencies of the radiated energy agree between classical theory and the Bohr model for large values of the quantum number n. Bohr's correspondence principle is verified for large orbits, where classical and quantum physics should agree.

By 1915, as Bohr's model gained widespread acceptance, the critics of the quantum concept were finding it harder to gain an audience. Bohr had demonstrated the necessity of Planck's quantum constant in understanding atomic structure, and Einstein's conception of the photoelectric effect was generally accepted as well. The *assumption* of quantized angular momentum $L_n = n\hbar$ led to the quantization of other quantities r, v, and E. We collect the following three equations here for easy reference.

Orbital radius $\qquad r_n = \dfrac{4\pi\epsilon_0\hbar^2}{me^2} n^2 \qquad$ (4.24)

Velocity $\qquad v_n = \dfrac{n\hbar}{mr_n} \qquad$ (4.22b)

Energy $\qquad E_n = -\dfrac{e^2}{8\pi\epsilon_0 a_0 n^2} \qquad$ (4.25)

4.5 Successes and Failures of the Bohr Model

As we briefly mentioned in the previous section, the Bohr atomic model was a first step in understanding the structure of the atom. We will discuss the correct description of the hydrogen atom in Chapter 7 after we introduce quantum theory in Chapter 6. Wavelength measurements for the atomic spectrum of hydrogen are precise and exhibit a small disagreement with the Bohr model results just presented. These disagreements can be corrected by looking more carefully at our original assumptions, one of which was to assume an infinite nuclear mass.

Reduced Mass Correction

Figure 4.17 Because the nucleus does not actually have an infinite mass, the electron and nucleus rotate about a common center of mass that is located very near the nucleus.

The electron and hydrogen nucleus actually revolve about their mutual center of mass as shown in Figure 4.17. This is a two-body problem, and our previous analysis should be in terms of r_e and r_M instead of just r. A straightforward analysis derived from classical mechanics shows that this two-body problem can be reduced to an equivalent one-body problem in which the motion of a particle of mass μ_e moves in a central force field around the center of mass. The only change required in the results of Section 4.4 is to replace the electron mass m_e by its **reduced mass** μ_e where

Reduced mass

$$\mu_e = \frac{m_e M}{m_e + M} = \frac{m_e}{1 + \dfrac{m_e}{M}} \tag{4.36}$$

and M is the mass of the nucleus (see Problem 53). In the case of the hydrogen atom, M is the proton mass, and the correction for the hydrogen atom is $\mu_e = 0.999456\, m_e$. This difference can be measured experimentally. The Rydberg constant for infinite nuclear mass, R_∞, defined in Equation (4.29), should be replaced by R, where

$$R = \frac{\mu_e}{m_e} R_\infty = \frac{1}{1 + \dfrac{m_e}{M}} R_\infty = \frac{\mu_e e^4}{4\pi c \hbar^3 (4\pi\epsilon_0)^2} \tag{4.37}$$

The Rydberg constant for hydrogen is $R_H = 1.096776 \times 10^7 \text{ m}^{-1}$.

Example 4.8

Calculate the wavelength for the $n_u = 3 \to n_\ell = 2$ transition (called the H_α line) for the atoms of hydrogen, deuterium, and tritium.

Strategy We use Equation (4.30) but with R_∞ replaced by the Rydberg constant expressed in Equation (4.37). In order to use Equation (4.37) we will need the masses for hydrogen, deuterium, and tritium.

Solution The following masses are obtained by subtracting the electron mass from the atomic masses given in Appendix 8.

$$\text{Proton} = 1.007276 \text{ u}$$

$$\text{Deuteron} = 2.013553 \text{ u}$$

Triton (tritium nucleus) $= 3.015500$ u

The electron mass is $m_e = 0.0005485799$ u. The Rydberg constants are

$$R_H = \frac{1}{1 + \dfrac{0.0005486}{1.00728}} R_\infty = 0.99946 R_\infty \qquad \text{Hydrogen}$$

$$R_D = \frac{1}{1 + \dfrac{0.0005486}{2.01355}} R_\infty = 0.99973 R_\infty \qquad \text{Deuterium}$$

$$R_T = \frac{1}{1 + \dfrac{0.0005486}{3.01550}} R_\infty = 0.99982 R_\infty \qquad \text{Tritium}$$

The calculated wavelength for the H_α line is

$$\frac{1}{\lambda} = R\left(\frac{1}{2^2} - \frac{1}{3^2}\right) = 0.13889R$$

$$\lambda(H_\alpha, \text{hydrogen}) = 656.47 \text{ nm}$$

$$\lambda(H_\alpha, \text{deuterium}) = 656.29 \text{ nm}$$

$$\lambda(H_\alpha, \text{tritium}) = 656.23 \text{ nm}$$

Deuterium was discovered when two closely spaced spectral lines of hydrogen near 656.4 nm were observed in 1932. These proved to be the H_α lines of atomic hydrogen and deuterium.

The Bohr model may be applied to any single-electron atom (hydrogen-like) even if the nuclear charge is greater than 1 proton charge $(+e)$, for example He^+ and Li^{++}. The only change needed is in the calculation of the Coulomb force, where e^2 is replaced by Ze^2 to account for the nuclear charge of $+Ze$. Bohr applied his model to the case of singly ionized helium, He^+. The Rydberg Equation (4.30) now becomes

$$\frac{1}{\lambda} = Z^2 R\left(\frac{1}{n_\ell^2} - \frac{1}{n_u^2}\right) \tag{4.38}$$

where the Rydberg constant is given by Equation (4.37). We emphasize that Equation (4.38) is valid only for single-electron atoms (H, He^+, Li^{++}, and so on) and does not apply to any other atoms (for example He, Li, Li^+). Charged atoms, such as He^+, Li^+, and Li^{++}, are called *ions*.

In his original paper of 1913, Bohr predicted the spectral lines of He^+ although they had not yet been identified in the lab. He showed that certain lines (generally ascribed to hydrogen) that had been observed by Pickering in stellar spectra, and by Fowler in vacuum tubes containing both hydrogen and helium, could be identified as singly ionized helium. Bohr showed that the wavelengths predicted for He^+ with $n_\ell = 4$ are almost identical to those of H for $n_\ell = 2$, except that He^+ has *additional lines* between those of H (see Problem 35). The correct explanation of this fact by Bohr gave credibility to his model.

Example 4.9

Calculate the shortest wavelength that can be emitted by the Li^{++} ion.

Strategy The shortest wavelength occurs when the electron changes from the highest state (unbound, $n_u = \infty$) to the lowest state ($n_\ell = 1$). We use Equation (4.38) to calculate the wavelength.

Solution Equation (4.38) gives

$$\frac{1}{\lambda} = (3)^2 R\left(\frac{1}{1^2} - \frac{1}{\infty}\right) = 9R$$

$$\lambda = \frac{1}{9R} = 10.1 \text{ nm}$$

When we let $n_u = \infty$, we have what is known as the *series limit*, which is the shortest wavelength possibly emitted for each of the named series.

Other Limitations

As the level of precision increased in optical spectrographs, it was observed that each of the lines, originally believed to be single, actually could be resolved into two or more lines. Arnold Sommerfeld adapted the special theory of relativity (assuming some of the electron orbits were elliptical) to Bohr's hypotheses and was able to account for some of the "splitting" of spectral lines. Subsequently it has been found that other factors (especially the electron's *spin*, or *intrinsic angular momentum*) also affect the fine structure of spectral lines.

It was soon observed that external magnetic fields (the Zeeman effect) and external electric fields (the Stark effect) applied to the radiating atoms affected the spectral lines, splitting and broadening them. Although classical electromagnetic theory could quantitatively explain the (normal) Zeeman effect (see Chapter 7), it was unable to account for the Stark effect; for this the quantum model of Bohr and Sommerfeld was necessary.

Although the Bohr model was a great step forward in the application of the new quantum theory to understanding the tiny atom, it soon became apparent that the model had its limitations:

Limitations of Bohr model

1. It could be successfully applied only to single-electron atoms (H, He^+, Li^{++}, and so on).
2. It was not able to account for the intensities or the fine structure of the spectral lines.
3. Bohr's model could not explain the binding of atoms into molecules.

We discuss in Chapter 7 the full quantum mechanical theory of the hydrogen atom, which accounts for all of these phenomena. The Bohr model was an ad hoc theory to explain the hydrogen spectral lines. Although it was useful in the beginnings of quantum physics, we now know that the Bohr model does not correctly describe atoms. Despite its flaws, Bohr's model should not be denigrated. It was the first step from a purely classical description of the atom to the correct quantum explanation. As usually happens in such tremendous changes of understanding, Bohr's model simply did not go far enough—he retained too many classical concepts. Einstein, many years later, noted* that Bohr's achievement "appeared to me like a miracle and appears as a miracle even today."

4.6   Characteristic X-Ray Spectra and Atomic Number

By 1913 when Bohr's model was published, little progress had been made in understanding the structure of many-electron atoms. It was believed that the general characteristics of the Bohr-Rutherford atom would prevail. We discussed the production of x rays from the bombardment of various materials by electrons in Section 3.7. It was known that an x-ray tube with an anode made from a given element produced a continuous spectrum of bremsstrahlung x rays upon which are superimposed several peaks whose frequencies are characteristic of that element (see Figure 3.19).

Characteristic x-ray wavelengths

We can now understand these **characteristic x-ray wavelengths** by adopting Bohr's *electron shell* hypothesis. Bohr's model suggests that an electron shell based on the radius r_n can be associated with each of the principal quantum numbers n. Electrons with lower values of n are more tightly bound to the nucleus than those with higher values. The radii of the electron orbits increase in proportion to n^2 [Equation (4.24)]. A specific energy is associated with each value of n. We may assume that when we add electrons to a fully ionized many-electron atom, the inner

*P. A. Schillp, ed., *Albert Einstein, Philosopher-Scientist*, La Salle, IL: The Open Court, 1949.

shells (low values of n) are filled before the outer shells. We have not yet discussed how many electrons each shell contains or even why electrons tend to form shells, rather than a featureless glob, for example. Historically, the shells were given letter names: the $n = 1$ shell was called the K shell, $n = 2$ was the L shell, and so on. The shell structure of an atom is indicated in Figure 4.18. In heavy atoms with many electrons, we may suppose that several shells contain electrons. What happens when a high-energy electron in an x-ray tube collides with one of the K-shell electrons (we shall call these *K electrons*) in a target atom? If enough energy can be transferred to the K electron to dislodge it from the atom, the atom will be left with a vacancy in its K shell. The atom is most stable in its lowest energy state or *ground state*, so it is likely that an electron from one of the higher shells will change its state and fill the inner-shell vacancy at lower energy, emitting radiation as it changes its state. When this occurs in a heavy atom we call the electromagnetic radiation emitted an *x ray*, and it has the energy

$$E\,(\text{x ray}) = E_u - E_\ell \tag{4.39}$$

The process is precisely analogous to what happens in an excited hydrogen atom. The photon produced when the electron falls from the L shell into the K shell is called a K_α x ray; when it falls from the M shell into the K shell, the photon is called a K_β x ray. This scheme of x-ray identification is diagrammed in Figure 4.18. The relative positions of the energy levels of the various shells differ for each element, so the characteristic x-ray energies of the elements are simply the energy differences between the shells. The two strong peaks in the molybdenum spectrum of Figure 3.19 are the K_α and K_β x rays.

This simple description of the electron shells, which will be modified later in Chapters 7 and 8 by the full quantum mechanical treatment, was not understood by early 1913. The experimental field of x-ray detection was beginning to flourish (see Section 3.3), and the precise identification of the wavelengths of characteristic x rays was possible. In 1913 H. G. J. Moseley, working in Rutherford's Manchester laboratory, was engaged in cataloguing the characteristic x-ray spectra of a series of elements. He concentrated on the K- and L-shell x rays produced in an x-ray tube. Physicists in Rutherford's Manchester lab had already fully accepted the concept of the atomic number, although there was no firm experimental evidence for doing so. Most of the European physicists still believed that **atomic weight** A was the important factor, and the periodic table of elements was so structured. The **atomic number** Z is the number of protons in the nucleus. The makeup of the nucleus was unknown at the time, so Z was related to the positive charge of the nucleus.

Moseley compared the frequencies of the characteristic x rays with the then supposed atomic number of the elements and found empirically an amazing linear result when he plotted the atomic number Z versus the square root of the measured frequency as shown in Figure 4.19:

$$f_{K_\alpha} = \frac{3cR}{4}(Z-1)^2 \tag{4.40}$$

This result holds for the K_α x rays, and a similar result was found for the L shell. The data shown in Figure 4.19 are known as a *Moseley plot*. Moseley began his work in 1913 in Manchester and, after moving to Oxford late in 1913, completed the investigation in early 1914. Although it is clear that Bohr and Moseley discussed physics and even corresponded after Bohr left for Copenhagen, Moseley does not mention Bohr's model in his 1914 paper. Thus, it is not known whether Bohr's ideas had any influence on Moseley's work.

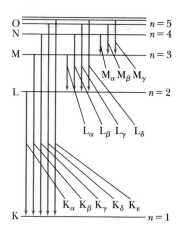

Figure 4.18 Historically, the stationary states were also given letter identifications: K shell ($n = 1$), L shell ($n = 2$), M shell ($n = 3$), and so on. The x rays emitted when an atom changes energy states are given different names depending on the initial and final states. The Greek letter subscripts indicate the value of Δn and the roman letters the value of n for the final state.

Significance of atomic number

Figure 4.19 Moseley's original data indicating the relationship between the atomic number Z and the characteristic x-ray frequencies. Notice the missing entries for elements $Z = 43$, 61, and 75, which had not yet been identified. There are also a few errors in the atomic number designations for the elements. © *From H. G. J. Moseley*, Philosophical Magazine *(6)*, **27**, *703 (1914).*

Using Bohr's model we can understand Moseley's empirical result, Equation (4.40). If a vacancy occurs in the K shell, there is still one electron remaining in the K shell. (We will see in Chapter 8 that, at most, two electrons can occupy the K shell.) An electron in the L shell will feel an effective charge of $(Z - 1)e$ due to $+Ze$ from the nucleus and $-e$ from the remaining K-shell electron, because the L-shell orbit is normally outside the K-shell orbit. The other electrons outside the K shell hardly affect the L-shell electron. The x ray produced when a transition occurs from the $n = 2$ to the $n = 1$ shell has the wavelength, from Equation (4.38), of

$$\frac{1}{\lambda_{K_\alpha}} = R(Z - 1)^2\left(\frac{1}{1^2} - \frac{1}{2^2}\right) = \frac{3}{4}R(Z - 1)^2 \qquad (4.41)$$

or

$$f_{K_\alpha} = \frac{c}{\lambda_{K_\alpha}} = \frac{3cR}{4}(Z - 1)^2 \qquad (4.42)$$

which is precisely the equation Moseley found describing the K_α-shell x rays. We can write Equation (4.41) in a more general form for the K series of x ray wavelengths:

$$\frac{1}{\lambda_K} = R(Z - 1)^2\left(\frac{1}{1^2} - \frac{1}{n^2}\right) = R(Z - 1)^2\left(1 - \frac{1}{n^2}\right) \qquad (4.43)$$

Moseley correctly concluded that the atomic number Z was the determining factor in the ordering of the periodic table, and this reordering was more consistent with chemical properties than one based on atomic weight. It put potassium ($Z = 19$, $A = 39.10$) after argon ($Z = 18$, $A = 39.95$) by atomic number rather than the reverse by atomic weight. Moseley concluded that the atomic number of an element should be identified with the number of positive units of electricity in the nucleus (that is, the number of protons). He tabulated all the atomic numbers between Al ($Z = 13$) and Au ($Z = 79$) and pointed out there were still three elements ($Z = 43, 61,$ and 75) yet to be discovered! The element promethium ($Z = 61$) was finally discovered around 1940.

Moseley's research helped put the Rutherford-Bohr model of the atom on a firmer footing. It clarified the importance of the electron shells for all the elements, not just for hydrogen. It also helped show that the atomic number was the significant factor in the ordering of the periodic table, not the atomic weight.

Henry G. J. Moseley (1887–1915), shown here working in 1910 in the Balliol-Trinity laboratory of Oxford University, was a brilliant young experimental physicist with varied interests. Unfortunately, he was killed in action at the young age of 27 during the English expedition to the Dardanelles. Moseley volunteered and insisted on combat duty in World War I, despite the attempts of Rutherford and others to keep him out of action. *University of Oxford, Museum of the History of Science/Courtesy AIP Niels Bohr Library.*

Example 4.10

Moseley found experimentally that the equation describing the frequency of the L_α spectral line was

$$f_{L_\alpha} = \frac{5}{36}cR(Z - 7.4)^2 \qquad (4.44)$$

How can the Bohr model explain this result? What is the general form for the L-series wavelengths λ_L?

Strategy We follow the general procedure that we used to find Equation (4.42). The L_α x ray results from a transition from the M shell ($n_u = 3$) to the L shell ($n_\ell = 2$). There may be several electrons in the L shell and two electrons in the K shell that shield the nuclear charge $+Ze$ from the M-shell electron making the transition to the L shell. Let's assume the effective charge that the electron sees is $+Z_{eff}e$. Then we can use Equation (4.38) to find both Z_{eff} and the general form for the λ_L series of wavelengths.

Solution We replace Z by Z_{eff} in Equation (4.38) and find

$$f_{L_\alpha} = \frac{c}{\lambda_{L_\alpha}} = cRZ_{eff}^2\left(\frac{1}{2^2} - \frac{1}{3^2}\right) \qquad (4.45)$$

$$f_{L_\alpha} = \frac{5cRZ_{eff}^2}{36}$$

According to Moseley's data the effective charge Z_{eff} must be $Z - 7.4$. This result is within the spirit of the Bohr model, which applied primarily to hydrogen-like atoms.

We rewrite Equation (4.45) to determine λ_L for the entire series:

$$\frac{1}{\lambda_L} = RZ_{eff}^2\left(\frac{1}{2^2} - \frac{1}{n^2}\right) = R(Z - 7.4)^2\left(\frac{1}{4} - \frac{1}{n^2}\right) \qquad (4.46)$$

James Franck (1882–1964), shown here on the left with Gustav Hertz in Tübingen, Germany, in 1926, came to America in 1935 to avoid Nazi persecution and became an important American scientist who trained many experimental physicists. Gustav Hertz (1887–1975), the nephew of Heinrich Hertz who discovered electromagnetic waves, worked in German universities and industrial labs before going to the Soviet Union in 1945. They received the Nobel Prize for Physics for the experiment named after them (Franck-Hertz experiment) in 1925. *AIP Emilio Segrè Visual Archives.*

4.7   Atomic Excitation by Electrons

All the evidence for the quantum theory discussed so far has involved quanta of electromagnetic radiation (photons). In particular, the Bohr model explained experimental optical spectra of certain atoms. Spectroscopic experiments were typically performed by exciting the elements, for example in a high-voltage discharge tube, and then examining the emission spectra.

The German physicists James Franck and Gustav Hertz decided to study electron bombardment of gaseous vapors in order to study the phenomenon of ionization. They set out in 1914 explicitly to study the possibility of transferring a part of an electron's kinetic energy to an atom. If such measurements were possible, they would provide a distinctive new technique for studying atomic structure.

An experimental arrangement similar to that used by Franck and Hertz is shown in Figure 4.20. This particular arrangement is one actually used in a typical undergraduate physics laboratory experiment. Electrons are emitted thermionically from a hot cathode (filament) and are then accelerated by an electric field with its intensity determined by a variable (0- to 45-V) power supply. After passing through a grid consisting of wire mesh, the electrons are subjected to a decelerating voltage (typically 1.5 V) between grid and anode (collector). If the electrons have greater than 1.5 eV after passing through the grid they will have enough energy to reach the collector and be registered as current in an extremely sensitive ammeter (called an *electrometer*). A voltmeter measures the accelerating voltage V. The experiment consists of measuring the current I in the electrometer as a function of V.

The accelerating electrons pass through a region containing mercury (Hg) vapor (a monatomic gas). Franck and Hertz found that as long as the accelerating voltage V was below about 5 V (that is, the maximum kinetic energy of the electrons was below 5 eV), the electrons apparently did not lose energy. The electron current registered in the electrometer continued to increase as V increased. However, as the accelerating voltage increased above 5 V, there was a sudden drop in the current (see Figure 4.21, which was constructed using data taken by students performing this experiment). As the accelerating voltage continued to increase above 5 V, the current increased again, but suddenly dropped again above 10 V. Franck and Hertz first interpreted this behavior as the onset of ionization of

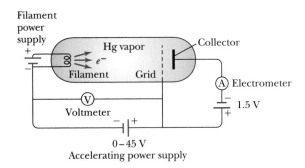

Figure 4.20 Schematic diagram of apparatus used in an undergraduate physics laboratory for the Franck-Hertz experiment. The hot filament produces electrons, which are accelerated through the mercury vapor toward the grid. A decelerating voltage between grid and collector prevents the electrons from registering in the electrometer unless the electron has a certain minimum energy.

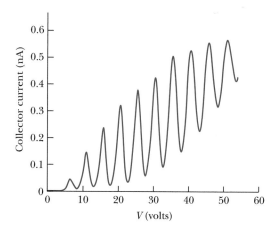

Figure 4.21 Data from an undergraduate student's Franck-Hertz experiment using apparatus similar to that shown in Figure 4.20. The energy difference between peaks is about 5 V, but the first peak is not at 5 V because of the work function differences of the metals used for the filament and grid.

the Hg atom; that is, an atomic electron is given enough energy to remove it from the Hg, leaving the atom ionized. They later realized that the Hg atom was actually being excited to its first excited state.

We can explain the experimental results of Franck and Hertz within the context of Bohr's picture of *quantized atomic energy levels*. In the most popular representation of atomic energy states, we say that the atom when all the electrons are in their lowest possible energy states is the **ground state**. We define this energy E_0 to be zero. The first quantized energy state above the ground state is called the first excited state, and it has energy E_1. The energy difference $E_1 - 0 = E_1$ is called the **excitation energy** of the state E_1. We show the position of one electron in an energy-level diagram of Hg in Figure 4.22 in both the ground state and first excited state. The first excited state of Hg is at an excitation energy of 4.88 eV. As long as the accelerating electron's kinetic energy is below 4.88 eV, no energy can be transferred to Hg because not enough energy is available to excite an electron to the next energy level in Hg. The Hg atom is so much more massive than the electron that almost no kinetic energy is transferred to the recoil of the Hg atom; the collision is *elastic*. The electron can only bounce off the Hg atom and continue along a new path with about the same kinetic energy. If the electron gains at least 4.88 eV of kinetic energy from the accelerating potential, it can transfer 4.88 eV to an electron in Hg, promoting it to the first excited state. This is an *inelastic* collision. A bombarding electron that has lost energy in an inelastic collision then has too little energy (after it passes the grid) to reach the collector. Above 4.88 V, the current dramatically drops because the inelastically scattered electrons no longer reach the collector.

When the accelerating voltage is increased to 7 or 8 V, even electrons that have already made an inelastic collision have enough remaining energy to reach the collector. Once again the current increases with V. However, when the accelerating voltage reaches 9.8 V, the electrons have enough energy to excite two Hg atoms in successive inelastic collisions, losing 4.88 eV in each (2×4.88 eV $= 9.76$ eV). The current drops sharply again. As we see in Figure 4.21, even with student apparatus it is possible to observe several successive excitations as the accelerating voltage is increased. Notice that the energy differences between peaks are typically 4.9 eV. The first peak does not occur at 4.9 eV because of the difference in the work functions between the dissimilar metals used as cathode and anode. Other highly excited states in Hg can also be excited in an inelastic collision, but

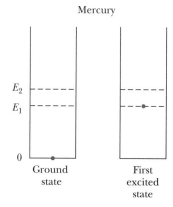

Figure 4.22 A valence electron is shown in the ground state of mercury on the left. On the right side the electron has been elevated to the first excited state after a bombarding electron inelastically scattered from the mercury atom.

the probability of exciting them is much smaller than that for the first excited state. Franck and Hertz, however, were able to detect them.

The Franck-Hertz experiment convincingly proved the quantization of atomic electron energy levels. The bombarding electron's kinetic energy can change only by certain discrete amounts determined by the atomic energy levels of the mercury atom. They performed the experiment with gases of several other elements and obtained similar results.

Conceptual Example 4.11

Would it be experimentally possible to observe radiation emitted from the first excited state of Hg after it was produced by an electron collision?

Solution If the collision of the bombarding electron with the mercury atom is elastic, mercury will be left in its ground state. If the collision is inelastic, however, the mercury atom will end up in its excited state at 4.9 eV (see Figure 4.22). The mercury atom will not exist long in its first excited state and should decay quickly ($\sim 10^{-8}$ s) back to the ground state.

Franck and Hertz even considered this possibility and looked for x rays. They observed no radiation emitted when the electron's kinetic energy was below about 5 V, but as soon as the current dropped as the voltage went past 5 V, indicating excitation of Hg, an emission line of wavelength 254 nm (ultraviolet) was observed. Franck and Hertz set $E = 4.88$ eV $= hf = (hc)/\lambda$ and showed that the value of h determined from $\lambda = 254$ nm was in good agreement with values of Planck's constant determined by other means.

We have learned in this chapter about the Rutherford-Bohr concept of the atom. Rutherford showed that the atom consisted of an object with most of the mass in the positively charged nucleus. Electrons apparently orbited the nucleus. Bohr was able to derive the important Rydberg equation by proposing his quantized shell model of the atom and explaining how electrons can have stable orbits around the nucleus. The experiment of Franck and Hertz confirmed the quantized shell behavior. Nevertheless, it was clear that Bohr's model was primarily effective for hydrogen-like atoms and that a full and complete description for the majority of the atomic elements was lacking. Before pursuing that in Chapters 6–8, we must first return to investigate the wave properties of matter in Chapter 5, where even more surprises await us.

Summary

Rutherford proposed a model of the atom consisting of a very massive, compact (relative to the size of the atom), positively charged nucleus surrounded by electrons. His assistants, Geiger and Marsden, performed scattering experiments with energetic alpha particles and showed that the number of backward-scattered α particles could be accounted for only if the model were correct. The relation between the impact parameter b and scattering angle θ for Coulomb scattering is

$$b = \frac{Z_1 Z_2 e^2}{8\pi\epsilon_0 K} \cot\frac{\theta}{2} \qquad (4.6)$$

Rutherford's equation for the number of particles scattered at angle θ is

$$N(\theta) = \frac{N_i n t}{16}\left(\frac{e^2}{4\pi\epsilon_0}\right)^2 \frac{Z_1^2 Z_2^2}{r^2 K^2 \sin^4(\theta/2)} \quad (4.13)$$

where the dependence on charges $Z_1 e$ and $Z_2 e$, the kinetic energy K, the target thickness t, and the scattering angle θ were verified experimentally. The classical planetary atomic model predicts the rapid demise of the atom because of electromagnetic radiation.

Niels Bohr was able to derive the empirical Rydberg formula for the wavelengths of the optical spectrum of hydrogen by using his "general assumptions." This led to the quantization of various physical parameters of the hydrogen atom, including the radius, $r_n = n^2 a_0$, where $a_0 = 0.53 \times 10^{-10}$ m, and the energy, $E_n = -E_0/n^2$, where $E_0 = 13.6$ eV. The Rydberg equation

$$\frac{1}{\lambda} = R\left(\frac{1}{n_\ell^2} - \frac{1}{n_u^2}\right)$$

gives the wavelengths, where n_ℓ and n_u are the quantum numbers for the lower and upper stationary states, respectively. The Bohr model could explain the optical spectra of hydrogen-like atoms such as He^+ and Li^{++}, but could not account for the characteristics of many-electron atoms. This indicated that the model was incomplete and only approximate. Bohr's correspondence principle relates quantum theories to classical ones in the limit of large quantum numbers.

By examining the characteristic x-ray spectra of the chemical elements, Moseley proved the fundamental significance of the atomic number. We can derive the empirical Moseley relation

$$f_{K_\alpha} = \frac{3cR}{4}(Z-1)^2 \quad (4.40)$$

from the structure of the atom proposed by Rutherford, together with Bohr's model of hydrogen-like energy levels.

Another way of studying atomic structure is by using electron scattering rather than photon or optical methods. Franck and Hertz were able to confirm the quantized structure of the atom and determine a value of Planck's constant h in good agreement with other methods.

Questions

1. Thomson himself was perhaps the biggest critic of the model referred to as "plum pudding." He tried for years to make it work. What experimental data could he not predict? Why couldn't he make the planetary model of Rutherford-Bohr work?

2. Does it seem fortuitous that most of the successful physicists who helped unravel the secrets of atomic structure (Thomson, Rutherford, Bohr, Geiger, and Moseley) worked either together or in close proximity in England? Why do you suppose we don't hear names of physicists working on this idea in other European countries or the United States?

3. Could the Rutherford scattering of α particles past $90°$ be due to scattering from electrons collected together (say $100\ e^-$) in one place over a volume of diameter 10^{-15} m? Explain.

4. In an intense electron bombardment of the hydrogen atom, significant electromagnetic radiation is produced in all directions upon decay. Which emission line would you expect to be most intense? Why?

5. Why are peaks due to higher-lying excited states in the Franck-Hertz experiment not more observable?

6. As the voltage increases above 5 V in the Franck-Hertz experiment, why doesn't the current suddenly jump back up to the value it had below 5 V?

7. Using Hg gas in the Franck-Hertz experiment, approximately what range of voltages would you expect for the first peak? Explain.

8. When are photons likely to be emitted in the Franck-Hertz experiment?

9. Is an electron most strongly bound in an H, He^+, or Li^{++} atom? Explain.

10. Why do we refer to atoms as being in the "ground" state or "stationary"? What does an "excited" state mean?

11. What lines would be missing for hydrogen in an absorption spectrum? What wavelengths are missing for hydrogen in an emission spectrum?

12. Why can't the Bohr model be applied to the neutral He atom? What difficulties do you think Bohr had in modifying his model for He?

13. Describe how the hydrogen atom might absorb a photon of energy less than 13.6 eV. Describe a process by which a 9.8-eV photon might be absorbed. What about a 15.2-eV photon?

Problems

Note: The more challenging problems are indicated by a shaded box.

4.1 The Atomic Models of Thomson and Rutherford

1. In Thomson's plum-pudding model, devise an atomic composition for carbon that consists of a pudding of charge $+6e$ along with six electrons. Try to configure a system where the charged particles move only about points in stable equilibrium.

2. How large an error in the velocity do we make by treating the velocity of a 7.7-MeV α particle non-relativistically?

3. In Example 4.1, show that the electron's velocity must be $v'_e \approx 2v_\alpha$ in order to conserve energy and linear momentum.

4. Thomson worked out many of the calculations for multiple scattering. If we find an average scattering angle of $1°$ for α scattering, what would be the probability that from multiple scattering the α particle could scatter by as much as $80°$? The probability for large-angle scattering is $\exp(-(\theta/\langle\theta\rangle)^2)$. Geiger and Marsden found that about 1 in 8000 α particles were deflected past $90°$. Can multiple scattering explain the experimental results of Geiger and Marsden? Explain.

4.2 Rutherford Scattering

5. Calculate the impact parameter for scattering a 7.7-MeV α particle from gold at an angle of (a) $1°$ and (b) $90°$.

6. A beam of 8.0-MeV α particles scatters from a thin gold foil. What is the ratio of the number of α particles scattered to angles greater than $1°$ to the number scattered to angles greater than $2°$?

7. For aluminum ($Z = 13$) and gold ($Z = 79$) targets, what is the ratio of α-particle scattering at any angle for equal numbers of scattering nuclei per unit area?

8. What fraction of 5-MeV α particles will be scattered through angles greater than $6°$ from a gold foil ($Z = 79$, density $= 19.3$ g/cm^3) of thickness 10^{-8} m?

9. In an experiment done by scattering 5.5-MeV α particles on a thin gold foil, students find that 10,000 α particles are scattered at an angle greater than $50°$. (a) How many of these α particles will be scattered greater than $90°$? (b) How many will be scattered between $70°$ and $80°$?

10. Students want to construct a scattering experiment using a powerful source of 5.5-MeV α particles to scatter from a gold foil. They want to be able to count 1 particle/s at $50°$, but their detector is limited to a maximum count rate of 2000 particles/s. Their detector subtends a small angle. Will their experiment work without modifying the detector if the other angle they want to measure is $6°$?

11. The nuclear radii of aluminum and gold are approximately $r = 3.6$ fm and 7.0 fm, respectively. The radii of protons and alpha particles are 1.3 fm and 2.6 fm, respectively. (a) What energy α particles would be needed in head-on collisions for the nuclear surfaces to just touch? (This is about where the nuclear force becomes effective.) (b) What energy protons would be needed? In both (a) and (b) perform the calculation for both aluminum and gold.

12. Consider the scattering of an alpha particle from the positively charged part of the Thomson plum-pudding model. Let the kinetic energy of the α particle be K (nonrelativistic) and let the atomic radius be R. (a) Assuming that the maximum transverse Coulomb force acts on the α particle for a time $\Delta t = 2R/v$ (where v is the initial speed of the α particle), show that the largest scattering angle we can expect from a single atom is

$$\theta = \frac{2Z_2 e^2}{4\pi\epsilon_0 KR}$$

(b) Evaluate θ for an 8.0-MeV α particle scattering from a gold atom of radius ~ 0.1 nm.

13. Using the results of the previous problem, (a) find the average scattering angle of a 10-MeV α particle from a gold atom ($R \approx 10^{-10}$ m) for the positively charged part of the Thomson model. (b) How does this compare with the scattering from the electrons?

4.3 The Classical Atomic Model

14. The radius of a hydrogen nucleus is believed to be about $r = 1.2 \times 10^{-15}$ m. (a) If an electron rotates around the nucleus at that radius, what would be its speed according to the planetary model? (b) What would the total mechanical energy be? (c) Are these reasonable?

15. Assume that the nucleus is composed of electrons and that the protons are outside. (a) If the size of an atom were about 10^{-10} m, what would be the speed of a proton? (b) What would be the total mechanical energy? (c) What is wrong with this model?

16. Calculate the speed and radial acceleration for an electron in the hydrogen atom. Do the same for the Li^{++} ion.

17. What is the total mechanical energy for a ground-state electron in H, He$^+$, and Li^{++} atoms? For which atom is the electron most strongly bound? Why?

18. Calculate the time, according to classical laws, it would take the electron of the hydrogen atom to radiate its energy and crash into the nucleus. [*Hint*: The

radiated power P is given by $(1/4\pi\epsilon_0)(2Q^2/3c^3)$ $(d^2\vec{r}/dt^2)^2$ where Q is the charge, c the speed of light, and $\vec{r}$ the position vector of the electron from the center of the atom.]

4.4 The Bohr Model of the Hydrogen Atom

19. The Ritz combination rules express relationships between observed frequencies of the emission optical spectra. Explain one of the more important ones:

$$f(K_\alpha) + f(L_\alpha) = f(K_\beta)$$

where K_α and K_β refer to the Lyman series and L_α to the Balmer series of hydrogen (Figure 4.18).

20. Calculate the angular momentum in $kg \cdot m^2/s$ for the lowest electron orbit in the hydrogen atom. Compare the result with Planck's constant h.

21. Use the known values of ϵ_0, h, m, and e and calculate the following to five significant figures: hc (in $eV \cdot nm$), $e^2/4\pi\epsilon_0$ (in $eV \cdot nm$), mc^2 (in keV), a_0 (in nm), and E_0 (in eV).

22. What is the speed (ratio of v/c) of the electron in the first three Bohr orbits of the H atom?

23. A hydrogen atom in an excited state absorbs a photon of wavelength 434 nm. What were the initial and final states of the hydrogen atom?

24. A hydrogen atom in an excited state emits a photon of wavelength 95 nm. What are the initial and final states of the hydrogen atom?

25. What is the calculated binding energy of the electron in the ground state of (a) deuterium, (b) He^+, and (c) Be^{+++}?

26. The **isotope shift** of spectral lines refers to the shift in wavelengths (or frequencies) due to the different masses of given elements. Find the isotope shifts for each of the four visible Balmer series wavelengths for deuterium and tritium compared to hydrogen.

27. Find the isotope shift (see previous problem) of the ground-state energy for deuterium and tritium compared to the ground-state energy of hydrogen. Express the answer in eV.

28. Describe the visible absorption spectra for (a) a hydrogen atom and (b) an ionized helium atom, He^+.

29. A hydrogen atom exists in an excited state for typically 10^{-8} s. How many revolutions would an electron make in an $n = 3$ state before decaying?

30. Electromagnetic radiation of wavelength 100 nm is incident upon the ground-state hydrogen atom at rest. What is the highest state to which hydrogen can be excited?

31. A muonic atom consists of a muon ($m = 106$ MeV/c^2) in place of an electron. For the muon in a hydrogen atom, what is (a) the smallest radius and (b) the binding energy of the muon in the ground state? (c) Calculate the series limit of the wavelength for the first three series.

32. Positronium is an atom composed of an electron and a positron ($m = m_e$, $Q = +e$). Calculate the distance between the particles and the energy of the lowest energy state of positronium. (*Hint*: what is the reduced mass of the two particles? See Problem 53.)

33. (a) Find the Bohr radius of the positronium atom described in the previous problem. (b) Find the wavelength for the transition from $n_u = 2$ to $n_\ell = 1$ for positronium.

34. What is the difference in the various Bohr radii r_n for the hydrogen atom (a) between r_1 and r_2, (b) between r_5 and r_2, (c) between r_5 and r_6, and (d) between r_{10} and r_{11}?

4.5 Successes and Failures of the Bohr Model

35. Compare the Balmer series of hydrogen with the series where $n_\ell = 4$ for the ionized helium atom He^+. What is the difference between the wavelengths of the L_α and L_β line of hydrogen and the $n_u = 6$ and 8 of He^+? Is there a wavelength of the Balmer series very similar to any wavelength values where $n_\ell = 4$ in He^+?

36. Calculate the Rydberg constant for the single-electron (hydrogen-like) ions of helium, potassium, and uranium. Compare each of them with R_∞ and determine the percentage difference.

37. In 1896 Pickering found lines from the star Zeta Puppis that had not been observed on Earth. Bohr showed in 1913 that the lines were due to He^+. Show that an equation giving these wavelengths is

$$\frac{1}{\lambda} = R\left(\frac{1}{n_\ell^2} - \frac{1}{n_u^2}\right)$$

What value should the Rydberg constant R have in this case?

4.6 Characteristic X-Ray Spectra and Atomic Number

38. What wavelengths for the L_α lines did Moseley predict for the missing $Z = 43, 61$, and 75 elements? (See Example 4.10.)

39. If the resolution of a spectrograph is $\Delta\lambda = 10^{-12}$ m, would it be able to separate the K_α lines for platinum and gold? Explain.

40. Determine the correct equation to describe the K_β frequencies measured by Moseley. Compare that with Moseley's equation for K_α frequencies. Does the result agree with the data in Figure 4.19? Explain.

41. Calculate the K_α and K_β wavelengths for He and Li.

42. (a) Calculate the ratio of K_α wavelengths for uranium and carbon. (b) Calculate the ratio of L_α wavelengths for tungsten and calcium.

43. Calculate the three longest wavelengths and the series limit for the molybdenum atom.

44. An unknown element is used as a target in an x-ray tube. Measurements show that the characteristic spectral lines with the longest wavelengths are 0.155 nm and 0.131 nm. What is the element? (*Hint:* you will find the answer to Problem 40 to be useful.)

4.7 Atomic Excitation by Electrons

45. If an electron of 40 eV had a head-on collision with a Hg atom at rest, what would be the kinetic energy of the recoiling Hg atom? Assume an elastic collision.

46. In the Franck-Hertz experiment, explain why the small potential difference between the grid and collector plate is useful. Redraw the data of Figure 4.21 the way you believe the data would be without this small retarding potential.

47. Calculate the value of Planck's constant determined by Franck and Hertz when they observed the 254-nm ultraviolet radiation using Hg vapor.

48. Consider an element having excited states at 3.6 eV and 4.6 eV used as a gas in the Franck-Hertz experiment. Assume that the work functions of the materials involved cancel out. List all the possible peaks that *might* be observed with electron scattering up to an accelerating voltage of 18 V.

General Problems

49. The redshift measurements of spectra from magnesium and iron are important in understanding distant galaxies. What are the K_α and L_α wavelengths for magnesium and iron?

50. In the early 1960s the strange optical emission lines from starlike objects that also were producing tremendous radio signals confused scientists. Finally, in 1963 Maarten Schmidt of the Mount Palomar observatory discovered that the optical spectra were just those of hydrogen, but redshifted because of the tremendous velocity of the object with respect to Earth. The object was moving away from Earth at a speed of 50,000 km/s! Compare the wavelengths of the normal and redshifted spectral lines for the K_α and K_β lines of the hydrogen atom.

51. A beam of 8-MeV α particles scatters from a gold foil of thickness 0.4 μm. (a) What fraction of the α particles is scattered between 1° and 2°? (b) What is the ratio of α particles scattered through angles greater than 1° to the number scattered through angles greater than 10°; greater than 90°?

52. In Rutherford scattering we noted that angular momentum is conserved. The angular momentum of the incident α particle relative to the target nucleus is mv_0b where m is the mass, v_0 is the initial velocity of the α particle, and b is the impact parameter. Start with $\vec{L} = \vec{r} \times \vec{p}$ and show that angular momentum is conserved, and the magnitude is given by mv_0b along the entire path of the α particle while it is scattered by the Coulomb force from a gold nucleus.

53. The proton (mass M) and electron (mass m) in a hydrogen atom actually rotate about their common center of mass as shown in Figure 4.17. The distance $r = r_e + r_M$ is still defined to be the electron-nucleus distance. Show that Equation (4.24) is only modified by substituting for m by

$$\mu = \frac{m}{1 + m/M}$$

54. In Bohr's Assumption D, he assumed the mean value K of the kinetic energy of the electron-nucleus system to be $nhf_{orb}/2$ where f_{orb} is the orbital frequency of the electron around the nucleus. Calculate f_{orb} in the ground state in the following ways: (a) Use $f_{classical}$ in Equation (4.34). (b) Use Equation (4.33a), but first determine v and r. (c) Show that the mean value K is equal to the absolute value of the electron-nucleus system total energy and that this is 13.6 eV. Use this value of K to determine f_{orb} from the relation for K stated above.

55. Show that the quantization of angular momentum $L = n\hbar$ follows from Bohr's Assumption D that the mean value K of the kinetic energy of the electron-nucleus system is given by $K = nhf_{orb}/2$. Assume a circular orbit.

56. (a) Calculate the energies of the three lowest states of positronium. (b) Determine the wavelengths of the K_α, K_β, L_α, and L_β transitions.

Wave Properties of Matter and Quantum Mechanics I

I thus arrived at the following overall concept which guided my studies: for both matter and radiations, light in particular, it is necessary to introduce the corpuscle concept and the wave concept at the same time.

Louis de Broglie, 1929

We regard quantum mechanics as a complete theory for which the fundamental physical and mathematical hypotheses are no longer susceptible of modification.

**Werner Heisenberg and Max Born,
paper delivered to the Solvay Congress, 1927**

Chapter 3 presented compelling evidence that light (electromagnetic radiation) must be particle-like in order to explain phenomena such as the photoelectric effect and Compton scattering. The emission and absorption of photons in atoms allow us to understand the optical spectra of hydrogen atoms.

In this chapter we discuss so many new surprising results that an overview of them here is in order. For example, we already know that photons, as electromagnetic radiation, demonstrate wavelike properties. The only way we can interpret certain experimental observations is to conclude that wavelike properties are also exhibited by "particles" of matter. We begin the chapter by discussing experiments which prove that photons, in the form of x rays, behave as waves when passing through crystals. De Broglie's suggestion that particles may also behave as waves was verified by the electron-scattering experiments of Davisson and Germer.

We then present a short review of wave phenomena, including a description of the localization of a particle in terms of a collection of waves. Physicists in the first part of the twentieth century had considerable difficulty understanding how wavelike and particle-like properties can occur in nature in the same entity. We now face the same hurdle. Niels Bohr's principle of complementarity convinces us that *both* wavelike and particle-like properties are needed to give a complete description of matter (electrons, protons, and so on) and radiation (photons). We shall see that certain physical observables can only be expressed in terms of probabilities, with those probabilities determined by using wave functions $\Psi(x, t)$.

Max von Laue (1879–1960) was born, educated, and worked most of his life in Germany. After studying at Strasburg, Göttingen, and Munich, he received his doctorate from the University of Berlin in 1903 where he studied under Max Planck. He subsequently worked at several German universities and did his Nobel Prize–winning work on x-ray diffraction at Munich. He spent most of his productive career as a theoretical physicist in Berlin, where he had considerable influence on the development of scientific research in Germany. *Photogravure by Gen. Stab. Lit. Anst., courtesy AIP Emilio Segrè Visual Archives, Weber and Physics Today Collections.*

Heisenberg's uncertainty principle plays a major role in our understanding of particle-like and wavelike behavior. This principle prohibits the precise, simultaneous knowledge of both momentum and position or of both energy and time. We will see that no experiment requires us to utilize both wave and particle properties *simultaneously*. Although modern quantum theory is applicable primarily at the atomic level, there are many macroscopic observations of its effects.

5.1   X-Ray Scattering

Following Röntgen's discovery of x rays in 1895, intense efforts were made to determine the nature and origin of the new penetrating radiation. Charles Barkla (Nobel Prize, 1917) made many x-ray measurements at Liverpool University during the early 1900s and is given credit for discovering that each element emits x rays of characteristic wavelengths and that x rays exhibit properties of polarization.

By 1912 it became clear that x rays were a form of electromagnetic radiation and must therefore have wave properties. However, because it had proved difficult to refract or diffract x rays as easily as visible light, it was suggested that their wavelengths must be much shorter than those of visible light. Max von Laue (1879–1960), a young theoretical physicist at the University of Munich, became interested in the nature of x rays primarily because of the presence at Munich of Röntgen and the theorist Arnold Sommerfeld (1868–1951), who would later play an important role in understanding atomic structure. Wilhelm Wien (1864–1928) and Sommerfeld, among others, estimated the wavelength of an x ray to be between 10^{-10} and 10^{-11} m. Knowing the distance between atoms in a crystal to be about 10^{-10} m, Laue made the brilliant suggestion that x rays should scatter from the atoms of crystals. He suggested that, if x rays were a form of electromagnetic radiation, interference effects should be observed. From the study of optics, we know that wave properties are most easily demonstrated when the sizes of apertures or obstructions are about equal to or smaller than the wavelength of the light. We use gratings in optics to separate light by diffraction into different wavelengths. Laue suggested that crystals might act as three-dimensional gratings, scattering the waves and producing observable interference effects.

Max von Laue (Nobel Prize for Physics, 1914) designed the experiment and convinced two experimental physicists at Munich, Walter Friedrich and Paul Knipping, to perform the measurement. A schematic diagram of the transmission Laue process is shown in Figure 5.1, along with one of Friedrich and Knipping's earliest experimental results. When they rotated the crystals, the positions and intensities of the diffraction maxima were shown to change. Laue was able to perform the complicated analysis necessary to prove that x rays were scattered as waves from a three-dimensional crystal grating. Though the primary purpose of Laue's proposal was to prove the wave nature of x rays, he ended up also demonstrating the lattice structure of crystals, which led to the origin of solid-state physics and the development of modern electronics.

Laue proved wave nature of x rays and emphasized lattice structure of crystals

Two English physicists, William Henry Bragg and his son, William Lawrence Bragg, fully exploited the wave nature of x rays and simplified Laue's analysis. W. L. Bragg pointed out in 1912 that each of the images surrounding the bright central spot of the Laue photographs could be interpreted as the reflection of the incident x-ray beam from a unique set of planes of atoms within the crystal. Each dot in the pattern corresponds to a different set of planes in the crystal (see Figure 5.1b).

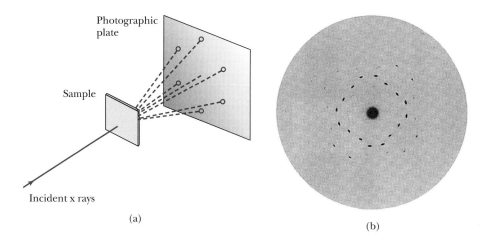

Photographic plate

Sample

Incident x rays

(a)

(b)

Figure 5.1 (a) Schematic diagram of Laue diffraction transmission method. A wide range of x-ray wavelengths scatters from a crystal sample. The x rays constructively interfere from certain planes, producing dots. (b) One of the first results of Friedrich and Knipping in 1912 showing the symmetric placement of *Laue dots* of x-ray scattering from ZnS. The analysis of these results by Laue, although complex, convincingly proved that x rays are waves. *Photo from W. Friedrich, P. Knipping, and M. Laue, Sitzungsberichte der Bayerischen Akademie der Wissenschaften, 303–322 und 5 Tafeln, (1912), reprinted in Max von Laue, Gesammelte Schriften und Vorträge, Band 1, Braunschweig: Vieweg (1961).*

Is x-ray scattering from atoms within crystals consistent with what we know from classical physics? From classical electromagnetic theory we know that the oscillating electric field of electromagnetic radiation polarizes an atom, causing the positively charged nucleus and negatively charged electrons to move in opposite directions. The result is an asymmetric charge distribution, or electric dipole. The electric dipole oscillates at the same frequency as the incident wave and in turn reradiates electromagnetic radiation at the same frequency, but in the form of spherical waves. These spherical waves travel throughout the matter and, in the case of crystals, may constructively or destructively interfere as the waves pass through different directions in the crystal.

If we consider x rays scattered from a simple rock salt crystal (NaCl, shown in Figure 5.2), we can, by following the Bragg simplification, determine conditions necessary for constructive interference. We will study solids in Chapter 10, but for now note that the atoms of crystals like NaCl form lattice planes,

William Lawrence Bragg (1890–1971) (left) and William Henry Bragg (1862–1942) (right) were a son-father team, both of whom were educated at Cambridge. The father spent 22 years at the University of Adelaide in Australia, where his son was born. Both father and son initially studied mathematics, but eventually changed to physics. The father was a skilled experimenter, and the son was able to conceptualize physical problems and express them mathematically. They did their important work on x-ray crystallography in 1912–14 while the father was at the University of Leeds and the son was a graduate student at Cambridge working under J. J. Thomson. Both physicists had long and distinguished careers, with the son being director of the famous Cavendish Laboratory at Cambridge from 1938 to 1953. W. Lawrence Bragg received his Nobel Prize at age 25. *Courtesy Edgar Fahs Smith Memorial Collection, Department of Special Collections, University of Pennsylvania Library.*

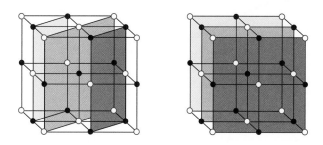

Figure 5.2 The crystal structure of NaCl (rock salt) showing two of the possible sets of lattice planes (Bragg planes).

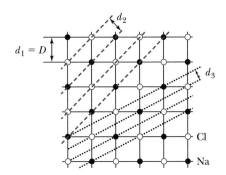

Figure 5.3 Top view of NaCl (cubic crystal), indicating possible lattice planes. D is the interatomic spacing and d is the distance between lattice planes.

called **Bragg planes**. We can see from Figure 5.3 that it is possible to have many Bragg planes in a crystal, each with different densities of atoms. Figure 5.4 shows an incident plane wave of monochromatic x rays of wavelength λ scattering from two adjacent planes. There are two conditions for constructive interference of the scattered x rays:

Conditions for constructive interference

1. The angle of incidence must equal the angle of reflection of the outgoing wave.
2. The difference in path lengths ($2d \sin \theta$) shown in Figure 5.4 must be an integral number of wavelengths.

We will not prove condition 1 but will assume it.* It is referred to as the *law of reflection* ($\theta_{\text{incidence}} = \theta_{\text{reflection}}$), although the effect is actually due to diffraction and interference. Condition 2 will be met if

Bragg's law

$$n\lambda = 2d \sin \theta \qquad (n = \text{integer}) \tag{5.1}$$

as can be seen from Figure 5.4, where D is the interatomic spacing (distance between atoms) and d is the distance between lattice planes. Equation (5.1) was first presented by W. L. Bragg in 1912 after he learned of Laue's results. The integer n is called the *order of reflection*, following the terminology of ruled diffraction gratings in optics. Equation (5.1) is known as **Bragg's law** and is useful for determining

*See L. R. B. Elton and D. F. Jackson, *American Journal of Physics* **34**, 1036 (1966), for a proof.

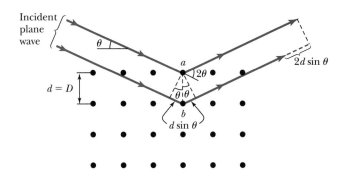

Figure 5.4 Schematic diagram illustrating x-ray scattering from Bragg lattice planes. The path difference of the two waves illustrated is $2d \sin \theta$. Notice that the actual scattering angle from the incident wave is 2θ.

Figure 5.5 Schematic diagram of Bragg spectrometer. X rays are produced by electron bombardment of metal target. The x rays are collimated by lead, scatter from a crystal, and are detected as a function of the angle 2θ.

Figure 5.6 A computer graphic of the DNA double helix is shown. This complex structure was understood only after hundreds of x-ray diffraction photos were carefully studied. Rosalind Franklin, while working at King's College in London in the early 1950s, produced the images of the DNA molecule that helped Watson and Crick unravel the DNA structure. Franklin died in 1958 at the age of 37, four years before the Nobel Prize was awarded to Watson and Crick. *Copyright Nelson Max/LLNL, Peter Arnold, Inc.*

either the wavelength of x rays or the interplanar spacing d of the crystal if λ is already known.

W. H. Bragg and W. L. Bragg (who shared the 1915 Nobel Prize) constructed an apparatus similar to that shown in Figure 5.5, called a *Bragg spectrometer*, and scattered x rays from several crystals. The intensity of the diffracted beam is determined as a function of scattering angle by rotating the crystal and the detector. The Braggs' studies opened up a whole new area of research that continues today.

The Bragg and Laue methods are complementary. In the Bragg method, the crystal is normally placed so that a certain set of planes produces constructive interference. The Bragg method is best for measuring x-ray wavelengths, and monochromatic x rays usually are used. The Laue method emphasizes x-ray transmission and is best for studying the crystal structure. This technique of studying crystals is widespread and many incident wavelengths are normally used simultaneously in order to obtain many dots. These techniques tell us almost everything we know about the structures of solids, liquids, and even complex molecules such as DNA (see Figure 5.6).

If a single large crystal is not available, then many small crystals may be used. If these crystals are ground into a powdered form, the small crystals will then each have random orientations. Because of random orientations, when a beam of x rays passes through the powdered crystal, the dots become a series of rings. A schematic diagram of the **powder technique** is shown in Figure 5.7a along with the film arrangement to record powder photographs in Figure 5.7b. The lines indicated in part (b) are sections of rings called the *Debye-Scherrer pattern*, named after the discoverers. Figure 5.7c is a sequence of four photographs, each with an increasingly larger number of crystals, which indicates the progression from the Laue dots to the rings characteristic of the powder photographs.

Powder technique

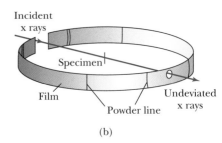

Powder line

Incident
x rays   Powder
specimen

Photographic
plate

(a)

Incident
x rays

Specimen

Film   Powder line   Undeviated
x rays

(b)

Figure 5.7 (a) Diagram showing the experimental arrangement of producing powder photographs from random-oriented crystals. (b) Film arrangement to record powder photographs. (c) The four photos show a progression of x-ray photographs for fluorite from a single crystal (clearly showing dots), through a few crystals, to a large number of crystals, which gives the rings the characteristic of an ideal powder photograph. *(a) and (b) from N. F. M. Henry, H. Lipson, and W. A. Wooster,* The Interpretation of X-ray Diffraction Photographs, *London: MacMillan (1960). (c) From H. S. Lipson,* Crystals and X-Rays, *London: Wykeham (1970).*

(c)

Example 5.1

X rays scattered from rock salt (NaCl) are observed to have an intense maximum at an angle of 20° from the incident direction. Assuming $n = 1$ (from the intensity), what must be the wavelength of the incident radiation?

Strategy We will use Equation (5.1) to find λ, but we need to know d, the lattice spacing, and the angle θ. Notice that the angle between the incident beam and scattered wave for constructive interference is always 2θ (see Figures 5.4 and 5.5), and because $2\theta = 20°$, we have $\theta = 10°$. We can use the density of NaCl to help find d, because the volume taken up by one atom is d^3.

Solution In Section 4.1 we showed that

$$\frac{\text{Number of molecules}}{\text{Volume}} = \frac{N_A \rho}{M}$$

where N_A is Avogadro's number, ρ is the density, and M is the gram-molecular weight. For NaCl, $\rho = 2.16 \text{ g/cm}^3$ and $M = 58.5 \text{ g/mol}$.

$$\frac{N_A \rho}{M} = \frac{\left(6.02 \times 10^{23} \dfrac{\text{molecules}}{\text{mol}}\right)\left(2.16 \dfrac{\text{g}}{\text{cm}^3}\right)}{58.5 \dfrac{\text{g}}{\text{mol}}}$$

$$= 2.22 \times 10^{22} \frac{\text{molecules}}{\text{cm}^3}$$

$$= 4.45 \times 10^{22} \frac{\text{atoms}}{\text{cm}^3}$$

$$= 4.45 \times 10^{28} \frac{\text{atoms}}{\text{m}^3}$$

Because NaCl has a cubic array, we take d as the distance between Na and Cl atoms, so we have a volume of d^3 per atom.

$$\frac{1}{d^3} = 4.45 \times 10^{28} \frac{\text{atoms}}{\text{m}^3}$$

$$d = 2.82 \times 10^{-10} \text{ m} = 0.282 \text{ nm}$$

This technique of calculating the lattice spacing works for only a few cases because of the variety of crystal structures, many of which are non-cubic.

We use Equation (5.1) to find λ.

$$\lambda = \frac{2d \sin \theta}{n} = \frac{(2)(0.282 \text{ nm})(\sin 10°)}{1} = 0.098 \text{ nm}$$

which is a typical x-ray wavelength. NaCl is a useful crystal for determining x-ray wavelengths and for calibrating experimental apparatus.

The previous example shows the value of x-ray crystallography and its tremendous usefulness. The technique pioneered by physicists in the first part of the twentieth century continues to be useful to many scientists in varied fields today.

5.2   De Broglie Waves

By 1920 it was established that x rays were electromagnetic radiation that exhibited wave properties. X-ray crystallography and its usefulness in studying the crystalline structure of atoms and molecules was being established. But a detailed understanding of the atom was still lacking. Many physicists believed that a new, more general theory was needed to replace the rudimentary Bohr model of the atom. An essential step in this development was made by a young French graduate student, Prince Louis V. de Broglie, who began studying the problems of the Bohr model in 1920.

De Broglie was well versed in the work of Planck, Einstein, and Bohr. He was aware of the duality of nature expressed by Einstein in which matter and energy were not independent but were in fact interchangeable. De Broglie was particularly struck by the fact that photons (electromagnetic radiation) had both wave (x-ray crystallography) and corpuscular (photoelectric effect) properties. The concept of waves is needed to understand interference and diffraction (Section 5.1), but localized corpuscles are needed to explain phenomena like the photoelectric effect (Section 3.6) and Compton scattering (Section 3.8). If electromagnetic radiation must have *both wave and particle properties*, then why should material particles not have both wave and particle properties as well? According to de Broglie, the symmetry of nature encourages such an idea, and no laws of physics prohibit it.

When de Broglie presented his new hypothesis in a doctoral thesis to the University of Paris in 1924, it aroused considerable interest. De Broglie used Einstein's special theory of relativity together with Planck's quantum theory to establish the wave properties of particles. His fundamental relationship is the prediction

$$\lambda = \frac{h}{p} \tag{5.2}$$

De Broglie wavelength of a particle

That is, the wavelength to be associated with a particle is given by Planck's constant divided by the particle's momentum.

De Broglie was guided by the concepts of phase and group velocities of waves (see Section 5.4) to arrive at Equation (5.2). Recall that for a photon $E = pc$, and $E = hf$, so that

$$hf = pc = p\lambda f$$

$$h = p\lambda$$

After serving in World War I, Prince Louis de Broglie (1892–1987) resumed his studies toward a doctoral degree at the University of Paris in 1924, where he reported his concept of matter waves as part of his doctoral dissertation. De Broglie spent his life in France where he enjoyed much success as an author and teacher. *AIP Niels Bohr Library, W. F. Meggers Collection.*

CAVENDISH LABORATORY

*B*efore the 1870s most of our scientific knowledge resulted from the research of people working in their own private laboratories. William Thomson (who would later become Lord Kelvin) established a laboratory at the University of Glasgow in the 1840s, and in the 1860s efforts began at both Oxford and Cambridge to build physical laboratories. In 1871 James Clerk Maxwell was called from his Scottish home to become the first Cavendish Professor at Cambridge University. Maxwell began planning and supervising the construction of the laboratory on Free School Lane in central Cambridge with an unexpected fervor while he gave regular lectures to students. The most important work of the day was to demonstrate the existence of Maxwell's electromagnetic waves, but they were "scooped" by Heinrich Hertz in Germany. Maxwell's successor was Lord Rayleigh, who published 50 papers during his five years at Cavendish before returning to his estate farm where he made most of his discoveries (including the noble gases) at his private laboratory.

The appointment of the young J. J. Thomson at age 28 as Cavendish Professor in 1884 was the beginning of a long and fruitful era in atomic physics. The discovery of the electron in 1897, the arrival of the young Ernest Rutherford from New Zealand as a student, and the early work of C. T. R. Wilson that led to the development of the cloud chamber all helped the Cavendish Laboratory expand, prosper, and grow in stature under Thomson's leadership. Thomson's 35-year leadership was remarkable in many ways, particularly in the manner he stepped down in 1919 upon the opportunity of attracting Rutherford back to Cavendish to be the next Professor.

During Rutherford's 19-year reign, the Cavendish became the most renowned center of science in the world. It attracted the best students, researchers, and visitors from all over the world. Rutherford was a team leader, and he surrounded himself with a collection of young physicists whom he called "his boys." By the end of the Rutherford era in 1937, the laboratory was moving into new directions with particle accelerators and cryogenic labs.

World War II would change the face of the Cavendish forever. Physicists spread out to perform wartime research, particularly on the development of the atomic bomb and the development of radar that were to play large roles in the allied victory. William Lawrence Bragg returned to Cambridge as Cavendish Professor to succeed Rutherford in 1937, and the field of x-ray crystallography flourished. The Cavendish scientists have had an uncanny ability to choose productive research areas. It has been said that the fields of molecular biology and radio astronomy started at the Cavendish in the late 1940s, and Bragg must be given credit for the foresight to support these fledgling subjects in the face of "Big Science" in the United States. Bragg's tenure as Cavendish Professor ended in 1953 just when Watson and Crick succeeded in discovering the DNA structure. Bragg also supported J. A. Ratcliffe and Martin Ryle, who had worked on radar at the Cavendish during the war, to construct the first radio telescope. This effort led to the discovery of quasars and pulsars.

When Sir Nevill Mott succeeded Bragg as Cavendish Professor in 1954, the lab made a turn toward solid state physics. Mott had worked on collision theory and nuclear problems in the 1930s, but eventually turned to theoretical investigations of electronic systems. Brian Josephson did his pioneering theoretical work (see Chapter 10) on the supercurrent through a tunnel barrier while a student, graduating in 1964 with his Ph.D. In 1974 the Cavendish moved to a new site in West Cambridge. Condensed matter physics now accounts for the greater part of research at the Cavendish, but the groups in radio astronomy and high-energy physics are still important. The Cavendish Laboratory has set a standard that other laboratories can only hope to emulate.

We end with a list of Nobel Prizes awarded to those who did their most important work at the Cavendish Laboratory. The asterisks (for example, Rutherford and Rayleigh) indicate Nobel Prizes awarded primarily for work done elsewhere to people who are still widely associated with the Cavendish Laboratory.

Figure A Upper left, the old Cavendish Laboratory on Free School Lane in Cambridge. The original building is to the left of the gate. The first four Cavendish professors: James Clerk Maxwell, upper right; Lord Rayleigh, bottom left; and Sir J. J. Thomson (left) and Lord Rutherford, bottom right.
From Cambridge University Press.

Cavendish Laboratory Nobel Prizes

1904	Physics	Lord Rayleigh*	Density of gases, discovery of argon
1906	Physics	Sir J. J. Thomson	Investigations of electricity in gases
1908	Chemistry	Lord Rutherford*	Element disintegration
1915	Physics	Sir William Lawrence Bragg	X-ray analysis of crystals
1917	Physics	Charles G. Barkla	Secondary x rays
1922	Chemistry	Francis W. Aston	Isotopes discovery
1927	Physics	Charles T. R. Wilson	Cloud chamber
1928	Physics	Sir Owen W. Richardson	Thermionic emission
1935	Physics	Sir James Chadwick	Neutron discovery
1937	Physics	Sir George P. Thomson	Electron diffraction
1947	Physics	Sir Edward V. Appleton*	Upper atmosphere investigations
1948	Physics	Lord Patrick M. S. Blackett	Discoveries in nuclear physics
1951	Physics	Sir John D. Cockcroft and Ernest T. S. Walton	Nuclear transmutation
1962	Physiology or Medicine	Francis H. C. Crick and James D. Watson	DNA discoveries
1962	Chemistry	Max Perutz and Sir John Kendrew	Structures of globular proteins
1973	Physics	Brian D. Josephson	Supercurrent in tunnel barriers
1974	Physics	Sir Martin Ryle and Antony Hewish	Radio astrophysics, pulsars
1977	Physics	Sir Nevill F. Mott	Magnetic and disordered systems
1978	Physics	P. L. Kapitsa*	Low-temperature physics
1982	Chemistry	Sir Aaron Klug	Nucleic acid–protein complexes

and

$$\lambda = \frac{h}{p} \tag{5.3}$$

De Broglie extended this relation for photons to all particles. Particle waves were **Matter waves** called **matter waves** by de Broglie, and the wavelength expressed in Equation (5.2) is now called the **de Broglie wavelength** of a particle.

Calculate the de Broglie wavelength of (a) a tennis ball of mass 57 g traveling 25 m/s (about 56 mph) and (b) an electron of energy 50 eV.

Strategy The calculation for both of these wavelengths is a straightforward application of Equation (5.2).

Solution (a) For the tennis ball, $m = 0.057$ kg, so

$$\lambda = \frac{h}{p} = \frac{6.63 \times 10^{-34}\, \text{J} \cdot \text{s}}{(0.057 \text{ kg})(25 \text{ m/s})} = 4.7 \times 10^{-34} \text{ m}$$

(b) For the electron, it is more convenient to use eV units, so we rewrite the wavelength λ as

$$\lambda = \frac{h}{p} = \frac{h}{\sqrt{2mK}} = \frac{hc}{\sqrt{2(mc^2)K}}$$

$$\lambda = \frac{1240 \text{ eV} \cdot \text{nm}}{\sqrt{(2)(0.511 \times 10^6 \text{ eV})(50 \text{ eV})}} = 0.17 \text{ nm}$$

Note that because the kinetic energy of the electron is so small, we have used a nonrelativistic calculation. Calculations in modern physics are normally done using eV units, both because it is easier and also because eV values are more appropriate for atoms and nuclei (MeV, GeV) than are joules. The values of hc and some masses can be found inside the front cover.

How can we show whether such objects as the tennis ball or the electron in the previous example exhibit wavelike properties? The best way is to pass the objects through a slit having a width of the same dimension as the object's wavelength. We expect it to be virtually impossible to demonstrate interference or diffraction for the tennis ball, because we cannot find a slit as narrow as 10^{-34} m. It is unlikely we will ever be able to demonstrate the wave properties of the tennis ball. But the de Broglie wavelength of the 50-eV electron, about 0.2 nm, is large enough that we should be able to demonstrate its wave properties. Because of their small mass, electrons can have a small momentum and in turn a large wavelength $(\lambda = h/p)$. Electrons offer our best chance of observing effects due to matter waves.

Bohr's Quantization Condition

One of Bohr's assumptions concerning his hydrogen atom model was that the angular momentum of the electron-nucleus system in a stationary state is an integral multiple of $h/2\pi$. Let's now see if we can predict this result using de Broglie's result. Represent the electron as a standing wave in an orbit around the proton. This standing wave will have nodes and be an integral number of wavelengths. We show an example of this in Figure 5.8. In order for it to be a correct standing wave, we must have

$$n\lambda = 2\pi r$$

where r is the radius of the orbit. Now we use the de Broglie relation for the wavelength and obtain

Figure 5.8 A schematic diagram of waves in an electron orbit around a nucleus. An integral number of wavelengths fits in the orbit. Note that the electron does not "wiggle" around the nucleus. The displacement from the dashed line represents its wave amplitude.

$$2\pi r = n\lambda = n\frac{h}{p}$$

The angular momentum of the electron in this orbit is $L = rp$, so we have, using the above relation,

$$L = rp = \frac{nh}{2\pi} = n\hbar$$

We have arrived at Bohr's quantization assumption by simply applying de Broglie's wavelength for an electron in a standing wave. This result seemed to justify Bohr's assumption. De Broglie's wavelength theory for particles was a crucial step toward the new quantum theory, but experimental proof was lacking. As we will see in the next section, this was soon to come.

5.3   Electron Scattering

In 1925, a laboratory accident led to experimental proof for de Broglie's wavelength hypothesis. C. Davisson and L. H. Germer of Bell Telephone Laboratories (now part of Lucent Technologies) were investigating the properties of metallic surfaces by scattering electrons from various materials when a liquid air bottle exploded near their apparatus. Because the nickel target they were currently using was at a high temperature when the accident occurred, the subsequent breakage of their vacuum system caused significant oxidation of the nickel. The target had been specially prepared and was rather expensive, so they tried to repair it by, among other procedures, prolonged heating at various high temperatures in hydrogen and under vacuum to deoxidize it.

A simple diagram of the Davisson-Germer apparatus is shown in Figure 5.9. Upon putting the refurbished target back in place and continuing the experiments, Davisson and Germer found a striking change in the way electrons were scattering from the nickel surface. They had previously seen a smooth variation of intensity with scattering angle, but the new data showed large increases for certain energies at a given scattering angle. Davisson and Germer were so puzzled by their new data that after a few days they cut open the tube in order to examine the nickel target. They found that the high temperature had modified the polycrystalline structure of the nickel. The many small crystals of the original target had been changed into a few large crystals as a result of the heat treatment.

Clinton J. Davisson (1881–1958) is shown here in 1928 on the right looking at the electronic diffraction tube held by Lester H. Germer (1896–1971). Davisson received his undergraduate degree at the University of Chicago and his doctorate at Princeton. They performed their work at Bell Telephone Laboratory located in New York City. Davisson received the Nobel Prize in Physics in 1937. *AIP Emilio Segrè Visual Archives.*

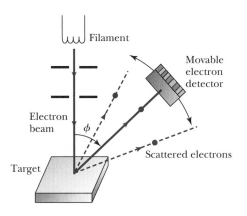

Figure 5.9 Schematic diagram of Davisson-Germer experiment. Electrons are produced by the hot filament, accelerated, and focused onto the target. Electrons are scattered at an angle ϕ into a detector, which is movable. The distribution of electrons is measured as a function of ϕ. The entire apparatus is located in a vacuum.

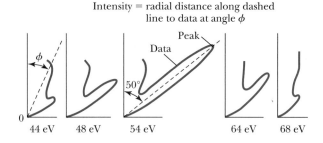

Figure 5.10 Davisson and Germer data for scattering of electrons from Ni. The peak $\phi = 50°$ builds dramatically as the energy of the electron nears 54 eV. *From C. J. Davisson, Franklin Institute Journal* **205**, *597–623 (1928).*

Davisson surmised it was this new crystal structure of nickel—the arrangement of atoms in the crystals, not the structure of the atoms—that had caused the new intensity distributions. Some 1928 experimental results of Davisson and Germer for 54-eV electrons scattered from nickel are shown in Figure 5.10. The scattered peak occurs for $\phi = 50°$.

The electrons were apparently being diffracted much like x rays, and Davisson, being aware of de Broglie's results, found that the Bragg law applied to their data as well. Davisson and Germer were able to vary the scattering angles for a given wavelength and vary the wavelength (by changing the electron accelerating voltage and thus the momentum) for a given angle.

The relationship between the incident electron beam and the nickel crystal is shown in Figure 5.11. In the Bragg law, 2θ is the angle between the incident and exit beams. Therefore, $\phi = \pi - 2\theta = 2\alpha$. Because $\sin \theta = \cos(\phi/2) = \cos \alpha$, we have for the Bragg condition, $n\lambda = 2d \cos \alpha$. However, d is the lattice plane spacing and is related to the interatomic distance D by $d = D \sin \alpha$ so that

$$n\lambda = 2d \sin \theta = 2d \cos \alpha = 2D \sin \alpha \cos \alpha$$

$$n\lambda = D \sin 2\alpha = D \sin \phi \tag{5.4}$$

or

$$\lambda = \frac{D \sin \phi}{n} \tag{5.5}$$

For nickel the interatomic distance is $D = 0.215$ nm. If the peak found by Davisson and Germer at 50° was $n = 1$, then the electron wavelength should be

$$\lambda = (0.215 \text{ nm}) (\sin 50°) = 0.165 \text{ nm}$$

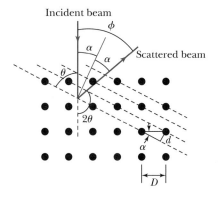

Figure 5.11 The scattering of electrons by lattice planes in a crystal. This figure is useful to compare the scattering relations $n\lambda = 2d \sin \theta$ and $n\lambda = D \sin \phi$ where θ and ϕ are the angles shown, D = interatomic spacing, and d = lattice plane spacing.

Example 5.3

Determine the de Broglie wavelength for a 54-eV electron used by Davisson and Germer.

Strategy We shall use the de Broglie wavelength Equation (5.2) to determine the wavelength λ. We need to find the momentum of a 54-eV electron, but because the energy is so low, we can do a nonrelativistic calculation. We shall do a general calculation for the wavelength of any electron accelerated over a voltage of V_0.

Solution We write the kinetic energy K.E. in terms of the final momentum of the electron and the voltage V_0 across which the electron is accelerated.

$$\frac{p^2}{2m} = \text{K.E.} = eV_0 \tag{5.6}$$

We find the momentum from this equation to be $p = \sqrt{(2m)(eV_0)}$. The de Broglie wavelength from Equation (5.2) is now

$$\lambda = \frac{h}{p} = \frac{hc}{pc} = \frac{hc}{\sqrt{(2mc^2)(eV_0)}}$$

$$= \frac{1240\ \text{eV}\cdot\text{nm}}{\sqrt{(2)(0.511\times10^6\ \text{eV})(eV_0)}}$$

$$= \frac{1.226\ \text{nm}\cdot\text{V}^{1/2}}{\sqrt{V_0}} \tag{5.7}$$

where the constants h, c, and m have been evaluated and V_0 is the voltage. For $V_0 = 54$ V, the wavelength is

$$\lambda = \frac{1.226\ \text{nm}\cdot\text{V}^{1/2}}{\sqrt{54\ \text{V}}} = 0.167\ \text{nm}$$

We note that the value of the de Broglie wavelength 0.167 nm found in the previous example is in good agreement with that experimentally found (0.165 nm) by Davisson and Germer for the peak at 50°. This is an important result and shows that electrons have wavelike properties.

Shortly after Davisson and Germer reported their experiment, George P. Thomson (1892–1975), son of J. J. Thomson, reported seeing the effects of electron diffraction in transmission experiments. The first target was celluloid, and soon after that gold, aluminum, and platinum were used. The randomly oriented polycrystalline sample of SnO_2 produces rings, as shown in Figure 5.12. Davisson and Thomson received the Nobel Prize in 1937 for their investigations, which clearly showed that particles exhibited wave properties. In the next few years hydrogen and helium atoms were also shown to exhibit wave diffraction. An important modern measurement technique uses diffraction of neutrons to study the crystal and molecular structure of biologically important substances. All these experiments are consistent with the de Broglie hypothesis for the wavelength of a particle with mass.

Example 5.4

In introductory physics, we learned that a particle (ideal gas) in thermal equilibrium with its surroundings has a kinetic energy of $3kT/2$. Calculate the de Broglie wavelength for (a) a neutron at room temperature (300 K) and (b) a "cold" neutron at 77 K (liquid nitrogen).

Strategy In both of these cases we will use Equation (5.2) to find the de Broglie wavelength. First, we will need to determine the momentum, and we note in both cases the energies of the particles will be so low that we can perform a nonrelativistic calculation. Neutrons have a rest energy of almost 1000 MeV, and their kinetic energies at these temperatures will be quite low (0.026 eV at 300 K).

Solution We begin by finding the de Broglie wavelength of the neutron in terms of the temperature.

$$\frac{p^2}{2m} = \text{K.E.} = \frac{3}{2}kT \tag{5.8}$$

$$p = \sqrt{3mkT}$$

$$\lambda = \frac{h}{p} = \frac{h}{\sqrt{3mkT}} = \frac{hc}{\sqrt{3(mc^2)kT}}$$

$$= \frac{1}{T^{1/2}} \frac{1240 \text{ eV} \cdot \text{nm}}{\sqrt{3(938 \times 10^6 \text{ eV})(8.62 \times 10^{-5} \text{ eV/K})}}$$

It again has been convenient to use eV units.

$$\lambda = \frac{2.52}{T^{1/2}} \text{ nm} \cdot \text{K}^{1/2}$$

$$\lambda(300 \text{ K}) = \frac{2.52 \text{ nm} \cdot \text{K}^{1/2}}{\sqrt{300 \text{ K}}} = 0.145 \text{ nm} \tag{5.9}$$

$$\lambda(77 \text{ K}) = \frac{2.52 \text{ nm} \cdot \text{K}^{1/2}}{\sqrt{77 \text{ K}}} = 0.287 \text{ nm}$$

These wavelengths are thus suitable for diffraction by crystals. "Supercold" neutrons, used to produce even larger wavelengths, are useful because extraneous electric and magnetic fields do not affect neutrons nearly as much as electrons.

(a)

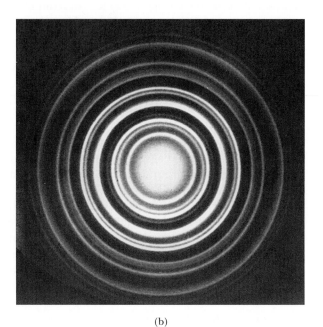

(b)

Figure 5.12 Examples of transmission electron diffraction photographs. (a) Produced by scattering 120-keV electrons on the quasicrystal $Al_{80}Mn_{20}$. (b) An early pattern from polycrystalline SnO_2. Notice that the dots in (a) indicate that the sample was a crystal whereas the rings in (b) indicate that a randomly oriented sample (or powder) was used. *(a) Courtesy of David Follstaedt, Sandia National Laboratory. (b) From G. P. Thomson and W. Cochrane,* Theory and Practice of Electron Diffraction, *London: Macmillan (1939).*

5.4   Wave Motion

Because particles exhibit wave behavior, as shown in the last section for electron diffraction, it must be possible to formulate a wave description of particle motion. This is an essential step in our progress toward understanding the behavior of matter—the quantum theory of physics. Our development of quantum theory will be based heavily on waves, so now we digress briefly to review the elementary physics of wave motion, which we shall ultimately apply to particles.

In elementary physics, we study waves of several kinds, including sound waves and electromagnetic waves (including light). The simplest form of wave has a sinusoidal form; at a fixed time (say, $t = 0$) its spatial variation looks like

$$\Psi(x, t)|_{t=0} = A \sin\left(\frac{2\pi}{\lambda} x\right) \qquad (5.10)$$

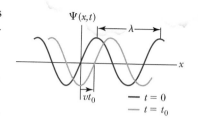

as shown in Figure 5.13. The function $\Psi(x, t)$ represents the *instantaneous amplitude* or **displacement** of the wave as a function of position x and time t. In the case of a traveling wave moving down a string, Ψ is the displacement of the string from equilibrium; in the case of a sound wave, Ψ is the displacement of the air molecules; and in the case of electromagnetic radiation, Ψ is the magnitude of the electric field $\vec{E}$ or magnetic field $\vec{B}$. The maximum displacement A is normally called the **amplitude**, but a better term for a harmonic wave such as we are considering may be **harmonic amplitude**.

Figure 5.13 Wave form of a wave moving to the right at speed v shown at $t = 0$ and $t = t_0$.

As time increases, the position of the wave will change, so the general expression for the wave is

$$\Psi(x, t) = A \sin\left[\frac{2\pi}{\lambda}(x - vt)\right] \qquad (5.11) \qquad \textbf{Wave form}$$

The position at time $t = t_0$ is also shown in Figure 5.13. The **wavelength** λ is defined to be the distance between points in the wave with the same phase, for example, positive wave crests. The **period** T is the time required for a wave to travel a distance of one wavelength λ. Because the velocity [actually phase velocity, see Equation (5.17)] of the wave is v, we have $\lambda = vT$. The frequency $f \,(= 1/T)$ of a harmonic wave is the number of times a crest passes a given point (a complete *cycle*) per second. A traveling wave of the type described by Equation (5.11) satisfies the wave equation*

$$\frac{\partial^2 \Psi}{\partial x^2} = \frac{1}{v^2}\frac{\partial^2 \Psi}{\partial t^2} \qquad (5.12)$$

If we use $\lambda = vT$, we can rewrite Equation (5.11) as

$$\Psi(x, t) = A \sin\left[2\pi\left(\frac{x}{\lambda} - \frac{t}{T}\right)\right] \qquad (5.13)$$

We can write Equation (5.13) more compactly by defining the **wave number** k and **angular frequency** ω by

Wave number and angular frequency

$$k \equiv \frac{2\pi}{\lambda} \quad \text{and} \quad \omega = \frac{2\pi}{T} \qquad (5.14)$$

Equation (5.13) then becomes

$$\Psi(x, t) = A \sin(kx - \omega t) \qquad (5.15)$$

This is the mathematical description of a sine curve traveling in the positive x direction that has a displacement $\Psi = 0$ at $x = 0$ and $t = 0$. A similar wave traveling in the negative x direction would have the form

$$\Psi(x, t) = A \sin(kx + \omega t) \qquad (5.16)$$

---

*The derivation of the wave equation is presented in most introductory physics textbooks for a wave on a string, although it is often an optional section and might have been skipped. It would be worthwhile for the student to review its derivation now, especially the use of the partial derivatives.

The **phase velocity** v_{ph} is the velocity of a point on the wave that has a given phase (for example, the crest) and is given by

Phase velocity
$$v_{ph} = \frac{\lambda}{T} = \frac{\omega}{k} \tag{5.17}$$

Phase constant

If the wave does not have $\Psi = 0$ at $x = 0$ and $t = 0$, we can describe the wave using a **phase constant**:

$$\Psi(x, t) = A \sin(kx - \omega t + \phi) \tag{5.18}$$

For example, if $\phi = \pi/2$, Equation (5.18) can be written

$$\Psi(x, t) = A \cos(kx - \omega t) \tag{5.19}$$

Principle of superposition

Observation of many different kinds of waves has established the general result that when two or more waves traverse the same region, they act independently of each other. According to the **principle of superposition**, we add the displacements of all waves present. A familiar example is the superposition of two sound waves of nearly equal frequencies: The phenomenon of beats is observed. Examples of superposition are shown in Figure 5.14. The net displacement depends on the harmonic amplitude, the phase, and the frequency of each of the individual waves. When we add waves at a given position and time, we simply add their instantaneous displacements. This can lead to constructive and interference effects like we saw in x-ray scattering in Section 5.1.

In quantum theory (or *quantum mechanics* as it is sometimes called to reflect its differences from *classical mechanics*), we will soon learn that we will use waves to represent a moving particle. How can we do that? In Figure 5.14 we see that when two waves are added together, we obtain regions of relatively large (and small) displacement. If we add many waves of different amplitudes and frequencies in

Wave packet

particular ways, it is possible to obtain what is called a **wave packet**. The important property of the wave packet is that its net amplitude differs from zero only over a small region Δx as shown in Figure 5.15. We can localize the position of a particle in a particular region by using a wave packet description (see Problem 63 for a calculation of this effect).

Let us examine in detail the superposition of two waves. Assume both waves have the same harmonic amplitude A but different wave numbers (k_1 and k_2) and angular frequencies (ω_1 and ω_2). The superposition of the two waves is the sum

$$
\begin{aligned}
\Psi(x, t) &= \Psi_1(x, t) + \Psi_2(x, t) \\
&= A \cos(k_1 x - \omega_1 t) + A \cos(k_2 x - \omega_2 t) \tag{5.20} \\
&= 2A \cos\left[\frac{1}{2}(k_1 - k_2)x - \frac{1}{2}(\omega_1 - \omega_2)t\right]\cos\left[\frac{1}{2}(k_1 + k_2)x - \frac{1}{2}(\omega_1 + \omega_2)t\right] \\
&= 2A \cos\left(\frac{\Delta k}{2}x - \frac{\Delta \omega}{2}t\right)\cos(k_{av}x - \omega_{av}t) \tag{5.21}
\end{aligned}
$$

where $\Delta k = k_1 - k_2$, $\Delta \omega = \omega_1 - \omega_2$, $k_{av} = (k_1 + k_2)/2$, and $\omega_{av} = (\omega_1 + \omega_2)/2$. We display similar waves in Figure 5.14a–d, where the heavy solid line indicates the sum of the two waves. In Figure 5.14f the blue dashed line indicates an envelope, which denotes the maximum displacement of the combined waves. The combined (or summed) wave Ψ oscillates within this envelope with the wave number

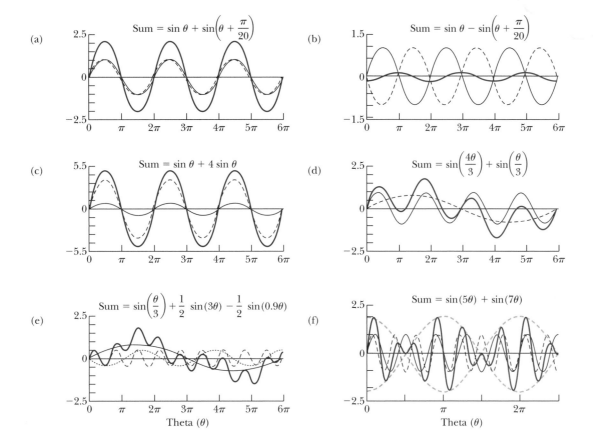

Figure 5.14 Superposition of waves. The heavy blue line is the resulting wave. (a) Two waves of equal frequency and amplitude that are almost in phase. The result is a larger wave. (b) As in (a) but the two waves are almost out of phase. The result is a smaller wave. (c) Superposition of two waves with the same frequency, but different amplitudes. (d) Superposition of two waves of equal amplitude but different frequencies. (e) Superposition of three waves of different amplitudes and frequencies. (f) Superposition of two waves of almost the same frequency over many wavelengths, indicating the phenomenon of beats. The blue dashed line indicates an envelope that denotes the maximum displacement of the combined waves.

k_av and angular frequency ω_av. The envelope is described by the first cosine factor of Equation (5.21), which has the wave number $\Delta k/2$ and angular frequency $\Delta\omega/2$. The individual waves Ψ_1 and Ψ_2 each move with their own phase velocity: ω_1/k_1 and ω_2/k_2. The combined wave has a phase velocity $\omega_\text{av}/k_\text{av}$. When combining many more than two waves, one obtains a pulse, or wave packet, which moves at the **group velocity**, as shown later. Only the group velocity, which describes the speed of the envelope ($u_\text{gr} = \Delta\omega/\Delta k$), is important when dealing with wave packets.

In contrast to the pulse or wave packet, the combination of only two waves is not localized in space. However, for purposes of illustration, we can identify a "localized region" $\Delta x = x_2 - x_1$ where x_1 and x_2 represent two consecutive points where the envelope is zero (or maximum, see Figure 5.14f). The term $\Delta k \cdot x/2$ in Equation (5.21) must be different by a phase of π for the values x_1 and x_2, because

Figure 5.15 An idealized wave packet localized in space over a region Δx is the superposition of many waves of different amplitudes and frequencies.

$x_2 - x_1$ represents only one half of the wavelength of the envelope confining the wave.

$$\frac{1}{2}\Delta k\, x_2 - \frac{1}{2}\Delta k\, x_1 = \pi$$

(5.22)

$$\Delta k(x_2 - x_1) = \Delta k\, \Delta x = 2\pi$$

Similarly, for a given value of x we can determine the time Δt over which the wave is localized and obtain

$$\Delta\omega\, \Delta t = 2\pi$$

(5.23)

The results of Equations (5.22) and (5.23) can be generalized for a case in which many waves form the wave packet (see Problem 63). The equations, $\Delta k\, \Delta x = 2\pi$ and $\Delta\omega\, \Delta t = 2\pi$, are significant because they tell us that in order to know precisely the position of the wave packet envelope (Δx small), we must have a large range of wave numbers (Δk large). Similarly, to know precisely when the wave is at a given point (Δt small), we must have a large range of frequencies ($\Delta\omega$ large). Equation (5.23) is the origin of the bandwidth relation, which is important in electronics. A particular circuit component must have a large bandwidth $\Delta\omega$ in order for its signal to respond in a short time Δt.

If we are to treat particles as matter waves, we have to be able to describe the particle in terms of waves. An important aspect of a particle is its localization in space. That is why it is so important to form the wave packet that we have been discussing. We extend Equation (5.20) by summing over many waves with possibly different wave numbers, angular frequencies, and amplitudes.

$$\Psi(x, t) = \sum_i A_i \cos(k_i x - \omega_i t)$$

(5.24)

Fourier series and integral   Such a result is called a **Fourier series**. When dealing with a continuous spectrum, it may be desirable to extend Equation (5.24) to the integral form called a **Fourier integral**.

$$\Psi(x, t) = \int \tilde{A}(k)\cos(kx - \omega t)\, dk$$

(5.25)

The amplitudes A_i and $\tilde{A}(k)$ may be functions of k. The use of Fourier series and Fourier integrals is at a more advanced level of mathematics than we want to pursue now.* We can, however, indicate their value by one important example.

Gaussian Wave Packet   Gaussian wave packets are often used to represent the position of particles, as illustrated in Figure 5.16, because the associated integrals are relatively easy to evaluate. At a given time t, say $t = 0$, a Gaussian wave can be expressed as

Gaussian function

$$\Psi(x, 0) = \psi(x) = Ae^{-\Delta k^2 x^2}\cos(k_0 x)$$

(5.26)

where Δk expresses the range of wave numbers used to form the wave packet. The $\cos(k_0 x)$ term describes the wave oscillating inside the envelope bounded by the (Gaussian) exponential term $e^{-\Delta k^2 x^2}$. The intensity distribution $\mathcal{I}(k)$ for the wave numbers leading to Equation (5.26) is shown in Figure 5.16a. There is a high probability of a particular measurement of k being within one standard deviation of the

---

*See John D. McGervey, *Introduction to Modern Physics*, Chap. 4, Orlando, FL: Academic Press (1983).

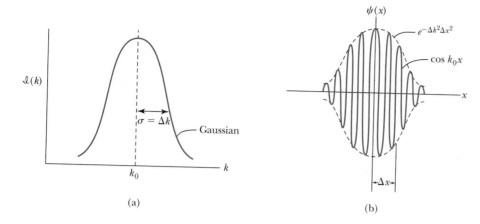

Figure 5.16 The form of the probability distribution or intensity $\mathcal{L}(k)$ shown in (a) is taken to have a Gaussian shape with a standard deviation of Δk [determined when the function $\exp[-(k - k_0)^2/2\sigma^2]$ has $k = k_0 \pm \Delta k$ and $\Delta k = \sigma$, the standard deviation]. This $\mathcal{L}(k)$ leads to $\psi(x)$, as shown in (b). The envelope for $\psi(x)$ is described by the $\exp(-\Delta k^2 x^2)$ term with the oscillating term $\cos(k_0 x)$ contained by the envelope. At the given time $t = 0$, the wave packet (particle) is localized to the area $x \approx 0 \pm \Delta x$ with wave numbers $k \approx k_0 \pm \Delta k$.

mean value k_0. The function $\psi(x)$ is shown in Figure 5.16b. For simplicity, let the constant A be 1. There is a good probability of finding the particle within the values of $x = 0$ [$\psi(x) = 1$] and $x = \Delta x/2$ [$\psi(x) = \exp((-\Delta k^2 \Delta x^2)/4)$]. Roughly, the value of $\psi(x)$ at the position $x = \Delta x/2$ is about 0.6 (see Figure 5.16b), so we have

$$e^{-\Delta k^2 \Delta x^2/4} \approx 0.6$$

We take the logarithm of both sides and find

$$-\frac{\Delta k^2 \Delta x^2}{4} \approx 0.5 \qquad \text{or} \qquad \Delta k\, \Delta x \approx 1.4 \tag{5.27}$$

This has been a rough calculation, and the result depends on the assumptions we have made. A more detailed calculation gives $\Delta k\, \Delta x = 1/2$. The important point is that with the Gaussian wave packet, we have arrived at a result similar to Equation (5.22), namely, that the product $\Delta k\, \Delta x$ is about unity. The localization of the wave packet over a small region Δx to describe a particle requires a large range of wave numbers; that is, Δk is large. Conversely, a small range of wave numbers cannot produce a wave packet localized within a small distance.

To complete our study of waves and the representation of particles by wave packets, we must be convinced that the superposition of waves is actually able to describe particles. We found earlier for the superposition of two waves that the group velocity, $u_{\text{gr}} = \Delta\omega/\Delta k$, represented the motion of the envelope. We can generalize this for the case of the wave packet and find that the wave packet moves with the group velocity u_{gr} given by

$$u_{\text{gr}} = \frac{d\omega}{dk} \tag{5.28}$$

Because the wave packet consists of many wave numbers, we should remember to evaluate this derivative at the center of the wave packet (that is, $k = k_0$).

For a de Broglie wave, we have $E = hf$ and $p = h/\lambda$. We can rewrite both of these equations in terms of $\hbar$.

$$E = hf = \hbar(2\pi f) = \hbar\omega \tag{5.29}$$

$$p = \frac{h}{\lambda} = \hbar\frac{2\pi}{\lambda} = \hbar k \tag{5.30}$$

where we have used the relations $\omega = 2\pi f$ and $k = 2\pi/\lambda$. If we multiply the denominator and numerator in Equation (5.28) by $\hbar$, we have

$$u_{gr} = \frac{d\omega}{dk} = \frac{d(\hbar\omega)}{d(\hbar k)} = \frac{dE}{dp}$$

We use the relativistic relation $E^2 = p^2c^2 + m^2c^4$ and its derivative to find

$$2E\,dE = 2pc^2\,dp$$

or

$$u_{gr} = \frac{dE}{dp} = \frac{pc^2}{E} \tag{5.31}$$

This is the velocity of a particle of momentum p and total energy E. Thus, it is plausible to assume that the group velocity of the wave packet can be associated with the velocity of a particle.

The phase velocity of a wave is represented by

$$v_{ph} = \lambda f = \frac{\omega}{k} \tag{5.32}$$

so that $\omega = kv_{ph}$.

Then, the group velocity is related to the phase velocity by

$$u_{gr} = \frac{d\omega}{dk} = \frac{d}{dk}(v_{ph}k) = v_{ph} + k\frac{dv_{ph}}{dk} \tag{5.33}$$

Thus, the group velocity may be greater or less than the phase velocity. A medium is called *nondispersive* when the phase velocity is the same for all frequencies and $u_{gr} = v_{ph}$. An example is electromagnetic waves in vacuum. Water waves are a good example of waves in a dispersive medium. When one throws a rock in a still pond, the envelope of the waves moves more slowly than the individual waves moving outward (Figure 5.17).

Dispersion plays an important role in the shape of wave packets. For example, in the case of the Gaussian wave packet shown in Figure 5.16 at $t = 0$, the wave packet will spread out as time progresses. A packet that is highly localized at one time will have its waves added together in a considerably different manner at another time due to the superposition of the waves.

Does the preceding discussion agree with our classical ideas? Consider a particle of mass m moving nonrelativistically with speed v. The phase velocity of this particle, if we treat it as a de Broglie wave, can be found by

$$v_{ph} = \lambda f = \frac{h}{p}\frac{E}{h} = \frac{E}{p} = \frac{p^2/2m}{p} = \frac{p}{2m} = \frac{mv}{2m} = \frac{v}{2} \tag{5.34}$$

The phase velocity is half of the particle's velocity, so the particle does not move with the phase velocity.

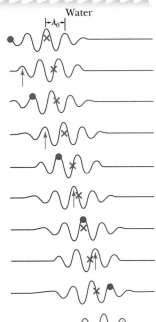

Figure 5.17 Progression with time of wave packet for which $u_{gr} = v_{ph}/2$. Note how the individual wave (arrow and dot alternately) moves through the wave packet (symbol ×) with time.

Example **5.5**

We just saw that the speed of a nonrelativistic particle of mass m is not equal to its phase velocity. Show that the particle speed is equal to the group velocity.

Strategy We can use the relation for the group velocity in Equation (5.28) or (5.31). Either should work and using both equations will be instructive.

Solution First, we look at Equation (5.31) for our nonrelativistic particle:

$$u_{gr} = \frac{pc^2}{E} = \frac{mvc^2}{mc^2} = v \qquad (5.35)$$

In order to use Equation (5.28) we utilize the results in Equations (5.29) and (5.30) for ω and k.

$$u_{gr} = \frac{d\omega}{dk} = \frac{d(E/\hbar)}{d(p/\hbar)} = \frac{dE}{dp} = \frac{d}{dp}\frac{p^2}{2m} = \frac{2p}{2m} = v$$

We agree that the particle, when acting as a wave, moves with the group velocity, not the phase velocity.

Example **5.6**

Newton showed that deep-water waves have a phase velocity of $\sqrt{g\lambda/2\pi}$. Find the group velocity of such waves and discuss the motion.

Strategy We use Equation (5.33) to relate the group and phase velocities, but first we need to find the phase velocity v_{ph} in terms of k.

Solution If we use $\lambda = 2\pi/k$, we have

$$v_{ph} = \sqrt{\frac{g\lambda}{2\pi}} = \sqrt{\frac{g}{k}} = \sqrt{g}k^{-1/2}$$

Now we can take the necessary derivative for Equation (5.33).

$$u_{gr} = \sqrt{\frac{g}{k}} + k\frac{d}{dk}\left(\sqrt{g}k^{-1/2}\right) = \sqrt{\frac{g}{k}} + k\sqrt{g}\left(-\frac{1}{2}k^{-3/2}\right)$$

$$= \sqrt{\frac{g}{k}} - \frac{1}{2}\sqrt{\frac{g}{k}} = \frac{1}{2}\sqrt{\frac{g}{k}} = \frac{1}{2}v_{ph}$$

The group velocity is determined to be one half of the phase velocity. Such an effect can be observed by throwing a rock in a still pond. As the circular waves move out, the individual waves seem to run right through the wave crests and then disappear (see Figure 5.17).

5.5 Waves or Particles?

By this point it is not unusual to be a little confused. We have learned that electromagnetic radiation behaves sometimes as waves (as in interference and diffraction) and other times as particles (as in the photoelectric and Compton effects). We have been presented evidence in this chapter that particles also behave as waves (electron diffraction). Can all this really be true? If a particle is a wave, what is waving? In the preceding section we learned that, at least mathematically, we can describe particles by using wave packets. Can we represent matter as waves and particles simultaneously? And can we represent electromagnetic radiation as waves and particles simultaneously? We must answer these questions about the **wave-particle duality** before proceeding with our study of quantum theory.

Double-Slit Experiment with Light To better understand the differences and similarities of waves and particles, we analyze Young's double-slit diffraction experiment, which is studied in detail in introductory physics courses to show the

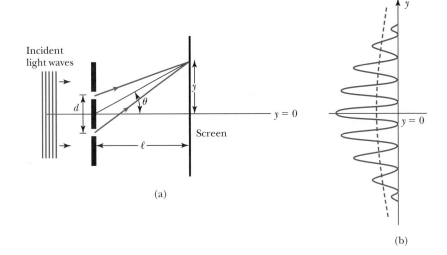

Figure 5.18 (a) Schematic diagram of Young's double-slit experiment. This experiment is easily performed with a laser as the light source (and $\ell \gg d$, where d = distance between slit centers). (b) The solid line indicates the interference pattern due to both slits. If either of the slits is covered, single-slit diffraction gives the result shown in the dashed curve.

interference character of light. Figure 5.18a shows a schematic diagram of the experiment. This experiment is easily performed with the use of a low-power laser. With both slits open, a nice interference pattern is observed, with bands of maxima and minima. When one of the slits is covered, this interference pattern is changed, and a rather broad peak is observed (see Figure 5.18b). Thus, we conclude that the double-slit interference pattern is due to light passing through *both* slits—a wave phenomenon.

However, if the light intensity is reduced, and we observe the pattern on a screen, we learn that the light arriving on the screen produces flashes at various points, entirely representative of particle behavior! If we take pictures of the screen after varying lengths of time, the pictures will look like those shown in Figure 5.19. Eventually the interference pattern characteristic of wave behavior emerges. There is therefore no contradiction in this experiment. If we want to know the precise location of the light (photon), we must use the particle description and not the wave description.

Figure 5.19 Computer simulation of Young's double-slit interference experiment for light or electrons. This calculation was performed for slit width $a = 4\lambda$ and slit distance $d = 20\lambda$. The four pictures are for increasing number of counts: 20, 100, 500, 4000. The interference pattern clearly has emerged for 500 counts. *Simulation and photos courtesy of Julian V. Noble.*

(a) 20 counts

(b) 100 counts

(c) 500 counts

(d) ~4000 counts

Electron Double-Slit Experiment Now let us examine a similar double-slit experiment that uses electrons rather than light. If matter also behaves as waves, then shouldn't the same experimental results be obtained if we use electrons rather than light? The answer is yes, and physicists did not doubt the eventual result. This experiment is not as easy to perform as the similar one with light. The difficulty arises in constructing slits narrow enough to exhibit wave phenomena. This requires $\lambda \sim a$, where a is the slit width. For light of $\lambda = 600$ nm, slits can be produced mechanically. However, for electrons of energy 50 keV, $\lambda = 5 \times 10^{-3}$ nm, which is smaller than a hydrogen atom (~ 0.1 nm). Nevertheless, C. Jönsson* of Tübingen, Germany, succeeded in 1961 in showing double-slit interference effects for electrons (Figure 5.20) by constructing very narrow slits and using relatively large distances between the slits and the observation screen. Copper slits were made by electrolytically depositing copper on a polymer strip printed on silvered glass plates. This experiment demonstrated that precisely the same behavior occurs for both light (waves) and electrons (particles). We have seen similar behavior previously from the Debye-Scherrer rings produced by the diffraction of x rays (waves) and electrons (particles).

*C. Jönsson, *American Journal of Physics* **42**, 4 (1974), translation of *Zeitschrift für Physik* **161**, 454 (1961).

Figure 5.20 Demonstration of electron interference using two slits similar in concept to Young's double-slit experiment for light. This result by Claus Jönsson clearly shows that electrons exhibit wave behavior (see also Example 5.7). *Reprinted with permission from C. Jönsson*, American Journal of Physics *42, 4 (1974). © 1974 American Association of Physics Teachers.*

Example 5.7

In the experiment by Jönsson, 50-keV electrons impinged on slits of width 500 nm separated by a distance of 2000 nm. The observation screen was located 350 mm beyond the slits. What was the distance between the first two maxima?

Strategy The equation specifying the orders of maxima and the angle θ from incidence is (see Figure 5.18) given by

$$d \sin \theta = n\lambda \qquad (5.36)$$

The first maximum is of order $n = 0$ and occurs for $\theta = 0$. The next maximum, at angle θ, occurs for $n = 1$:

$$\sin \theta = \frac{\lambda}{d} = \frac{\lambda}{2000 \text{ nm}}$$

We need to determine the wavelength λ in order to find the angle θ. Once we know θ, we can find the distance between the two maxima on the observation screen.

Solution We have already calculated the wavelength for electrons of energy eV_0 in Equation (5.7).

$$\lambda = \frac{1.226 \text{ nm} \cdot V^{1/2}}{\sqrt{50 \times 10^3 \text{ V}}} = 5.48 \times 10^{-3} \text{ nm}$$

Because 50 keV may be too high an energy for a nonrelativistic calculation such as that done in Equation (5.7), we had better perform a relativistic calculation to be certain. We first find the momentum and insert that into $\lambda = h/p$.

$$(pc)^2 = E^2 - E_0^2 = (K + E_0)^2 - E_0^2$$
$$= (50 \times 10^3 \text{ eV} + 0.511 \times 10^6 \text{ eV})^2$$
$$- (0.511 \times 10^6 \text{ eV})^2$$
$$= (0.231 \times 10^6 \text{ eV})^2$$

Now we can determine the wavelength.

$$\lambda = \frac{h}{p} = \frac{hc}{pc} = \frac{1240 \text{ eV} \cdot \text{nm}}{0.231 \times 10^6 \text{ eV}} = 5.36 \times 10^{-3} \text{ nm}$$

Therefore, we find the more accurate relativistic value to be somewhat less (2%) than the nonrelativistic value. Now we can determine the angle.

$$\sin \theta = \frac{5.36 \times 10^{-3} \text{ nm}}{2000 \text{ nm}} = 2.68 \times 10^{-6}$$

The distance of the first maximum along the screen is $y = \ell \tan \theta$, but for such a small angle, $\sin \theta \approx \tan \theta$.

$$y = \ell \tan \theta \approx \ell \sin \theta = 350 \text{ mm}(2.68 \times 10^{-6})$$
$$= 9.38 \times 10^{-4} \text{ mm} \frac{10^6 \text{ nm}}{\text{mm}} = 938 \text{ nm}$$

Such a diffraction pattern is too small to be viewed by the naked eye. Jönsson magnified the pattern by a series of electronic lenses and then observed a fluorescent screen with a 10-power optical microscope in order to see the diffraction pattern as expected.

Figure 5.21 An attempt to measure the slit an electron passes through in the double-slit experiment. A powerful light source scatters a photon from the electron, and the scattered photon is observed. The motion of the electron is affected.

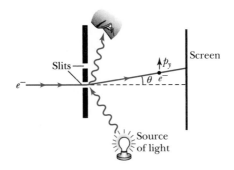

Another *Gedanken* Experiment If we were to cover one of the slits in the preceding Jönsson experiment, the double-slit interference pattern would be destroyed—just as it was when light was used. But our experience tells us the electron is a particle, and we know that it can go through only one of the slits. Let's devise a *gedanken* experiment, shown in Figure 5.21, to determine which slit the electron went through. We set up a light shining on the double slit and use a powerful microscope to look at the region. After the electron passes through one of the slits, light bounces off the electron; we observe the reflected light, so we know which slit the electron came through.

In order to do this experiment, we need to use light having wavelength narrower than the slit separation d, in order to determine which slit the electron went through. We use a subscript "ph" to denote variables for light (photon). Therefore, we have $\lambda_{ph} < d$. The momentum of the photon is

$$p_{ph} = \frac{h}{\lambda_{ph}} > \frac{h}{d}$$

For us to show the interference effects for the electrons passing through the slits, the electrons must also have a wavelength on the order of the slit separation d, $\lambda_{el} \sim d$. The momentum of the electrons will be on the order of

$$p_{el} = \frac{h}{\lambda_{el}} \sim \frac{h}{d}$$

The difficulty is that the momentum of the photons used to determine which slit the electron went through is sufficiently great to strongly modify the momentum of the electron itself, thus changing the direction of the electron! The attempt to identify which slit the electron is passing through will in itself change the interference pattern. We will take a closer look at this experiment in Section 5.6. In trying to determine which slit the electron went through, we are examining the particle-like behavior of the electron. When we are examining the interference pattern of the electron, we are using the wavelike behavior of the electron.

Bohr resolved this dilemma by pointing out that the particle-like and wavelike aspects of nature are *complementary*. Both are needed—they just can't be observed simultaneously.

Principle of complementarity

Bohr's principle of complementarity: *It is not possible to describe physical observables simultaneously in terms of both particles and waves.*

Physical observables

Physical observables are those quantities such as position, velocity, momentum, and energy that can be experimentally measured. In any given instance we must use either the particle description or the wave description. Usually the choice is clear. The interference pattern of the double-slit experiment suggests that the light (or electron) had to go through both slits, and we must use the wave description. In our description of nature, we cannot describe phenomena by displaying both particle and wave behavior at the same time.

By the use of the principle of complementarity, we can solve the **wave-particle duality** problem, which has been plaguing us. It is not unusual for students to feel uncomfortable with this solution, because it cannot be directly proven. However, as a "principle" and not a "law," the complementarity principle does seem to describe nature, and, as such, we use it. We must pay close attention to the fact that we do not use waves and particles simultaneously to describe a particular phenomenon. Experiments dictate what actually happens in nature, and we must draw up a set of rules to describe our observations. These rules naturally lead to a probability interpretation of experimental observations. If we set up a series of small detectors along the screen in the electron double-slit experiment, we can speak of the probability of the electron being detected by one of the detectors. The interference pattern can guide us in our probability determinations. But once the electron has been registered by one of the detectors, the probability of its being seen in the other detectors is zero. Matter and radiation propagation is described by wavelike behavior, but matter and radiation interact (that is, undergo creation/annihilation or detection) as particles.

Solution of wave-particle duality

5.6   Uncertainty Principle

In Section 5.4, when we discussed the superposition of waves, we learned that in order to localize a wave packet over a small region Δx, we had to use a large range Δk of wave numbers. For the case of two waves, we found in Equation (5.22) that $\Delta k \, \Delta x = 2\pi$. If we examine a Gaussian wave packet closely, we would find that the product $\Delta k \, \Delta x = 1/2$. The minimum value of the product $\Delta k \, \Delta x$ is obtained when Gaussian wave packets are used.

In Section 5.4 we learned that it is impossible to measure simultaneously, with no uncertainty, the precise values of k and x for the same particle. The wave number k may be rewritten as

$$k = \frac{2\pi}{\lambda} = \frac{2\pi}{h/p} = p\frac{2\pi}{h} = \frac{p}{\hbar} \tag{5.37}$$

and

$$\Delta k = \frac{\Delta p}{\hbar} \tag{5.38}$$

so that, in the case of the Gaussian wave packet,

$$\Delta k \, \Delta x = \frac{\Delta p}{\hbar} \, \Delta x = \frac{1}{2}$$

or

$$\Delta p \, \Delta x = \frac{\hbar}{2} \tag{5.39}$$

for Gaussian wave packets.

The relationship in Equation (5.39) was first presented in 1927 by the German physicist Werner Heisenberg, who won the Nobel Prize for Physics in 1932. This uncertainty applies in all three dimensions, so we should put a subscript on Δp to indicate the x direction Δp_x. Heisenberg's **uncertainty principle** can therefore be written

Heisenberg uncertainty principle for p_x and x

$$\Delta p_x \, \Delta x \geq \frac{\hbar}{2} \qquad (5.40)$$

which establishes limits on the simultaneous knowledge of the values of p_x and x.* The limits on Δp_x and Δx represent the lowest possible limits on the uncertainties in knowing the values of p_x and x, no matter how good an experimental measurement is made. It is possible to have a greater uncertainty in the values of p_x and x, but it is not possible to know them with more precision than allowed by the uncertainty principle. The uncertainty principle does not apply to the products of Δp_z and Δx or to that of Δp_y and Δz. The value of $\Delta p_z \, \Delta x$ can be zero. Equation (5.40) is true not only for specific waves such as water or sound, but for matter waves as well. It is a consequence of the de Broglie wavelength of matter. If we want to know the position of a particle very accurately, then we must accept a large uncertainty in the momentum of the particle. Similarly, if we want to know the precise value of a particle's momentum, it is not possible to specify the particle's location very precisely. The uncertainty principle represents another sharp digression with classical physics, where it is assumed that it is possible to specify simultaneously and precisely both the particle's position and momentum. Because of the small value of $\hbar$, the uncertainty principle becomes important only on the atomic level.

Consider a particle whose location is known within a width of ℓ along the x axis. We then know the position of the particle to within a distance $\Delta x \leq \ell / 2$. The uncertainty principle specifies that Δp is limited by

$$\Delta p \geq \frac{\hbar}{2 \, \Delta x} \geq \frac{\hbar}{\ell} \qquad (5.41)$$

Because $p = mv$, we have $\Delta p = m \Delta v$, and

$$\Delta v = \frac{\Delta p}{m} \geq \frac{\hbar}{m\ell} \qquad (5.42)$$

These results have some interesting implications. For example, consider a particle with low energy. What is the minimum kinetic energy such a particle can have? We can use nonrelativistic equations, so we have $K = p^2/2m$. Equation (5.41) indicates there is an uncertainty in the momentum, so we can assume the minimum value of the momentum will be at least as large as its uncertainty and $p_{min} \geq \Delta p$ to find the minimum value of the kinetic energy K_{min}.

$$K_{min} = \frac{p_{min}^2}{2m} \geq \frac{(\Delta p)^2}{2m} \geq \frac{\hbar^2}{2m\ell^2} \qquad (5.43)$$

Note that this equation indicates that if we are uncertain as to the exact position of a particle, for example an electron somewhere inside an atom of diameter ℓ, the particle can't have zero kinetic energy.

Werner Heisenberg (1901–1976) was born in Germany where he spent his entire career at various universities including Munich, Leipzig, and Berlin. He was appointed director of the Kaiser Wilhelm Institute in Berlin in 1942, the highest scientific position in Germany. After World War II Heisenberg spent much of his effort supporting research and opportunities for young physicists and speaking out against the atom bomb. *AIP Emilio Segrè Visual Archives.*

*In some representations of the uncertainty principle, the factor $\frac{1}{2}$ is absent. Our form represents the lower limit of uncertainty.

Example 5.8

Calculate the momentum uncertainty of (a) a tennis ball constrained to be in a fence enclosure of length 35 m surrounding the court, and (b) an electron within the smallest diameter of a hydrogen atom.

Strategy We will use Equation (5.40) to find Δp_x. The position uncertainty Δx is approximately half of the enclosure.

Solution (a) If we insert the uncertainty of the location of the tennis ball, $\Delta x = (35 \text{ m})/2$, into Equation (5.40), we have

$$\Delta p_x \geq \frac{1}{2} \frac{\hbar}{\Delta x} = \frac{1.05 \times 10^{-34} \text{ J} \cdot \text{s}}{2(35 \text{ m})/2} = 3 \times 10^{-36} \text{ kg} \cdot \text{m/s}$$

We will have no problem specifying the momentum of the tennis ball!

(b) The diameter of the hydrogen atom in its lowest energy state (smallest radius) is $2a_0$. The uncertainty Δx is half the diameter or equal to the radius, $\Delta x = a_0$.

$$\Delta x = a_0 = 0.529 \times 10^{-10} \text{ m}$$

$$\Delta p_x \geq \frac{1}{2} \frac{\hbar}{\Delta x} = \frac{1.05 \times 10^{-34} \text{ J} \cdot \text{s}}{2(0.529 \times 10^{-10} \text{ m})}$$

$$= 1 \times 10^{-24} \text{ kg} \cdot \text{m/s}$$

Example 5.9

Treat the hydrogen atom as a one-dimensional entity of length $2a_0$ and determine the electron's minimum kinetic energy.

Strategy We will use the uncertainty principle to determine $K_{\min}$. Equation (5.43) gives us the minimum kinetic energy for a particle known to be located within a distance ℓ.

Solution Equation (5.43) gives

$$K_{\min} = \frac{\hbar^2}{2m\ell^2} = \frac{(\hbar c)^2}{2mc^2\ell^2}$$

$$= \frac{(197 \text{ eV} \cdot \text{nm})^2}{(2)(0.511 \times 10^6 \text{ eV})(2 \times 0.0529 \text{ nm})^2} = 3.4 \text{ eV}$$

A calculation considering three dimensions would give a result about twice this value. This simple calculation gives a reasonable value for the kinetic energy of the ground state electron of the hydrogen atom.

Energy-Time Uncertainty Principle Equation (5.40) is not the only form of the uncertainty principle. We can find another form by using Equation (5.23) from our study of wave motion. When we superimposed two waves to form a wave packet we found $\Delta\omega \, \Delta t = 2\pi$. If we evaluate this same product using Gaussian packets, we will find

$$\Delta\omega \, \Delta t = \frac{1}{2} \tag{5.44}$$

just as we did for the product $\Delta k \, \Delta x$. A relationship like this is easy to understand. If we are to localize a wave packet in a small time Δt (instead of over an infinite time as for a single wave), we must include the frequencies of many waves in order to have them cancel everywhere but over the time interval Δt. Because $E = hf$, we have for each wave

$$\Delta E = h \, \Delta f = h \frac{\Delta\omega}{2\pi} = \hbar \, \Delta\omega$$

Therefore

$$\Delta\omega = \frac{\Delta E}{\hbar} \quad \text{and} \quad \Delta\omega\,\Delta t = \frac{\Delta E}{\hbar}\Delta t = \frac{1}{2}$$

We can therefore obtain another form of Heisenberg's uncertainty principle:

Heisenberg uncertainty principle for energy and time

$$\Delta E\,\Delta t \geq \frac{\hbar}{2} \tag{5.45}$$

Other *conjugate variables* similar to p_x and x in Equation (5.40) also form uncertainty principle relations. The product of conjugate variables (like p_x and x or E and t) must have the same dimensions as Planck's constant. Conjugate variable pairs include the angular momentum L and angle θ, as well as the rotational inertia I and angular velocity ω. Similar uncertainty relations can be written for them.

We once again must emphasize that the uncertainties expressed in Equations (5.40) and (5.45) are intrinsic. They are not due to our inability to construct better measuring equipment. No matter how well we can measure, no matter how accurate an instrument we build, and no matter how long we measure, we can never do any better than the uncertainty principle allows. Many people, including Einstein, have tried to think of situations in which it is violated, but they have not succeeded. At the 1927 Solvay conference Bohr and Einstein had several discussions about the uncertainty principle. Every morning at breakfast Einstein would present a new *gedanken* experiment that would challenge the uncertainty principle. In his careful, deliberate manner, Bohr would refute each objection. Eventually Einstein conceded—he could not provide a valid example of contradiction. These discussions continued off and on into the 1930s, because Einstein had difficulty accepting the idea that the quantum theory could give a complete description of physical phenomena. He believed that quantum theory could give a statistical description of a collection of particles but could not describe the motion of a single particle. Einstein presented several paradoxes to support his ideas. Bohr was able to analyze each paradox and present a reasonable answer. Bohr stressed his complementarity principle, which precludes a simultaneous explanation in terms of waves and particles, as well as Heisenberg's uncertainty principle.

Bohr and Einstein discussions

Figure 5.22 Niels Bohr's coat of arms was designed in 1947 when he was awarded the Danish Order of the Elephant. This award was normally given only to royalty and foreign presidents. Bohr chose the Chinese yin-yang symbol because it stands for the two opposing but inseparable elements of nature. The translation of the Latin motto is "Opposites are complementary." It was hung near the king's coat of arms in the church of Frederiksborg Castle at Hillerod. *AIP Niels Bohr Library, Margarethe Bohr Collection.*

Let's return to the previous discussion of determining which slit an electron passes through in the double-slit experiment (see Figure 5.21). We again shine light on the electrons passing through the slits and look with a powerful microscope. This time we will use the uncertainty principle and make a more detailed calculation. Photons from the shining light bounce off the electron as the electron passes through one of the slits. Photons then scatter into the microscope where we observe them. We must be able to locate the electron's position in y to at least within $\Delta y < d/2$ (where d is the distance between the two slits) in order to know which slit each electron went through. If the position of the electron is uncertain to less than $d/2$, then according to the uncertainty principle, the electron's momentum must be uncertain to at least $\Delta p_y > \hbar/d$. Just by scattering photons off the electrons in order to know which slit the electron went through, we introduce an uncertainty in the electron's momentum. This uncertainty has been caused by the measurement itself.

Consider an electron originally moving in a particular direction; let us choose $\theta = 0$ for convenience. By scattering the photon from the electron we now have an uncertainty in the angle θ due to the "kick" given the electron by the photon in the measurement process. The uncertainty in the electron's angle due to a possible momentum change along the y axis is $\Delta\theta = \Delta p_y/p$, but because $p = h/\lambda$, we have

$$\Delta\theta = \frac{\Delta p_y}{p} = \frac{(\Delta p_y)\lambda}{h} = \frac{(\hbar)\lambda}{hd} = \frac{\lambda}{2\pi d}$$

According to Equation (5.36) the first interference maximum will be at $\sin\theta = \lambda/d$ and the first minimum at $\sin\theta = \lambda/2d$. For small angle scattering, $\sin\theta \approx \theta$, and the angle of the first minimum is $\theta_{min} \approx \lambda/2d$. Note that the position of the first minimum is on the same order as our uncertainty in $\Delta\theta$, so the interference pattern is washed out. If we insist on identifying the electrons as particles and knowing which slit the electrons pass through, the wave characteristics of the electron disappear. We cannot simultaneously treat the electron as both a particle and a wave. This limitation seems to be a fundamental characteristic of the laws of nature. Only the smallness of Planck's constant h keeps us from encountering this limitation in everyday life.

Niels Bohr tried to turn this limitation into a philosophical principle. When he was awarded the Danish Order of the Elephant he featured on his coat of arms (see Figure 5.22) the Chinese yin-yang symbol, which stands for the two opposing but inseparable elements in nature. The Latin motto on the center of the coat of arms means "Opposites are complementary."

Example 5.10

Calculate the minimum kinetic energy of an electron that is localized within a typical nuclear radius of 6×10^{-15} m.

Strategy Let's assume the minimum electron momentum is equal to that determined by the uncertainty principle for an electron constrained within the distance Δx equal to the nuclear radius ($\Delta x = \pm r$). We can then determine the minimum electron energy from the minimum momentum.

Solution Given that $\Delta x \approx r = 6 \times 10^{-15}$ m, we have

$$\Delta p \geq \frac{\hbar}{2\Delta x} = \frac{6.58 \times 10^{-16}\ \text{eV} \cdot \text{s}}{1.2 \times 10^{-14}\ \text{m}}$$

$$\geq (5.48 \times 10^{-2}\ \text{eV} \cdot \text{s/m})\left(\frac{3 \times 10^8\ \text{m/s}}{c}\right)$$

$$\geq 1.64 \times 10^7\ \text{eV}/c$$

Because we assumed that the momentum p is at least as large as the uncertainty in p, we have

$$p \approx \Delta p \geq 1.64 \times 10^7\ \text{eV}/c$$

Because we don't yet know the electron's energy, let's be careful and calculate it relativistically.

$$E^2 = (pc)^2 + E_0^2$$

$$= \left[\left(1.64 \times 10^7\ \frac{\text{eV}}{c}\right)c\right]^2 + (0.511\ \text{MeV})^2$$

$$= (16.4\ \text{MeV})^2 + (0.511\ \text{MeV})^2$$

$$E = 16.4\ \text{MeV}$$

$$\text{K.E.} = E - E_0 = 16.4\ \text{MeV} - 0.51\ \text{MeV}$$

$$= 15.9\ \text{MeV}$$

Note that because $E \gg E_0$, a relativistic calculation was needed.

Conceptual Example 5.11

We found in the last example that if an electron is confined within the size of a nuclear radius, the uncertainty principle suggests that the minimum kinetic energy of the electron must have a minimum value of about 16 MeV. What does this indicate about the possibility of electrons existing within the nucleus?

Solution The value of 16 MeV for the electron's kinetic energy is larger than that observed for electrons emitted from nuclei in beta decay. We conclude that electrons are not confined within the nucleus. Electrons emitted from the nucleus (during beta decay) must actually be created when they are emitted.

Example 5.12

An atom in an excited state normally remains in that state for a very short time ($\sim 10^{-8}$ s) before emitting a photon and returning to a lower energy state. The "lifetime" of the excited state can be regarded as an uncertainty in the time Δt associated with a measurement of the energy of the state. This, in turn, implies an "energy width," namely, the corresponding energy uncertainty ΔE. Calculate (a) the characteristic "energy width" of such a state and (b) the uncertainty ratio of the frequency $\Delta f / f$ if the wavelength of the emitted photon is 300 nm.

Strategy (a) We use the uncertainty principle, Equation (5.45), to determine ΔE because we know Δt.

(b) We can determine Δf from the energy uncertainty ΔE by using $E = hf$: $\Delta E = h\,\Delta f$. We can determine the frequency by $f = c/\lambda$.

Solution (a) Equation (5.45) gives

$$\Delta E \geq \frac{\hbar}{2\,\Delta t} = \frac{6.58 \times 10^{-16}\ \text{eV} \cdot \text{s}}{(2)(10^{-8}\ \text{s})} = 3.3 \times 10^{-8}\ \text{eV}$$

This is a small energy, but many excited energy states have such energy widths. For stable ground states, $\tau = \infty$, and $\Delta E = 0$. For excited states in the nucleus, the lifetimes can be as short as 10^{-20} s (or shorter) with energy widths of 100 keV (or more).

(b) The frequency is found to be

$$f = \frac{c}{\lambda} = \frac{3 \times 10^8\ \text{m/s}}{300 \times 10^{-9}\ \text{m}} = 10^{15}\ \text{Hz} \tag{5.46}$$

The uncertainty Δf is

$$\Delta f = \frac{\Delta E}{h} = \frac{3.3 \times 10^{-8}\ \text{eV}}{4.136 \times 10^{-15}\ \text{eV} \cdot \text{s}} = 8 \times 10^6\ \text{Hz} \tag{5.47}$$

The uncertainty ratio of the frequency $\Delta f / f$ is

$$\frac{\Delta f}{f} = \frac{8 \times 10^6\ \text{Hz}}{10^{15}\ \text{Hz}} = 8 \times 10^{-9}$$

Modern instruments are capable of measuring ratios approaching 10^{-15}, or 1 Hz in a frequency of 10^{15} Hz! Experimental physicists have managed to improve this ratio by a factor of 100 every three years over the last two decades. The experimental limitations are considerably better than needed to measure the energy widths.

5.7 Probability, Wave Functions, and the Copenhagen Interpretation

We learned in elementary physics that the instantaneous wave intensity of electromagnetic radiation (light) is $\epsilon_0 c E^2$ where E is the electric field. Thus the probability of observing light is proportional to the square of the electric field. In the double-slit light experiment we can be assured that the electric field of the light wave is relatively large at the bright spots on the screen and small in the region of the dark places.

If Young's double-slit experiment is performed with very low intensity levels of light, individual flashes can be seen on the observing screen. We show a simulation of the experiment in Figure 5.19. After only 20 flashes (Figure 5.19a) we cannot make any prediction as to the eventual pattern. But we still know that the *probability* of observing a flash is proportional to the square of the electric field. We now briefly review this calculation that is normally given in introductory physics courses. If the distance from the central ray along the screen we are observing in an experiment like that depicted in Figure 5.18a is denoted by y, the probability for the photon to be found between y and $y + dy$ is proportional to the intensity of the wave (E^2) times dy. For Young's double-slit experiment, the value of the electric field $\vec{E}$ produced by the two interfering waves is large where the flash is likely to be observed and small where it is not likely to be seen. By counting the number of flashes we relate the energy flux I (called the intensity) of the light to the number flux, N per unit area per unit time, of photons having energy hf. In the wave description, we have $I = \epsilon_0 c \langle E^2 \rangle$ (where E is the electric field), and in what appears to be the particle description, $I = Nhf$. The flux of photons N, or the probability P of observing the photons, is proportional to the average value of the square of the electric field $\langle E^2 \rangle$.

How can we interpret the probability of finding the electron in the wave description?

First, let's remember that the localization of a wave can be accomplished by using a wave packet. We used a function $\Psi(x, t)$ to denote the superposition of many waves to describe the wave packet. We call this function $\Psi(x, t)$ the **wave function**. In the case of light, we know that the electric field $\vec{E}$ and magnetic field $\vec{B}$ satisfy a wave equation. In electrodynamics either $\vec{E}$ or $\vec{B}$ serves as the wave function Ψ. For particles (say electrons) a similar behavior occurs. In this case the wave function $\Psi(x, t)$ determines the probability, just as the flux of photons N arriving at the screen and the electric field $\vec{E}$ determined the probability in the case of light.

Wave function

For matter waves having a de Broglie wavelength, it is the wave function Ψ that determines the likelihood (or probability) of finding a particle at a particular position in space at a given time. The value of the wave function Ψ has no physical significance itself, and as we will see later, it can have a **complex** value (containing both real and imaginary numbers). The quantity $|\Psi|^2$ is called the **probability density** and represents the probability of finding the particle in a given unit volume at a given instant of time.

Probability density

In general, $\Psi(x, y, z, t)$ is a complex quantity and depends on the spatial coordinates x, y, and z as well as time t. The complex nature will be of no concern to us: we use Ψ times its complex conjugate Ψ^* when finding probabilities. We are interested here in only a single dimension y along the observing screen and for a given time t. In this case $\Psi^* \Psi \, dy = |\Psi|^2 \, dy$ is the probability of observing an electron in the interval between y and $y + dy$ at a given time, and we call this $P(y) \, dy$.

$$P(y) \, dy = |\Psi(y, t)|^2 \, dy \qquad (5.48)$$

Because the electron has to have a probability of unity of being observed *somewhere* along the screen, we integrate the probability density over all space by integrating over y from $-\infty$ to ∞. This process is called **normalization**.

Normalization

$$\int_{-\infty}^{\infty} P(y) \, dy = \int_{-\infty}^{\infty} |\Psi(y, t)|^2 \, dy = 1 \qquad (5.49)$$

Max Born (Nobel Prize, 1954), one of the founders of the quantum theory, first proposed this probability interpretation of the wave function in 1926. The determination of the wave function $\Psi(x, t)$ will be discussed in much more detail in the next chapter.

The use of wave functions $\Psi(x, y, z, t)$ rather than the classical positions $x(t)$, $y(t), z(t)$ represents a clean break between classical and modern physics. Physicists have developed a set of rules and procedures in the quantum theory to determine physical observables like position, momentum, and energy (see Section 6.2).

The Copenhagen Interpretation

Erwin Schrödinger and Werner Heisenberg worked out independent and separate mathematical models for the quantum theory in 1926. We shall examine Schrödinger's theory in Chapter 6, because it is somewhat easier to understand and is based on waves. Paul Dirac reported his relativistic quantum theory in

Max Born (1882–1970) was born a German in what is now Poland. After studying at several European universities he received his degree in 1907 from the University of Göttingen. After visiting several universities and serving in World War I, he became a professor at Göttingen in 1921 where he did his most important work on the statistical meaning of the new quantum theory (Nobel Prize in Physics, 1954). He and his student, Werner Heisenberg, collaborated on the matrix mechanics version of quantum mechanics. Born, a Jew, was forced to emigrate from Germany in 1933, and after visiting Italy, Cambridge, and India, he settled at the University of Edinburgh in 1936, from which he retired in 1953. *AIP Emilio Segrè Visual Archives.*

1928. Today there is little disagreement about the mathematical formalism of quantum theory. That is not the case regarding its interpretation.

We want to examine the *Copenhagen interpretation*, because it is the mainstream interpretation of quantum theory. Werner Heisenberg announced his uncertainty principle in early 1927 while he was a lecturer in Bohr's Institute of Theoretical Physics. At first Bohr, the mentor, thought Heisenberg's uncertainty principle was too narrow, and he pointed out a mistake in Heisenberg's paper concerning a *gedanken* experiment about a gamma-ray microscope used by Heisenberg to prove his point. Heisenberg, the 25-year-old rising star, strongly objected at first to Bohr's opinion and refused Bohr's suggestion to withdraw his paper on the uncertainty principle. Bohr and Heisenberg had many discussions in 1927 formulating the interpretation of quantum mechanics now known as the "Copenhagen interpretation," "Copenhagen school," or sometimes unkindly as "Copenhagen orthodoxy." It was strongly supported by Max Born and Wolfgang Pauli (profiled in Chapter 8).

There are various formulations of the interpretation, but it is generally based on the following:

1. The uncertainty principle of Heisenberg
2. The complementarity principle of Bohr
3. The statistical interpretation of Born, based on probabilities determined by the wave function

Together these three concepts form a logical interpretation of the physical meaning of quantum theory. According to the Copenhagen interpretation, physics depends on the outcomes of measurement. Consider a single electron passing through the two-slit experiment. We can determine precisely where the electron hits the screen by noting a flash. The Copenhagen interpretation rejects arguments about where the electron was between the times it was emitted in the apparatus (and subsequently passed through the two slits) and when it flashed on the screen. The measurement process itself randomly chooses one of the many possibilities allowed by the wave function, and the wave function instantaneously changes to represent the final outcome. Bohr argued that it is not the task of physics to find out how nature is, because we can never understand the quantum world or assign physical meaning to the wave function. Bohr and Heisenberg argued that measurement outcomes are the only reality in physics.

Many physicists objected (and some still do!) to the Copenhagen interpretation for widely varying reasons. One of the basic objections is to its nondeterministic nature. They also object to the vague measurement process that converts probability functions into nonprobabilistic measurements. Famous physicists who objected to the Copenhagen interpretation were Albert Einstein, Max Planck, Louis de Broglie, and Erwin Schrödinger. Einstein and Schrödinger never accepted the Copenhagen interpretation. Einstein was particularly bothered by the reliance upon probabilities, and he wrote Born in 1926 that "God does not throw dice." But it is fair to say that the great majority of physicists today accept the Copenhagen interpretation as the primary interpretation of quantum mechanics. In the past decade physicists have used feedback systems to demonstrate that quantum indeterminism can be reduced by guiding the outcome of a probabilistic quantum process toward a deterministic outcome.*

---

*See, for example, J. M. Geremia, J. K. Stockton, and H. Mabuchi, *Science* **304**, 270 (2004).

Several paradoxes have been proposed by physicists to refute the Copenhagen interpretation. They include the famous Schrödinger cat paradox,* the Einstein-Podolsky-Rosen paradox,[†] and Bell's theorem (or inequality).[‡] Space does not allow us to describe these paradoxes (see Problems 50, 51, and 52). A Princeton University graduate student, Hugh Everett III, announced an alternate interpretation to the Copenhagen view in 1957. In Everett's "Many Worlds" interpretation, the concept of parallel universes is invoked—in itself such a weird idea that it has not gained wide acceptance, but it overcomes some objections to the Copenhagen interpretation. Nevertheless, the Copenhagen interpretation remains the favored interpretation.

5.8   Particle in a Box

Let's now consider the situation of a particle of mass m trapped in a one-dimensional box of width ℓ. We have already used the uncertainty principle in Equation (5.43) to calculate the minimum kinetic energy of such a particle. Now let's determine the possible energies of such a particle. Because of our discussion in the previous section we want to use the wave nature of the particle in this determination.

First, what is the most probable location of the particle in the state with the lowest energy at a given time, say $t = 0$, so that $\Psi(x, 0) = \Psi(x)$? In order to find the probable location, we will treat the particle as a sinusoidal wave. The particle cannot be physically outside the confines of the box, so the amplitude of the wave motion must vanish at the walls and beyond. In the language of the wave function, its probability of being outside is zero, so the wave function must vanish outside. The wave function must be continuous, and the probability distribution can have only one value at each point in the box. In order for the probability to vanish at the walls, we must have an integral number of half wavelengths $\lambda/2$ fit into the box. Note that all the possible waves shown in Figure 5.23 fit this requirement.

The requirement of an integral number of half wavelengths $\lambda/2$ means that

$$\frac{n\lambda}{2} = \ell \quad \text{or} \quad \lambda_n = \frac{2\ell}{n} \quad (n = 1, 2, 3, \ldots) \tag{5.50}$$

The possible wavelengths are quantized, and the wave shapes will have $\sin(n\pi x/\ell)$ factors. If we treat the problem nonrelativistically and assume there is no potential energy, the energy E of the particle is

$$E = \text{K.E.} = \frac{1}{2}mv^2 = \frac{p^2}{2m} = \frac{h^2}{2m\lambda^2}$$

If we insert the values for λ_n, we have

$$E_n = \frac{h^2}{2m}\left(\frac{n}{2\ell}\right)^2 = n^2\frac{h^2}{8m\ell^2} \quad (n = 1, 2, 3, \ldots) \tag{5.51}$$

---

*Schrödinger published an essay in 1935, "The Present Situation in Quantum Mechanics," in which he described a thought experiment where a cat in a closed box either lived or died according to whether a quantum event occurred. The paradox was that it was not possible to know whether the cat was dead or alive until an observer opened the box, an apparent contradiction to the intuitive notion that the cat is either alive or dead at any moment.

[†]A. Einstein, B. Podolsky, and N. Rosen, Can quantum-mechanical description of physical reality be considered complete? *Physical Review* **47**, 777 (1935).

[‡]J. S. Bell, On the Einstein Podolsky Rosen paradox, *Physics* **1**, 195 (1964).

Figure 5.23 Possible ways of fitting waves into a one-dimensional box of length ℓ. The left side shows the wave functions for the four lowest energy values. The right side shows the corresponding probability distributions.

Therefore, the possible energies of the particle are quantized, and each of these energies E_n is a possible energy level. Note that the lowest energy is $E_1 = h^2/8m\ell^2$. Because we assumed the potential energy to be zero, E_n is also equal to the kinetic energy. We previously found in Equation (5.43) a value for $K_{\min} = \hbar^2/2m\ell^2$, which differs by a factor $1/\pi^2$ from the value of E_1. Why the difference? Equation (5.51) is based on the wave theory of physics, and as we shall see in the next chapter, it is a better calculation than the result in Equation (5.43), which used approximations.

The probability of observing the particle between x and $x + dx$ in each state is $P_n\, dx \propto |\psi_n(x)|^2\, dx$. Notice that $E_0 = 0$ is not a possible state, because $n = 0$ corresponds to $\psi_0 = 0$. The lowest energy level is therefore E_1, with a probability density $P_1 \propto |\psi_1(x)|^2$, shown in Figure 5.23. The most probable location for the particle in the lowest energy state is in the middle of the box.

This particle-in-a-box model is more important than it might seem. It is our first application of what we call "quantum theory" or "quantum mechanics." Notice how the quantization of energy arises from the need to fit a whole number of half-waves into the box and how we obtained the corresponding probability densities of each of the states. The concept of energy levels, as first discussed in the Bohr model, has surfaced in a natural way by using waves. The procedure followed is the same as finding the allowed modes of standing waves inside the box. We can use all the results that we learned about waves in elementary physics.

Example 5.13

Find the quantized energy levels of an electron constrained to move in a one-dimensional atom of width 0.1 nm.

Strategy We previously found the minimum kinetic energy of an electron in a similar situation in Example 5.9.

In the present case we want to use quantum theory, so we use Equation (5.51) for the energy levels.

Solution We use Equation (5.51) and insert the appropriate values for m and ℓ.

$$E_n = n^2 \frac{h^2}{8m\ell^2} = n^2 \frac{h^2 c^2}{8mc^2\ell^2}$$

$$= n^2 \frac{(1239.8 \text{ eV} \cdot \text{nm})^2}{(8)(0.511 \times 10^6 \text{ eV})(0.1 \text{ nm})^2}$$

$$= n^2(38 \text{ eV})$$

The first three energy levels are $E_1 = 38$ eV, $E_2 = 152$ eV, and $E_3 = 342$ eV.

Summary

Max von Laue suggested the scattering of x rays from matter, thereby firmly establishing the wave nature of x rays and the lattice structure of crystals. W. H. Bragg and W. L. Bragg exploited the wave behavior of x rays by utilizing x-ray scattering to determine the spacing d between crystal planes according to Bragg's law:

$$n\lambda = 2d \sin \theta \qquad (5.1)$$

In an important conceptual leap, de Broglie suggested that particles might also exhibit wave properties, with a wavelength λ determined by their momentum:

$$\lambda = \frac{h}{p} \qquad \text{de Broglie wavelength} \qquad (5.2)$$

Davisson and Germer, and G. P. Thomson independently, demonstrated the wave characteristics of particles by diffracting low-energy electrons from crystals.

Particles may be described using waves by representing them as wave packets, the superposition of many waves of different amplitudes and frequencies. The group velocity $u_{gr} = d\omega/dk$ represents the speed of the particle described by the wave packet.

Niels Bohr proposed a principle of complementarity, stating that it is not possible to describe physical behavior simultaneously in terms of both particles and waves. We must use either one form of description or the other, thus resolving (or avoiding) the wave-particle duality problem.

We describe particles exhibiting wave behavior by using wave functions Ψ, which in general may be complex-valued functions of space and time. The probability of observing a particle between x and $x + dx$ at time t is $|\Psi(x, t)|^2 \, dx$.

Werner Heisenberg pointed out that it is not possible to know simultaneously both the exact momentum and position of a particle or to know its precise energy at a precise time. These relationships

$$\Delta p_x \, \Delta x \geq \hbar/2 \qquad (5.40)$$

$$\Delta E \, \Delta t \geq \hbar/2 \qquad (5.45)$$

are called *Heisenberg's uncertainty principle* and are consistent with Bohr's complementarity principle. No experiment, regardless of how clever, can measure p, x, E, and t better than the uncertainties expressed in Equations (5.40) and (5.45). The mainstream interpretation of quantum theory is the Copenhagen interpretation, which depends on the uncertainty principle, the complementarity principle, and the statistical interpretation.

Energy levels naturally arise when a particle in a box is considered to have wave behavior.

Questions

1. In 1900, did it seem clear that x rays were electromagnetic radiation? Give reasons for your answer. Was it important to perform further experiments to verify the characteristics of x rays?
2. In the early 1900s it was found that x rays were more difficult to refract or diffract than visible light. Why did this lead researchers to suppose that the wavelengths of x rays were shorter rather than longer than those of light?
3. What determines whether a given photon is an x ray? Could an x ray have a wavelength longer than ultraviolet light?
4. For a single crystal, transmission x-ray scattering will produce dots. However, if there are randomly oriented crystals, as in powder, concentric rings appear. Explain the difference qualitatively.
5. How many particles do you think might be shown experimentally to exhibit wavelike properties? List at least four and discuss possible experiments.
6. Why are neutrons more widely used than protons for studying crystal structure? What about using a hydrogen atom?
7. Why is it important to use "cold" neutrons for studying crystal structure? How could one obtain "cold" neutrons?
8. It has been said that many experimental discoveries are made as a result of accidents (for example, that of Davisson and Germer). This statement may have some truth,

but what traits and abilities must good scientists possess to take advantage of their accidents?

9. Images taken with transmission electron microscopes are taken by passing very high energy electrons (40–100 keV) through matter. Why are the images always in black and white (or any two colors)?

10. Are the following phenomena wave or particle behaviors? Give your reasoning. (a) Television picture, (b) rainbows, (c) football sailing through goal posts, (d) telescope observing the moon, (e) police radar.

11. The experiment by Jönsson that showed the wavelike properties of electrons passing through a double slit is considered a pedagogically interesting experiment but not a landmark experiment. Why do you suppose this is true?

12. Can you think of an experiment other than those mentioned in this chapter that might show the wavelike properties of particles? Discuss it.

13. Why doesn't the uncertainty principle restriction apply between the variables p_z and x?

14. How does the uncertainty principle apply to a known stable atomic system that apparently has an infinite lifetime? How well can we know the energy of such a system?

15. According to the uncertainty principle, can a particle having a kinetic energy of exactly zero be confined somewhere in a box of length ℓ? Explain.

16. What is similar about the conjugate variable pairs (p, x), (E, t), (L, θ), and (I, ω)?

17. What are the dimensions of the wave function $\Psi(x, t)$ that describes matter waves? Give your reasoning.

18. Soon after their discovery, Davisson and Germer were using their experimental technique to point out new crystal structures of nickel. Do you think they were justified? Explain how you think their results allowed them to make such statements.

19. Albert Einstein was a dissenter to the Copenhagen Interpretation and what it represented until the day he died. In a letter to Max Born in December 1926, Einstein wrote "The theory yields a lot, but it hardly brings us any closer to the secret of the Old One. In any case I am convinced that God does not throw dice." What do you think Einstein meant by this statement? Who or what is the Old One?

20. It has been said that the energy-time version of the uncertainty principle allows a violation of the conservation of energy. The argument is that the uncertainty ΔE allows the possibility that we may not know that energy conservation is violated during a small time Δt. Discuss arguments pro and con concerning this possibility.

Problems

Note: The more challenging problems are indicated by a shaded box.

5.1 X-Ray Scattering

1. X rays scattered from a crystal have a first-order diffraction peak at $\theta = 15°$. At what angle will the second- and third-order peaks appear?

2. X rays of wavelength 0.186 nm are scattered from NaCl. What is the angular separation between first- and second-order diffraction peaks? Assume scattering planes that are parallel to the surface.

3. Potassium chloride is a crystal with lattice spacing of 0.314 nm. The first peak for Bragg diffraction occurs at 12°. What energy x rays were diffracted? What other order peaks can be observed ($\theta \leq 90°$)?

4. A cubic crystal with interatomic spacing of 0.24 nm is used to select γ rays of energy 100 keV from a radioactive source containing a continuum of energies. If the incident beam is normal to the crystal, at what angle do the 100-keV γ rays appear?

5.2 De Broglie Waves

5. Calculate the de Broglie wavelength of a 3.0-kg rock thrown with a speed of 6 m/s into a pond. Is this

wavelength similar to that of the water waves produced? Explain.

6. Calculate the de Broglie wavelength of a nitrogen molecule in the atmosphere on a hot summer day (37°C). Compare this with the diameter (less than 1 nm) of the molecule.

7. Transmission electron microscopes that use high-energy electrons accelerated over a range from 40 to 100 kV are employed in many areas including the study of biological samples (like a virus) and nanoscience research and development (alloy particles and carbon nanotubes, for example). What would be the spatial limitation for this range of electrons? It is often true that resolution is limited by the optics of the lens system, not by the intrinsic limitation due to the de Broglie wavelength.

8. A 3-MV transmission electron microscope has been in operation at Osaka University in Japan for several years. The higher energy electrons allow for deeper sample penetration and extremely high resolution. What is the resolution limit for these electrons?

9. Work out Example 5.2b strictly using SI units of m, J, kg, and so on, and compare with the method of the example using eV units.

10. Assume that the total energy E of an electron greatly exceeds its rest energy. If a photon has a wavelength equal to the de Broglie wavelength of the electron, what is the photon's energy? What if $E = 2E_0$ for the electron?

11. Determine the de Broglie wavelength of a particle of mass m and kinetic energy K. Do this for both (a) a relativistic and (b) a nonrelativistic particle.

12. The Stanford Linear Accelerator can accelerate electrons to an energy of 50 GeV. What is the de Broglie wavelength of these electrons? What fraction of a proton's diameter ($d \approx 2 \times 10^{-15}$ m) can such a particle probe?

13. Find the kinetic energy of (a) photons, (b) electrons, (c) neutrons, and (d) α particles that have a de Broglie wavelength of 0.15 nm.

14. Find the de Broglie wavelength of neutrons in equilibrium at the temperatures (a) 5.0 K and (b) 0.010 K.

15. An electron initially at rest is accelerated across a potential difference of 3.00 kV. What are its wavelength, momentum, kinetic energy, and total energy?

16. What is the wavelength of an electron with kinetic energy (a) 40 eV, (b) 400 eV, (c) 4.0 keV, (d) 40 keV, (e) 0.40 MeV, and (f) 4.0 MeV? Which of these energies are most suited for study of the NaCl crystal structure?

17. Calculate the de Broglie wavelength of (a) an oxygen (O_2) molecule darting around the room at 480 m/s, (b) a bacterium of mass 1.5×10^{-15} kg moving at a speed of 1.0×10^{-6} m/s.

18. What is the de Broglie wavelength of the 1.0-TeV protons accelerated in the Fermilab Tevatron accelerator? These high-energy protons are needed to probe elementary particles (see Chapter 14).

5.3 Electron Scattering

19. In an electron-scattering experiment, an intense reflected beam is found at $\phi = 32°$ for a crystal with an interatomic distance of 0.23 nm. What is the lattice spacing of the planes responsible for the scattering? Assuming first-order diffraction, what are the wavelength, momentum, kinetic energy, and total energy of the incident electrons?

20. Davisson and Germer performed their experiment with a nickel target for several electron bombarding energies. At what angles would they find diffraction maxima for 48-eV and 64-eV electrons?

21. A beam of 2-keV electrons incident on a crystal is refracted and observed (by transmission) on a screen 35 cm away. The radii of three concentric rings on the screen, all corresponding to first-order diffraction, are 2.1 cm, 2.3 cm, and 3.2 cm. What is the lattice-plane spacing corresponding to each of the three rings?

22. A beam of thermal neutrons (K.E. = 0.025 eV) scatters from a crystal with interatomic spacing 0.45 nm. What is the angle of the first-order Bragg peak?

5.4 Wave Motion

23. Generating plants in some power systems drop 10% of their load when the AC frequency changes by 0.3 Hz from the standard of 60 Hz. How often must the reading be monitored in order for the automatic operating system to be able to take corrective action? Let the time between measurements be at least half that determined by the bandwidth relation.

24. Consider electrons of kinetic energy 5 eV and 500 keV. For each electron, find the de Broglie wavelength, particle speed, phase velocity (speed), and group velocity (speed).

25. A wave, propagating along the x direction according to Equation (5.11), has a maximum displacement of 3.0 cm at $x = 0$ and $t = 0$. The wave speed is 4.0 cm/s, and the wavelength is 7.0 cm. (a) What is the frequency? (b) What is the wave's amplitude at $x = 10$ cm and $t = 13$ s?

26. A wave of wavelength 4.0 cm has a wave speed of 4.2 cm/s. What is its (a) frequency, (b) period, (c) wave number, and (d) angular frequency?

27. Two waves are traveling simultaneously down a long Slinky. They can be represented by $\Psi_1(x, t) = 0.0030 \sin(6.0x - 300t)$ and $\Psi_2(x, t) = 0.0030 \sin(7.0x - 250t)$. Distances are measured in meters and time in seconds. (a) Write the expression for the resulting wave. (b) What are the phase and group velocities? (c) What is Δx between two adjacent zeros of Ψ? (d) What is $\Delta k \, \Delta x$?

28. A wave packet describes a particle having momentum $p = mv$. Show that the group velocity is βc and the phase velocity is c/β (where $\beta = v/c$). How can the phase velocity physically be greater than c?

29. For waves in shallow water the phase velocity is about equal to the group velocity. What is the dependence of the phase velocity on the wavelength?

30. Find the group and phase velocities of 8-MeV protons and 8-MeV electrons (see Problem 28).

31. Use Equation (5.25) with $\tilde{A}(k) = A_0$ for the range of $k = k_0 - \Delta k/2$ to $k_0 + \Delta k/2$, and $\tilde{A}(k) = 0$ elsewhere, to determine $\Psi(x, 0)$, that is, at $t = 0$. Sketch the envelope term, the oscillating term, and $|\Psi(x, 0)|^2$. Approximately what is the width Δx over the full-width-half-maximum part of $|\Psi(x, 0)|^2$? What is the value of $\Delta k \, \Delta x$?

32. Using Equation (5.31), show that u_{gr} correctly represents the velocity of the particle both relativistically and classically.

5.5 Waves or Particles?

33. Light of intensity I_0 passes through two sets of apparatus. One contains one slit and the other two slits. The slits have the same width. What is the ratio of the outgoing intensity amplitude for the central peak for the two-slit case compared to the single slit?

34. Design a double-slit electron-scattering experiment using 1.0-keV electrons that will provide the first maximum at an angle of 1°. What will be the slit separation d?

35. You want to design an experiment similar to the one done by Jönsson that does not require magnification of the interference pattern in order to be seen. Let the two slits be separated by 2000 nm. Assume that you can discriminate visually between maxima that are as little as 0.3 mm apart. You have at your disposal a lab that allows the screen to be placed 80 cm away from the slits. What energy electrons will you require? Do you think such low-energy electrons will represent a problem? Explain.

5.6 Uncertainty Principle

36. A proton is confined in a uranium nucleus of radius 8×10^{-15} m. Determine the proton's minimum kinetic energy according to the uncertainty principle if the proton is confined to a one-dimensional box that has length equal to the nuclear diameter.

37. A neutron is confined in a deuterium nucleus (deuteron) of diameter 2×10^{-15} m. Use the energy-level calculation of a one-dimensional box to calculate the neutron's minimum kinetic energy. What is the neutron's minimum kinetic energy according to the uncertainty principle?

38. What is the ratio uncertainty of the velocities $(\Delta v/v)$ of (a) an electron and (b) a proton confined to a one-dimensional box of length 1.7 nm?

39. Show that the uncertainty principle can be expressed in the form $\Delta L \, \Delta \theta \geq \hbar/2$, where θ is the angle and L the angular momentum. For what uncertainty in L will the angular position of a particle be completely undetermined?

40. Some physics theories indicate that the lifetime of the proton is about 10^{36} years. What would such a prediction say about the energy of the proton?

41. What is the bandwidth $\Delta \omega$ of an amplifier for radar if it amplifies a pulse of width 2.4 μs?

42. Find the minimum uncertainty in the speed of a bacterium having mass 3.0×10^{-15} kg if we know the position of the bacterium to within 1.0 micron, that is, to about its own size.

43. An atom in an excited state of 4.7 eV emits a photon and ends up in the ground state. The lifetime of the excited state is 1.0×10^{-13} s. (a) What is the energy uncertainty of the emitted photon? (b) What is the spectral line width (in wavelength) of the photon?

44. An electron microscope is designed to resolve objects as small as 0.14 nm. What energy electrons must be used in this instrument?

45. Rayleigh's criterion is used to determine when two objects are barely resolved by a lens of diameter d. The angular separation must be greater than θ_R where

$$\theta_R = 1.22 \frac{\lambda}{d}$$

In order to resolve two objects 4000 nm apart at a distance of 20 cm with a lens of diameter 5 cm, what energy (a) photons or (b) electrons should be used? Is this consistent with the uncertainty principle?

46. Calculate the de Broglie wavelength of a 5.5-MeV α particle emitted from an ^{241}Am nucleus. Could this particle exist inside the ^{241}Am nucleus (diameter $\approx 1.6 \times 10^{-14}$ m)?

47. Show that the minimum energy of a simple harmonic oscillator is $\hbar \omega/2$. What is the minimum energy in joules for a mass of 2.3 g oscillating on a spring with a spring constant of 8.2 N/m?

5.7 Probability, Wave Functions, and the Copenhagen Interpretation

48. The wave function of a particle in a one-dimensional box of width L is $\psi(x) = A \sin(\pi x/L)$. If we know the particle must be somewhere in the box, what must be the value of A?

49. The Fifth Solvay Congress, held in Brussels in October 1927, was dedicated to the quantum theory. A photo taken of the famous participants is often reproduced. Identify at least 10 participants and discuss what their contributions were to quantum physics, either experimentally or theoretically.

50. Write a cogent description of the Schrödinger cat paradox. Discuss variations of the paradox and the current status of its experimental verification.

51. Write a cogent description of the Einstein-Podolsky-Rosen paradox. Discuss variations of the paradox and the current status of its experimental verification.

52. Write a cogent description of the Bell inequality. Discuss variations and the current status of its experimental verification.

53. Summarize the discussions that Einstein and Bohr had at the 1927 Solvay Congress. List at least three objections that Einstein had to the Copenhagen interpretation of quantum mechanics and give Bohr's explanation.

5.8 Particle in a Box

54. Write down the normalized wave functions for the first three energy levels of a particle of mass m in a one-dimensional box of width L. Assume there are equal probabilities of being in each state.

55. A particle in a one-dimensional box of length L has a kinetic energy much greater than its rest energy. What is the ratio of the following energy levels E_n: E_2/E_1, E_3/E_1, E_4/E_1? How do your answers compare with the nonrelativistic case?

General Problems

56. Consider a wave packet having the product $\Delta p\, \Delta x = \hbar$ at a time $t = 0$. What will be the width of such a wave packet after the time $m(\Delta x)^2/\hbar$?

57. Analyze the Gaussian wave packet carefully and show that $\Delta k\, \Delta x = 1/2$. You must justify the assumptions you make concerning uncertainties in k and x. Take the Gaussian form given in Equation (5.26). (*Hint:* the linear "spread" of the wave packet Δx is given by one standard deviation, at which point the probability amplitude ($|\Psi|^2$) has fallen to one half its peak value.)

58. Most of the particles known to physicists are unstable. For example, the lifetime of the neutral pion, π^0, is about 10^{-16} s. Its mass is 135 MeV/c^2. What is the energy width of the π^0 in its ground state?

59. The range of the nuclear strong force is believed to be about 1.2×10^{-15} m. An early theory of nuclear physics proposed that the particle that "mediates" the strong force (similar to the photon mediating the electromagnetic force) is the pion. Assume that the pion moves at the speed of light in the nucleus, and calculate the time Δt it takes to travel between nucleons. Assume that the distance between nucleons is also about 1.2×10^{-15} m. Use this time Δt to calculate the energy ΔE for which energy conservation is violated during the time Δt. This ΔE has been used to estimate the mass of the pion. What value do you determine for the mass? Compare this value with the measured value of 135 MeV/c^2 for the neutral pion.

60. The planes of atoms in a cubic crystal lie parallel to the surface, 0.80 nm apart. X rays having wavelength 0.50 nm are directed at an angle θ to the surface. (a) For what values of θ will there be a strong reflection? (b) What energy electrons could give the same result?

61. Aliens visiting Earth are fascinated by baseball. They are so advanced that they have learned how to vary $\hbar$ to make sure that a pitcher cannot throw a strike with any confidence. Assume the width of the strike zone is 0.38 m, the speed of the baseball is 35 m/s, the mass of the baseball is 145 g, and the ball travels a distance of 18 m. What value of $\hbar$ is required? (*Hint:* there are two uncertainties here: the width of the strike zone and the transverse momentum of the pitched ball.)

62. Neutrons from nuclear reactors are used in neutron diffraction experiments to measure interplanar spacings of a crystal lattice. The interplanar spacing can be measured as an indication of strain in the sample. Neutrons are particularly useful because they are less destructive than x rays and are able to penetrate deep into the sample. Their magnetic moment allows their use to study magnetic properties of matter. To study a particular polycrystalline sample with a planar spacing of 0.156 nm, a detector is mounted at an angle of 26° from the incident neutron beam. What energy neutrons from the reactor must be used in this experiment? (An accelerator-based neutron spallation source is expected to be in operation in 2006 at Oak Ridge National Laboratory.)

63. Use a computer program to produce a wave packet using the function $\psi_n = A_n \cos 2\pi n x$ where the integer n ranges from 9 to 15. Let the amplitude $A_{12} = 1$ with the amplitudes A_n decreasing symmetrically by 1/2, 1/3, 1/4 on either side of A_{12} (for example, $A_{10} = 1/3$ and $A_{15} = 1/4$). (a) Plot the wave packet $\psi = \sum_n \psi_n$ versus x and each wave ψ_n over a wide enough range in x to see repeatable behavior for the wave packet. (b) Where is the wave packet centered? Over what value of x is the wave packet repeated?

64. Most elementary particles (see Chapter 14) are not stable, and physicists have measured their mean lifetime τ. Consider the uncertainty that this places on their mass-energy. The energy spread Γ is the full width of the particle's energy distribution at half its maximum value. (a) If we relate $\tau = \Delta t$ and $\Gamma = 2\, \Delta E$, what is the relation $\Gamma\tau$ in terms of the uncertainty principle? (b) What is the energy spread in the mass-energy of the following particles with their mean lifetimes in parentheses: neutron (887 s), charged pion π^- or π^+ (2.6×10^{-8} s), and upsilon (10^{-20} s)?

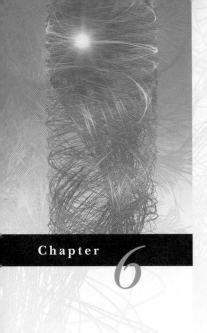

Quantum Mechanics II

I think it is safe to say that no one understands quantum mechanics. Do not keep saying to yourself, if you can possibly avoid it, "But how can it be like that?" because you will get "down the drain" into a blind alley from which nobody has yet escaped. Nobody knows how it can be like that.

Richard Feynman

Those who are not shocked when they first come across quantum mechanics cannot possibly have understood it.

Niels Bohr

As we discussed in the previous chapter, tremendous progress was made during the 1920s to correct the deficiencies of Bohr's atomic model. The origination of the quantum theory, also called **quantum mechanics**, is generally credited to Werner Heisenberg and Erwin Schrödinger, whose answers were clothed in very different mathematical formulations. Heisenberg (along with Max Born and Pascual Jordan) presented the *matrix formulation* of quantum mechanics in 1925 and 1926. The mathematical tools necessary to introduce matrix mechanics are not intrinsically difficult, but would require too lengthy an exposition for us to study them here. The other solution, proposed in 1926 by Schrödinger, is called *wave mechanics*; its mathematical framework is similar to the classical wave descriptions we have already studied in elementary physics. Paul Dirac and Schrödinger himself (among others) later showed that the matrix and wave mechanics formulations give identical results and differ only in their mathematical form. We shall study only the wave theory of Schrödinger here.

We discussed in the previous chapter the Copenhagen interpretation of quantum theory and the lack of universal agreement among physicists. Quantum theory is indeed a complex subject, and its probabilistic nature is contrary to the direct cause and effect seen in classical physics. We will do what thousands before us have done: "Shut up and calculate!"* In this chapter we will determine wave

---

*Many people credit this to Richard Feynman, but David Mermin (*Physics Today*, May 2004) says that this quote should not be attributed to Feynman. Certainly many students studying quantum physics have heard similar phrases uttered by their professors.

functions for some simple potentials and use these wave functions to predict the values of physical observables such as position and energy. We will see that particles are able to tunnel through potential barriers to exist in places that are not allowed by classical physics. Nuclear α decay and electronic tunnel diodes are examples of tunneling we will discuss.

6.1 The Schrödinger Wave Equation

The Austrian physicist Erwin Schrödinger (Nobel Prize, 1933) was presenting a seminar at the University of Zurich in November 1925 on de Broglie's wave theory for particles when Peter Debye suggested that there should be a wave equation. Within a few weeks Schrödinger had found a suitable wave equation based on what he knew about geometrical and wave optics.

In our previous study of elementary physics, we learned that Newton's laws, especially the second law of motion, govern the motion of particles. We need a similar set of equations to describe the wave motion of particles; that is, we need a **wave equation** that is dependent on the potential field (for example, the Coulomb or strong force field) that the particle experiences. We can then find the wave function Ψ (discussed in Chapter 5) that will allow us to calculate the probable values of the particle's position, energy, momentum, and so on.

We point out that although our procedure is similar to that followed in classical physics, we will no longer be able to calculate and specify the *exact* position, energy, and momentum simultaneously. Our calculations now must be consistent with the uncertainty principle and the notion of probability. It will take time and experience to get used to these new ideas (see Bohr's quote at the beginning of the chapter), and we will strive to give you that experience in this chapter.

There are several possible paths by which we could plausibly obtain the **Schrödinger wave equation**. Because none of the methods is actually a derivation, we prefer to present the equation and indicate its usefulness. Its ultimate correctness rests on its ability to explain and describe experimental results. The Schrödinger wave equation in its **time-dependent** form for a particle of energy E moving in a potential V in one dimension is

$$i\hbar\frac{\partial\Psi(x,\,t)}{\partial t} = -\frac{\hbar^2}{2m}\frac{\partial^2\Psi(x,\,t)}{\partial x^2} + V\Psi(x,\,t) \qquad (6.1)$$

Time-dependent Schrödinger wave equation

Erwin Schrödinger (1887–1961) was an Austrian who worked at several European universities before fleeing Nazism in 1938 and accepting a position at the University of Dublin where he remained until his retirement in 1956. His primary work on the wave equation was performed during the period he was in Zurich from 1920 to 1927. Schrödinger worked in many fields including philosophy, biology, history, literature, and language. *AIP Emilio Segrè Visual Archives.*

where $i = \sqrt{-1}$ is an imaginary number and we have used partial derivatives. Both the potential V and wave function Ψ may be functions of space and time, $V(x,\,t)$ and $\Psi(x,\,t)$.

The extension for Equation (6.1) into three dimensions is fairly straightforward.

$$i\hbar\frac{\partial\Psi}{\partial t} = -\frac{\hbar^2}{2m}\left(\frac{\partial^2\Psi}{\partial x^2} + \frac{\partial^2\Psi}{\partial y^2} + \frac{\partial^2\Psi}{\partial z^2}\right) + V\Psi(x,\,y,\,z,\,t) \qquad (6.2)$$

We will restrict ourselves to the one-dimensional form until Section 6.5.

Let's compare Equation (6.1) with the classical wave equation given by

$$\frac{\partial^2\Psi(x,\,t)}{\partial x^2} = \frac{1}{v^2}\frac{\partial^2\Psi(x,\,t)}{\partial t^2} \qquad (6.3)$$

In Equation (6.3) the wave function may be as varied as the amplitude of a water wave, a guitar-string vibration, or even the electric field $\vec{E}$ or magnetic field $\vec{B}$. Notice that the classical wave equation contains a second-order time derivative, whereas the Schrödinger wave equation contains only a first-order time derivative. This already gives us some idea that we are dealing with a somewhat different phenomenon.

Because the time-dependent Schrödinger Equation (6.1) is such a departure from our known physical laws, there is no derivation for it. We need *new* physical principles. Despite the fact that the Schrödinger wave equation has not been derived, it is still a useful tool because it describes experimental results. In science, and especially in physics, the test of theoretical calculation is that it agrees with what we observe. In most of the remainder of this chapter, we shall apply the Schrödinger wave equation to several simple situations in order to illustrate its usefulness.

Example 6.1

The wave equation must be linear in order that we can use the superposition principle to form wave packets using two or more waves. Prove that the wave function in Equation (6.1) is linear by showing that it is satisfied for the wave function

$$\Psi(x, t) = a\Psi_1(x, t) + b\Psi_2(x, t)$$

where a and b are constants, and Ψ_1 and Ψ_2 describe two waves each satisfying Equation (6.1).

Strategy We take the derivatives needed for Equation (6.1) and insert them in a straightforward manner. If Equation (6.1) is satisfied, then the wave function is linear.

Solution We take each of the derivatives needed for Equation (6.1).

$$\frac{\partial \Psi}{\partial t} = a\frac{\partial \Psi_1}{\partial t} + b\frac{\partial \Psi_2}{\partial t}$$

$$\frac{\partial \Psi}{\partial x} = a\frac{\partial \Psi_1}{\partial x} + b\frac{\partial \Psi_2}{\partial x}$$

$$\frac{\partial^2 \Psi}{\partial x^2} = a\frac{\partial^2 \Psi_1}{\partial x^2} + b\frac{\partial^2 \Psi_2}{\partial x^2}$$

We insert these derivatives into Equation (6.1) to yield

$$i\hbar\left(a\frac{\partial \Psi_1}{\partial t} + b\frac{\partial \Psi_2}{\partial t}\right) = -\frac{\hbar^2}{2m}\left(a\frac{\partial^2 \Psi_1}{\partial x^2} + b\frac{\partial^2 \Psi_2}{\partial x^2}\right)$$
$$+ V(a\Psi_1 + b\Psi_2)$$

Rearrangement of this equation gives

$$a\left(i\hbar\frac{\partial \Psi_1}{\partial t} + \frac{\hbar^2}{2m}\frac{\partial^2 \Psi_1}{\partial x^2} - V\Psi_1\right)$$
$$= -b\left(i\hbar\frac{\partial \Psi_2}{\partial t} + \frac{\hbar^2}{2m}\frac{\partial^2 \Psi_2}{\partial x^2} - V\Psi_2\right)$$

Because Ψ_1 and Ψ_2 each satisfy Equation (6.1), the quantities in parentheses are identically zero, and therefore Ψ is also a solution.

In Section 5.4 we discussed wave motion and the formation of wave packets from waves. For a wave of wave number k and angular frequency ω moving in the $+x$ direction, the wave function is

$$\Psi(x, t) = A \sin(kx - \omega t + \phi) \tag{5.18}$$

Equation (5.18) is not the most general form of a wave function, which may include both sines and cosines. Our wave function is also not restricted to being real. Only the physically measurable quantities must be real, and Equation (6.1) already has an imaginary number. A more general form of wave function is

$$\Psi(x, t) = Ae^{i(kx - \omega t)} = A[\cos(kx - \omega t) + i\sin(kx - \omega t)] \tag{6.4}$$

which also describes a wave moving in the $+x$ direction. In general the amplitude A may also be complex.

Example 6.2

Show that $Ae^{i(kx-\omega t)}$ satisfies the time-dependent Schrödinger wave equation.

Strategy We take appropriate derivatives needed for Equation (6.1) and insert them into Equation (6.1) to see if it is satisfied.

Solution

$$\frac{\partial \Psi}{\partial t} = -i\omega Ae^{i(kx-\omega t)} = -i\omega \Psi$$

$$\frac{\partial \Psi}{\partial x} = ik\Psi$$

$$\frac{\partial^2 \Psi}{\partial x^2} = i^2 k^2 \Psi = -k^2 \Psi$$

We insert these results into Equation (6.1) to yield

$$i\hbar(-i\omega\Psi) = -\frac{\hbar^2}{2m}(-k^2\Psi) + V\Psi$$

$$\left(\hbar\omega - \frac{\hbar^2 k^2}{2m} - V\right)\Psi = 0$$

If we use $E = hf = \hbar\omega$ and $p = \hbar k$, we obtain

$$\left(E - \frac{p^2}{2m} - V\right)\Psi = 0$$

which is zero in our nonrelativistic formulation, because $E = K + V = p^2/2m + V$. Thus $e^{i(kx-\omega t)}$ appears to be an acceptable solution at this point.

We showed in Example 6.2 that $e^{i(kx-\omega t)}$ represents an acceptable solution to the Schrödinger wave equation. It is not true that all functions of $\sin(kx - \omega t)$ and $\cos(kx - \omega t)$ are solutions. We show this in the following example.

Example 6.3

Determine whether $\Psi(x, t) = A\sin(kx - \omega t)$ is an acceptable solution to the time-dependent Schrödinger wave equation.

Strategy We again take the derivatives needed for Equation (6.1) and insert them into the equation to see if it is satisfied.

Solution

$$\frac{\partial \Psi}{\partial t} = -\omega A\cos(kx - \omega t)$$

$$\frac{\partial \Psi}{\partial x} = kA\cos(kx - \omega t)$$

$$\frac{\partial^2 \Psi}{\partial x^2} = -k^2 A\sin(kx - \omega t) = -k^2 \Psi$$

After we insert these relations into Equation (6.1), we have

$$-i\hbar\omega\cos(kx - \omega t) = \left(\frac{\hbar^2 k^2}{2m} + V\right)\Psi$$

$$= \left(\frac{\hbar^2 k^2}{2m} + V\right)A\sin(kx - \omega t) \quad (6.5)$$

(not true)

This equation is generally *not* satisfied for all x and t, and $A\sin(kx - \omega t)$ is, therefore, not an acceptable wave function. This function is, however, a solution to the classical wave equation [Equation (6.3)].

Normalization and Probability

We begin by reviewing the probability interpretation of the wave function that we discussed in Section 5.7. The probability $P(x)\,dx$ of a particle being between x and $x + dx$ was given in Equation (5.48).

$$P(x)\ dx = \Psi^*(x, t)\Psi(x, t)\ dx \tag{6.6}$$

The probability of the particle being between x_1 and x_2 is given by

Probability
$$P = \int_{x_1}^{x_2} \Psi^*\Psi\ dx \tag{6.7}$$

The wave function must also be normalized so that the probability of the particle being somewhere on the x axis is 1.

Normalization
$$\int_{-\infty}^{\infty} \Psi^*(x, t)\Psi(x, t)\ dx = 1 \tag{6.8}$$

Example 6.4

Consider a wave packet formed by using the wave function $Ae^{-\alpha|x|}$, where A is a constant to be determined by normalization. Normalize this wave function and find the probabilities of the particle being between 0 and $1/\alpha$, and between $1/\alpha$ and $2/\alpha$.

Strategy This wave function is sketched in Figure 6.1. We will use Equation (6.8) to normalize Ψ. Then we will find the probability by using the limits in the integration of Equation (6.7).

Solution If we insert the wave function into Equation (6.8), we have

$$\int_{-\infty}^{\infty} A^2 e^{-2\alpha|x|}\ dx = 1$$

Because the wave function is symmetric about $x = 0$, we can integrate from 0 to ∞, multiply by 2, and drop the absolute value signs on $|x|$.

$$2\int_{0}^{\infty} A^2 e^{-2\alpha x}\ dx = 1 = \frac{2A^2}{-2\alpha}e^{-2\alpha x}\Big|_{0}^{\infty}$$

$$1 = \frac{-A^2}{\alpha}(0 - 1) = \frac{A^2}{\alpha}$$

The coefficient $A = \sqrt{\alpha}$, and the normalized wave function Ψ is

$$\Psi = \sqrt{\alpha}\,e^{-\alpha|x|}$$

We use Equation (6.7) to find the probability of the particle being between 0 and $1/\alpha$, where we again drop the absolute signs on $|x|$ because x is positive.

$$P = \int_{0}^{1/\alpha} \alpha e^{-2\alpha x}\ dx$$

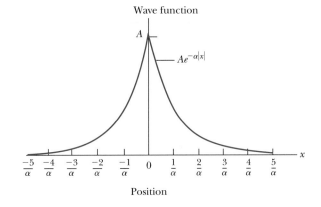

Wave function

Figure 6.1 The wave function $Ae^{-\alpha|x|}$ is plotted as a function of x. Note that the wave function is symmetric about $x = 0$.

The integration is similar to the previous one.

$$P = \frac{\alpha}{-2\alpha}e^{-2\alpha x}\Big|_{0}^{1/\alpha} = -\frac{1}{2}(e^{-2} - 1) = 0.432$$

The probability of the particle being between $1/\alpha$ and $2/\alpha$ is

$$P = \int_{1/\alpha}^{2/\alpha} \alpha e^{-2\alpha x}\ dx$$

$$P = \frac{\alpha}{-2\alpha}e^{-2\alpha x}\Big|_{1/\alpha}^{2/\alpha} = -\frac{1}{2}(e^{-4} - e^{-2}) = 0.059$$

We conclude that the particle is much more likely to be between 0 and $1/\alpha$ than between $1/\alpha$ and $2/\alpha$. This is to be expected, given the shape of the wave function shown in Figure 6.1.

The wave function $e^{i(kx-\omega t)}$ represents a particle under zero net force (constant V) moving along the x axis. There is a problem with this wave function, because if we try to normalize it, we obtain an infinite result for the integral. This occurs because there is a finite probability for the particle to be anywhere along the x axis. Over the entire x axis, these finite probabilities add up, when integrated, to infinity. The only other possibility is a zero probability, but that is not an interesting physical result. Because this wave function has precise k and ω values, it represents a particle having a definite energy and momentum. According to the uncertainty principle, because $\Delta E = 0$ and $\Delta p = 0$, we must have $\Delta t = \infty$ and $\Delta x = \infty$. We cannot know where the particle is at any time. We can still use such wave functions if we restrict the particle to certain positions in space, such as in a box or in an atom. We can also form wave packets from such functions in order to localize the particle.

Properties of Valid Wave Functions

Besides the Schrödinger wave equation, there are certain properties (sometimes called **boundary conditions**) that an acceptable wave function Ψ must also satisfy:

1. In order to avoid infinite probabilities, Ψ must be finite everywhere.
2. In order to avoid multiple values of the probability, Ψ must be single valued.   **Boundary conditions**
3. For finite potentials, Ψ and $\partial\Psi/\partial x$ must be continuous. This is required because the second-order derivative term in the wave equation must be single valued. (There are exceptions to this rule when V is infinite.)
4. In order to normalize the wave functions, Ψ must approach zero as x approaches $\pm\infty$.

Solutions for Ψ that do not satisfy these properties do not generally correspond to physically realizable circumstances.

Time-Independent Schrödinger Wave Equation

In many cases (and in most of the cases discussed here), the potential will not depend explicitly on time. The dependence on time and position can then be separated in the Schrödinger wave equation. Let

$$\Psi(x, t) = \psi(x)f(t) \tag{6.9}$$

We insert this $\Psi(x, t)$ into Equation (6.1) and obtain

$$i\hbar\psi(x)\frac{\partial f(t)}{\partial t} = -\frac{\hbar^2 f(t)}{2m}\frac{\partial^2\psi(x)}{\partial x^2} + V(x)\psi(x)f(t)$$

We divide by $\psi(x)f(t)$ to yield

$$i\hbar\frac{1}{f(t)}\frac{df(t)}{dt} = -\frac{\hbar^2}{2m}\frac{1}{\psi(x)}\frac{d^2\psi(x)}{dx^2} + V(x) \tag{6.10}$$

The left side of Equation (6.10) depends only on time, and the right side depends only on spatial coordinates. We have changed the partial derivatives to ordinary derivatives, because each side depends on only one variable. It follows that each side must be equal to a constant (which we label B), because one variable may change independently of the other. We integrate the left side of Equation (6.10) in an effort to determine the value of B.

$$ i\hbar \frac{1}{f} \frac{df}{dt} = B $$

$$ i\hbar \int \frac{df}{f} = \int B \, dt $$

We integrate both sides and find

$$ i\hbar \ln f = Bt + C $$

where C is an integration constant that we may choose to be 0. Therefore

$$ \ln f = \frac{Bt}{i\hbar} $$

From this equation we determine f to be

$$ f(t) = e^{Bt/i\hbar} = e^{-iBt/\hbar} \tag{6.11} $$

If we compare this function for $f(t)$ to the free-particle wave function that has the time dependence $e^{-i\omega t}$, we see that $B = \hbar\omega = E$. This is a general result.

We now have, from Equation (6.10),

$$ i\hbar \frac{1}{f(t)} \frac{df(t)}{dt} = E \tag{6.12} $$

Time-independent Schrödinger wave equation

$$ -\frac{\hbar^2}{2m} \frac{d^2\psi(x)}{dx^2} + V(x)\psi(x) = E\psi(x) \tag{6.13} $$

Equation (6.13) is known as the **time-independent Schrödinger wave equation**, and it is a fundamental equation in quantum mechanics.

Equation (6.11) can be rewritten as

$$ f(t) = e^{-i\omega t} \tag{6.14} $$

and the wave function $\Psi(x, t)$ becomes

$$ \Psi(x, t) = \psi(x)e^{-i\omega t} \tag{6.15} $$

We will restrict our attention for the present to time-independent potentials in one space dimension. Many important and useful results can be obtained from this nonrelativistic and one-dimensional form of quantum mechanics, because usually only the spatial part of the wave function $\psi(x)$ is needed. Therefore, we need only use Equation (6.13), the time-independent form of the Schrödinger wave equation.

Let's examine the probability density $\Psi^*\Psi$ discussed in Section 5.7. For the case of Equation (6.15), where the potential does not depend on time, we have

$$ \Psi^*\Psi = \psi^2(x)(e^{i\omega t}e^{-i\omega t}) $$

$$ \Psi^*\Psi = \psi^2(x) \tag{6.16} $$

The probability distributions are constant in time. We have seen in introductory physics the phenomenon called *standing waves* (for example, oscillations of strings fixed at both ends). Such standing waves can be formed from traveling waves moving in opposite directions. In quantum mechanics, we say the system is in a **stationary state**.

Example 6.5

Consider a metal where there are free electrons, and the potential is zero. What mathematical form does the wave function $\psi(x)$ take?

Strategy This is our first attempt at solving Equation (6.13), the time-independent Schrödinger wave equation. We let $V(x) = 0$ in Equation (6.13) and try to solve the differential equation for $\psi(x)$.

Solution If we let $V(x) = 0$ in Equation (6.13), we have

$$-\frac{\hbar^2}{2m}\frac{d^2\psi(x)}{dx^2} = E\psi(x)$$

We drop the x dependence on $\psi(x)$ and rewrite this equation as

$$\frac{d^2\psi}{dx^2} = -\frac{2mE}{\hbar^2}\psi = -k^2\psi$$

We have seen the differential equation $d^2\psi/dx^2 = -k^2\psi$ several times in calculus and in introductory physics. It occurs in small angle oscillations for pendula and in simple harmonic motion. If the energy E is positive, then k^2 is real, and the wave function solution is sinusoidal [$\psi(x) = A\sin kx + B\cos kx$]. However, for negative energy E, then k^2 is negative, and k is imaginary. An exponential wave function [$\psi(x) = Ce^{ikx}$] is appropriate.

Comparison of Classical and Quantum Mechanics We can gain insight by looking briefly at the similarities and differences between classical and quantum mechanics. Newton's second law ($\vec{F} = d\vec{p}/dt$) and Schrödinger's wave equation are both differential equations. They are both postulated to explain certain observed behavior, and experiments show they are successful. Newton's second law can be derived from the Schrödinger wave equation, so the latter is the more fundamental. Newton's laws may seem more fundamental—because they describe the precise values of the system's parameters, whereas the wave equation only produces wave functions that give probabilities—but by now we know from the uncertainty principle that it is not possible to simultaneously know precise values of both position and momentum and of both energy and time. Classical mechanics only appears to be more precise because it deals with macroscopic phenomena. The underlying uncertainties in macroscopic measurements are just too small to be significant.

An interesting parallel between classical mechanics and wave mechanics can be made by considering ray optics and wave optics. Throughout the 1700s, scientists argued about which of the optics formulations was the more fundamental; Newton favored ray optics. Finally, it was shown early in the 1800s that wave optics was needed to explain the observed phenomena of diffraction and interference. Ray optics is a good approximation as long as the wavelength of the radiation is much smaller than the dimensions of the apertures and obstacles it passes. Rays of light are characteristic of particle-like behavior. However, in order to describe interference phenomena, wave optics is required. Similarly for macroscopic objects, the de Broglie wavelength is so small that wave behavior is not apparent. However, advances in instrumentation and experimentation made it possible to observe behavior at the atomic level, and eventually the wave descriptions and quantum mechanics were required to understand all the data. Classical mechanics is a good macroscopic approximation and is accurate enough in the limit of large quantum numbers, but as far as we know now, there is only one correct theory and that is quantum mechanics.

6.2   Expectation Values

In order to be useful, the wave equation formalism must be able to determine values of the measurable quantities like position, momentum, and energy. In this section we will discuss how the wave function is able to provide this information. We will do this here in only one dimension, but the discussion can be extended to three dimensions as well. We will also evaluate the values of the physical quantities for a given time t, because in general the whole system, including the values of the physical quantities, evolves with time.

Consider a measurement of the position x of a particular system (for example, the position of a particle in a box—see Section 5.8). If we make three measurements of the position, we are likely to obtain three different results. Nevertheless, if our method of measurement is inherently accurate, then there is some physical significance to the average of our measured values of x. Moreover, the precision of our result improves as more measurements are made. In quantum mechanics we use wave functions to calculate the expected result of the average of many measurements of a given quantity. We call this result the **expectation value**; the expectation value of x is denoted by $\langle x \rangle$. Any measurable quantity for which we can calculate the expectation value is called a **physical observable**. The expectation values of physical observables (for example, position, linear momentum, angular momentum, and energy) must be real, because the experimental results of measurements are real.

Expectation value

Physical observables

Let's first review the method of determining average values. Consider a particle that is constrained to move along the x axis. If we make many measurements of the particle, we may find the particle N_1 times at x_1, N_2 times at x_2, N_i times at x_i, and so forth. The average value of x, denoted by $\bar{x}$ [or $(x)_{av}$], is then

$$\bar{x} = \frac{N_1 x_1 + N_2 x_2 + N_3 x_3 + N_4 x_4 + \cdots}{N_1 + N_2 + N_3 + N_4 + \cdots} = \frac{\sum_i N_i x_i}{\sum_i N_i}$$

We can change from discrete to continuous variables by using the probability $P(x, t)$ of observing the particle at a particular x. The previous equation then becomes

$$\bar{x} = \frac{\int_{-\infty}^{\infty} x P(x) \, dx}{\int_{-\infty}^{\infty} P(x) \, dx} \tag{6.17}$$

In quantum mechanics we must use the probability distribution given in Equation (6.6), $P(x) \, dx = \Psi^*(x, t) \Psi(x, t) \, dx$, to determine the average or expectation value. The procedure for finding the expectation value $\langle x \rangle$ is similar to that followed in Equation (6.17):

$$\langle x \rangle = \frac{\int_{-\infty}^{\infty} x \Psi^*(x, t) \Psi(x, t) \, dx}{\int_{-\infty}^{\infty} \Psi^*(x, t) \Psi(x, t) \, dx} \tag{6.18}$$

If the wave function is normalized, the denominator becomes 1. The expectation value is then given by

$$\langle x \rangle = \int_{-\infty}^{\infty} x \Psi^*(x, t) \Psi(x, t) \; dx \qquad (6.19)$$

If the wave function has not been normalized, then Equation (6.18) should be used.

The same general procedure can be used to find the expectation value of any function $g(x)$ for a normalized wave function $\Psi(x, t)$.

$$\langle g(x) \rangle = \int_{-\infty}^{\infty} \Psi^*(x, t) g(x) \Psi(x, t) \; dx \qquad (6.20)$$

We emphasize again that the wave function can provide only the expectation value of a given function $g(x)$ that can be written as a function of x. It cannot give us the value of each individual measurement. When we say the wave function provides a complete description of the system, we mean that the expectation values of the physical observables can be determined.

Any knowledge we might have of the simultaneous values of the position x and momentum p must be consistent with the uncertainty principle. To find the expectation value of p, we first need to represent p in terms of x and t. As an example, let's consider once more the wave function of the free particle, $\Psi(x, t) = e^{i(kx - \omega t)}$. If we take the derivative of $\Psi(x, t)$ with respect to x, we have

$$\frac{\partial \Psi}{\partial x} = \frac{\partial}{\partial x} \left[e^{i(kx - \omega t)} \right] = ik e^{i(kx - \omega t)} = ik\Psi$$

But because, $k = p/\hbar$ this becomes

$$\frac{\partial \Psi}{\partial x} = i \frac{p}{\hbar} \Psi$$

After rearrangement, this yields

$$p[\Psi(x, t)] = -i\hbar \frac{\partial \Psi(x, t)}{\partial x}$$

An **operator** is a mathematical operation that transforms one function into another. For example, an operator, denoted by $\hat{A}$, transforms the function $f(x)$ by $\hat{A}f(x) = g(x)$. In the previous wave function equation, the quantity $-i\hbar(\partial/\partial x)$ is operating on the function $\Psi(x, t)$ and is called the *momentum operator* $\hat{p}$, where the ^ sign over the letter p indicates an operator.

$$\hat{p} = -i\hbar \frac{\partial}{\partial x} \qquad (6.21)$$

The existence of the momentum operator is not unique. Each of the physical observables has an associated operator that is used to find that observable's expectation value. In order to compute the expectation value of some physical observable A, the operator $\hat{A}$ must be placed between Ψ^* and Ψ so that it *operates* on $\Psi(x, t)$ in the order shown:

$$\langle A \rangle = \int_{-\infty}^{\infty} \Psi^*(x, t) \hat{A} \Psi(x, t) \; dx \qquad (6.22)$$

Thus, the expectation value of the momentum becomes

$$\langle p \rangle = -i\hbar \int_{-\infty}^{\infty} \Psi^*(x, t) \frac{\partial \Psi(x, t)}{\partial x} \, dx \tag{6.23}$$

The position x is its own operator. Operators for observables that are functions of both x and p can be constructed from x and $\hat{p}$.

Now let's take the time derivative of the free-particle wave function.

$$\frac{\partial \Psi}{\partial t} = \frac{\partial}{\partial t}\left[e^{i(kx-\omega t)}\right] = -i\omega e^{i(kx-\omega t)} = -i\omega\Psi$$

We substitute $\omega = E/\hbar$, and then rearrange to find

$$E[\Psi(x, t)] = i\hbar \frac{\partial \Psi(x, t)}{\partial t} \tag{6.24}$$

We call the quantity operating on $\Psi(x, t)$ the *energy operator*.

$$\hat{E} = i\hbar \frac{\partial}{\partial t} \tag{6.25}$$

It is used to find the expectation value $\langle E \rangle$ of the energy.

$$\langle E \rangle = i\hbar \int_{-\infty}^{\infty} \Psi^*(x, t) \frac{\partial \Psi(x, t)}{\partial t} \, dx \tag{6.26}$$

Although we have found the momentum and energy operators for only the free-particle wave functions, they are general results. We shall have occasion later to use these operators to determine the physical observables (position, momentum, and energy, for example) and compare with experimental results.

Example 6.6

Use the momentum and energy operators with the conservation of energy to produce the Schrödinger wave equation.

Strategy We first find the energy E as the sum of the kinetic and potential energies. Our treatment is entirely nonrelativistic. We want to use the operator functions, so we write the kinetic energy in terms of momentum and keep the total energy E.

Solution The energy is

$$E = K + V = \frac{p^2}{2m} + V \tag{6.27}$$

We allow the operators of both sides of this equation to act on the wave function. The left side gives

$$\hat{E}\Psi = i\hbar \frac{\partial \Psi}{\partial t} \tag{6.28}$$

The application of the operators on the right side of Equation (6.27) on Ψ gives

$$\left[\frac{1}{2m}(\hat{p})^2 + V\right]\Psi = \frac{1}{2m}\left(-i\hbar \frac{\partial}{\partial x}\right)^2 \Psi + V\Psi$$

$$= -\frac{\hbar^2}{2m} \frac{\partial^2 \Psi}{\partial x^2} + V\Psi$$

Notice that the operator $(\hat{p})^2$ implies two successive applications of the $\hat{p}$ operator, not the algebraic square of one $\hat{p}$ operator. Now we set the previous equation equal to Equation (6.28) and obtain

$$i\hbar \frac{\partial \Psi}{\partial t} = -\frac{\hbar^2}{2m} \frac{\partial^2 \Psi}{\partial x^2} + V\Psi \tag{6.29}$$

which is the time-dependent Schrödinger wave equation, Equation (6.1). It should be noted that this example is not a determination of the Schrödinger wave equation, but rather a verification of the consistency of the definitions.

6.3   Infinite Square-Well Potential

We have thus far established the time-independent Schrödinger wave equation and have discussed how the wave functions can be used to determine the physical observables. Now we would like to find the wave function for several possible potentials and see what we can learn about the behavior of a system having those potentials. In the process of doing this we will find that some observables, including energy, have quantized values. We begin by exploring the simplest such system—that of a particle trapped in a box with infinitely hard walls that the particle cannot penetrate. This is the same physical system as the particle in a box we presented in Section 5.8, but now we present the full quantum-mechanical solution.

The potential, called an *infinite square well*, is shown in Figure 6.2 and is given by

$$V(x) = \begin{cases} \infty & x \le 0,\ x \ge L \\ 0 & 0 < x < L \end{cases} \qquad (6.30)$$

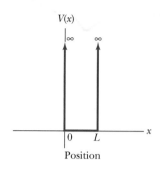

Figure 6.2 Infinite square-well potential. The potential is $V = \infty$ everywhere except the region $0 < x < L$, where $V = 0$.

The particle is constrained to move only between $x = 0$ and $x = L$, where the particle experiences no forces. Although the infinite square-well potential is simple, we will see that it is useful because many physical situations can be approximated by it. We will also see that requiring the wave function to satisfy certain boundary conditions leads to energy quantization. We will use this fact to explore energy levels of simple atomic and nuclear systems.

As we stated previously, most of the situations we encounter allow us to use the time-independent Schrödinger wave equation. Such is the case here. If we insert $V = \infty$ in Equation (6.13), we see that the only possible solution for the wave function is $\psi(x) = 0$. Therefore, there is zero probability for the particle to be located at $x \le 0$ or $x \ge L$. Because the kinetic energy of the particle must be finite, the particle can never penetrate into the region of infinite potential. However, when $V = 0$, Equation (6.13) becomes, after rearranging,

$$\frac{d^2\psi}{dx^2} = -\frac{2mE}{\hbar^2}\psi = -k^2\psi$$

where we have used Equation (6.13) with $V = 0$ and let the wave number $k = \sqrt{2mE/\hbar^2}$. A suitable solution to this equation that satisfies the properties given in Section 6.1 is

$$\psi(x) = A\sin kx + B\cos kx \qquad (6.31)$$

The wave function must be continuous, which means that $\psi(x) = 0$ at both $x = 0$ and $x = L$ as already discussed. The proposed solution in Equation (6.31) therefore must have $B = 0$ in order to have $\psi(x = 0) = 0$. If $\psi(x = L) = 0$, then $A\sin(kL) = 0$, and because $A = 0$ leads to a trivial solution, we must have

$$kL = n\pi \qquad (6.32)$$

where n is a positive integer. The value $n = 0$ leads to $\psi = 0$, a physically uninteresting solution, and negative values of n do not give different physical solutions than the positive values. The wave function is now

$$\psi_n(x) = A\sin\left(\frac{n\pi x}{L}\right) \qquad (n = 1, 2, 3, \ldots) \qquad (6.33)$$

The property that $d\psi/dx$ must be continuous is not satisfied in this case, because of the infinite step value of the potential at $x = 0$ and $x = L$, but we were warned of this particular situation in Section 6.1, and it creates no problem. We normalize our wave function over the total distance $-\infty < x < \infty$.

$$\int_{-\infty}^{\infty} \psi_n^*(x)\psi_n(x)\,dx = 1$$

Substitution of the wave function yields

$$A^2 \int_0^L \sin^2\left(\frac{n\pi x}{L}\right) dx = 1$$

This is a straightforward integral (with the help of integral tables, see Appendix 3) and gives $L/2$, so that $A^2(L/2) = 1$ and $A = \sqrt{2/L}$.

The normalized wave function becomes

$$\psi_n(x) = \sqrt{\frac{2}{L}} \sin\left(\frac{n\pi x}{L}\right) \qquad (n = 1, 2, 3, \ldots) \tag{6.34}$$

These wave functions are identical to the ones obtained for a vibrating string with its ends fixed that are studied in elementary physics. The application of the boundary conditions here corresponds to fitting standing waves into the box. It is not a surprise to obtain standing waves in this case, because we are considering time-independent solutions. Because $k_n = n\pi/L$ from Equation (6.32), we have

$$k_n = \frac{n\pi}{L} = \sqrt{\frac{2mE_n}{\hbar^2}}$$

Notice the subscript n on k_n and E_n denoting that they depend on the integer n and have multiple values. This equation is solved for E_n to yield

Quantized energy levels
$$E_n = n^2 \frac{\pi^2 \hbar^2}{2mL^2} \qquad (n = 1, 2, 3, \ldots) \tag{6.35}$$

The possible energies E_n of the particle are quantized, and the integer n is a quantum number. Notice that the results for the quantized energy levels in Equation (6.35) are identical to those obtained in Equation (5.51) of Section 5.8, when we treated a particle in a one-dimensional box as a wave. The quantization of the energy occurs in a natural way from the application of the boundary conditions (standing waves) to possible solutions of the wave equation. Each wave function $\psi_n(x)$ has associated with it a unique energy E_n. In Figure 6.3 we show the wave function ψ_n, probability density $|\psi_n|^2$, and energy E_n for the lowest three values of n (1, 2, 3).

The lowest energy level given by $n = 1$, is called the *ground state*, and its energy is given by

$$E_1 = \frac{\pi^2 \hbar^2}{2mL^2}$$

Note that the lowest energy cannot be zero because we have ruled out the possibility of $n = 0$ ($\psi_0 = 0$). Classically, the particle can have zero or any positive

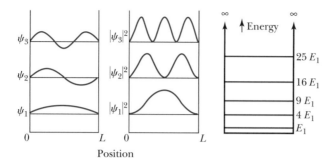

Figure 6.3 Wave functions ψ_n, probability densities $|\psi_n|^2$, and energy levels E_n for the lowest quantum numbers for the infinite square-well potential.

energy. If we calculate E_n for a macroscopic object in a box (for example, a tennis ball in a tennis court), we will obtain a very small number for E_1. Adjacent energy levels would be so close together that we could not measure their differences. Macroscopic objects must have very large values of n.

Classically, the particle has equal probability of being anywhere inside the box. The classical probability density (see Section 6.2) is $P(x) = 1/L$ (for $0 < x < L$, zero elsewhere) in order for the probability to be 1 for the particle to be in the box. According to Bohr's correspondence principle (see Section 4.4), we should obtain the same probability in the region where the classical and quantum results should agree, that is, for large n. The quantum probability density is $(2/L)\sin^2(k_n x)$. For large values of n, there will be many oscillations within the box. The average value of $\sin^2\theta$ over one complete cycle is $1/2$. The average value of $\sin^2\theta$ over many oscillations is also $1/2$. Therefore, the quantum probability approaches $1/L$ in this limit, in agreement with the classical result.

Example 6.7

Show that the wave function $\Psi_n(x, t)$ for a particle in an infinite square well corresponds to a standing wave in the box.

Strategy We have just found the wave function $\psi_n(x)$ in Equation (6.34). According to Equation (6.14), we can obtain $\Psi_n(x, t)$ by multiplying the wave function $\psi_n(x)$ by $e^{-i\omega_n t}$.

Solution The product of $\psi_n(x)$ from Equation (6.34) and $f(t) = e^{-i\omega_n t}$ gives

$$\Psi_n(x, t) = \sqrt{\frac{2}{L}}\sin(k_n x)e^{-i\omega_n t}$$

We can write $\sin(k_n x)$ as

$$\sin(k_n x) = \frac{e^{ik_n x} - e^{-ik_n x}}{2i}$$

so that the wave function* becomes

$$\Psi_n(x, t) = \sqrt{\frac{2}{L}}\frac{e^{i(k_n x - \omega_n t)} - e^{-i(k_n x + \omega_n t)}}{2i}$$

This is the equation of a standing wave for a vibrating string, for example. It is the superposition of a wave traveling to the right with a wave traveling to the left. They interfere to produce a standing wave of angular frequency ω_n.

*The imaginary number i should be of no concern, because the probability values are determined by a product of $\psi^*\psi$, which gives a real number.

Example 6.8

Determine the expectation values for x, x^2, p, and p^2 of a particle in an infinite square well for the first excited state.

Strategy The first excited state corresponds to $n = 2$, because $n = 1$ corresponds to the lowest energy state or the ground state. Equation (6.34) gives us the wave function that we need to find the expectation values given in Section 6.2.

Solution The wave function for this case, according to Equation (6.34), is

$$\psi_2(x) = \sqrt{\frac{2}{L}} \sin\left(\frac{2\pi x}{L}\right)$$

The expectation value $\langle x \rangle_{n=2}$ is

$$\langle x \rangle_{n=2} = \frac{2}{L}\int_0^L x \sin^2\left(\frac{2\pi x}{L}\right) dx = \frac{L}{2}$$

We evaluate all these integrations by looking up the integral in Appendix 3. As we expect, the average position of the particle is in the middle of the box ($x = L/2$), even though the actual probability of the particle being there is zero (see $|\psi|^2$ in Figure 6.3).

The expectation value $\langle x^2 \rangle_{n=2}$ of the square of the position is given by

$$\langle x^2 \rangle_{n=3} = \frac{2}{L}\int_0^L x^2 \sin^2\left(\frac{2\pi x}{L}\right) dx = 0.32L^2$$

The value of $\sqrt{\langle x^2 \rangle_{n=2}}$ is $0.57L$, larger than $\langle x \rangle_{n=2} = 0.5L$. Does this seem reasonable? (*Hint:* look again at the shape of the wave function in Figure 6.3.)

The expectation value $\langle p \rangle_{n=2}$ is determined by using Equation (6.23).

$$\langle p \rangle_{n=2} = (-i\hbar)\frac{2}{L}\int_0^L \sin\left(\frac{2\pi x}{L}\right)\left[\frac{d}{dx}\sin\left(\frac{2\pi x}{L}\right)\right] dx$$

which reduces to

$$\langle p \rangle_{n=2} = -\frac{4i\hbar}{L^2}\int_0^L \sin\left(\frac{2\pi x}{L}\right)\cos\left(\frac{2\pi x}{L}\right) dx = 0$$

Because the particle is moving left as often as right in the box, the average momentum is zero.

The expectation value $\langle p^2 \rangle_{n=2}$ is given by

$$\langle p^2 \rangle_{n=2} = \frac{2}{L}\int_0^L \sin\left(\frac{2\pi x}{L}\right)\left(-i\hbar\frac{d}{dx}\right)\left(-i\hbar\frac{d}{dx}\right)\sin\left(\frac{2\pi x}{L}\right) dx$$

$$= (-i\hbar)^2\frac{2}{L}\int_0^L \sin\left(\frac{2\pi x}{L}\right)\left(\frac{2\pi}{L}\frac{d}{dx}\right)\cos\left(\frac{2\pi x}{L}\right) dx$$

$$= -(-\hbar^2)\frac{8\pi^2}{L^3}\int_0^L \sin\left(\frac{2\pi x}{L}\right)\sin\left(\frac{2\pi x}{L}\right) dx$$

$$= \frac{4\pi^2\hbar^2}{L^2}$$

This value can be compared with E_2 [Equation (6.35)]:

$$E_2 = \frac{4\pi^2\hbar^2}{2mL^2} = \frac{\langle p^2 \rangle_{n=2}}{2m}$$

which is correct, because nonrelativistically we have $E = p^2/2m + V$ and $V = 0$.

Example 6.9

A typical diameter of a nucleus is about 10^{-14} m. Use the infinite square-well potential to calculate the transition energy from the first excited state to the ground state for a proton confined to the nucleus. Of course, this is only a rough calculation for a proton in a nucleus.

Strategy In order to find the transition energy between the ground and first excited energy states, we use Equation (6.35) to find E_1 and E_2.

Solution The energy of the ground state, from Equation (6.35), is

$$E_1 = \frac{\pi^2\hbar^2c^2}{2mc^2L^2} = \frac{1}{mc^2}\frac{\pi^2(197.3 \text{ eV}\cdot\text{nm})^2}{2(10^{-5} \text{ nm})^2}$$

$$= \frac{1}{mc^2}(1.92 \times 10^{15} \text{ eV}^2)$$

The mass of the proton is $938.3 \text{ MeV}/c^2$, which gives

$$E_1 = \frac{1.92 \times 10^{15} \text{ eV}^2}{938.3 \times 10^6 \text{ eV}} = 2.0 \text{ MeV}$$

The first excited state energy is found [again from Equation (6.35)] to be $E_2 = 4E_1 = 8$ MeV, and the transition energy is $\Delta E = E_2 - E_1 = 6$ MeV. This is a reasonable value for protons in the nucleus.

If we had done a calculation, similar to that in the previous example, for an electron in the nucleus, we would find energies on the order of 10^4 MeV, much larger than the rest energy of the electron. A correct relativistic treatment is necessary, and it would give electron energies significantly less than 10^4 MeV but still much larger than those of electrons actually observed being emitted from the nucleus in β decay. Such reasoning indicates that electrons do not exist inside the nucleus.

6.4 Finite Square-Well Potential

We gained some experience in the last section in dealing with the time-independent Schrödinger wave equation. Now we want to look at a more realistic potential—one that is not infinite. The finite square-well potential is similar to the infinite one, but we let the potential be V_0 rather than infinite in the region $x \leq 0$ and $x \geq L$.

$$V(x) = \begin{cases} V_0 & x \leq 0 & \text{region I} \\ 0 & 0 < x < L & \text{region II} \\ V_0 & x \geq L & \text{region III} \end{cases} \qquad (6.36)$$

The three regions of the potential are shown in Figure 6.4. We will consider a particle of energy $E < V_0$ that classically is bound inside the well. We will find that quantum mechanics *allows the particle to be outside the well*. We set the potential $V = V_0$ in the time-independent Schrödinger Equation (6.13) for regions I and III outside the square well. This gives

$$-\frac{\hbar^2}{2m}\frac{1}{\psi}\frac{d^2\psi}{dx^2} = E - V_0 \qquad \text{regions I, III} \qquad (6.37)$$

We rewrite this using $\alpha^2 = 2m(V_0 - E)/\hbar^2$, a positive constant.

$$\frac{d^2\psi}{dx^2} = \alpha^2\psi$$

The solution to this differential equation has exponentials of the form $e^{\alpha x}$ and $e^{-\alpha x}$. In the region $x > L$, we can reject the positive exponential term, because it would become infinite as $x \rightarrow \infty$. Similarly, the negative exponential can be rejected for $x < 0$. The wave functions become

$$\psi_I(x) = Ae^{\alpha x} \qquad \text{region I, } x < 0 \qquad (6.38)$$

$$\psi_{III}(x) = Be^{-\alpha x} \qquad \text{region III, } x > L \qquad (6.39)$$

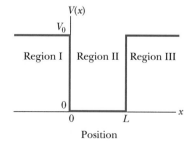

Region I   Region II   Region III

Position

Figure 6.4 A finite square-well potential has the value V_0 everywhere except $0 < x < L$, where $V = 0$. The three regions I, II, and III are indicated.

Figure 6.5 The energy levels E_n and wave functions ψ_n for the lowest quantum numbers for the finite square-well potential. Notice that ψ extends past $x < 0$ and $x > L$, where classically the particle is forbidden. *From R. Eisberg and R. Resnick*, Quantum Physics of Atoms, Molecules, Solids, Nuclei, and Particles, *2nd ed., New York: Wiley (1985).*

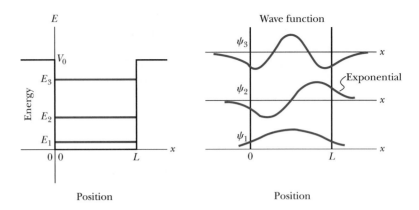

Inside the square well, where the potential V is zero, the wave equation becomes

$$\frac{d^2\psi}{dx^2} = -k^2\psi$$

where $k = \sqrt{(2mE)/\hbar^2}$. Instead of a sinusoidal solution, we can write it as

$$\psi_{II} = Ce^{ikx} + De^{-ikx} \qquad \text{region II, } 0 < x < L \qquad (6.40)$$

We now want to satisfy the boundary-value properties listed in Section 6.1. We have already made sure that all but properties 2 and 3 have been satisfied. The wave functions are finite everywhere, even at infinity. In order for the wave functions to be single valued, we must have $\psi_I = \psi_{II}$ at $x = 0$ and $\psi_{II} = \psi_{III}$ at $x = L$. Both ψ and $\partial\psi/\partial x$ must be continuous at $x = 0$ and $x = L$. We will not perform these tedious procedures here, but the results for the wave functions are presented graphically in Figure 6.5.

The application of the boundary conditions leads to quantized energy values E_n and to particular wave functions $\psi_n(x)$. One remarkable result is that the particle has a finite probability of being outside the square well, as indicated by Figure 6.5. Notice that the wave functions join smoothly at the edges of the well and approach zero exponentially outside the well.

What other differences can we easily discern between the infinite and finite square well? For example, by examination of Figures 6.5 and 6.3, we can see that the de Broglie wavelength is larger for the finite square well because the waves extend past the square well. This in turn leads to a smaller momentum and lower energy levels. The number of energy levels will, of course, be limited because of the potential height V_0 (see Figure 6.5). When $E > V_0$ the particle is unbound, a situation that will be discussed in Section 6.7.

The occurrence of the particle outside the square well is clearly prohibited classically, but it occurs naturally in quantum mechanics. Note that because of the exponential decrease of the wave functions ψ_I and ψ_{III}, the probability of the particle penetrating a distance greater than $\delta x \approx 1/\alpha$ begins to decrease markedly.

$$\delta x \approx \frac{1}{\alpha} = \frac{\hbar}{\sqrt{2m(V_0 - E)}} \qquad (6.41)$$

We call δx the *penetration depth*. However, later we will find values of δx as large as $10/\alpha$ and $20/\alpha$ for electrons tunneling through semiconductors (Example 6.14) and for nuclear alpha decay (Example 6.17), respectively. The fraction of particles that successfully tunnel through in these cases is exceedingly small, but the results have important applications, especially in electronics.

It should not be surprising to find that the penetration distance that violates classical physics is proportional to Planck's constant $\hbar$. This result is also consistent with the uncertainty principle because in order for the particle to be in the barrier region, the uncertainty ΔE of the energy must be very large. According to the uncertainty principle ($\Delta E \, \Delta t \geq \hbar/2$), this can occur for only a very short period of time Δt.

6.5 Three-Dimensional Infinite-Potential Well

In order to use quantum theory to solve the atomic physics problems that we shall face in Chapters 7 and 8, it is necessary to extend the Schrödinger equation to three dimensions. This is easily accomplished with the operator notation already developed in Section 6.2. After obtaining the three-dimensional equation, we shall use it to study the problem of a three-dimensional infinite-potential well.

We anticipate that there will be time-independent solutions, so we shall start with the time-independent Schrödinger wave equation. The wave function ψ must be a function of all three spatial coordinates, that is, $\psi = \psi(x, y, z)$. We could just directly modify Equation (6.13) to three dimensions, but there is a simple method to arrive at the Schrödinger equation. We begin with the conservation of energy.

$$E = K + V = \frac{p^2}{2m} + V$$

We multiply this equation times the wave function ψ, which gives

$$\frac{p^2}{2m}\psi + V\psi = E\psi \tag{6.42}$$

We now use Equation (6.21) to express p^2 as an operator to act on ψ. But because $p^2 = p_x{}^2 + p_y{}^2 + p_z{}^2$, we must apply the momentum operator in all three dimensions.

$$\hat{p}_x\psi = -i\hbar\frac{\partial\psi}{\partial x}$$

$$\hat{p}_y\psi = -i\hbar\frac{\partial\psi}{\partial y}$$

$$\hat{p}_z\psi = -i\hbar\frac{\partial\psi}{\partial z}$$

The application of $\hat{p}^2$ in Equation (6.42) gives

$$-\frac{\hbar^2}{2m}\left(\frac{\partial^2\psi}{\partial x^2} + \frac{\partial^2\psi}{\partial y^2} + \frac{\partial^2\psi}{\partial z^2}\right) + V\psi = E\psi \tag{6.43}$$

Time-independent Schrödinger wave equation in three dimensions

This is the time-independent Schrödinger wave equation in three dimensions.

You may recognize the expression in parentheses as the Laplacian operator in mathematics. It is usually written with the shorthand notation

$$\nabla^2 = \frac{\partial^2}{\partial x^2} + \frac{\partial^2}{\partial y^2} + \frac{\partial^2}{\partial z^2} \tag{6.44}$$

With this notation, we can write

$$-\frac{\hbar^2}{2m}\nabla^2\psi + V\psi = E\psi \tag{6.45}$$

Example 6.10

Consider a free particle inside a box with lengths L_1, L_2, and L_3 along the x, y, and z axes, respectively, as shown in Figure 6.6. The particle is constrained to be inside the box. Find the wave functions and energies. Then find the ground energy and wave function and the energy of the first excited state for a cube of sides L.

Strategy We employ some of the same strategies to solve this problem as we used for the one-dimensional case. First, because we are considering the walls of the box to be absolutely closed, they are infinite potential barriers, and the wave function ψ must be zero at the walls and beyond. We expect to see standing waves similar to Equation (6.31).

But how should we write the wave function so as to properly include the x, y, and z dependence of the wave function? In this case the mathematics will follow from the physics. The particle is free within the box. Therefore, the x-, y-, and z-dependent parts of the wave function must be independent of each other. Inside the box $V = 0$, so the wave equation we must solve is

$$-\frac{\hbar^2}{2m}\nabla^2\psi = E\psi \tag{6.46}$$

It is therefore reasonable to try a wave function of the form

$$\psi(x, y, z) = A \sin(k_1 x)\sin(k_2 y)\sin(k_3 z) \tag{6.47}$$

where A is a normalization constant. The quantities k_i ($i = 1, 2, 3$) are determined by applying the appropriate boundary conditions. To find the energies, we substitute the wave function into the Schrödinger equation and solve for E.

Solution The condition that $\psi = 0$ at $x = L_1$ requires that $k_1 L_1 = n_1\pi$ or $k_1 = n_1\pi/L_1$. The values for the k_i are

$$k_1 = \frac{n_1\pi}{L_1} \qquad k_2 = \frac{n_2\pi}{L_2} \qquad k_3 = \frac{n_3\pi}{L_3} \tag{6.48}$$

where n_1, n_2, and n_3 are integers. Not surprisingly, we have found that in three dimensions, it is necessary to use *three* quantum numbers to describe the physical state.

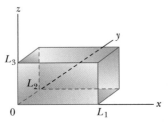

Figure 6.6 A three-dimensional box that contains a free particle. The potential is infinite outside the box, so the particle is constrained to be inside the box.

In order to find the energies using Equation (6.43), we first need to take the appropriate derivatives of the wave function. We do this first for the variable x.

$$\frac{\partial\psi}{\partial x} = \frac{\partial}{\partial x}[A \sin(k_1 x)\sin(k_2 y)\sin(k_3 z)]$$

$$= k_1 A \cos(k_1 x)\sin(k_2 y)\sin(k_3 z)$$

$$\frac{\partial^2\psi}{\partial x^2} = \frac{\partial}{\partial x}[k_1 A \cos(k_1 x)\sin(k_2 y)\sin(k_3 z)]$$

$$= -(k_1)^2 A \sin(k_1 x)\sin(k_2 y)\sin(k_3 z)$$

$$= -k_1^2\psi$$

The derivatives for y and z are similar, and Equation (6.43) becomes

$$\frac{\hbar^2}{2m}(k_1^2 + k_2^2 + k_3^2)\psi = E\psi$$

This gives

$$E = \frac{\hbar^2}{2m}(k_1^2 + k_2^2 + k_3^2)$$

We substitute the values of k_i from Equation (6.48) in this equation to obtain

$$E = \frac{\pi^2 \hbar^2}{2m}\left(\frac{n_1^2}{L_1^2} + \frac{n_2^2}{L_2^2} + \frac{n_3^2}{L_3^2}\right) \qquad (6.49)$$

The allowed energy values depend on the values of the three quantum numbers n_1, n_2, and n_3.

Let us now consider the special case of the *cubical* box, with $L_1 = L_2 = L_3 = L$. The energy values of Equation (6.49) can be expressed as

$$E = \frac{\pi^2 \hbar^2}{2mL^2}(n_1^2 + n_2^2 + n_3^2) \qquad (6.50)$$

For the ground state we have $n_1 = n_2 = n_3 = 1$, so the ground state energy is

$$E_{gs} = \frac{3\pi^2 \hbar^2}{2mL^2} \qquad (6.51)$$

and the ground state wave function is

$$\psi_{gs} = A \sin\left(\frac{\pi x}{L}\right)\sin\left(\frac{\pi y}{L}\right)\sin\left(\frac{\pi z}{L}\right) \qquad (6.52)$$

What is the energy of the first excited state? Higher values of the quantum numbers n_i correspond to higher energies; therefore, it is logical to try something like $n_1 = 2, n_2 = 1$, and $n_3 = 1$. But we could just as well assign quantum numbers $n_1 = 1, n_2 = 2, n_3 = 1$ to the first excited state, or $n_1 = 1, n_2 = 1, n_3 = 2$. In each of these cases the total energy is

$$E_{1st} = \frac{\pi^2 \hbar^2}{2mL^2}(2^2 + 1^2 + 1^2) = \frac{3\pi^2 \hbar^2}{mL^2}$$

In physics we say that a given state is **degenerate** when there is more than one wave function for a given energy. We have this situation in Example 6.10, where all three possible wave functions for the first excited state have the same energy. The degeneracy in this case is a result of the symmetry of the cube. If the box had sides of three different lengths, we say the *degeneracy is removed*, because the three quantum numbers in different orders (211, 121, 112) would result in three different energies. Degeneracy is not a new phenomenon. It also occurs in classical physics, for example, in planetary motion, where orbits with different eccentricities may have the same energy. Degeneracy results from particular properties of the potential energy function that describes the system. A *perturbation* of the potential energy can remove the degeneracy. Energy levels can be split (and the degeneracy removed) by applying external magnetic fields (Zeeman effect, Section 7.4) and external electric fields (Stark effect, discussed in the Chapter 8 Special Topic, "Rydberg Atoms").

Degenerate state

6.6 Simple Harmonic Oscillator

Because of their common occurrence in nature, we now want to examine simple harmonic oscillators. In introductory physics you studied the case of a mass oscillating in one dimension on the end of a spring. Consider a spring having spring constant* κ that is in equilibrium at $x = x_0$. The restoring force (see Figure 6.7a) along the x direction is $F = -\kappa(x - x_0)$, and the potential energy stored in the spring is $V = \kappa(x - x_0)^2/2$ (see Figure 6.7b). The resulting motion is called *simple harmonic motion* (SHM), and the equations describing it are well known.

Besides springs and pendula (small oscillations), many phenomena in nature can be approximated by SHM, for example, diatomic molecules and atoms in a solid lattice of atoms. Systems can also be approximated by SHM in a general way. As an example, consider a lattice in which the force on the atoms depends on the distance x from some equilibrium position x_0. If we expand the potential in a Taylor series in terms of the distance $(x - x_0)$ from equilibrium, we obtain

---

*We let κ be the spring constant in this section rather than the normal k to avoid confusion with the wave number. It is important to note the context in which variables like k and κ are used, because either might be used as wave number or spring constant.

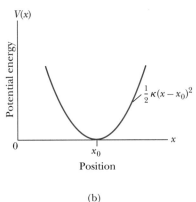

Figure 6.7 (a) The restoring force for a spring having a spring constant κ is $F = -\kappa(x - x_0)$. (b) The potential energy has the form $\kappa(x - x_0)^2/2$.

(a)

(b)

$$V(x) = V_0 + V_1(x - x_0) + \tfrac{1}{2}V_2(x - x_0)^2 + \cdots \tag{6.53}$$

where V_0, V_1, and V_2 are constants, and we have kept only the three lowest terms of the series, because $(x - x_0) \approx 0$ for small excursions from the equilibrium position x_0. At $x = x_0$ we have a minimum of the potential, so $dV/dx = 0$ at $x = x_0$. This requires that $V_1 = 0$, and if we redefine the zero of potential energy to require $V_0 = 0$, then the lowest term of the potential $V(x)$ is

$$V(x) = \tfrac{1}{2}V_2(x - x_0)^2$$

This is the origin of the $V = \kappa x^2/2$ potential energy term that occurs so often. Near the equilibrium position a parabolic form as displayed in Figure 6.8 may approximate many potentials.

We want to study the quantum description of simple harmonic motion by inserting a potential $V = \kappa x^2/2$ (we let $x_0 = 0$, see Figure 6.9a) into Equation (6.13), the time-independent Schrödinger wave equation.

$$\frac{d^2\psi}{dx^2} = -\frac{2m}{\hbar^2}\left(E - \frac{\kappa x^2}{2}\right)\psi = \left(-\frac{2mE}{\hbar^2} + \frac{m\kappa x^2}{\hbar^2}\right)\psi \tag{6.54}$$

If we let

$$\alpha^2 = \frac{m\kappa}{\hbar^2} \tag{6.55a}$$

and

$$\beta = \frac{2mE}{\hbar^2} \tag{6.55b}$$

Figure 6.8 Many potentials in nature can be approximated near their equilibrium position by the simple harmonic potential (black dashed curve). Such is the case here for the potential energy $V(r)$ of a diatomic molecule near its equilibrium position r_0 (blue curve).

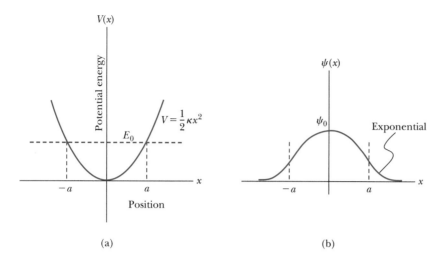

then

$$\frac{d^2\psi}{dx^2} = (\alpha^2 x^2 - \beta)\psi \tag{6.56}$$

Before discussing the solution of Equation (6.56), let us first examine what we can learn about the problem qualitatively. Because the particle is confined to the potential well, centered at $x = 0$, it has zero probability of being at $x = \infty$. Therefore, $\psi(x) \to 0$ as $x \to \infty$.

What is the lowest energy level possible for the harmonic oscillator? Is $E = 0$ possible? If $E = 0$, then $x = 0$ and $V = 0$ in order to allow $E \geq V$. But if E and V are zero, then the kinetic energy $K = 0$, and the momentum $p = 0$. Simultaneously having both $x = 0$ and $p = 0$ (that is, both x and p are known exactly) violates the uncertainty principle. Therefore, the minimum energy E cannot be zero. In fact, the energy levels must all be positive, because $E > V \geq 0$. The state having the lowest energy, denoted here by E_0, will have an energy like that shown in Figure 6.9a, and the wave function ψ_0 for that state will most likely be a simple wave fitting inside the region defined by the potential (see Figure 6.9b). Let $E_0 = V_0 = \kappa a^2/2$. The distances $\pm a$ define the classical limits of the particle, but we know from the previous section that the particle has a small probability of being outside the potential well dimensions of $\pm a$. Therefore, the wave function will not be zero at $x = \pm a$, but will have a finite value that decreases rapidly to zero on the other side of the barrier. Thus a plausible guess for the lowest-order wave function ψ_0 is like that shown in Figure 6.9b. We shall find the minimum energy E_0, called the *zero-point energy*, in the following example.

Example 6.11

Estimate the minimum energy of the simple harmonic oscillator allowed by the uncertainty principle.

Strategy In introductory physics you learned that the average kinetic energy is equal to the average potential energy for simple harmonic oscillators over the range of motion (from $-x$ to $+x$), and both the average potential and kinetic energies are equal to one half the total energy. By relating the mean square deviation values to the uncertainty values Δx and Δp, we will determine the minimum energy.

222   Chapter 6   Quantum Mechanics II

Solution The energies are related by

$$K_{av} = \frac{1}{2}E = \frac{1}{2}\kappa(x^2)_{av} = \frac{1}{2m}(p^2)_{av}$$

The mean value of x is zero, but the mean value of $(x^2)_{av}$ is the mean square deviation $(\Delta x)^2$. Similarly, $(p^2)_{av} = (\Delta p)^2$. From the previous equation, we therefore have the energy $E = \kappa(\Delta x)^2 = (\Delta p)^2/m$ and, as a result, we must have $\Delta x = \Delta p/\sqrt{m\kappa}$. From the uncertainty principle we have $\Delta p \Delta x \geq \hbar/2$, so the minimum value of $\Delta x = \hbar/(2\Delta p)$. Now we have for the lowest energy E_0

$$E = \kappa(x^2)_{av} = \kappa(\Delta x)^2 = \kappa\left(\frac{\Delta p}{\sqrt{m\kappa}}\right)\left(\frac{\hbar}{2\Delta p}\right)$$

$$E_0 = \frac{\hbar}{2}\sqrt{\frac{\kappa}{m}} = \frac{\hbar\omega}{2}$$

Our estimate for the zero-point energy of the harmonic oscillator is $\hbar\omega/2$. This agrees with the zero-point energy found by more rigorous means.

The zero-point energy is not just a curious oddity. For example, the zero-point energy for ^{4}He is large enough to prevent liquid ^{4}He from freezing at atmospheric pressure, no matter how cold the system, even near 0 K.

The wave function solutions ψ_n for Equation (6.56) are

$$\psi_n = H_n(x)e^{-\alpha x^2/2} \tag{6.57}$$

where $H_n(x)$ are polynomials of order n, and n is an integer ≥ 0. The functions $H_n(x)$ are related by a constant to the *Hermite polynomial functions* tabulated in many quantum mechanics books. The first few values of ψ_n and $|\psi_n|^2$ are shown in Figure 6.10. In contrast to the particle in a box, where the oscillatory wave function is a sinusoidal curve, in this case the oscillatory behavior is due to the polynomial, which dominates at small x. The exponential tail is provided by the Gaussian function, which dominates at large x.

The energy levels are given by

$$E_n = (n + \tfrac{1}{2})\hbar\sqrt{\kappa/m} = (n + \tfrac{1}{2})\hbar\omega \tag{6.58}$$

where $\omega^2 = \kappa/m$, and ω is the classical angular frequency. From Equation (6.58) we see that the *zero-point energy* E_0 is

$$E_0 = \frac{1}{2}\hbar\omega \tag{6.59}$$

Notice that this result for E_0 is precisely the value found in Example 6.11 by using the uncertainty principle. The uncertainty principle is responsible for the minimum energy of the simple harmonic oscillator. In Section 5.6 we mentioned that the minimum value (that is, the equality sign) of the uncertainty principle is found for Gaussian wave packets. We note here that the wave functions for the simple harmonic oscillators are of just the Gaussian form (see Figure 6.10). The minimum energy E_0 allowed by the uncertainty principle, sometimes called the *Heisenberg limit*, is found for the ground state of the simple harmonic oscillator.

Finally, let us compare the motion as described by classical and quantum theory. We recall the classical motion of the mass at the end of a spring. The speed is greatest as it passes through its equilibrium position. The speed is lowest (zero) at the two ends (compressed or extended positions of the spring), when the mass stops and reverses direction. Classically, the probability of finding the mass is greatest at the ends of motion and smallest at the center (that is, proportional to the amount of time the mass spends at each position). The classical probability is shown by the black dashed line in Figure 6.11.

The quantum theory probability density for the lowest energy state (ψ_0^2, see Figure 6.10) is contrary to the classical one. The largest probability for this lowest energy state is for the particle to be at the center. We are not surprised to see such a marked difference between classical and quantum predictions (see Section 4.4).

(a)

Wave functions

$$\psi_3(x) = \left(\frac{\alpha}{\pi}\right)^{1/4} \frac{1}{\sqrt{3}} \left(\sqrt{\alpha}\,x\right)\left(2\alpha x^2 - 3\right) e^{-\alpha x^2/2}$$

$$\psi_2(x) = \left(\frac{\alpha}{\pi}\right)^{1/4} \frac{1}{\sqrt{2}} \left(2\alpha x^2 - 1\right) e^{-\alpha x^2/2}$$

$$\psi_1(x) = \left(\frac{\alpha}{\pi}\right)^{1/4} \sqrt{2\alpha}\; x e^{-\alpha x^2/2}$$

$$\psi_0(x) = \left(\frac{\alpha}{\pi}\right)^{1/4} e^{-\alpha x^2/2}$$

(b)

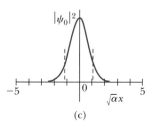

(c)

Figure 6.10 Results for simple harmonic oscillator potential. (a) The energy levels for the lowest four energy states are shown with the corresponding wave functions listed. (b) The wave functions for the four lowest energy states are displayed. Notice that even quantum numbers have symmetric $\psi_n(x)$, and the odd quantum numbers have antisymmetric $\psi_n(x)$. (c) The probability densities $|\psi_n|^2$ for the lowest four energy states are displayed.

However, from the correspondence principle we would expect the classical and quantum probabilities to be similar as the quantum number n becomes very large. In Figure 6.11 we show ψ_n^2 for the case of $n = 10$, and we see that the average probabilities become similar. As n continues to increase, the peaks and valleys of the quantum probabilities are hardly observable, and the average value approaches the classical result.

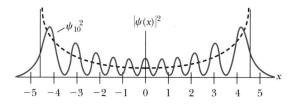

Figure 6.11 The probability distribution $|\psi_{10}|^2$ for the $n = 10$ state is compared with the classical probability (dashed line). As n increases, the two probability distributions become more similar.

Example 6.12

Normalize the ground state wave function ψ_0 for the simple harmonic oscillator and find the expectation values $\langle x \rangle$ and $\langle x^2 \rangle$.

Strategy We can use Equation (5.49) for the normalization process. Let's assume that all we know about the wave function ψ_0 is the form given in Equation (6.57). $H_0(x)$ has no dependence on x, so we take it to be a constant A. We find the expectation values as discussed in Section 6.2.

Solution If we let $H_0(x) = A$, the ground state wave function becomes

$$\psi_0(x) = Ae^{-\alpha x^2/2}$$

The normalization determines A.

$$\int_{-\infty}^{\infty} \psi_0^*(x)\psi_0(x) \, dx = 1$$

$$A^2 \int_{-\infty}^{\infty} e^{-\alpha x^2} \, dx = 1$$

$$2A^2 \int_{0}^{\infty} e^{-\alpha x^2} \, dx = 1$$

We determine this integral with the help of integral tables (see Appendix 6), with the result

$$2A^2 \left(\frac{1}{2}\sqrt{\frac{\pi}{\alpha}} \right) = 1$$

$$A^2 = \sqrt{\frac{\alpha}{\pi}}$$

$$A = \left(\frac{\alpha}{\pi} \right)^{1/4}$$

For the ground state wave function, this gives

$$\psi_0(x) = \left(\frac{\alpha}{\pi} \right)^{1/4} e^{-\alpha x^2/2} \tag{6.60}$$

This is precisely the wave function given in Figure 6.10 and is of the Gaussian form.

The expectation value of x is

$$\langle x \rangle = \int_{-\infty}^{\infty} \psi_0^*(x)\, x \psi_0(x) \, dx$$

$$= \sqrt{\frac{\alpha}{\pi}} \int_{-\infty}^{\infty} x e^{-\alpha x^2} \, dx$$

The value of $\langle x \rangle$ must be zero, because we are integrating an odd function of x over symmetric limits from $-\infty$ to $+\infty$ (see Appendix 6). Both classical and quantum mechanics predict the average value of x to be zero because of the symmetric nature of the potential, $\kappa x^2/2$.

The expectation value $\langle x^2 \rangle$, however, must be positive, because x^2 is never negative.

$$\langle x^2 \rangle = \int_{-\infty}^{\infty} \psi_0^*(x)\, x^2 \psi_0(x) \, dx$$

$$= \sqrt{\frac{\alpha}{\pi}} \int_{-\infty}^{\infty} x^2 e^{-\alpha x^2} \, dx$$

$$= 2\sqrt{\frac{\alpha}{\pi}} \int_{0}^{\infty} x^2 e^{-\alpha x^2} \, dx$$

This integral can be found in a table of integrals (see Appendix 6), and the result is

$$\langle x^2 \rangle = 2\sqrt{\frac{\alpha}{\pi}} \left(\frac{\sqrt{\pi}}{4\alpha^{3/2}} \right) = \frac{1}{2\alpha}$$

Inserting the value of the constant α from Equation (6.55a) gives

$$\langle x^2 \rangle = \frac{\hbar}{2\sqrt{m\kappa}}$$

Because $\omega = \sqrt{\kappa/m}$, we have

$$\langle x^2 \rangle = \frac{\hbar}{2m\omega} \tag{6.61}$$

In Example 6.11 we argued that

$$(x^2)_{av} = (\Delta x)^2 = \frac{E_0}{\kappa}$$

and showed that $E_0 = \hbar\omega/2$, the minimum energy allowed by the uncertainty principle. We can now see that these results are consistent, because

$$\langle x^2 \rangle = (x^2)_{av} = \frac{E_0}{\kappa} = \frac{\hbar\omega}{2\kappa} = \frac{\hbar\omega}{2m\omega^2} = \frac{\hbar}{2m\omega}$$

as we determined in Equation (6.61).

6.7 Barriers and Tunneling

The methods we have already seen for studying particles in potential wells can be applied to the problem of a particle approaching a potential barrier. As you will see, potential barriers are real physical phenomena, and our results have immediate applications.

Potential Barrier with $E > V_0$

Consider a particle of energy E approaching a potential barrier of height V_0 for $0 < x < L$. The potential elsewhere is zero. First, let us consider the case where the particle's energy is $E > V_0$ as shown in Figure 6.12. Classically we know the particle would pass the barrier, moving with reduced velocity in the region of V_0 ($K = mv^2/2 = E - V_0$, rather than $K = mv^2/2 = E$). On the other side of the barrier, where $V = 0$, the particle will have its original velocity again. According to quantum mechanics, the particle will behave differently because of its wavelike character. In regions I and III (where $V = 0$) the wave numbers are

$$k_\mathrm{I} = k_\mathrm{III} = \frac{\sqrt{2mE}}{\hbar} \qquad \text{where } V = 0 \qquad (6.62a)$$

In the barrier region, however, we have

$$k_\mathrm{II} = \frac{\sqrt{2m(E - V_0)}}{\hbar} \qquad \text{where } V = V_0 \qquad (6.62b)$$

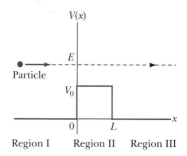

Figure 6.12 A particle having energy E approaches a potential barrier of width L and height V_0 with $E > V_0$. The one-dimensional space is divided into three regions as shown.

We consider an analogy with optics. When light in air penetrates another medium (for example, glass), the wavelength changes because of the index of refraction. Some of the light will be reflected, and some will be transmitted into the medium. Because we must consider the wave behavior of particles interacting with potential barriers, we might expect similar behavior. The wave function will consist of an incident wave, a reflected wave, and a transmitted wave (see Figure 6.13). These wave functions can be determined by solving the Schrödinger wave equation, subject to appropriate boundary conditions. The difference from classical wave theories is that the wave function allows us to compute only probabilities.

Classical mechanics allows *no* reflection if $E > V_0$ and *total* reflection for $E < V_0$. Quantum mechanics predicts almost total *transmission* for $E \gg V_0$ and almost complete *reflection* for $E \ll V_0$. In the regime where E is comparable to V_0, unusual nonclassical phenomena may appear.

The potentials and the Schrödinger equation for the three regions are as follows:

Region I ($x < 0$) $V = 0$ $\dfrac{d^2\psi_\mathrm{I}}{dx^2} + \dfrac{2m}{\hbar^2}E\psi_\mathrm{I} = 0$

Region II ($0 < x < L$) $V = V_0$ $\dfrac{d^2\psi_\mathrm{II}}{dx^2} + \dfrac{2m}{\hbar^2}(E - V_0)\psi_\mathrm{II} = 0$

Region III ($x > L$) $V = 0$ $\dfrac{d^2\psi_\mathrm{III}}{dx^2} + \dfrac{2m}{\hbar^2}E\psi_\mathrm{III} = 0$

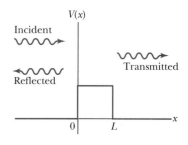

Figure 6.13 The incident particle in Figure 6.12 can be either transmitted or reflected.

The wave functions obtained for these equations are

Region I ($x < 0$)	$\psi_I = Ae^{ik_1x} + Be^{-ik_1x}$	(6.63a)
Region II ($0 < x < L$)	$\psi_{II} = Ce^{ik_{II}x} + De^{-ik_{II}x}$	(6.63b)
Region III ($x > L$)	$\psi_{III} = Fe^{ik_1x} + Ge^{-ik_1x}$	(6.63c)

We assume that we have incident particles coming from the left moving along the $+x$ direction. In this case the term Ae^{ik_1x} in region I represents the incident particles. The term Be^{-ik_1x} represents the reflected particles moving in the $-x$ direction. In region III there are no particles initially moving along the $-x$ direction, so the only particles present must be those transmitted through the barrier. Thus $G = 0$, and the only term in region III is Fe^{ik_1x}. We summarize these wave functions:

Incident wave	$\psi_I(\text{incident}) = Ae^{ik_1x}$	(6.64a)
Reflected wave	$\psi_I(\text{reflected}) = Be^{-ik_1x}$	(6.64b)
Transmitted wave	$\psi_{III}(\text{transmitted}) = Fe^{ik_1x}$	(6.64c)

The probability of particles being reflected or transmitted is determined by the ratio of the appropriate $\psi^*\psi$. The probabilities are

Probability of reflection

$$R = \frac{|\psi_I(\text{reflected})|^2}{|\psi_I(\text{incident})|^2} = \frac{B^*B}{A^*A} \tag{6.65}$$

Probability of transmission

$$T = \frac{|\psi_{III}(\text{transmitted})|^2}{|\psi_I(\text{incident})|^2} = \frac{F^*F}{A^*A} \tag{6.66}$$

where R and T are reflection and transmission probabilities, respectively. Because the particles must be either reflected or transmitted, we must have $R + T = 1$; the probability of the wave being either reflected or transmitted has to be unity.

The values of R and T are found by applying the properties (boundary conditions) of Section 6.1 as $x \rightarrow \pm \infty$, $x = 0$, and $x = L$. These conditions will result in relationships between the coefficients A, B, C, D, and F. We will not go through the long algebra steps here, but the result for the transmission probability is

$$T = \left[1 + \frac{V_0^2 \sin^2(k_{II}L)}{4E(E - V_0)} \right]^{-1} \tag{6.67}$$

Notice that there is a situation in which the transmission probability is 1. This occurs when $k_{II}L = n\pi$, where n is an integer. It is possible for particles moving along the $+x$ direction to be reflected both at $x = 0$ and $x = L$. Their path difference back toward the $-x$ direction is $2L$. When $2L$ equals an integral number of the wavelengths inside the potential barrier, the reflected wave functions are precisely out of phase and will completely cancel.

Figure 6.14 A particle having energy E approaches a potential barrier of height V_0 with $E < V_0$. Classically, the particle will be reflected.

Potential Barrier with $E < V_0$

Now we consider the situation where classically the particle does not have enough energy to surmount the potential barrier, $E < V_0$. We show the situation in Figure 6.14. In the classical situation, the particle cannot penetrate the barrier because its kinetic energy ($K = E - V_0$) is negative. The particle is reflected at $x = 0$ and returns. The quantum mechanical result, however, is one of the most remarkable features of modern physics, and there is ample experimental proof of its existence. There is a small, but finite, probability that the particle can penetrate the barrier and even emerge on the other side. Such a surprising result requires a careful inspection of the wave functions. Fortunately, there are only a few changes to the equations already presented, and they occur in region II. The wave function in region II becomes $\psi_{II} = Ce^{\kappa x} + De^{-\kappa x}$ where

$$\kappa = \frac{\sqrt{2m(V_0 - E)}}{\hbar} \tag{6.68}$$

The parameter κ is a positive, real number, because $V_0 > E$. The application of the boundary conditions will again relate the coefficients of the wave functions.

The equations for the reflection and transmission probabilities of Equations (6.65) and (6.66) are unchanged, but the results will be modified by changing $ik_{II} \to \kappa$. Quantum mechanics allows the particle to actually be on the other side of the potential barrier despite the fact that all the incident particles came in from the left moving along the $+x$ direction (Figure 6.15). This effect is called **tunneling**. The result for the transmission probability in this case is

Tunneling

$$T = \left[1 + \frac{V_0^2 \sinh^2(\kappa L)}{4E(V_0 - E)} \right]^{-1} \tag{6.69}$$

Note that the sine function in Equation (6.67) has been replaced by the hyperbolic sine (sinh). When $\kappa L \gg 1$, the transmission probability equation (6.67) reduces to

$$T = 16 \frac{E}{V_0} \left(1 - \frac{E}{V_0} \right) e^{-2\kappa L} \tag{6.70}$$

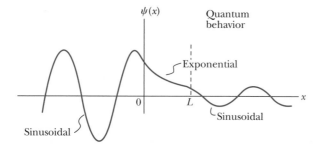

$\psi(x)$

Quantum behavior

Exponential

Sinusoidal

Sinusoidal

0   L   x

Figure 6.15 According to quantum mechanics, the particle approaching the potential barrier of Figure 6.14 may actually pass into the barrier and has a small probability of tunneling through the barrier and emerging at $x = L$. The particle may also be reflected at each boundary.

Conceptual Example 6.13

Which is more effective in preventing tunneling, the barrier potential height or the barrier width?

Solution The probability of penetration is dominated by the exponentially decreasing term. The exponential factor in Equation (6.70) depends linearly on the barrier width but only on the square root of the potential barrier height ($\kappa \sim \sqrt{V_0 - E}$). Thus, the width of the barrier is more effective than the potential height in preventing tunneling. It comes as no surprise that tunneling is observed only at the smallest distances on the atomic scale.

Example 6.14

In a particular semiconductor device, electrons that are accelerated through a potential of 5 V attempt to tunnel through a barrier of width 0.8 nm and height 10 V. What fraction of the electrons are able to tunnel through the barrier if the potential is zero outside the barrier?

Strategy We use either Equation (6.69) or (6.70) to calculate the tunneling probability, depending on the value of κL. We need to know V_0, κ, L, and E. We are given L and the fact that the potential barrier has $V_0 = 10$ eV and is zero outside the barrier. We determine from the accelerating voltage that the energy E of the electrons is $K = 5$ eV. We find the value of κ from a variation of Equation (6.62b) where we let $ik_{\mathrm{II}} \rightarrow \kappa$.

Solution We determine κ by using the mass of the electron and the appropriate energies.

$$\kappa = \frac{\sqrt{2m(V_0 - E)}}{\hbar}$$

$$= \frac{\sqrt{2(0.511 \times 10^6 \text{ eV}/c^2)(10 \text{ eV} - 5 \text{ eV})}}{6.58 \times 10^{-16} \text{ eV} \cdot \text{s}}$$

$$= \frac{3.43 \times 10^{18} \text{ s}^{-1}}{c} = \frac{3.43 \times 10^{18} \text{ s}^{-1}}{3 \times 10^8 \text{ m/s}} = 1.15 \times 10^{10} \text{ m}^{-1}$$

The value of $\kappa L = (1.15 \times 10^{10} \text{ m}^{-1})(0.8 \times 10^{-9} \text{ m}) = 9.2$, which might be considered to be much greater than 1, so we can try Equation (6.70). Let's calculate the transmission probability using both equations. The approximate Equation (6.70) gives

$$T = 16\left(\frac{5 \text{ eV}}{10 \text{ eV}}\right)\left(1 - \frac{5 \text{ eV}}{10 \text{ eV}}\right)e^{-18.4} = 4.1 \times 10^{-8}$$

The more accurate Equation (6.69) gives

$$T = \left[\frac{1 + (10 \text{ eV})^2 \sinh^2(9.2)}{4(5 \text{ eV})(5 \text{ eV})}\right]^{-1} = 4.1 \times 10^{-8} \quad (6.71)$$

The approximate equation, valid when $\kappa L \gg 1$, works well in this case.

Example 6.15

Consider Equation (6.70) and let all the factors multiplying the exponential term be denoted by M:

$$T = 16\frac{E}{V_0}\left(1 - \frac{E}{V_0}\right)e^{-2\kappa L} = Me^{-2\kappa L} \quad (6.72)$$

Consider the value of M and assign an average value. Calculate the probability of the electron tunneling through the barrier of the previous example.

Strategy We first consider a range of values E/V_0 and estimate an average value. Then we use Equation (6.72) to determine the tunneling probability and compare with the value found in Equation (6.71) of Example 6.14.

Solution The maximum value of

$$M = 16\frac{E}{V_0}\left(1 - \frac{E}{V_0}\right)$$

will be $M = 4$ when $E/V_0 = 0.5$. The values of M are symmetric around $E/V_0 = 0.5$, and $M = 1.4$ when $E/V_0 = 0.1$. A typical value of M might be 2, so Equation (6.72) becomes

$$T \approx 2e^{-2\kappa L} \quad (6.73)$$

We emphasize that Equation (6.73) is only an estimate to give an order of magnitude value for the tunneling probability.

For the values in Example 6.14, $\kappa L = 9.2$, and we have

$$T \approx 2e^{-2(9.2)} = 2e^{-18.4} = 2 \times 10^{-8}$$

As we expected, Equation (6.73) gives a result that agrees in order of magnitude with Example 6.14.

A simple argument based on the uncertainty principle helps us understand tunneling. Inside the barrier region (where $0 < x < L$), the wave function ψ_{II} is dominated by the $e^{-\kappa x}$ term, and $|\psi_{\mathrm{II}}|^2 \approx e^{-2\kappa x}$, so that over the interval $\Delta x = \kappa^{-1}$, the probability density of observing the particle has decreased markedly ($e^{-2} = 0.14$). Because $\Delta p \Delta x \geq \hbar$, we have $\Delta p \geq \hbar/\Delta x = \hbar \kappa$. The minimum kinetic energy in this interval must be

$$K_{min} = \frac{(\Delta p)^2}{2m} = \frac{\pi^2 \kappa^2}{2m} = V_0 - E$$

where we have substituted for κ in the last step. The violation allowed by the uncertainty principle (K_{min}) is equal to the *negative* kinetic energy required! The particle is allowed by quantum mechanics and the uncertainty principle to penetrate into a classically forbidden region.

Let us return briefly to our analogy with wave optics. If light passing through a glass prism reflects from an internal surface with an angle greater than the critical angle, total internal reflection occurs as seen in Figure 6.16a. However, the electromagnetic field is not exactly zero just outside the prism. If we bring another prism very close to the first one, experiment shows that the electromagnetic wave (light) appears in the second prism (see Figure 6.16b). The situation is analogous to the tunneling described here. This effect was observed by Newton and can be demonstrated with two prisms and a laser. The intensity of the second light beam decreases exponentially as the distance between the two prisms increases.*

*See D. D. Coon, *American Journal of Physics* **34**, 240 (1966).

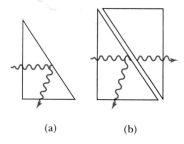

(a)   (b)

Figure 6.16 (a) A light wave will be totally reflected inside a prism if the reflection angle is greater than the critical angle. (b) If a second prism is brought close to the first, there is a small probability for the wave to pass through the air gap and emerge in the second prism.

Example 6.16

Consider a particle of kinetic energy K approaching the step function of Figure 6.17 from the left, where the potential barrier steps from 0 to V_0 at $x = 0$. Find the penetration distance Δx, where the probability of the particle penetrating into the barrier drops to $1/e$. Calculate the penetration distance for a 5-eV electron approaching a step barrier of 10 eV.

Strategy We use the results of this section to find the wave functions in the two regions $x < 0$ and $x > 0$.

$$\psi_I = A e^{ikx} + B e^{-ikx} \qquad x < 0$$
$$\psi_{II} = C e^{\kappa x} + D e^{-\kappa x} \qquad x > 0$$

where

$$k = \frac{\sqrt{2mE}}{\hbar} = \frac{\sqrt{2mK}}{\hbar}$$

$$\kappa = \frac{\sqrt{2m(V_0 - E)}}{\hbar}$$

Because the wave function ψ_{II} must go to zero when $x \to \infty$, the coefficient $C = 0$, so we have

$$\psi_{II} = D e^{-\kappa x} \qquad x > 0$$

The probability distribution for $x > 0$ is given by $|\psi_{II}|^2$. We need to find the value of x when the probability has dropped to e^{-1}. Let's call this distance ℓ.

Solution For the penetration distance ℓ, we have

$$e^{-1} = \frac{\psi_{II}^2 (x = \ell)}{\psi_{II}^2 (x = 0)} = e^{-2\kappa \ell}$$

$V(x)$

V_0

E

Particle

0

x

Position

Figure 6.17 A particle of energy E approaches a potential barrier from the left. The step potential is $V = 0$ for $x < 0$ and $V = V_0$ for $x > 0$.

From this equation we have $1 = 2\kappa \ell$, and the penetration distance becomes

$$\ell = \frac{1}{2\kappa} = \frac{\hbar}{2\sqrt{2m(V_0 - E)}}$$

This is the result we needed.

Now we find the penetration distance for the $E = K = $ 5-eV electron.

$$\ell = \frac{\hbar c}{2\sqrt{2mc^2(V_0 - E)}}$$

$$= \frac{197.3 \text{ eV} \cdot \text{nm}}{2\sqrt{2(0.511 \times 10^6 \text{ eV})(10 \text{ eV} - 5 \text{ eV})}} = 0.044 \text{ nm}$$

Electrons do not penetrate very far into the classically forbidden region.

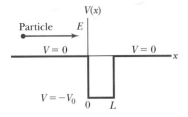

Figure 6.18 A particle of energy E approaches a potential well from the left. The potential is $V = 0$ everywhere except between $0 < x < L$, where $V = -V_0$.

Potential Well

Consider a particle of energy $E > 0$ passing through the potential well region (Figure 6.18), rather than into a potential barrier. Let $V = -V_0$ in the region $0 < x < L$ and zero elsewhere. Classically, the particle would speed up passing the well region, because $K = mv^2/2 = E + V_0$. According to quantum mechanics, reflection and transmission may occur, but the wavelength inside the potential well is smaller than outside. When the width of the potential well is precisely equal to half-integral or integral units of the wavelength, the reflected waves may be out of phase or in phase with the original wave, and cancellations or resonances may occur. The reflection/cancellation effects can lead to almost pure transmission or pure reflection for certain wavelengths. For example, at the second boundary ($x = L$) for a wave passing to the right, the wave may reflect and be out of phase with the incident wave. The effect would be a cancellation inside the well.

Alpha-Particle Decay

The phenomenon of tunneling explains the alpha-particle decay of heavy, radioactive nuclei. Many nuclei heavier than lead are natural emitters of alpha particles, but their emission rates vary over a factor of 10^{13}, whereas their energies tend to range only from 4 to 8 MeV. Inside the nucleus, an alpha particle feels the strong, short-range attractive nuclear force as well as the repulsive Coulomb force. The shape of the potential well is shown in Figure 6.19. The nuclear force dominates inside the nuclear radius r_N, and the potential can be approximated by a square well. However, outside the nucleus, the Coulomb force dominates. The so-called Coulomb potential energy barrier of Figure 6.19 can be several times the typical kinetic energy K (~5 MeV) of an alpha particle.

The alpha particle therefore is trapped inside the nucleus. Classically, it does not have enough energy to surmount the Coulomb potential barrier. According to quantum mechanics, however, the alpha particle can "tunnel" through the barrier. The widely varying rates of alpha emission from radioactive nuclei can be explained by small changes in the potential barrier (both height and width). A small change in the barrier can manifest itself greatly in the transmission probability, because of the exponential behavior in $e^{-2\kappa L}$.

Figure 6.19 An α particle of energy E_α is trapped inside a heavy nucleus by the large nuclear potential. Classically, it can never escape, but quantum mechanically it may tunnel through and escape.

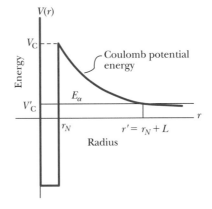

Example 6.17

Consider the α-particle emission from a ^{238}U nucleus, which emits a 4.2-MeV α particle. We represent the potentials as shown in Figure 6.19. The α particle is contained inside the nuclear radius of $r_N \approx 7 \times 10^{-15}$ m. Find the barrier height and the distance the α particle must tunnel and use a square-top potential to calculate the tunneling probability.

Strategy We shall calculate the barrier height [$V_C\ (r = r_N)$ in Figure 6.19] by calculating the Coulomb potential between an α particle and the remainder of the uranium nucleus for a separation of the nuclear radius, 7×10^{-15} m. We find the tunneling distance by setting the kinetic energy equal to the Coulomb potential and then use Equation (6.70) to determine the tunneling probability.

Solution The Coulomb potential is

$$V_C = \frac{Z_1 Z_2 e^2}{4\pi\epsilon_0 r_N}$$

$$= \frac{2(90)(1.6 \times 10^{-19}\ \text{C})^2 (9 \times 10^9\ \text{N} \cdot \text{m}^2 / \text{C}^2)}{7 \times 10^{-15}\ \text{m}}$$

$$\times \frac{10^{-6}\ \text{MeV}}{1.6 \times 10^{-19}\ \text{J}}$$

$$= 37\ \text{MeV}$$

We determine the distance r' through which the α particle must tunnel by setting $K = V_C\ (r = r')$ at that distance (see Figure 6.19). Because $K = 4.2$ MeV, we have

$$4.2\ \text{MeV} = \frac{Z_1 Z_2 e^2}{4\pi\epsilon_0 r'}$$

We solve this equation for r':

$$r' = \frac{37\ \text{MeV}}{4.2\ \text{MeV}} r_N = 6.2 \times 10^{-14}\ \text{m} = 62\ \text{fm}$$

where we have used the values above for V_C and r_N.

We make a simple, but rough, approximation of a square-top potential where $V = 37$ MeV for 7 fm $< r <$ 62 fm. We then find

$$\kappa = \frac{\sqrt{2m(V - E)}}{\hbar}$$

$$= \frac{\sqrt{2(3727\ \text{MeV}/c^2)(37\ \text{MeV} - 4.2\ \text{MeV})}}{6.58 \times 10^{-22}\ \text{MeV} \cdot \text{s}}$$

$$= 2.5 \times 10^{15}\ \text{m}^{-1}$$

where the mass of the α particle is 3727 MeV/c^2. The barrier width L is the difference between r' and r_N.

$$L = r' - r_N$$

$$= 62\ \text{fm} - 7\ \text{fm} = 55\ \text{fm}$$

The value of $\kappa L = (2.5 \times 10^{15}\ \text{m}^{-1})(55 \times 10^{-15}\ \text{m}) = 138$. Because $kL \gg 1$, we use Equation (6.70) to calculate the tunneling probability.

$$T = 16\left(\frac{4.2\ \text{MeV}}{37\ \text{MeV}}\right)\left(1 - \frac{4.2\ \text{MeV}}{38\ \text{MeV}}\right)e^{-275}$$

$$= 1.6 e^{-275} = 6 \times 10^{-120}$$

which is an extremely small number.

Our assumption of a square-top potential of the full height and full width is unrealistic. A closer approximation to the potential shown in Figure 6.19 would be a square-top potential of only half the maximum Coulomb potential (18 MeV rather than 37 MeV) and a barrier width of only half L (28 fm rather than 55 fm). If we use 18 MeV in the calculation of κ we obtain 1.7×10^{15} m^{-1}. The tunneling probability now becomes

$$T = 16\frac{4.2\ \text{MeV}}{18\ \text{MeV}}\left(1 - \frac{4.2\ \text{MeV}}{18\ \text{MeV}}\right)$$

$$\times \exp[-2(1.7 \times 10^{15}\ \text{m}^{-1})(2.8 \times 10^{-14}\ \text{m})]$$

$$= 2.9 \exp(-95) = 1.6 \times 10^{-41}$$

This still seems like a very low probability, but let us see if we can determine how much time it takes the α particle to tunnel out. If the α particle has a kinetic energy of 4.2 MeV, its speed is determined nonrelativistically by

$$K = \frac{1}{2}mv^2$$

$$v = \sqrt{\frac{2K}{m}} = \sqrt{\frac{2(4.2\ \text{MeV})}{3727\ \text{MeV}/c^2}} = 0.047c = 1.4 \times 10^7\ \text{m/s}$$

The diameter of the nucleus is about 1.4×10^{-14} m, so it takes the α particle $(1.4 \times 10^{-14}\ \text{m})/(1.4 \times 10^7\ \text{m/s}) \approx 10^{-21}$ s to cross. The α particle must make many traverses back and forth across the nucleus before it can escape. According to our probability calculation it must make about 10^{41} attempts, so we estimate the α particle may tunnel through in about 10^{20} s. The half-life of a ^{238}U nucleus is 4.5×10^9 y or about 10^{17} s. Our rough estimate does not seem all that bad.

Special Topic

SCANNING PROBE MICROSCOPES

Heinrich Rohrer (left, 1933–) and Gerd Binnig (1947–) received the Nobel Prize for Physics in 1986 for their design of the scanning tunneling microscope. The Swiss Rohrer was educated at the Swiss Federal Institute of Technology in Zurich and joined the IBM Research Laboratory in Zurich in 1963. The German Binnig received his doctorate from the University of Frankfurt (Germany) in 1978 and then joined the same IBM Research Laboratory. He moved to the IBM Physics Group in Munich in 1984. *Courtesy of IBM Zurich Research Laboratory.*

Scanning probe microscopes allow a computer-generated contour map of a surface, atom by atom (see Figure 1.7). They consist of two types, **scanning tunneling microscopes** (STM) and **atomic force microscopes** (AFM), which have revolutionized the imaging of atomic surfaces. Gerd Binnig and Heinrich Rohrer (Nobel Prize, 1986) invented the STM in the early 1980s at the IBM Research Laboratory in Zurich, Switzerland. Later while Binnig was on leave at Stanford University and IBM's Almaden Research Center, he thought up the concept of the AFM, which he developed in 1985 with Christoph Gerber of IBM and Calvin Quate of Stanford.

In the most common form of the STM a constant bias voltage of appropriate polarity is applied between the atoms of a tip and the sample to be examined (Figure A). As the tip runs over the surface of the sample just an atom's diameter away, electrons attracted to the needle tunnel across the gap, and the sensitivity of the tunneling current to the gap distance is the key to the STM capability. This tunneling current can be as small as a few pA (10^{-12} A), and a change in the tunneling gap of only 0.4 nm can cause a factor of 10^4 in the tunneling current. A highly sensitive feedback mechanism regulates the position of the tip to maintain a steady current of electrons. The resulting up and down movements of the tip effectively trace the contours of the sample atoms, producing something like a topographic map. The resulting image (path of the tip) is shown by the solid black line in Figure A.

The AFM depends on the interatomic forces between the tip and sample atoms as shown in Figure B. In some systems, the sample atoms are scanned horizontally while the sample is moved up and down to keep the force between the tip and sample atoms constant. The interatomic forces cause the very sensitive cantilever to bend. A laser is reflected off the end of the cantilever arm into an optical sensor, and the feedback signal from this sensor controls the sample height, giving the topography of the atomic surface. The tip is scanned over the surface for a constant cantilever deflection and a constant interatomic force between tip and atom.

The interaction between tip and sample is much like that of a record player stylus moving across the record but is about a million times more sensitive. The optical feedback system prevents the tip from actually damaging or distorting the sample atoms. Cantilevers having spring constants as small as 0.1 N/m have been microfabricated from silicon and silicon compounds. The cantilever lateral dimensions are on the order of 100 μm with thicknesses of about 1 μm. In comparison the spring constant of a piece of household aluminum foil 4 mm by 1 mm is about 1 N/m. The tapered tips may have an end dimension of only 50 nm. The tracking forces felt by the cantilever can be as small as 10^{-9} N.

Advantages of the AFM compared to the STM are that a conducting surface is not required and neither special sample preparation nor expensive vacuum equipment is needed. Because the AFM works for both

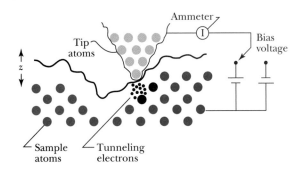

Figure A Highly schematic diagram of the scanning tunneling microscope process. Electrons, represented in the figure as small dots, tunnel across the gap between the atoms of the tip and sample. A feedback system that keeps the tunneling current constant causes the tip to move up and down, tracing the contours of the sample atoms.

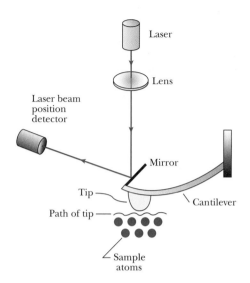

Laser

Lens

Laser beam
position
detector

Mirror

Tip

Path of tip

Cantilever

Sample
atoms

Figure B Highly schematic diagram of the atomic force microscope. A feedback signal from the detection of the laser beam
reflecting off the mirror that is mounted on the cantilever provides
a signal to move the sample atoms up or down to keep the cantilever force constant. The movement of the sample atoms traces
the contours of the sample atoms.

Figure C These three photos, taken with an STM, show xenon
atoms placed on a nickel surface. The xenon atoms are 0.16 nm
high and adjacent xenon atoms are 0.5 nm apart (the vertical scale
has been exaggerated). The small force between the STM tip and an
atom is enough to drag one xenon atom at a time across the nickel.
The nickel atoms are represented by the black-and-white stripes on
the horizontal surface. See also Figure 1.7. The image is magnified
about 5 million times. *Courtesy of International Business Machines.*

Figure D An atomic force microscope scan of a stamper used to
mold compact disks. The numbers given are in nm. The bumps on
this metallic mold stamp out 60-nm-deep holes in tracks that are
1.6 μm apart in the optical disks. *Photo courtesy of Digital Instruments/
Veeco Metrology Group, Santa Barbara, CA.*

insulators and conductors, it can be used for ceramics,
polymers, optical surfaces, and biological specimens.
However, special vibrational insulation is needed for
AFM to dampen air currents and even human voices.
Figure 1.7 is a photo indicating individual atoms that
was taken using an STM. Those atoms can be individually moved as shown in Figure C.

In addition to providing individual atomic images,
the STM and AFM open tremendous possibilities for
investigation of a variety of atomic surface features.
The topographical images produced by these microscopes allow scientists to examine the gross features of
a sample, such as the flatness of materials, grain structures, and the breakup of thin films. Industrial applications of this technology include the inspections of
magnetic bit shapes, integrated circuit topography, lu-
bricant thicknesses, optical disk stampers (Figure D),
and measurement of line widths on integrated circuit
masks. Biological applications of AFM include the imaging of amino acids, DNA, proteins, and even leaf
sections from a plant. The AFM has been used to observe the polymerization of fibrin, a blood-clotting
protein. Real-time imaging of biological samples offers
incredible possibilities, for example, the attachment of

the AIDS virus onto cell membranes. Now that STM
and AFM instruments are both available commercially,
new applications for this revolutionary technology are
being developed that include such diverse techniques
as developing x-ray mirrors by moving atoms to increase reflectivity, "nanoengineering" electronics to
improve performance, and repairing organic molecules on an atomic basis.

Tunnel Diode An extremely useful application of tunneling is that of a tunnel diode, which is a special kind of semiconductor. In a tunnel diode, electrons may pass from one region through a junction into another region. We can depict the behavior by considering a potential barrier over the region of the junction, which may be only 10 nm wide. Both positive and negative bias voltages may be applied to change the barrier height to allow the electrons to tunnel either way through the barrier. In a normal semiconductor junction, the electrons (and holes) diffuse through, a relatively slow process. In a tunnel diode, the electrons tunnel through quite rapidly when the tunneling probability is relatively high. Because the applied bias voltage can be changed rapidly, a tunnel diode is an extremely fast device and, as such, has important uses in switching circuits and high-frequency oscillators.

Summary

Werner Heisenberg and Erwin Schrödinger developed modern quantum theory in the 1920s. The time-dependent Schrödinger wave equation for the wave function $\Psi(x, t)$ is expressed as

$$i\hbar \frac{\partial \Psi(x, t)}{\partial t} = -\frac{\hbar^2}{2m} \frac{\partial^2 \Psi(x, t)}{\partial x^2} + V\Psi(x, t) \quad (6.1)$$

The time-independent form for the spatial dependence (in one dimension) of $\psi(x)$, where $\Psi(x, t) = \psi(x)e^{-iEt/\hbar}$, is

$$-\frac{\hbar^2}{2m} \frac{d^2\psi(x)}{dx^2} + V(x)\psi(x) = E\psi(x) \quad (6.13)$$

Certain properties of Ψ and $\partial\Psi/\partial x$ lead to quantized behavior. The wave function $\Psi(x, t)$ must be finite, single valued, and continuous; $\partial\Psi/\partial x$ must be continuous. The wave function must be normalized for use in determining probabilities.

Average values of the physical observables are determined by calculating the expectation values using the wave functions. The expectation value of a function $g(x)$ is found from

$$\langle g(x)\rangle = \int_{-\infty}^{\infty} \Psi^*(x, t)g(x)\Psi(x, t)\, dx \quad (6.20)$$

To find the expectation values of the momentum and energy, we need to know the appropriate operators. In these two cases the operators are

$$\hat{p} = -i\hbar \frac{\partial}{\partial x} \quad (6.21)$$

$$\hat{E} = i\hbar \frac{\partial}{\partial t} \quad (6.25)$$

and the expectation values $\langle p\rangle$ and $\langle E\rangle$ are

$$\langle p\rangle = -i\hbar \int_{-\infty}^{\infty} \Psi^*(x, t)\frac{\partial \Psi(x, t)}{\partial x}\, dx \quad (6.23)$$

$$\langle E\rangle = i\hbar \int_{-\infty}^{\infty} \Psi^*(x, t)\frac{\partial \Psi(x, t)}{\partial t}\, dx \quad (6.26)$$

The infinite square-well potential is a particularly simple application of the Schrödinger wave equation, and it leads to quantized energy levels and quantum numbers. The three-dimensional infinite square-well potential leads to the concept of degenerate states, different physical states with the same energy.

The simple harmonic oscillator, where the potential is $V(x) = \kappa x^2/2$, is an important application of the Schrödinger wave equation because it approximates many complex systems in nature but is exactly soluble. The energy levels of the simple harmonic oscillator are $E_n = (n + \frac{1}{2})\hbar\omega$, where $n = 0$ represents the ground state energy $E_0 = \hbar\omega/2$. The fact that the minimum energy is not zero—that the oscillator exhibits zero-point motion—is a consequence of the uncertainty principle.

Finite potentials lead to the possibility of a particle entering a region that is classically forbidden, where $V_0 > E$ (negative kinetic energy). This quantum process is called tunneling and is studied by considering various potential barrier shapes. Important examples of quantum tunneling are alpha decay and tunnel diodes. Tunneling is consistent with the uncertainty principle and occurs only for short distances.

Questions

1. Why can we use the nonrelativistic form of the kinetic energy in treating the structure of the hydrogen atom?
2. How do you reconcile the fact that the probability density for the ground state of the quantum harmonic oscillator (Figure 6.10c) has its peak at the center and its minima at its ends, whereas the classical harmonic oscillator's probability density (Figure 6.11) has a minimum at the center and peaks at each end? If you do this experiment with an actual mass and spring, what experimental result for its position distribution would you expect to obtain? Why?
3. Notice for the finite square-well potential that the wave function ψ is not zero outside the well despite the fact that $E < V_0$. Is it possible classically for a particle to be in a region where $E < V_0$? Explain this result.
4. In a given tunnel diode the pn junction (see Chapter 11) width is fixed. How can we change the time response of the tunnel diode most easily? Explain.
5. A particle in a box has a first excited state that is 3 eV above its ground state. What does that tell you about the box?
6. Does the wavelength of a particle change after it tunnels through a barrier as shown in Figure 6.15? Explain.

7. Can a particle be observed while it is tunneling through a barrier? What would its wavelength, momentum, and kinetic energy be while it tunnels through the barrier?
8. Is it easier for an electron or a proton of the same energy to tunnel through a given potential barrier? Explain.
9. Can a wave packet be formed from a superposition of wave functions of the type $e^{i(kx-\omega t)}$? Can it be normalized?
10. Given a particular potential V and wave function Ψ, how could you prove that the given Ψ is correct? Could you determine an appropriate E if the potential is independent of time?
11. Compare the infinite square-well potential with the finite one. Where is the Schrödinger wave equation the same? Where is it different?
12. Tunneling can occur for an electron trying to pass through a very thin tunnel diode. Can a baseball tunnel through a very thin window? Explain.
13. For the three-dimensional cubical box, the ground state is given by $n_1 = n_2 = n_3 = 1$. Why is it not possible to have one $n_i = 1$ and the other two equal to zero?

Problems

Note: The more challenging problems are indicated by a shaded box.

6.1 The Schrödinger Wave Equation

1. Try to normalize the wave function $e^{i(kx-\omega t)}$. Why can't it be done over all space? Explain why it is not possible.
2. (a) In what direction does a wave of the form $A\sin(kx - \omega t)$ move? (b) What about $B\sin(kx + \omega t)$? (c) Is $e^{i(kx-\omega t)}$ a real number? (d) In what direction is the wave in (c) moving? Explain.
3. Show directly that the trial wave function $\Psi(x, t) = Ae^{i(kx-\omega t)}$ satisfies Equation (6.1).
4. Normalize the wave function $Ae^{i(kx-\omega t)}$ in the region $x = 0$ to a.
5. Normalize the wave function $Are^{-r/\alpha}$ from $r = 0$ to ∞ where α and A are constants.
6. Property 2 of the boundary conditions for wave functions specifies that Ψ must be continuous in order to avoid discontinuous probability values. Why can't we have such probabilities?
7. Consider the wave function $Ae^{-\alpha|x|}$ that we used in Example 6.4. (a) How does this wave function satisfy the boundary conditions of Section 6.1? (b) What can we conclude about this wave function? (c) If the

wave function is unacceptable as is, how could it be fixed?

6.2 Expectation Values

8. A set of measurements has given the following result for the measurement of x (in some units of length): 3.4, 3.9, 5.2, 4.7, 4.1, 3.8, 3.9, 4.7, 4.1, 4.5, 3.8, 4.5, 4.8, 3.9, and 4.4. Find the average value of x, called $\bar{x}$ or $\langle x \rangle$, and average value of x^2, called $\langle x^2 \rangle$. Show that the standard deviation of x, given by

$$\sigma = \sqrt{\frac{\sum (x_i - \bar{x})^2}{N}}$$

where x_i is the individual measurement and N is the number of measurements, is also given by $\sigma = \sqrt{\langle x^2 \rangle - \langle x \rangle^2}$. Find the value of σ for the set of data.
9. If the potential V is independent of time, show that the expectation value for x is independent of time.
10. A wave function ψ equals $A(e^{ix} + e^{-ix})$ in the region $-\pi < x < \pi$ and zero elsewhere. Normalize the wave function and find the probability of the particle being between $x = 0$ and $\pi/8$.

11. A wave function has the value $A \sin x$ between 0 and π, but zero elsewhere. Normalize the wave function and find the probability that the particle is between $x = 0$ and $\pi/4$.

6.3 Infinite Square-Well Potential

12. Find an equation for the difference between adjacent energy levels ($\Delta E_n = E_{n+1} - E_n$) for the infinite square-well potential. Calculate ΔE_1, ΔE_8, and ΔE_{800}.

13. Determine the average value of $\psi_n^2(x)$ inside the well for the infinite square-well potential for $n = 1, 5, 20$, and 100. Compare these averages with the classical probability of detecting the particle inside the box.

14. A particle in an infinite square-well potential has a ground state energy of 4.3 eV. (a) Calculate and sketch the energies of the next three levels, and (b) sketch the wave functions on top of the energy levels.

15. We can approximate an electron moving in a nanowire as a one-dimensional infinite square-well potential. Let the wire be 2 μm long. The nanowire is cooled to a temperature of 13 K, and we assume the electron's average kinetic energy is that of gas molecules at this temperature ($= 3/2\ kT$). (a) What are the three lowest possible energy levels of the electrons? (b) What is the approximate quantum number of electrons moving in the wire?

16. An electron moves with a speed $v = 10^{-4}c$ inside a one-dimensional box ($V = 0$) of length 48.5 nm. The potential is infinite elsewhere. The particle may not escape the box. What approximate quantum number does the electron have?

17. For the infinite square-well potential, find the probability that a particle in its ground state is in each third of the one-dimensional box: $0 \le x \le L/3$, $L/3 \le x \le 2L/3$, $2L/3 \le x \le L$.

18. Repeat the previous problem using the first excited state.

19. Repeat Example 6.9 for an electron inside the nucleus. Assume nonrelativistic equations and find the transition energy for an electron. (See Example 6.9 for an interpretation of the result.)

20. What is the minimum energy of (a) a proton and (b) an α particle trapped in a one-dimensional region the size of a uranium nucleus (radius = 7×10^{-15} m)?

21. An electron is trapped in an infinite square-well potential of width 0.5 nm. If the electron is initially in the $n = 4$ state, what are the various photon energies that can be emitted as the electron jumps to the ground state?

6.4 Finite Square-Well Potential

22. Consider a finite square-well potential well of width 3×10^{-15} m that contains a particle of mass 2 GeV/c^2. How deep does this potential well need to be to contain three energy levels? (Except for the energy levels, this situation approximates a deuteron.)

23. Compare the results of the infinite and finite square-well potentials. (a) Are the wavelengths longer or shorter for the finite square well compared with the infinite well? (b) Using physical arguments, do you expect the energies (for a given quantum number n) to be (i) larger or (ii) smaller for the finite square well than for the infinite square well? (c) Why will there be a finite number of bound energy states for the finite potential?

24. Apply the boundary conditions to the finite square-well potential at $x = 0$ to find the relationships between the coefficients A, C, and D and the ratio C/D.

25. Apply the boundary conditions to the finite square-well potential at $x = L$ to find the relationship between the coefficients B, C, and D and the ratio C/D.

6.5 Three-Dimensional Infinite-Potential Well

26. Find the energies of the second, third, fourth, and fifth levels for the three-dimensional cubical box. Which are degenerate?

27. Write the possible (unnormalized) wave functions for each of the first four excited energy levels for the cubical box.

28. Find the normalization constant A for the ground state wave function for the cubical box in Equation (6.52).

29. Complete the derivation of Equation (6.49) by substituting the wave function given in Equation (6.47) into Equation (6.46). What is the origin of the three quantum numbers?

6.6 Simple Harmonic Oscillator

30. In Figure 6.9 we showed a plausible guess for the wave function ψ_0 for the lowest energy level E_0 of the simple harmonic oscillator. Explain the shape of the wave function and explain why it is a maximum at $x = 0$ and not zero when $E = V_0$.

31. What is the energy level difference between adjacent levels $\Delta E_n = E_{n+1} - E_n$ for the simple harmonic oscillator? What is ΔE_0, ΔE_2, and ΔE_{50}? How many possible energy levels are there?

32. The wave function for the first excited state ψ_1 for the simple harmonic oscillator is $\psi_1 = Axe^{-\alpha x^2/2}$. Normalize the wave function to find the value of the constant A. Determine $\langle x \rangle$, $\langle x^2 \rangle$, and $\Delta x = \sqrt{\langle x^2 \rangle - \langle x \rangle^2}$.

33. An atom of mass 3.32×10^{-26} kg oscillates in one dimension at a frequency of 10^{13} Hz. What are its effective force constant and quantized energy levels?

34. One possible solution for the wave function ψ_n for the simple harmonic oscillator is

$$\psi_n = A(2\alpha x^2 - 1)e^{-\alpha x^2/2}$$

where A is a constant. What is the value of the energy level E_n?

35. What would you expect for $\langle p \rangle$ and $\langle p^2 \rangle$ for the ground state of the simple harmonic oscillator? (*Hint*: use symmetry and energy arguments.)

36. Show that the energy of a simple harmonic oscillator in the $n = 1$ state is $3\hbar\omega/2$ by substituting the wave function $\psi_1 = Axe^{-\alpha x^2/2}$ directly into the Schrödinger equation.

37. A H_2 molecule can be approximated by a simple harmonic oscillator having spring constant $k = 1.1 \times 10^3$ N/m. Find (a) the energy levels, and (b) the possible wavelengths of photons emitted when the H_2 molecule decays from the third excited state eventually to the ground state.

6.7 Barriers and Tunneling

38. The creation of elements in the early universe and in stars involves protons tunneling through nuclei. Find the probability of the proton tunneling through ^{12}C. Let the temperature of the star containing the proton and carbon be 12,000 K.

39. Compare the wavelength of a particle when it passes a barrier of height (a) $+V_0$ (see Figure 6.12) and (b) $-V_0$ where $E > |V_0|$ (see Figure 6.18). Calculate the momentum and kinetic energy for both cases.

40. (a) Calculate the transmission probability of an α particle of energy $E = 5$ MeV through a Coulomb barrier of a heavy nucleus that is approximated by a square barrier with $V_0 = 15$ MeV and barrier width $L = 1.3 \times 10^{-14}$ m. Also, calculate the probability (b) by doubling the potential barrier height and (c) by using the original barrier height but doubling the barrier width. Compare all three probabilities.

41. Consider a particle of energy E trapped inside the potential well shown in the accompanying figure. Sketch a possible wave function inside and outside the potential well. Explain your sketch.

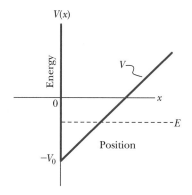

Problem 41 A potential well is infinite for $x < 0$ but increases linearly from $V = -V_0$ at $x = 0$.

42. When a particle of energy E approaches a potential barrier of height V_0, where $E \gg V_0$, show that the reflection coefficient is about $\{[V_0 \sin(kL)]/2E\}^2$.

43. Let 11.0-eV electrons approach a potential barrier of height 3.8 eV. (a) For what barrier thickness is there

no reflection? (b) For what barrier thickness is the reflection a maximum?

44. A 1-eV electron has a 10^{-4} probability of tunneling through a 2.5-eV potential barrier. What is the probability of a 1-eV proton tunneling through the same barrier?

45. An electron is attempting to tunnel through a square barrier potential. (a) Draw a potential function with zero potential on either side of a square-top potential similar to Figure 6.12. Draw the wave function before, after, and inside the barrier. (b) Let the barrier be twice as wide and repeat part (a). (c) Let the barrier be about twice as tall as in (a) and repeat (a). Do not perform calculations; make estimates only.

46. Use the approximate Equation (6.73) to estimate the probability of (a) a 1.4-eV electron tunneling through a 6.4-eV-high barrier of width 2.8 nm, and (b) a 4.4-MeV α particle tunneling through a uranium nucleus where the potential barrier is 19.2 MeV and 6.7 fm wide. (c) Discuss whether the approximation was valid for these two cases. Explain.

General Problems

47. A particle of mass m is trapped in a three-dimensional rectangular potential well with sides of length L, $L/\sqrt{2}$, and $2L$. Inside the box $V = 0$, outside $V = \infty$. Assume that

$$\psi = A \sin(k_1 x)\sin(k_2 y)\sin(k_3 z)$$

inside the well. Substitute this wave function into the Schrödinger equation and apply appropriate boundary conditions to find the allowed energy levels. Find the energy of the ground state and first four excited levels. Which of these levels are degenerate?

48. For a region where the potential $V = 0$, the wave function is given by $\sqrt{2/\alpha} \sin(3\pi x/\alpha)$. Calculate the energy of this system.

49. Consider the semi-infinite-well potential in which $V = \infty$ for $x \leq 0$, $V = 0$ for $0 < x < L$, and $V = V_0$ for $x > L$. (a) Show that possible wave functions are $A \sin kx$ inside the well and $Be^{-\kappa x}$ for $x > L$, where $k = \sqrt{2mE}/\hbar$ and $\kappa = \sqrt{2m(V_0 - E)}/\hbar$. (b) Show that the application of the boundary conditions gives $\kappa \tan(kL) = -k$.

50. Assume that $V_0 = \hbar^2/2mL^2$ and show that the ground state energy of a particle in the semi-infinite well of the previous problem is given by $0.04\hbar^2/2mL^2$.

51. Prove that there are a limited number of bound solutions for the semi-infinite well.

52. Use the semi-infinite-well potential to model a deuteron, a nucleus consisting of a neutron and a proton. Let the well width be 3.5×10^{-15} m and $V_0 - E = 2.2$ MeV. Determine the energy E. How many excited states are there, and what are their energies?

53. The wave function for the $n = 2$ state of a simple harmonic oscillator is $A(1 - 2\alpha x^2)e^{-\alpha x^2/2}$. (a) Show that its energy level is $5\hbar\omega/2$ by substituting the wave function into the Schrödinger equation. (b) Find $\langle x \rangle$ and $\langle x^2 \rangle$.

54. A particle is trapped inside an infinite square-well potential between $x = 0$ and $x = L$. Its wave function is a superposition of the ground state and first excited state. Its wave function is given by

$$\psi(x) = \frac{1}{2}\psi_1(x) + \frac{\sqrt{3}}{2}\psi_2(x)$$

Show that the wave function is normalized.

55. The *Morse potential* is a good approximation for a real potential to describe diatomic molecules. It is given by $V(r) = D(1 - e^{-a(r-r_e)})^2$ where D is the molecular dissociation energy, and r_e is the equilibrium distance between the atoms. For small vibrations, $r - r_e$ is small, and $V(r)$ can be expanded in a Taylor series to reduce to a simple harmonic potential. Find the lowest term of $V(r)$ in this expansion and show that it is quadratic in $(r - r_e)$.

56. Show that the vibrational energy levels E_v for the Morse potential of the previous problem are given by

$$E_v = \hbar\omega\left(n + \frac{1}{2}\right) - \frac{\hbar^2\omega^2}{4D}\left(n + \frac{1}{2}\right)^2$$

where

$$\omega = a\sqrt{\frac{2D}{m_r}}$$

and n is the vibrational quantum number, m_r is the reduced mass, and $E_v \ll D$. Find the three lowest energy levels for KCl where $D = 4.42$ eV, and $a = 7.8$ nm^{-1}.

57. Consider a particle of mass m trapped inside a two-dimensional square box of sides L aligned along the x and y axes. Show that the wave function and energy levels are given by

$$\psi(x, y) = \frac{2}{L}\sin\frac{n_x\pi x}{L}\sin\frac{n_y\pi y}{L}$$

$$E = \frac{\hbar^2\pi^2}{2mL^2}(n_x^2 + n_y^2)$$

Plot the first six energy levels and give their quantum numbers.

58. Make a sketch for each of the following situations for both the infinite square-well and finite square-well potentials in one dimension: (a) the four lowest energy levels, and (b) the probability densities for the four lowest states. (c) Discuss the differences between the two potentials and why they occur.

59. Two nanowires are separated by 1.3 nm as measured by STM. Inside the wires the potential energy is zero, but between the wires the potential energy is greater than the electron's energy by only 0.9 eV. Estimate the probability that the electron passes from one wire to the other.

60. The WKB approximation is useful to obtain solutions to the one-dimensional time-independent Schrödinger equation in cases where $E > V(x)$ and the potential $V(x)$ changes slowly and gradually with x. In this case the wavelength $\lambda(x)$ varies with x because of the $V(x)$ dependence on x. (a) Argue that we can write the wavelength as

$$\lambda(x) = \frac{h}{\sqrt{2m[E - V(x)]}}$$

for a particle of mass m in a potential $V(x)$. (b) By considering the number of oscillations that can be fit into a distance dx, show that the following equation is valid, where n is an integer and represents the number of standing waves that fit inside the potential well.

$$2\int\sqrt{2m[E - V(x)]}\,dx = nh \qquad \text{where } n \text{ is an integer}$$
(6.74)

This is the WKB approximation. (*Hint:* the equation

$$\int\frac{dx}{\lambda(x)} = \frac{n}{2}$$

might be helpful.)

61. Use the WKB approximation of Equation (6.74) in the previous problem with the potential shown in the accompanying figure.

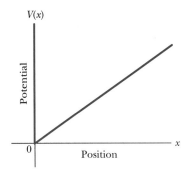

Problem 61 The potential $V(x) = \infty$ for $x \leq 0$ and $V(x) = Ax$ for $x > 0$.

$V(x) = \infty$ for $x \leq 0$, and $V(x) = Ax$ for $x > 0$. (a) Find the quantized energy values, and (b) sketch the wave functions on top of the $V(x)$ function for the three lowest states.

The Hydrogen Atom

The atom of modern physics can be symbolized only through a partial differential equation in an abstract space of many dimensions. All its qualities are inferential; no material properties can be directly attributed to it. That is to say, any picture of the atom that our imagination is able to invent is for that very reason defective. An understanding of the atomic world in that primary sensuous fashion . . . is impossible.

Werner Heisenberg

Because it is the simplest atom, the hydrogen atom has been the object of more experimentation and study than any other atom. The hydrogen line spectra discussed in Chapter 3 and the Bohr model of the hydrogen atom in Chapter 4 are examples of hydrogen studies that resulted in significant breakthroughs in physics. In Chapter 6 we studied the Schrödinger equation and its application to several model systems. We now have the tools to apply quantum mechanics to real physical systems, which we will do in the next few chapters. Our first major subject is atomic physics, and we naturally begin by applying the Schrödinger equation to the hydrogen atom. We will find that several quantum numbers are needed in order to explain experimental results. Although we generally confine ourselves to the hydrogen atom in this chapter, we occasionally digress, for example for the Stern-Gerlach experiment in Section 7.4, in order to incorporate an important experimental result that we need for our understanding of the hydrogen atom. Two sections in this chapter (Sections 7.2 and 7.6) are advanced topics and may be skipped without losing continuity.

7.1 Application of the Schrödinger Equation to the Hydrogen Atom

The hydrogen atom is the first system we shall consider that requires the full complexity of the three-dimensional Schrödinger equation. To a good approximation the potential energy of the electron-proton system is electrostatic:

$$V(r) = -\frac{e^2}{4\pi\epsilon_0 r} \tag{7.1}$$

We rewrite the three-dimensional time-independent Schrödinger Equation (6.43) as

$$-\frac{\hbar^2}{2m}\frac{1}{\psi(x,y,z)}\left[\frac{\partial^2\psi(x,y,z)}{\partial x^2}+\frac{\partial^2\psi(x,y,z)}{\partial y^2}+\frac{\partial^2\psi(x,y,z)}{\partial z^2}\right]=E-V(r) \quad (7.2)$$

As discussed in Chapter 4, the correct mass value m to be used is the reduced mass μ of the proton-electron system. We can also study other hydrogen-like (called *hydrogenic*) atoms such as He$^+$ or Li^{++} by inserting the appropriate reduced mass μ and by replacing e^2 in Equation (7.1) with Ze^2, where Z is the atomic number.

We note that the potential $V(r)$ in Equation (7.2) depends only on the distance r between the proton and electron. The potential in this case is the central force—perhaps the most important in quantum mechanics. To take advantage of the radial symmetry, we transform to spherical polar coordinates. The transformation is given in Figure 7.1, where the relationships between the Cartesian coordinates x, y, z and the spherical polar coordinates r, θ, ϕ are shown. The transformation of Equation (7.2) into spherical polar coordinates is straightforward. After inserting the Coulomb potential into the transformed Schrödinger equation, we have

Schrödinger equation in spherical coordinates

$$\frac{1}{r^2}\frac{\partial}{\partial r}\left(r^2\frac{\partial\psi}{\partial r}\right)+\frac{1}{r^2\sin\theta}\frac{\partial}{\partial\theta}\left(\sin\theta\frac{\partial\psi}{\partial\theta}\right)+\frac{1}{r^2\sin^2\theta}\frac{\partial^2\psi}{\partial\phi^2}+\frac{2\mu}{\hbar^2}(E-V)\psi=0$$

$$(7.3)$$

The wave function ψ is now a function of r, θ, ϕ [$\psi(r,\theta,\phi)$], but we will write it simply as ψ for brevity. In the terminology of partial differential equations, Equation (7.3) is separable, meaning a solution may be found as a product of three functions, each depending on only one of the coordinates r, θ, ϕ. (This is exactly analogous to our separating the time-dependent part of the Schrödinger equation solution as $e^{-iEt/\hbar}$.) Let us try a solution of the form

$$\psi(r,\theta,\phi)=R(r)f(\theta)g(\phi) \quad (7.4)$$

This substitution allows us to separate the partial differential in Equation (7.3) into three separate differential equations, each depending on one coordinate: $r, \theta,$ or ϕ.

We have a good idea from Chapter 6 what to expect the results to look like. For each of the three differential equations we must apply appropriate boundary conditions on the functions $R(r), f(\theta),$ and $g(\phi)$. This will lead to three quantum numbers, one for each of the three separate differential equations. Notice that there is one quantum number for each dimension of motion available; recall that in Chapter 6 we obtained one quantum number for one-dimensional motion and three quantum numbers for three-dimensional motion.

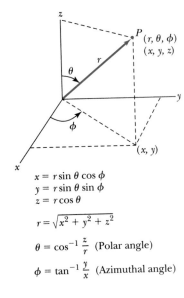

$$x=r\sin\theta\cos\phi$$
$$y=r\sin\theta\sin\phi$$
$$z=r\cos\theta$$

$$r=\sqrt{x^2+y^2+z^2}$$

$$\theta=\cos^{-1}\frac{z}{r} \quad \text{(Polar angle)}$$

$$\phi=\tan^{-1}\frac{y}{x} \quad \text{(Azimuthal angle)}$$

Figure 7.1 Relationship between spherical polar coordinates (r, θ, ϕ) and Cartesian coordinates (x, y, z).

7.2 Solution of the Schrödinger Equation for Hydrogen

The first step is to substitute the trial solution, Equation (7.4), into Equation (7.3). Then we can separate the resulting equation into three equations: one for $R(r)$, one for $f(\theta)$, and one for $g(\phi)$. The solutions to those equations will then allow us to understand the structure of the hydrogen atom, in the ground state and in the excited states as well.

Separation of Variables

Starting with Equation (7.4), we find the necessary derivatives to be

$$\frac{\partial \psi}{\partial r} = fg \frac{\partial R}{\partial r} \qquad \frac{\partial \psi}{\partial \theta} = Rg \frac{\partial f}{\partial \theta} \qquad \frac{\partial^2 \psi}{\partial \phi^2} = Rf \frac{\partial^2 g}{\partial \phi^2} \tag{7.5}$$

We substitute these results into the Schrödinger Equation (7.3) and find

$$\frac{fg}{r^2} \frac{\partial}{\partial r}\left(r^2 \frac{\partial R}{\partial r} \right) + \frac{Rg}{r^2 \sin \theta} \frac{\partial}{\partial \theta}\left(\sin \theta \frac{\partial f}{\partial \theta} \right) + \frac{Rf}{r^2 \sin^2 \theta} \frac{\partial^2 g}{\partial \phi^2} + \frac{2\mu}{\hbar^2}(E - V)\, Rfg = 0$$

$$\tag{7.6}$$

Next we multiply both sides of Equation (7.6) by $r^2 \sin^2 \theta / Rfg$ and rearrange to have

$$-\frac{\sin^2 \theta}{R} \frac{\partial}{\partial r}\left(r^2 \frac{\partial R}{\partial r} \right) - \frac{2\mu}{\hbar^2} r^2 \sin^2 \theta (E - V) - \frac{\sin \theta}{f} \frac{\partial}{\partial \theta}\left(\sin \theta \frac{\partial f}{\partial \theta} \right) = \frac{1}{g} \frac{\partial^2 g}{\partial \phi^2} \tag{7.7}$$

Look closely at Equation (7.7). Notice that only the variables r and θ (and their functions R and f) appear on the left side, whereas only ϕ and its function g appear on the right side. We have achieved a separation of variables, completely isolating ϕ. What does this mean? The left side of the equation cannot change as ϕ changes, because it does not contain ϕ or any function depending on ϕ. Similarly, the right side cannot change with either r or θ. The only way for this to be true is for each side of Equation (7.7) to be equal to a constant. We choose now to let this constant have the value $-m_\ell^2$ so that we can more easily introduce a new quantum number in Equation (7.15). If we set the constant $-m_\ell^2$ equal to the right side of Equation (7.7), we have

$$\frac{1}{g} \frac{\partial^2 g}{\partial \phi^2} = -m_\ell^2$$

or, after rearranging,

$$\frac{d^2 g}{d\phi^2} = -m_\ell^2 g \tag{7.8} \qquad \textbf{Azimuthal equation}$$

Notice that because ϕ is the only variable, we have replaced the partial derivative with the ordinary derivative. Because the angle ϕ in spherical coordinates corresponds to the azimuth angle in astronomy, Equation (7.8) is traditionally referred to as the **azimuthal equation**. This is simply the equation of a harmonic oscillator that we have studied in introductory physics, and the solutions for $g(\phi)$ will take the form of sines and cosines or exponential functions. We find it convenient to choose a solution of the form $e^{im_\ell \phi}$.

One may easily verify by direct substitution that $e^{im_\ell \phi}$ satisfies Equation (7.8) for any value of m_ℓ. However, in order to have a physically valid solution for any value of ϕ, it is necessary that the solution be single valued, that is, $g(\phi) = g(\phi + 2\pi)$. This means, for example, that $g(\phi = 0) = g(\phi = 2\pi)$, which requires that $e^0 = e^{2\pi i m_\ell}$. The only way for this to be true is for m_ℓ to be zero or an integer (either positive or negative). The quantum number m_ℓ is therefore restricted to be zero or a positive or negative integer. If the sign on the right-hand side of Equation (7.8) were positive rather than negative, the solution would not be physically realized. It could not be normalized and would not be single valued in ϕ. We shall defer further discussion of solutions for Equation (7.8) until later. For now it is sufficient to realize that readily obtainable solutions exist.

Now we set the left side of Equation (7.7) equal to the constant $-m_\ell^2$ and rearrange to have

$$\frac{1}{R}\frac{\partial}{\partial r}\left(r^2\frac{\partial R}{\partial r}\right) + \frac{2\mu r^2}{\hbar^2}(E - V) = \frac{m_\ell^2}{\sin^2\theta} - \frac{1}{f\sin\theta}\frac{\partial}{\partial\theta}\left(\sin\theta\frac{\partial f}{\partial\theta}\right) \quad (7.9)$$

Notice that we have again achieved a successful separation of variables, with everything depending on r on the left side and everything depending on θ on the right side. We can set each side of Equation (7.9) equal to a constant, which this time we call $\ell(\ell + 1)$. Doing so with each side of the equation in succession yields (after more rearrangement) the two equations

Radial equation

$$\frac{1}{r^2}\frac{d}{dr}\left(r^2\frac{dR}{dr}\right) + \frac{2\mu}{\hbar^2}\left[E - V - \frac{\hbar^2}{2\mu}\frac{\ell(\ell + 1)}{r^2}\right]R = 0 \quad (7.10)$$

and

Angular equation

$$\frac{1}{\sin\theta}\frac{d}{d\theta}\left(\sin\theta\frac{df}{d\theta}\right) + \left[\ell(\ell + 1) - \frac{m_\ell^2}{\sin^2\theta}\right]f = 0 \quad (7.11)$$

where, after separation, we have again replaced the partial derivatives with the ordinary ones.

The process of separation of variables is now complete. The original Schrödinger equation has been separated into three ordinary second-order differential equations [Equations (7.8), (7.10), and (7.11)], each containing only one variable.

Solution of the Radial Equation

Equation (7.10), appropriately called the *radial equation*, is another differential equation for which solutions are well known. It is called the **associated Laguerre equation** after the French mathematician Edmond Nicolas Laguerre (1834–1886). The solutions R to this equation that satisfy the appropriate boundary conditions are called *associated Laguerre functions*. We shall consider these solutions in some detail in Section 7.6. We can obtain some idea of how the ground-state wave function looks if we assume that the ground state has the lowest possible quantum number $\ell = 0$ of the system. We will soon see that this requires the value $m_\ell = 0$. Notice that $\ell = 0$ greatly simplifies the radial wave Equation (7.10) to be

$$\frac{1}{r^2}\frac{d}{dr}\left(r^2\frac{dR}{dr}\right) + \frac{2\mu}{\hbar^2}(E - V)R = 0 \quad (7.12)$$

The derivative of the bracketed expression in the first term of Equation (7.12) yields two terms by using the derivative product rule. We write out both of those terms and insert the Coulomb potential energy, Equation (7.1), to find

$$\frac{d^2R}{dr^2} + \frac{2}{r}\frac{dR}{dr} + \frac{2\mu}{\hbar^2}\left(E + \frac{e^2}{4\pi\epsilon_0 r}\right)R = 0 \quad (7.13)$$

Those students with some experience in solving differential equations will recognize that an exponential solution is required. We try a solution having the form

$$R = Ae^{-r/a_0}$$

where A is a normalization constant and a_0 is a constant with the dimension of length (we shall see that it was no accident that we chose the constant a_0!). It is reasonable to try to verify the trial solution by inserting it into the radial equation (7.13). The first and second derivatives are

$$\frac{dR}{dr} = -\frac{1}{a_0}R \qquad \frac{d^2R}{dr^2} = \frac{R}{a_0^2}$$

We insert these derivatives into Equation (7.13) and rearrange terms to yield

$$\left(\frac{1}{a_0^2} + \frac{2\mu}{\hbar^2}E\right) + \left(\frac{2\mu e^2}{4\pi\epsilon_0\hbar^2} - \frac{2}{a_0}\right)\frac{1}{r} = 0 \qquad (7.14)$$

By the same reasoning that we applied in the separation of variables method, the only way for Equation (7.14) to be satisfied for *any* value of r is for *each* of the two expressions in parentheses to be equal to zero. We set the second expression in parentheses equal to zero and solve for a_0 to find

$$a_0 = \frac{4\pi\epsilon_0\hbar^2}{\mu e^2}$$

We see that a_0 is in fact equal to the Bohr radius [see Equation (4.24)]! Now we set the first bracketed term in Equation (7.14) equal to zero and solve for E to find

$$E = -\frac{\hbar^2}{2\mu a_0^2} = -E_0$$

Again this is equal to the Bohr result, with E_0 having the value 13.6 eV.

Because we are not prepared to deal with the full scope of the associated Laguerre functions in this book, we shall not consider higher energy states here but will summarize some of the key results.

Introduction of Quantum Numbers The full solution to the radial wave equation requires (not surprisingly) the introduction of a quantum number, which we shall call n, such that n is a positive integer (but not zero). Equation (7.11), which we shall call the *angular equation*, was first solved by the famous mathematician Adrien-Marie Legendre (1752–1833). It is well known in the theory of differential equations as the **associated Legendre equation**. Application of the appropriate boundary conditions (see Section 6.1) to Equations (7.10) and (7.11), a process too lengthy to present here, leads to the following restrictions on the quantum numbers ℓ and m_ℓ:

$$\ell = 0, 1, 2, 3, \ldots$$
$$m_\ell = -\ell, -\ell + 1, \ldots, -2, -1, 0, 1, 2, \ldots, \ell - 1, \ell \qquad (7.15)$$

That is, the quantum number ℓ must be zero or a positive integer, and the quantum number m_ℓ must be a positive or negative integer, or zero, subject to the restriction that $|m_\ell| \le \ell$. The choice of $\ell(\ell + 1)$ as the constant for Equation (7.9) provides us with the succinct results in Equation (7.15). There is a further restriction that the quantum number ℓ can only take on values less than n. The consequences of this, along with a full consideration of allowed sets of the three quantum numbers n, ℓ, and m_ℓ, will be explained in Section 7.3. Let us note, however, that the predicted energy levels turn out to be

$$E_n = -\frac{E_0}{n^2}$$

in agreement with the Bohr result.

With the introduction of the quantum numbers, we are prepared to present the wave functions. The first few radial wave functions $R_{n\ell}$ are listed in Table 7.1, where $a_0 =$ Bohr radius $= 0.53 \times 10^{-10}$ m. Note the subscripts on R specify the values of n and ℓ.

Solution of the Angular and Azimuthal Equations

We now return to the azimuthal Equation (7.8). We note that its solutions can be expressed in exponential form as $e^{im_\ell\phi}$ or $e^{-im_\ell\phi}$. But because the angular equation also contains the quantum number m_ℓ, solutions to the angular and azimuthal equations are linked. It is customary to group these solutions together into functions called the **spherical harmonics** $Y(\theta, \phi)$, defined as

Spherical harmonics

$$Y(\theta, \phi) = f(\theta)g(\phi) \tag{7.16}$$

The $f(\theta)$ part of the $Y(\theta, \phi)$ is always a polynomial function of $\sin\theta$ and $\cos\theta$ of order ℓ. See Table 7.2 for a listing of the normalized spherical harmonics up to $\ell = 3$.

Example 7.1

Show that the spherical harmonic $Y_{11}(\theta, \phi)$ satisfies the angular Equation (7.11).

Strategy We insert the value for $Y_{11}(\theta, \phi)$ into Equation (7.11) and let the values $\ell = 1$ and $m_\ell = 1$. Because $Y(\theta, \phi) = f(\theta)g(\phi)$ [see Equation (7.16)], and θ and ϕ are independent variables, we will be able to separate the constants and variable ϕ from the factors involving θ.

Solution We first write the value of $Y_{11}(\theta, \phi)$ from Table 7.2 and separate the factor involving θ from all the other factors.

$$Y_{11}(\theta, \phi) = -\frac{1}{2}\sqrt{\frac{3}{2\pi}}\sin\theta \, e^{i\phi} = A\sin\theta$$

where the term A includes no factors involving θ.

After inserting the values of ℓ, m_ℓ, and $Y_{11}(\theta, \phi)$ into Equation (7.11), we have

$$\frac{1}{\sin\theta}\frac{d}{d\theta}\left[A\sin\theta\frac{d(\sin\theta)}{d\theta}\right] + \left(2 - \frac{1}{\sin^2\theta}\right)A\sin\theta = 0$$

We divide by A and take the derivative inside the square bracket and find

$$\frac{1}{\sin\theta}\frac{d}{d\theta}(\sin\theta\cos\theta) + 2\sin\theta - \frac{1}{\sin\theta} = 0$$

We take the final derivative and rewrite the resulting term in square brackets.

$$\frac{1}{\sin\theta}\left[\cos^2\theta - \sin^2\theta\right] + 2\sin\theta - \frac{1}{\sin\theta} = 0$$

$$\frac{1}{\sin\theta}\left[1 - 2\sin^2\theta\right] + 2\sin\theta - \frac{1}{\sin\theta} = 0$$

$$\frac{1}{\sin\theta} - 2\sin\theta + 2\sin\theta - \frac{1}{\sin\theta} = 0$$

$$0 = 0$$

So we indeed found that Equation (7.11) is satisfied by $Y_{11}(\theta, \phi)$.

Table 7.1		Hydrogen Atom Radial Wave Functions
n	ℓ	$R_{n\ell}(r)$
1	0	$\dfrac{2}{(a_0)^{3/2}}e^{-r/a_0}$
2	0	$\left(2-\dfrac{r}{a_0}\right)\dfrac{e^{-r/2a_0}}{(2a_0)^{3/2}}$
2	1	$\dfrac{r}{a_0}\dfrac{e^{-r/2a_0}}{\sqrt{3}(2a_0)^{3/2}}$
3	0	$\dfrac{1}{(a_0)^{3/2}}\dfrac{2}{81\sqrt{3}}\left(27-18\dfrac{r}{a_0}+2\dfrac{r^2}{a_0^2}\right)e^{-r/3a_0}$
3	1	$\dfrac{1}{(a_0)^{3/2}}\dfrac{4}{81\sqrt{6}}\left(6-\dfrac{r}{a_0}\right)\dfrac{r}{a_0}e^{-r/3a_0}$
3	2	$\dfrac{1}{(a_0)^{3/2}}\dfrac{4}{81\sqrt{30}}\dfrac{r^2}{a_0^2}e^{-r/3a_0}$

Table 7.2		Normalized Spherical Harmonics $Y(\theta,\phi)$
ℓ	m_ℓ	$Y_{\ell m_\ell}$
0	0	$\dfrac{1}{2\sqrt{\pi}}$
1	0	$\dfrac{1}{2}\sqrt{\dfrac{3}{\pi}}\cos\theta$
1	± 1	$\mp\dfrac{1}{2}\sqrt{\dfrac{3}{2\pi}}\sin\theta\,e^{\pm i\phi}$
2	0	$\dfrac{1}{4}\sqrt{\dfrac{5}{\pi}}(3\cos^2\theta-1)$
2	± 1	$\mp\dfrac{1}{2}\sqrt{\dfrac{15}{2\pi}}\sin\theta\cos\theta\,e^{\pm i\phi}$
2	± 2	$\dfrac{1}{4}\sqrt{\dfrac{15}{2\pi}}\sin^2\theta\,e^{\pm 2i\phi}$
3	0	$\dfrac{1}{4}\sqrt{\dfrac{7}{\pi}}(5\cos^3\theta-3\cos\theta)$
3	± 1	$\mp\dfrac{1}{8}\sqrt{\dfrac{21}{\pi}}\sin\theta(5\cos^2\theta-1)\,e^{\pm i\phi}$
3	± 2	$\dfrac{1}{4}\sqrt{\dfrac{105}{2\pi}}\sin^2\theta\cos\theta\,e^{\pm 2i\phi}$
3	± 3	$\mp\dfrac{1}{8}\sqrt{\dfrac{35}{\pi}}\sin^3\theta\,e^{\pm 3i\phi}$

The probability density for the electron in the hydrogen atom is given by $\psi^*\psi$. Therefore, the spherical harmonics together with the radial wave function R will determine the overall shape of the probability density for the various quantum states. The total wave function $\psi(r, \theta, \phi)$ depends on the quantum numbers n, ℓ, and m_ℓ. We can now write the wave function as

$$\psi_{n\ell m_\ell}(r, \theta, \phi) = R_{n\ell}(r)Y_{\ell m_\ell}(\theta, \phi) \tag{7.17}$$

where we indicate by the subscripts that $R_{n\ell}(r)$ depends only on n and ℓ, and $Y_{\ell m_\ell}(\theta, \phi)$ depends only on ℓ and m_ℓ. We shall look at these wave functions again in Section 7.6.

Example 7.2

Show that the hydrogen wave function ψ_{211} is normalized.

Strategy We refer to Equation (6.8) in Chapter 6 where we normalized the wave function in one dimension. Now we want to normalize the wave function in three dimensions in spherical polar coordinates. The normalization condition is

$$\int \psi_{n\ell m_\ell}^*\psi_{n\ell m_\ell}\, d\tau = 1 = \int \psi_{211}^*\psi_{211}r^2 \sin\theta\, dr\, d\theta\, d\phi \tag{7.18}$$

where $d\tau = r^2 \sin\theta\, dr\, d\theta\, d\phi$ is the volume element. We look up the wave function ψ_{211} using Tables 7.1 and 7.2.

$$\psi_{211} = R_{21}Y_{11} = \left[\frac{r}{a_0}\frac{e^{-r/2a_0}}{\sqrt{3}(2a_0)^{3/2}}\right]\left[\frac{1}{2}\sqrt{\frac{3}{2\pi}}\sin\theta e^{i\phi}\right]$$

Solution We insert the wave function ψ_{211} into Equation (7.18), insert the integration limits for r, θ, and ϕ, and do the integration. First we find $\psi_{211}^*\psi_{211}$:

$$\psi_{211}^*\psi_{211} = \frac{1}{64\pi a_0^5}r^2 e^{-r/a_0}\sin^2\theta$$

where we have combined factors. The normalization condition from Equation (7.18) becomes

$$\int \psi_{211}^*\psi_{211}r^2 \sin\theta\, dr\, d\theta\, d\phi$$

$$= \frac{1}{64\pi a_0^5}\int_0^\infty r^4 e^{-r/a_0}\, dr \int_0^\pi \sin^3\theta\, d\theta \int_0^{2\pi} d\phi$$

$$= \frac{1}{64\pi a_0^5}[24a_0^5]\left[\frac{4}{3}\right][2\pi]$$

$$= 1$$

We have not shown all the steps in the integration, but we have shown the results of each integration in each of the square brackets. The integrals needed are in Appendix 3. The wave function is indeed normalized.

7.3 Quantum Numbers

The three quantum numbers obtained from solving Equation (7.3) are

n **Principal quantum number**

ℓ **Orbital angular momentum quantum number**

m_ℓ **Magnetic quantum number**

Boundary conditions were given in Section 6.1.   Their values are obtained by applying the boundary conditions to the wave function $\psi(r, \theta, \phi)$ as discussed in Section 6.1. The boundary conditions require that the wave functions have acceptable properties, including being single valued and finite. The restrictions imposed by the boundary conditions are

$$n = 1, 2, 3, 4, \ldots \qquad\qquad \text{Integer}$$

$$\ell = 0, 1, 2, 3, \ldots, n - 1 \qquad \text{Integer} \qquad (7.19)$$

$$m_\ell = -\ell, -\ell + 1, \ldots, 0, 1, \ldots, \ell - 1, \ell \quad \text{Integer}$$

These three quantum numbers must be integers. The orbital angular momentum quantum number must be less than the principal quantum number, $\ell < n$, and the magnitude of the magnetic quantum number (which may be positive or negative) must be less than or equal to the orbital angular momentum quantum number, $|m_\ell| \leq \ell$. We can summarize these conditions as

$$n > 0$$

$$\ell < n \qquad\qquad (7.20)$$

$$|m_\ell| \leq \ell$$

The lowest value of n is 1, and for $n = 1$, we must have $\ell = 0$, $m_\ell = 0$. For $n = 2$, we may have $\ell = 0$, $m_\ell = 0$ as well as $\ell = 1$, $m_\ell = -1, 0, +1$.

Conceptual Example 7.3

What are the possible quantum numbers for the state $n = 4$ in atomic hydrogen?

Solution We want to apply the restrictions for the quantum numbers given in Equations (7.19) and (7.20). If $n = 4$, then the possible values of ℓ are $\ell = 0, 1, 2, 3$, because $\ell_{\max} = n - 1$. For each value of ℓ, m_ℓ goes from $-\ell$ to $+\ell$. We show the results in tabular form.

n	ℓ	m_ℓ
4	0	0
4	1	$-1, 0, 1$
4	2	$-2, -1, 0, 1, 2$
4	3	$-3, -2, -1, 0, 1, 2, 3$

As yet these quantum numbers may seem to have little physical meaning. Let us examine each of them more carefully and try to find classical analogies where possible.

Principal Quantum Number n

The principal quantum number n results from the solution of the radial wave function $R(r)$ in Equation (7.4). Because the radial equation includes the potential energy $V(r)$, it is not surprising to find that the boundary conditions on $R(r)$ quantize the energy E. The result for this quantized energy is

$$E_n = \frac{-\mu}{2}\left(\frac{e^2}{4\pi\epsilon_0\hbar}\right)^2 \frac{1}{n^2} = -\frac{E_0}{n^2} \qquad (7.21)$$

which is precisely the value found in Chapter 4 from the Bohr theory [Equations (4.25) and (4.26)]. So far, the energy levels of the hydrogen atom depend only on the principal quantum number n. The negative value of the energy E indicates that the electron and proton are bound together.

It is perhaps surprising that the total energy of the electron does not depend on the angular momentum. However, a similar situation occurs for planetary motion, where the energy depends on the semimajor axis of the elliptical planetary orbits and not on the eccentricity of the orbits. This peculiarity occurs for the solar system and the hydrogen atom because both the gravitational and Coulomb forces are central; they also both have inverse-square-law dependences on distance.

Orbital Angular Momentum Quantum Number ℓ

The orbital angular momentum quantum number ℓ is associated with the $R(r)$ and $f(\theta)$ parts of the wave function. The electron-proton system has orbital angular momentum as the particles pass around each other. Classically, this orbital angular momentum is $\vec{L} = \vec{r} \times \vec{p}$ with magnitude $L = mv_{\text{orbital}}r$, where v_{orbital} is the orbital velocity, perpendicular to the radius. The quantum number ℓ is related to the magnitude of the orbital angular momentum L by

$$L = \sqrt{\ell(\ell + 1)}\,\hbar \tag{7.22}$$

This curious dependence of L on ℓ [$L^2 \sim \ell(\ell + 1)$ rather than ℓ^2] is a wave phenomenon—it results from the application of the boundary conditions on $\psi(r, \theta, \phi)$. We will present a justification for it later in this section. The quantum result disagrees with the elementary Bohr theory of the hydrogen atom, where $L = n\hbar$. This is most obvious in an $\ell = 0$ state, where $L = \sqrt{0(1)}\,\hbar = 0$. Based on these results, we will have to discard Bohr's semiclassical "planetary" model of electrons orbiting a nucleus.

We show in Figure 7.2 several classical orbits corresponding to the same total energy. For an electron in an atom, the energy depends on n; for planetary motion, the energy depends on the semimajor axis. Do not take the elliptical orbits literally for electrons; only probability functions can describe the electron positions, which must be consistent with the uncertainty principle. We say that a certain energy level is *degenerate* with respect to ℓ when the energy is independent of the value of ℓ (see Section 6.5). For example, the energy for an $n = 3$ level is the same for all possible values* of ℓ ($\ell = 0, 1, 2$).

It is customary to use letter names for the various ℓ values. These are

$\ell =$	0	1	2	3	4	5 ...
Letter $=$	s	p	d	f	g	h ...

Figure 7.2 Various possible electron (or planetary) classical orbits. The energy depends only on the principal quantum number n and not on the angular momentum of the hydrogen atom. There is a finite probability for an $\ell = 0$ electron to be present within the nucleus. Of course, none of the planets has $\ell = 0$, and (obviously) they do not pass through the sun.

These particular letter designations for the first four values resulted from empirical visual observations from early experiments: *sharp, principal, diffuse,* and *fundamental*. After $\ell = 3$ (f state), the letters follow alphabetical order.

Atomic states are normally referred to by their n number and ℓ letter. Thus a state with $n = 2$ and $\ell = 1$ is called a $2p$ state. Examples of other various atomic states are $1s$ ($n = 1$, $\ell = 0$), $2s$ ($n = 2$, $\ell = 0$), $4d$ ($n = 4$, $\ell = 2$), and $6g$ ($n = 6$, $\ell = 4$). A state such as $2d$ is not possible, because this refers to $n = 2$ and $\ell = 2$. Our boundary conditions require $n > \ell$.

---

*This statement is true for single-electron atoms like hydrogen. We will learn later in Chapter 8 that for many-electron atoms (atoms with more than 1 electron), electrons with lower ℓ values lie lower in energy for a given n value.

Magnetic Quantum Number m_ℓ

The orbital angular momentum quantum number ℓ determines the magnitude of the angular momentum $\vec{L}$, but because $\vec{L}$ is a vector, it also has a direction. Classically, because there is no torque in the hydrogen atom system in the absence of external fields, the angular momentum $\vec{L}$ is a constant of the motion and is conserved. The solution to the Schrödinger equation for $f(\theta)$ specified that ℓ must be an integer, and therefore the magnitude of $\vec{L}$ is quantized.

The angle ϕ is a measure of the rotation about the z axis. The solution for $g(\phi)$ specifies that m_ℓ is an integer *and* related to the z component of the angular momentum $\vec{L}$.

$$L_z = m_\ell \hbar \tag{7.23}$$

The relationship of L, L_z, ℓ, and m_ℓ is displayed in Figure 7.3 for the value $\ell = 2$. The magnitude of L is fixed $\left[L = \sqrt{\ell(\ell + 1)}\hbar = \sqrt{6}\hbar\right]$. Because L_z is quantized, only certain orientations of $\vec{L}$ are possible, each corresponding to a different m_ℓ (and therefore L_z). This phenomenon is called **space quantization** because only certain orientations of $\vec{L}$ are allowed in space.

Space quantization

We can ask whether we have established a preferred direction in space by choosing the z axis. The choice of the z axis is completely arbitrary unless there is an external magnetic field to define a preferred direction in space. It is customary to choose the z axis to be along $\vec{B}$ if there is a magnetic field. This is why m_ℓ is called the *magnetic quantum number*.

Will the angular momentum be quantized along the x and y axes as well? The answer is that quantum mechanics allows $\vec{L}$ to be quantized along only one direction in space. Because we know the magnitude of $\vec{L}$, the knowledge of a second component would imply a knowledge of the third component as well because of the relation $L^2 = L_x^2 + L_y^2 + L_z^2$. The following argument shows that this would violate the Heisenberg uncertainty principle: If all three components of $\vec{L}$ were known, then the direction of $\vec{L}$ would also be known. In this case we would have a precise knowledge of one component of the electron's position in space, because the electron's orbital motion is confined to a plane perpendicular to $\vec{L}$. But confinement of the electron to that plane means that the electron's

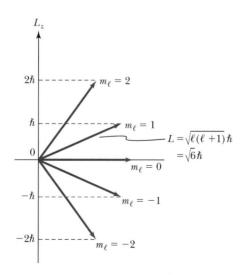

Figure 7.3 Schematic diagram of the relationship between $\vec{L}$ and L_z with the allowed values of m_ℓ.

momentum component along $\vec{L}$ is *exactly* zero. This simultaneous knowledge of the same component of position and momentum is forbidden by the uncertainty principle.

Only the magnitude $\left|\vec{L}\right|$ and L_z may be specified simultaneously. The values of L_x and L_y must be consistent with $L^2 = L_x^2 + L_y^2 + L_z^2$ but cannot be specified individually. Physicists refer to the known values of L and L_z as "sharp" and the unknown L_x and L_y as "fuzzy." The angular momentum vector $\vec{L}$ *never* points in the z direction (see Figure 7.3) because $L = \sqrt{\ell(\ell+1)}\hbar$ and $\left|\vec{L}\right| > \left|L_z\right|_{max} = \ell\hbar$. Our results from solving the Schrödinger equation for the hydrogen atom are consistent with the uncertainty principle.

The space quantization just mentioned is an experimental fact. The values of L_z range from $-\ell$ to $+\ell$ in steps of 1, for a total of $2\ell + 1$ allowed values. Because there is nothing special about the three directions x, y, and z, we expect the average of the angular momentum components squared in the three directions to be the same, $\langle L_x^2 \rangle = \langle L_y^2 \rangle = \langle L_z^2 \rangle$. The average value of $\langle L^2 \rangle$ is equal to three times the average value of the square of any one of the components, so we choose the z component, $\langle L^2 \rangle = 3\langle L_z^2 \rangle$. To find the average value of L_z^2, we just have to sum all the squares of the quantum numbers for L_z and divide by the total number, $2\ell + 1$.

$$\langle L^2 \rangle = 3\langle L_z^2 \rangle = \frac{3}{2\ell + 1} \sum_{m_\ell=-\ell}^{\ell} m_\ell^2 \hbar^2 = \ell(\ell + 1)\hbar^2 \tag{7.24}$$

where we have used a math table for the summation result. This rather simple argument to explain the $\ell(\ell + 1)$ dependence for the expectation value of L^2 (rather than using a sophisticated quantum-mechanical calculation) was originally due to Richard Feynman and simplified by P. W. Milonni.*

*P. W. Milonni, *American Journal of Physics* **58**, 1012 (1990).

Example 7.4

What is the degeneracy of the $n = 3$ level? That is, how many different states are contained in the energy level $E_3 = -E_0/9$?

Strategy This is a good opportunity to review the quantum numbers we have been discussing. The energy eigenvalues for atomic hydrogen depend only on the principal quantum number n (in the absence of a magnetic field). For each value of n, there can be n different orbital angular momentum ℓ states ($\ell = 0, 1, \ldots, n - 1$). For each value of ℓ, there are $2\ell + 1$ different magnetic quantum states ($m_\ell = -\ell, -\ell + 1, \ldots, 0, 1, \ldots, +\ell$).

Solution To find the total degeneracy for $n = 3$ we have to add all the possibilities.

n	ℓ	m_ℓ	$2\ell + 1$
3	0	0	1
3	1	$-1, 0, 1$	3
3	2	$-2, -1, 0, 1, 2$	5
		Total = 9	

The $n = 3$ level is degenerate (in the absence of a magnetic field) because all nine states have the same energy but different quantum numbers. Their wave functions, however, are quite different. You may notice that, in general, the degeneracy is n^2 (see Problem 16).

7.4 Magnetic Effects on Atomic Spectra— The Normal Zeeman Effect

It was shown as early as 1896 by the Dutch physicist Pieter Zeeman that the spectral lines emitted by atoms placed in a magnetic field broaden and appear to split. The splitting of an energy level into multiple levels in the presence of an external magnetic field is called the **Zeeman effect**. When a spectral line is split into three lines, it is called the *normal* Zeeman effect. But more often a spectral line is split into more than three lines; this effect is called the *anomalous* Zeeman effect. The normal Zeeman effect, discussed here, can be understood by considering the atom to behave like a small magnet. We will return to our discussion of the anomalous Zeeman effect, which is more complicated, in Section 8.3. By the 1920s considerable **fine structure** of atomic spectral lines from hydrogen and other elements had been observed. Fine structure refers to the splitting of a spectral line into two or more closely spaced lines.

As a rough model, think of an electron circulating around the nucleus as a circular current loop. The current loop has a magnetic moment $\mu = IA$ where A is the area of the current loop and the current $I = dq/dt$ is simply the electron charge ($q = -e$) divided by the period T for the electron to make one revolution ($T = 2\pi r/v$). Thus

$$\mu = IA = \frac{q}{T}A = \frac{(-e)\pi r^2}{2\pi r/v} = \frac{-erv}{2} = -\frac{e}{2m}L \tag{7.25}$$

where $L = mvr$ is the magnitude of the orbital angular momentum. Both the magnetic moment $\vec{\mu}$ and angular momentum $\vec{L}$ are vectors so that

$$\vec{\mu} = -\frac{e}{2m}\vec{L} \tag{7.26}$$

The relationship between $\vec{\mu}$ and $\vec{L}$ is displayed in Figure 7.4.

In the absence of an external magnetic field to align them, the magnetic moments $\vec{\mu}$ of atoms point in random directions. In classical electromagnetism, if a magnetic dipole having a magnetic moment $\vec{\mu}$ is placed in an external magnetic field, the dipole will experience a torque $\vec{\tau} = \vec{\mu} \times \vec{B}$ tending to align the dipole with the magnetic field. The dipole also has a potential energy V_B in the field given by

$$V_B = -\vec{\mu} \cdot \vec{B} \tag{7.27}$$

According to classical physics, if the system can change its potential energy, the magnetic moment will align itself with the external magnetic field to minimize energy.

Note the similarity with the case of the spinning top in a gravitational field. A child's spinning top is said to precess about the gravitational field; that is, the axis (around which the spinning top is rotating) itself rotates about the direction of the gravitational force (vertical). The gravitational field is not parallel to the angular momentum, and the force of gravity pulling down on the spinning top results in a ***precession*** of the top about the field direction. Precisely the same thing happens with the magnetic moment of an atom in a magnetic field. The angular momentum is aligned with the magnetic moment, and the torque between $\vec{\mu}$ and $\vec{B}$ causes a precession of $\vec{\mu}$ about the magnetic field (see Figure 7.5), not an

The Dutch physicist Pieter Zeeman (1865–1943) studied at the University of Leiden under the famous physicists H. Kamerlingh Onnes and H. A. Lorentz and received his degree in 1890. While at Leiden he showed that atomic spectral lines were split under the influence of an applied magnetic field. After this discovery he left Leiden in 1897 to go to the University of Amsterdam where he remained until 1935. He shared the 1902 Nobel Prize for Physics with his mentor Lorentz. *AIP Emilio Segrè Visual Archives, W. F. Meggers Collection.*

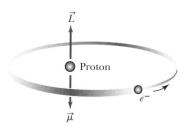

Figure 7.4 Representation of the orbital angular momentum $\vec{L}$ and magnetic moment $\vec{\mu}$ of the hydrogen atom due to the electron orbiting the proton. The directions of $\vec{L}$ and $\vec{\mu}$ are opposite because of the negative electron charge.

alignment. The magnetic field establishes a preferred direction in space along which we customarily define the z axis. Then we have

$$\mu_z = \frac{e\hbar}{2m} m_\ell = -\mu_B m_\ell \tag{7.28}$$

Bohr magneton　where $\mu_B = e\hbar/2m$ is a unit of magnetic moment called a **Bohr magneton**. Because of the quantization of L_z and the fact that $L = \sqrt{\ell(\ell+1)}\,\hbar > m_\ell \hbar$, we cannot have $|\vec{\mu}| = \mu_z$; the magnetic moment cannot align itself exactly in the z direction. Just like the angular momentum $\vec{L}$, the magnetic moment $\vec{\mu}$ has only certain allowed quantized orientations. Note also that in terms of the Bohr magneton, $\vec{\mu} = -\mu_B \vec{L}/\hbar$.

Example 7.5

Determine the precessional frequency of an atom having magnetic moment $\vec{\mu}$ in an external magnetic field $\vec{B}$. This precession is known as the *Larmor precession*.

Strategy We have already seen that the torque $\vec{\tau}$ is equal to $\vec{\mu} \times \vec{B}$, but we also know from classical mechanics that the torque is $d\vec{L}/dt$. The torque in Figure 7.5 is perpendicular to $\vec{\mu}$, $\vec{L}$, and $\vec{B}$ and is out of the page. This must also be the direction of the change in momentum $d\vec{L}$ as seen in Figure 7.5. Thus $\vec{L}$ and $\vec{\mu}$ precess about the magnetic field. The Larmor frequency ω_L is given by $d\phi/dt$.

Solution The magnitude of $d\vec{L}$ is given by $L \sin\theta\, d\phi$ (see Figure 7.5), so ω_L is given by

$$\omega_L = \frac{d\phi}{dt} = \frac{1}{L \sin\theta}\frac{dL}{dt} \tag{7.29}$$

We now insert the magnitude of $L = 2m\mu/e$ from Equation (7.26). The value of dL/dt, the magnitude of $\vec{\mu} \times \vec{B}$, can be determined from Figure 7.5 to be $\mu B \sin\theta$. Equation (7.29) becomes

$$\omega_L = \left(\frac{e}{2m\mu \sin\theta}\right)\mu B \sin\theta = \frac{eB}{2m} \tag{7.30}$$

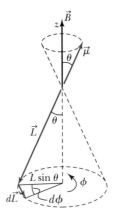

Figure 7.5 An atom having magnetic moment $\vec{\mu}$ feels a torque $\vec{\tau} = \vec{\mu} \times \vec{B}$ due to an external magnetic field $\vec{B}$. This torque must also be equal to $d\vec{L}/dt$. The vectors $\vec{\mu}$ and $\vec{L}$ are antiparallel, so the vector $d\vec{L}/dt$ must be perpendicular to $\vec{\mu}$, $\vec{B}$, and $\vec{L}$. As shown in the figure, $d\vec{L}/dt$ requires both $\vec{\mu}$ and $\vec{L}$ to precess (angle ϕ) about the magnetic field $\vec{B}$.

What about the energy of the orbiting electron in a magnetic field? It takes work to rotate the magnetic moment away from $\vec{B}$. With $\vec{B}$ along the z direction, we have from Equation (7.27)

$$V_B = -\mu_z B = +\mu_B m_\ell B \tag{7.31}$$

The potential energy is thus quantized according to the *magnetic quantum number* m_ℓ; each (degenerate) atomic level of given ℓ is split into $2\ell + 1$ different energy states according to the value of m_ℓ. The energy degeneracy of a given $n\ell$ level

(a)

(b)

Figure 7.6 The normal Zeeman effect. (a) An external magnetic field removes the degeneracy of a $2p$ level and reveals the three different energy states. (b) There are now transitions with three different energies between an excited $2p$ level and the $1s$ ground state in atomic hydrogen. The energy ΔE has been grossly exaggerated along the energy scale.

is removed by a magnetic field (see Figure 7.6a). If the degenerate energy of a state is given by E_0, the three different energies in a magnetic field B for a $\ell = 1$ state is

m_ℓ	Energy
1	$E_0 + \mu_B B$
0	E_0
−1	$E_0 - \mu_B B$

Conceptual Example 7.6

What is the lowest $n\ell$ state in the hydrogen atom that has a degeneracy of 5?

Solution We want to find the lowest energy $n\ell$ state that has five m_ℓ states. That will be true for a $\ell = 2$ state, because

$2\ell + 1 = 5$. The lowest possible $\ell = 2$ state will be $3d$, because $n > \ell$ is required.

Example 7.7

What is the value of the Bohr magneton? Use that value to calculate the energy difference between the $m_\ell = 0$ and $m_\ell = +1$ components in the $2p$ state of atomic hydrogen placed in an external field of 2 T.

Strategy To find the Bohr magneton we insert the known values of e, $\hbar$, and m into the equation for μ_B [see text after Equation (7.28)]. The energy difference is determined from Equation (7.31).

Solution The Bohr magneton is determined to be

$$\mu_B = \frac{e\hbar}{2m}$$

$$= \frac{(1.602 \times 10^{-19}\,\text{C})(1.055 \times 10^{-34}\,\text{J·s})}{2(9.11 \times 10^{-31}\,\text{kg})}$$

$$\mu_B = 9.27 \times 10^{-24}\,\text{J/T} \tag{7.32}$$

The international system of units has been used (T = tesla for magnetic field). The energy splitting is determined from Equation (7.31) (see also Figure 7.6a)

$$\Delta E = \mu_B B \, \Delta m_\ell \qquad (7.33)$$

where $\Delta m_\ell = 1 - 0 = 1$. Hence, we have

$$\Delta E = (9.27 \times 10^{-24}\,\text{J/T})(2\,\text{T}) = 1.85 \times 10^{-23}\,\text{J}$$
$$= 1.16 \times 10^{-4}\,\text{eV}$$

An energy difference of 10^{-4} eV is easily observed by optical means.

Otto Stern (1888–1969) was born in a part of Germany (now in Poland), where he was educated. He worked in several universities until he left Germany in 1933 to avoid Nazi persecution and emigrated to the United States. He was educated and trained as a theorist, but changed to experimentation when he began his molecular beam experiments in 1920 at the University of Frankfurt with Walter Gerlach. He continued his distinguished career in Hamburg and later at Carnegie Institute of Technology in Pittsburgh. He received the Nobel Prize in 1943. *AIP Emilio Segrè Visual Archives, Segrè Collection.*

The splitting of spectral lines, called the normal Zeeman effect, can be partially explained by the application of external magnetic fields (see Figure 7.6). When a magnetic field is applied, the 2p level of atomic hydrogen is split into three different energy states (Figure 7.6a) with the energy difference given by Equation (7.33). A transition for an electron in the excited 2p level to the 1s ground state results in three different energy transitions as shown (greatly exaggerated) in Figure 7.6b. The energy differences between the three spectral lines shown in Figure 7.6b are quite small and were first observed by Pieter Zeeman in 1896. The application of external magnetic fields eliminates much of the energy degeneracy, because quantized states that previously had the same energy now have slight differences. When electrons make the transition between these states, the photons absorbed or produced have more widely varying energies. We will see in Section 7.6 that the selection rule for m_ℓ does not allow more than three different spectral lines in the normal Zeeman effect (see Problem 33).

Efforts were begun in the 1920s to detect the effects of space quantization (m_ℓ) by measuring the energy difference ΔE as in Example 7.7. In 1922 O. Stern and W. Gerlach reported the results of an experiment that clearly showed evidence for space quantization. If an external magnetic field is inhomogeneous—for example, if it is stronger at the south pole than at the north pole—then there will be a net force on a magnet placed in the field as well as a torque. This force is represented in Figure 7.7, where the net force on $\vec{\mu}$ (direction of S to N in bar magnet) is different for different orientations of $\vec{\mu}$ in the inhomogeneous magnetic field $\vec{B}$.

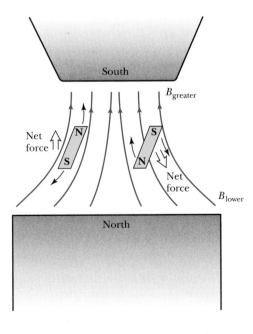

Figure 7.7 An inhomogeneous magnetic field is created by the smaller south pole. Two bar magnets representing atomic magnetic moments have $\vec{\mu}$ in opposite directions. Because the force on the top of the bar magnets is greater than that on the bottom, there will be a net translational force on the bar magnets (atoms).

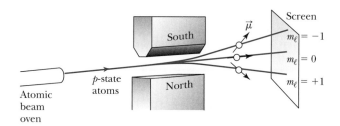

Figure 7.8 Schematic diagram of expected result of Stern-Gerlach experiment if atoms in a p state are used. Three patterns of atoms, due to $m_\ell = \pm 1, 0$, are expected to be observed on the screen. The magnet poles are arranged to produce a magnetic field gradient as shown in Figure 7.7. The experiment performed by Stern and Gerlach reported only two lines, not three (see Section 7.5).

Now if we pass an atomic beam of particles in the $\ell = 1$ state through a magnetic field along the z direction, we have from Equation (7.31) that $V_B = -\mu_z B$, and the force on the particles is $F_z = -(dV_B/dz) = \mu_z(dB/dz)$. There will be a different force on each of the three possible m_ℓ states. A schematic diagram of the Stern-Gerlach experiment is shown in Figure 7.8. The $m_\ell = +1$ state will be deflected down, the $m_\ell = -1$ state up, and the $m_\ell = 0$ state will be undeflected.

Stern and Gerlach performed their experiment with silver atoms and observed two distinct lines, not three. This was clear evidence of space quantization, although the number of m_ℓ states is always odd $(2\ell + 1)$ and should have produced an odd number of lines if the space quantization were due to the magnetic quantum number m_ℓ.

Example 7.8

In 1927 T. E. Phipps and J. B. Taylor of the University of Illinois reported an important experiment similar to the Stern-Gerlach experiment but using hydrogen atoms instead of silver. This was done because hydrogen is the simplest atom, and the separation of the atomic beam in the inhomogeneous magnetic field would allow a clearer interpretation. The atomic hydrogen beam was produced in a discharge tube having a temperature of 663 K. The highly collimated beam passed along the x direction through an inhomogeneous field (of length 3 cm) having an average value of 1240 T/m along the z direction. If the magnetic moment of the hydrogen atom is 1 Bohr magneton, what is the separation of the atomic beam?

Strategy The force can be found from the potential energy of Equation (7.31).

$$F_z = -\frac{dV}{dz} = \mu_z \frac{dB}{dz}$$

The acceleration of the hydrogen atom along the z direction is $a_z = F_z/m$. The separation of the atom along the z direction due to this acceleration is $d = a_z t^2/2$. The time that the atom spends within the inhomogeneous field is $t = \Delta x/v_x$ where Δx is the length of the inhomogeneous field, and v_x is the constant speed of the atom within the field. The separation d is therefore found from

$$d = \frac{1}{2}a_z t^2 = \frac{1}{2}\left(\frac{F_z}{m}\right)t^2 = \frac{1}{2m}\left(\mu_z \frac{dB}{dz}\right)\left(\frac{\Delta x}{v_x}\right)^2$$

We know all the values needed to determine d except the speed v_x, but we do know the temperature of the hydrogen gas. The average energy of the atoms collimated along the x direction is $\frac{1}{2}m\langle v_x^2 \rangle = \frac{3}{2}kT$.

Solution We calculate $\langle v_x^2 \rangle$ to be

$$v_x^2 = \frac{3kT}{m} = \frac{3(1.38 \times 10^{-23}\,\text{J/T})(663\,\text{K})}{1.67 \times 10^{-27}\,\text{kg}}$$

$$= 1.64 \times 10^7\,\text{m}^2/\text{s}^2$$

The separation d of the one atom is now determined to be

$$d = \frac{1}{2(1.67 \times 10^{-27}\,\text{kg})}(9.27 \times 10^{-24}\,\text{J/T})(1240\,\text{T/m})$$

$$\times \frac{(0.03\,\text{m})^2}{(1.64 \times 10^7\,\text{m}^2/\text{s}^2)} = 0.19 \times 10^{-3}\,\text{m}$$

Phipps and Taylor found only two distinct lines as did Stern and Gerlach for silver atoms, and the separation of the lines from the central ray with no magnetic field was 0.19 mm as we just calculated! The total separation of the two lines (one deflected up and one down) was 0.38 mm. The mystery still remained as to why there were only two lines.

7.5   Intrinsic Spin

It was clear by the early 1920s that there was a problem with space quantization and the number of lines observed in the Stern-Gerlach experiment. Wolfgang Pauli was the first to suggest that a fourth quantum number (after n, ℓ, m_ℓ) assigned to the electron might explain the anomalous optical spectra discussed in the previous section. His reasoning for four quantum numbers was based on relativity, where there are four coordinates—three space and one time. The physical significance of this fourth quantum number was not made clear.

In 1925 Samuel Goudsmit and George Uhlenbeck, two young physics graduate students in Holland, proposed that *the electron must have an intrinsic angular momentum* and therefore a magnetic moment (because the electron is charged). Classically, this corresponds in the planetary model to the fact that the Earth rotates on its own axis as it orbits the sun. However, this simple classical picture runs into serious difficulties when applied to the spinning charged electron. In order to achieve the angular momentum needed, Paul Ehrenfest showed that the surface of the spinning electron (or electron cloud) would have to be moving at a velocity greater than the speed of light! If such an intrinsic angular momentum exists, we must regard it as a *purely quantum-mechanical result* (see Problems 44 and 45).

Intrinsic spin quantum number

In order to explain experimental data, Goudsmit and Uhlenbeck proposed that the electron must have an **intrinsic spin quantum number** $s = 1/2$. The spinning electron reacts similarly to the orbiting electron in a magnetic field. Therefore, we should try to find quantities analogous to the angular momentum variable L, L_z, ℓ, and m_ℓ. By analogy, there will be $2s + 1 = 2(1/2) + 1 = 2$ components of the spin angular momentum vector $\vec{S}$. Thus the **magnetic spin quantum number** m_s has only two values, $m_s = \pm 1/2$. The electron's spin will be oriented either "up" or "down" in a magnetic field (see Figure 7.9), and the electron can never be spinning with its *magnetic moment* μ_s exactly along the z axis (the direction of the external magnetic field $\vec{B}$).

Magnetic spin quantum number

For each atomic state described by the three quantum numbers (n, ℓ, m_ℓ) discussed previously, there are now two distinct states, one with $m_s = +1/2$ and one with $m_s = -1/2$. These states will be degenerate in energy unless the atom is in an external magnetic field. In a magnetic field these states will have different energies due to an energy separation like that of Equation (7.33). We say the *splitting* of these energy levels by the magnetic field has removed the energy degeneracy.

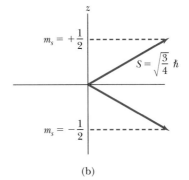

Figure 7.9 (a) A purely classical schematic of the intrinsic spin angular momentum, $\vec{S}$, of a spinning electron. (b) The quantization of $\vec{S}$, which can have only two positions in space relative to z (direction of external magnetic field). The z component of $\vec{S}$ is $S_z = \pm\hbar/2$.

(a)   (b)

The **intrinsic spin angular momentum** vector $\vec{S}$ has a magnitude of $|\vec{S}| = \sqrt{s(s+1)}\hbar = \sqrt{3/4}\hbar$. The magnetic moment is $\vec{\mu}_s = -(e/m)\vec{S}$, or $-2\mu_B\vec{S}/\hbar$. The fact that the coefficient of $\vec{S}/\hbar$ is $-2\mu_B$ rather than $-\mu_B$ as with the orbital angular momentum $\vec{L}$ is a consequence of the theory of relativity (Dirac equation), and we will not pursue the matter further here. This numerical factor relating the magnetic moment to each angular momentum vector is called the **gyromagnetic ratio**. It is designated by the letter g with the appropriate subscript (ℓ or s), so that $g_\ell = 1$ and $g_s = 2$. In terms of the gyromagnetic ratios, then,

Intrinsic spin angular momentum vector

$$\vec{\mu}_\ell = -\frac{g_\ell \mu_B \vec{L}}{\hbar} = -\frac{\mu_B \vec{L}}{\hbar} \tag{7.34a}$$

Gyromagnetic ratios

$$\vec{\mu}_s = -\frac{g_s \mu_B \vec{S}}{\hbar} = -2\frac{\mu_B \vec{S}}{\hbar} \tag{7.34b}$$

The z component of $\vec{S}$ is $S_z = m_s\hbar = \pm\hbar/2$.

We can now understand why the experiment of Stern and Gerlach produced only two distinct lines. If the atoms were in a state with $\ell = 0$, there would be no splitting due to m_ℓ. However, there is still space quantization due to the intrinsic spin that would be affected by the inhomogeneous magnetic field. The same arguments used previously for $\vec{\mu}_\ell$ (we now use the subscript ℓ to indicate the magnetic moment due to the orbiting electron and the subscript s to indicate the magnetic moment due to intrinsic spin) can now be applied to $\vec{\mu}_s$, and the potential energy, Equation (7.27), becomes

$$V_B = -\vec{\mu}_s \cdot \vec{B} = +\frac{e}{m}\vec{S} \cdot \vec{B} \tag{7.35}$$

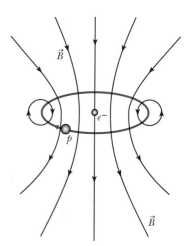

If we look at the hydrogen atom in the frame of the orbiting electron, we have the classical result shown in Figure 7.10. This classical picture indicates that the orbiting proton creates a magnetic field at the position of the electron. Therefore, even without an external magnetic field, the electron will feel the effects of an internal magnetic field, and Equation (7.35) predicts an energy difference depending on whether the electron's spin is up or down. Many levels are effectively split into two different states called *doublets*.

The relativistic quantum theory proposed by P. A. M. Dirac in 1928 showed that the intrinsic spin of the electron *required* a fourth quantum number as a consequence of the theory of relativity.

Figure 7.10 The hydrogen atom in the frame of reference of the electron. In this case, the orbiting proton creates a magnetic field at the position of the electron.

Example 7.9

How many distinctly different states (and therefore wave functions) exist for the $4d$ level of atomic hydrogen?

Strategy In the absence of an applied magnetic field, the fourth quantum number for intrinsic spin makes the degeneracy of the nth quantum level $2n^2$. The number of states has increased significantly.

Solution For the $4d$ level ($n = 4$, $\ell = 2$) there are $2\ell + 1 = 5$ different values of m_ℓ. For each of these values of m_ℓ (-2, -1, 0, 1, 2), there are two m_s states ($\pm 1/2$). Therefore there are 10 different possible individual states for a $4d$ level of atomic hydrogen.

HYDROGEN AND THE 21-CM LINE TRANSITION

B oth the proton and the electron in the hydrogen atom have intrinsic spin. We can think of each of these spinning particles as behaving like a small magnet. We show a schematic picture of the hydrogen atom in Figure A in two cases: one with the proton and electron spins (magnetic moments) aligned and one opposed. We know from introductory physics that when the spins are opposed, the system (the hydrogen atom in this case) is in a lower energy state. We mentioned this effect in our discussion of Equation (7.35). All you have to do is to hold two bar magnets in your hands and see in which configuration the bar magnets want to be when you bring them close. They will want to have their magnetic fields in opposite directions, in effect just like that shown on the right side of Figure A. When the electron changes its spin from aligned to opposed, the process is called a *spin-flip transition*.

During the German occupation of the Netherlands in World War II, Jan Oort (1900–1992), the great Dutch astronomer, wanted to find a radio spectral line that would allow him to detect objects more than a few thousand lightyears away. He wanted a spectral line that would allow the Doppler shift of the line to be easily observed, and he suggested to his graduate student, H. C. van de Hulst, that hydrogen might be a good candidate. In 1944 Hulst decided that a transition between the parallel (aligned) and antiparallel (opposed) configurations of the $n = 1$ line in atomic hydrogen might work. The parallel configuration (magnetic moments

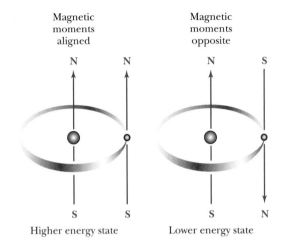

Figure A The state ($n = 1$) for the hydrogen atom when the proton and electron spins are aligned (left side) is a slightly higher energy state than when they are opposed (right side). When the electron spin changes from the aligned to the opposed state (called a *spin-flip transition*), a photon having a characteristic 21-cm wavelength is emitted.

are aligned) has an energy of only 5.9×10^{-6} eV greater than the antiparallel one. Such small energy differences are called *hyperfine structures*. When the hydrogen atom is excited into the parallel state and eventually decays to the antiparallel ground state, the photon emitted has a frequency of 1420 MHz and wavelength of 21.1 cm. This is the origin of the 21-cm line transition of atomic hydrogen, which was first detected in 1951.

The *interstellar medium* is the space between stars that contains a diffuse medium of gas and dust. About 90% of this medium is in the form of hydrogen

---

7.6 Energy Levels and Electron Probabilities

We are now in a position to discuss a complete description of the hydrogen atom. Every possible state of the hydrogen atom has a distinct wave function that is completely specified by four quantum numbers: (n, ℓ, m_ℓ, m_s). In many cases the energy differences associated with the quantum numbers m_ℓ and m_s are insignificant (that is, the states are nearly degenerate), and we can describe the states adequately by n and ℓ alone: for example, $1s$, $2p$, $2s$, $3d$, and so on. Generally, capital letters (that is, S, P, D) are used to describe the orbital angular momentum of (atomic and molecular). The Milky Way Galaxy contains neutral hydrogen gas with a density of about 1 atom/cm^3 and temperatures of 100 K or greater. This is enough hydrogen so that about every 500 years a given hydrogen atom collides with another and one of the hydrogen atoms is excited into the higher energy spin-aligned state in $n = 1$. Somewhat later (perhaps 20 million or so years!), the state jumps back to the ground state and emits the electromagnetic radiation of wavelength 21 cm. There is so much hydrogen in interstellar space that even with the rare collisions and long decay time, there are still many transitions to observe. The narrow line width of the excited state allows precise Doppler shift measurements to be made of the emitted radiation. This allows the velocity of the interstellar gas to be measured, including various rotations of galaxies. Estimated distances to gas clouds allow astronomers to map the distribution of matter in the galaxy. The 21-cm line radiation is the best way to map the structure of our Milky Way Galaxy.

In 1959 Philip Morrison and Giuseppe Cocconi suggested that the best frequency to search for signals from intelligent extraterrestrials is 1420 MHz, corresponding to the 21-cm line of the hydrogen atom. The search for extraterrestrial intelligence (SETI) has been ongoing since 1960 when Frank Drake pointed a radio telescope to the sky to search for evidence of the 21-cm line from Tau Ceti and Epsilon Eridani, two stars that are relatively close and sunlike. The first signals from Earth seeking contact were sent by the large radio telescope at Arecibo, Puerto Rico, in 1974. Three spacecraft, *Pioneer 10*, *Voyagers 1* and *2*, are exiting the solar

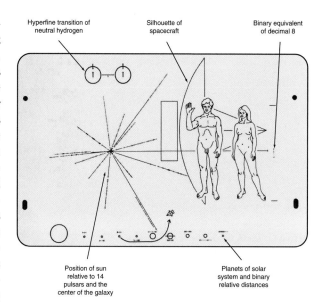

Figure B This plaque with a pictorial message from mankind is on board the *Pioneer 10* spacecraft (and *Pioneer 11*). Astronomers Carl Sagan and Frank Drake designed the plaque (23 cm by 15 cm) that was first launched on *Pioneer 10* in 1972. The explanations of some of the symbolism on the plaque are shown here, but are not on the plaque. The *Pioneer 10* left the solar system in the 1980s and was estimated in 2004 to be 13×10^9 km from Earth. It was last heard from in 2002, and its power source is now too weak to transmit to us. *NASA*.

system and have messages on board for extraterrestrials (see Figure B). Note that the 21-cm line transition is prominently displayed in the upper left in Figure B. SETI continues today with no positive results.

atomic states and lowercase letters (that is, *s*, *p*, *d*) are used to describe those for individual electrons. For hydrogen it makes little difference because each state has only a single electron, and we will use either specification.

We show an energy-level diagram in Figure 7.11 for hydrogen in the absence of an external magnetic field. The energy levels are degenerate with respect to ℓ, m_ℓ, and m_s, but in a magnetic field this degeneracy is removed. For heavier atoms with several electrons, the degeneracy is removed—either because of internal magnetic fields within the atom or because the average potential energy due to the nucleus plus electrons is non-Coulombic. In atoms with $Z > 1$ the smaller ℓ values tend to lie at a lower energy level for a given n (see Section 8.1). For example, in sodium or potassium, the energy states are $E(4S) < E(4P) < E(4D) < E(4F)$.

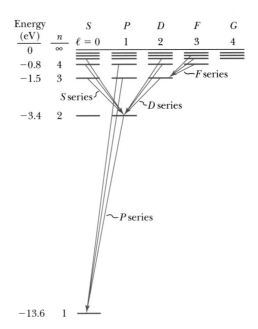

Figure 7.11 Energy-level diagram of hydrogen atom with no external magnetic field. Also shown are allowed photon transitions between some levels.

For hydrogen, the energy levels depend only on the principal quantum number n and are predicted with great accuracy by the Bohr theory.

We have previously learned that atoms emit characteristic electromagnetic radiation when they make transitions to states of lower energy. An atom in its ground state cannot emit radiation; it can absorb electromagnetic radiation, or it can gain energy through inelastic bombardment by particles, especially electrons. The atom will then have one or more of its electrons transferred to a higher energy state.

Selection Rules

We can use the wave functions obtained from the solution of the Schrödinger equation to calculate transition probabilities for the electron to change from one state to another. The results of such calculations show that electrons absorbing or emitting photons are much more likely to change states when $\Delta\ell = \pm 1$. Such transitions are called **allowed**. Other transitions, with $\Delta\ell \neq \pm 1$, are theoretically possible but occur with much smaller probabilities and are called **forbidden transitions**. There is no selection rule restricting the change Δn of the principal quantum number. The selection rule for the magnetic quantum number is $\Delta m_\ell = 0, \pm 1$. The magnetic spin quantum number m_s can (but need not) change between $1/2$ and $-1/2$. We summarize the selection rules for allowed transitions:

Allowed and forbidden transitions

$$\Delta n = \text{anything}$$

Selection rules

$$\Delta\ell = \pm 1 \qquad (7.36)$$

$$\Delta m_\ell = 0, \pm 1$$

Some allowed transitions are diagrammed in Figure 7.11. Notice that there are no transitions shown for $3P \to 2P$, $3D \to 2S$, and $3S \to 1S$ because those transitions violate the $\Delta\ell = \pm 1$ selection rule.

If the orbital angular momentum of the atom changes by $\hbar$ when absorption or emission of radiation takes place, we must still check that all conservation laws are obeyed. What about the conservation of angular momentum? The only external effect on the atom during the absorption or emission process is that due to the photon being absorbed or emitted. If the state of the atom changes, then the photon must possess energy, linear momentum, and angular momentum. The selection rule $\Delta \ell = \pm 1$ strongly suggests that the photon carries one unit ($\hbar$) of angular momentum. By applying quantum mechanics to Maxwell's equations, it is possible to show* that electromagnetic radiation is quantized into photons having energy $E = hf$ and intrinsic angular momentum of $\hbar$.

Example 7.10

Which of the following transitions for quantum numbers (n, ℓ, m_ℓ, m_s) are allowed for the hydrogen atom, and for those allowed, what is the energy involved?

(a) $(2, 0, 0, 1/2) \rightarrow (3, 1, 1, 1/2)$
(b) $(2, 0, 0, 1/2) \rightarrow (3, 0, 0, 1/2)$
(c) $(4, 2, -1, -1/2) \rightarrow (2, 1, 0, 1/2)$

Strategy We want to compare $\Delta \ell$ and Δm_ℓ with the selection rules of Equation (7.36). If allowed, the energies may be obtained from Equation (7.21) with $E_0 = 13.6$ eV.

Solution

(a) $\Delta \ell = +1, \Delta m_\ell = 1$; allowed.

$$\Delta E = E_3 - E_2 = -13.6 \text{ eV}\left(\frac{1}{3^2} - \frac{1}{2^2}\right)$$

$$= 1.89 \text{ eV, corresponding to absorption of a 1.89-eV photon}$$

(b) $\Delta \ell = 0, \Delta m_\ell = 0$; not allowed, because $\Delta \ell \neq \pm 1$.

(c) $\Delta \ell = -1, \Delta m_\ell = 1$; allowed. Notice that $\Delta n = -2$ and $\Delta m_s = +1$ does not affect whether the transition is allowed.

$$\Delta E = E_2 - E_4 = -13.6 \text{ eV}\left(\frac{1}{2^2} - \frac{1}{4^2}\right)$$

$$= -2.55 \text{ eV, corresponding to emission of a 2.55-eV photon}$$

Probability Distribution Functions

In the Bohr theory of the hydrogen atom, the electrons were pictured as orbiting around the nucleus in simple circular (or elliptical) orbits. The position vector $\vec{r}$ of the electron was well defined. In the wave picture of the atom, we must use wave functions to calculate the probability distributions† of the electrons. The "position" of the electron is therefore spread over space and is not well defined. The distributions can be found by examining the separable wave functions $R(r)$, $f(\theta)$, and $g(\phi)$. The $g(\phi)$ distribution is simplest because it leads to uniform probability—all values of ϕ are equally likely. It is easy to see why. Because the azimuthal part of the wave function is always of the form $e^{im_\ell \phi}$, the probability density $\psi^*\psi$ contains a corresponding factor of $(e^{im_\ell \phi})^* e^{im_\ell \phi} = e^{-im_\ell \phi + im_\ell \phi} = e^0 = 1$. (Remember that ψ^* means we take the complex conjugate of ψ.)

We may use the radial wave function $R(r)$ to calculate radial probability distributions of the electron (that is, the probability of the electron being at a given r). As discussed in Section 5.7, the probability of finding the electron in a differential volume element $d\tau$ is

$$dP = \psi^*(r, \theta, \phi)\, \psi(r, \theta, \phi)\, d\tau \tag{7.37}$$

*See Leonard Schiff, *Quantum Mechanics*, 3rd ed., New York: McGraw-Hill (1968) for a discussion of both the semiclassical and quantum treatment of radiation.

†It may be useful at this time to review Section 5.7, where the relationships between probability and wave functions were discussed.

We are interested in finding the probability $P(r)\,dr$ of the electron being between r and $r + dr$. The differential volume element in spherical polar coordinates is

$$d\tau = r^2 \sin\theta\,dr\,d\theta\,d\phi$$

Therefore,

$$P(r)\,dr = r^2 R^*(r)R(r)\,dr \int_0^\pi |f(\theta)|^2 \sin\theta\,d\theta \int_0^{2\pi} |g(\phi)|^2\,d\phi \qquad (7.38)$$

We are integrating over θ and ϕ, because we are only interested in the radial dependence. If the integrals over $f(\theta)$ and $g(\phi)$ have already been normalized to unity, the probability of finding the electron between r and $r + dr$ reduces to

Radial probability

$$P(r)\,dr = r^2 |R(r)|^2\,dr \qquad (7.39)$$

The radial probability density is

$$P(r) = r^2 |R(r)|^2 \qquad (7.40)$$

This probability density depends only on n and ℓ. In Figure 7.12 we display both $R(r)$ and $P(r)$ for the lowest-lying states of the hydrogen atom.

Radial wave functions $(R_{n\ell})$

Radial probability distribution $(P_{n\ell})$

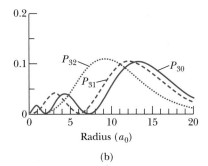

Radius (a_0)

Radius (a_0)

(a)

(b)

Figure 7.12 (a) The radial wave function $R_{n\ell}(r)$ plotted as a function of radius (in units of Bohr radius a_0) for several states of the hydrogen atom. (b) The radial probability distribution $P_{n\ell}$, which gives the probability of the electron being between r and $r + dr$.

Example 7.11

Find the most probable radius for the electron of a hydrogen atom in the $1s$ and $2p$ states.

Strategy To find the maximum and minimum of a function we take the derivative of the function with respect to the variable and set the derivative equal to zero. To find the most probable radial value we take the derivative of the probability density $P(r)$ with respect to r and set it equal to zero.

Solution $1s$ **state:**

$$\frac{d}{dr}P(r) = 0 = \frac{d}{dr}\left(\frac{4e^{-2r/a_0}}{a_0^3}r^2\right)$$

$$0 = \frac{4}{a_0^3}\left(-\frac{2}{a_0}r^2 + 2r\right)e^{-2r/a_0}$$

$$\frac{2r^2}{a_0} = 2r$$

$$r = a_0 \quad \text{Most probable radius}$$
$$\text{for } 1s \text{ state electron} \qquad (7.41)$$

$2p$ state:

$$\frac{d}{dr}\left[\frac{e^{-r/a_0}}{3(2a_0)^3}\frac{r^4}{a_0^2}\right] = 0$$

$$\frac{1}{24a_0^5}\left(-\frac{r^4}{a_0} + 4r^3\right) = 0$$

$$\frac{r^4}{a_0} = 4r^3$$

$$r = 4a_0 \quad \text{Most probable radius}$$
$$\text{for } 2p \text{ state electron} \qquad (7.42)$$

Notice that the most probable radii for the $1s$ and $2p$ states agree with the Bohr radii. This occurs only for the largest possible ℓ value for each n (see Problem 36).

Example 7.12

Calculate the average orbital radius of a $1s$ electron in the hydrogen atom.

Strategy To find the average value, we shall find the expectation value.

Solution The expectation (or average) value of r is (see Section 6.2)

$$\langle r \rangle = \int \psi^*(r, \theta, \phi)r\psi(r, \theta, \phi)\, d\tau = \int rP(r)\, dr$$

where we have again integrated over θ and ϕ.

$$\langle r \rangle = \int_0^\infty \frac{4}{a_0^3}e^{-2r/a_0}r^3\, dr$$

We look up this integral in Appendix 3 and find

$$\int_0^\infty r^3 e^{-2r/a_0}\, dr = \frac{3a_0^4}{8}$$

so that

$$\langle r \rangle = \frac{4}{a_0^3}\frac{3a_0^4}{8} = \frac{3}{2}a_0 \qquad \text{For the } 1s \text{ state electron}$$

Therefore, the average electron radius in the $1s$ state is larger than the most probable value, the Bohr radius. We can see that this result is reasonable by examining the radial probability distribution for the $1s$ state displayed in Figure 7.12. The maximum (or most probable) value occurs at a_0, but the average is greater than a_0 because of the shape of the "tail" of the distribution.

Example 7.13

What is the probability of the electron in the $1s$ state of the hydrogen atom being at a radius greater than the Bohr radius a_0?

Strategy In order to find the probability, we integrate the radial probability distribution from $r = a_0$ to ∞, because $P(r)$ is already normalized (that is, it has a unit probability of being somewhere between 0 and ∞).

Solution

$$\text{Probability} = \int_{a_0}^{\infty} P(r)\, dr$$

$$= \frac{4}{a_0^3} \int_{a_0}^{\infty} e^{-2r/a_0}\, r^2\, dr$$

We look up the indefinite integral in Appendix 3 and evaluate to find the result

$$\text{Probability} = \frac{4}{a_0^3}\left(\frac{5}{4}\, a_0^3 e^{-2}\right) = 5e^{-2} = 0.68$$

The probability of the electron being outside the Bohr radius in a $1s$ state is greater than 50%. This explains why we found $\langle r \rangle_{1s} = 1.5\, a_0$. This result is consistent with the shape of the $1s$ curve in Figure 7.12b.

The probability distributions for the $\ell = 0$ state electrons are spherically symmetric because the wave functions have no θ or ϕ dependence (see Table 7.2). For $\ell > 0$ the distributions are interesting because of the $f(\theta)$ dependence. For example, consider a p orbital. Referring to Table 7.2, we see that there are two possibilities for the angular part of the wave function. If $\ell = 1$ and $m_\ell = 0$, Y_{10} will be a factor in the wave function, and therefore $\cos^2 \theta$ will be a factor in the probability density $\psi^*\psi$. In this case the probability density will be highest near $0°$ and $180°$, that is, near the $+z$ axis and $-z$ axis. The other possible combinations for the quantum numbers of an electron in a p orbital are $\ell = 1$ and $m_\ell = \pm 1$. The wave function will contain $Y_{1\pm1}$, and therefore $\sin^2 \theta$ will be a factor in the probability density $\psi^*\psi$. The probability is highest at $\theta = 90°$, that is, in the xy plane. The probability distributions seen in Figure 7.13 are consistent with this analysis.

When we look at d orbitals, the situation becomes a bit more complicated, but a similar analysis will allow us to see at what angles θ the probability density is maximized. For the $\ell = 2$, $m_\ell = 0$ state, we see that $|Y_{20}|^2$ must have a maximum around $\theta = 0°$ and $180°$. Once again, these results are shown in Figure 7.13. Similarly, $|Y_{2\pm2}|^2$ (corresponding to the $\ell = 2$, $m_\ell = \pm 2$ states) has a maximum in the xy plane. For the $\ell = 2$, $m_\ell = \pm 1$ states, we find a factor $\sin^2 \theta \cos^2 \theta$ coming from $|Y_{2\pm1}|^2$. For these states the probability maxima are at $\theta = 45°$ and $135°$.

It is interesting to consider in which state (for a given n) the electron is closest to the origin. We can calculate $\langle r \rangle$ for the $2s$ and $2p$ states (see Problem 40) and find that the $2p$ average radius is smaller. However, because the $P(r)$ for the $2s$ state has two maxima, one with $r < a_0$, the electron in the $2s$ state will actually spend more time very close to the nucleus than will one in the $2p$ state. This effect can be seen in Figure 7.12, where the radial distribution for $P(r)$ in the $2s$ state extends farther out than that for $2p$, but there is a secondary maximum for $P(r)$ for the $2s$ state near a_0.

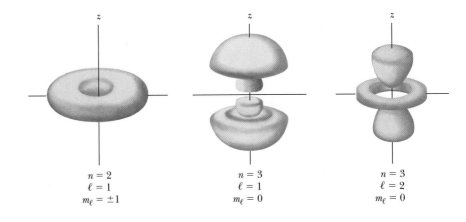

Figure 7.13 Pictorial representation of the probability density $|\psi(r, \theta, \phi)|^2$ for the hydrogen atom for three different electron states. There is axial symmetry about the z axis in each case. The shaded regions indicate those parts of space with the highest probability densities.

$n = 2$
$\ell = 1$
$m_\ell = \pm 1$

$n = 3$
$\ell = 1$
$m_\ell = 0$

$n = 3$
$\ell = 2$
$m_\ell = 0$

There are still minor corrections to be made to our model of the hydrogen atom. For example, in Section 8.2 of the next chapter we shall discuss the spin-orbit coupling. Each effect we cover has a smaller and smaller effect on the energy levels. Also in Chapter 8 we will continue our study of the subfield of physics called *atomic physics* and consider multielectron atoms.

Summary

The Schrödinger wave theory is applied to atomic physics, beginning with the hydrogen atom. The application of the boundary conditions leads to three quantum numbers:

n Principal quantum number
ℓ Orbital angular momentum quantum number
m_ℓ Magnetic quantum number

with the values and restrictions (all are integers)

$$n = 1, 2, 3, 4, \ldots \qquad n > 0$$

$$\ell = 0, 1, 2, 3, \ldots, n - 1 \qquad \ell < n$$

$$m_\ell = -\ell, -\ell + 1, \ldots, 0, 1, \ldots, \ell - 1, \ell \qquad |m_\ell| \le \ell$$
$$(7.19)$$

The energy of the electron-proton system is quantized and depends to the first order only on n. The orbital angular momentum L is quantized by $L = \sqrt{\ell(\ell + 1)}\hbar$ and not by $n\hbar$ as in the Bohr theory. We use letter names $s, p, d, f, g, h, \ldots$ to indicate the ℓ value for a given electron.

The z component of $\vec{L}$ is quantized, and $L_z = m_\ell\hbar$. This is referred to as *space quantization*, because $\vec{L}$ can only have certain orientations in space. In the absence of a magnetic field, the energy is degenerate with respect to ℓ and m_ℓ. In an external magnetic field each $n\ell$ level is split into $2\ell + 1$ different energy states (normal Zeeman effect).

In order to explain increasingly complex atomic spectra, Goudsmit and Uhlenbeck introduced a fourth quantum number. This quantum number s is related to the electron's intrinsic angular momentum, commonly referred to as spin. The electron spin quantum number is $s = 1/2$, and the values of the magnetic spin quantum number m_s are $\pm1/2$. Stern and Gerlach observed in 1922 the effects of intrinsic spin, although at the time it was confused with orbital angular momentum.

The selection rules for allowed transitions for a change from one state to another are

$$\Delta n = \text{anything}$$

$$\Delta \ell = \pm 1 \qquad (7.36)$$

$$\Delta m_\ell = 0, \pm 1$$

The probability of finding an electron between r and $r + dr$ is $P(r)\,dr = r^2|R(r)|^2\,dr$ where $R(r)$ is the radial wave function.

Questions

1. Do the radial wave functions depend on m_ℓ? Explain your reasons.
2. Would the radial wave functions be different for a potential $V(r)$ other than the Coulomb potential? Explain.
3. For what energy levels in the hydrogen atom will we not find $\ell = 2$ states?
4. What are the differences and similarities between atomic levels, atomic states, and atomic spectral lines? When do spectral lines occur?
5. What are the differences and similarities between the quantization of angular momentum in the Bohr model and the Schrödinger theory?
6. Can the magnetic moment of an atom line up exactly with an external magnetic field? Explain.
7. What are the possible magnetic quantum numbers for an f state?
8. List all the reasons why you think a fourth quantum number (intrinsic spin) might have helped explain complex optical spectra in the early 1920s.
9. Is it possible for the z component of the orbital magnetic moment to be zero, but not the orbital angular momentum? Explain.
10. A close examination of the spectral lines coming from starlight can be used to determine the star's magnetic field. Explain how this is possible.
11. If a hydrogen atom in the $2p$ excited state decays to the $1s$ ground state, explain how the following properties are conserved: energy, linear momentum, and angular momentum.

Problems

Note: The more challenging problems are indicated by a shaded box.

7.1 Application of the Schrödinger Equation to the Hydrogen Atom

1. Assume that the electron in the hydrogen atom is constrained to move only in a circle of radius a in the xy plane. Show that the separated Schrödinger equation for ϕ becomes

$$\frac{1}{a^2}\frac{d^2\psi}{d\phi^2} + \frac{2m}{\hbar^2}|E|\psi = 0$$

   where ϕ is the angle describing the position on the circle. This is similar to the Bohr assumption.

2. Solve the equation in the previous problem for ψ. Find the allowed energies and angular momenta.

3. After separating the Schrödinger equation using $\psi = R(r)f(\theta)g(\phi)$, the equation for ϕ is

$$-\frac{1}{g}\frac{d^2g}{d\phi^2} = k^2$$

   where k = constant. Solve for $g(\phi)$ in this equation and apply the appropriate boundary conditions. Show that k must be 0 or a positive or negative integer ($k = m_\ell$, the magnetic quantum number).

4. Using the transformation equations between Cartesian coordinates and spherical polar coordinates given in Figure 7.1, transform the Schrödinger Equation (7.2) from Cartesian to spherical coordinates as given in Equation (7.3).

7.2 Solution of the Schrödinger Equation for Hydrogen

5. Show that the radial wave function R_{20} for $n = 2$ and $\ell = 0$ satisfies Equation (7.13). What energy E results? Is this consistent with the Bohr model?

6. Show that the radial wave function R_{21} for $n = 2$ and $\ell = 1$ satisfies Equation (7.10). What energy results? Is this consistent with the Bohr model?

7. Show that the radial wave function R_{21} for $n = 2$ and $\ell = 1$ is normalized.

8. The normalized wave function ψ for the ground state of hydrogen is given by

$$\psi_{100}(r, \theta, \phi) = \frac{1}{\sqrt{\pi a_0^3}} e^{-r/a_0}$$

   Show that the wave function is normalized over all space.

7.3 Quantum Numbers

9. List all the possible quantum numbers (n, ℓ, m_ℓ) for the $n = 6$ level in atomic hydrogen.

10. For a $3p$ state give the possible values of $n, \ell, m_\ell, L, L_z, L_x$, and L_y.

11. Write down all the wave functions for the $3p$ level of hydrogen. Identify the wave functions by their quantum numbers. Use the solutions in Tables 7.1 and 7.2.

12. Prove that $\langle L^2 \rangle = \ell(\ell + 1)\hbar^2$ by actually performing the summation for Equation (7.24).

13. What is the degeneracy of the $n = 6$ shell of atomic hydrogen considering (n, ℓ, m_ℓ) and no magnetic field?

14. Draw for a $3d$ state all the possible orientations of the angular momentum vector $\vec{L}$. What is $L_x^2 + L_y^2$ for the $m_\ell = -1$ component?

15. What is the smallest value that ℓ may have if $\vec{L}$ is within $3°$ of the z axis?

16. Prove that the degeneracy of an atomic hydrogen state having principal quantum number n is n^2 (ignore the spin quantum number).

17. Write out the hydrogen wave functions $\psi_{n\ell m_\ell}$ for $n\ell m_\ell$ values of $(2, 1, -1)$, $(2, 1, 0)$, and $(3, 2, -1)$.

18. Show that the hydrogen wave functions ψ_{200} and ψ_{21-1} are normalized. If the integrals required are not in Appendix 3, consult a table of integrals.

7.4 Magnetic Effects on Atomic Spectra— The Normal Zeeman Effect

19. Calculate the possible z components of the orbital angular momentum for an electron in a $4p$ state.

20. For hydrogen atoms in a $4d$ state, what is the *maximum* difference in potential energy between atoms when placed in a magnetic field of 2.5 T? Ignore intrinsic spin.

21. Show that the wavelength difference between adjacent transitions in the normal Zeeman effect is given approximately by

$$\Delta\lambda = \frac{\lambda_0^2 \mu_B B}{hc}$$

22. For hydrogen atoms in a d state sketch the orbital angular momentum with respect to the z axis. Use units of $\hbar$ along the z axis and calculate the allowed angles of μ_ℓ with respect to the z axis.

23. For a hydrogen atom in the $6f$ state, what is the minimum angle between the orbital angular momentum vector and the z axis?

24. The red line of the Balmer series in hydrogen ($\lambda = 656.5$ nm) is observed to split into three different spectral lines with $\Delta\lambda = 0.04$ nm between two adjacent lines when placed in a magnetic field B. What is the value of B if $\Delta\lambda$ is due to the energy splitting between two adjacent m_ℓ states?

25. A hydrogen atom in an excited $5f$ state is in a magnetic field of 3 T. How many different energy states can the electron have in the $5f$ subshell? (Ignore the magnetic spin effects.) What is the energy of the $5f$ state in the absence of a magnetic field? What will be the energy of each state in the magnetic field?

26. The magnetic field in a Stern-Gerlach experiment varies along the vertical direction as $dB_z/dz = 20$ T/cm. The horizontal length of the magnet is 7.1 cm, and the speed of the silver atoms averages 925 m/s. The mass of the silver atoms is 1.8×10^{-25} kg. Show that the z component of its magnetic moment is 1 Bohr magneton. What is the separation of the two silver atom beams as they leave the magnet?

27. An experimenter wants to separate silver atoms in a Stern-Gerlach experiment by at least 1 cm (a large separation) as they exit the magnetic field. To heat the silver she has an oven that can reach 1000°C and needs to order a suitable magnet. What should be the magnet specifications?

7.5 Intrinsic Spin

28. In an external magnetic field, can the electron spin vector $\vec{S}$ point in the direction of $\vec{B}$? Draw a diagram with $\vec{B} = B_0\hat{k}$ showing $\vec{S}$ and S_z.

29. Using all four quantum numbers (n, ℓ, m_ℓ, m_s), write down all possible sets of quantum numbers for the $5f$ state of atomic hydrogen. What is the total degeneracy?

30. Using all four quantum numbers (n, ℓ, m_ℓ, m_s), write down all possible sets of quantum numbers for the $6d$ state of atomic hydrogen. What is the total degeneracy?

31. The 21-cm line transition for atomic hydrogen results from a spin-flip transition for the electron in the parallel state of the $n = 1$ state. What temperature in interstellar space gives a hydrogen atom enough energy $(5.9 \times 10^{-6}$ eV) to excite another hydrogen atom in a collision?

32. Prove that the total degeneracy for an atomic hydrogen state having principal quantum number n is $2n^2$.

7.6 Energy Levels and Electron Probabilities

33. Show that, for transitions between any two n states of atomic hydrogen, no more than three different spectral lines can be obtained for the normal Zeeman effect.

34. Find whether the following transitions are allowed, and if they are, find the energy involved and whether the photon is absorbed or emitted for the hydrogen atom:
(a) $(5, 2, 1, \frac{1}{2}) \rightarrow (5, 2, 1, -\frac{1}{2})$
(b) $(4, 3, 0, \frac{1}{2}) \rightarrow (4, 2, 1, -\frac{1}{2})$
(c) $(5, 2, -2, -\frac{1}{2}) \rightarrow (1, 0, 0, -\frac{1}{2})$
(d) $(2, 1, 1, \frac{1}{2}) \rightarrow (4, 2, 1, \frac{1}{2})$

35. In Figure 7.12, the radial distribution function $P(r)$ for the $2s$ state of hydrogen has two maxima. Find the values of r (in terms of a_0) where these maxima occur.

36. Find the most probable radial position for the electron of the hydrogen atom in the $2s$ state. Compare this value with that found for the $2p$ state in Example 7.11.

37. Sketch the probability function as a function of r for the $2s$ state of hydrogen. At what radius is the position probability equal to zero?

38. Calculate the probability of an electron in the ground state of the hydrogen atom being inside the region of the proton (radius $\approx 1 \times 10^{-15}$ m). (*Hint:* note that $r \ll a_0$.)

39. Calculate the probability that an electron in the ground state of the hydrogen atom can be found between $0.95a_0$ and $1.05a_0$.

40. Find the expectation value of the radial position for the electron of the hydrogen atom in the $2s$ and $2p$ states.

41. Calculate the probability of an electron in the $2s$ state of the hydrogen atom being inside the region of the proton (radius $\approx 1 \times 10^{-15}$ m). Repeat for a $2p$ electron. (*Hint:* note that $r \ll a_0$.)

42. Find the most probable radial position of an electron in the $3d$ state of the hydrogen atom.

43. What is the probability that an electron in the $3d$ state is located at a radius greater than a_0?

General Problems

44. Assume the following (incorrect!) classical picture of the electron intrinsic spin. Take the electrical energy of the electron to be equal to its mass energy concentrated into a spherical shell of radius R.

$$\frac{e^2}{4\pi\epsilon_0 R} = mc^2$$

Calculate R (called the *classical electron radius*). Now let this spherical shell rotate and calculate the tangential speed v along the sphere's equator in order to obtain the electron intrinsic spin.

$$\text{Angular momentum} = I\omega = I\frac{v}{R} = \frac{\hbar}{2}$$

where I = moment of inertia of a spherical shell = $2mR^2/3$. Is the value of v obtained in this manner consistent with the theory of relativity? Explain.

45. As in the previous problem, we want to calculate the speed of the rotating electron. Now let's assume that the diameter of the electron is equal to the Compton wavelength of an electron. Calculate v and comment on the result.

46. Consider a hydrogen-like atom such as He^+ or Li^{++} that has a single electron outside a nucleus of charge $+Ze$. (a) Rewrite the Schrödinger equation with the new Coulomb potential. (b) What change does this new potential have on the separation of variables? (c) Will the radial wave functions be affected? Explain. (d) Will the spherical harmonics be affected? Explain.

47. For the previous problem find the wave function ψ_{100}.

48. Consider a hydrogen atom in the $3p$ state. (a) At what radius is the electron probability equal to zero? (b) At what radius will the electron probability be a maximum? (c) For $m_\ell = 1$, at what angles θ will the electron probability be equal to zero? What about for $m_\ell = -1$?

49. Consider a "muonic atom," which consists of a proton and a negative muon, symbol μ^-. Compute the ground state energy following the methods used for the hydrogen atom.

50. The lifetime of the excited component of the $n = 1$ state (parallel spins) that produces the 21-cm line transition is approximately 10^7 years. What is the energy line width of this state?

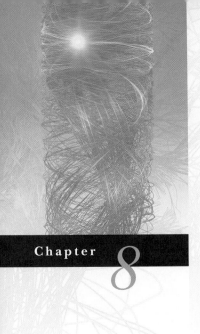

Atomic Physics

What distinguished Mendeleev was not only genius, but a passion for the elements. They became his personal friends; he knew every quirk and detail of their behavior.

J. Bronowski

W e began our study of atomic physics in the previous chapter with a study of the hydrogen atom. Now we examine more complex atoms with multiple electrons. Physicists and chemists have been studying the properties of the elements for centuries. We know much about atomic sizes, chemical behavior, ionization energies, magnetic moments, and spectroscopic properties, including x-ray spectra. In 1869 the Russian chemist Dmitri Mendeleev arranged the just over 60 known elements into a periodic table that systematized many of their chemical properties. His table generally had the elements arranged in order of atomic weight. When he put the elements in rows, a definite pattern appeared, but only if he left vacancies. Based on his systematization of elements known at the time, Mendeleev was able to predict several hitherto unknown elements. His result was initially looked upon with some skepticism, but after the discovery of three of the predicted elements, gallium (in 1875), scandium (1879), and germanium (1886), the value of Mendeleev's periodic table was widely accepted. The elucidation of the underlying physical basis of his (empirical) periodic table became one of the outstanding goals of physics. This goal was finally attained by the end of the 1920s and was one of the significant achievements of quantum mechanics. We shall also discover how even a qualitative understanding of atomic structure allows us to explain some of the physical and chemical properties of the elements.

Dmitri Ivanovich Mendeleev (1834–1907), born the 14th child of an educated family in Siberia, entered the University of St. Petersburg (Russia) at age 16 where he eventually earned his doctorate. He remained a chemist at St. Petersburg for most of his life and announced his periodic table there in 1869. He successfully predicted the properties of several elements that were eventually discovered: gallium (1875), scandium (1879), and germanium (1886). *AIP Emilio Segrè Visual Archives, W. F. Meggers Collection.*

8.1 Atomic Structure and the Periodic Table

We now have a good basis for understanding the hydrogen atom. How do we proceed to understand atoms with more than one electron? The obvious procedure is to add one more electron (helium atom) to the Schrödinger equation and solve for the wave functions. We soon run into formidable mathematical problems. Not only do we now have a nucleus with charge $+2e$ attracting two electrons, we also have the interaction of the two electrons repelling one another. The energy levels obtained previously for the single electron in the hydrogen atom will be changed because of these new interactions. In general the problem of many-electron atoms cannot be solved exactly with the Schrödinger equation because of the complex potential interactions. Modern computers have allowed us to make great progress, and numerical calculations can be carried out with great precision for various models. We will see in this section that we can understand many experimental results without actually computing the wave functions of many-electron atoms. We can learn a great deal about atoms by carefully applying the boundary conditions and selection rules.

The rise of quantum physics was accompanied by a vast accumulation of precise atomic spectroscopic data for optical frequencies. In order to understand these data, Wolfgang Pauli (Nobel Prize, 1945) proposed in 1925 his famous exclusion principle:

> **Pauli exclusion principle:** No two electrons in an atom may have the same set of quantum numbers (n, ℓ, m_ℓ, m_s).

This principle has far-reaching implications. We can use it to describe in a reasonable, precise fashion the organization of atomic electrons to form the elements. Pauli's exclusion principle applies to all particles of half-integer spin, which are called *fermions*, and can be generalized to include particles in the nucleus, where it is crucial to nuclear structure because neutrons and protons are both fermions. You will learn more about the properties of fermions in Chapter 9.

The atomic electron structure* leading to the observed ordering of the periodic table can be understood by the application of two rules:

> 1. The electrons in an atom tend to occupy the lowest energy levels available to them.
> 2. Only one electron can be in a state with a given (complete) set of quantum numbers (Pauli exclusion principle).

(8.1)

Let us apply these rules to the first few atoms in the periodic table. Hydrogen has quantum numbers (n, ℓ, m_ℓ, m_s) equal to $(1, 0, 0, \pm 1/2)$ when it is in its lowest energy state (ground state). In the absence of a magnetic field, the $m_s = 1/2$ state is degenerate with the $m_s = -1/2$ state. In neutral helium the quantum numbers must be different for the two electrons, so if the quantum numbers are $(1, 0, 0, 1/2)$ for the first electron, those for the second electron must be $(1, 0, 0, -1/2)$. Direct experimental evidence shows that the two electrons in the He atom have their spins antialigned (spin angular momentum opposed) rather than aligned. This supports the Pauli exclusion principle. These two electrons form a rather strong bond with their spin angular momentum antialigned. We speak of two electrons having the same quantum numbers n, ℓ, m_ℓ but with their spin angular momentum antialigned ($m_s = +1/2$ and $m_s = -1/2$) as being *paired*, and the total spin of the pair is zero.

The principal quantum number n has also been given letter codes:

$$
\begin{array}{llll}
n = 1 & 2 & 3 & 4 \ldots \\
\text{Letter} = \text{K} & \text{L} & \text{M} & \text{N} \ldots
\end{array}
$$

(8.2)

*The use of hydrogenic quantum numbers for other atoms implies a hydrogen-like central field for the outer electrons of these atoms.

Wolfgang Pauli (1900–1958) was born in Austria, studied at Munich under Arnold Sommerfeld, and spent brief periods at Göttingen (with Max Born), Copenhagen (with Niels Bohr), and Hamburg before accepting an appointment at Zurich in 1925 where he remained, except for brief periods at American universities including Princeton University during World War II. Pauli was a brilliant theoretical physicist who formulated the exclusion principle named after him, proposed a quantum spin number for the electron, and recognized the existence of the neutrino to explain nuclear beta decay. He received the Nobel Prize in 1945 for discovering the exclusion principle. *Photo taken by S. A. Goudsmidt, AIP Niels Bohr Library.*

Electron shells

Electron subshells

Because the binding energies depend mainly on n, the electrons for a given n are said to be in **shells**. We speak of the K shell, L shell, and so on (recall from Chapter 4 that this was nomenclature used to describe Moseley's x-ray results). The $n\ell$ descriptions are called **subshells**. We have 1s, 2p, 3d subshells. Both electrons in the He atom are in the K shell and 1s subshell (which is a shell in itself). We use a superscript to denote the number of electrons in each subshell. The hydrogen atom description is $1s^1$ or 1s (the superscript 1 is sometimes omitted), and the helium atom is $1s^2$.

The next atom in the table is lithium. The K shell has no more space because only two electrons are allowed. The next shell is the L shell ($n = 2$), and the possible subshells are 2s and 2p. Rule 1 says the electrons will occupy the state with the lowest energy. Remember that semiclassically the 2s state (with zero angular momentum) has an orbit through the nucleus, whereas the 2p state has a more nearly circular orbit. An electron in the 2p subshell (Li) will experience a $+3e$ nuclear charge, but the positive nuclear charge will be partially screened* by the two electrons in the 1s shell. The effective charge that the 2p electron sees (or feels) will therefore be $Z_{\text{eff}} \approx +1e$. The 2s electron, on the other hand, spends more time than a 2p electron actually passing near the nucleus; hence the effective charge it experiences will be $Z_{\text{eff}} > +1e$. Therefore, an electron in the 2s subshell will experience a more attractive potential than a 2p electron and will thus lie lower in energy. The electronic structure of Li is $1s^2 2s^1$. The third electron has the quantum numbers $(2, 0, 0, \pm 1/2)$.

How many electrons may be in each subshell in order to not violate the Pauli exclusion principle?

	Total
For each m_ℓ: two values of m_s	2
For each ℓ: $(2\ell + 1)$ values of m_ℓ	$2(2\ell + 1)$

Thus each $n\ell$ subshell can have $2(2\ell + 1)$ electrons. The 1s, 2s, 3s, 4s subshells can have only two electrons. The 2p, 3p, 4p subshells can have up to six. The 3d, 4d, 5d subshells can have up to ten, and so on.

We can now describe the electronic configurations of many-electron atoms. Although there are effects due to internal magnetic fields, in the absence of external magnetic fields the m_ℓ and m_s quantum numbers do not affect the atom's total energy. Thus, the different states available within the same subshell are nearly degenerate. For a qualitative understanding we need only refer to $n\ell$.

The filling of electrons in an atom generally proceeds until each subshell is full. When a subshell has its maximum number of electrons, we say it is *closed* or *filled*. Electrons in outer shells with lower ℓ values spend more time inside the (inner) closed shells. Classically, we understand this result, because the lower ℓ values have more elliptical orbits than the higher ℓ values. The electrons with higher ℓ values are therefore more shielded from the nuclear charge $+Ze$, feel less Coulomb attraction, and lie higher in energy than those with lower ℓ values. For a given n the subshells fill in the order $s, p, d, f, g, \ldots$ This shielding effect becomes so pronounced that the 4s subshell actually fills before the 3d subshell even though it has a larger n. This happens often as the higher-lying shells fill with electrons. Experimental evidence shows that the order of subshell filling given in Table 8.1 is generally correct. Some important variations from this order produce

---

* "Screened" in this case means the electron will react to both the $+3e$ nucleus charge and $-2e$ electron charge within its own orbit.

Table 8.1	Order of Electron Filling in Atomic Subshells			
n	ℓ	Subshell	Subshell Capacity	Total Electrons in All Subshells
1	0	$1s$	2	2
2	0	$2s$	2	4
2	1	$2p$	6	10
3	0	$3s$	2	12
3	1	$3p$	6	18
4	0	$4s$	2	20
3	2	$3d$	10	30
4	1	$4p$	6	36
5	0	$5s$	2	38
4	2	$4d$	10	48
5	1	$5p$	6	54
6	0	$6s$	2	56
4	3	$4f$	14	70
5	2	$5d$	10	80
6	1	$6p$	6	86
7	0	$7s$	2	88
5	3	$5f$	14	102
6	2	$6d$	10	112

Figure 8.1 Approximate energy ordering of the subshells for the outermost electron in an atom. This representation assumes that the given subshell is receiving its first electron and that all lower subshells are full and all higher subshells are empty.

the rare earth lanthanides and actinides. A schematic diagram of the subshell energy levels is shown in Figure 8.1.

One nomenclature for identifying atoms is $_ZX$ where Z is the atomic number of the element (the number of protons), and X is the chemical symbol that identifies the element. The Z notation is superfluous because every element has a unique Z. For example, $_8O$ and O stand for the same element, because oxygen always has $Z = 8$. In Chapter 12 we will discuss isotopes of elements in which the mass number of the element varies because the number of neutrons in the nucleus is different. Note that in a neutral atom, the number of electrons is equal to Z.

Example 8.1

Give the electron configuration and the $n\ell$ value of the last electrons in the subshell (called *valence* electrons) for the following atoms: $_{11}$Na, $_{18}$Ar, $_{20}$Ca, $_{35}$Br.

Strategy We can use the rules of Equation (8.1), the shell and subshell nomenclature, and the results just discussed about electron subshell ordering. We utilize the ordering given in Figure 8.1 and Table 8.1.

Solution $_{11}$Na: Sodium has 11 electrons, so we start filling them in the correct order. Two electrons go in the $1s$ subshell,

two go in the $2s$ subshell, and six go in the $2p$ subshell. That gives us ten electrons in the filled $1s^2 2s^2 2p^6$ subshells, which is called the *core* because it is filled. According to the order of filling given in Table 8.1 and Figure 8.1, the extra electron must be in the $3s$ subshell with $n = 3$, $\ell = 0$. The electronic configuration is then $1s^2 2s^2 2p^6 3s^1$. The chemical properties of Na are determined almost exclusively by the one extra electron outside the core. The core is rather inert with the orbital and intrinsic angular momenta of the electrons paired to zero.

$_{18}$**Ar:** From Table 8.1 we see that 18 electrons complete the $3p$ subshell, so the 18th and last electron has $n = 3$, $\ell = 1$, and the electronic configuration is $1s^2 2s^2 2p^6 3s^2 3p^6$. Argon has completely closed subshells and no extra (valence) electrons. This is the reason argon, one of the inert gases, is chemically inactive.

$_{20}$**Ca:** After Ar the next two electrons go into the $4s$ subshell, so $n = 4$, $\ell = 0$, and the electronic configuration for calcium is $1s^2 2s^2 2p^6 3s^2 3p^6 4s^2$. There is a rather large energy gap between the $3p$ subshell and the $4s$ and $3d$ subshells (see Figure 8.1). The two electrons in the $4s$ subshell are situated rather precariously outside the inert core of Ar and can react strongly with other atoms.

$_{35}$**Br:** One more electron added to $_{35}$Br finishes the $4p$ subshell and makes the strongly inert gas krypton. The last electron in $_{35}$Br has $n = 4$ and $\ell = 1$, and the electronic configuration of the last few subshells is $3p^6 4s^2 3d^{10} 4p^5$. Bromine badly needs one more electron to complete its subshell and is very active chemically with a high electron affinity—searching for that last electron to fill its $4p$ subshell.

It is now relatively easy to understand the structure of the periodic table shown in Figure 8.2. The ordering of electrons into subshells follows from the two rules of Equation (8.1). In Figure 8.2 the horizontal groupings are according to separate subshells. Atomic electron configurations are often denoted by only the last subshell, and all previous subshells are assumed to be filled. In Figure 8.2 only the last unfilled subshell configurations are shown within the element box. In some cases a smooth order does not occur. For example, $_{40}$Zr has $5s^2 4d^2$, that is, the $5s$ subshell is filled, but the next element, $_{41}$Nb, has the structure $5s^1 4d^4$. An electron has been taken from the $5s$ subshell and placed in the $4d$ subshell with an additional electron to make a total of four electrons in the $4d$ subshell. Several such unusual cases occur as the atomic number increases. These details reflect the complex electron-electron interactions in a system of many particles.

Groups and periods   Let us briefly review some of the special arrangements of the periodic table. The vertical columns (or **groups**) have similar chemical and physical properties. This occurs because they have the same valence electron structure—that is, they have the same number of electrons in an ℓ orbit and can form similar chemical bonds. The horizontal rows are called **periods**, and they correspond to filling of the subshells. For example, in the fourth row the $4s$ subshell is filled first with 2 electrons, next the $3d$ subshell is filled with 10 electrons, and finally the $4p$ subshell is filled with 6 electrons. The fourth row consists of 18 elements and the filling of the $4s$, $3d$, and $4p$ subshells.

In order to compare some properties of elements we show the ionization energies of elements in Figure 8.3 and atomic radii in Figure 8.4. (The ionization energy is the energy required to remove the weakest bound electron, forming a positive ion.) The electrical conductivity and resistivity also show subshell effects. Good electrical conductors need free electrons that are only weakly bound to their nuclei. In Chapter 10 we shall see similar patterns in superconducting properties. The differences according to subshells are remarkable.

Inert Gases

The last group of the periodic table is the inert gases. They are unique in that they all have closed subshells. For all inert gases except helium the closed subshell is a p subshell. They have no valence electrons, and the p subshell is tightly bound. These elements therefore are chemically inert. They do not easily form chemical bonds with other atoms. They have zero net spin, large ionization energy (Figure 8.3), and poor electrical conductivity. Their boiling points are quite low, and at room temperature they are monatomic gases, because their atoms interact so weakly with each other.

Periodic Table of Elements

Figure 8.2 The atomic number and element symbol are given in the top of each box. The electron configuration for each element is specified by giving the values of the principal quantum numbers n, the angular momentum quantum numbers ℓ (s, p, d, or f), and the number of electrons outside closed shells. The configuration of some of the closed shells is given on the left.

Alkalis

Hydrogen and the alkali metals (Li, Na, K, and so on) form Group 1 of the periodic table. They have a single s electron outside an inert core. This electron can be easily removed, so the alkalis easily form positive ions with a charge $+1e$. Therefore, we say that their *valence* is $+1$. Figure 8.3 shows that the alkali metals

Figure 8.3 The ionization energies of the elements are shown versus the atomic numbers. The element symbols are shown for the peaks and valleys with the subshell closure in parentheses where appropriate. When a single electron is added to the p and d subshells, the ionization energy significantly decreases, indicating the shell effects of atomic structure.

have the lowest ionization energies. The drop in ionization energies between the inert gases and the alkalis is precipitous. The alkali metals are relatively good electrical conductors, because the valence electrons are free to move around from one atom to another.

Alkaline Earths

The alkaline earths are in Group 2 of the periodic table. These elements (Be, Mg, Ca, Sr, and so on) have two s electrons in their outer subshell, and although these subshells are filled, the s electrons can extend rather far from the nucleus and can be relatively easily removed. The alkali metals and alkaline earths have the largest atomic radii (Figure 8.4), because of their loosely bound s electrons. The ionization energies (Figure 8.3) of the alkaline earths are also low, but their electrical conductivity is high. The valence of these elements is $+2$, and they are rather active chemically.

Figure 8.4 Atomic radii from "covalent" data determined from bond lengths in the molecules of chemical compounds. The smallest radii occur when the subshells are filled. *From Darrell D. Ebbing, General Chemistry, 3rd ed., Houghton Mifflin (1990).*

Halogens

Immediately to the left of the inert gases, Group 17 is one electron short of having a filled outermost subshell. These elements (F, Cl, Br, I, and so on) all have a valence of -1 and are chemically very active. They form strong ionic bonds (for example, NaCl) with the alkalis (valence $+1$) by gaining the electron given up easily by the alkali atom. In effect, a compound such as NaCl consists of Na^+ and Cl^- ions strongly bound by their mutual Coulomb interaction. The groups to the immediate left of the halogens have fewer electrons in the p shell. In Figure 8.4 it is apparent that the radii of the p subshell decrease as electrons are added. A more stable configuration occurs in the p subshell as it is filled, resulting in a more tightly bound atom.

Transition Metals

The three rows of elements in which the $3d$, $4d$, and $5d$ subshells are being filled are called the *transition elements* or *transition metals*. Their chemical properties are similar—primarily determined by the s electrons, rather than by the d subshell being filled. This occurs because the s electrons, with higher n values, tend to have greater radii than the d electrons. The filling of the $3d$ subshell leads to some important characteristics for elements in the middle of the period. These elements (for example Fe, Co, and Ni) have d-shell electrons with unpaired spins (as dictated by Hund's rules, see Section 8.2). The spins of neighboring atoms in a crystal lattice align themselves, producing large magnetic moments and the ferromagnetic properties of these elements (see Section 10.4). As the d subshell is filled, the electron spins eventually pair off, and the magnetic moments, as well as the tendency for neighboring atoms to align spins, are reduced.

Lanthanides

The lanthanides ($_{58}$Ce to $_{71}$Lu), also called the *rare earths*, all have similar chemical properties. This occurs because they all have the outside $6s^2$ subshell completed while the smaller $4f$ subshell is being filled. The ionization energies (see Figure 8.3) are similar for the lanthanides. As occurs in the $3d$ subshell, the electrons in the $4f$ subshell often have unpaired electrons. These unpaired electrons align themselves, and because there can be so many electrons in the f subshell, large magnetic moments may occur. The large orbital angular momentum ($\ell = 3$) also contributes to the large ferromagnetic effects that some of the lanthanides have. The element holmium can have an extremely large internal magnetic field at low temperatures, much larger than even that of iron.

Actinides

The actinides ($_{90}$Th to $_{103}$Lr) are similar to the lanthanides in that inner subshells are being filled while the $7s^2$ subshell is complete. It is difficult to obtain reliable chemical data for these elements, because they are all radioactive. A few actinide isotopes can be kept in significant quantities because they have longer half-lives. Examples of these are thorium-232, uranium-235, and uranium-238, which occur naturally, and neptunium-237 and plutonium-239, which are produced in the laboratory.

Special Topic

RYDBERG ATOMS

Rydberg atoms are highly excited atoms with their outermost electron at a high energy level, very near ionization. They are named after Johannes Rydberg, who developed the empirical relation bearing his name that produces the correct wavelengths of the hydrogen atom [Equation (3.13)]. Rydberg atoms appear somewhat like hydrogen atoms because the highly excited electron is in such an extreme orbit that it stays well outside the orbits of the other electrons. A Rydberg atom of atomic number Z has an electron well outside a positive core of charge $+e$ [Z protons and $(Z-1)$ electrons], just like the hydrogen atom. In principle any atom can become a Rydberg atom.

Rydberg atoms are gigantic, as much as 100,000 times larger than normal atoms. Despite being in such a highly excited energy state, they are surprisingly long lived because the selection rules given in Equation (7.36) do not allow them to easily decay to lower energy levels (especially because of their high ℓ values). Their lifetime can be as long as a second, which is over a million times the lifetime of a normal excited atom. On the atomic scale, these long-lived Rydberg atoms live almost forever.

We recall from Chapter 4 that the energy levels of the hydrogen atom are given by $-E_0/n^2$ and the radius is given by $n^2 a_0$, where $E_0 = 13.6$ eV and $a_0 = 5.3 \times 10^{-11}$ m. Rydberg atoms have been observed in radio astronomy measurements from outer space with n values near 400, but those produced in the laboratory are rarely larger than 100 and are more commonly studied near 30. Note that a Rydberg atom acting like hydrogen and having $n = 400$ would have a diameter of $10^5 \times 10^{-10}$ m or $10\,\mu$m, an incredibly large atom! A transition from $n = 401 \rightarrow 400$ results in a 4×10^{-7} eV photon emission having a wavelength near 3 m, a radio wave.

Rydberg atoms can be made in the laboratory by bombarding gaseous atoms with charged particles. A revolution in their study came about, however, from the use of tunable lasers (see Chapter 10), which allows specific states to be excited by transferring a laser photon of precise energy to an electron. The density of atoms must be kept low because a collision between Rydberg atoms and normal atoms may quickly lead to de-excitation. The reason Rydberg atoms are so easily found (relatively speaking, of course) in interstellar space is because once created, a Rydberg atom has a poor chance of colliding with another atom.

The German physicist Johannes Stark discovered in 1913 the effect, named after him, that atomic spectral lines are split when subjected to a strong, external electric field. The most dramatic, and most useful, property of Rydberg atoms is due to this *Stark effect*. Because of their large n values, Rydberg atoms are highly de-

---

Conceptual Example 8.2

Copper and silver have the two highest electrical conductivities. Explain how the electronic configurations of copper and silver can account for their very high electrical conductivities.

Solution We need to refer to Figure 8.2 to investigate their electron configurations. We see that $_{28}$Ni has the structure $3d^8 4s^2$, but $_{29}$Cu has the structure $3d^{10} 4s^1$ and the next element, $_{30}$Zn, has $3d^{10} 4s^2$. Copper is unique in that one electron from the $4s$ subshell has changed to the $3d$ subshell. The remaining $4s$ electron is very weakly bound—in fact it is almost free.

Something similar happens to $_{47}$Ag in the next period. The elements on either side have completed the $4d^{10}$ subshell, and for $_{47}$Ag the $5s$ electron is only weakly bound. The elements $_{41}$Nb through $_{45}$Rh have an unpaired $5s$ electron, but incomplete $4d$ subshells, and so less screening—their $5s$ electrons are less free to wander than that of $_{47}$Ag.

Copper and silver have one very weakly bound electron outside a closed subshell core. The electron is practically free and is able to move around easily in the metal.

generate. Remember that two states are *degenerate* when they have different quantum numbers but have the same energy. Many states can have the same high value of n but have different values of ℓ and m_ℓ. In highly degenerate Rydberg atoms, the Stark effect is significant because the splitting of the many energy levels varies linearly with the electric field as shown in Figure A. It requires only a weak electric field to either ionize or change the energy level of a Rydberg atom.

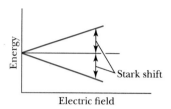

Figure A The thin black line represents the degenerate energy level, and the blue lines represent the maximum energy shift for a given electric field. Because of the large degeneracy, states may have many of the energies between the extremes.

Applications

Rydberg atoms are used for fundamental atomic measurements, including studies of the correspondence principle (see Section 4.4), because of their high n values. Detectors sensitive to electromagnetic radiation from the infrared to microwave wavelengths could be built using the very small differences between atomic energy levels in Rydberg atoms. Such long-wavelength radiation (and low frequency) is difficult to detect using normal atoms because of the larger energy differences between adjacent energy states. The use of electric fields allows the energy-level differences to be fine-tuned for Rydberg atoms. Such a detector, for example, might be useful in astronomy where suitable detectors are difficult to find.

Rydberg atoms may be useful in developing lasers because their high n values can serve as long-lived upper level metastable states (see Chapter 10). The use of Rydberg atoms has also been suggested for the separation of isotopes for uranium enrichment. A laser could be used to promote the atoms of one particular isotope, but not the others, to moderate atomic excitation, while a second laser excites the atoms of all other isotopes to highly excited states. These other atoms could then be isolated by ionization, leaving the single isotope.

8.2 Total Angular Momentum

If an atom has an orbital angular momentum and a spin angular momentum due to one or more of its electrons, we expect that, as is true classically, these angular momenta combine to produce a **total angular momentum**. We saw previously, in Section 7.5, that an interaction between the orbital and spin angular momenta in one-electron atoms causes splitting of energy levels into doublets, even in the absence of external magnetic fields. In this section we examine how the orbital and spin angular momenta combine and see how this results in energy-level splitting.

Single-Electron Atoms

We initially discuss only atoms having a single electron outside an inert core (for example, the alkalis). For an atom with orbital angular momentum $\vec{L}$ and spin angular momentum $\vec{S}$, the total angular momentum $\vec{J}$ is given by

$$\vec{J} = \vec{L} + \vec{S}$$

(8.3)   **Total angular momentum**

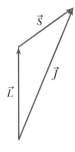

$j = \ell + s$
$= 1 + \frac{1}{2} = \frac{3}{2}$

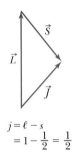

$j = \ell - s$
$= 1 - \frac{1}{2} = \frac{1}{2}$

Figure 8.5 When forming the total angular momentum from the orbital and spin angular momenta, the addition must be done vectorially, $\vec{J} = \vec{L} + \vec{S}$. We show schematically the addition of $\vec{L}$ and $\vec{S}$ with $\ell = 1$ and $s = 1/2$ to form vectors $\vec{J}$ with quantum numbers $j = 1/2$ and $3/2$.

Spin-orbit coupling

Because L, L_z, S, and S_z are quantized, the total angular momentum and its z component J_z are also quantized. If j and m_j are the appropriate quantum numbers for the single electron we are considering, quantized values of J and J_z are, in analogy with the single electron of the hydrogen atom,

$$J = \sqrt{j(j+1)}\,\hbar \tag{8.4a}$$

$$J_z = m_j\hbar \tag{8.4b}$$

Because m_ℓ is integral and m_s is half-integral, m_j will always be half-integral. Just as the value of m_ℓ ranges from $-\ell$ to ℓ, the value of m_j ranges from $-j$ to j, and therefore j will be half-integral.

The quantization of the magnitudes of $\vec{L}$, $\vec{S}$, and $\vec{J}$ are all similar.

$$L = \sqrt{\ell(\ell+1)}\,\hbar$$
$$S = \sqrt{s(s+1)}\,\hbar \tag{8.5}$$
$$J = \sqrt{j(j+1)}\,\hbar$$

The total angular momentum quantum number for the single electron can only have the values

$$j = \ell \pm s \tag{8.6}$$

which, because $s = 1/2$, can only be $\ell + 1/2$ or $\ell - 1/2$ (but j must be $1/2$ if $\ell = 0$). The relationships of $\vec{J}$, $\vec{L}$, and $\vec{S}$ are shown in Figure 8.5. For an ℓ value of 1, the quantum number j is $3/2$ or $1/2$, depending on whether $\vec{L}$ and $\vec{S}$ are aligned or antialigned. The notation commonly used to describe these states is

$$nL_j \tag{8.7}$$

where n is the principal quantum number, j is the total angular momentum quantum number, and L is an uppercase letter (S, P, D, and so on) representing the orbital angular momentum quantum number.

In Section 7.5 we briefly mentioned that the single electron of the hydrogen atom can feel an internal magnetic field $\vec{B}_{\text{internal}}$ due to the proton, because in the rest system of the electron, the proton appears to be circling the electron (see Figure 7.10). A careful examination of this effect shows that the spins of the electron and the orbital angular momentum interact, an effect called **spin-orbit coupling**. As usual the dipole potential energy $V_{s\ell}$ is equal to $-\vec{\mu}_s \cdot \vec{B}_{\text{internal}}$. The spin magnetic moment is proportional to $-\vec{S}$, and $\vec{B}$ internal is proportional to $\vec{L}$, so that $V_{s\ell} \sim \vec{S} \times \vec{L} = SL \cos \alpha$, where α is the angle between $\vec{S}$ and $\vec{L}$. The result of this effect is to make the states with $j = \ell - 1/2$ slightly lower in energy than for $j = \ell + 1/2$, because α is smaller when $j = \ell + 1/2$. The same applies for the atom when placed in an external magnetic field. The same effect leads us to accept j and m_j as better quantum numbers han m_ℓ and m_s, even for single-electron atoms like hydrogen. We mean "better" because j and m_j are more directly related to a physical observable. A given state having a definite energy can no longer be assigned a definite L_z and S_z, but it can have a definite J_z. The wave functions now depend on n, ℓ, j, and m_j. The spin-orbit interaction splits the $2P$ level into two states, $2P_{3/2}$ and $2P_{1/2}$, with $2P_{1/2}$ being lower in energy. There are additional relativistic effects, not discussed here, that give corrections to the spin-orbit effect.

In the absence of an external magnetic field, the total angular momentum has a fixed magnitude and a fixed z component. Remember that only J_z can be known;

Figure 8.6 (a) The vectors $\vec{L}$ and $\vec{S}$ precess around $\vec{J}$. The total angular momentum $\vec{J}$ can have a fixed value in only one direction in space—not shown in this figure. (b) However, with an external magnetic field $\vec{B}_{ext}$ along the z axis, $\vec{J}$ will precess around the z direction (J_z is fixed), and both $\vec{L}$ and $\vec{S}$ precess around $\vec{J}$. We have shown the case where $\vec{L}$ and $\vec{S}$ are aligned.

the uncertainty principle forbids J_x or J_y from being known at the same time as J_z. The vectors $\vec{L}$ and $\vec{S}$ will precess around $\vec{J}$ (see Figure 8.6a). In an external magnetic field, however, $\vec{J}$ will precess about $\vec{B}_{ext}$, while $\vec{L}$ and $\vec{S}$ still precess about $\vec{J}$ as shown in Figure 8.6b. The motion of $\vec{L}$ and $\vec{S}$ then becomes quite complicated.

Optical spectra are due to transitions between different energy levels. We have already discussed transitions for the hydrogen atom in Section 7.6 and gave the rules listed in Equation (7.36). For single-electron atoms, we now add the selection rules for Δj. The restriction of $\Delta\ell = \pm 1$ will require $\Delta j = \pm 1$ or 0. The allowed transitions for a single-electron atom are

$$\Delta n = \text{anything} \qquad \Delta\ell = \pm 1$$
$$\Delta m_j = 0, \pm 1 \qquad \Delta j = 0, \pm 1 \tag{8.8}$$

Single-electron atom allowed transitions

The selection rule for Δm_j follows from our results for Δm_ℓ in Equation (7.36) and from the result that $m_j = m_\ell + m_s$, where m_s is not affected.

Figure 7.11 presented an energy-level diagram for hydrogen showing many possible transitions. Figure 8.7 is a highly exaggerated portion of the hydrogen energy-level diagram for $n = 2$ and $n = 3$ levels showing the spin-orbit splitting.

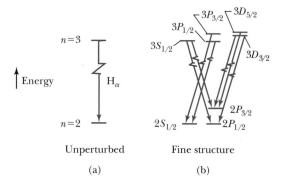

Unperturbed

(a)

Fine structure

(b)

Figure 8.7 (a) The unperturbed H_α line is shown due to a transition between the $n = 3$ and $n = 2$ shells of the hydrogen atom. (b) The more detailed level structure (not to scale) of the hydrogen atom leads to optical fine structure. The spin-orbit interaction splits each of the $\ell \neq 0$ states.

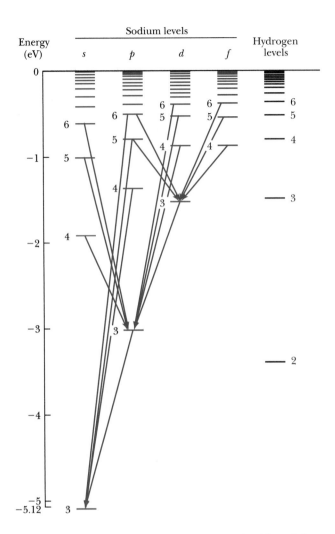

Figure 8.8 The energy-level diagram of sodium (a single electron outside an inert core) is compared to that of hydrogen. Coulomb effects cause the lower ℓ states of sodium to be lower than the corresponding levels of hydrogen. Several allowed transitions are shown for sodium.

All of the states (except for the s states) are split into doublets. What appeared in Figure 7.11 to be one transition is now actually seven different transitions. The splitting is quite small, but measurable—typically on the order of 10^{-5} eV in hydrogen. For example, the splitting between the $2P_{3/2}$ and $2P_{1/2}$ levels has been found to be 4.5×10^{-5} eV.

We show in Figure 8.8 the energy levels of a single-electron atom, sodium, compared with those of hydrogen. The single electron in sodium is $3s^1$, and the energy levels of sodium should be similar to those of $n = 3$ and above for hydrogen. However, the strong attraction of the electrons with small ℓ values causes those energy levels to be considerably lower than for higher ℓ. Notice in Figure 8.8 that the $5f$ and $6f$ energy levels of sodium closely approach the hydrogen energy levels, but the $3s$ energy level of sodium is considerably lower. The transitions between the energy levels of sodium displayed in Figure 8.8 are consistent with the selection rules of Equation (8.8).

The fine splitting of the levels for different j is too small to be seen in Figure 8.8. Nevertheless these splittings are important, and they are easily detected in the optical spectrum of sodium. The energy levels $3P_{3/2}$ and $3P_{1/2}$ are separated by about 2×10^{-3} eV, for example. This splits the $3p \rightarrow 3s$ (~2.1 eV) optical line into a doublet: the famous yellow sodium doublet, with $\lambda = 589.0$ nm and 589.6 nm (see also Example 8.8).

Example 8.3

Show that an energy difference of 2×10^{-3} eV for the $3p$ subshell of sodium accounts for the 0.6-nm splitting of a spectral line at 589.3 nm.

Strategy The wavelength λ of a photon is related to the energy of a transition by

$$E = \frac{hc}{\lambda}$$

For a small splitting, we approximate ΔE by using a differential:

$$dE = \frac{-hc}{\lambda^2} d\lambda$$

Then, letting $\Delta E = dE$ and $\Delta \lambda = d\lambda$ and taking absolute values yields

$$|\Delta E| = \frac{hc}{\lambda^2}|\Delta \lambda| \quad \text{or} \quad |\Delta \lambda| = \frac{\lambda^2}{hc}|\Delta E|$$

Solution We insert the values of λ, ΔE, and hc to obtain

$$|\Delta \lambda| = \frac{(589.3 \text{ nm})^2 (2 \times 10^{-3} \text{ eV})}{1.240 \times 10^3 \text{ eV} \cdot \text{nm}} = 0.6 \text{ nm}$$

The value of 0.6 nm agrees with the experimental measurement for sodium.

Many-Electron Atoms

The interaction of the various spins and angular momenta becomes formidable for more than two electrons outside an inert core. Various empirical rules help in applying the quantization results to such atoms. The best-known rule set is called Hund's rules, introduced in 1925 by the German physicist Friedrich Hund (1896–1997), who is known mainly for his work on the electronic structure of atoms and molecules. We consider here the case of two electrons outside a closed shell (for example, helium and the alkaline earths).*

The order in which a given subshell is filled is governed by Hund's rules:

Rule 1. The total spin angular momentum S should be maximized to the extent possible without violating the Pauli exclusion principle.

Rule 2. Insofar as rule 1 is not violated, L should also be maximized.

Rule 3. For atoms having subshells less than half full, J should be minimized.

Hund's rules

For example, the first five electrons to occupy a d subshell should all have the same value of m_s. This requires that each one has a different m_ℓ (because the allowed m_ℓ values are $-2, -1, 0, 1, 2$). By rule 2 the first two electrons to occupy a d subshell should have $m_\ell = 2$ and $m_\ell = 1$ *or* $m_\ell = -2$ and $m_\ell = -1$.

Besides the spin-orbit interaction already discussed, there are now spin-spin and orbital-orbital interactions. There are also effects due to the spin of the nucleus that lead to *hyperfine* structure (see Special Topic, "Hydrogen and the 21-cm Line Transition," in Chapter 7), but the nuclear effects are much smaller than the ones we are presently considering. For the two-electron atom, we label the electrons 1 and 2 so that we have $\vec{L}_1, \vec{S}_1$ and $\vec{L}_2, \vec{S}_2$. The total angular momentum $\vec{J}$ is the vector sum of the four angular momenta:

$$\vec{J} = \vec{L}_1 + \vec{L}_2 + \vec{S}_1 + \vec{S}_2 \tag{8.9}$$

*See H. G. Kuhn, *Atomic Spectra*, 2nd ed., New York: Academic Press (1969), or H. E. White, *Introduction to Atomic Spectra*, New York: McGraw-Hill (1934) for further study.

There are two schemes, called **LS coupling** and **jj coupling**, for combining the four angular momenta to form *J*. We shall discuss these next. The decision of which scheme to use depends on relative strengths of the various interactions. We shall see that *jj* coupling predominates for heavier elements.

LS Coupling

The *LS* coupling scheme, also called *Russell-Saunders coupling*, is used for most atoms when the magnetic field is weak. The orbital angular momenta $\vec{L}_1$ and $\vec{L}_2$ combine to form a total orbital angular momentum $\vec{L}$ and similarly for $\vec{S}$:

$$\vec{L} = \vec{L}_1 + \vec{L}_2 \tag{8.10}$$

$$\vec{S} = \vec{S}_1 + \vec{S}_2 \tag{8.11}$$

Then $\vec{L}$ and $\vec{S}$ combine to form the total angular momentum:

$$\vec{J} = \vec{L} + \vec{S} \tag{8.12}$$

One of Hund's rules states that the electron spins combine to make $\vec{S}$ a maximum. Physically, this occurs because of the mutual repulsion of the electrons, which want to be as far away from each other as possible in order to have the lowest energy. If two electrons in the same subshell have the same m_s, they must then have different m_ℓ, normally indicating different spatial distributions. Similarly, the lowest energy states normally occur with a maximum $\vec{L}$. We can understand this physically, because the electrons would revolve around the nucleus in the same direction if aligned, thus staying as far apart as possible. If $\vec{L}_1$ and $\vec{L}_2$ were antialigned, the electrons would pass each other more often, and therefore would tend to have a higher interaction energy.

For the case of two electrons in a single subshell, the total spin angular momentum quantum number* may be $S = 0$ or 1 depending on whether the spins are antiparallel or parallel. For a given value of L, there are $2S + 1$ values of J, because J goes from $L - S$ to $L + S$ (for $L > S$). For $L < S$ there are fewer than $2S + 1$ possible values of J (see Examples 8.4, 8.5, and 8.7). The value of $2S + 1$ is called the **multiplicity** of the state.

Multiplicity

The notation nL_j discussed before for a single-electron atom becomes

Spectroscopic symbols

$$n^{2S+1}L_J \tag{8.13}$$

The letters and numbers used in this notation are called *spectroscopic* or *term symbols*. For two electrons we have **singlet** states ($S = 0$) and **triplet** states ($S = 1$), which refer to the multiplicity $2S + 1$. Recall that a single-electron state (with $s = 1/2$) is a doublet, with $2s + 1 = 2$.

Consider two electrons: One is in the $4p$ and one is in the $4d$ subshell. For the atomic states shown in Table 8.2, we have the following possibilities: $S_1 = 1/2$, $S_2 = 1/2$, $L_1 = 1$, and $L_2 = 2$. A schematic diagram showing the relative energies of these states appears in Figure 8.9. The spin-spin interaction breaks the unperturbed state into the singlet and triplet states. The Coulomb effect, due to the electrons, orders the highest L value for each of these states to be lowest in

---

*It is customary to use capital letters L, S, and J for the angular momentum quantum numbers of many-electron atoms. This can lead to confusion, because we are accustomed to thinking, for example, $S = |\vec{S}|$. To avoid confusion, remember that the magnitude of an angular momentum vector is always some number times $\hbar$, but the new angular momentum quantum numbers, L, S, and J are simply integers or half-integers.

Table 8.2   Spectroscopic Symbols for Two Electrons: One in 4p and One in 4d

S	L	J	Spectroscopic Symbol
0 (singlet)	1	1	4^1P_1
	2	2	4^1D_2
	3	3	4^1F_3
1 (triplet)	1	2	4^3P_2
		1	4^3P_1
		0	4^3P_0
1 (triplet)	2	3	4^3D_3
		2	4^3D_2
		1	4^3D_1
1 (triplet)	3	4	4^3F_4
		3	4^3F_3
		2	4^3F_2

energy. Finally, the spin-orbit splitting causes the lowest J value to be lowest in energy ($\vec{L}$ and $\vec{S}$ antialigned).

As an example of the optical spectra obtained from two-electron atoms, we consider the energy-level diagram of magnesium in Figure 8.10. The most obvious characteristic of this figure is that we have separated the energy levels according

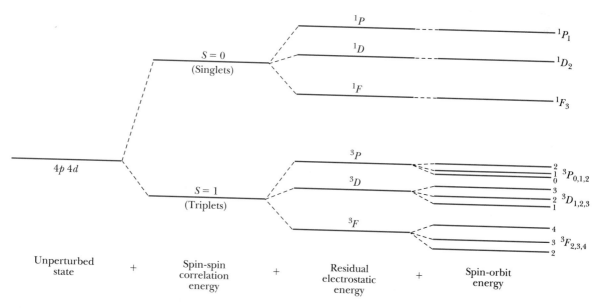

Figure 8.9 Schematic diagram indicating the increasing fine-structure splitting due to different effects. This case is for an atom having two valence electrons, one in the 4p and the other in the 4d state. The energy is not to scale. *From R. B. Leighton,* Principles of Modern Physics, *New York: McGraw-Hill (1959), p. 261.*

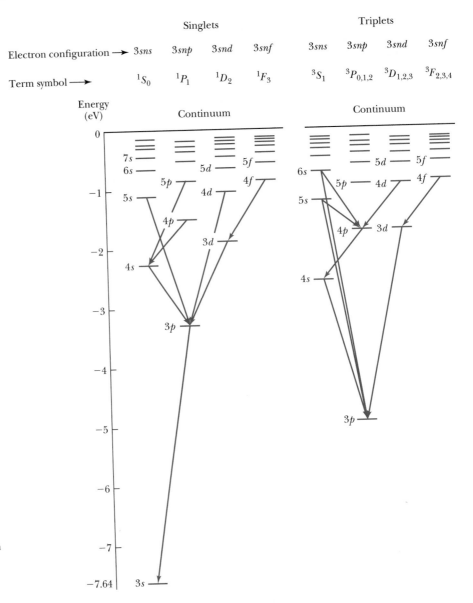

Figure 8.10 Energy-level diagram for magnesium (two-electron atom) with one electron in the 3s subshell and the other electron excited into the $n\ell$ subshell indicated. The singlet and triplet states are separated, because transitions between them are not allowed by the $\Delta S = 0$ selection rule. Several allowed transitions are indicated.

to whether they are $S = 0$ or $S = 1$. This is because *allowed* transitions must have $\Delta S = 0$, and no allowed transitions are possible between singlet and triplet states. This does not mean that it is impossible for such transitions to occur. Transitions that are not allowed, called *forbidden* transitions, occur, but with much lower probability than allowed transitions.

The two electrons in magnesium outside the closed $2p$ subshell are $3s^2$. Therefore, the ground state of magnesium is 3^1S_0 ($S = 0$, $L = 0$, and $J = 0$). The 3^3S_1 state ($S = 1$, $J = 1$) cannot exist because $m_s = 1/2$ for both electrons in order to have $S = 1$, and this is forbidden by the exclusion principle. A $^3S_1$ state is allowed if one of the electrons is in a higher n shell. The energy-level diagram of Figure 8.10 is generated by one electron remaining in the $3s$ subshell while the other electron is promoted to the subshell indicated on the diagram. The allowed transitions (for the LS coupling scheme) are

$$\Delta L = \pm 1 \qquad \Delta S = 0$$
$$\Delta J = 0, \pm 1 \qquad (J = 0 \rightarrow J = 0 \text{ is forbidden})$$
(8.14)

A magnesium atom excited to the $3s3p$ triplet state has no lower triplet state to which it can decay. The only state lower in energy is the $3s^2$ ground state, which is singlet. Such an excited triplet state may exist for a relatively long time ($\gg 10^{-8}$ s) before it finally decays to the ground state as a forbidden transition. Such a $3s3p$ triplet state is called **metastable**, because it lives for such a long time on the atomic scale.

Metastable states

jj Coupling

This coupling scheme predominates for the heavier elements, where the nuclear charge causes the spin-orbit interactions to be as strong as the forces between the individual $\vec{S}_i$ and the individual $\vec{L}_i$. The coupling order becomes

$$\vec{J}_1 = \vec{L}_1 + \vec{S}_1$$
(8.15a)

$$\vec{J}_2 = \vec{L}_2 + \vec{S}_2$$
(8.15b)

and then

$$\vec{J} = \sum_i \vec{J}_i$$
(8.16)

The spectroscopic or term notation is also used to describe the final states in this coupling scheme.

Conceptual Example 8.4

What are the total angular momentum and the spectroscopic notation for the ground state of helium?

Solution The two electrons for helium are both $1s$ electrons. Because helium is a light atom, we use the LS coupling scheme. We have $L_1 = 0$ and $L_2 = 0$, and therefore $L = 0$. We can have $S = 0$ or 1 for two electrons, but not in the same subshell. The spins must be antialigned and $S = 0$. Therefore $J = 0$ also. We can write the ground state spectroscopic symbol for helium as $1\,^1S_0$.

Example 8.5

Consider two electrons in an atom with orbital quantum numbers $\ell_1 = 1$ and $\ell_2 = 2$. Use LS coupling and find all possible values for the total angular momentum quantum numbers for $\vec{J}$.

Strategy We first find all the ways $\vec{L}_1$ and $\vec{L}_2$ combine to form the total orbital angular momentum $\vec{L}$. We find all the

possible vectors for the spin angular momentum $\vec{S}$, which will be $S = 0$ or 1, because the two electrons can only be aligned or antialigned, and $s = 1/2$. Then we add $\vec{L}$ and $\vec{S}$ to find the quantum numbers for $\vec{J}$.

Solution The total orbital angular momentum quantum number ranges from $|\ell_1 - \ell_2|$ to $|\ell_1 + \ell_2|$, so we have

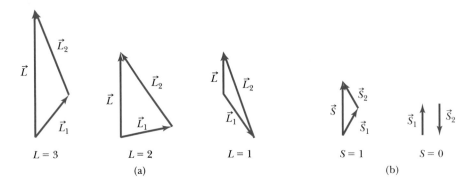

$L = 3$ $L = 2$ $L = 1$ $S = 1$ $S = 0$

(a)   (b)

Figure 8.11 (a) Two electrons having orbital angular momentum quantum numbers of 1 and 2 combine to form L values of 1, 2, 3. (b) Two electrons having spin angular quantum numbers of 1/2 and −1/2 form S values of 0 and 1.

values of L that are 1, 2, and 3. We show the vectors for $\vec{L}$ in Figure 8.11a. We also show in Figure 8.11b how $\vec{S}_1$ and $\vec{S}_2$ form to have S values of 0 and 1. Now $\vec{J} = \vec{L} + \vec{S}$ and the range of quantum numbers for J range from $|L - S|$ to $|L + S|$, so we have values of $J = 0, 1, 2, 3, 4$.

Example 8.6

Consider the values $L = 3$ and $S = 1$ in the last example, and choose the minimum value of J. Draw the coupling of the vectors for the case of no magnetic field. Show the precession of the vectors.

Strategy We showed in Figure 8.6a how the vectors combine in LS coupling. In this case we have the situation where L and S are antialigned, because we are considering only the minimum value of J.

Solution We are using the situation in Example 8.5 where $\vec{L}_1$ and $\vec{L}_2$ are aligned to form the maximum value of $\vec{L}$, and $\vec{S}_1$ and $\vec{S}_2$ are aligned to form the maximum value of $\vec{S}$. But the value of J is a minimum. The vectors $\vec{L}$ and $\vec{S}$ both precess about $\vec{J}$. The vector $\vec{J}$ precesses randomly around the z axis; only the component J_z is fixed in space. We show the result in Figure 8.12.

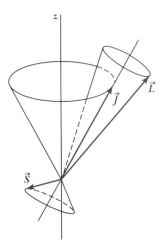

Figure 8.12 $\vec{S}$ and $\vec{L}$ are antialigned and form $\vec{J} = \vec{L} + \vec{S}$. Both $\vec{S}$ and $\vec{L}$ precess about $\vec{J}$, while $\vec{J}$ precesses randomly about the z axis.

Example 8.7

What are the L, S, and J values for the first few excited states of helium?

Strategy The lowest excited states of helium must be $1s^12s^1$ or $1s^12p^1$—that is, one electron is promoted to either the $2s^1$ or $2p^1$ subshell. It turns out that *all* excited states of

helium are single-electron states, because to excite both electrons requires more than the ionization energy. We expect the excited states of $1s^12s^1$ to be lower than those of $1s^12p^1$, because the subshell $2s^1$ is lower in energy than the $2p^1$ subshell.

Solution The possibilities are

$$1s^1 2s^1 \qquad L = 0$$

$$\text{If } S = 0, \text{ then } J = 0$$

$$\text{If } S = 1, \text{ then } J = 1$$

with $S = 1$ being lowest in energy. The lowest excited state is $^3S_1$ and then comes $^1S_0$.

$$1s^1 2p^1 \qquad L = 1$$

$$\text{If } S = 0, \text{ then } J = 1$$

$$\text{If } S = 1, \text{ then } J = 0, 1, 2$$

The state $^3P_0$ has the lowest energy of these states, followed by $^3P_1$, $^3P_2$, and $^1P_1$. The energy-level diagram for helium is shown in Figure 8.13.

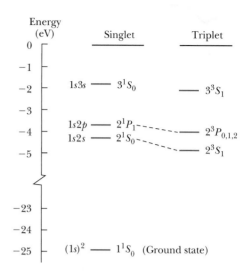

Figure 8.13 The low-lying atomic states of helium are shown. The ground state ($^1S_0$) is some 20 eV below the grouping of the lowest excited states. The level indicated by $^3P_{0,1,2}$ is actually three states ($^3P_0$, $^3P_1$, $^3P_2$), but the separations are too small to be indicated.

Example 8.8

If the spin-orbit splitting of the $3P_{3/2}$ and $3P_{1/2}$ states of sodium is 0.002 eV, what is the internal magnetic field causing the splitting?

Strategy The potential energy due to the spin magnetic moment is

$$V = -\vec{\mu}_s \cdot \vec{B} \tag{8.17}$$

By analogy with Equations (7.34), the z component of the total magnetic moment is

$$\mu_z = -g_z \left(\frac{e\hbar}{2m} \right) \frac{J_z}{\hbar} \tag{8.18}$$

where we have used the gyromagnetic ratio $g_s = 2$, because this splitting is actually due to spin. We associate the energy splitting, denoted ΔE, with the potential energy, $\Delta E = V$.

We determine the value of μ_z in Equation (8.18) and insert this into Equation (8.17) to find B, because we are given the energy splitting ΔE.

Solution The difference in spins between the $3P_{3/2}$ and $3P_{1/2}$ states is $\hbar$ so that

$$\Delta E = g_s \left(\frac{e\hbar}{2m} \right) \frac{\hbar}{\hbar} B = \frac{e\hbar}{m} B$$

Then

$$B = \frac{m \Delta E}{e\hbar} = \frac{(9.11 \times 10^{-31}\, \text{kg})(0.002\, \text{eV})}{(1.6 \times 10^{-19}\, \text{C})(6.58 \times 10^{-16}\, \text{eV} \cdot \text{s})}$$

$$= 17\, \text{T}$$

This is a large magnetic field.

Example 8.9

What are the possible energy states for atomic carbon?

Strategy The element carbon has two $2p$ subshell electrons outside the closed $2s^2$ subshell. Both electrons have

$\ell = 1$, so we can have $L = 0$, 1, or 2 using the LS coupling scheme. The spin angular momentum is $S = 0$ or 1.

We list the possible states:

S	L	J	Spectroscopic Notation	
0	0	0	$^1S_0$	
	1	1	$^1P_1$	Not allowed
	2	2	$^1D_2$	
1	0	1	$^3S_1$	Not allowed
	1	0, 1, 2	$^3P_{0,1,2}$	
	2	1, 2, 3	$^3D_{1,2,3}$	Not allowed

Solution The $^3S_1$ state is not allowed by the Pauli exclusion principle, because both electrons in the $2p^2$ subshell would have $m_s = +1/2$ and $m_\ell = 0$. Similarly, the $^3D_{1,2,3}$ states are not allowed, because both electrons would have $m_s = +1/2$ and $m_\ell = 1$. According to Hund's rules, the triplet states $S = 1$ will be lowest in energy, so the ground state will be one of the $^3P_{0,1,2}$ states. The spin-orbit interaction then indicates the $^3P_0$ state to be the ground state; the others are excited states.

The fact that the $^1P_1$ state is not allowed is a result of the antisymmetrization of the wave function, which we have not discussed. This rule, which requires electrons to have anti-symmetric wave functions, is basically an extension of the Pauli exclusion principle for this example, and it allows the states in which the m_ℓ of the two electrons are equal to combine only with $S = 0$ states. The m_ℓ values for the electrons forming the $S = 1$ state must be unequal. This rule is the theoretical basis for Hund's rules, described previously. It precludes the ^1P, ^3S, and ^3D states from existing for $2p^2$ electrons. The states with one electron having $m_\ell = 1$, $m_s = -1/2$ and the other having $m_\ell = 0$, $m_s = -1/2$ still exist, but they can be included in the $^1D_2$ state, for example, because $m_s = 0$ and $m_\ell = 2, 1, 0, -1, -2$.

The low-lying excited states of carbon are then $^3P_1$, $^3P_2$, $^1D_2$, and $^1S_0$.

8.3   Anomalous Zeeman Effect

In Section 7.4 we discussed the normal Zeeman effect and showed that the splitting of an optical spectral line into three components in the presence of an external magnetic field could be understood by considering the interaction $(\vec{\mu}_\ell \cdot \vec{B}_{\text{ext}})$ of the orbital angular momentum magnetic moment m_ℓ and the external magnetic field. Soon after the discovery of this effect by Zeeman in 1896, it was found that often more than the three closely spaced optical lines were observed. This observation was called the *anomalous* Zeeman effect. We are now able to explain both Zeeman effects. We shall see that the anomalous effect depends on the effects of electron intrinsic spin.

The interaction that splits the energy levels in an external magnetic field $\vec{B}_{\text{ext}}$ is still caused by the $\vec{\mu} \cdot \vec{B}$ interaction. However, the magnetic moment is due not only to the orbital contribution $\vec{\mu}_\ell$; it also depends on the spin magnetic moment $\vec{\mu}_s$. The $2J + 1$ degeneracy (due to m_J) for a given total angular momentum state J is removed by the effect of the external magnetic field. If the external field $\vec{B}_{\text{ext}}$ is small in comparison with the internal magnetic field (say $B_{\text{ext}} < 0.1$ T), then $\vec{L}$ and $\vec{S}$ (using the LS coupling scheme) precess about $\vec{J}$ while $\vec{J}$ precesses *slowly* about $\vec{B}_{\text{ext}}$.

We can see this more easily by calculating $\vec{\mu}$ in terms of $\vec{L}$, $\vec{S}$, and $\vec{J}$. The total magnetic moment $\vec{\mu}$ is

$$\vec{\mu} = \vec{\mu}_\ell + \vec{\mu}_s \tag{8.19}$$

$$= -\frac{e}{2m}\vec{L} - \frac{e}{m}\vec{S} \tag{8.20}$$

where $\vec{\mu}_\ell$ is obtained from Equation (7.26) and $\vec{\mu}_s$ from Section 7.5.

$$\vec{\mu} = -\frac{e}{2m}(\vec{L} + 2\vec{S}) = -\frac{e}{2m}(\vec{J} + \vec{S}) \tag{8.21}$$

The vectors $-\vec{\mu}$ and $\vec{J}$ are along the same direction only when $S = 0$. We show schematically in Figure 8.14 what is happening. The vector $\vec{B}_{\text{ext}}$ defines the z direction. We plot $-\vec{\mu}$ instead of $+\vec{\mu}$ in order to emphasize the relationship between $\vec{\mu}$ and $\vec{J}$. In a weak magnetic field the precession of $\vec{\mu}$ around $\vec{J}$ is much faster than the precession of $\vec{J}$ around $\vec{B}_{\text{ext}}$. Therefore, we first find the average $\vec{\mu}_{\text{av}}$ about $\vec{J}$ and then find the interaction energy of $\vec{\mu}_{\text{av}}$ with $\vec{B}_{\text{ext}}$. We leave this as an exercise for the student (see Problem 31). The result is

$$V = \frac{e\hbar B_{\text{ext}}}{2m} g m_J = \mu_B B_{\text{ext}} g m_J \qquad (8.22)$$

where μ_B is the Bohr magneton and

$$g = 1 + \frac{J(J+1) + S(S+1) - L(L+1)}{2J(J+1)} \qquad (8.23)$$

is a dimensionless number called the **Landé g factor**. The magnetic total angular momentum numbers m_J range from $-J$ to J in integral steps. The external field $\vec{B}_{\text{ext}}$ splits each state J into $2J + 1$ equally spaced levels separated by $\Delta E = V$, with V determined in Equation (8.22), each level being described by a different m_J.

In addition to the previous selection rules [Equation (8.14)] for photon transitions between energy levels, we must now add one for m_J:

$$\Delta m_J = \pm 1, 0 \qquad (8.24)$$

but $m_{J_1} = 0 \rightarrow m_{J_2} = 0$ is forbidden when $\Delta J = 0$.

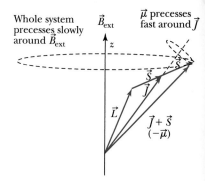

Figure 8.14 Relationships between $\vec{S}, \vec{L}, \vec{J}$, and $\vec{\mu}$ are indicated. The vector $\vec{B}_{\text{ext}}$ is in the z direction. The magnetic moment $\vec{\mu}$ precesses fast around $\vec{J}$ as $\vec{J}$ precesses more slowly around the weak $\vec{B}_{\text{ext}}$. *After J. D. McGervey*, Introduction to Modern Physics, *New York: Academic Press (1983), p. 329.*

Conceptual Example 8.10

Show that the normal Zeeman effect should be observed for transitions between the $^1D_2$ and $^1P_1$ states.

Strategy Because $2S + 1 = 1$ for both states, then $S = 0$ and $J = L$. The g factor from Equation (8.23) is equal to 1 (as it will always be for $S = 0$). The $^1D_2$ state splits into five equally spaced levels, and the $^1P_1$ state splits into three (see Figure 8.15). We use the selection rules from Equations (8.14) and (8.24) to determine which transitions are allowed.

Solution We start with every level in the $^1D_2$ state and determine using the selection rules which transitions to levels in the $^1P_1$ state are allowed. We show in Figure 8.15 that there are only nine allowed transitions. The other transitions are disallowed by the selection rule for Δm_J. Even though there are nine different transitions, there are only three different energies for emitted or absorbed photons, because transition energies labeled 1, 3, 6 are identical, as are 2, 5, 8, and also 4, 7, 9. Thus the three equally spaced transitions of the normal Zeeman effect are observed whenever $S = 0$.

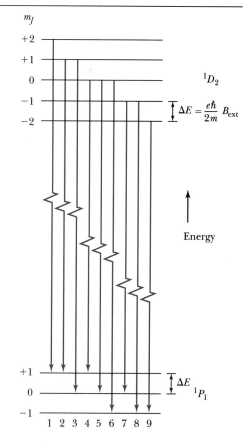

Figure 8.15 Examples of transitions for the normal Zeeman effect. The nine possible transitions are labeled, but there are only three distinctly different energies because the split energy levels are equally spaced (ΔE) for both the $^1D_2$ and $^1P_1$ states.

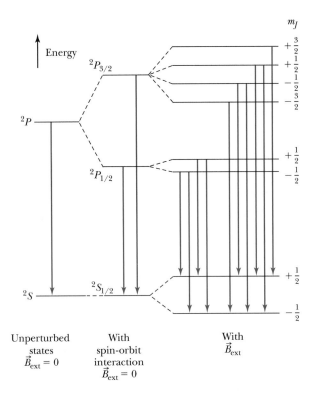

Figure 8.16 Schematic diagram of anomalous Zeeman effect for sodium (energy levels not to scale). With $\vec{B}_{ext} = 0$ for the unperturbed states, there is only one transition. With the spin-orbit interaction splitting the 2P state into two states, there are two possible transitions when $\vec{B}_{ext} = 0$. Finally, $\vec{B}_{ext}$ splits J into $2J + 1$ components, each with a different m_J. The energy splitting ΔE for each major state is different because $\Delta E = g m_J \, (e\hbar 2m) B_{ext}$ and the Landé g factor for $g[\Delta E(^2S_{1/2})] > g[\Delta E(^2P_{3/2})] > g[\Delta E(^2P_{1/2})]$. All allowed transitions are shown.

The anomalous Zeeman effect is a direct result of intrinsic spin. Let us consider transitions between the ^2P and ^2S states of sodium as shown in Figure 8.16. In a completely unperturbed state the $^2P_{3/2}$ and $^2P_{1/2}$ states are degenerate. However, the internal spin-orbit interaction splits them, with $^2P_{1/2}$ being lower in energy. The $^2S_{1/2}$ state is not split by the spin-orbit interaction because $\ell = L = 0$.

When sodium is placed in an external magnetic field, all three states are split into $2J + 1$ levels with different m_J (see Figure 8.16). The appropriate Landé g factors are

$$^2S_{1/2} \qquad g = 1 + \frac{\frac{1}{2}(\frac{1}{2} + 1) + \frac{1}{2}(\frac{1}{2} + 1)}{2 \cdot \frac{1}{2}(\frac{1}{2} + 1)} = 2$$

$$^2P_{1/2} \qquad g = 1 + \frac{\frac{1}{2}(\frac{1}{2} + 1) + \frac{1}{2}(\frac{1}{2} + 1) - 1(1 + 1)}{2 \cdot \frac{1}{2}(\frac{1}{2} + 1)} = 0.67$$

$$^2P_{3/2} \qquad g = 1 + \frac{\frac{3}{2}(\frac{3}{2} + 1) + \frac{1}{2}(\frac{1}{2} + 1) - 1(1 + 1)}{2 \cdot \frac{3}{2}(\frac{3}{2} + 1)} = 1.33$$

All g factors are different, and the energy-splitting ΔE calculated using Equation (8.22) for the three states are different. The allowed transitions using the selection rules are shown in Figure 8.16. There are four different energy transitions for $^2P_{1/2} \rightarrow {}^2S_{1/2}$ and six different energy transitions for $^2P_{3/2} \rightarrow {}^2S_{1/2}$.

If the external magnetic field is increased, then $\vec{L}$ and $\vec{S}$ precess too rapidly about $\vec{B}_{ext}$ and our averaging procedure for $\vec{\mu}$ around $\vec{B}_{ext}$ breaks down. In that case, the equations developed in this section are incorrect. This occurrence, called the *Paschen-Back effect*, must be analyzed differently. We will not pursue this calculation further.*

*See H. E. White, *Introduction to Atomic Spectra*, New York: McGraw-Hill (1934) for more information.

Summary

The Pauli exclusion principle states that no two electrons in an atom may have the same set of quantum numbers (n, ℓ, m_ℓ, m_s). Because electrons normally occupy the lowest energy state available, the Pauli exclusion principle may be used to produce the periodic table and understand many properties of the elements.

The total angular momentum $\vec{J}$ is the vector sum of $\vec{L}$ and $\vec{S}$, $\vec{J} = \vec{L} + \vec{S}$. The coupling of $\vec{S}$ and $\vec{L}$, called the spin-orbit interaction, leads to lower energies for smaller values of J. For two or more electrons in an atom we can couple the $\vec{L}_i$ and $\vec{S}_i$ of the valence electrons by either LS or jj coupling. The spectroscopic notation for an atomic state is $n^{2S+1}L_J$.

The allowed transitions now have

$$\Delta L = \pm 1 \qquad \Delta S = 0$$
$$\Delta J = 0, \pm 1 \qquad (J = 0 \rightarrow J = 0 \text{ is forbidden}) \tag{8.14}$$

The anomalous Zeeman effect is explained by the removal of the $2J + 1$ degeneracy when an atom is placed in a weak magnetic field. Each state has a different m_J, which has the selection rule for transitions of $\Delta m_J = \pm 1, 0$ (with exceptions). The normal Zeeman effect (three spectral lines) occurs when $S = 0$.

Questions

1. Explain in terms of the electron shell configuration why it is dangerous to throw sodium into water.
2. Why are the inert gases in gaseous form at room temperature?
3. Which groups of elements have the best and which the poorest electrical conductivities? Explain.
4. Why are the elements with good electrical conductivities also generally good thermal conductors?
5. Boron, carbon, and aluminum are not part of the alkalis or alkaline earths, yet they are generally good electrical conductors. Explain.
6. The alkali metals have the lowest ionization energies (Figure 8.3), yet they have the largest atomic radii (Figure 8.4). Is this consistent? Explain.
7. List four compounds that you believe should be strongly bound. Explain why.

8. Explain why the transition metals have good thermal and electrical conductivities.
9. Why do the alkaline earths have low resistivities?
10. Why is there no spin-orbit splitting for the ground state of hydrogen?
11. Is it possible for both atoms in a hydrogen molecule to be in the $(1, 0, 0, -1/2)$ state? Explain.
12. Discuss in your own words the differences between $\vec{L}$ and ℓ, between m_ℓ and m_s, and between J_z and m_j.
13. Why do Rydberg atoms live so long? *Hint:* Consider the selection rules and the values of their quantum numbers.
14. Discuss whether the atomic state represented by $3^2D_{7/2}$ exists, and give reasons.

Problems

Note: The more challenging problems are indicated by a shaded box.

8.1 Atomic Structure and the Periodic Table

1. A lithium atom has three electrons. Allow the electrons to interact with each other and the nucleus. Label each electron's spin and angular momentum. List all the possible interactions.
2. For all the elements through neon list the electron descriptions of these elements in their ground state using $n\ell$ notation (for example, helium is $1s^2$).

3. How many subshells are in the following shells: L, N, and O?
4. What electron configuration would you expect $(n\ell)$ for the first excited state of argon and krypton?
5. Using Table 8.1 and Figure 8.2 write down the electron configuration ($n\ell$ notation) of the following elements: potassium, vanadium, selenium, zirconium, samarium, and uranium.
6. Use Figure 8.2 to list all the (a) inert gases, (b) alkalis, and (c) alkaline earths.
7. The $3s$ state of Na has an energy of -5.14 eV. Determine the effective nuclear charge.

8. List the quantum numbers (n, ℓ, m_ℓ, m_s) for all the electrons in a nitrogen atom.

9. What atoms have the configuration (a) $1s^2 2s^2 2p$, (b) $1s^2 2s^2 2p^6 3s$, (c) $3s^2 3p^6$?

10. What are the electronic configurations for the ground states of the elements Pd, Hf, and Sb?

11. What atoms have the configuration (a) $4s^2 4p^3$, (b) $4p^6 4d^{10} 5s$, and (c) $5s^2 5p^6 4f^{11}$?

8.2 Total Angular Momentum

12. If the zirconium atom ground state has $S = 1$ and $L = 3$, what are the permissible values of J? Write the spectroscopic notation for these possible values of S, L, and J. Which one of these is likely to represent the ground state?

13. Using the information in Table 8.2, determine the ground state spectroscopic symbol for gallium.

14. List all the elements through calcium that you would expect not to have a spin-orbit interaction that splits the ground state energy. Explain.

15. For the hydrogen atom in the $3d$ excited state find the possible values of ℓ, m_ℓ, j, s, m_s, and m_j. Give the term notation for each possible configuration.

16. What are S, L, and J for the following states: $^1S_0$, $^2D_{5/2}$, $^5F_1$, $^3F_4$?

17. What are the possible values of J_z for the $8^2G_{7/2}$ state?

18. (a) What are the possible values of J_z for the $6^2F_{7/2}$ state? (b) Determine the minimum angle between the total angular momentum vector and the z axis for this state.

19. Explain why the spectroscopic term symbol for lithium in the ground state is $^2S_{1/2}$.

20. What is the spectroscopic term symbol for aluminum in its ground state? Explain.

21. The $4P$ state in potassium is split by its spin-orbit interaction into the $4P_{3/2}$ ($\lambda = 766.41$ nm) and $4P_{1/2}$ ($\lambda = 769.90$ nm) states (the wavelengths are for the transitions to the ground state). Calculate the spin-orbit energy splitting and the internal magnetic field causing the splitting.

22. Draw the energy-level diagram for the states of carbon discussed in Example 8.9. Draw lines between states that have allowed transitions and list ΔL, ΔS, and ΔJ.

23. An $n = 2$ shell (L shell) has a $2s$ state and two $2p$ states split by the spin-orbit interaction. Careful measurements of the K_α x-ray ($n = 2 \rightarrow n = 1$) transition reveal only two spectral lines. Explain.

24. What is the energy difference between a spin-up state and spin-down state for an electron in an s state if the magnetic field is 1.7 T?

25. Which of the following elements can have either (or both) singlet and triplet states and which have neither: He, Al, Ca, Sr? Explain.

26. If the minimum angle between the total angular momentum vector and the z axis is 32.3° (in a single-electron atom), what is the total angular momentum quantum number?

27. Use the Biot-Savart law to find the magnetic field in the frame of an electron circling a nucleus of charge Ze. If the velocity of the electron around the nucleus is $\vec{v}$ and the position vector of the proton with respect to the electron is $\vec{r}$, show that the magnetic field at the electron is

$$\vec{B} = \frac{Ze}{4\pi\epsilon_0} \frac{\vec{L}}{mc^2 r^3}$$

where m is the electron mass and $\vec{L}$ is the angular momentum, $\vec{L} = m\vec{r} \times \vec{v}$.

28. Using the internal magnetic field of the previous problem show that the potential energy of the spin magnetic moment μ_s interacting with $\vec{B}_{\text{internal}}$ is given by

$$V_{s\ell} = \frac{Ze^2}{4\pi\epsilon_0} \frac{\vec{S} \cdot \vec{L}}{m^2 c^2 r^3}$$

There is an additional factor of $1/2$ to be added from relativistic effects called the *Thomas factor*.

29. The difference between the $2P_{3/2}$ and $2P_{1/2}$ doublet in hydrogen due to the spin-orbit splitting is 4.5×10^{-5} eV. (a) Compare this with the potential energy given in the previous problem. (b) Compare this with a more complete calculation giving the potential energy term V as

$$V = -\frac{Z^4 \alpha^4}{2n^3} mc^2 \left(\frac{2}{2j+1} - \frac{3}{4n} \right)$$

where α is the fine-structure constant, $\alpha = 1/137$.

8.3 Anomalous Zeeman Effect

30. For which L and S values does an atom exhibit the normal Zeeman effect? Does this apply to both ground and excited states? Can an atom exhibit both the normal and anomalous Zeeman effects?

31. Derive Equations (8.22) and (8.23). First find the average value of $\vec{\mu}$ and $\vec{J}$. Use

$$\vec{\mu}_{\text{av}} = (\vec{\mu} \cdot \hat{J}) \frac{\vec{J}}{\vec{J} \cdot \vec{J}} \quad \text{and} \quad V = -\vec{\mu}_{\text{av}} \cdot \vec{B}$$

[Remember $|\vec{J}|^2 = J(J+1)\hbar^2$.]

32. In the early 1900s the normal Zeeman effect was useful to determine the electron's e/m if Planck's constant was assumed known. Calcium is an element that exhibits the normal Zeeman effect. The difference between adjacent components of the spectral lines is observed to be 0.013 nm for $\lambda = 422.7$ nm when calcium is placed in a magnetic field of 1.5 T. From these data calculate the value of $e\hbar/m$ and compare with the accepted value today. Calculate e/m assuming the known value of $\hbar$.

33. Calculate the Landé g factor for an atom with a single (a) s electron, (b) p electron, (c) d electron.

34. An atom with the states $^2G_{9/2}$ and $^2H_{11/2}$ is placed in a weak magnetic field. Draw the energy levels and indicate the possible allowed transitions between the two states.

35. Repeat the previous problem for $^3P_1$ and $^3D_2$ states.

36. With no magnetic field, the spectral line representing the transition from the $^2P_{1/2}$ state to the $^2S_{1/2}$ state in sodium has the wavelength 589.76 nm (see Figure 8.16). This is one of the two strong yellow lines in sodium. Calculate the difference in wavelength between the shortest and longest wavelength between these two states when placed in a magnetic field of 0.50 T.

37. When sodium in the $^2P_{3/2}$ state is placed in a magnetic field of 0.50 T, the energy level splits into four levels (see Figure 8.16). Calculate the energy difference between these levels.

General Problems

38. (a) Write down the configurations for the ground states of magnesium and aluminum. (b) What are the *LS* coupling quantum numbers for the outside subshell electrons? Write the spectroscopic symbol for each atom.

39. (a) Write down the configurations for the ground states of the ionized ions Sc^- and Al^-. (b) What are the *LS* coupling quantum numbers for the outside subshell electrons? Write the spectroscopic symbol for each atom.

Statistical Physics

Ludwig Boltzmann, who spent much of his life studying statistical mechanics, died in 1906 by his own hand. Paul Ehrenfest, carrying on his work, died similarly in 1933. Now it is our turn to study statistical mechanics. Perhaps it will be wise to approach the subject cautiously.

David L. Goodstein (States of Matter, Mineola, New York: Dover, 1985)

Statistics are important in a number of areas of physics, one of which is atomic physics. The position of the electron in the hydrogen atom was described in Section 7.6 in terms of a probability distribution. Another important application of the concept of probability to atomic physics arises for transitions between atomic states. Like most atomic systems, the hydrogen atom may be in the ground state or in one of a number of excited states. An atom in an excited state is likely to make a transition to a lower energy state and eventually to the ground state. The question that arises logically is: just how likely are each of the allowed transitions? One may also ask: what are the relative probabilities of finding an atom in any particular state? These are not simple questions, and the answers require knowledge of the wave functions for each of the states involved. We shall not attempt to solve these problems here. We should simply be aware of the fact that transitions between quantum states must usually be described in probabilistic terms because there is no simple causal mechanism we can use to track the electron(s) from one level to another.

There is a simpler way to begin to understand how statistics and probability theory are used in physics. Historically, the need for these mathematical tools became apparent to those studying problems in heat and thermodynamics. Accordingly, we begin this chapter with a review of how probability and statistics found their way into physics in the nineteenth century. The important results of Maxwell and Boltzmann are outlined and used to derive some of the basic laws of the kinetic theory of gases. Then we look at how quantum statistics differ from classical statistics. The last two sections deal with applications of quantum statistics (Fermi-Dirac and Bose-Einstein) to a number of problems in modern physics. The use of statistics in quantum systems, particularly in solids, will be continued in more detail in Chapter 10.

9.1   Historical Overview

At the beginning of the nineteenth century, physicists were generally not inclined to use probability and statistics to describe physical processes. Indeed, there was a strong tendency to view the universe as a machine run by strict, unvarying, deterministic laws. This mechanistic view was due in large part to Newton and to the successful run of Newtonian physics through the eighteenth and nineteenth centuries. New mathematical methods developed by Lagrange around 1790 and Hamilton around 1840 added significantly to the computational power of Newtonian mechanics. They enabled physicists to describe extremely complicated physical systems with relatively simple second-order differential equations.

Probably the most extreme defender of the absolute power of classical mechanics (not to mention the fact that he was one of its greatest practitioners) was Pierre-Simon de Laplace (1749–1827). Laplace held that it should be possible in principle to have perfect knowledge of the physical universe. Such knowledge would come from measuring precisely at one time the position and velocity of every particle of matter and then applying Newton's physical laws. Because of the absolute immutability of Newton's laws, this knowledge could be extended indefinitely into the future and the past, even to the creation of the universe. We realize now that Heisenberg's uncertainty principle (Section 5.6) creates serious problems for Laplace's position. A famous but perhaps apocryphal story demonstrates the conviction of the defenders of the mechanistic view. Laplace once presented the principle of perfect knowledge just described to Emperor Napoleon. After hearing it in some detail, Napoleon asked where God fit into this mechanistic system. Laplace is supposed to have answered, "I have no need of this hypothesis." That was a rather extreme statement for anyone to make in the early nineteenth century, but it should indicate the great sense of confidence felt by those true believers in the Newtonian program.

In fairness to Laplace we must point out that he did make major contributions to the theory of probability. He may have been inspired to study probability by his understanding of the practical limits of measurement in classical mechanics. His 1812 treatise *Théorie analytique des probabilités* was the standard reference on the subject for much of that century.

The development of statistical physics in the nineteenth century was tied closely to the development of thermodynamics (which in turn was driven by the Industrial Revolution). In 1800 the common view of heat was as a material substance known as *caloric*, a fluid that could flow through bodies to effect changes in temperature. In 1798 Benjamin Thompson (Count Rumford) put forward the idea that what we call *heat* is merely the motion of individual particles in a substance. Rumford's idea was essentially correct, but it was not accepted quickly. His work was too sketchy and qualitative, although it did plant the seeds that later blossomed in the work of Maxwell and Boltzmann. In 1822 Joseph Fourier published his theory of heat, which was the first truly mathematical treatise on the subject. It was not statistical in nature, but it provided a quantitative basis for later work.

The concept of energy is central to modern thermodynamics. Of supreme importance in the history of energy is the work of James Prescott Joule (1818–1889). Joule is probably best known for his experiment demonstrating the mechanical equivalent of heat, first performed in about 1843. In that experiment (Figure 9.1) a falling weight was used to turn a paddle wheel through water. Joule showed conclusively that the energy of the falling weight was transferred to internal energy in the water.

Figure 9.1 A schematic drawing of J. P. Joule's paddle wheel apparatus used to determine the mechanical equivalent of heat. Falling weights were used to turn paddle wheels through water (in the center drum), which raised the temperature of the water. Joule concluded that a weight of 772 lb (mass = 350.2 kg) must fall through a distance of 1 ft (30.48 cm) in order to raise the temperature of 1 lb of water (2.2 kg) by 1°F ($\frac{5}{9}$°C).

James Clerk Maxwell did the great work of bringing the mathematical theories of probability and statistics to bear on the physical problems of thermodynamics. Using the understanding of energy gained by Joule and others, Maxwell derived expressions for the distributions of velocities and speeds of molecules in an ideal gas and showed that these distributions can be used to derive the observed macroscopic phenomena (see Sections 9.2–9.4). In spite of the great importance of this work, Maxwell is probably best known for his synthesis of electromagnetic theory (Maxwell's equations). There are some similarities between Maxwell's electromagnetic and thermodynamic theories. They are both highly mathematical (Maxwell was one of the foremost mathematicians of his day). They are also very mechanical, in keeping with the Newtonian tradition. In his electromagnetic treatises Maxwell set up elaborate molecular vortices to explain the electric and magnetic properties of matter. In thermodynamics he believed that all relevant properties were due to the motions of individual molecules. These similarities are important, because the great and immediate success of Maxwell's electromagnetic theory played a role in winning scientists over to the statistical view of thermodynamics.

But it was not really until the early twentieth century that statistical mechanics won the day. In 1905 Einstein (perhaps in his spare time after developing special relativity and his theory of the photoelectric effect!) published a theory of Brownian (random) motion, a theory that helped support the view that atoms are real. Experiments done several years later by Perrin confirmed Einstein's results. Soon after this came Bohr's successful atomic theory, and then the quantum theory we have developed in the previous chapters of this book.

We conclude this section with a brief discussion of the philosophy of statistical physics. Some people have difficulty accepting some of the uses of probability and statistics because of what they perceive as a lack of determinism or causality. It is perhaps difficult to reconcile this indeterminism with the fairly strict determinism still found in much of physics. Can the probabilistic laws governing the behavior of atoms and subatomic particles really be that different from our everyday experience of cause and effect? No less a figure than Einstein worried about this. He said, "God does not play dice," implying that some causal mechanism is at work that is simply beyond the scope of our experiments. This philosophical discussion is an interesting and ongoing one, but it should be made clear that statistical physics is necessary regardless of the ultimate nature of physical reality. This is true for three reasons. First, even as simple a problem as determining the outcome of the toss of a coin is so complex that it is most often useful to reduce it to statistical terms. Air resistance, rotational dynamics, and restitution problems make predicting the outcome of a coin toss a formidable mechanics problem. Other problems, particularly in quantum theory, are even more challenging. Second, when the number of particles is large (for example, on the order of 10^{23} particles for a modest sample of an ideal gas), it is highly impractical to study individual particles if one is more interested in the overall behavior of a system of particles (for example, the pressure, temperature, or specific heat of an ideal gas). This is made clear in subsequent sections of this chapter. Third, as Heisenberg showed, uncertainties are inherent in physics, and they are of significant size in atomic and subatomic systems. Therefore, it appears that statistics will always be an important part of physics.

9.2   Maxwell Velocity Distribution

As Laplace pointed out, we could, in principle, know everything about an ideal gas by knowing the position and instantaneous velocity of every molecule. This actually entails knowing six parameters per molecule, because three numbers are needed to specify the position (x, y, z) and three more to specify the velocity (v_x, v_y, v_z). Many relevant physical quantities must depend on one or more of these six parameters. In physics we sometimes think of these parameters as the components of a six-dimensional **phase space**.

Phase space

Maxwell focused on the three velocity components because he was most interested in the properties of ideal gases. The velocity components of the molecules of an ideal gas are clearly more important than the (random) instantaneous positions, because the energy of a gas should depend only on the velocities and not on the instantaneous positions. The crucial question for Maxwell was: what is the distribution of velocities for an ideal gas at a given temperature? Let us define a **velocity distribution function** $f(\vec{v})$ (see Appendix 5) such that

$$f(\vec{v}) \, d^3\vec{v} = \text{the probability of finding a particle with}$$
$$\text{velocity between } \vec{v} \text{ and } \vec{v} + d^3\vec{v}$$

Velocity distribution function

where $d^3\vec{v} = dv_x \, dv_y \, dv_z$. Note that because $\vec{v}$ is a vector quantity, the preceding statement implies three separate conditions. The vector $\vec{v}$ has components v_x, v_y, and v_z. Therefore $f(\vec{v}) \, d^3\vec{v}$ is the probability of finding a particle with v_x between v_x and $v_x + dv_x$, with v_y between v_y and $v_y + dv_y$, *and* with v_z between v_z and $v_z + dv_z$. We can think of the distribution function $f(\vec{v})$ as playing a role analogous to the probability density $\Psi^*\Psi$ in quantum theory.

Maxwell was able to prove* that the probability distribution function is proportional to $\exp(-\frac{1}{2} mv^2/kT)$, where m is the molecular mass, v is the molecular speed, k is Boltzmann's constant, and T is the absolute temperature. Therefore, we may write

$$f(\vec{v}) \, d^3\vec{v} = C \exp(-\tfrac{1}{2}\beta mv^2) \, d^3\vec{v} \tag{9.1}$$

where C is a proportionality factor and $\beta \equiv (kT)^{-1}$. (Don't confuse this parameter β in thermal physics with $\beta = v/c$ from relativity.) We can easily rewrite

---

*There are numerous ways of demonstrating that the distribution is proportional to $\exp(-\frac{1}{2} mv^2/kT)$. See for example Daniel Schroeder, *Thermal Physics*, Addison-Wesley (1999), pp. 220–223.

Equation (9.1) in terms of the three velocity components, because $v^2 = v_x^2 + v_y^2 + v_z^2$. Then

$$f(\vec{v}) \, d^3\vec{v} = C \exp\left(-\tfrac{1}{2}\beta m v_x^2 - \tfrac{1}{2}\beta m v_y^2 - \tfrac{1}{2}\beta m v_z^2\right) d^3\vec{v} \qquad (9.2)$$

Equation (9.2) can be rewritten in turn as the product of three factors, each of which contains one of the three velocity components. Let us define them as

$$g(v_x) \, dv_x \equiv C' \exp\left(-\tfrac{1}{2}\beta m v_x^2\right) dv_x$$
$$g(v_y) \, dv_y \equiv C' \exp\left(-\tfrac{1}{2}\beta m v_y^2\right) dv_y \qquad (9.3)$$
$$g(v_z) \, dv_z \equiv C' \exp\left(-\tfrac{1}{2}\beta m v_z^2\right) dv_z$$

with a new constant $C' = C^{1/3}$. Equations (9.3) give us the distributions of the three velocity components.

In order to perform any useful calculations using the distributions of Equations (9.3), we will need to know the value of the constant C'. Now $g(v_x) \, dv_x$ is simply the probability that the x component of a gas molecule's velocity lies between v_x and $v_x + dv_x$. If we sum (or integrate) $g(v_x) \, dv_x$ over all possible values of v_x, the result must be 1, because every molecule has a velocity component v_x somewhere in this range. This is just the same process of normalization that we have followed throughout our study of quantum theory in Chapters 5–7. Performing the integral (see Appendix 6) yields

$$\int_{-\infty}^{\infty} g(v_x) \, dv_x = C'\left(\frac{2\pi}{\beta m}\right)^{1/2} = 1 \qquad (9.4)$$

Then

$$C' = \left(\frac{\beta m}{2\pi}\right)^{1/2}$$

and

$$g(v_x) \, dv_x = \left(\frac{\beta m}{2\pi}\right)^{1/2} \exp\left(-\frac{1}{2}\beta m v_x^2\right) dv_x \qquad (9.5)$$

With this distribution we can calculate the mean value of v_x (see Appendix 5):

$$\overline{v_x} = \int_{-\infty}^{\infty} v_x g(v_x) \, dv_x = C' \int_{-\infty}^{\infty} v_x \exp\left(-\frac{1}{2}\beta m v_x^2\right) dv_x = 0 \qquad (9.6)$$

because v_x is an odd function (see Appendix 6). This result makes sense physically, because in a random distribution of velocities one would expect the velocity components to be distributed evenly around the peak at $v_x = 0$ (Figure 9.2).

Similarly, the mean value of v_x^2 is

$$\overline{v_x^2} = C' \int_{-\infty}^{\infty} v_x^2 \exp\left(-\frac{1}{2}\beta m v_x^2\right) dv_x$$
$$= 2C' \int_{0}^{\infty} v_x^2 \exp\left(-\frac{1}{2}\beta m v_x^2\right) dv_x$$
$$\overline{v_x^2} = \left(\frac{\beta m}{2\pi}\right)^{1/2} \frac{\sqrt{\pi}}{2} \left(\frac{2}{\beta m}\right)^{3/2} = \frac{1}{\beta m} = \frac{kT}{m} \qquad (9.7)$$

Of course there is nothing special about the x direction (the gas makes no distinction among x, y, and z), so the results for the x, y, and z velocity components

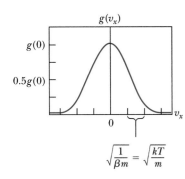

Figure 9.2 The Maxwell velocity distribution as a function of one velocity dimension (v_x). Notice that at $v_x = \sqrt{1/\beta m} = \sqrt{kT/m}$ we have $g(v_x) = g(0)e^{-1/2} \approx 0.607 g(0)$.

are identical. The three components may be used together to find the mean translational kinetic energy of a molecule:

$$\overline{K} = \frac{1}{2}\,\overline{mv^2} = \frac{1}{2}\,m\left(\overline{v_x^2} + \overline{v_y^2} + \overline{v_z^2}\right) = \frac{1}{2}\,m\left(\frac{3kT}{m}\right) = \frac{3}{2}\,kT \qquad (9.8)$$

We have just confirmed one of the principal results of kinetic theory! The fact that we have done so from purely statistical considerations is good evidence of the validity of this statistical approach to thermodynamics.

Example 9.1

Find the mean translational kinetic energy of (a) a single ideal gas molecule, and (b) a mole of ideal gas, both at room temperature 293 K.

Strategy The mean kinetic energy of a single molecule is given by Equation (9.8). For an entire mole, it's necessary to multiply the result by Avogadro's number N_A, the number of atoms in a mole.

Solution (a) For a single molecule, Equation (9.8) gives

$$\overline{K} = \tfrac{3}{2}kT = \tfrac{3}{2}(1.38 \times 10^{-23}\,\text{J/K})\,(293\,\text{K}) = 6.07 \times 10^{-21}\,\text{J}$$

Because of the small size of this energy, it may be useful to convert to units of electron volts.

$$\overline{K} = 6.07 \times 10^{-21}\,\text{J} \times \frac{1\,\text{eV}}{1.60 \times 10^{-19}\,\text{J}} = 0.038\,\text{eV}$$

The mean molecular energy at room temperature is about $1/25$ eV. This is a good number to remember, because this computation applies for *any* ideal gas at room temperature.

(b) For an entire mole, the mean kinetic energy is a factor of N_A larger:

$$\overline{K} = 6.07 \times 10^{-21}\,\text{J} \times N_A$$

$$= (6.07 \times 10^{-21}\,\text{J})(6.02 \times 10^{23}) = 3650\,\text{J}$$

9.3 Equipartition Theorem

The results of Section 9.2 can be extended into a rather general statement relating the internal energy of a thermodynamic system to its temperature. According to Equation (9.7)

$$\tfrac{1}{2}\,m\overline{v_x}^2 = \tfrac{1}{2}\,kT \qquad (9.9)$$

Similarly, an average energy $\frac{1}{2}kT$ is associated with each of the other two velocity components, producing a net average translational kinetic energy of $\frac{3}{2}kT$ per molecule. In a monatomic gas such as helium or argon, virtually all of the gas's energy is in this form. But consider instead a diatomic gas, such as oxygen (O_2). If we think of this molecule as two oxygen atoms connected by a massless rod, it is clear that this molecule can also have *rotational* kinetic energy. The crucial question is: how much rotational energy is there and how is it related to temperature?

The answer to this question is provided by the equipartition theorem, which we state without proof.

Equipartition theorem: *In equilibrium a mean energy of $\frac{1}{2}kT$ per molecule is associated with each independent quadratic term in the molecule's energy.*

Equipartition theorem

We shall see that those independent quadratic terms may be quadratic in coordinate, velocity component, angular velocity component, or anything else that when

Degree of freedom

squared is proportional to energy. Each independent phase space coordinate is called a **degree of freedom** for the system.

For example, in a monatomic ideal gas (such as helium), each molecule has a kinetic energy

$$K = \tfrac{1}{2} m v^2 = \tfrac{1}{2} m(v_x^2 + v_y^2 + v_z^2)$$

There are three independent phase space coordinates (v_x, v_y, and v_z). Thus there are three degrees of freedom, and the equipartition theorem predicts a mean kinetic energy of $3(\tfrac{1}{2} kT) = \tfrac{3}{2} kT$ per molecule.

How can we check our calculation of $\tfrac{3}{2} kT$? This is done by measuring helium's heat capacity at constant volume. In a gas of N helium molecules the total internal energy should be

$$U = N\overline{E} = \tfrac{3}{2} NkT$$

The heat capacity at constant volume is $C_V = (\partial U / \partial T)_V$, or

$$C_V = \tfrac{3}{2} Nk$$

We will use the standard notation of uppercase C_V for the heat capacity of a general amount (such as N molecules) of a substance, and lowercase c_V when describing the heat capacity for 1 mole. For 1 mole, $N = N_A$, Avogadro's number, and

$$c_V = \tfrac{3}{2} N_A k = \tfrac{3}{2} R = 12.5 \text{ J/K} \tag{9.10}$$

where we have used the fact that $N_A k = R = 8.31$ J/K, the ideal gas constant.

The measured molar heat capacity of He is very close to this value. In Table 9.1 the molar heat capacities of a number of gases are listed. For monatomic gases the measured heat capacities match the value predicted in Equation (9.10).

Table 9.1 Molar Heat Capacities for Selected Gases at 15°C and 1 Atmosphere

Gas	c_V (J/K)	c_V/R
Ar	12.5	1.50
He	12.5	1.50
CO	20.7	2.49
H_2	20.4	2.45
HCl	21.4	2.57
N_2	20.6	2.49
NO	20.9	2.51
O_2	21.1	2.54
Cl_2	24.8	2.98
CO_2	28.2	3.40
CS_2	40.9	4.92
H_2S	25.4	3.06
N_2O	28.5	3.42
SO_2	31.3	3.76

However, the heat capacities of diatomic gases are significantly greater. We can explain this by considering the rigid rotator model of the oxygen molecule described earlier and shown in Figure 9.3, and by applying the equipartition theorem to it. The molecule is free to rotate about either the x or y axis, and the corresponding rotational energies can be written in terms of rotational inertia and angular velocity components as $\frac{1}{2}I_x\omega_x^2$ and $\frac{1}{2}I_y\omega_y^2$. Each of these is quadratic in angular velocity, so the equipartition theorem tells us that we should add $2(\frac{1}{2}kT) = kT$ per molecule to the translational kinetic energy, for a total of $\frac{5}{2}kT$. Stated another way, there are five degrees of freedom (three translational and two rotational), so the energy per molecule is $5(\frac{1}{2}kT) = \frac{5}{2}kT$.

Why don't we include rotations about the z axis? The answer lies in quantum theory. In the quantum theory of the rigid rotator the allowed energy levels are

$$E = \frac{L^2}{2I} = \frac{\hbar^2 \ell(\ell + 1)}{2I}$$

where I is the rotational inertia and ℓ is a quantum number equal to zero or a positive integer. Notice that this result is consistent with the quantization of angular momentum given previously in Chapters 7 and 8. Here the kinetic energy of a rotator with angular momentum L is equal to $L^2/(2I)$. Consider again our diatomic molecule modeled in Figure 9.3. In studying atomic physics in Chapter 8 we saw that the vast majority of the mass of an atom is confined to a nucleus that is four or five orders of magnitude smaller than the whole atom. Therefore the diatomic molecule's rotational inertia (I_z) about the axis connecting the two atoms is orders of magnitude smaller than I_x and I_y. A small value of I_z in the denominator of the energy equation given above leads to a high energy, relative to that obtained with I_x or I_y and comparable quantum numbers. Thus, when the rotational energy is relatively low and small quantum numbers are required, only rotations about the x and y axes are allowed.*

Therefore, oxygen should have an average kinetic energy of $\frac{5}{2}kT$ per molecule, which translates to a molar heat capacity of $\frac{5}{2}R$. The measured molar heat capacity of O_2 is reasonably close to this value. Look again at the measured heat capacities in Table 9.1. For most diatomic gases the measured heat capacities are similar to the value predicted by the equipartition theorem, $\frac{5}{2}R$ per mole. Some of the diatomic gases (principally Cl_2) do not match up as well, though. Why not? The physics of the two-atom molecule is evidently not as simple as we have pictured it. From our study of atomic physics we know that atoms are complex systems and that the quantum theory must be used to replace classical mechanics when studying atomic systems. Further, we have not even begun to discuss the nature of molecular bonding; the solid massless rod joining the two atoms is undoubtedly a crude approximation. Given all these approximations it is a wonder that the equipartition theorem serves as well as it does!

In some circumstances it is a better approximation to think of atoms connected to each other by a massless spring rather than a rigid rod. How many degrees of freedom does this add? One may be tempted to say just one, because of the potential energy $\frac{1}{2}\kappa(r - r_0)^2$, where κ is the spring's force constant, r the separation between atoms, and r_0 the equilibrium separation between atoms. But another degree of freedom is associated with the vibrational velocity (dr/dt), because the vibrational kinetic energy is $\frac{1}{2}m(dr/dt)^2$. This makes sense because in classical physics the average kinetic energy is equal to the average potential

Rigid rotator model

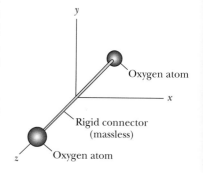

Figure 9.3 The rigid rotator model of the O_2 molecule, with the two oxygen atoms connected by a rigid, massless rod along the z axis.

---

*A more complete description of why quantum theory overrules the classical equipartition theorem in this case is given by Clayton A. Gearhart, *American Journal of Physics* **64**, 995–1000 (1996).

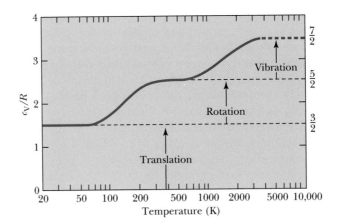

Figure 9.4 Molar heat capacity c_V as a function of temperature for H_2, a typical diatomic gas. The heat capacity c_V equals $3R/2$ at low temperatures, rises to $5R/2$ at higher temperatures as the rotational mode is excited, and finally approaches $7R/2$ when the molecule dissociates at very high temperatures.

energy for a harmonic oscillator (see Problem 47). If this molecule is still free to rotate, there is now a total of seven degrees of freedom: three translational, two rotational, and two vibrational. The resulting molar heat capacity is $\frac{7}{2}R$.

Is this vibrational mode ever a factor? Yes, it is, but high temperatures are normally required to excite the vibrational modes in diatomic gases. In fact, the heat capacities of diatomic gases are temperature dependent, indicating that the different degrees of freedom are "turned on" at different temperatures. One of the more striking examples is H_2 (see Figure 9.4). At temperatures just above its boiling point (about 20 K), the molar heat capacity is about $\frac{3}{2}R$. At just under 100 K the rotational mode is excited, and c_V then increases gradually until it reaches $\frac{5}{2}R$, where it remains from about 250 K up to about 1000 K. At that point the vibrational mode becomes effective (also gradually). We can see that c_V is beginning to approach $\frac{7}{2}R$, but the curve terminates at this point because the molecule dissociates.

Apparently it is difficult to excite the vibrational mode in a diatomic gas. In a polyatomic gas, we may choose to think of the molecule as consisting of a number of masses (atoms) connected by springs. In classical mechanics such systems have a frequency corresponding to each "normal mode" of oscillation, with the number of normal modes increasing as the number of masses in the system increases. Therefore polyatomic systems may have a number of different vibration frequencies, each turning on at a different temperature. This can cause the vibrational spectra of polyatomic molecules to be quite complex.

Let us briefly turn our attention to the thermal molecular motion in a solid. Now the vibrational mode will be of supreme importance, for the atoms in a solid are not free to translate or rotate. As an example, we consider a sample of pure copper, the atoms of which are arranged in a face-centered cubic lattice, as shown in Figure 9.5. How many degrees of freedom are in this system? It is possible to think of each atom as a three-dimensional harmonic oscillator. Just as described earlier, a one-dimensional harmonic oscillator has two degrees of freedom, one coming from the kinetic energy and one from the potential energy. Therefore there are *six* degrees of freedom for our three-dimensional oscillator, and the molar heat capacity is $6(\frac{1}{2}R) = 3R$. The experimental value of molar heat capacity is almost exactly $3R$ for copper near room temperature. It is also observed that, near room temperature, the molar heat capacity increases slightly with increasing temperature. As we will see in Section 9.6, this can be attributed to the conduction electrons, which have been neglected in our consideration of the equipartition theorem.

Figure 9.5 The lattice structure of copper, an example of the face-centered cubic lattice.

Consider the two gases HF and Ne, both at a (room) temperature of 300 K. Compare the average translational kinetic energy and total kinetic energy of the two types of molecules.

Solution The HF and Ne molecules have about the same molecular mass, about 20 u, where recall 1 u = 1 atomic mass unit = 1.66054×10^{-27} kg. According to the equipartition theorem, the molecules' translational kinetic energies are the same, $\frac{3}{2}kT$ per molecule or $\frac{3}{2}RT$ per mole. However, the diatomic HF should also have two rotational degrees of freedom, so its total kinetic energy is predicted to be about $\frac{3}{2}RT + RT = \frac{5}{2}RT$ per mole. The monatomic Ne gas has no rotational degrees of freedom, so its total kinetic energy is $\frac{3}{2}RT$ per mole.

9.4 Maxwell Speed Distribution

Let us return to the Maxwell velocity distribution in the form

$$f(\vec{v}) \, d^3\vec{v} = C \exp(-\tfrac{1}{2}\beta m v^2) \, d^3\vec{v} \qquad (9.1)$$

where we know that

$$C = (C')^3 = \left(\frac{\beta m}{2\pi}\right)^{3/2} \qquad (9.11)$$

Even though $f(\vec{v})$ is a function of the speed (v) and not the velocity ($\vec{v}$), this is still a velocity distribution because the probability equation, Equation (9.1), contains the differential velocity element. It will be useful for us to turn this velocity distribution into a speed distribution $F(v)$ where, by the usual definition,

$F(v) \, dv$ = the probability of finding a particle with speed between v and $v + dv$

As we shall soon see, it is *not* possible simply to assume that $f(\vec{v}) = F(v)$.

This is just the kind of problem for which the phase space concept is useful. Consider the analogous problem in normal three-dimensional (x, y, z) space. Suppose there exists some distribution of particles $f(x, y, z)$, as in Figure 9.6a. A particle at the point (x, y, z) is a distance $r = (x^2 + y^2 + z^2)^{1/2}$ from the origin, and $\vec{r}$ is a position vector directed from the origin to the point (x, y, z). Then $f(x, y, z) \, d^3\vec{r}$ = the probability of finding a particle between $\vec{r}$ and $\vec{r} + d^3\vec{r}$ with $d^3\vec{r} = dx \, dy \, dz$. Now let us change to a radial distribution $F(r)$, such that

$F(r) \, dr$ = the probability of finding a particle between r and $r + dr$

The space between r and $r + dr$ is a spherical shell. Therefore this problem is simply one of counting all of the particles in a spherical shell of radius r and thickness dr. The volume $(d^3\vec{r})$ of a spherical shell is $4\pi r^2 \, dr$. Thus we may write

$$F(r) \, dr = f(x, y, z) 4\pi r^2 \, dr \qquad (9.12)$$

Returning to our problem of obtaining a speed distribution $F(v)$ from a velocity distribution $f(\vec{v})$, we see that all we have to do is count the number of particles in a spherical shell in velocity space. Simply replace the coordinates x, y, and z with the velocity space coordinates (that is, the velocity components) v_x, v_y, and v_z (Figure 9.6b). The speed $v = (v_x^2 + v_y^2 + v_z^2)^{1/2}$ is the velocity space analog of

(a)

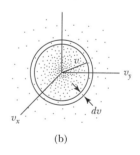

(b)

Figure 9.6 (a) A distribution of particles in three-dimensional space. The distribution function $f(r)$ is proportional to the number of particles in a spherical shell, between r and $r + dr$. (b) A similar distribution in three-dimensional velocity space. This shows that the speed distribution $f(v)$ must be proportional to the number of particles found in a spherical shell in velocity space between v and $v + dv$.

radius $r = (x^2 + y^2 + z^2)^{1/2}$. The preceding analysis indicates that the "volume" of our spherical shell in velocity space is $4\pi v^2 \, dv$, and the desired speed distribution is $F(v)$ where

$$F(v) \, dv = f(\vec{v})4\pi v^2 \, dv \qquad (9.13)$$

Using Equation (9.1) we obtain the Maxwell speed distribution:

Maxwell speed distribution

$$F(v) \, dv = 4\pi C \exp\left(-\tfrac{1}{2}\beta mv^2\right)v^2 \, dv \qquad (9.14)$$

A graph of a typical Maxwell speed distribution is shown in Figure 9.7. There is a qualitative similarity between this distribution and some of the radial probability distributions we encountered in Chapter 7 (see for example Figure 7.12). There is a quantitative distinction, however, in that the quadratic (v^2) argument in the exponential in Equation (9.14) differs from the linear (r) power found throughout Section 7.6. Notice also that the speed distribution is qualitatively different from the velocity distribution in that it is not symmetric about its peak. It is simple to show that $F(v)$ approaches zero in the limiting cases of very high and very low speeds (Problem 8). Because it was derived from purely classical considerations, the Maxwell speed distribution gives a nonzero probability of finding a particle with a speed greater than c (Problem 5). We know particle speeds cannot exceed the speed of light. Therefore, Equation (9.14) is valid only in the classical limit. This presents no serious problems in using the distribution because (you should convince yourself of this) the predicted probability of $v > c$ is extremely low at any reasonable temperature. Further, the other formulas we use in the kinetic theory of ideal gases (for example, K.E. $= \tfrac{1}{2}mv^2$) already restrict us to the classical limit.

The asymmetry of the distribution curve leads to an interesting result: the most probable speed v^*, the mean speed $\bar{v}$, and the root-mean-square speed v_{rms} are all slightly different from each other. The most probable speed v^* corresponds to the peak of the curve. Thus

$$\frac{d}{dv}\left[4\pi C \exp\left(-\frac{1}{2}\beta mv^2 \right)v^2 \right]\Bigg|_{v=v^*} = 0$$

$$\exp\left(-\frac{1}{2}\beta mv^{*2} \right)(2v^*) - \frac{1}{2}\beta m(2v^*)\exp\left(-\frac{1}{2}\beta mv^{*2} \right)v^{*2} = 0$$

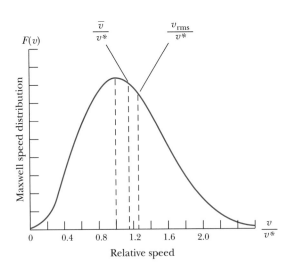

Figure 9.7 The Maxwell speed distribution, expressed in terms of the most probable speed v^*. Note the positions of $\bar{v}$ and v_{rms} relative to v^*.

Solving for v^*:

$$v^* = \sqrt{\frac{2}{\beta m}} = \sqrt{\frac{2kT}{m}}$$

(9.15)   **Most probable speed v^***

A curious corollary of this result is that the kinetic energy of a molecule moving with the most probable speed v^* is

$$K^* = \tfrac{1}{2}mv^{*2} = kT$$

(9.16)

which is exactly two thirds of the mean kinetic energy.

Now the mean speed is

$$\bar{v} = \int_0^\infty vF(v)\,dv = 4\pi C \int_0^\infty v^3 \exp\left(-\frac{1}{2}\beta mv^2\right) dv$$

See Appendix 6 for an evaluation of this integral. The result is

$$\bar{v} = 4\pi C\left[\frac{1}{2(\frac{1}{2}\beta m)^2}\right] = 8\pi \left(\frac{\beta m}{2\pi}\right)^{3/2}\left(\frac{1}{\beta m}\right)^2$$

$$\bar{v} = \frac{4}{\sqrt{2\pi}}\sqrt{\frac{kT}{m}}$$

(9.17)   **Mean speed $\bar{v}$**

Comparing the results (9.15) and (9.17), we see that

$$\frac{\bar{v}}{v^*} = \sqrt{\frac{4}{\pi}} \approx 1.13$$

(9.18)

In other words, $\bar{v}$ is about 13% greater than v^* *at any temperature*.

We define the **root-mean-square (rms) speed** v_{rms} to be

Root-mean-square speed

$$v_{\text{rms}} \equiv \left(\overline{v^2}\right)^{1/2}$$

(9.19)

In order to find the root-mean-square speed, it is first necessary to calculate $\overline{v^2}$:

$$\overline{v^2} = \int_0^\infty v^2 F(v)\,dv = 4\pi C \int_0^\infty v^4 \exp\left(-\frac{1}{2}\beta mv^2\right) dv$$

$$= 4\pi C\left[\frac{3\sqrt{\pi}}{8(\frac{1}{2}\beta m)^{5/2}}\right] = \frac{3}{\beta m} = \frac{3kT}{m}$$

Then

$$v_{\text{rms}} = \left(\overline{v^2}\right)^{1/2} = \sqrt{\frac{3kT}{m}}$$

(9.20)

As before, it is instructive to compare this result with the most probable speed:

$$\frac{v_{\mathrm{rms}}}{v^*} = \sqrt{\frac{3}{2}} \approx 1.22 \tag{9.21}$$

The rms speed is about 22% greater than the most probable speed for any temperature. Notice that the result in Equation (9.20) is again in keeping with our basic law of kinetic theory, namely

$$\overline{K} = \tfrac{1}{2}m\overline{v^2} = \tfrac{3}{2}kT$$

Finally, we calculate the standard deviation of the molecular speeds (see Appendix 5):

$$\sigma_v = (\overline{v^2} - \overline{v}^2)^{1/2} = \left(\frac{3kT}{m} - \frac{8kT}{\pi m}\right)^{1/2} = \left[3 - \left(\frac{8}{\pi}\right)\right]^{1/2}\left(\frac{kT}{m}\right)^{1/2} \approx 0.48v^* \tag{9.22}$$

The result expressed in Equation (9.22) indicates that σ_v increases in proportion to $\sqrt{T}$. The distribution of molecular speeds therefore widens somewhat as temperature increases.

To summarize, here are the three particular speeds that are important in the Maxwell speed distribution:

- The most probable speed v^* is at the peak of the Maxwell speed distribution.
- The mean speed $\overline{v}$ is the weighted average of all speeds in the distribution.
- The rms speed v_{rms} is the speed associated with the mean kinetic energy.

Example 9.3

Compute the mean molecular speed $\overline{v}$ in the light gas hydrogen (H_2) and the heavy gas radon (Rn), both at room temperature 293 K. (Assume this is the longest-lived radon isotope, which has a mass of 222 u.) Compare the results.

Strategy The mean speed is given by Equation (9.17) as

$$\overline{v} = \frac{4}{\sqrt{2\pi}}\sqrt{\frac{kT}{m}}$$

With the temperatures the same, we need to use the appropriate molecular masses to complete the computation. The lighter mass (hydrogen) will have the higher average speed.

Solution The mass of the hydrogen molecule is twice that of a hydrogen atom (neglecting the small binding energy), or $2(1.008\ \mathrm{u}) = 2.02\ \mathrm{u}$. Thus the average molecular speed of hydrogen is

$$\overline{v} = \frac{4}{\sqrt{2\pi}}\sqrt{\frac{kT}{m}}$$

$$= \frac{4}{\sqrt{2\pi}}\sqrt{\frac{(1.38 \times 10^{-23}\,\mathrm{J/K})(293\ \mathrm{K})}{(2.02\ \mathrm{u})(1.66 \times 10^{-27}\,\mathrm{kg/u})}} = 1750\ \mathrm{m/s}$$

The average molecular speed of radon is

$$\overline{v} = \frac{4}{\sqrt{2\pi}}\sqrt{\frac{kT}{m}}$$

$$= \frac{4}{\sqrt{2\pi}}\sqrt{\frac{(1.38 \times 10^{-23}\,\mathrm{J/K})(293\ \mathrm{K})}{(222\ \mathrm{u})(1.66 \times 10^{-27}\,\mathrm{kg/u})}} = 167\ \mathrm{m/s}$$

The hydrogen molecule is more than 10 times faster, on average. That's to be expected, because its mass is more than 100 times lighter. Most other gases have molecular masses that fall between these two extremes, so their speeds should be between the two values we computed here.

Example 9.4

What fraction of the molecules in an ideal gas in equilibrium have speeds within $\pm 1\%$ of v^*?

Strategy Recall that $F(v)\,dv$ is the probability of finding a particle with speed between v and $v + dv$. In principle we could integrate the distribution $F(v)$ in Equation (9.14) from the limits $0.99v^*$ to $1.01v^*$.

$$P(\pm 1\%) = \int_{0.99v^*}^{1.01v^*} F(v)\,dv \qquad (9.23)$$

Unfortunately, the indefinite integral cannot be done in closed form. We can obtain an approximate solution by calculating $F(v^*)$ and multiplying by $dv \approx \Delta v = 0.02v^*$.

Solution The product of $F(v^*)(0.02v^*)$ gives us our probability

$$P(\pm 1\%) \approx F(v^*)(0.02v^*)$$

$$\approx 4\pi C \exp\left(-\frac{1}{2}\beta m v^{*2}\right) v^{*2}(0.02v^*)$$

$$\approx 4\pi\left(\frac{\beta m}{2\pi}\right)^{3/2} e^{-1}(0.02)\left(\frac{2}{\beta m}\right)^{3/2}$$

$$\approx \frac{4}{\sqrt{\pi}} e^{-1}(0.02) \approx 0.017$$

Students with computer programming experience are encouraged to do the integration in Equation (9.23) numerically and compare the result with this approximation. A numerical integration, though still an approximation, should yield a more accurate result.

Conceptual Example 9.5

Use the shape of the Maxwell speed distribution curve (Figure 9.7) to explain why $v_{rms} > \bar{v} > v^*$.

Solution The speed distribution curve shown in Figure 9.7 is not symmetric. It could best be described as lopsided, with the wider part of the curve to the right of v^*. This means that there are more molecules with speeds greater than v^* than molecules with speeds less than v^*. The computation of the mean speed $\bar{v}$ is a weighted average. With more molecules having speeds above v^* than below, this result is $\bar{v} > v^*$.

The root-mean-square speed v_{rms} comes from a similar average, but one that uses the average *square* of the speed. In such a computation, higher speeds are weighted even more heavily, because the squares of numbers rise more rapidly than the numbers themselves. For example, the number 11 is 1.1 times larger than the number 10, but comparing the squares of these numbers, 121 (11^2) is 1.21 times larger than 100 (11^2). This is why $v_{rms} > \bar{v}$.

9.5 Classical and Quantum Statistics

In the previous section we mentioned that the Maxwell speed distribution is only valid in the classical (that is, nonrelativistic) limit, and also that this fact only prohibits the use of the Maxwell distribution at exceptionally high temperatures. A more severe restriction on the use of the Maxwell distribution comes from quantum theory. You may have already wondered why it has not been necessary to apply quantum theory to the ideal gas systems we have been studying to this point. After all, the particles involved are of molecular and atomic size. The important point is that an ideal gas is dilute. The molecules of the gas are so far apart that they can be considered not to interact with each other. When collisions between molecules do occur, they can be considered totally elastic, and therefore they have no effect on the distributions and mean values calculated in the previous sections of this chapter.

James Clerk Maxwell (1831–1879) was one of the greatest physicists of the nineteenth century. Born and educated in Scotland, he developed a theory that united electricity and magnetism, stating them succinctly in the set of four Maxwell's equations. Maxwell collaborated extensively with other physicists of the day, particularly William Thomson (Lord Kelvin). Maxwell developed a successful kinetic theory of gases and laid the foundation for what is now statistical mechanics in his 1871 work *Theory of Heat*. Maxwell was keenly interested in the second law of thermodynamics and proposed a thought experiment, now known as Maxwell's demon, as a challenge to the statistical nature of the second law of thermodynamics. *AIP Niels Bohr Library*.

But what happens when matter is in the liquid or solid state, where the density of matter is generally several orders of magnitude higher than the density of gases? In liquids and solids the assumption of a collection of noninteracting particles may no longer be valid. If molecules, atoms, or subatomic particles are packed closely together, the Pauli exclusion principle (see Chapter 8) prevents two particles with identical wave functions from sharing the same space. This limits the allowed energy states of any particle subject to the Pauli principle, and therefore the distribution of energies for a system of particles will undoubtedly be affected too.

There is another fundamental difference between classical and quantum energy states. In classical physics there is no restriction on particle energies, but in quantum systems only certain energy values are allowed. This also affects the overall distribution of energies.

Classical Distributions

Because energy levels are of such fundamental importance in quantum theory, it will be useful for us to rewrite the results of Section 9.4 in terms of energy rather than velocity. The Maxwell speed distribution was given by

$$F(v) \, dv = 4\pi C \exp\left(-\tfrac{1}{2}\beta m v^2\right) v^2 \, dv \tag{9.14}$$

For a monatomic gas the energy is all translational kinetic energy. Thus

$$E = \tfrac{1}{2} m v^2$$

$$dE = mv \, dv$$

$$dv = \frac{dE}{mv} = \frac{dE}{\sqrt{2mE}} \tag{9.24}$$

With Equation (9.24) the speed distribution of Equation (9.14) can be turned into an energy distribution (see Problem 19):

$$F(v) \, dv = F(E) \, dE \tag{9.25}$$

where

Maxwell-Boltzmann energy distribution

$$F(E) = \frac{8\pi C}{\sqrt{2}m^{3/2}} \exp(-\beta E) E^{1/2} \tag{9.26}$$

Two factors in Equation (9.26) contain E. The factor $E^{1/2}$ can be traced back to the phase space analysis done in Section 9.4. It is a feature of this particular kind of distribution (that is, the distribution of molecular speeds in an ideal gas, for which the energy of a molecule is $\tfrac{1}{2} m v^2$). The factor $\exp(-\beta E)$ is of more fundamental importance. Boltzmann showed that the statistical factor $\exp(-\beta E)$ is a characteristic of any classical system, regardless of how quantities other than molecular speeds may affect the energy of a given state. Thus we define the **Maxwell-Boltzmann factor** for classical systems as

Maxwell-Boltzmann factor

$$F_{\text{MB}} = A \exp(-\beta E) \tag{9.27}$$

where A is a normalization constant. The energy distribution for a classical system will have the form

$$n(E) = g(E) F_{\text{MB}} \tag{9.28}$$

where, in the usual sense, $n(E)$ is a distribution such that $n(E)\ dE$ represents the number of particles with energies between E and $E + dE$. The function $g(E)$, known as the **density of states**, is the number of states available per unit energy range. It makes sense that $n(E)$ is proportional to $g(E)$. The density of states can therefore be thought of as an essential element in all distributions, and we shall keep this in mind as we develop the quantum distributions. The factor F_{MB} tells us the relative probability that an energy state is occupied at a given temperature.

Density of states

Quantum Distributions

Now we turn our attention from classical to quantum distribution functions. In quantum theory, particles are described by wave functions. Identical particles cannot be distinguished from one another if there is a significant overlap of their wave functions. It is this characteristic of indistinguishability that makes quantum statistics different from classical statistics.

To illustrate this point consider the following example. Suppose that we have a system of just two particles, each of which has an equal probability (0.5) of being in either of two energy states. If the particles are distinguishable (call them A and B), then the possible configurations we may measure are as follows:

State 1	State 2
AB	
A	B
B	A
	AB

These four configurations are equally likely; therefore the probability of each is one fourth (0.25). However, if the two particles are indistinguishable, then our probability table changes:

State 1	State 2
XX	
X	X
	XX

Now there are only three equally likely configurations, each having a probability of one third (~0.33).

It turns out that two kinds of quantum distributions are needed. This is because some particles obey the Pauli exclusion principle and others do not. As mentioned earlier in this section, the Pauli principle has a significant impact on how energy states can be occupied and therefore on the corresponding energy distribution. It is easy to tell whether a particle will obey the Pauli principle: we determine the spin of the particle. Particles with half-integer spins obey the Pauli principle and are known collectively as **fermions**; those with zero or integer spins do not obey the Pauli principle and are known as **bosons**. Protons, neutrons, and electrons are fermions. Photons and pions are bosons. Also, atoms and molecules consisting of an even number of fermions must be bosons when considered as a whole, because their total spin will be zero or an integer. Similarly, those (relatively few) atoms and molecules made up of an odd number of fermions will be fermions.

Fermions and bosons

Austrian physicist Ludwig Boltzmann (1844–1906) worked independently of Maxwell on developing the laws governing the statistical behavior of classical particles. Boltzmann is remembered most for his work on the statistical nature of entropy, and he supported the notion that the equipartition theorem is a fundamental part of statistical physics and thermodynamics. On Boltzmann's Vienna tombstone is carved his famous formula for entropy: $S = k \log W$, where W is the number of possible ways a state can be configured and k is the constant that was named in Boltzmann's honor. *University of Vienna, courtesy AIP Niels Bohr Library.*

We state here without proof the Fermi-Dirac distribution, which is valid for fermions:

$$n(E) = g(E)F_{FD} \tag{9.29}$$

where

Fermi-Dirac distribution

$$F_{FD} = \frac{1}{B_1 \exp(\beta E) + 1} \tag{9.30}$$

Similarly the Bose-Einstein distribution, valid for bosons, is

$$n(E) = g(E)F_{BE} \tag{9.31}$$

where

Bose-Einstein distribution

$$F_{BE} = \frac{1}{B_2 \exp(\beta E) - 1} \tag{9.32}$$

In each case B_i ($i = 1$ or 2) is a normalization factor, and $g(E)$ is the density of states appropriate for a particular physical situation. Notice that the Fermi-Dirac and Bose-Einstein distributions look very similar; they differ only by the normalization constant and by the sign attached to the 1 in the denominator. This sign difference causes a significant difference in the properties of bosons and fermions, as will become evident in Sections 9.6 and 9.7. It is also important to see that both the Fermi-Dirac and Bose-Einstein distributions reduce to the classical Maxwell-Boltzmann distribution when $B_i \exp(\beta E)$ is much greater than 1* (in that case the normalization constant $A = 1/B_i$). This means that the Maxwell-Boltzmann factor $A \exp(-\beta E)$ is much less than 1 (that is, the probability that a particular energy state will be occupied is much less than 1). This is consistent with our earlier use of Maxwell-Boltzmann statistics for a dilute, noninteracting system of particles. See Table 9.2 for a summary of the properties of the three distribution functions.

A comparison of the three distribution functions is shown in Figure 9.8, with each one graphed as a function of energy. The normalization constants for the distributions (A for the Maxwell-Boltzmann, B_1 for the Fermi-Dirac, and B_2 for the Bose-Einstein) depend on the physical system being considered. For convenience, we set them all equal to 1 for this comparison. Notice that the Bose-Einstein factor F_{BE} is higher than the Fermi-Dirac factor F_{FD} at any given energy. Mathematically, this is due to the difference between +1 and −1 in the denominators of the two functions. Physically, the higher value of F_{BE} results from the fact that bosons do not obey the Pauli exclusion principle, so more bosons are allowed to fill lower energy states. Another thing to notice in Figure 9.8 is that the three graphs coincide at high energies—the classical limit. That is why Maxwell-Boltzmann statistics may be used in the classical limit, regardless of whether the particles in the system are fermions or bosons.

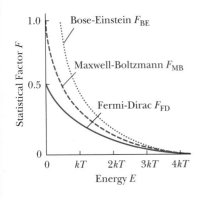

Figure 9.8 A comparison of the three distribution functions, each drawn as a function of energy over the same range. The normalization constants A, B_1, and B_2 have been set equal to 1 for convenience. The Bose-Einstein distribution is higher than the Fermi-Dirac distribution, because bosons do not obey the Pauli principle. At high energies, the three distributions are close enough so that the classical Maxwell-Boltzmann distribution can be used to replace either quantum distribution.

---

*This happens at high temperatures and low densities. A good rule of thumb is to compare the interparticle spacing with the average de Broglie wavelength. If the interparticle spacing is much greater than the de Broglie wavelength, then Maxwell-Boltzmann statistics are fairly accurate. Otherwise one should use the quantum statistics.

Table 9.2   Classical and Quantum Distributions

Distributors	Properties of the Distribution	Examples	Distribution Function
Maxwell-Boltzmann	Particles are identical but distinguishable	Ideal gases	$F_{MB} = A \exp(-\beta E)$
Bose-Einstein	Particles are identical and indistinguishable with integer spin	Liquid ^{4}He, photons	$F_{BE} = \dfrac{1}{B_2 \exp(\beta E) - 1}$
Fermi-Dirac	Particles are identical and indistinguishable with half-integer spin	Electron gas	$F_{FD} = \dfrac{1}{B_1 \exp(\beta E) + 1}$

Example 9.6

In a gas of atomic hydrogen (assume that the Maxwell-Boltzmann distribution is valid), what is the relative number of atoms in the ground state and first excited state at 293 K (room temperature), 5000 K (the temperature at the surface of a star), and 10^6 K (a temperature in the interior of a star)?

Strategy The desired ratio is

$$\frac{n(E_2)}{n(E_1)} = \frac{g(E_2)}{g(E_1)} \exp\left[\beta(E_1 - E_2)\right]$$

In the ground state ($n = 1$) of hydrogen there are two possible configurations for the electron, that is, $g(E_1) = 2$. There are eight possible configurations in the first excited state (see Chapter 7), so $g(E_2) = 8$. For atomic hydrogen $E_1 - E_2 = -10.2$ eV. Therefore

$$\frac{n(E_2)}{n(E_1)} = 4 \exp\left[\beta(-10.2 \text{ eV})\right] = 4 \exp(-10.2 \text{ eV}/kT)$$

for a given temperature T. We need to insert numerical values for each of the temperatures.

Solution The desired numerical results are

$$\frac{n(E_2)}{n(E_1)} = 4 \exp(-404) \approx 10^{-175} \qquad \text{for } T = 293 \text{ K}$$

$$= 4 \exp(-23.7) \approx 2 \times 10^{-10} \qquad \text{for } 5000 \text{ K}$$

$$= 4 \exp(-0.118) \approx 3.55 \qquad \text{for } 10^6 \text{ K}$$

Notice that at very high temperatures ($T \gg 10^6$ K), the exponential factor approaches 1, so the ratio $n(E_2)/n(E_1)$ approaches 4, the ratio of the densities of states. In fact, atomic hydrogen cannot exist at such high temperatures (10^6 K or greater). The electrons dissociate from the nuclei to form a state of matter known as a *plasma*.

9.6   Fermi-Dirac Statistics

Introduction to Fermi-Dirac Theory

The Fermi-Dirac distribution, as expressed in Equations (9.29) and (9.30), provides the basis for our understanding of the behavior of a collection of fermions. Let us examine the distribution in some detail before applying it to the problem of electrical conduction.

First we consider the role of the factor B_1. In principle one may compute B_1 for a particular physical situation by integrating $n(E)\, dE$ over all allowed energies. Because the parameter $\beta\ (= 1/kT)$ is contained in F_{FD}, it is clear that B_1 should be temperature dependent. It is possible to express this temperature dependence as

$$B_1 = \exp(-\beta E_F) \qquad (9.33)$$

where E_F is called the **Fermi energy**. We can then rewrite the Fermi-Dirac factor more conveniently as

$$F_{FD} = \frac{1}{\exp[\beta(E - E_F)] + 1} \qquad (9.34)$$

Equation (9.34) shows us an important fact about the Fermi energy: when $E = E_F$, the exponential term is 1, and therefore $F_{FD} = \frac{1}{2}$ (exactly). In fact it is common to define the Fermi energy as the energy at which $F_{FD} = \frac{1}{2}$.

Consider now the temperature dependence of F_{FD}, as expressed in Equation (9.34). In the limit as $T \to 0$, it is seen (Problem 30) that

$$F_{FD} = \begin{cases} 1 & \text{for } E < E_F \\ 0 & \text{for } E > E_F \end{cases} \qquad (9.35)$$

The physical basis for Equation (9.35) is easily understood. At $T = 0$, fermions occupy the lowest energy levels available to them. They cannot all be in the lowest level, because that would violate the Pauli principle. Rather, fermions will fill all the available energy levels up to a particular energy (the Fermi energy). Near $T = 0$ there is little chance that thermal agitation will kick a fermion to an energy greater than E_F.

As the temperature increases from $T = 0$, more and more fermions may be in higher energy levels. The Fermi-Dirac factor "smears out" from the sharp step function in Figure 9.9a to a smoother curve, shown in Figure 9.9b. It is sometimes useful to consider a **Fermi temperature**, defined as $T_F \equiv E_F/k$. A plot of F_{FD} at $T = T_F$ is shown in Figure 9.9c. When $T \ll T_F$ the step function approximation for F_{FD} in Equation (9.35) is reasonably accurate. When $T \gg T_F$, F_{FD} approaches a simple decaying exponential (see Figure 9.9d). This is just as one would expect, for at sufficiently high temperatures we expect Maxwell-Boltzmann statistics to be reasonably accurate [see Equation (9.27) and the subsequent discussion in Section 9.5].

Figure 9.9 The Fermi-Dirac factor F_{FD} at various temperatures: (a) $T = 0$, (b) $T > 0$, (c) $T = T_F = E_F/k$, and (d) $T \gg T_F$. At $T = 0$ the Fermi-Dirac factor is a step function. As the temperature increases, the step is gradually rounded. Finally, at very high temperatures, the distribution approaches the simple decaying exponential of the Maxwell-Boltzmann distribution.

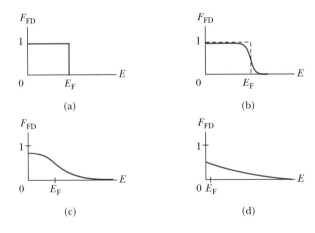

We can apply the Fermi-Dirac theory to the problem of understanding electrical conduction in metals. For comparison we will first present a brief review of the classical theory. A more complete description of the classical theory can be found in most good introductory physics texts.

Classical Theory of Electrical Conduction

In 1900 Paul Drude developed a theory of electrical conduction in an effort to explain the observed conductivity of metals. His model assumed that the electrons in a metal existed as a gas of free particles. This is a fair assumption, because in a conductor the outermost electron(s)—which may average up to several per atom—are so weakly bound to an atom that they may be stripped away easily by even a very weak electric field. In the Drude model the metal is thought of as a lattice of positive ions with a gas of electrons free to flow through it. Just as in the ideal gases we considered earlier in this chapter, the electrons have a thermal kinetic energy proportional to temperature. The mean speed of an electron at room temperature can be calculated [using Equation (9.17)] to be about 10^5 m/s. But remember that the velocities of the particles in a gas (electrons, in this case) are directed randomly. Therefore there will be no net flow of electrons unless an electric field is applied to the conductor.

When an electric field is applied, the negatively charged electrons flow in the direction opposite that of the field. According to Drude, their flow is severely restricted by collisions with the lattice ions. Based on several simple assumptions from classical mechanics, Drude was able to show that the current in a conductor should be linearly proportional to the applied electric field. That is consistent with Ohm's law, a well-known experimental fact. The principal success of Drude's theory was that it did predict Ohm's law.

Unfortunately, the numerical predictions of the theory were not so successful. One important prediction was that the electrical conductivity could be expressed* as

$$\sigma = \frac{ne^2\tau}{m} \tag{9.36}$$

where n is the number density of conduction electrons, e is the electron charge, τ is the average time between electron-ion collisions, and m is the electronic mass. It is possible to measure n by the Hall effect (see Chapter 11). The parameter τ is not as easy to measure, but it can be estimated using transport theory. The best estimates of τ, when combined with the other parameters in Equation (9.36), produced a value of σ that is about one order of magnitude too small for most conductors. The Drude theory is therefore incorrect in this prediction.

A restatement of Equation (9.36) will show another deficiency of the classical theory. The mean time between collisions τ should be simply the mean distance ℓ traveled by an electron between collisions (called the **mean free path**) divided by the mean speed $\bar{v}$ of electrons. That is,

Mean free path

$$\tau = \ell/\bar{v}$$

Then the electrical conductivity can be expressed as

$$\sigma = \frac{ne^2\ell}{m\bar{v}} \tag{9.37}$$

Paul Adrien Maurice Dirac (1902–1984) had a Swiss father and English mother and was born in England. While still a young man in 1932, he was named Lucasian Professor of Mathematics at Cambridge, the Chair once held by Newton and currently held by Stephen Hawking. Dirac was one of the pioneers in applying quantum mechanics to solve problems that could not be understood classically. He predicted the existence of antimatter before it was observed experimentally. Along with Fermi, he developed the correct quantum-mechanical laws (Fermi-Dirac statistics) governing the statistics of half-integer spin particles, now known as fermions. *A. B. Örtzells Tryckeri, courtesy AIP Niels Bohr Library.*

*See the introductory text, R. Serway and J. Jewett, *Physics for Scientists and Engineers*, 6th ed., Pacific Grove, CA: Brooks/Cole (2004), pp. 841–842.

Equation (9.17) shows that the mean speed is proportional to the square root of the absolute temperature. Hence, according to the Drude model, the conductivity should be proportional to $T^{-1/2}$. But for most conductors the conductivity is very nearly proportional to T^{-1} except at very low temperatures, where it no longer follows a simple relation. Clearly the classical model of Drude has failed to predict this important experimental fact.

Finally, there is the problem of calculating the electronic contribution to the heat capacity of a solid conductor. As discussed in Section 9.3, the heat capacity of a solid can be almost completely accounted for by considering the six degrees of freedom in the lattice vibrations. This gives a molar heat capacity of $6(\frac{1}{2}R) = 3R$. According to the equipartition theorem, we should add another $3(\frac{1}{2}R) = \frac{3}{2}R$ for the heat capacity of the electron gas, giving a total of $\frac{9}{2}R$. This is not consistent with experimental results. The electronic contribution to the heat capacity depends on temperature and is typically only about $0.02R$ per mole at room temperature. Clearly a different theory is needed to account for the observed values of electrical conductivity and heat capacity as well as the temperature dependence of the conductivity.

Quantum Theory of Electrical Conduction

To obtain a better solution we have to turn to quantum theory. It is necessary to understand how the electron energies are distributed in a conductor. Electrons are fermions, and therefore we need to use the Fermi-Dirac distribution described earlier in this section. The real problem we face is to find $g(E)$, the number of allowed states per unit energy. The question is: what energy values should we use? Let us retain the "free electron" assumption of the Drude model and use the results obtained in Chapter 6 for a three-dimensional infinite square-well potential (which corresponds physically to a cubic lattice of ions). The allowed energies are

$$E = \frac{h^2}{8mL^2}(n_1^2 + n_2^2 + n_3^2) \qquad (9.38)$$

where L is the length of a side of the cube and n_i are the integer quantum numbers. For the time being we will not worry about the effects an elevated temperature might have on the distribution of energies. We will solve the problem at $T = 0$ first and consider the effect of temperature later. It turns out that the distribution at room temperature is very much like the $T = 0$ distribution anyway because we are in the $T \ll T_F$ regime (for example, $T_F \approx 80{,}000$ K for copper—see Table 9.4).

Equation (9.38) can be rewritten as

$$E = r^2 E_1 \qquad (9.39)$$

where $r^2 = n_1^2 + n_2^2 + n_3^2$ and $E_1 = h^2/8mL^2$. The parameter r is the "radius" of a sphere in phase space and is a dimensionless quantity—it should not be confused with a radius in Euclidean space. Note that E_1 is just a constant, not the ground state energy. (It is actually one third of the ground state energy.) We have defined r in this way in order to construct a geometric solution to the problem of counting the number of allowed quantum states per unit energy. Think of the n_i as the "coordinates" of a three-dimensional number space, as in Figure 9.10. The number of allowed states up to "radius" r (or up to energy $E = r^2 E_1$) is directly related to the spherical "volume" $\frac{4}{3}\pi r^3$. The exact number of states up to radius r is

$$N_r = (2)\left(\frac{1}{8}\right)\left(\frac{4}{3}\pi r^3\right) \qquad (9.40)$$

Cube of unit "volume"

Figure 9.10 The three-dimensional number space used to count the number of allowed states within a sphere of radius $r = \sqrt{n_1^2 + n_2^2 + n_3^2}$. We may also think of this as counting the number of unit $(1 \times 1 \times 1)$ cubes within this octant of space.

The extra factor of 2 is due to spin degeneracy: for each set of quantum numbers there may be two electrons, one with spin up and one with spin down. The factor of 1/8 is necessary because we are restricted to positive quantum numbers, and therefore to one octant of the three-dimensional number space. Equations (9.39) and (9.40) can be used to express N_r as a function of E:

$$N_r = \frac{1}{3}\pi\left(\frac{E}{E_1}\right)^{3/2} \tag{9.41}$$

At $T = 0$ the Fermi energy is the energy of the highest occupied energy level, as we saw in Figure 9.9a. If there are a total of N electrons, then

$$N = \frac{1}{3}\pi\left(\frac{E_F}{E_1}\right)^{3/2}$$

Table 9.3 Free-Electron Number Densities for Selected Elements at $T = 300$ K

Element	N/V ($\times 10^{28}$ m^{-3})	Element	N/V ($\times 10^{28}$ m^{-3})
Cu	8.47	Mn (α)	16.5
Ag	5.86	Zn	13.2
Au	5.90	Cd	9.27
Be	24.7	Hg (78 K)	8.65
Mg	8.61	Al	18.1
Ca	4.61	Ga	15.4
Sr	3.55	In	11.5
Ba	3.15	Sn	14.8
Nb	5.56	Pb	13.2
Fe	17.0		

From N. W. Ashcroft and N. D. Mermin, *Solid State Physics*, Philadelphia: Saunders College (1976).

Table 9.4   **Fermi Energies ($T = 300$ K), Fermi Temperatures, and Fermi Velocities for Selected Metals**

Element	E_F (eV)	T_F ($\times 10^4$ K)	u_F ($\times 10^6$ m/s)
Li	4.74	5.51	1.29
Na	3.24	3.77	1.07
K	2.12	2.46	0.86
Rb	1.85	2.15	0.81
Cs	1.59	1.84	0.75
Cu	7.00	8.16	1.57
Ag	5.49	6.38	1.39
Au	5.53	6.42	1.40
Be	14.3	16.6	2.25
Mg	7.08	8.23	1.58
Ca	4.69	5.44	1.28
Sr	3.93	4.57	1.18
Ba	3.64	4.23	1.13
Nb	5.32	6.18	1.37
Fe	11.1	13.0	1.98
Mn	10.9	12.7	1.96
Zn	9.47	11.0	1.83
Cd	7.17	8.68	1.62
Hg	7.13	8.29	1.58
Al	11.7	13.6	2.03
Ga	10.4	12.1	1.92
In	8.63	10.0	1.74
Tl	8.15	9.46	1.69
Sn	10.2	11.8	1.90
Pb	9.47	11.0	1.83
Bi	9.90	11.5	1.87
Sb	10.9	12.7	1.96

From N. W. Ashcroft and N. D. Mermin, *Solid State Physics*, Philadelphia: Saunders College (1976).

Solving for E_F:

$$E_F = E_1 \left(\frac{3N}{\pi} \right)^{2/3} = \frac{h^2}{8m} \left(\frac{3N}{\pi L^3} \right)^{2/3} \tag{9.42}$$

Equation (9.42) is useful because the ratio N/L^3 is well known for most conductors: it is simply the number density of conduction electrons, a quantity easily measured using the Hall effect, which is discussed in detail in Chapter 11.

Example 9.7

Calculate the Fermi energy and Fermi temperature for copper.

Strategy Equation (9.42) can be used to compute the Fermi energy, provided the number density of conduction electrons is known. That number for copper is given in Table 9.3. (See also Problem 27.)

Solution The number density of conduction electrons in copper is given by Table 9.3 as 8.47×10^{28} m^{-3}. Using this value of N/L^3 in Equation (9.42) gives

$$E_F = \frac{(6.626 \times 10^{-34} \text{ J} \cdot \text{s})^2}{8(9.11 \times 10^{-31} \text{ kg})} \left[\frac{3(8.47 \times 10^{28} \text{ m}^{-3})}{\pi} \right]^{2/3}$$

$$= 1.13 \times 10^{-18} \text{ J} = 7.03 \text{ eV}$$

Within rounding errors, this result is equivalent to that given in Table 9.4.

$$T_F = \frac{E_F}{k} = \frac{7.03 \text{ eV}}{8.62 \times 10^{-5} \text{ eV/K}} = 8.16 \times 10^4 \text{ K}$$

Fermi energies and Fermi temperatures for other common conductors are listed in Table 9.4. Note that E_F changes little between $T = 0$ and room temperature.

The density of states can be calculated by differentiating Equation (9.41) with respect to energy:

$$g(E) = \frac{dN_r}{dE} = \frac{\pi}{2} E_1^{-3/2} E^{1/2}$$

This result can be expressed more conveniently in terms of E_F rather than E_1. Using Equation (9.42) for E_1 we find

$$g(E) = \frac{\pi}{2} \left(E_F^{-3/2} \frac{3N}{\pi} \right) E^{1/2} = \frac{3N}{2} E_F^{-3/2} E^{1/2} \qquad (9.43)$$

The distribution of electronic energies is then given by Equation (9.29). Because we are considering the $T = 0$ case, it is possible to use the step function form of the Fermi-Dirac factor (Equation 9.35). Therefore at $T = 0$ we have

$$n(E) = \begin{cases} g(E) & \text{for } E < E_F \\ 0 & \text{for } E > E_F \end{cases} \qquad (9.44)$$

With the distribution function $n(E)$ the mean electronic energy can be calculated easily:

$$\overline{E} = \frac{1}{N} \int_0^\infty E n(E) \, dE = \frac{1}{N} \int_0^{E_F} E g(E) \, dE$$

$$= \frac{1}{N} \int_0^{E_F} \left(\frac{3N}{2} \right) E_F^{-3/2} E^{3/2} \, dE$$

$$= \frac{3}{2} E_F^{-3/2} \int_0^{E_F} E^{3/2} \, dE = \frac{3}{5} E_F \qquad (9.45)$$

This is a reasonable result, considering the shapes of the curves in Figure 9.11.

We can now proceed to find the electronic contribution to the heat capacity of a conductor. Recall that the general expression for heat capacity at constant volume is $C_V = \partial U/\partial T$, where U is the internal energy of the system in question. By Equation (9.45)

$$U = N\overline{E} = \tfrac{3}{5} N E_F$$

Figure 9.11 (a) The density of states $g(E) = (3N/2)E_F^{-3/2}E^{1/2}$ and (b) distribution function $n(E) = g(E)F_{FD}$ for an electron gas. The function $n(E)$ is shown at $T = 0$ (dashed line) and $T = 300$ K (solid line).

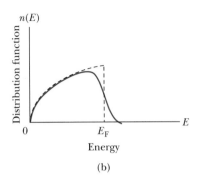

at $T = 0$. The important question for determining the heat capacity is how U increases with temperature. Because the energy levels are filled up to E_F, we expect that only those electrons within about kT of E_F will be able to absorb thermal energy and jump to a higher state. Therefore the fraction of electrons capable of participating in this thermal process is on the order of kT/E_F. The exact number of electrons depends on temperature, because the shape of the curve $n(E)$ changes with temperature (see Figure 9.11). In general we can say that

$$U = \frac{3}{5}NE_F + \alpha N\frac{kT}{E_F}kT \tag{9.46}$$

where α is a constant >1 due to the shape of the distribution curve. Therefore the electronic contribution to the heat capacity is

$$C_V = \frac{\partial U}{\partial T} = 2\alpha Nk^2\frac{T}{E_F}$$

As with ideal gases it is common to express this result as a molar heat capacity c_V (see Section 9.3). For 1 mole we have $N = N_A$. Then using $N_A k = R$ (the ideal gas constant) and $E_F = kT_F$, we find that

$$c_V = 2\alpha R\frac{T}{T_F} \tag{9.47}$$

Arnold Sommerfield used the correct distribution $n(E)$ at room temperature and found a value* for α of $\pi^2/4$. With the value $T_F = 80,000$ K for copper, we obtain $c_V \approx 0.02R$, which is just what is measured experimentally! The increase in c_V with increasing temperature is also seen experimentally. The quantum theory has proved to be a success in a case where the classical theory failed. The small value of the electronic contribution to the heat capacity can be attributed to the unique nature of the Fermi-Dirac distribution, and in particular to the Pauli exclusion principle, which severely restricts the number of electrons that may participate in heat absorption.

Turning our attention to the electrical conductivity, we can make some improvement by replacing the mean speed $\bar{v}$ in Equation (9.37) with what is called **Fermi speed**   a **Fermi speed** u_F, defined from $E_F = \frac{1}{2}mu_F^2$. We can justify this change by noting that conduction electrons in a metal are those most loosely bound to their atoms. Therefore these electrons must be at the highest energy level. At room

---

*See C. Kittel, *Introduction to Solid State Physics*, 6th ed., New York: Wiley (1986).

temperature the highest energy level is close to the Fermi energy, as discussed previously. This means that we should use

$$u_F = \sqrt{\frac{2E_F}{m}} \approx 1.6 \times 10^6 \text{ m/s}$$

for copper. Unfortunately, this is an even higher speed than $\bar{v}$, which seems undesirable because the value we calculated for σ using $\bar{v}$ was already too low. It turns out that there is also a problem with the classical value Drude used for the mean free path ℓ. He thought that because the ions were so large and occupied so much space in a solid, the mean free path could be no more than several tenths of a nanometer. But in quantum theory the ions are not hard spheres, and the electrons can be thought of as waves. Taking into account the wavelike properties of electrons, the mean free path can be longer than Drude estimated. Einstein calculated the value of ℓ to be on the order of 40 nm in copper at room temperature. This gives a conductivity of

$$\sigma = \frac{ne^2\ell}{mu_F} \approx 6 \times 10^7 \ \Omega^{-1} \cdot \text{m}^{-1}$$

which is just right.

Finally, let us consider the temperature dependence of the electrical conductivity. We have already replaced $\bar{v}$ in Equation (9.37) with the Fermi speed u_F, which is nearly temperature independent. However, as the temperature of a conductor is increased, ionic vibrations become more severe. Thus electron-ion collisions will become more frequent, the mean free path will become smaller, and the conductivity will be reduced. According to elementary transport theory (this result is shown in many introductory physics texts*) the mean free path is inversely proportional to the cross-sectional area of these ionic scatterers. Let us assume that the ions are harmonic oscillators. The energy of a harmonic oscillator is proportional to the square of its vibration amplitude. But the effective cross-sectional area should also be proportional to the square of the vibration amplitude. Therefore, the mean free path is inversely proportional to vibration energy (and temperature, because thermal energy is proportional to temperature). We may summarize this sequence of proportions by saying that

$$\sigma \propto \ell \propto r^{-2} \propto U^{-1} \propto T^{-1}$$

In other words, the electrical conductivity varies inversely with temperature. This is another success for the quantum theory, because electrical conductivity is observed to vary inversely with temperature for most pure metals.

*See the introductory text, R. Serway and J. Jewett, *Physics for Scientists and Engineers*, 6th ed., Pacific Grove, CA: Brooks/Cole, 2004, p. 658.

Example 9.8

Use the Fermi theory to compute the electronic contribution to the molar heat capacity of (a) copper and (b) silver, each at temperature $T = 293$ K. Express the results as a function of the molar gas constant R.

Strategy From the Fermi theory the molar heat capacity is $c_V = 2\alpha RT/T_F$. Recall that we can use $\alpha = \pi^2/4$ at room temperature ($T = 293$ K), so

$$c_V = 2\left(\frac{\pi^2}{4}\right)R\frac{T}{T_F} = \frac{\pi^2 RT}{2T_F}$$

Solution (a) Inserting numerical values (from Table 9.4, $T_F = 8.16 \times 10^4$ K for copper),

$$c_V = \frac{\pi^2(293 \text{ K})}{2(8.16 \times 10^4 \text{ K})} R = 0.0177R$$

for copper, which rounds to the value $0.02R$ quoted in the text.

(b) For silver, we simply replace the Fermi temperature with 6.38×10^4 K (Table 9.4):

$$c_V = \frac{\pi^2(293 \text{ K})}{2(6.38 \times 10^4 \text{ K})} R = 0.0227R$$

We see that the electronic contribution to the molar heat capacity is not much different for these two good conductors, and in each case it is small compared with the lattice contribution (Section 9.3).

9.7   Bose-Einstein Statistics

Like Fermi-Dirac statistics, Bose-Einstein statistics can be used to solve problems that are beyond the scope of classical physics. In this section we will concentrate on two examples: a derivation of the Planck formula for blackbody radiation and an investigation of the properties of liquid helium.

Blackbody Radiation

You may wish to review Section 3.5 on blackbody radiation. We will use the ideal blackbody described in that section: a nearly perfectly absorbing cavity that emits a spectrum of electromagnetic radiation. The problem is to find the intensity of the emitted radiation as a function of temperature and wavelength [Equation (3.23)].

$$\mathcal{I}(\lambda, T) = \frac{2\pi c^2 h}{\lambda^5} \frac{1}{e^{hc/\lambda kT} - 1} \tag{3.23}$$

Of course, in quantum theory we must begin with the assumption that the electromagnetic radiation is really a collection of photons of energy hc/λ. Recall that photons are bosons with spin 1. Our approach to this problem is to use the Bose-Einstein distribution to find how the photons are distributed by energy, and then use the relationship $E = hc/\lambda$ to turn the energy distribution into a wavelength distribution. The desired temperature dependence should already be included in the Bose-Einstein factor [see Equation (9.32)].

As in the previous section, the key to the problem is being able to find the density of states $g(E)$. In fact, it is possible to model the photon gas just as we did the electron gas: a collection of free particles within a three-dimensional infinite potential well. We cannot use Equation (9.38) for the energy states, however, for we are now dealing with massless particles. To solve this problem it is necessary to recast the solution to the particle-in-a-box problem in terms of momentum states rather than energy states. For a free particle of mass m the energy is $p^2/2m$. We may rewrite Equation (9.38) in terms of momentum:

$$p = \sqrt{p_x^2 + p_y^2 + p_z^2} = \frac{h}{2L} \sqrt{n_1^2 + n_2^2 + n_3^2} \tag{9.48}$$

The energy of a photon is pc, so

$$E = \frac{hc}{2L} \sqrt{n_1^2 + n_2^2 + n_3^2} \tag{9.49}$$

We proceed now to calculate the density of states as we did in Section 9.6. Think of the n_i as the coordinates of a number space and define $r^2 = n_1^2 + n_2^2 + n_3^2$. The number of allowed energy states within "radius" r is

$$N_r = 2\left(\tfrac{1}{8}\right)\left(\tfrac{4}{3}\pi r^3\right) \tag{9.50}$$

where the factor $1/8$ again comes from the restriction to positive values of n_i. This time the factor of 2 is due to the fact that there are two possible photon polarizations. Note that energy is proportional to r, namely,

$$E = \frac{hc}{2L}r \tag{9.51}$$

Thus, we can rewrite N_r in terms of E:

$$N_r = \frac{8\pi L^3}{3h^3 c^3}E^3 \tag{9.52}$$

The density of states $g(E)$ is

$$g(E) = \frac{dN_r}{dE} = \frac{8\pi L^3}{h^3 c^3}E^2 \tag{9.53}$$

The energy distribution is as always the product of the density of states and the appropriate statistical factor, in this case the Bose-Einstein factor:

$$n(E) = g(E)F_{\mathrm{BE}}$$

$$= \frac{8\pi L^3}{h^3 c^3}E^2\frac{1}{e^{E/kT} - 1} \tag{9.54}$$

Notice that the normalization factor B_2 in the Bose-Einstein factor has been set equal to unity. This is because we have a non-normalized collection of photons. As photons are absorbed and emitted by the walls of the cavity, the number of photons is not constant. The distribution in Equation (9.54) will serve to provide a relative number of photons at different energies, but it is impossible to normalize to a particular number of photons.*

The next step is to convert from a number distribution to an energy density distribution $u(E)$. To do this it is necessary to multiply by the factor E/L^3 (that is, energy per unit volume):

$$u(E) = \frac{En(E)}{L^3} = \frac{8\pi}{h^3 c^3}E^3\frac{1}{e^{E/kT} - 1}$$

For all photons in the range E to $E + dE$

$$u(E)\,dE = \frac{8\pi}{h^3 c^3}\frac{E^3\,dE}{e^{E/kT} - 1} \tag{9.55}$$

Using $E = hc/\lambda$ and $|dE| = (hc/\lambda^2)\,d\lambda$,[†] we find

$$u(\lambda,\,T)\,d\lambda = \frac{8\pi hc}{\lambda^5}\frac{1}{e^{hc/\lambda kT} - 1} \tag{9.56}$$

---

*For more detail on the derivation of Equation (9.54), see F. Reif, *Fundamentals of Statistical and Thermal Physics*, New York: McGraw-Hill (1965), pp. 339–340.

[†]The negative sign is dropped because it would be meaningless in a distribution in which a (positive) number representing a probability is required. Physically the negative sign means that energy increases as wavelength decreases, but that fact is not relevant here.

Satyendra Nath Bose (1894–1974), who first applied the distribution that bears his name (along with Einstein's) to the quantum statistics of photons. After this first contact with Einstein, Bose collaborated with other European scientists on applications of quantum theory. He served for over 30 years as a professor of physics, first at the University of Dacca (in his native Bengal region, now Bangladesh) and then at Calcutta University. His other work includes collaboration with the Indian physicist M. N. Saha on the ionization of gases in stars. Although shunned by some European physicists early in his career, he eventually won their respect and high honors, including election as a Fellow of the Royal Society of London in 1958. *Indian National Council of Science Museums, courtesy AIP Emilio Segrè Visual Archives.*

In the SI system, multiplying by a constant factor $c/4$ is required* to change the energy density [$u(\lambda, T)$ is energy per unit volume per unit wavelength inside the cavity] to a spectral intensity [$\mathscr{I}(\lambda, T)$ is power per unit area per unit wavelength for radiation emitted from the cavity]:

$$\mathscr{I}(\lambda, T) = \frac{2\pi c^2 h}{\lambda^5} \frac{1}{e^{hc/\lambda kT} - 1} \tag{9.57}$$

which is identical with Equation (3.23).

Planck did not use the Bose-Einstein distribution to derive his radiation law. Nevertheless, it is an excellent example of the power of the statistical approach to see that such a fundamental law can be derived with relative ease. This problem was first solved in 1924 by the young Indian physicist Satyendra Nath Bose (for whom the boson is named). Einstein's name was added to the distribution because he helped Bose publish his work in the West and later applied the distribution to other problems. It is remarkable that Bose did this work prior to the full development of quantum mechanics, and in particular prior to the elucidation of the concept of spin in quantum theory.

Liquid Helium

Helium is an element with a number of remarkable properties. It has the lowest boiling point of any element (4.2 K at 1 atmosphere pressure) and has no solid phase at normal pressures. Liquid helium has been studied extensively since its discovery in 1908 by Heike Kamerlingh Onnes. It continues to be used by experimental physicists as a cryogenic (cooling) device. Many effects can be observed only at extremely low temperatures, and liquid helium can be used to cool materials to 4.2 K and lower. In Chapter 10 we will examine low-temperature methods in more detail, along with our study of superconductivity, perhaps the best-known low-temperature phenomenon.

Liquid helium is also quite interesting in its own right, especially at temperatures somewhat less than its boiling point. In 1924 Kamerlingh Onnes (along with J. Boks) measured the density of liquid helium as a function of temperature and obtained the curve shown in Figure 9.12. W. Keesom and K. Clausius measured the specific heat of liquid helium as a function of temperature in 1932. Their results are shown in Figure 9.13.

Figure 9.12 Density and number density versus temperature for liquid helium. Notice the sharp drop in density above the lambda point, $T = 2.17$ K. *From F. London, Superfluids, New York: Dover (1964). Reprinted with permission.*

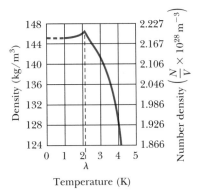

*The factor c is as may be expected dimensionally, and the extra factor of $1/4$ comes from a geometrical analysis. For a detailed calculation see J. J. Brehm and W. J. Mullin, *Introduction to the Structure of Matter*, New York: Wiley (1989), p. 80. See also our Chapter 3, Problem 30.

Figure 9.13 Specific heat of liquid helium as a function of temperature. The characteristic shape of this curve gives the transition point of 2.17 K the name "lambda point." *From F. London,* Superfluids, *New York: Dover (1964). Reprinted with permission.*

Heike Kamerlingh Onnes (1853–1926), Dutch physicist who performed the first important studies in low-temperature physics. After studying for a time under the German physicists Bunsen and Kirchhoff, he returned to Holland and eventually earned a physics chair at the University of Leiden. There, Onnes was the first to liquefy helium in 1908, an advance that allowed him to discover the phenomenon of superconductivity in 1911. For this work he was awarded the 1913 Nobel Prize in Physics. Along with J. D. van der Waals and H. A. Lorentz, Onnes helped advance the cause of Dutch physics. *Burndy Library, courtesy AIP Niels Bohr Library.*

It should be clear that something extraordinary is happening at about 2.17 K. That temperature is commonly referred to as the *critical temperature* (T_c), *transition temperature*, or simply the *lambda point*, a name derived from the shape of the curve in Figure 9.13. Indeed, visual inspection of the liquid helium is sufficient to demonstrate that something special happens at the lambda point. As the temperature is reduced from 4.2 K toward the lambda point, the liquid boils vigorously. But at 2.17 K the boiling suddenly stops and the liquid becomes quite calm. At lower temperatures evaporation from the surface continues, but there is no further boiling.

We call what happens at 2.17 K a transition from the **normal phase** to the **superfluid phase**. The word *superfluid* comes from yet another peculiar property of liquid helium below the lambda point. The flow rate of liquid helium through a capillary tube as a function of temperature is shown in Figure 9.14. The rate of flow increases dramatically as the temperature is reduced (Figure 9.15). Another way of describing this phenomenon is to say that the superfluid has a lower viscosity

Figure 9.14 The rate of capillary flow in liquid helium II (superfluid) and liquid helium I (normal liquid). The capillary flow rate increases dramatically with decreasing temperature below the lambda point. *From M. Zemansky,* Temperatures Very Low and Very High, *New York: Dover (1964). Reprinted with permission.*

Special Topic

SUPERFLUID ^{3}He

$\mathcal{S}$uperfluidity in ^{3}He was discovered in the early 1970s. This makes it a relatively recent discovery, compared with superfluidity in ^{4}He. There are good reasons, both experimental and theoretical, for the delay. The isotope ^{3}He accounts for only 0.00013% of naturally occurring helium. This is not enough to produce significant amounts of ^{3}He for experimental purposes. Since World War II it has been possible to obtain ^{3}He through the radioactive decay of tritium, which is produced in nuclear reactors. Also recall that the superfluid transition in ^{3}He occurs at just 2.7 mK, a temperature not reached until the 1950s. Even then many physicists believed that a superfluid phase in ^{3}He was impossible because the Bose condensation responsible for the superfluid phase of ^{4}He cannot occur in ^{3}He (a fermion).

Superfluid ^{3}He was finally discovered in 1971 by Douglas Osheroff, Robert Richardson, and David Lee at Cornell University. As is sometimes the case in science, they made the discovery while studying something else: the magnetic properties of solid ^{3}He. (^{3}He, like ^{4}He, has a solid phase at extremely high pressures.) Liquid ^{3}He was present in the experiment only as a refrigeration ingredient. The researchers saw a sudden change in the rate at which the pressure in their apparatus changed, and they were ultimately able to show that a transition from normal ^{3}He to superfluid ^{3}He was responsible.

Subsequent studies showed something even more remarkable about ^{3}He: it has three distinct superfluid phases. Two of them are shown in Figure A. The third

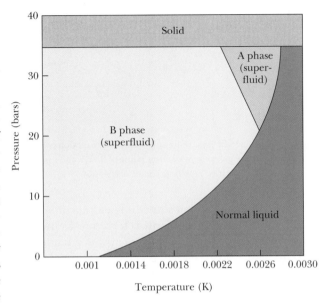

Figure A Phase diagram for ^{3}He with zero applied magnetic field. In this case there are two superfluid phases (A and B), and phase A exists only at very high pressures (1 bar = 10^5 Pa $\approx$ 1 atm). *From N. D. Mermin and D. M. Lee, "Superfluid Helium 3,"* Scientific American *235, 68 (1976). Image copyright © George V. Kelvin. Used with permission.*

phase can be produced only in the presence of a magnetic field (Figure B). This raises an interesting point about ^{3}He: with three spin 1/2 particles in its nucleus, it has a net magnetic moment. The magnetic behavior of ^{3}He, along with the fermion/boson distinction discussed in the text, is largely responsible for the different behavior of the two isotopes. It also helps us distinguish between the fermion pairing mechanism

---

Creeping film than the normal fluid. The viscosity can be so low that superfluid helium has been observed to form what is called a **creeping film** and flow up and over the (vertical) walls of its container. But there is a curious note to add regarding the viscosity of liquid helium. If one tries to measure the viscosity by, for example, measuring the drag on a metal plate as it is passed over the surface of the liquid, the result is just about what one would get with a normal fluid, even at temperatures somewhat below the lambda point. In other words, there appears to be a contradiction between the results of this experiment and the capillary flow experiment just described.

A solution to this conundrum is provided by a theory proposed by Fritz London in 1938. London claimed that liquid helium below the lambda point is part in superfluid ^{3}He and the fermion pairing mechanism in superconductors (Section 10.5). The pairs of ^{3}He atoms form in such a way as to have a net spin (the magnetic moments of the two atoms tend to align with each other) and net angular momentum (they also tend to revolve around one another), but electron pairs in a superconductor have zero spin and zero angular momentum. However, the fact that physicists had previously considered the possibility of a superconducting state with nonzero angular momentum helped them quickly understand the nonzero angular momentum state of superfluid ^{3}He when it was observed.

How do we characterize the three superfluid phases? In the A_1 phase, which exists only in the presence of an applied magnetic field and just below the superfluid transition temperature (Figure B), the magnetic moments align with the applied field, and the mutual pair revolution just described takes place in a plane parallel to the applied field. The A and B phases are more difficult to describe. Let it suffice to say that they correspond to various superpositions of the allowed fermion wave functions. Although the A and B phases do not depend on the existence of an external magnetic field, their behavior when such a field is imposed is quite interesting. The superfluid is then said to exhibit *anisotropic* behavior. That is, the properties of the superfluid are dependent upon the direction one considers with respect to the magnetic field. For example, the superfluid flow rate is quite different when measured parallel to the applied field than when measured in a plane perpendicular to the applied field.

Recently physicists have been interested in studying the vortex motion of superfluid ^{3}He. Upon rotation the superfluid tends to break into separate vor-

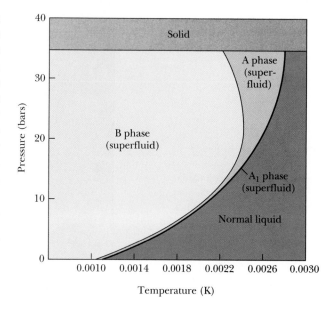

Figure B Phase diagram for ^{3}He in the presence of a strong magnetic field. The third superfluid phase (A_1) appears in a very narrow band on the pressure-temperature diagram. *From N. D. Mermin and D. M. Lee, "Superfluid Helium 3,"* Scientific American *235, 68 (1976). Image copyright © George V. Kelvin. Used with permission.*

tices—small regions of rotation with rotation patterns distinct from those in neighboring regions. The interesting thing about this discovery is that the vortex patterns are similar to those observed in superconductors. This new connection between superconductivity and superfluidity in ^{3}He continues to be the subject of careful study. In 1996 Osheroff, Richardson, and Lee were awarded the Nobel Prize in Physics for their discovery of superfluid ^{3}He.

---

superfluid and part normal. The superfluid component increases as the temperature is reduced from the lambda point, approaching 100% superfluid as the temperature approaches absolute zero. It has been shown that the fraction F of helium atoms in the superfluid state follows fairly closely the relation

$$F = 1 - \left(\frac{T}{T_c}\right)^{3/2} \tag{9.58}$$

The two-fluid model can explain our viscosity paradox, if we assume that only the superfluid part participates in capillary flow, and that enough normal fluid

Figure 9.15 (a) Liquid helium above the lambda point shows vigorous boiling. The suspended vessel contains holes in the bottom that are so fine that they do not allow passage of the normal fluid. (b) Liquid helium below the lambda point no longer boils, and the superfluid is dripping through the small holes in the bottom of the vessel. *Reproduced by kind permission of the Department of Physics, Clarendon Laboratory, Oxford.*

(a)   (b)

is always present to retard the motion of a metal plate dragged across its surface.

With two protons, two neutrons, and two electrons, the helium atom is a boson and therefore subject to Bose-Einstein statistics. How does this help explain the behavior of superfluid helium? One of the most important properties of bosons is that they are not subject to the Pauli exclusion principle. There is no limit to the number of bosons that may be in the same quantum state. Superfluid liquid helium is referred to as a **Bose-Einstein condensation** into a single state, in this case the superfluid state. All the particles in a Bose-Einstein condensate are in the same quantum state, which is not forbidden for bosons. As we saw in the case of the Fermi electron gas, fermions must "stack up" into their energy states, no more than two per energy state. Therefore, such a condensation process is not possible with fermions. As a striking demonstration of this fact, consider the isotope ^{3}He. With one less neutron than the more common ^{4}He, this isotope is a fermion. Liquid ^{3}He undergoes a superfluid transition at 2.7 mK (a temperature almost a factor of 1000 lower than T_c in ^{4}He!). The superfluid mechanism for the fermion ^{3}He is radically different than the Bose-Einstein condensation described earlier. It is more like the electron (fermion) pairing that one finds in the superconducting transition, which will be described in Section 10.5. Although they are very similar chemically (as two isotopes of the same element usually are), ^{3}He and ^{4}He are radically different when it comes to superfluidity, simply because one is a fermion and the other a boson. (See also Special Topic, "Superfluid ^{3}He.")

Bose-Einstein statistics may be used to estimate T_c for the superfluid phase of ^{4}He. Rather than derive a new density of states function for bosons, let us take advantage of the density of states function we have already developed for fermions. Using that result [Equation (9.43)] and substituting for the constant E_F yields

Bose-Einstein condensation

$$g_{\text{FD}}(E) = \frac{\pi}{2}\left(\frac{h^2}{8mL^2}\right)^{-3/2} E^{1/2} \tag{9.59}$$

where the FD subscript indicates a fermion distribution. The only difference for bosons is that they do not obey the Pauli principle, and therefore the density of states for bosons should be less by a factor of 2. Dividing by 2 and rearranging, we find

$$g_{\text{BE}}(E) = \frac{2\pi V}{h^3}(2m)^{3/2}E^{1/2} \tag{9.60}$$

with $V \equiv L^3$. Of course, the mass m is now the mass of a helium atom.

The number distribution $n(E)$ is now

$$n(E) = g_{\text{BE}}(E)\,F_{\text{BE}}$$

$$= \frac{2\pi V}{h^3}(2m)^{3/2}\,E^{1/2}\,\frac{1}{B_2 e^{E/kT} - 1} \tag{9.61}$$

In a collection of N helium atoms the normalization condition is

$$N = \int_0^\infty n(E)\,dE$$

$$= \frac{2\pi V}{h^3}(2m)^{3/2}\int_0^\infty \frac{E^{1/2}}{B_2 e^{E/kT} - 1}\,dE \tag{9.62}$$

With the substitution $u = E/kT$, this reduces to

$$N = \frac{2\pi V}{h^3}(2mkT)^{3/2}\int_0^\infty \frac{u^{1/2}}{B_2 e^u - 1}\,du \tag{9.63}$$

We do not know the value of B_2, but we will proceed using the minimum allowed value $B_2 = 1$. Then the integral in Equation (9.63), when evaluated (as in Appendix 7), has a value of 2.315. Because we used the minimum value of B_2, this result will correspond to the maximum value of N. In other words

$$N \leq \frac{2\pi V}{h^3}(2mkT)^{3/2}(2.315) \tag{9.64}$$

Rearranging, we find

$$T \geq \frac{h^2}{2mk}\left[\frac{N}{2\pi V(2.315)}\right]^{2/3} \tag{9.65}$$

Equation (9.65) can be evaluated numerically, because N/V is simply the number density of liquid helium in the normal state (2.11×10^{28} m^{-3}). The result is

$$T \geq 3.06\text{ K} \tag{9.66}$$

The value 3.06 K is an estimate of T_c, because our analysis has predicted that this is the minimum temperature at which we may expect to find a normal Bose-Einstein distribution. At lower temperatures there is condensation into the superfluid state. Our result is a bit off, because we used a density of states derived for a noninteracting gas rather than a liquid. Still, we have come within 1 kelvin of the correct value of T_c. As you would expect, calculations using a density of states constructed especially for a liquid produce even better results.

Example 9.9

For a gas of nitrogen (N_2) at room temperature (293 K) and 1 atmosphere pressure, calculate the Maxwell-Boltzmann constant A and thereby show that Bose-Einstein statistics can be replaced by Maxwell-Boltzmann statistics in this case.

Strategy The normalization condition for a gas of N molecules is

$$N = \int_0^\infty n(E) \, dE$$

where the distribution function is $n(E) = g(E)F_{MB}$. Equation (9.60) gives the density of states for this boson gas:

$$g(E) = \frac{2\pi V}{h^3}(2m)^{3/2} E^{1/2}$$

and the Maxwell-Boltzmann factor is $F_{MB} = A \exp(-E/kT)$. The solution will require doing the normalization integral and solving for the constant A.

Solution Putting all the factors into the normalization integral:

$$N = \int_0^\infty n(E) \, dE$$

$$= \frac{2\pi V}{h^3}(2m)^{3/2}A \int_0^\infty E^{1/2} \exp(-E/kT) \, dE$$

The integral is a standard definite integral, which yields

$$N = \frac{2\pi V}{h^3}(2m)^{3/2}A \frac{\sqrt{\pi}}{2}(kT)^{3/2}$$

$$= \frac{V}{h^3}(2\pi mkT)^{3/2}A$$

Therefore

$$A = \frac{h^3 N}{V}(2\pi mkT)^{-3/2}$$

Under normal conditions (atmospheric pressure and room temperature) the number density of nitrogen gas is $N/V = 2.50 \times 10^{25}$ m^{-3}. Plugging this into our result for A along with the molecular mass of N_2 and $T = 293$ K yields the value $A = 1.8 \times 10^{-7}$. Because this is much less than unity, the use of Maxwell-Boltzmann statistics is justified (see the discussion at the end of Section 9.5).

Bose-Einstein Condensation in Gases

For years physicists attempted to demonstrate Bose-Einstein condensation in gases. They were hampered, however, by the strong Coulomb interactions among the gas particles. This kept researchers from obtaining the low temperatures and high densities needed to produce the condensate.

Finally in 1995 success was achieved by a group led by Eric Cornell and Carl Wieman, working at Boulder, Colorado. Forming a Bose-Einstein condensate from a gas requires the achievement of extremely low temperatures, which this group was able to do in a two-step process. First, they used laser cooling (see the Special Topic box in Chapter 2) to cool their gas of ^{87}Rb atoms to about 1 mK. Then they used a magnetic trap to cool the gas to a temperature of about 20 nK (2.0×10^{-8} K). In their magnetic trap the researchers varied the magnetic field at radio frequencies in such a way that atoms with higher speeds and further from the center are driven away. What remained was an extremely cold, dense cloud. At temperatures of about 170 nK, the rubidium cloud was observed to pass through a transition, from a gas with a normal, broad velocity distribution to an extremely narrow one. It was also observed that the fraction of atoms in the condensed state increased as the temperature was lowered further, just as in a superfluid. Figure 9.16 shows how a Bose-Einstein condensate forms as a gas is cooled.

Later in 1995 Bose condensation in a gas was confirmed, this time with sodium gas, by a group led by Wolfgang Ketterle at MIT. For sodium the transition to the condensed state was observed to begin at a relatively warm 2 μK.

Figure 9.16 This series of graphs shows sequentially how a Bose-Einstein condensate forms as a sample of atoms is cooled. In these three-dimensional representations, the two horizontal axes represent velocity components, and the vertical axis represents the number of atoms having those velocities. Thus, a higher and sharper peak around the center of the graph (at zero velocity) shows the condensation of atoms into a small space. The field of view of each of the three frames is about 200 μm by 270 μm, and the absolute temperatures represented in the three frames are 200 nK, 100 nK, and 20 nK. *Images courtesy of Eric Cornell and Mike Matthews, JILA Research Team, University of Colorado at Boulder.*

For their important discoveries Cornell, Ketterle, and Wieman were awarded the 2001 Nobel Prize in Physics.

Today physicists are striving to understand the properties of these and other Bose condensates and reconcile their observations with quantum theory. On the more practical side, some physicists are exploring possible applications of Bose-Einstein condensation. One such application is an *atomic laser*, first demonstrated by Ketterle and others in 1996. The idea is that the condensate will produce a coherent beam of atoms analogous to the way a laser produces a coherent beam of photons. As of 2005 the condensate and resulting atomic beam are too small to have much practical value, but there's hope for future improvements.

Summary

By the end of the nineteenth century, the work of Maxwell and Boltzmann had made it clear that statistics could be useful in describing physical processes. In particular, Maxwell's statistical distribution for the velocities of molecules in an ideal gas is $f(\vec{v})$, where

$$f(\vec{v}) \, d^3\vec{v} = C \exp\left(-\tfrac{1}{2}\beta m v_x^2 - \tfrac{1}{2}\beta m v_y^2 - \tfrac{1}{2}\beta m v_z^2\right) d^3\vec{v} \tag{9.2}$$

The corresponding speed distribution $F(v)$ is

$$F(v) \, dv = 4\pi C \exp\left(-\tfrac{1}{2}\beta m v^2\right) v^2 \, dv \tag{9.14}$$

where $C = (\beta m/2\pi)^{3/2}$ and $\beta = (kT)^{-1}$. The speed distribution can be used to predict many of the observed properties of ideal gases. Computing the mean kinetic energy of a monatomic ideal gas molecule using the Maxwell velocity distribution yields

$$\overline{K} = \tfrac{3}{2}kT \tag{9.8}$$

This result is in accord with the equipartition theorem, which states that in equilibrium a mean energy of $\frac{1}{2}kT$ is associated with each independent quadratic term in a molecule's energy. The equipartition theorem can be applied to the rotational and vibrational modes of a diatomic molecule, and it can thereby be used to compute the heat capacities of diatomic gases at various temperatures.

Under conditions of higher densities and lower temperatures the particles' wave functions overlap, and the indistinguishability of the particles becomes a factor. As a result it is necessary to use quantum statistics: Fermi-Dirac statistics for fermions and Bose-Einstein statistics for bosons. These distributions differ because fermions obey the Pauli exclusion principle and bosons do not. The statistical factors associated with the classical (Maxwell-Boltzmann) and quantum distributions are given by

$$F_{MB} = A \exp(-\beta E) \tag{9.27}$$

$$F_{FD} = \frac{1}{B_1 \exp(\beta E) + 1} \tag{9.30}$$

$$F_{BE} = \frac{1}{B_2 \exp(\beta E) - 1} \tag{9.32}$$

where A, B_1, and B_2 are normalization factors. In each case the distribution function $n(E)$ can be expressed as the product of the density of states $g(E)$ [where $g(E)$ is defined as the number of states available per unit energy range] and the appropriate statistical factor.

Fermi-Dirac statistics are needed in order to predict the correct behavior of conduction electrons in a metal. Using the Fermi-Dirac distribution, one can find the electrical conductivity (and the correct temperature dependence thereof) as well as the electronic contribution to the specific heat of a metal.

Bose-Einstein statistics can be used to derive the Planck law for blackbody radiation:

$$\mathscr{I}(\lambda, T) = \frac{2\pi c^2 h}{\lambda^5} \frac{1}{e^{hc/\lambda kT} - 1} \tag{9.57}$$

Bose-Einstein statistics also help us understand some of the properties of liquid helium, which undergoes a transition to the superfluid state at the temperature $T_c = 2.17$ K. The extraordinarily low viscosity of a superfluid is due to the fact that the molecules in a superfluid obey Bose-Einstein statistics. In recent years, Bose-Einstein condensation has been observed in a variety of new materials.

Questions

1. How relevant is the Heisenberg uncertainty principle in frustrating Laplace's goal of determining the behavior of an essentially classical system of particles (say an ideal gas) through knowledge of the motions of individual particles?

2. Why might the measured molar heat capacity of Cl_2 not match the prediction of the equipartition theorem as well as that of O_2?

3. In a diatomic gas why should it be more difficult to excite the vibrational mode than the rotational mode?

4. What is the physical significance of the root-mean-square speed in an ideal gas?

5. For room temperature and 1 atmosphere pressure compare v^* for N_2 and O_2. Do the same comparison for $\bar{v}$ and v_{rms}.

6. An insulated container filled with an ideal gas moves through field-free space with a constant velocity. Describe the effect this has on the Maxwell velocity distributions and Maxwell speed distribution.

7. Maxwell conceived of a device (later called *Maxwell's demon*) that could use measurements of individual molecular speeds in a gas to separate faster molecules from slower ones. This device would operate a trap door between two (initially identical) compartments, allowing only faster molecules to pass one way and slower molecules the other way, thus creating a temperature imbalance. Then the temperature difference could be used to run a heat engine, and the result would be the production

of mechanical work with no energy input. Of course this would violate the laws of thermodynamics. Discuss reasons why you think Maxwell's demon cannot work.

8. If the distribution function $f(q)$ for some physical property q is even, it follows that $\bar{q} = 0$. Does it also follow that the most probable value $q^* = 0$?

9. Which of the following act as fermions and which as bosons: hydrogen atoms, deuterium atoms, neutrinos, muons, table tennis balls?

10. Theorists tend to believe that free quarks (with charge $\pm\frac{2}{3}e$ and $\pm\frac{1}{3}e$) do not exist, but the question is by no means decided. If free quarks are found to exist, what can you say about the distribution function they would obey?

11. Explain why you expect $\bar{E}$ to be greater than $\frac{1}{2}E_F$ [Equation (9.45)].

12. How would the behavior of metals be different if electrons were bosons rather than fermions?

13. What would happen to the Planck distribution and the behavior of liquid helium if we let $h \rightarrow 0$ in the Bose-Einstein density of states [Equation (9.53)]?

14. If only the superfluid component of liquid helium flows through a very fine capillary, is it possible to use capillary flow to separate completely the superfluid component of a sample of liquid helium from the normal component?

15. Why is $B_2 = 1$ the minimum allowed value in the integral in Equation (9.63)?

Problems

Note: The more challenging problems are indicated by a shaded box.

9.2   Maxwell Velocity Distribution

1. (a) Use Equation (9.5) to show that the one-dimensional rms speed is

$$v_{x\,\text{rms}} = \left(\overline{v_x^2}\right)^{1/2} = \left(\frac{kT}{m}\right)^{1/2}$$

(b) Show that Equation (9.5) can be rewritten as

$$g(v_x)\,dv_x = (2\pi)^{-1/2}(v_{x\,\text{rms}})^{-1}\exp\left(-\tfrac{1}{2}v_x^2/v_{x\,\text{rms}}^2\right)\,dv_x$$

2. The result of Problem 1 can be used to estimate the relative probabilities of various velocities. Pick a small interval $\Delta v_x = 0.002 v_{x\,\text{rms}}$. For 1 mole of an ideal gas compute the number of molecules within the range Δv_x centered at (a) $v_x = 0.01 v_{x\,\text{rms}}$, (b) $v_x = 0.20 v_{x\,\text{rms}}$, (c) $v_x = v_{x\,\text{rms}}$, (d) $v_x = 5v_{x\,\text{rms}}$, (e) $v_x = 100 v_{x\,\text{rms}}$.

3. Consider an ideal gas enclosed in a spectral tube. When a high voltage is placed across the tube, many atoms are excited, and all excited atoms emit electromagnetic radiation at characteristic frequencies. According to the Doppler effect, the frequencies observed in the laboratory depend on the velocity of the emitting atom. The nonrelativistic Doppler shift of radiation emitted in the x direction is $f = f_0(1 + v_x/c)$. The resulting wavelengths observed in the spectroscope are spread to higher and lower values due to the (respectively) lower and higher frequencies, corresponding to negative and positive values of v_x. We say that the spectral line has been *Doppler broadened* (this is what allows us to see the lines easily in the spectroscope, because the Heisenberg uncertainty principle does not cause significant line broadening in atomic transitions). (a) What is the mean frequency of the radiation observed in the spectroscope? (b) To get an idea of how much the spectral line is broadened at particular temperatures, derive an expression for the standard deviation of frequencies, defined to be

$$\text{Standard deviation} = \left[\overline{(f - f_0)^2}\right]^{1/2}$$

Your result should be a function of f_0, T, and constants. (c) Use your results from (b) to estimate the fractional line width, defined by the ratio of the standard deviation to f_0, of hydrogen (H_2) gas at $T = 293$ K. Repeat for a gas of atomic hydrogen at the surface of a star, with $T = 5500$ K.

9.3   Equipartition Theorem

4. Consider the model of the diatomic gas oxygen (O_2) shown in Figure 9.3. (a) Assuming the atoms are point particles separated by a distance of 0.85 nm, find the rotational inertia I_x for rotation about the x axis.

(b) Now compute the rotational inertia of the molecule about the z axis, assuming almost all of the mass of each atom is in the nucleus, a nearly uniform solid sphere of radius 3.0×10^{-15} m. (c) Compute the rotational energy associated with the first ($\ell = 1$) quantum level for a rotation about the x axis. (d) Using the energy you computed in (c), find the quantum number ℓ needed to reach that energy level with a rotation about the z axis. Comment on the result in light of what the equipartition theorem predicts for diatomic molecules.

9.4   Maxwell Speed Distribution

5. Using the Maxwell speed distribution, (a) write an integral expression for the number of molecules in an ideal gas that would have speed $v > c$ at $T = 293$ K. (b) Explain why the numerical result of the expression you found in (a) is negligible.

6. Use a computer to explore the numerical value of the definite integral you constructed in the previous problem.

7. It is important for nuclear engineers to know the thermal properties of neutrons in a nuclear reactor. Assuming that a gas of neutrons is in thermal equilibrium, find $\bar{v}$ and v^* for neutrons at (a) 300 K and (b) 2000 K.

8. Show that the Maxwell speed distribution function $F(v)$ approaches zero by carefully taking the limit as $v \to 0$ and as $v \to \infty$.

9. Find v^* for N_2 gas in air (a) on a cold day at $T = -5°C$ and (b) on a hot day at $T = 30°C$.

10. For an ideal gas O_2 at $T = 293$ K find the two speeds v that satisfy the equation $2F(v) = F(v^*)$. Which of the two speeds you found is closer to v^*? Does this make sense?

11. For the ideal gas Ar at $T = 293$ K, use a computer to show that

$$\int_0^\infty F(v)\,dv = 1$$

and thereby verify that $C = (\beta m/2)^{3/2}$.

12. Consider the ideal gas H_2 at $T = 293$ K. Use a numerical integration program on a computer to find the fraction of molecules with speeds in the following ranges: (a) 0 to 1 m/s, (b) 0 to 100 m/s, (c) 0 to 1000 m/s, (d) 1000 m/s to 2000 m/s, (e) 2000 m/s to 5000 m/s, (f) 0 to 5000 m/s.

13. (a) Find v_{rms} for H_2 gas and N_2 gas, both at $T = 293$ K. (b) Considering your answers to part (a), discuss why our atmosphere contains nitrogen but not hydrogen.

14. (a) Find the total translational kinetic energy of 1 mole of argon atoms at $T = 273$ K. (b) Would your

answer be the same or different for 1 mole of oxygen (O_2) molecules? Explain.

9.5 Classical and Quantum Statistics

15. Using the Maxwell-Boltzmann energy distribution Equation (9.26), (a) find the mean translational kinetic energy of an ideal gas and (b) compare your results with $\frac{1}{2} m \bar{v}^2$ and $\frac{1}{2} m \overline{v^2}$.

16. From the Maxwell-Boltzmann energy distribution, find the most probable energy E^*. Plot $F(E)$ versus E and indicate the position of E^* on your plot.

17. Inside a certain kind of star there are regions in which approximately one hydrogen atom per million is in the first excited level ($n = 2$). The other atoms can be assumed to be in the $n = 1$ level. Use this information to estimate the temperature there, assuming that Maxwell-Boltzmann statistics are valid. (*Hint:* In this case the density of states depends on the number of possible quantum states available on each level, which is 8 for $n = 2$ and 2 for $n = 1$.)

18. One way to decide whether Maxwell-Boltzmann statistics are valid is to compare the de Broglie wavelength λ of a typical particle with the average interparticle spacing d. If $\lambda \ll d$ then Maxwell-Boltzmann statistics are generally acceptable. (a) Using de Broglie's relation $\lambda = h/p$, show that

$$\lambda = \frac{h}{(3mkT)^{1/2}}$$

(b) Use the fact that $N/V = 1/d^3$ to show that the inequality $\lambda \ll d$ can be expressed as

$$\frac{N}{V} \frac{h^3}{(3mkT)^{3/2}} \ll 1$$

(c) Use the result of (b) to determine whether Maxwell-Boltzmann statistics are valid for Ar gas at room temperature (293 K) and for the conduction electrons in pure silver at $T = 293$ K.

19. Use Equation (9.24) to turn the Maxwell speed distribution, Equation (9.14), into an energy distribution [Equations (9.25) and (9.26)].

20. Consider an atom with a magnetic moment μ and a total spin of 1/2. The atom is placed in a uniform magnetic field of magnitude B at temperature T. (a) Assuming Maxwell-Boltzmann statistics are valid at this temperature, find the ratio of atoms with spins aligned with the field to those aligned opposite the field. (b) Evaluate numerically with $B = 6$ T, for $T = 77$ K, $T = 273$ K, and $T = 600$ K.

9.6 Fermi-Dirac Statistics

21. The Fermi energy can be defined as the energy at which the Fermi factor $F_{FD} = 0.5$. Using this definition, show that the constant B_1 in Equation (9.30) is equal to $\exp(-\beta E_F)$ and that

$$F_{FD} = \frac{1}{\exp[\beta(E - E_F)] + 1}$$

In other words, verify Equations (9.33) and (9.34).

22. At $T = 0$, what fraction of electrons have energy $E < \bar{E}$?

23. Silver has exactly one conduction electron per atom. (a) Using the density of silver (1.05×10^4 kg/m³) and the mass of 107.87 g/mol, find the density of conduction electrons in silver. (b) At what temperature is $A = 1$ for silver (where A is the normalization constant in the Maxwell-Boltzmann distribution)? (c) At what temperature is $A = 10^{-3}$?

24. What fraction of the electrons in a good conductor have energies between $0.95 E_F$ and E_F at $T = 0$?

25. Using the data in Problem 23, compute (a) E_F and (b) u_F for silver.

26. The Fermi energy for gold is 5.51 eV at $T = 293$ K. (a) Find the average energy of a conduction electron at that temperature. (b) Compute the temperature at which the average kinetic energy of an ideal gas molecule would equal the average energy you found in (a). (c) Comment on the relative temperatures in (a) and (b).

27. The density of pure copper is 8.92×10^3 kg/m³, and its molar mass is 63.546 grams. Using the experimental value of the conduction electron density, 8.47×10^{28} m⁻³, compute the number of conduction electrons per atom.

28. Aluminum has a density of 2.70×10^3 kg/m³ at a temperature of 293 K, and its molar mass is 26.98 g. (a) Compute the number of aluminum atoms per unit volume at that temperature. (b) Use the fact that $E_F = 11.63$ eV for aluminum at 293 K to find the number density of free electrons. (c) Combine your results from (a) and (b) to estimate the number of conduction electrons per atom—the valence number for aluminum.

29. Compute the Fermi speed for (a) Sr ($E_F = 3.93$ eV) and (b) Zn ($E_F = 9.47$ eV).

30. Verify that Equation (9.35) is valid in the limit $T \to 0$.

31. Show that in general (that is, $T \neq 0$) the energy distribution of N electrons in a conductor with Fermi energy E_F at temperature T is

$$n(E) = \frac{3N}{2} E_F^{-3/2} \frac{E^{1/2}}{\exp[\beta(E - E_F)] + 1}$$

32. Use the result of Problem 31 with copper ($E_F = 7.0$ eV) to sketch $n(E)$ at (a) $T = 0$, (b) $T = 293$ K, and (c) $T = 1500$ K.

33. Use numerical integration of the function given in Problem 31 to verify that

$$\int_0^\infty n(E)\, dE = N$$

Choose the parameters $T = 300$ K and use $E_F = 7.00$ eV for copper.

34. Use numerical integration of the function given in Problem 31 to find the fraction of conduction electrons with energies between 6.00 eV and 7.00 eV in copper at $T = 293$ K. Comment on your results.

35. In a neutron star the entire star's mass has collapsed essentially to nuclear density. For a neutron star with radius 10 km and mass 4.50×10^{30} kg, find the Fermi energy of the neutrons.

36. Consider a collection of fermions at $T = 293$ K. Find the probability that a single-particle state will be occupied if that state's energy is (a) 0.1 eV less than E_F; (b) equal to E_F; (c) 0.1 eV greater than E_F.

37. Suppose you have an ideal gas of fermions at room temperature (293 K). How large must $E - E_F$ be for Fermi-Dirac and Maxwell-Boltzmann statistics to agree to within 1%? Do you think the agreement is within 1% for ideal gases under normal conditions?

9.7 Bose-Einstein Statistics

38. Consider the problem of photons in a spherical cavity at temperature T, as described in Section 9.7. (a) For the entire collection of photons, what is the number density (number of photons per unit volume)? (b) Evaluate your result from (a) numerically at $T = 500$ K and $T = 5500$ K (the approximate temperature of the sun's surface).

39. Use numerical integration on a computer to verify the output of the definite integral in Equation (9.63).

General Problems

40. Use the method described in Appendix 6 to evaluate the integral

$$\int_0^\infty x^n \exp(-ax^2)\, dx$$

in terms of the constant a for $n = 3$, 4, and 5.

41. Assume that air is an ideal gas under a uniform gravitational field, so that the potential energy of a molecule of mass m at altitude z is mgz. Show that the distribution of molecules varies with altitude as given by the distribution function $f(z)\, dz = C_z \exp(-\beta mgz)\, dz$ and that the normalization constant $C_z = kT/mg$. This distribution is referred to as the *law of atmospheres*.

42. Use the law of atmospheres (Problem 41) to compare the air densities at sea level, Denver (altitude 1610 m), and the summit of Mt. Rainier (altitude 4390 m), assuming the same temperature 273 K in each case.

43. Consider the law of atmospheres (Problem 41). First show that the pressure difference ΔP corresponding to an altitude change Δz is approximately

$$\Delta P = -\rho g\, dz = -\frac{Nmg}{V}\, dz$$

Next, assume that the temperature is constant over small altitude changes and then show that $P \approx P_0 \exp(-\beta mgz)$ where P_0 is the pressure at $z = 0$.

44. Consider a thin-walled, fixed-volume container of volume V that holds an ideal gas at constant temperature T. It can be shown by dimensional analysis that the number of particles striking the walls of the container per unit area per unit time is given by $n\bar{v}/4$, where as usual n is the particle number density. The container has a small hole of area A in its surface through which the gas can leak slowly. Assume that A is much less than the surface area of the container. (a) Assuming that the pressure inside the container is much greater than the outside pressure (so that no gas will leak from the outside back in), estimate the time it will take for the pressure inside to drop to half the initial value. Your answer should contain A, V, and the mean molecular speed $\bar{v}$. (b) Obtain a numerical result for a spherical container with a diameter of 40 cm containing air at 293 K, if there is a circular hole of diameter 1.0 mm in the surface.

45. For the situation described in Problem 44, show that the speed distribution of the escaping molecules is proportional to $v^3 \exp(-\frac{1}{2}\beta mv^2)$ and that the mean energy of the escaping molecules is $2kT$.

46. (a) What density of conduction electrons in copper is needed in order for the Maxwell-Boltzmann normalization constant to be $A = 1$ at $T = 293$ K? (b) Repeat the calculation for neutrons at the same temperature. (c) Repeat the calculation for He gas at the same temperature.

47. For a (classical) simple harmonic oscillator with fixed total energy E, find the mean value of kinetic energy $\bar{K}$ and the mean value of potential energy $\bar{V}$. Show that $\bar{K} = \bar{V} = E/2$.

48. During World War II, physicists developed methods to separate the uranium-235 and -238 isotopes. One method involved converting the uranium metal to a gas, UF_6, and then allowing the gas to diffuse through a porous barrier, with the lighter gas diffusing faster. What is the difference in the speeds of the two UF_6 gas species at room temperature?

49. Find the number density N/V for Bose-Einstein condensation to occur in helium at room temperature (293 K). Compare your answer with the number density for an ideal gas at room temperature at 1 atmosphere pressure.

50. Bose-Einstein condensates have reached temperatures of 20 nK. Treating rubidium as an ideal gas, find the rms speed of a rubidium atom at that temperature.

51. The ^{40}Ar isotope of argon (its most common form) is a boson, like ^{4}He. (a) Following the methods of Section 9.7, estimate the temperature at which argon should become a Bose-Einstein condensate. Use a number density 2.5×10^{28} m^{-3} for argon. (b) Why isn't argon observed to become a Bose-Einstein condensate? (*Hint:* The freezing point of argon is 84 K.)

52. In one experiment done by Cornell and Wieman, a Bose-Einstein condensate contained 2000 rubidium-87 atoms within a volume of about 10^{-15} m^3. Estimate the temperature at which Bose-Einstein condensation should have occurred.

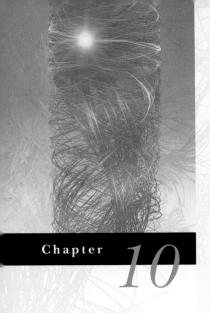

Molecules and Solids

The secret of magnetism, now explain that to me! There is no greater secret, except love and hate.

Johann Wolfgang von Goethe

In Chapters 7 and 8, you learned about the properties of individual atoms. This chapter builds on that knowledge to find out what happens when those atoms join together to form molecules and solids.

Beginning with the simplest combination of atoms, we first consider the diatomic molecule. Molecular spectra are quantized, just like the atomic spectra you've seen previously, and the quantum principles you've already learned help explain the observed spectra. One important application of the study of atomic and molecular energy levels is the use of stimulated emission in lasers. In Section 10.2 the working mechanisms of several kinds of lasers are explained, and some of their many applications, including holography, are discussed.

The remainder of the chapter is devoted to a study of solids. Solids can have many different crystal structures, although in some cases they lack long-range order altogether. Thermal and magnetic properties of solids (including thermal expansion, thermal conductivity, and magnetic susceptibility) can be explained by utilizing quantum theory.

Superconductivity is a remarkable phenomenon observed in many solid materials at low temperatures. We present the development of superconducting materials, the theory of superconductivity, and the prospects for future basic research in this field. The final section of the chapter addresses some of the developed and proposed applications of superconductivity. Zero-resistance electrical circuits and transmission lines, magnetic levitation, magnetic resonance imaging, and high-field magnets are just a few of the applications being studied. The amazing properties of superconductors give rise to many more exciting possibilities for the future.

10.1 Molecular Bonding and Spectra

How are atoms held together to form molecules? The attractive forces between atoms in molecules must be due to the Coulomb force, because the Coulomb force is the only one that has both the strength and long range necessary to bind atoms at the distances observed (normally on the order of 10^{-10} m). All atoms have both positive charges (in the nucleus) and negative charges (electrons) needed to provide attractive forces, though not all atoms readily form molecules. Because every neutral atom contains both kinds of charge, the net force of one atom on another must be a combination of attractive and repulsive forces. It is the combination of attractive and repulsive forces that creates a stable molecular structure.

As an example let us consider the bonding in a diatomic molecule (the mechanisms are similar, though sometimes more complex, in multiatom molecules). We will find it useful to look at molecular binding using potential energy V. Recall that in conservative systems, force is related to potential energy by $F = -dV/dr$, where r is the distance of separation. This means that, in cases in which the magnitude of the force decreases with increasing distance (as it does with Coulomb forces), we should associate a negative slope ($dV/dr < 0$) with repulsive forces and a positive slope ($dV/dr > 0$) with attractive forces. An approximation of the force felt by one atom in the vicinity of another atom may then be written

$$V = \frac{A}{r^n} - \frac{B}{r^m} \qquad (10.1)$$

where A and B are positive constants that depend on the types of atoms involved and n and m are small, positive numbers. You might expect n and m to be equal to 1 because the forces are Coulomb, but because of the complicated shielding effects of the various electron shells, this is not so.

As shown in Figure 10.1, the sum of attractive and repulsive potentials produces a potential well that provides a stable equilibrium for total energy $E < 0$. The exact shape of the curve depends on the parameters A, B, n, and m from Equation (10.1). You should convince yourself that $n > m$ is required in order to produce the potential well shown in Figure 10.1. Vibrations are excited thermally, so the exact level of E depends on temperature, with E at the bottom of the potential well when $T = 0$. Once a pair of atoms is joined, one would then have to supply enough energy to raise the total energy of the system to zero (effectively taking the second atom to $r = \infty$) in order to separate the molecule into two neutral atoms. These potential energy curves typically have a minimum value, and the corresponding value of r is an *equilibrium separation*.

The amount of energy required to separate the two atoms completely is known as the **binding energy** for that molecule. The binding energy is roughly equal to the depth of the potential well shown in Figure 10.2. The binding energy and well depth may not be exactly equal, however, because by the Heisenberg uncertainty principle, the ground state energy cannot lie exactly at the bottom of the well.

Molecular Bonds

The bonding mechanism for a particular molecule depends principally on the electronic structure of the atoms involved. The simplest of these bonding mechanisms, the **ionic bond**, typically occurs when the two atoms involved are easily ionized. For example, sodium, which has the electronic configuration $1s^2 2s^2 2p^6 3s^1$, readily gives up its $3s$ electron to become Na^+, while chlorine, with electronic configuration $1s^2 2s^2 2p^6 3s^2 3p^5$, readily gains an electron to become Cl^-. Notice that both Na^+ and Cl^- have filled electronic shells. The Na^+ and Cl^- ions are electrostatically attracted to form the NaCl molecule.

In a **covalent bond**, the atoms are not as easily ionized. In order to form a molecule, they must somehow share their outer shell electrons. Diatomic molecules

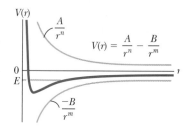

Figure 10.1 Attractive and repulsive potentials (lightly shaded lines) experienced by one atom in the vicinity of another atom. The sum of attractive and repulsive potentials is represented by the solid blue line. Bound states can exist for a total energy $E < 0$.

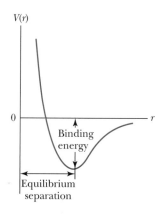

Figure 10.2 The net potential energy curve, showing the equilibrium separation and binding energy.

Binding energy

Ionic bond

Covalent bond

ℓ E_{rot}

5 ——— $15E_1$

4 ——— $10E_1$

3 ——— $6E_1$

Energy

2 ——— $3E_1$

1 ——— E_1
0 ——— 0

Figure 10.3 Energy levels for a rigid rotator. The spacing between levels increases with increasing energy. The levels are given in terms of the $\ell = 1$ energy level E_1, where $E_1 = \hbar^2/I$.

formed by the combination of two identical atoms (H_2, N_2, O_2, and so on, sometimes referred to as *homopolar* molecules) tend to be covalent, because neither atom is more likely than the other to gain or lose an electron. Larger molecules (like organic molecules) are formed principally with covalent bonds.

Several more exotic bonding mechanisms are described only briefly here. The *van der Waals bond* is a relatively weak bond found mostly in liquids and solids at low temperatures. One common example of the van der Waals bond in action is in graphite, a form of pure carbon. In graphite, the carbon atoms form two-dimensional sheets, held together by strong covalent bonds. The much weaker van der Waals bond holds together adjacent sheets of carbon atoms. As a result, one layer of atoms slides over the next layer with little friction, which you see for example when the graphite in your pencil slides easily over paper. The *hydrogen bond* is important in holding together many organic molecules. In a *metallic bond* essentially free valence electrons may be shared by a number of atoms.

Rotational States

Regardless of the types of molecular bonds present, we can learn much about the properties of molecules by studying how molecules absorb, emit, and scatter electromagnetic radiation. This kind of study is referred to broadly as *molecular spectroscopy*. Let us begin by considering a simple two-atom molecule, such as N_2. As we saw in Section 9.3 on the equipartition theorem, we can model this molecule in different ways, depending on the physical situation. For example, one may think of the N_2 molecule as two N atoms held together with a massless, rigid rod. This is known as the *rigid rotator* model. Because these are purely rotational modes of motion, the quantum theory of angular momentum can be used to determine those energy levels. In a purely rotational system, the kinetic energy is expressed in terms of the angular momentum L and the rotational inertia I as

$$E_{\text{rot}} = \frac{L^2}{2I}$$

The angular momentum is quantized (see Section 7.3) in the form

$$L = \sqrt{\ell(\ell + 1)}\,\hbar \tag{7.22}$$

where ℓ is the angular momentum quantum number. Then the energy levels are given by

$$E_{\text{rot}} = \frac{\hbar^2 \ell(\ell + 1)}{2I} \tag{10.2}$$

Energy levels of a simple rotational state

The rotational inertia I is constant for a rigid rotator, and therefore E_{rot} varies only as a function of the quantum number ℓ, as shown in Figure 10.3.

Example 10.1

Estimate the value of E_{rot} for the lowest rotational energy state of N_2.

Strategy If we consider the nitrogen atoms to be point masses (each with mass m) separated by a distance R (see Figure 10.4), then the rotational inertia about an axis passing

through the center of the molecule and perpendicular to the line joining the atoms is

$$I = m\left(\frac{R}{2}\right)^2 + m\left(\frac{R}{2}\right)^2 = \frac{mR^2}{2}$$

Figure 10.4 Schematic of a diatomic molecule (N_2) as a rigid rotator, with an equilibrium separation R between atomic centers. (Center of mass is indicated as c.m.)

With this numerical estimate of rotational inertia I, the energy levels follow from Equation (10.2).

Solution For nitrogen, $m = 2.33 \times 10^{-26}$ kg. If we assume $R \approx 10^{-10}$ m in this order-of-magnitude estimate, then

$$I \approx \frac{(2 \times 10^{-26}\ \text{kg})\,(10^{-20}\ \text{m}^2)}{2} = 10^{-46}\ \text{kg}\cdot\text{m}^2$$

For $\ell = 1$

$$E_{\text{rot}} = \frac{2\hbar^2}{2I} = \frac{\hbar^2}{I} \approx \frac{10^{-68}\ \text{J}^2\cdot\text{s}^2}{10^{-46}\ \text{kg}\cdot\text{m}^2} = 10^{-22}\ \text{J} \approx 10^{-3}\ \text{eV}$$

Remember, however, that this is just the energy of the *lowest* state. The energy of the $\ell = 4$ state [with $\ell(\ell+1) = 20$] is ten times higher, or about 0.01 eV. This energy is still at least two orders of magnitude less than the energy associated with an electronic transition in hydrogen (the kind that produces visible photons). Therefore we can expect photons generated by transitions between adjacent rotational states to be in the *infrared* or *microwave* portion of the spectrum.

Vibrational States

In addition to rotation, there is also the possibility that a vibrational energy mode will be excited. As we saw in Section 9.3, there will be no thermal excitation of this mode in a diatomic gas at ordinary temperatures. However, it is possible to stimulate vibrations in molecules using electromagnetic radiation. As a crude model of vibration in a diatomic molecule, we again assume that the two atoms are point masses connected by a massless spring, as in Figure 10.5. Then the atoms can execute simple harmonic motion, just as in classical physics. The energy levels must be those of a quantum-mechanical oscillator (see Section 6.6), namely

$$E_{\text{vibr}} = (n + \tfrac{1}{2})\hbar\omega \tag{10.3}$$

where ω is the natural (classical) angular frequency of the oscillator.

One way of estimating E_{vibr} is to estimate ω from purely classical considerations. From classical mechanics, the frequency of a two-particle oscillator is

$$\omega = \sqrt{\frac{\kappa}{\mu}} \tag{10.4}$$

where $\mu = m_1 m_2/(m_1 + m_2)$ is the reduced mass of the system and κ is a force (spring) constant. Now we can estimate κ by assuming that the force holding the masses together is Coulomb. If the bond were purely ionic, then the point masses would be point charges $+e$ and $-e$. Thus we can compute the force constant κ (force per unit distance):

$$\kappa = \left|\frac{dF}{dr}\right| \approx \left|\frac{d}{dr}\left(\frac{e^2}{4\pi\epsilon_0 r^2}\right)\right| = \frac{e^2}{2\pi\epsilon_0 r^3} \tag{10.5}$$

and

$$\omega \approx \sqrt{\frac{e^2}{2\pi\epsilon_0 \mu r^3}} \tag{10.6}$$

Figure 10.5 A model of a diatomic molecule with the two masses m_1 and m_2 connected by a massless spring with force constant κ.

If we use $r \approx 10^{-10}$ m, Equation (10.5) yields a force constant of $\kappa \approx 460$ N/m. Then $\omega = \sqrt{\kappa/\mu} \approx 2.0 \times 10^{14}$ s^{-1} (or $f \approx 3.2 \times 10^{13}$ Hz), and Equation (10.3)

Table 10.1	Fundamental Vibrational Frequencies and Effective Force Constants for Some Diatomic Molecules	
Molecule	Frequency (Hz), $n = 0$ to $n = 1$	Force Constant (N/m)
HF	8.72×10^{13}	970
HCl	8.66×10^{13}	480
HBr	7.68×10^{13}	410
HI	6.69×10^{13}	320
CO	6.42×10^{13}	1860
NO	5.63×10^{13}	1530

From G. M. Barrow, *The Structure of Molecules*, New York: Benjamin (1963).

gives an energy $E_{\text{vibr}} \approx 0.2$ eV for the $n = 1$ vibrational level of N_2. Actual values of the fundamental vibrational frequencies and effective force constants are shown in Table 10.1.

Example 10.2

(a) Given that the spacing between vibrational energy levels of the HCl molecule is 0.36 eV, calculate the effective force constant. (b) Find the classical temperature associated with this difference between vibrational energy levels in HCl.

Strategy (a) Because $\kappa = \mu\omega^2$ we first need to find μ and ω. The reduced mass μ is given by $\mu = m_1 m_2/(m_1 + m_2)$. We also know that $\Delta E = hf = \hbar\omega$, so $\omega = \Delta E/\hbar$. At the lowest level $n = 0$ and $E_{\text{vibr}} = \hbar\omega/2$.

(b) Two degrees of freedom are associated with a one-dimensional oscillator, one from the kinetic energy and one from the potential (see Section 9.3). Therefore we can say that

$$\Delta E = \hbar\omega = 2\left(\frac{kT}{2}\right) = kT$$

(where k is Boltzmann's constant), and so $T = \Delta E/k$.

Solution (a) If we assume the most common chlorine-35 isotope, the reduced mass is

$$\mu = \frac{m_1 m_2}{m_1 + m_2} = \frac{(34.97 \text{ u})(1.008 \text{ u})}{34.97 \text{ u} + 1.008 \text{ u}}$$

$$= 0.9798 \text{ u} = 1.63 \times 10^{-27} \text{ kg}$$

Then

$$\omega = \frac{\Delta E}{\hbar} = \frac{0.36 \text{ eV}}{6.58 \times 10^{-16} \text{ eV} \cdot \text{s}} = 5.47 \times 10^{14} \text{ rad/s}$$

Putting together the calculated numerical values, the force constant is given by Equation (10.4) as

$$\kappa = \mu\omega^2 = (1.63 \times 10^{-27} \text{ kg})(5.47 \times 10^{14} \text{ rad/s})^2$$

$$= 490 \text{ N/m}$$

which is quite close to the value in Table 10.1.

(b) Using the numerical value of ΔE,

$$T = \frac{\Delta E}{k} = \frac{0.36 \text{ eV}}{8.62 \times 10^{-5} \text{ eV/K}} = 4200 \text{ K}$$

In order to excite this vibrational state, we need a temperature of 4200 K. This is why vibrational levels are not thermally excited at ordinary temperatures.

Vibration and Rotation Combined

In a real molecular system it is possible to excite the rotational and vibrational modes of motion simultaneously. We combine the previous results to produce the total energy of our simple vibration-rotation system:

$$E = E_{\text{rot}} + E_{\text{vibr}} = \frac{\hbar^2 \ell(\ell + 1)}{2I} + \left(n + \frac{1}{2}\right)\hbar\omega \qquad (10.7)$$

A diatomic molecule that has been stimulated to an excited state will, as in atomic systems, emit a photon upon decaying to a lower energy state. Generally, then, it is possible to observe a wide spectrum of emitted photons, corresponding to various rotational and vibrational transitions.

One outstanding characteristic of emission spectra can be deduced by examining Equation (10.7). Because the vibrational energies are spaced at regular intervals ($E_{\text{vibr}} = \frac{1}{2}\hbar\omega, \frac{3}{2}\hbar\omega, \dots$), emission features due to vibrational transitions appear at regular intervals. This is also the case for rotational features, though for a different reason. Consider, for example, a transition from the $\ell + 1$ state to the ℓ state. The photon produced by that transition will have an energy

$$E_{\text{ph}} = \frac{\hbar^2}{2I}[(\ell + 1)(\ell + 2) - \ell(\ell + 1)]$$

$$= \frac{\hbar^2}{2I}[\ell^2 + 3\ell + 2 - \ell^2 - \ell] = \frac{\hbar^2}{I}(\ell + 1) \qquad (10.8)$$

Now we see an emission-spectrum spacing that varies with ℓ. Specifically, the higher the starting energy level, the greater the photon energy for a transition with $\Delta\ell = -1$. This is consistent with the energy-level spacings shown in Figure 10.3. Because the photon energy increases linearly with the quantum number ℓ, photon energies increase at regular intervals.

As our estimates have indicated, vibrational energies are typically greater than rotational energies by as much as an order of magnitude. This significant energy difference, along with the different spacing characteristic noted earlier, results in the **band spectrum** shown in Figure 10.6. We see an evenly spaced vibrational spectrum with a more closely spaced rotational spectrum superimposed on each vibrational line. **Band spectrum**

In any molecular spectrum the positions and intensities of the observed bands are governed by the rules of quantum mechanics. We note two features in particular. First, the relative intensities of the bands are due to different transition probabilities. The probabilities of transitions from an initial state (for example, $n = 5$) to each possible final state ($n = 4, 3, \dots$) are not necessarily the same. The spectral lines vary in brightness, with more intense lines corresponding to more probable transitions. Second, some transitions are forbidden by the

Wavelength →

Figure 10.6 Typical section of the emission spectrum of a diatomic molecule. Equally spaced groups of lines correspond to the equal spacings between vibrational levels. The structure within each group is due to transitions between rotational levels.

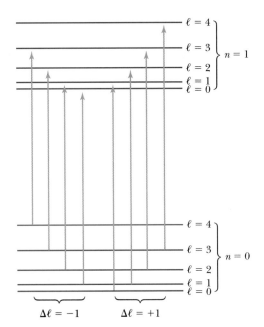

Figure 10.7 A schematic diagram of the absorptive transitions between adjacent vibrational states ($n = 0$ to $n = 1$) in a diatomic molecule.

quantum-mechanical selection rule that requires $\Delta\ell = \pm 1$. This must be so, because upon emission the photon carries away its intrinsic angular momentum of one quantum unit ($\hbar$).

An interesting application of the $\Delta\ell = \pm 1$ selection rule is found in the study **Absorption spectra** of **absorption spectra**. When electromagnetic radiation is incident upon a collection of a particular kind of molecule (for example, in a closed gas cell), molecules can absorb photons and make transitions to a higher vibrational state only if the rotational state changes by $\Delta\ell = \pm 1$. A schematic of the allowed transitions between two vibrational states is shown in Figure 10.7. Because ΔE increases linearly with ℓ as in Equation (10.8), one expects to see absorption bands at regular intervals of energy (or frequency, which is proportional to energy). This is evident in the absorption spectrum of HCl (Figure 10.8). The regular spacing between the peaks can be used to compute the rotational inertia I (Problem 15a). The missing peak in the center corresponds to the forbidden $\Delta\ell = 0$ transition. Therefore that central frequency f is just

$$f = \frac{1}{2\pi}\sqrt{\frac{k}{\mu}}$$

the frequency of a vibrational transition from $n = 1$ to $n = 0$.

Figure 10.8 An absorption spectrum of a diatomic molecule, HCl. The spacing is regular, as predicted in Equation (10.7), and each line is split due to the small mass difference between the chlorine isotopes ^{35}Cl and ^{37}Cl. The missing central peak at $f = 8.65 \times 10^{13}$ Hz corresponds to the forbidden $\Delta\ell = 0$ transition.

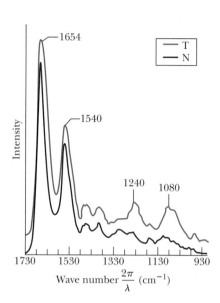

Figure 10.9 FTIR spectrum comparing tumor cells (T) and normal cells (N) from human lungs. Absorption in the infrared region is shown as a function of wave number $(2\pi/\lambda)$, so the wavelengths range from 36 μm to 68 μm. The FTIR method makes peak locations and intensities distinct enough so that this could be a useful diagnostic tool. *From E. Benedetti et al.,* Applied Spectroscopy *44, 1276–1280 (1990).*

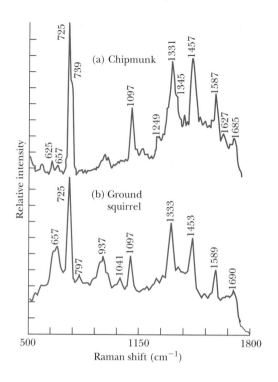

Figure 10.10 Raman spectra of eye pigments from (a) chipmunk and (b) ground squirrel. The "Raman shift" is the difference in wave number, $2\pi/\lambda$, corresponding to the frequency difference in Equation (10.10). These spectra are used to study metabolic and photochemical generation of lens pigments. *From S. Nie et al.,* Applied Spectroscopy *44, 571–575 (1990).*

Physicists and chemists have developed extremely sophisticated equipment and data reduction methods for the sole purpose of studying molecular spectra. One of the most popular methods is known as *Fourier transform infrared* (FTIR) *spectroscopy.* Most Fourier transform spectrometers use an interferometric system based on the Michelson interferometer (see Chapter 2) in order to make precise determinations of photon wavelengths. The Fourier transform is used to analyze the spectrum. In Fourier analysis, a spectrum can be decomposed into an infinite series of sine and cosine functions. With the proper knowledge of spectral characteristics, random and instrumental noise can be greatly reduced in order to produce a "clean" spectrum, that is, one with a high signal-to-noise ratio (see Figures 10.9 and 10.10).

It is not necessary that an incoming photon's energy precisely match the transition energy ΔE. If a photon of energy greater than ΔE is absorbed by a molecule, the excess energy may be released in the form of a scattered photon of lower energy. This process is known as **Raman scattering**. It is possible to examine the spectrum of Raman-scattered photons and learn some of the properties of the molecules being studied. In Raman scattering the angular momentum selection

Raman scattering

rule becomes $\Delta \ell = \pm 2$ because of the second photon involved.* Consider a transition from a state ℓ to a state $\ell + 2$. The rotational part of the transition energy is

$$\Delta E_{\mathrm{rot}} = \frac{\hbar^2}{2I}\left[(\ell + 2)(\ell + 3) - \ell(\ell + 1)\right]$$

$$= \frac{\hbar^2}{I}(2\ell + 3) \tag{10.9}$$

Suppose an incoming photon with energy hf is Raman-scattered and the scattered photon has energy hf'. Then the frequency of the scattered photon can be found in terms of the relevant rotational variables:

$$hf' = hf - \Delta E_{\mathrm{rot}} = hf - \frac{\hbar^2}{I}(2\ell + 3)$$

Thus

$$f' = f - \frac{\hbar}{2\pi I}(2\ell + 3) \tag{10.10}$$

Raman spectroscopy can be a useful tool in determining rotational properties (specifically ℓ and I) of molecules. It is also used to study polyatomic systems, in which the analysis of molecular properties is not so straightforward.

Raman spectroscopy is also used to study the vibrational properties of liquids and solids. In this case, the difference in energy between the incoming and Raman-shifted photon corresponds to a vibrational energy in the sample being illuminated. For molecules in solution, the set of observed energy differences provides a unique "fingerprint," allowing material identification. Solid samples also have characteristic vibrational spectra. Although vibrations in a crystalline material correspond to waves extending over many atomic sites, they still exhibit evidence of quantization. The quanta of vibration are called phonons, which we will discuss in more detail in Section 10.5.

Spontaneous emission

(a)

Stimulated emission

(b)

Figure 10.11 Schematic of (a) spontaneous emission, with $E_2 - E_1 = hf$ and (b) stimulated emission in a two-level system. In stimulated emission one photon triggers the emission of a second photon.

10.2 Stimulated Emission and Lasers

The emission of photons by molecules as described in Section 10.1 is known as **spontaneous emission**. A molecule in an excited state will decay to a lower energy state and emit a photon spontaneously, without any stimulus from the outside (Figure 10.11a). The laws of quantum mechanics do not allow us to say when the transition will occur. Because the process is probabilistic, the best we can do is calculate the mean lifetime of an excited state or the probability that a spontaneous transition will occur within a given amount of time. As a consequence, one can expect the phase of the emitted photon to be random. We note as an aside that the mean lifetime can often be estimated from the width of the emission spectrum line. If a spectral line has a width ΔE, then Heisenberg's uncertainty

---

*Of course $\Delta \ell = 0$ is also possible in a two-photon process. The $\Delta \ell = 0$ case is known as *Rayleigh scattering*. Quantum theory predicts that Rayleigh scattering becomes more likely (relative to Raman scattering) at shorter wavelengths. Rayleigh scattering of photons by the atmosphere causes the sky to be blue for the most part and orange-red near sunrise and sunset, because shorter wavelength photons (blue) are Rayleigh-scattered more than longer wavelength photons (red).

principle, Equation (5.45), gives an upper-bound estimate of the lifetime of $\Delta t = \hbar/(2\,\Delta E)$. For example, suppose you observe an atomic state with $E = 0.24$ eV and $\Delta E = 2.1 \times 10^{-6}$ eV. Then the upper-bound lifetime is $\Delta t = \hbar/(2\,\Delta E) = 1.6 \times 10^{-10}$ s, a rather short lifetime.

Conceptual Example 10.3

Why do we say that $\Delta t = \hbar/(2\,\Delta E)$ is an upper-bound estimate of the lifetime of an atomic state? Why isn't that the exact or approximate value of the lifetime?

Solution The width of the spectral line ΔE presumably comes from a series of measurements, each having some inherent uncertainty, plus some deviation due to a lack of experimental precision. Therefore we can expect that ΔE will usually be larger than the minimum allowed by the Heisenberg uncertainty principle. A larger value of ΔE leads to a smaller Δt, because of the inverse nature of the relationship.

It is possible, however, to make the emission process occur in a more controlled way. Electromagnetic radiation (a photon) incident upon a molecule in an excited state can cause the inherently unstable system to decay to a lower state. This is called **stimulated emission**. An important feature of stimulated emission is that the photon emitted in the process tends to have the same phase and direction as the stimulating radiation. And if the incoming photon has the same energy as the emitted photon, the result will be *two* photons of the same wavelength and phase traveling in the same direction, because the incoming photon is not absorbed, but rather triggers emission of the second photon. The two photons (of the same wavelength and phase) are then said to be *coherent*. A schematic diagram of the stimulated emission process is shown in Figure 10.11b.

Stimulated emission

Stimulated emission is the fundamental physical process in the operation of the laser. A simple argument by Einstein from his 1917 paper "On the Quantum Theory of Radiation" explains in a rather straightforward way why we can expect stimulated emission to occur. It is a testament to Einstein's genius that he did this work, as he did in many other areas of physics, on purely theoretical grounds, long before it was confirmed in the laboratory.

Here, in brief, is Einstein's analysis. We shall consider transitions between two molecular states with energies E_1 and E_2 (where $E_1 < E_2$). The photon associated with either emission or absorption has an energy E_{ph} and frequency f, where $E_{ph} = hf = E_2 - E_1$. If stimulated emission (that is, a process in which incoming radiation causes a transition from E_2 to E_1) occurs, the *rate* of those transitions must be proportional to the number of molecules in the higher state (call this N_2) and the energy density of the incoming radiation [call this $u(f)$]. Therefore let us say that the rate at which stimulated transitions are made from E_2 to E_1 is $B_{21}N_2u(f)$, where B_{21} is a proportionality constant that depends on the quantum-mechanical probability of stimulated emission. Similarly, the probability that a molecule at E_1 will *absorb* a photon can be expressed as $B_{12}N_1u(f)$. We must also take into account the possibility that spontaneous emission will occur. The rate of spontaneous emission is independent of $u(f)$, however, and can be expressed simply as AN_2, where A is a constant related to the quantum-mechanical probability of spontaneous emission.

Once the system has reached equilibrium with the incoming radiation, the total number of downward transitions must equal the total number of upward transitions. Then

$$B_{21}N_2u(f) + AN_2 = B_{12}N_1u(f)$$

or, rearranging,

$$[B_{21}u(f) + A]N_2 = B_{12}u(f)N_1$$

In thermal equilibrium each of the N_i are proportional to their respective Boltzmann factors $e^{-E_i/kT}$. Therefore

$$[B_{21}u(f) + A]e^{-E_2/kT} = B_{12}u(f)e^{-E_1/kT} \tag{10.11}$$

In the classical limit $T \to \infty$ (see Section 9.5). Then $e^{-E_1/kT} = e^{-E_2/kT}$ and the energy density $u(f)$ becomes very large, so the factor A becomes insignificant. Thus we see that in the classical limit

$$B_{12} \approx B_{21} \equiv B \tag{10.12}$$

That is, the probability of stimulated emission is approximately equal to the probability of absorption. What this basically means is that if the transition from E_1 to E_2 (absorption) can occur, then we should also expect that stimulated emission will occur.

Finally, we can use Equation (10.11) to obtain another useful relationship. Solving Equation (10.11) for $u(f)$ yields

$$u(f) = \frac{A}{B_{12}e^{(E_2-E_1)/kT} - B_{21}} = \frac{A}{B_{12}e^{hf/kT} - B_{21}}$$

or, if we use Equation (10.12),

$$u(f) = \frac{A/B}{e^{hf/kT} - 1} \tag{10.13}$$

Equation (10.13) should look familiar because it closely resembles the Planck radiation law, Equation (3.23). In fact, when the Planck law is expressed in terms of frequency instead of wavelength, it is

$$u(f, T) = \frac{8\pi f^2}{c^3} \frac{hf}{e^{hf/kT} - 1} \tag{10.14}$$

Therefore by Equations (10.13) and (10.14) it is required that

$$\frac{A}{B} = \frac{8\pi hf^3}{c^3} \tag{10.15}$$

In other words, the stimulated emission probability coefficient B is proportional to the spontaneous emission probability coefficient A in equilibrium. We may interpret this result to mean that in a process for which the probability of spontaneous emission is high, the probability of stimulated emission will also be high.

The laser   Stimulated emission is the fundamental physical process in a **laser**. *Laser* is an acronym for "light amplification by the stimulated emission of radiation." One also hears of **masers**, in which microwaves are used instead of visible light. The first working maser was made by Charles H. Townes in 1954, and the first laser by Theodore H. Maiman in 1960. The schematic drawing of a helium-neon laser (Figure 10.12), the type commonly found in most university physics departments today, will help you to understand the process. The main body of the laser is a

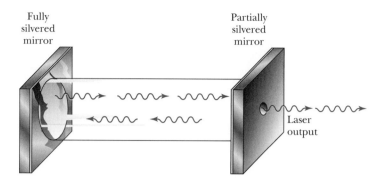

Fully
silvered
mirror

Partially
silvered
mirror

Laser
output

Figure 10.12 A schematic diagram of a He-Ne laser. The coherent photon beam escapes through the partially silvered mirror.

closed tube, filled with about a 9/1 ratio of helium and neon. Photons bouncing back and forth between the two mirrors are used to stimulate the transition in neon, which in turn produces more photons. Recall that photons produced by stimulated emission will be coherent, and thus the beam of photons that escapes through the partially silvered mirror will be the coherent beam we observe.

How are atoms put into the excited state in the first place? We cannot rely on the photons in the tube to do so; if we did, then there would be no net output of energy from the laser. Any photon produced by stimulated emission would have to be "used up" to excite another atom in order for the process to be continuous. A second potential difficulty is that there may be nothing to prevent spontaneous emission from atoms in the excited state. In that case the beam would not be coherent.

The way around both of those problems is to use a multilevel atomic system. (This will also work with molecular systems.) Consider first the **three-level system** shown in Figure 10.13. Atoms in the ground state (E_1) are *pumped* to a higher state (E_3) by some external source of energy. The atom then decays quickly to E_2. The key is that the transition from E_2 to E_1 must be forbidden, for example, by a $\Delta\ell = \pm 1$ selection rule. Then the state with energy E_2 is said to be *metastable*. Now a large number of atoms can exist for a relatively long time at E_2, where they are just waiting for a photon to come along and stimulate the transition to E_1. In normal operation many more atoms will be in the excited (metastable) state than in the ground state. This situation is, of course, contrary to what one finds under normal conditions. Therefore it is known as a *population inversion*. As we have described here, the population inversion is an essential feature of the operation of lasers.

There is a potential difficulty with the three-level system as we have described it. What happens to an atom after it has been returned to the ground state from E_2 by a stimulated emission? We would like the external power supply to return it immediately to E_3, but in practice it may take some time for this to happen, during which it is possible that a photon with energy $E_2 - E_1$ (just the energy that a photon created by stimulated emission would have) can be absorbed. That would be undesirable, for that photon would then be unavailable for stimulating another transition and for passing through the partially silvered mirror, as we expect all of them to do eventually. The result would be a much weaker beam, or perhaps none at all.

We can get over this last hurdle by using a **four-level system**, as shown in Figure 10.14. As before, atoms are pumped from the ground state to a higher state (now E_4), where they decay quickly to the metastable state E_3. The stimulated emission takes atoms from E_3 to E_2. The spontaneous transition from E_2 to E_1 is not forbidden, so the state E_2 will not exist long enough for a photon to be used up in kicking the system from E_2 to E_3. There is little chance that an atom in the

Charles Townes (1915–) was born in South Carolina and graduated from Furman University before earning his Ph.D. at Cal Tech. During World War II, he helped design radar bombing systems. After the war he joined the faculty of Columbia University and pursued research in microwave physics. Townes used his knowledge of microwaves to develop the concept of the maser, and he demonstrated the first maser in 1954. Along with Arthur Schawlow, he showed that masers could operate in infrared and optical wavelengths, leading to the development of the laser. *Ernest Orlando Lawrence Berkeley National Laboratory.*

Four-level system

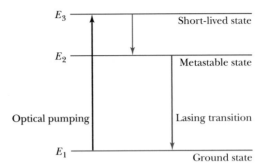

Figure 10.13 Transitions in a three-level laser. The lasing transition takes the system from the metastable state (E_2) to the ground state (E_1).

ground state will absorb a photon with energy $E_3 - E_2$, so the lasing process can proceed efficiently.

The red helium-neon laser uses transitions between energy levels in both helium and neon, as shown in Figure 10.15. The applied voltage excites a helium atom from its ground state to an excited level at approximately 20.61 eV. An excited helium atom will occasionally collide with a neon atom and transfer its excess energy to that atom, because 20.61 eV is very close to the energy needed to excite a neon atom from its ground state to the $2p^5 5s^1$ level (the metastable state). The lasing process then proceeds as shown in Figure 10.15, resulting in a coherent beam of red light at $\lambda = 632.8$ nm. There are other allowed transitions in neon, and some energy is invariably lost through those photons that do not participate in the lasing process. Green and orange helium-neon lasers, which are now available commercially, employ some of these transitions.

Tunable laser   In a **tunable laser** the wavelength of the emitted radiation can be adjusted over a range as wide as 200 nm. In the past, tunable lasers were made by using an organic dye as the lasing material. The organic compounds are chosen so that they have a great number of closely spaced energy levels. This is what makes stimulated emission over a range of wavelengths possible. Tuning is accomplished by changing the concentration of the dye and/or the length of the dye cell.* Semiconductor lasers (see Section 11.3) are now replacing dye lasers for many applications.

Free-electron laser   Another kind of tunable laser is the **free-electron laser**, shown schematically in Figure 10.16. This laser relies on the fact that charged particles, electrons in this case, emit electromagnetic radiation when accelerated. A series of magnets called

*For a more complete description see Eugene Hecht, *Optics*, 4th ed., San Francisco: Addison-Wesley (2002), pp. 597–601.

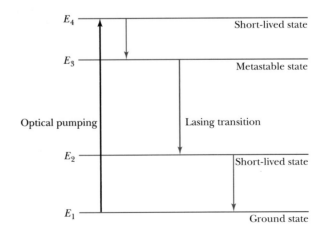

Figure 10.14 Transitions in a four-level laser. The lasing transition takes the system from the metastable state (E_3) to another short-lived state (E_2). The system then returns quickly to the ground state (E_1), so that photons cannot be reabsorbed in returning the system from E_2 to E_3.

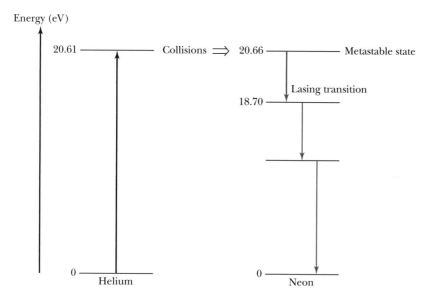

Figure 10.15 The energy levels and transitions in a helium-neon laser.

wigglers is used to accelerate a beam of electrons transversely as shown in the figure. The electron velocity is matched to the oscillations in the magnetic field in such a way that the emitted radiation is as coherent as in a gas-filled laser. Because the free electrons are not tied to atoms, they aren't dependent upon atomic energy levels and can be tuned to wavelengths well into the ultraviolet part of the spectrum. Several free-electron lasers in different countries now produce wavelengths of 100 nm and shorter. Electron-beam lasers have been made to produce outputs in the ultraviolet, visible, and infrared ranges. The exceptional fine-tuning that is possible with these devices, along with their relatively high efficiency, make electron-beam lasers and masers useful for a number of scientific applications, including spectroscopy, accelerator technology, radar, and fusion research.

Scientific Applications of Lasers

In recent years lasers have been used in a wide range of scientific applications. One obvious application is the use of the extremely coherent and nondivergent beam in making precise determinations of distances both large and small. Thanks in part to the improved accuracy of laser measurements, the speed of light in a vacuum has been *defined* to be $c = 299,792,458$ m/s. This definition of the speed of light has led to a redefinition of the meter. Formerly, the standard meter was a metal bar. Now the SI system defines the meter as the length of the path traveled by light in a vacuum during a time interval of $1/299,792,458$ of a second.

Figure 10.16 Schematic of the operation of a free-electron laser. The laser output is in the same direction (z in this drawing) as the injected electron beam. *From P. Sprangler and T. Coffey*, Physics Today, *Fig. 3, p. 48 (March 1984).*

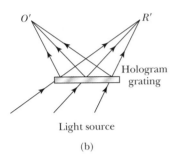

Figure 10.17 How a hologram is produced. (a) The production of interference fringes on a photographic plate, using interference of reference (*R*) and object (*O*) beams. (b) Illumination of the plate from the opposite side produces a real image of *O* at *O'*.

On the longer end of the distance scale, reflecting mirrors placed on the moon by Apollo astronauts have enabled scientists to reflect laser light from the moon's surface and thereby determine the Earth-to-moon distance to within 10 cm. Variations in this distance over the course of a month and a year have enabled us to better understand the orbital mechanics of the solar system. Lasers have been set up to measure land shifts near geologic fault lines in California, with the hope of better understanding and perhaps predicting earthquakes.

Pulsed lasers are now used in thin-film deposition. A short pulse of laser light incident upon the source material can create a thermal shock wave, expelling surface material from the target and onto a substrate. In this way many complex materials can be transferred to the substrate without chemical modification, yet with control on a nanometer scale. Such thin-layered materials are used in basic research to study the electronic properties of different materials and in the manufacture of integrated circuits, where smaller-sized circuit elements are desirable.

A potentially important use of lasers is in fusion research. The greatest problem in achieving controlled fusion in the laboratory is the difficulty in containing enough nuclei within a confined space at high temperature for a sufficient length of time in order for nuclei to fuse and thereby produce energy. In one scheme, known as *inertial confinement*, a pellet of deuterium and tritium would be induced into fusion by an intense burst of laser light coming simultaneously from many directions. Fusion will be discussed in Chapter 13.

Holography

The field of **holography** is one in which lasers have already been put to good use. The basic mechanism for constructing and viewing a hologram is shown in Figure 10.17. Consider laser light emitted by a **reference source** *R*. Through a simple combination of mirrors and lenses, this light can be made to strike both a piece of photographic plate and an object *O*. Because the laser light is coherent, the image on the film will essentially be an interference pattern, with the interfering beams coming from the reference source and object. After exposure this interference pattern is a hologram, and when the hologram is illuminated from the other side (as in Figure 10.17b), a real image of *O* is formed. The fact that not only intensity but also phase information is contained in the hologram means that a full three-dimensional image is formed. The other fascinating feature of holograms results from the fact that, if the lenses and mirrors are properly situated, light from virtually every part of the object will strike every part of the film, along with some of the reference beam. This means that each portion of the film contains enough information to reproduce practically the whole object! It is indeed possible to reconstruct a view of an object from a tiny piece of a hologram, although one generally loses some of the three-dimensional effect in that case. Figure 10.18 shows a close-up of a holographic plate and an image produced from it.

Transmission, reflection, and white light holograms

We have just described a **transmission hologram**, that is, one in which the reference beam is on the same side of the film as the object and the illuminating beam is on the opposite side. It is possible to make a **reflection hologram** by reversing the positions of the reference and illuminating beams, and if an overlaying series of holograms is made using several lasers of different colors, the result will be a **white light hologram**, in which the different colors contained in white light provide the colors seen in the image. This is what one commonly sees in magazine pictures and credit cards, because the casual observer does not have a laser handy for viewing a monochromatic hologram. With their complex interference patterns,

(a)   (b)

Figure 10.18 (a) A close-up view of a hologram, showing the interference pattern in the photographic plate. (b) The image produced by the hologram in (a). *From H. M. Smith*, Principles of Holography, *2nd ed., New York: Wiley (1975).*

holograms are difficult to counterfeit, and are therefore popular security tools on instruments such as credit cards and checks.

An interesting scientific application of holography is **interferometry**. Two holograms of the same object produced at different times can be used to detect small motion or growth that could not otherwise be seen. In Figure 10.19 we see a holographic image of a mushroom made seconds earlier superimposed on the mushroom itself. The interference fringes indicate growth patterns. Holographic interferometry is now widely used in industry to scan for imperfections, for example, in machined parts and computer chips.*

Holographic interferometry

Quantum Entanglement, Teleportation, and Information

Erwin Schrödinger employed the term "quantum entanglement" in two papers published in 1935 and 1936 to describe a strange correlation between two quantum systems. He was analyzing the Einstein-Podolsky-Rosen paradox that Einstein and others used in an attempt to refute the Copenhagen interpretation of quantum theory (see Chapter 5). Schrödinger considered the effects of entanglement

*See *The Industrial Physicist*, September 1997, pp. 37–39. (http://www.aip.org/tip/INPHFA/vol-3/iss-3/p37.pdf)

Figure 10.19 Interferometric observation of the growth of a live mushroom through a hologram. *Photo by T. Jeong.*

for quantum states acting across large distances, which Einstein referred to as "spooky action at a distance." John Bell showed in 1964 that the effects of quantum entanglement should be able to be observed experimentally (Bell's *inequality*). Subsequent experiments* showed that quantum entanglement is a real physical phenomenon, and entangled systems can interact across large distances.

No information can be transmitted through only quantum entanglement, but it is possible to transmit information using entangled systems in conjunction with classical information. This is known as **quantum teleportation**. Imagine a laser passing through a nonlinear crystal that produces a pair of entangled photons. Alice and Bob, who are spatially separated, try to transfer information about the photons. Alice doesn't know the properties of the photon entering her system, but she wants to send information about it to Bob. Measuring any physical property of the photon would disturb the quantum information because of the Heisenberg uncertainty principle. There are a number of ways she can transfer information. For example, a beam splitter can be used to produce two additional photons that can be used to trigger detectors. Alice can perform a manipulation on her quantum system and send that information over a classical information channel to Bob. He then arranges his part of the shared quantum system to detect information, for example the polarization status, about the unknown quantum state at his detector.

Several groups around the world, including those at the University of Innsbruck, University of Vienna, and NIST at Boulder, Colorado, have performed quantum teleportation experiments, sometimes over distances of more than a kilometer. Teleportation experiments have now been performed with both photons and atoms. Such experiments are needed in order to utilize quantum information to eventually construct quantum computers, but teleportation as in *Star Trek* is not likely to happen in our lifetime—if ever.

Other Laser Applications

Lasers have been important in medicine for years. They are used in surgery to make precise incisions in different kinds of body tissue. In addition to providing better precision than conventional means they often reduce bleeding, because the laser tends to help the blood coagulate while cutting. Lasers are used in a number of eye operations, particularly in retinal reattachment and in the treatment of glaucoma (excessive fluid pressure) by burning holes that allow small amounts of fluid to leak out and thereby reduce pressure. There have already been thousands of cases in which laser surgery has prevented blindness. Medical applications were not the goal of laser pioneers, who wanted to better understand atomic and optical physics. Arthur Schawlow, who worked with Townes in the 1950s to lay the foundation for masers and laser, wrote: "When we were working on the laser concepts, I had no idea there was such a thing as a detached retina. If we had been trying to prevent blindness, I do not think we would have concerned ourselves with amplification of stimulated emission by atoms. Research cannot always go directly toward the goals; you sometimes have to explore and hope something will come of it."

Lasers are used extensively for elective eye surgery to correct myopia (nearsightedness), a condition that arises when the eye focuses the image in front of

---

*For additional information, see "Quantum Teleportation" by Anton Zellinger, *Scientific American*, April 2000 and update in *The Edge of Physics*; "Rules for a Complex Quantum World" by Michael A. Nielsen, *Scientific American*, November 2002. Students may also use a search engine to find current information about quantum entanglement, quantum teleportation, quantum information, and quantum computers.

the retina. For decades this correction has been done using photorefractive keratectomy, in which the front surface of the cornea is flattened to shorten the eyeball and lengthen the eye's focal length. Now a more popular technique is laser-assisted in situ keratomileusis (LASIK), in which part of the cornea is cut into a flap and lifted, so that a laser can cut and reshape the underlying corneal tissue.

An application we see in everyday life is the scanning devices used by super-markets and other retailers. An optical scanner can analyze the reflection of a laser beam from the bar code of a packaged product. When properly analyzed, the bar code is translated by a microprocessor into the appropriate product and price information. A substantial amount of engineering has given laser scanners a high reliability nearly independent of the angular orientation of the code and the speed with which it is swept over the beam.

Another common application of lasers is in compact disk (CD) and digital video disk (DVD) players. Although the technology of these two devices is slightly different, the basic concept is the same. Laser light is directed toward disk tracks that contain encoded information. The reflected light is then sampled and turned into electronic signals that produce a digital output. Most commonly this takes the form of an audio and/or video signal, but a CD or DVD can store virtually any digitized file. Most CD players use a laser in the infrared part of the spectrum with a wavelength of around 800 nm. DVD players often use shorter wavelengths, and this helps them to store more information on a similar-sized disk.

Other applications of the laser are so numerous we cannot discuss them here. They include fiber-optic communications and laser printers.

10.3 Structural Properties of Solids

The remainder of this chapter and all of Chapter 11 fall under the general heading of **condensed matter physics**. Under this broad heading fall many kinds of research.* Probably the single most important subfield of condensed matter physics is the study of the electronic properties of solids. The fundamentals of electrical conductivity were covered in Chapter 9. In the last parts of this chapter and all of Chapter 11, you will learn about the remarkable properties of super-conductors and semiconductors. Other important work is devoted to under-standing the fundamental properties of the structure of solids and their bulk thermal and magnetic properties. It is these fundamental properties we shall study now, along with some of the many applications in science and engineering.

Condensed matter physics

The structures of different solids are quite varied. Many solids exhibit a **crys-tal structure**, in which the atoms are arranged in extremely regular, periodic patterns. Max von Laue proved the existence of crystal structures in solids in 1912, using the still popular method of x-ray diffraction (see Section 5.1). In an ideal crystal the same basic structural unit is repeated indefinitely throughout space. The set of points in space occupied by atomic centers is called a *lattice*. Figure 10.20 shows the principal lattice types in three dimensions. A perfect crystal is rare. Most solids are in a *polycrystalline* form, meaning that they are made up of many smaller crystals, whose size may be anything from a few atoms on a side to thousands. A solid lacking any significant lattice structure is called *amorphous*

Crystal structure

*Strictly speaking, condensed matter physics includes studies of solids and liquids. Because much more research is done on solids than on liquids, some people still refer to condensed matter physics as "solid state physics." To get an idea of just how significant the role of condensed matter physics is today, consider the fact that in a typical month the number of articles published in *Physical Review B*, the section of *Physical Review* devoted to condensed matter, is nearly equal to the number of articles published in *Physical Review A, C, D*, and *E* (devoted to the other major fields) *combined*.

Figure 10.20 Some of the crystal lattices found in solids.

(literally "without form"). Common glass is amorphous, and therefore amorphous materials are referred to colloquially as "glasses."

Why do the atoms of a solid arrange themselves in a particular crystal lattice? A qualitative answer to this question is that as the material is cooled and allowed to change from the liquid to the solid state, the atoms can each find a place relative to their neighbors that creates the minimum energy configuration. It is rather like an electron being captured by an atom "finding" its way to the ground state. This is why some solids that normally tend to form good crystals will become polycrystalline or amorphous if cooled too quickly from the liquid to the solid state. To give a more quantitative answer to the question of why solids form as they do, let us use as an example the structure of the sodium chloride crystal. The basic cubic structure of sodium chloride is shown in Figure 10.21. Sodium and chlorine easily ionize to form Na^+ and Cl^-, and we may think of the solid as a collection of (spherically symmetric) Na^+ and Cl^- ions alternating indefinitely in each of the three orthogonal directions. The spatial symmetry results because there is no preferred direction for bonding the spheres together. The fact that different atoms have different symmetries (consider the shapes of p and d orbitals illustrated in Section 7.6) suggests why crystal lattices take so many different forms, as illustrated in Figure 10.20.

In order for this to be a stable configuration, each ion must experience a net attractive potential energy, which we shall model as

$$V_{att} = -\frac{\alpha e^2}{4\pi\epsilon_0 r} \tag{10.16}$$

Figure 10.21 The NaCl crystal structure.

Madelung constant

where r is the nearest-neighbor distance (0.282 nm in NaCl). This looks like a normal Coulomb potential energy, except for the introduction of the constant α, known as the **Madelung constant**. V_{att} is the net potential energy of an ion in a lattice due to all the other ions in the lattice. The Madelung constant therefore depends on the type of crystal lattice. In the NaCl crystal, each ion has 6 nearest neighbors of the opposite kind of charge that supply an attractive potential. There-fore, the nearest-neighbor contribution to the Madelung constant is exactly 6. The next nearest neighbors are of *like* charge to the ion we are considering. There are 12 such like charges, each located a distance $\sqrt{2}\,r$ away. Those ions contribute $-12/\sqrt{2}$ to the Madelung constant. Next there are 8 ions of the opposite charge a distance $\sqrt{3}\,r$ away, which contribute $8/\sqrt{3}$ to the Madelung constant. Contin-uing this process, one has an infinite series:

$$\alpha = 6 - 12/\sqrt{2} + 8/\sqrt{3} - \cdots \approx 1.7476 \qquad (10.17)$$

In addition to the attractive potential of Equation (10. 16) there is a *repulsive* po-tential due to the Pauli exclusion principle and the overlap of electron shells. A good theoretical model of the repulsive potential is

$$V_{\text{rep}} = \lambda e^{-r/\rho} \qquad (10.18)$$

where λ and ρ are also constants of the particular lattice and compound. The ex-ponential factor in Equation (10.18) is a common feature of "screened" poten-tials. Because the value of $e^{-r/\rho}$ diminishes rapidly for $r > \rho$, the parameter ρ is roughly regarded as the *range* of the repulsive force. We shall see that the nu-merical value of ρ can be calculated from experimental data.

The net potential energy is the sum of attractive and repulsive potentials:

$$V = V_{\text{att}} + V_{\text{rep}} = -\frac{\alpha e^2}{4\pi\epsilon_0 r} + \lambda e^{-r/\rho} \qquad (10.19)$$

At the equilibrium position $(r = r_0)$, $F = -dV/dr = 0$. Thus

$$0 = \frac{\alpha e^2}{4\pi\epsilon_0 r_0^2} - \left(\frac{\lambda}{\rho}\right) e^{-r_0/\rho} \qquad (10.20a)$$

Therefore

$$e^{-r_0/\rho} = \frac{\rho \alpha e^2}{4\pi\epsilon_0 \lambda r_0^2} \qquad (10.20b)$$

and

$$V(r = r_0) = -\frac{\alpha e^2}{4\pi\epsilon_0 r_0}\left[1 - (\rho/r_0)\right] \qquad (10.21)$$

Typically the repulsive potential is very short range, so that the ratio ρ/r_0 is much less than 1. For example $\rho/r_0 = 0.11$ for NaCl, as shown in Example 10.4. This is important, because Equation (10.21) indicates that ρ/r_0 must be less than 1 in or-der for the net potential energy at $r = r_0$ to be negative, as is required.

Example 10.4

The **dissociation energy**, that is, the energy needed to break a NaCl crystal into individual sodium and chlorine atoms, is determined experimentally to be 764.4 kJ/mol at standard temperature and pressure (STP) (see Table 10.2). Use this value to calculate the range parameter ρ for NaCl.

Strategy First divide the experimental dissociation energy by Avogadro's number (the number of ion pairs per mole) to obtain a value of 1.269×10^{-18} J/ion pair. Then with the equilibrium position at $r = r_0$, we have $V(r = r_0) = -1.269 \times 10^{-18}$ J. With this numerical value of $V(r = r_0)$, Equation (10.21) can be solved for ρ/r_0.

Solution Solving Equation (10.21) for ρ/r_0 in terms of $V(r = r_0)$,

$$\frac{\rho}{r_0} = 1 + \frac{4\pi\epsilon_0 r_0 V(r = r_0)}{\alpha e^2}$$

$$= 1 + \frac{(0.282 \text{ nm}) \left(-1.269 \times 10^{-18} \text{ J}\right)}{(9 \times 10^9 \text{ N} \cdot \text{m}^2/\text{C}^2) (1.7476) (1.6 \times 10^{-19} \text{ C})^2}$$

$$= 0.112$$

Therefore $\rho = 0.112 r_0 = 0.0316$ nm, which is in agreement with the value listed in Table 10.2. Indeed, this shows that the repulsive potential is very short range.

Table 10.2	Properties of Salt Crystals with the NaCl Structure		
Salt Crystal	**Nearest-Neighbor Separation (nm)**	**Repulsive Range Parameter ρ (nm)**	**Dissociation Energy (kJ/mol)**
LiF	0.214	0.029	1014
LiCl	0.257	0.033	832.6
LiBr	0.275	0.034	794.5
LiI	0.300	0.037	743.8
NaF	0.232	0.029	897.5
NaCl	0.282	0.032	764.4
NaBr	0.299	0.033	726.7
NaI	0.324	0.035	683.2
KF	0.267	0.030	794.5
KCl	0.315	0.033	694.0
KBr	0.330	0.034	663.5
KI	0.353	0.035	627.5
RbF	0.282	0.030	759.3
RbCl	0.329	0.032	666.8
RbBr	0.345	0.034	638.8
RbI	0.367	0.035	606.6

From C. Kittel, *Introduction to Solid State Physics*, 5th ed., New York: Wiley (1976), p. 92.

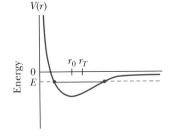

Figure 10.22 The asymmetry of the potential energy curve leads to thermal expansion. At $T = 0$ the mean spacing is r_0, but as T increases so does the mean spacing.

10.4   Thermal and Magnetic Properties of Solids

Thermal Expansion

One of the more ubiquitous properties of solids is **thermal expansion**, the tendency of a solid to expand as its temperature increases. A qualitative understanding can be obtained by studying Figure 10.22, the potential energy curve for an ion bound in a solid. We saw earlier in this chapter that this potential energy

curve is characteristic of the combined attractive and repulsive forces experienced by this ion (see Figure 10.1).

At $T = 0$, the ion is nearly "frozen solid" at $r = r_0$ because it has just the minimum energy possible (the zero-point energy, as described in Section 6.6). As the
temperature of the solid is increased from zero, the average energy of each ion
increases. The result is an oscillation between points r_1 ($<r_0$) and r_2 ($>r_0$). To a
first approximation the motion is simple harmonic, but it is not *exactly* so. The potential energy for a simple harmonic oscillator is of the form $V = \frac{1}{2}\kappa(r - r_0)^2$, a
function that is symmetric about the point $r = r_0$. The potential energy shown in
Figure 10.22 is not symmetric. The effect of this is that the mean lattice spacing
$r_T \approx (r_2 + r_1)/2$ is slightly greater than r_0 when $T > 0$ and increases gradually with
increasing temperature. The bulk effect of an increase in the mean lattice spacing is an overall expansion of the solid (Figure 10.23).

Now we shall develop a quantitative model of thermal expansion. Let $x =
r - r_0$, so that we will consider small oscillations of an ion about the equilibrium
position $x = 0$. A good model of the potential energy close to $x = 0$ is

$$V = ax^2 - bx^3 \tag{10.22}$$

where the x^3 term is responsible for the anharmonicity (that is, the deviation from
the standard harmonic oscillator) of the oscillation. The mean displacement $\langle x \rangle$
can be calculated using the Maxwell-Boltzmann distribution function $e^{-\beta V}$:

$$\langle x \rangle = \frac{\int_{-\infty}^{\infty} x e^{-\beta V}\,dx}{\int_{-\infty}^{\infty} e^{-\beta V}\,dx} \tag{10.23}$$

where we have used the usual notation $\beta = (kT)^{-1}$. In the numerator we can use
a Taylor expansion for the x^3 term because b is small:

$$x e^{-\beta V} = x e^{-\beta ax^2} e^{+\beta bx^3} = x e^{-\beta ax^2}(1 + \beta bx^3 + \cdots) \approx e^{-\beta ax^2}(x + \beta bx^4)$$

This allows us to easily evaluate the integral in the numerator of Equation
(10.23), because only the even (x^4) term survives integration from $-\infty$ to ∞ (see
Appendix 6):

$$\int_{-\infty}^{\infty} e^{-\beta ax^2}(x + \beta bx^4)\,dx = \int_{-\infty}^{\infty} e^{-\beta ax^2}\beta bx^4\,dx = \frac{3\sqrt{\pi}}{4} ba^{-5/2}\beta^{-3/2} \tag{10.24}$$

Because we are interested only in the first-order dependence on T, it is acceptable to approximate the denominator of Equation (10.23) by

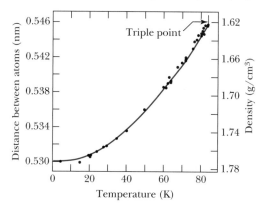

Figure **10.23** Thermal expansion of solid argon. The expansion is fairly linear except at very
low temperatures. *From O. G. Peter-
son, D. N. Batchelder, and R. O. Sim-
mons,* Physical Review **150**, 703
(1966).

$$\int_{-\infty}^{\infty} e^{-\beta V} \, dx \approx \int_{-\infty}^{\infty} e^{-\beta a x^2} \, dx = \left(\frac{\pi}{a\beta} \right)^{1/2} \tag{10.25}$$

Combining Equations (10.24) and (10.25), we obtain

$$\langle x \rangle = \frac{3b}{4a^2\beta} = \frac{3bkT}{4a^2} \tag{10.26}$$

Therefore thermal expansion is nearly linear with temperature in the classical limit. At very low temperatures the expansion is nonlinear, because the term calculated in Equation (10.26) vanishes as $T \to 0$. This is in agreement with experiment, as is seen in Figure 10.23.

Thermal Conductivity

Another important property of solids is their **thermal conductivity**, a measure of how well they transmit thermal energy. Materials that are good electrical conductors also tend to be good thermal conductors. Why should that be? Evidently the conduction electrons in a solid, because they are free to move around, are primarily responsible for the conduction of heat. We shall see, in fact, that the quantum theory of electrical conductivity developed in Chapter 9 is necessary in order to develop a good model for heat conduction.

Figure 10.24 A uniform rod of cross-sectional area A, used to illustrate thermal conductivity.

The standard way to define thermal conductivity is in terms of the flow of heat along a solid rod of uniform cross-sectional area A (Figure 10.24). It is found experimentally that the flow of heat per unit time along the rod is proportional to A and to the temperature gradient dT/dx. We define the thermal conductivity K to be the proportionality constant, so that

$$\frac{dQ}{dt} = -KA\frac{dT}{dx} \tag{10.27}$$

The negative sign in Equation (10.27) is due to the fact that heat flows in a direction opposite to the thermal gradient (that is, from hotter to colder).

In classical theory the thermal conductivity of an ideal free electron gas* is

$$K = \frac{n\bar{v}\ell c_V}{3N_A} \tag{10.28}$$

where n is the volume density of free electrons, $\bar{v}$ is the mean (thermal) speed, ℓ is the mean free path (see Chapter 9), and c_V is the molar heat capacity. Classically $c_V = \frac{3}{2}R = \frac{3}{2}N_A k$, so

$$K = \frac{1}{2}n\bar{v}\ell k \tag{10.29}$$

Because of their apparently close relationship, it is interesting to compare the thermal and electrical conductivities:

$$\frac{K}{\sigma} = \frac{\frac{1}{2}n\bar{v}\ell k}{n e^2 \ell / m\bar{v}} = \frac{m\bar{v}^2 k}{2e^2} \tag{10.30}$$

From classical thermodynamics the mean speed is (see Equation 9.17)

$$\bar{v} = \sqrt{\frac{8kT}{\pi m}}$$

*See, for example, F. Reif, *Statistical Physics*, New York: McGraw-Hill (1967), pp. 331–333.

Therefore

$$\frac{K}{\sigma} = \frac{4k^2 T}{\pi e^2} \tag{10.31}$$

Amazingly, then, the ratio K/σ is proportional to T. It is convenient to calculate the *constant* ratio

$$L \equiv \frac{K}{\sigma T} = \frac{4k^2}{\pi e^2} \tag{10.32}$$

Wiedemann-Franz law

Equation (10.32) is called the **Wiedemann-Franz law**, and the constant L is the **Lorenz number**. The numerical value of L is about $1.0 \times 10^{-8}\,\text{W} \cdot \Omega \cdot \text{K}^{-2}$. Experiments show that $K/\sigma T$ is indeed constant, but it has a numerical value about 2.5 times higher than predicted by Equation (10.32). Some experimental values are listed in Table 10.3.

Lorenz number

What is wrong with our analysis? The only problem is that we have used classical expressions for $\bar{v}$ and c_V. We should replace $\bar{v}$ with the Fermi speed u_F (because only electrons near the Fermi energy will be able to contribute to the conductivity) and replace $c_V = \frac{3}{2}R$ with the quantum-mechanical result [see Equation (9.47)]

$$c_V = \frac{\pi^2 R k T}{2 E_F} \tag{10.33}$$

so that Equation (10.28) can be rewritten

$$K = \frac{1}{3} \frac{n u_F \ell}{N_A} \frac{\pi^2 R k T}{2 E_F} \tag{10.34}$$

Note that $R = N_A k$ and $E_F = \frac{1}{2} m u_F^2$. Thus

$$K = \frac{n \ell \pi^2 k^2 T}{3 m u_F} \tag{10.35}$$

Table 10.3	Lorenz Number $L = K/\sigma T$ in Units of $10^{-8}\,\text{W} \cdot \Omega \cdot \text{K}^{-2}$ at Temperatures 273 K and 373 K	
Metal	$T = 273$ K	$T = 373$ K
Ag	2.31	2.37
Au	2.35	2.40
Cd	2.42	2.43
Cu	2.23	2.33
Ir	2.49	2.49
Mo	2.61	2.79
Pb	2.47	2.56
Pt	2.51	2.60
Sn	2.52	2.49
W	3.04	3.20
Zn	2.31	2.33

From C. Kittel, *Introduction to Solid State Physics*, 5th ed., New York: Wiley (1976), p. 178.

The proof of the validity of Equation (10.35) will be the correct quantum version of the Wiedemann-Franz law, that is, one in agreement with experimental results. As before, $L = K/\sigma T$, but now we use $\sigma = ne^2\ell/mu_\text{F}$:

$$L = \frac{K}{\sigma T} = \frac{n\ell\pi^2 k^2}{3mu_\text{F}} \frac{mu_\text{F}}{ne^2\ell}$$

Quantum Lorenz number

$$L = \frac{\pi^2 k^2}{3e^2} \tag{10.36}$$

The quantum-mechanically correct Lorenz number contains the same physical constants, k^2 and e^2, as the classical one, but strangely it is higher by a factor of $\pi^3/12$. This is just enough to bring the Lorenz number to a numerical value of $2.45 \times 10^{-8} \ \text{W} \cdot \Omega \cdot \text{K}^{-2}$, which is in agreement with experiment.

Magnetic Properties

The study of magnetic properties of solids constitutes an important subfield of solid state physics. Solids are characterized by their intrinsic magnetic moments (or lack thereof) and their responses to applied magnetic fields. Materials with a net magnetic moment even in zero applied magnetic field are called **ferromagnets**. Ferromagnets are sometimes referred to as "permanent magnets," although as we shall see later this is somewhat of a misnomer. In a **paramagnet** there is a net magnetic moment only in the presence of an applied field. The magnetic dipoles in a paramagnet tend to align themselves with the applied field. In a **diamagnet**, on the other hand, there is a (usually weak) tendency to have an induced magnetic moment opposite to the applied field. We shall consider each of these three principal kinds of materials separately.

Ferromagnets, paramagnets, and diamagnets

Magnetization

A useful quantity in studying magnetic materials is the **magnetization** M, which we define as the net magnetic moment per unit volume. Then the **magnetic susceptibility** χ is defined by

Magnetic susceptibility

$$\chi \equiv \frac{\mu_0 M}{B} \tag{10.37}$$

In other words, we may think of magnetic susceptibility as the induced magnetic moment per applied field, with a proportionality constant equal to the permeability constant μ_0. The magnetic susceptibility is positive for paramagnets and negative for diamagnets. Notice that χ is a dimensionless quantity (see Problem 43).

Today magnetic materials are found in many applications outside the research laboratory. The small size and extreme stability of magnetic domains (small regions of magnetization) in many materials makes them ideal for any device that requires data storage and retrieval. Computers, electronic instruments, and audio and videotapes all take advantage of magnetic materials in this way.

Diamagnetism

The behavior of a diamagnet may seem contrary to common sense, because in a diamagnet the magnetization opposes the applied field. However, one may think of the material as a whole responding to the applied field according to Faraday's law. A semiclassical model suggested by Feynman* helps explain the situation.

*Richard P. Feynman, Robert B. Leighton, and Matthew Sands, *The Feynman Lectures on Physics*, Vol. 2, Reading, MA: Addison-Wesley (1964), pp. 34-5 to 34-6.

Consider an electron orbiting counterclockwise in a circular orbit of radius r, as shown in Figure 10.25. Now suppose a magnetic field is applied gradually, directed out of the page. As the magnetic field increases from zero, there is an increasing magnetic flux upward through the atom's (circular) orbital path. Faraday's law states that the changing magnetic flux results in an induced electric field that is tangent to the electron's orbit, directed clockwise. Faraday's law gives the magnitude of the induced electric field:

$$E(2\pi r) = \left| \frac{d\Phi_B}{dt} \right| = \frac{d}{dt}(\pi r^2 B)$$

Therefore the induced electric field strength is

$$E = \frac{r}{2}\frac{dB}{dt}$$

This electric field produces a torque $\vec{\tau} = \vec{r} \times \vec{F} = \vec{r} \times (-e\vec{E})$, which has a magnitude reE ($\vec{r}$ and $\vec{E}$ are perpendicular) and a direction out of the page. Setting torque equal to the rate of change in angular momentum,

$$\tau = \frac{dL}{dt} = reE = \frac{er^2}{2}\frac{dB}{dt}$$

directed out of the page. Therefore, for a magnetic field going from 0 to B, directed out of the page, the angular momentum changes by an amount

$$\Delta L = \frac{er^2 B}{2}$$

with a direction $\Delta\vec{L}$ out of the page. This results in a magnetic moment changed by $\Delta\vec{\mu} = -\dfrac{e}{2m}\Delta\vec{L}$, which has a magnitude

$$\Delta\mu = \frac{e^2 r^2 B}{4m} \tag{10.38}$$

and a direction $\Delta\vec{\mu}$ *into* the page. The change in magnetic moment is opposite to the applied field, which is characteristic of diamagnetism.

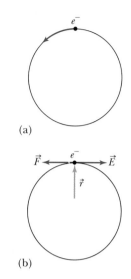

(a)

(b)

Figure 10.25 (a) Electron orbit used to illustrate diamagnetism. (b) The electron experiences a force in the applied magnetic field, resulting in a torque and a subsequent change in angular momentum.

Example 10.5

Estimate the size of the induced diamagnetic moment in a typical atom, assuming an orbital radius equal to the Bohr radius and an applied field of 2.0 T. Compare the result with the Bohr magneton $\mu_B = e\hbar/2m = 9.27 \times 10^{-24}$ J/T, a typical magnetic moment.

Strategy With the numerical values of the orbital radius and magnetic field known, the induced magnetic moment is given by Equation (10.38). Using SI units throughout, the units for change in magnetic moment will be in J/T.

Solution Using the numerical values provided along with the electron charge and mass,

$$\Delta\mu = \frac{e^2 r^2 B}{4m} = \frac{(1.60 \times 10^{-19}\ \text{C})^2\,(5.29 \times 10^{-11}\ \text{m})^2\,(2.0\ \text{T})}{4(9.11 \times 10^{-31}\ \text{kg})}$$

$$= 3.9 \times 10^{-29}\ \text{J/T}$$

The change in the magnetic moment is only about 4×10^{-6} as large as the Bohr magneton. This illustrates that diamagnetism is generally a weak effect.

Paramagnetism

In a paramagnet there exist unpaired magnetic moments that can be aligned by an external field. This is the case for rare earth elements and for many transition metals. The paramagnetic susceptibility χ is strongly temperature dependent. We can determine the temperature dependence by considering a collection of N unpaired magnetic moments per unit volume. At a given temperature there will be N^+ moments aligned parallel to the applied field and N^- moments aligned antiparallel to the applied field. The energy associated with a magnetic moment is $V = -\vec{\mu} \cdot \vec{B}$, so $V = -\mu B$ for a parallel alignment and $V = +\mu B$ for an antiparallel alignment. In the classical limit (that is, at ordinary temperatures) the distribution of magnetic moments will be governed by Maxwell-Boltzmann statistics, so that

$$N^+ = ANe^{\beta\mu B} \qquad \text{and} \qquad N^- = ANe^{-\beta\mu B} \qquad (10.39)$$

where A is a normalization constant, and as usual $\beta \equiv (kT)^{-1}$. Then the net magnetic moment (per unit volume) μ_{net} is

$$\mu_{net} = \mu(N^+ - N^-)$$
$$\mu_{net} = \mu AN(e^{\beta\mu B} - e^{-\beta\mu B}) \qquad (10.40)$$

Rather than directly calculating χ from μ_{net}, it is useful first to eliminate the constant A by considering $\bar{\mu}$, the mean magnetic moment per atom:

$$\bar{\mu} = \frac{\mu_{net}}{N} = \frac{\mu_{net}}{N^+ + N^-} = \frac{\mu AN(e^{\beta\mu B} - e^{-\beta\mu B})}{AN(e^{\beta\mu B} + e^{-\beta\mu B})}$$

$$\bar{\mu} = \mu\frac{e^{\beta\mu B} - e^{-\beta\mu B}}{e^{\beta\mu B} + e^{-\beta\mu B}} = \mu \tanh(\beta\mu B) \qquad (10.41)$$

Because we used the Maxwell-Boltzmann distribution, this expression should only be valid for $T \gg 0$. Note that this also means $\mu B \ll kT$, in which case $\tanh(\beta\mu B) = \tanh(\mu B/kT) \approx \mu B/kT$ and $\bar{\mu} = \mu^2 B/kT$. Therefore in the classical limit (note $M = N\bar{\mu}$)

Curie law

$$\chi = \frac{\mu_0 M}{B} = \frac{\mu_0 N\bar{\mu}}{B} = \frac{\mu_0 N\mu^2}{kT} \qquad (10.42)$$

Equation (10.42) is called the **Curie law** and is often simply stated as $\chi = C/T$, where $C = \mu_0 N\mu^2/k$ is a constant (the **Curie constant**) for a given paramagnetic material. Sample magnetization curves are shown in Figure 10.26, where the

Figure 10.26 An actual plot of magnetization for potassium chromate sulfate, a paramagnetic salt. Note the agreement with the Curie law for low magnetic fields. *From W. E. Henry, Physical Review* **88**, *559 (1952).*

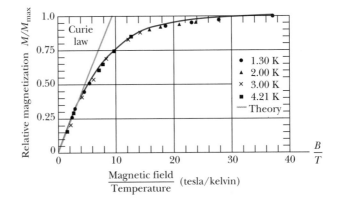

utility and the limitations of the Curie law can be seen. In Figure 10.26 it is clear that the M versus B curve is nearly linear over a wide range of magnetic fields. It is apparent that the Curie law breaks down at higher values of B, when the magnetization reaches a "saturation point" at which as many magnetic moments as possible have been aligned.

Example 10.6

Using the Bohr magneton $\mu_B = e\hbar/2m = 9.27 \times 10^{-24}\,\text{J/T}$ as a typical magnetic moment and $B = 0.50$ T, do the following:
(a) Find T at which $\mu B = 0.1kT$.
(b) For $\mu B = 0.1kT$ compare $\tanh(\beta\mu B)$ with $\beta\mu B$ and thereby check the suitability of this value of T as a "classical" temperature.
(c) Make the same comparison in the expression $\tanh(\beta\mu B) \approx \beta\mu B$ at $T = 100$ K.

Strategy With the numerical values of the magnetic moment and magnetic field given, the computations are all straightforward. For the comparisons called for in parts (b) and (c), the given values suggest using two significant figures.

Solution (a) We are given that $\mu B = 0.1kT$. We let $\mu = \mu_B$ and solve for T:

$$T = \frac{10\mu B}{k} = \frac{10(9.27 \times 10^{-24}\,\text{J/T})(0.5\,\text{T})}{1.38 \times 10^{-23}\,\text{J/K}} = 3.36\,\text{K}$$

(b) Now

$$\beta\mu B = \frac{\mu B}{kT} = 0.10$$

$$\tanh(\beta\mu B) = \tanh(0.10) = 0.10$$

to two significant digits. Therefore we conclude that the approximation is a good one for most purposes, even at this low temperature.
(c) Now

$$\beta\mu B = \frac{\mu\beta}{kT} = \frac{(9.27 \times 10^{-24}\,\text{J/T})(0.5\,\text{T})}{(1.38 \times 10^{-23}\,\text{J/K})(100\,\text{K})}$$

$$= 3.36 \times 10^{-3}$$

$$\tanh(\beta\mu B) = \tanh(3.36 \times 10^{-3}) = 3.36 \times 10^{-3}$$

so that $\beta\mu B$ and $\tanh(\beta\mu B)$ are the same to three significant figures. At these relatively high temperatures the approximation is an excellent one.

Ferromagnetism

Ferromagnetic materials are fairly rare. Of all the elements only five (Fe, Ni, Co, Gd, Dy) are ferromagnetic. A number of *compounds* are ferromagnetic, including some that do not contain any of these ferromagnetic elements. In the most powerful magnetic compounds (such as $Nd_2Fe_{14}B$) the magnetic field at the surface can exceed 1 T. In order to have a ferromagnet it is necessary to have not only unpaired spins, but also sufficient interaction between the magnetic moments so that, by their mutual interaction, a high degree of magnetic order is maintained. Opposing the maintenance of order is the continual randomizing effect of thermal motion, which contributes to the eventual "running down" of the magnetization of a ferromagnet. This phenomenon should not be surprising, in light of the second law of thermodynamics, which states that entropy (a statistical measure of disorder) evolves toward a maximum.

Sufficient thermal agitation at an elevated temperature can completely disrupt the magnetic order, to the extent that above a certain temperature (known as the **Curie temperature** T_C) a ferromagnet changes to a paramagnet. T_C for various ferromagnetic materials are listed in Table 10.4. At temperatures approaching T_C (from below), the magnetization of the ferromagnetic material begins to drop significantly as the thermal motion begins to disrupt the long-range magnetic order.

Curie temperature

Table 10.4 Selected Ferromagnets, with Curie Temperatures T_C	
Material	T_C (K)
Fe	1043
Co	1388
Ni	627
Gd	293
Dy	85
$CrBr_3$	37
Au_2MnAl	200
Cu_2MnAl	630
Cu_2MnIn	500
EuO	77
EuS	16.5
MnAs	318
MnBi	670
$GdCl_3$	2.2

From F. Keffer, *Handbuch der Physik* **18**, pt. 2, New York: Springer-Verlag (1966), and P. Heller, *Reports on Progress in Physics* **30**, pt. II, 731 (1967).

Ferromagnetic materials are used widely in scientific research and industry. Electrical generators use permanent magnets to take advantage of Faraday's law. Electric motors use permanent magnets in the reverse process, to turn electrical current into mechanical motion. Scientific uses include the wiggler magnets described in Section 10.2, low-field nuclear magnetic resonance, and Stern-Gerlach experiments (Section 7.4). Another application is in magnetically levitated transport systems (Section 10.6).

Antiferromagnetism and Ferrimagnetism

There are two more exotic kinds of magnetism. In **antiferromagnetic** materials, adjacent magnetic moments have opposing directions, as seen in Figure 10.27. The net effect is zero net magnetization below the ordering temperature (similar to T_C in ferromagnetic materials) called the *Neel temperature*, T_N. Above T_N, antiferromagnetic materials become paramagnetic. Negative ions seem to be most effective in providing the mechanism for antiparallel alignment, as in the antiferromagnetic materials MnO, FeO, MnS, MnF_2, and $NiCl_2$. In a **ferrimagnetic** substance a similar antiparallel alignment occurs, except that there are two different kinds of positive ions present. Thus the antiparallel moments do not precisely cancel, leaving a small net magnetization. The most common example of ferrimagnetic order is in magnetite, $FeO \cdot Fe_2O_3$.

10.5 Superconductivity

Superconductivity is perhaps the most remarkable phenomenon ever studied in solid state physics. Superconductivity is characterized by the absence of electrical resistance and the expulsion of magnetic flux from the superconductor.

For some time after its discovery over 90 years ago, superconductivity was a scientific curiosity with little potential for practical use. In recent years, particularly with the advent of high-temperature superconductors (that is, with temperatures exceeding 30 K), this has changed. Today superconductors are used in a wide variety of applications. Also, the prospect of even higher temperature superconductors makes it essential that physicists continue to study and learn more about their properties. We shall describe what is known today about superconductors in this section and consider the applications in Section 10.6.

Superconductivity is a special physical state characterized by two distinctive macroscopic features. The first of these is zero resistivity. Zero resistivity was first observed in solid mercury by the Dutch physicist Heike Kamerlingh Onnes in Leiden in 1911. He made this discovery shortly after liquefying helium, itself a significant accomplishment. (Remember that the boiling point of helium is 4.2 K at a pressure of 1 atm.) Kamerlingh Onnes could achieve temperatures approaching 1 K by reducing the pressure of the vapor surrounding his liquid helium. Today much lower temperatures are achieved by more sophisticated means. (See Special Topic, "Low-Temperature Methods.")

Figure 10.28 illustrates how the resistivity of a superconductor differs from that of a normal conductor. In a superconductor the resistivity drops abruptly to zero at what is called the *critical* (or *transition*) *temperature* T_c. It is important to realize that the resistivity is not merely very low; it really is zero. In the years 1956–1958 a group of British physicists led by S. C. Collins established a current in a superconducting ring and allowed it to flow with no external power source. It lasted until they were tired of watching it (about $2\frac{1}{2}$ years) with no detectable loss of current. Extrapolating from the uncertainty in their measuring instruments, they calculated that some current would remain after at least 100 million years (see Problem 58). The name "super" conductor is quite appropriate! By contrast, the current in a similar loop made of copper would be virtually gone in seconds, because although copper is considered an excellent conductor, it has some resistance even at extremely low temperatures.

Different superconductors have different transition temperatures, as seen in Table 10.5. Niobium is the transition-temperature champion of the pure elements, with $T_c = 9.25$ K. Notice that copper, silver, and gold, three of the four best conductors at room temperature, are not superconductors. This illustrates a rule of thumb that holds true with rare exceptions: The best conductors make the worst superconductors. Another interesting result is that superconducting behavior, like chemical behavior, tends to be similar within a given column of the periodic table. This should not be surprising, for it is the outermost electrons that are responsible for both chemical reactions and electrical conduction.

The second important macroscopic phenomenon associated with superconductivity is the **Meissner effect**, discovered by W. Meissner and R. Ochsenfeld in 1933. Succinctly stated, the Meissner effect is the complete expulsion of magnetic flux from within a superconductor. To do this it is necessary for the superconductor to generate currents, called *screening currents*. Just enough current is generated to expel the magnetic flux one tries to impose upon it. One can therefore view the superconductor as a perfect diamagnet, with a magnetic susceptibility $\chi = -1$. In Figure 10.29 we see a demonstration of the Meissner effect in which the induced currents within the superconductor create a magnetic field that opposes the field of a cubical magnet, thereby providing sufficient force to float it against gravity. There is an equilibrium position for the magnet, where the gravitational force on it is just balanced by the magnetic force at that distance from the superconductor.

The Meissner effect works only to a certain point, however. If a particular value of magnetic field (called the **critical field** B_c) is exceeded, magnetic flux does penetrate the material, and the superconductivity is lost until the magnetic field is reduced to below B_c. When the magnetic flux penetrates, the zero resistivity property is also lost, showing that zero resistivity and the Meissner effect do indeed go together in a superconductor.

Like the transition temperature, the critical field is also different for different superconductors. Superconductors with higher transition temperatures tend

Ferromagnetism

(a)

Antiferromagnetism

(b)

Ferrimagnetism

(c)

Figure 10.27 The alignment of spins in ferromagnets, antiferromagnets, and ferrimagnets.

Critical field

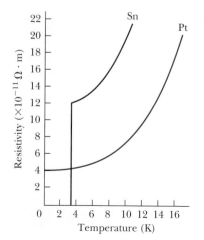

Figure 10.28 Resistivity of a normal conductor (platinum) and a superconductor (tin) at low temperature. The resistivity of tin drops dramatically to zero at its T_c, 3.7 K.

Table 10.5 Superconductivity Parameters of the Elements

Superconducting transition temperatures and critical fields

Upper number: Transition temperature in K

Lower number: Critical magnetic field at absolute zero in 10^{-4} tesla

Li	Be											B	C	N	O	F	Ne
Na	Mg											Al 1.18 / 105	Si	P	S	Cl	Ar
K	Ca	Sc	Ti 0.40 / 56	V 5.40 / 1408	Cr	Mn	Fe	Co	Ni	Cu	Zn 0.85 / 54	Ga 1.08 / 58	Ge	As	Se	Br	Kr
Rb	Sr	Y	Zr 0.61 / 47	Nb 9.25 / 2060	Mo 0.92 / 96	Tc 7.77 / 1410	Ru 0.49 / 69	Rh	Pd	Ag	Cd 0.52 / 28	In 3.41 / 282	Sn 3.72 / 505	Sb	Te	I	Xe
Cs	Ba	La 6.00 / 1046	Hf 0.13 / 13	Ta 4.47 / 829	W 0.02 / 1.15	Re 1.70 / 200	Os 0.66 / 70	Ir 0.11 / 16	Pt	Au	Hg 4.15 / 411	Tl 2.38 / 178	Pb 7.20 / 803	Bi	Po	At	Rn
Fr	Ra	Ac															

Ce	Pr	Nd	Pm	Sm	Eu	Gd	Tb	Dy	Ho	Er	Tm	Yb	Lu 0.1 / 350
Th 1.38 / 160	Pa 1.4	U	Np	Pu	Am	Cm	Bk	Cf	Es	Fm	Md	No	Lr

From B. W. Roberts, *Properties of Selected Superconductive Materials*, Supplement, NBS Technical Note 983, Washington, D.C.: U.S. Government Printing Office (1978).

Figure 10.29 Levitation of a cubical magnet over superconducting $YBa_2Cu_3O_7$ cooled to 77 K (well below its T_c of 93 K), illustrating the Meissner effect. Screening currents are generated in the superconductor, which provide a magnetic field to oppose the field of the magnet and thereby suspend it. © *Richard Megna, Fundamental Photographs, NYC.*

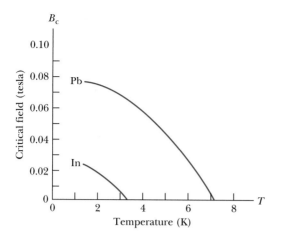

Figure 10.30 The temperature dependence of the critical fields of two superconductors. Note that B_c approaches zero as T approaches T_c (3.4 K for In and 7.2 K for Pb).

to have higher critical fields, although there are some exceptions. The critical field varies with temperature as shown in Figure 10.30. Just below T_c the critical field is low; that is, it takes very little magnetic field to eliminate the superconductivity. The fact that B_c for pure metals even at absolute zero is only on the order of 0.1 tesla is important, because one often encounters higher fields in laboratory situations. The critical field places a strict limit on how a particular superconductor can be used. Current-carrying wires generate magnetic fields, both inside and outside the wire. Therefore, if one wishes to use a superconducting wire to carry current without resistance, there will be a limit (known as the **critical current**) to the current that can be used. These effects severely limited the applications of superconductors for several decades, until superconductors with higher critical fields were discovered.

Critical current

The superconducting state we have just described is that of the pure metals Hg, Al, and many others. They are collectively known as **type I superconductors**. In **type II superconductors** (this category includes most superconducting alloys) there are two critical fields, a lower critical field B_{c1} and an upper critical field B_{c2}. Below B_{c1} and above B_{c2}, type II superconductors behave in the same manner as type I superconductors below and above B_c. Between B_{c1} and B_{c2}, however (known as the *vortex state*), there is a partial penetration of magnetic flux, as is shown in Figure 10.31, although the zero resistivity property is generally not lost. The good news is that B_{c2} can sometimes be very high, hundreds of a tesla or more. The bad

Type I and type II superconductors

(a) Type I

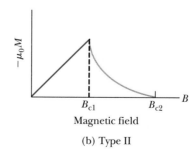

(b) Type II

Figure 10.31 A comparison of the temperature dependence of critical fields for (a) type I and (b) type II superconductors. In type I superconductors below B_c, we have $-\mu_0 M = B$, corresponding to $\chi = -1$, or complete expulsion of magnetic flux. Up to B_{c1}, the type II material expels all magnetic flux, and above B_{c2} it allows complete penetration. Between B_{c1} and B_{c2} (the vortex state) there is partial flux penetration.

Figure 10.32 A schematic of electron motion around a one-dimensional row of ions in a lattice, showing how the attraction of the two paired electrons varies as they propagate through the lattice.

news is that B_{c1} is seldom more than a few *hundredths* of one tesla. This must be kept in mind when considering applications of superconducting materials that depend on either the zero resistivity property or the Meissner effect.

The Meissner effect may remind you of a phenomenon from classical physics known as *Lenz's law*, which states that a changing magnetic flux generates a current in a conductor in such a way that the current produced will oppose the change in the original magnetic flux. In classical physics the current lasts only as long as the magnetic flux is changing (Faraday's law). You might expect that in a superconductor the current simply persists because of the zero resistivity property, but that is not what happens, and a simple experiment serves to demonstrate. One can impose a constant magnetic field on a material above its T_c, so that initially there is no current. If the material is then cooled to below T_c, the field is expelled instantly! These experimental results demonstrate that superconductors behave in ways that cannot be explained by classical physics.

What makes a superconductor display such unique properties? For years there **Isotope effect** were only vague guesses. In 1950 a phenomenon known as the **isotope effect** was discovered, and this eventually helped lead to a successful theory. Many superconductors follow the equation

$$M^{0.5}T_c = \text{constant} \qquad (10.43)$$

where M is the mass of the particular superconducting isotope. This means that T_c is just a bit higher for lighter isotopes. For example, a mercury sample with an average mass per atom of 199.5 u has $T_c = 4.185$ K, and a mercury sample with an average mass per atom of 203.4 u has $T_c = 4.146$ K. In other elements this general trend is followed, although for some the exponent differs slightly from 0.5.

The isotope effect indicates that the lattice ions are important in the superconducting state. This is at odds with the classical model of conduction, which leads one to believe that zero resistance can result only from zero interaction between electrons and lattice ions.

A successful theory* of superconductivity was developed in the mid-1950s by John Bardeen, Leon Cooper, and Robert Schrieffer (Nobel Prize, 1972) and is **BCS theory** referred to by their initials: **BCS**. The two principal features of the BCS theory of **Cooper pairs** superconductivity are that (1) electrons form pairs (**Cooper pairs**), which propagate throughout the lattice, and (2) such propagation is without resistance because the electrons move in resonance with the lattice vibrations. The lattice vi- **Phonons** brations are known as **phonons**, so called because, like *photons* of electromagnetic radiation, they represent quanta of energy. Hence the interaction described by **Electron-phonon** the BCS theory is known as the **electron-phonon interaction**.
interaction   How is it possible for two electrons to form a coherent pair? As a first approximation, consider the crude model shown in Figure 10.32, in which two electrons propagate in tandem along a single lattice row. Each of the two electrons

*J. Bardeen, L. N. Cooper, and J. R. Schrieffer, *Physical Review* **108**, 1175–1204 (1957).

experiences a net attraction toward the nearest positive ion. If the electrons get too close to each other in the region between ions, they are repelled by their mutual Coulomb force. There is a kind of balance between attraction and repulsion, just as we saw in our study of molecular bonding. In this way relatively stable electron pairs can be formed. The two fermions (electrons) combine to form a boson. Then the collection of these bosons in a bulk sample condense to form the superconducting state, similar to superfluid helium, the Bose condensate discussed in Section 9.7.

How can the zero resistivity property of superconductors be explained? Even at low temperatures there is some ionic motion. (Remember that a harmonic oscillator has a zero-point energy of $\frac{1}{2}\hbar\omega$.) That is why one would expect some resistance, even at the lowest temperatures. An electron moving through a wire should eventually collide with an ion. In this inelastic collision, electrical energy is converted to thermal energy and the wire heats up.

But if we neglect for a moment the second electron in the pair, we can understand how a single electron can travel between adjacent rows of ions without transfer of energy (Figure 10.33). The Coulomb attraction between the electron and ions causes a deformation of the lattice, which propagates along with the electron. This propagating wave is associated with phonon transmission, and the electron-phonon resonance allows the electron (along with its pair elsewhere in the lattice) to move without resistance. Keep in mind that this model is not as precise as the full, mathematically precise BCS theory. However, our model does illustrate the importance of long-range order in a superconductor, along with the wavelike nature of electrons, a purely quantum-mechanical phenomenon.

The complete BCS theory contains sophisticated mathematics and is based solidly on the foundations of the quantum theory. It also successfully predicts

Figure 10.33 Propagation of a single electron between rows of ions in a two-dimensional lattice.

several other observed phenomena. First, it predicts an isotope effect, with an exponent [Equation (10.43)] very close to 0.5. Second, it gives a critical field varying with temperature as

$$B_c(T) = B_c(0)\left[1 - \left(\frac{T}{T_c}\right)^2\right] \qquad (10.44)$$

which is in agreement with experiment (see Figure 10.30). It also accounts for the fact that the metals with higher resistivity at room temperature tend to be better superconductors. The BCS theory can be used to show that the magnetic flux through a superconducting ring is quantized in the form

Quantum fluxoid

$$\Phi = n\Phi_0 = \frac{nh}{2e} \qquad (10.45)$$

where n is an integer and $\Phi_0 = h/2e \approx 2.068 \times 10^{-15}$ Wb is known as the **quantum fluxoid**. (Remember that the weber, abbreviated Wb, is the SI unit for magnetic flux.) The quantization of magnetic flux, confirmed experimentally by B. S. Deaver, Jr., and W. M. Fairbank in 1961, is the basis for the Josephson junction, which will be discussed in Section 10.6.

Another correct prediction of the BCS theory concerns the energy gap (E_g) between the ground state (the superconducting state) and first excited state for conduction electrons. Electrons above the ground state can no longer move with zero resistance. Basically this means that E_g is the energy needed to break a Cooper pair apart, and the effect is that the larger the energy gap, the more stable the superconductor. The BCS theory predicts that

$$E_g(0) \approx 3.54 k T_c \qquad (10.46)$$

at $T = 0$ (see Table 10.6). This prediction is easily verified by studying the absorption of electromagnetic radiation by superconductors. Only photons with

Table 10.6   Energy Gaps in Superconductors at $T = 0$

Upper number: $E_g(0)$ in 10^{-4} eV
Lower number: $E_g(0)/k_B T_c$

												Al	
												3.5	Si
												3.4	
		V								Zn	Ga		
Sc	Ti	15.8	Cr	Mn	Fe	Co	Ni	Cu	2.3	3.3	Ge		
		3.4							3.2	3.5			
		Nb	Mo						Cd	In	Sn		
Y	Zr	30.3	2.7	Tc	Ru	Rh	Pd	Ag	1.4	10.6	11.2		
		3.80	3.4						3.2	3.6	3.5		
La		Ta							Hg	Tl	Pb		
19	Hf	13.9	W	Re	Os	Ir	Pt	Au	16.5	7.32	26.7		
2.3		3.6							4.6	3.6	4.3		

From N. W. Ashcroft and N. D. Mermin, *Solid State Physics*, Philadelphia: Saunders College (1976), p. 745.

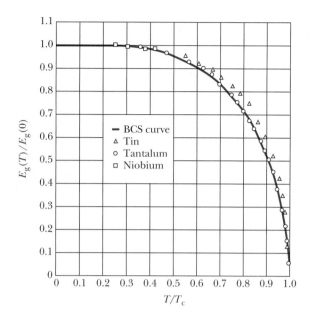

Figure 10.34 The superconductive energy gap as a function of temperature for various superconductors and according to the BCS theory. The agreement with theory is quite good for these pure metals. *From P. Townsend and J. Sutton*, Physical Review **128**, *591–595 (1962).*

energy greater than or equal to E_g are absorbed, an effect first observed by Michael Tinkham in 1960. At higher temperatures (just below T_c) BCS theory predicts

$$E_g(T) \approx 1.74 E_g(0)(1 - T/T_c)^{1/2} \qquad (10.47)$$

Again, the agreement with experiment is striking (Figure 10.34).

Example 10.7

Estimate the energy gap E_g for niobium at $T = 0$, and find the minimum photon wavelength needed to break the Cooper pair.

Strategy From Table 10.5 we see that $T_c = 9.25$ K for Nb. Therefore Equation (10.46) can be used to compute the energy gap. The photon wavelength can then be found using the usual formula relating energy and wavelength, $E = hc/\lambda$.

Solution From Equation (10.46),

$$E_g \approx 3.54(1.38 \times 10^{-23}\,\text{J/K})(9.25\,\text{K}) = 4.52 \times 10^{-22}\,\text{J}$$

$$= 2.82\,\text{meV}$$

This small energy corresponds to a photon wavelength of

$$\lambda = hc/E_g = 4.39 \times 10^{-4}\,\text{m}$$

This is in the far-infrared region of the electromagnetic spectrum. Photons with this wavelength or lower have sufficient energy to break the Cooper pair in niobium. Note that this estimate of E_g is close to the experimental value of 3.05 meV.

The Search for a Higher T_c

From the early days of research in superconductivity, physicists sought to achieve higher transition temperatures, because the high cost of cooling materials to extremely low temperatures more than offset the benefits of superconducting

Table 10.7	**Superconductivity Records Through the Years**		
Material	**Type**	T_c **(K)**	**Year of T_c Measurement**
Hg	Element	4.2	1911
Pb	Element	7.2	1913
Nb	Element	9.3	1930
Nb_3Sn	Alloy	18.1	1954
$Nb_3(Al_{0.75}Ge_{0.25})$	Intermetallic	20–21	1966
Nb_3Ga	Intermetallic	20.3	1971
Nb_3Ge	Intermetallic	23.2	1973
$Ba_xLa_{5-x}Cu_5O_{5(3-y)}$	Ceramic	30–35	1986
$(La_{0.9}Ba_{0.1})_2CuO_{4-\delta}$ (at 1 GPa pressure)	Ceramic	52.5	1986
$YBa_2Cu_3O_7$	Ceramic	90–95	1987
BiSrCaCuO	Ceramic	105–120	1988
TlBaCaCuO	Ceramic	110–125	1993
$HgBa_2Ca_2Cu_4O_{1+x}$	Ceramic	133	1994
$HgBa_2Ca_2Cu_3O_{8+x}$ (at 30 GPa pressure)	Ceramic	164	1994

From C. P. Poole, Jr., T. Datta, and H. A. Farach, *Copper Oxide Superconductors*, New York: Wiley Interscience (1988), p. 7.

properties, such as zero resistance. The cooling mechanism for superconductors has usually been a liquefied gas, such as helium. But liquid helium is very expensive to make,* and to keep anything at that temperature (4.2 K or lower) requires cumbersome and expensive insulation techniques. Further, helium gas itself is rare and expensive. It is found only in trace amounts in the atmosphere. (Helium is so light that its mean molecular speed is a significant fraction of escape speed from the Earth's surface; therefore it tends to leave the atmosphere.) Helium can be obtained from beneath the Earth's surface. It is found at the top of deposits of natural gas and in geological formations such as dolomite. Hydrogen, with a boiling point of 20 K, is a possible substitute, but its high flammability makes it undesirable for most uses. Nitrogen, on the other hand, is plentiful, safe, and relatively easy to liquefy and store. That is why 77 K, the boiling point of nitrogen at 1 atm pressure, was for many years the goal of physicists looking for higher temperature superconductors.

To put recent developments into perspective, it is instructive to look at the history of transition temperature increases (Table 10.7 and Figure 10.35). In 1911 Kamerlingh Onnes measured the transition temperature of mercury to be 4.2 K (it's just a coincidence that this is approximately the same as the boiling point of helium), and in 1930 Meissner found 9.3 K for niobium. The next significant

---

*A useful observation by B. S. Deaver was that in the 1970s liquid He cost about the same (for a given amount) as a fine scotch whiskey, while liquid N_2 cost about the same as milk. Today liquid He and liquid N_2 are somewhat cheaper by comparison.

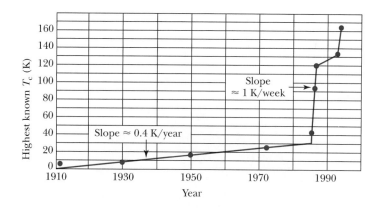

Figure 10.35 The highest known superconducting T_c by year from 1911 to 1994. Note the dramatic increase beginning in 1986. As of early 2005 there had been no further significant increases since 1994.

increase came in the 1950s with various niobium alloys. NbN, a compound used in many applications today for its stability and relatively high critical current, was found to have $T_c = 15$ K. In 1973 Berndt Matthias measured $T_c = 23.2$ K for Nb_3Ge, a record that stood until 1986. Nb_3Ge belongs to a class of compounds called **A-15**, which includes the relatively high-temperature superconductors Nb_3Sn, V_3Si, and others. This serves to illustrate that if you find one new superconductor, you have probably found several, because it is normally possible to substitute for one or more constituent atoms with another from the same column of the periodic table.

In 1986 the excitement began, when huge gains in T_c were suddenly realized. Georg Bednorz and Karl Alex Müller, working at the IBM Zurich research laboratory, found that a compound containing lanthanum, barium, copper, and oxygen had a transition temperature of at least 30 K. By playing the substitution game just described, Bednorz and Müller* were soon able to achieve a T_c of 40 K in $La_{1-x}Sr_xCuO$ with $x = 0.15$. Their discovery was revolutionary, certainly because they had nearly doubled the old record for T_c, but also because the essential ingredients in the new superconductors were copper and oxygen, which generally work against the formation of superconductors. For this discovery Bednorz and Müller were awarded the 1987 Nobel Prize for Physics.[†]

More significant gains in T_c came quickly. In early 1987 a group at the University of Houston led by Paul Chu more than doubled the record T_c again when they substituted yttrium for lanthanum and barium for strontium and changed the composition slightly. Chu reported a maximum T_c of about 93 K for $YBa_2Cu_3O_7$. The ratio of the three metallic elements has caused this compound to be referred to as "1-2-3." A T_c of 93 K was a fantastic advance because it surpasses the 77 K boiling point of nitrogen, achieving the long-sought goal!

Another useful property of the (type II) copper oxide superconductors is that they have extremely high upper critical fields. For $YBa_2Cu_3O_7$ the upper critical field $B_{c2} = 100$ T at 77 K. The value of B_{c2} at 0 K is so high that it cannot be measured, but by extrapolation it is taken to be about 300 T. This allows us to imagine future applications that use the Meissner effect or high currents. By the early twenty-first century, high-T_c yttrium−barium−copper oxide wire could be routinely produced with a critical current density of 10^9 A/m^2 in zero applied

---

*J. G. Bednorz and K. A. Müller, *Zeitschrift für Physik* **B64**, 189 (1986).

[†]This was the fastest such recognition in the history of the Nobel Prize in Physics.

Figure 10.36 Although almost 100 times smaller in cross section, four strands of American Superconductor's multifilamentary HTS wire (foreground) can transmit as much electrical current as conventional copper cable. *Courtesy of American Superconductor.*

magnetic field at 77 K and a critical current density of $2 \times 10^8 \, \text{A/m}^2$ in an applied magnetic field of 8 T at 77 K. For comparison, a normal conductor, #10 gauge copper wire, has a diameter of 2.59 mm and a recommended maximum current of 30 A (if insulated), or a maximum current density of $5.7 \times 10^6 \, \text{A/m}^2$. A critical current density of $10^9 \, \text{A/m}^2$ has been reported for newer mercury-based oxides at T = 110 K. More recently there have been reports of $10^{10} \, \text{A/m}^2$ critical currents at 77 K in specially prepared yttrium–barium–copper oxide tapes.

The copper oxide superconductors fall into a general category of materials called **ceramics**. Unfortunately, like most ceramic materials, they are extremely brittle and therefore are not easy to mold into convenient shapes. The copper oxide superconductors are relatively easy to make compared with the A-15 compounds or with the semiconductor wafers used in computers, but making them into shapes useful for practical applications is a challenge and requires innovative technology. It's now possible to make long, flexible wires, as illustrated in Figure 10.36. The wire shown is flexible enough to be bent into a curve with radius 10 cm, and it carries a critical current of 115 A at 77 K.

In early 1988 Chu and his group developed a superconducting oxide compound using bismuth and aluminum, $BiAl_{1-y}CaSrCoO_{7-d}$, which has $T_c =$ 114 K with $0 < y < 0.3$ and $d < 0.45$. By 1992 the T_c record reached 125 K in the

Figure 10.37 Superconducting transition temperatures in the thallium-based superconductor $Tl_2Ba_2Ca_{n-1}Cu_nO_{2n+4}$. The variation with n suggests that if n could be increased further, even higher transition temperatures could be achieved.

Relative number of copper oxide layers

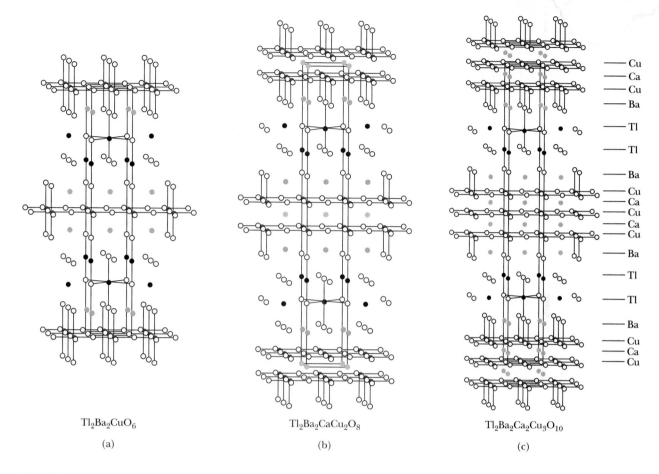

$Tl_2Ba_2CuO_6$

(a)

$Tl_2Ba_2CaCu_2O_8$

(b)

$Tl_2Ba_2Ca_2Cu_3O_{10}$

(c)

— Cu
— Ca
— Cu
— Ba

— Tl

— Tl

— Ba
— Cu
— Ca
— Cu
— Ca
— Cu
— Ba

— Tl

— Tl

— Ba

— Cu
— Ca
— Cu

Figure 10.38 Model of the thallium-based superconductor. Note the stacking of the copper oxide layers (oxygen is shown by open circles). *From C. C. Torardi et al., Science **240**, 631–634 (1988).*

thallium-based compound $Tl_2Ba_2Ca_{n-1}Cu_nO_{2n+4}$ with $n = 3$. The interesting thing is that there is a regular variation of T_c with n; $T_c = 80$ K for $n = 1$, $T_c = 110$ K for $n = 2$, and $T_c = 125$ K for $n = 3$. The curve shown in Figure 10.37 suggests that if we could make this thallium–copper oxide with $n = 3$, T_c could be increased further, perhaps indefinitely. In the structure (Figure 10.38) these higher values of n correspond to more stacked layers of copper and oxygen. So far it has not been possible to stack more than three. Also, thallium is discouraging because it is so toxic; it is a major component of rat poison and is dangerous to handle. A similar and slightly less toxic version that offers some hope is $Bi_2Sr_2Ca_{n-1}Cu_nO_x$, but its highest T_c is only about 110 K with $x = 1.6$.

In 1993 a higher T_c was achieved in mercury-based copper oxides. A T_c of 133 K was reported in the compound $HgBa_2Ca_2Cu_3O_{1+x}$ with x a small positive number. In 1994 the mercury-based compound $HgBa_2Ca_2Cu_3O_{8+x}$ was shown to have a superconducting transition temperature of 164 K at extremely high pressures—about 300,000 atm.*

The use of high-T_c superconductors is increasing in other applications for which long, flexible wires are unnecessary. For example, the wireless

*Physics Review B **50**(6), 4260–4263 (1994).

Special Topic

LOW-TEMPERATURE METHODS

Research in superconductivity and other low-temperature phenomena requires special laboratory equipment and techniques. There are well-established methods of achieving, maintaining, and measuring temperatures close to absolute zero. Here we introduce some of the hardware and techniques used by physicists who study materials at these very low temperatures.

Normally a very cold (cryogenic) liquid is used as a low-temperature bath. The sample being studied is immersed in the liquid, or else it is placed in good thermal contact with a material that does have contact with the bath. In designing low-temperature apparatus, one wishes to insulate the sample and bath from their room-temperature surroundings so that the sample does not absorb too much heat and thereby suffer an unwanted rise in temperature.

Heat can be transferred in any of three ways: conduction, convection, and radiation. The **dewar** flask, developed in 1892 by the low-temperature pioneer James Dewar (1842–1923), limits heat transfer by all three modes (see Figure A). The surfaces the liquid touches are designed to be poor conductors of heat. The cold liquid is insulated from convective transfer by surrounding it with a vacuum jacket. If the dewar is metallic, the shiny outer surface tends to limit heat transfer by radiation. Otherwise the surface is

Figure B Schematic drawing of a double-dewar apparatus. The outer dewar is maintained at 77 K with liquid nitrogen, and the inner dewar contains liquid helium. The inner dewar may be connected to a pump if temperatures less than 4.2 K are required.

"silvered," that is, coated with a thin layer of reflective material.

Figure B shows a schematic drawing of a double dewar apparatus typically used for very low temperature work. The dewar containing liquid helium, which has a temperature of 4.2 K at a pressure of 1 atm, is surrounded by a second dewar containing liquid nitrogen at 77 K. The logic of this device is that much of the heat that enters from the surroundings is absorbed by the liquid nitrogen. This causes the (relatively inexpensive) liquid nitrogen to boil away, rather than the liquid helium.

One may achieve temperatures lower than 4.2 K by pumping the vapor above the helium bath. The remaining liquid and vapor then cool by adiabatic expansion. As the liquid cools and the vapor pressure decreases, it becomes increasingly difficult to maintain the lower pressure by pumping. Therefore, there is a practical temperature limit of about 1 K with this method. In fact, at 0.7 K the creeping film (discussed in Section 9.7) extracts heat from the container walls, vaporizes, and gives up this heat to the bath by condensation. This presents a more serious limitation to the pumping method.

In 1926 W. F. Giauque and P. Debye (working independently) developed the idea of **adiabatic demagnetization** for achieving lower temperatures. Adiabatic

Figure A Schematic drawing of the dewar flask. The silvered inner surfaces reduce radiation, while the vacuum jacket reduces conduction of heat from the outside.

demagnetization is the second step in a two-step procedure. First, a paramagnetic salt is put into a vessel containing helium gas, which in turn is connected thermally to the helium bath. This is done with the dewar between the poles of an electromagnet. When the electromagnet is switched on, the magnetic dipoles in the salt align with the field. This magnetization takes place at a fixed temperature in the helium bath, and therefore this first step is referred to as *isothermal magnetization*. The key to understanding what is about to happen is to consider the entropy of the salt. Because the alignment of dipoles corresponds to a more ordered system (at the same temperature), the salt's entropy decreases. This entropy reduction is permitted by the second law of thermodynamics, because the gas surrounding the salt transfers some heat to the helium bath and thereby generates an entropy increase that at least offsets the entropy decrease just described.

Now the helium gas in contact with the salt is pumped away, so that no further heat transfer takes place. The magnet is removed, so that the magnetic dipoles resume random orientations. The demagnetization of the salt takes place without heat transfer (that is, adiabatically), and therefore it is this step that is properly called the adiabatic demagnetization. What is the result? The disordering of the magnetic dipoles surely corresponds to an entropy increase. But with no heat transfer possible, the total entropy of the salt should be constant. Therefore there must be a corresponding entropy *decrease*, and the only way this can occur is for the temperature of the salt to decrease. Figure C illustrates the entropy-temperature relationship in both steps of the process. This procedure can be repeated until (given the insulation constraints of the apparatus) no further heat can be removed from the salt. Temperatures of less than 1 K are routinely achieved by adiabatic demagnetization.

Today the most popular device for maintaining temperatures below 1 K is the ^{3}He-^{4}He dilution refrigerator, which also uses the principle of entropy exchange. Fritz London suggested this process in 1951, and the technology was developed in the 1960s and 1970s. This ^{3}He-^{4}He dilution is an effective method of cooling below 1 K, because ^{4}He at 1 K, in the superfluid state (see Section 9.7), has extremely little entropy, while ^{3}He still has a significant amount of entropy. The

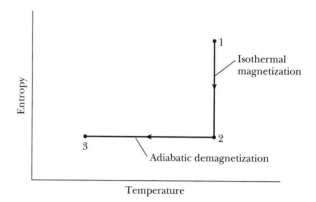

Figure C In isothermal magnetization a sample is taken from 1 to 2 along the vertical line of the entropy-temperature diagram, as the magnetic dipoles are aligned at a constant temperature. The path from 2 to 3 is adiabatic demagnetization. Because the total entropy is constant as the magnetic alignment vanishes, the temperature must drop.

entire process is too detailed to describe here. Let it suffice to say that adiabatic dilution serves the same purpose as the adiabatic demagnetization described earlier. Commercially available dilution refrigerators routinely reach temperatures on the order of 10^{-3} K.

Since the time of Kamerlingh Onnes, physicists have strived to cool materials to lower and lower temperatures. Exact temperature records depend on the type of material. Temperatures of about 10^{-6} K have been reached for a bulk sample. A research group at the Helsinki University of Technology in Finland has achieved temperatures of 10^{-7} K in silver nuclei. Bose-Einstein condensate gases (see Section 9.7) have the lowest temperatures on record, with a temperature of 500 pK (5×10^{-10} K) reported by a MIT research group in 2003.* In the past, lower temperatures have led to unexpected discoveries (for example, superconductivity and superfluids). Physicists today hope that with the achievement of even lower temperatures there will be new discoveries, along with new data to enhance our understanding of theoretical physics and materials science. For more information see F. Pobell, *Matter and Methods at Low Temperatures*, New York: Springer-Verlag (1995).

*A. Leanhardt et al., *Science* **301**, 1513–1515 (2003).

Figure 10.39 A schematic drawing of the C_{60} molecule. In the structures of the superconducting compounds based on C_{60}, atoms of the dopant are generally found on the inside of the "soccer ball" structure.

communications industry uses circuits with high-T_c components in electronic filters. These new filters are less noisy and consume less power than conventional, copper-based filters. Ceramic superconductors are also finding their way into nuclear magnetic resonance applications (see Section 10.6).

There is still much work to be done, both experimentally and theoretically, before we can understand the mechanism for superconductivity in the copper oxides. Generally they exhibit no isotope effect, which suggests that there may be a new, non-BCS mechanism at work. And while the advance in the highest T_c from 23 K to 164 K was spectacular, the mechanical properties of copper oxides make them less than desirable. Also, the greatest dream of low-temperature physicists has been to achieve a T_c not of 77 K but of 300 K (room temperature), thereby eliminating the need for cryogenic fluids. Whether this can be achieved in copper oxides or any other materials remains to be seen. If it is, there will likely be a Nobel Prize for the achievement, to add to the six Nobel Prizes already given in the field of superconductivity.

Superconducting Fullerenes

Another class of exotic superconductors was announced in 1991 by a research group at Bell Labs.* This class is based on the organic molecule C_{60}, which had been discovered just a few years earlier. The basic C_{60} molecule is called "buckminsterfullerene" (after the mathematician, architect, and general gadfly Buckminster Fuller) because of the large molecule's resemblance to Fuller's geodesic dome (Figure 10.39). The 1996 Nobel Prize in Chemistry was awarded to Robert F. Curl, Jr., Harold W. Kroto, and Richard E. Smalley for their 1985 discovery of fullerenes.

Although pure C_{60} is not superconducting, the addition of certain other elements can make it so. For example, when C_{60} is doped with the right amount of potassium, it forms the compound K_3C_{60} with a superconducting transition temperature of 18 K. When C_{60} is combined with thallium and rubidium, the T_c can be as high as 42.5 K. Although these transition temperatures are not as high as those achieved in the copper oxides, it is still encouraging whenever a new class of superconductors with a *relatively* high T_c is found. Further, each new discovery of this sort brings with it the potential for deeper understanding of the phenomenon of superconductivity, which in turn can lead to more experimental discoveries.

*D. R. Hoffman, Solid C_{60}, *Physics Today* **44**(11) 22–29 (November 1991).

10.6 Applications of Superconductivity

The remarkable properties of zero resistivity and the Meissner effect make super-conductors ideal for many applications. Some of these have been in use for years, while others may become feasible only if higher temperature superconductors become available. The cost of cryogenic systems and fluids must always be taken into account when one considers whether to use a superconductor.

Josephson Junctions

One of the earliest applications of superconductors was in a device known as a **Josephson junction**. In 1962 Brian Josephson predicted that electron pairs can tunnel from one superconductor through a thin layer of insulator into another superconductor. The superconductor/insulator/superconductor layer consti-tutes the Josephson junction. In the absence of any applied magnetic or electric field, a DC current will flow across the junction (the **DC Josephson effect**). When a DC voltage V is applied across the junction, the electron pair current across the junction oscillates with a frequency

$$f_j = \frac{2eV}{h} \tag{10.48}$$

This is the **AC Josephson effect**. Notice that the frequency of oscillation and the applied voltage are in the simple ratio $2e/h$. Because frequencies can be mea-sured to extremely high accuracy in these ranges ($f_j = 483.6$ GHz for $V = 1$ mV), Equation (10.48) provides a convenient way to measure and maintain voltage standards. The Josephson junction is used for just this purpose at the National Institute for Standards and Technology (NIST, formerly the National Bureau of Standards). With this device, the accuracy of the voltage standard is approxi-mately 1 part in 10^{10}.

It has been suggested that Josephson junctions could be used in integrated circuits, the heart of modern computers. These are superconducting devices, and therefore they would not be subject to the power losses that semiconductor-based circuits suffer. Studies throughout the 1970s showed that, given the costs of fab-rication and cooling, superconducting computers were simply not a competitive option. Needless to say, this view will have to be reevaluated whenever new and higher temperature superconductors become available.

Today Josephson junctions are used routinely in devices known as SQUIDs, superconducting quantum interference devices (Figure 10.40). The SQUID uses a pair of Josephson junctions in a current loop. It turns out that the current is ex-tremely sensitive to the magnetic flux applied to the loop. Therefore, SQUIDs are useful in measuring very small amounts of magnetic flux. They can be used to measure the quantum fluxoid $\Phi_0 = h/2e$ to within 1 part in 10^6. As ordinary mag-netometers, SQUIDs are capable of measuring magnetic fluctuations on the or-der of 10^{-13} T.

SQUIDs have been available commercially since the early 1970s. Since then many new applications for the SQUID have been found. For example, there are a number of biomedical applications, the most important of which is imaging soft tissues such as the brain. SQUIDs have also been used to examine materials for defects, search for explosives, detect the presence of bacteria, and search for oil.*

Brian D. Josephson (1940–) was born and raised in Cardiff, Wales. He earned his Ph.D. in physics at Cambridge in 1964. Josephson used the BCS theory to predict the superconducting tunneling cur-rents that bear his name. In recent years Josephson has turned his attention to the study of cognitive science, and in particular the inter-action of mind and matter. © *The Nobel Foundation.*

*A survey of applications can be found in *The Industrial Physicist* **4**(2), 20–23 (June 1998).

Figure 10.40 This is a close-up view of an yttrium–barium–copper oxide SQUID chip, manufactured by Conductus, Inc. This highly sophisticated chip contains 15 layers of material, of which three are superconducting. *Photo courtesy Conductus, Inc.*

Example 10.8

A uniform magnetic field is perpendicular to a loop of radius 1 cm. Find the value of the magnetic field such that the magnetic flux through the loop is equal to Φ_0.

Strategy In general, flux of a magnetic field is a scalar product of the magnetic field with the area, or $\Phi_0 = BA \cos \theta$, where A is the area of the loop and θ is the angle between the field vector and a vector normal to the loop. Because the magnetic field is perpendicular to the loop, $\cos \theta = 1$, and so $\Phi_0 = BA$.

Solution Solving for the magnetic field and using the given size of the loop,

$$B = \Phi_0/A = (2.068 \times 10^{-15} \text{ T} \cdot \text{m}^2)/[\pi(10^{-2} \text{ m})^2]$$
$$= 6.58 \times 10^{-12} \text{ T}$$

This exceptionally small value is an indication of just how small the quantum fluxoid is.

Maglev

For many years, scientists and engineers have imagined using strong magnetic fields to levitate various transport systems, particularly trains. The idea is to make a more comfortable ride at higher speeds while reducing frictional losses. The availability of superconducting magnets has made magnetic levitation of trains (generically referred to as **maglev** systems) more feasible.

There are two common approaches to maglev, illustrated in Figure 10.41. In each system the car glides above a track "guideway." In an electrodynamic (EDS)

EDS

Electromagnets on the
guideway levitate the car.

EMS

Electromagnets on
the cars lift the cars.

Figure 10.41 The two types of magnetic levitation (maglev) systems. In the EDS system, cars are lifted by repulsion from below. In the EMS system, magnets below the guideway are attracted upward. *From http://www.monorails .org/tMspages/TPMagIntro.html.*

system (also called repulsive maglev), magnets on the guideway repel the car to lift it. In an electromagnetic (EMS) system (also called attractive maglev), magnets attached to the bottom of the car lie below the guideway and are attracted upward toward the guideway to lift the car. The EMS system does not require superconducting magnets. However, the design is one of unstable equilibrium, so the train's path has to be monitored continually and adjusted so that the train doesn't scrape the guideway. The EDS system is more stable, but lifting the train by repulsion requires superconducting magnets, which are presently more expensive.

Both EMS and EDS systems have been built around the world, and many more are proposed. A Japanese EDS system test train has run for many years and has reached a top speed of 550 km/h. An operational EMS system that opened in Shanghai, China, runs over a length of 30 km with a top speed of 430 km/h.

Generation and Transmission of Electricity

There exists a potential for significant energy savings if superconductors can be used in electrical generators and motors. Part of the savings would obviously come from cutting resistive losses, but there would be even more savings if the heavy iron cores used today could be replaced by lighter superconducting magnets. The largest generator used now can produce energy at a rate of about 1 GW, an upper limit due to the size of the iron core. Larger generators would make for better economies of scale.

Once electrical power is generated it must be transmitted through power lines to industries and homes. Superconducting transmission lines would save significant amounts of energy (approaching 5% in some cases), again because there would be no resistive loss. Another advantage of superconducting transmission is that expensive transformers would no longer have to be used to step up voltage for transmission and down again for use. Transformers are used at both ends of conventional lines because the energy loss rate is

$$P_{\text{lost}} = I^2 R = P_{\text{trans}}^2 R / V^2 \tag{10.49}$$

It is desirable to make V as large as possible for transmission, but this is not necessary if $R = 0$. Finally, the larger current densities possible in superconducting wire could make it possible to reduce the number and size of transmission lines.

Superconducting rings may be used for energy storage. Today, power plants are strained during "peak hours," when it is necessary to produce electrical energy at up to several times the average use rate. Superconducting storage rings would allow plants to produce at just the average rate, with extra energy generated during low usage hours and stored for use during peak hours. This storage option might also allow us to make better use of other forms of energy, especially solar and wind.

Other Scientific and Medical Applications

There are other ways in which superconductors are already used in scientific research. Magnets used to confine plasma in fusion research are superconducting. Magnets for large particle accelerators, including existing ones at Fermilab and the European Organization for Nuclear Research (CERN) and the planned Large Hadron Collider (LHC) at CERN, must be superconducting in order to produce the large fields required. This is feasible using existing superconducting technology and liquid helium. However, higher-T_c magnets can improve the operation of large particle accelerators in two ways. First, the cost of operating is much lower with liquid nitrogen than with liquid helium. Second, recent work has shown that some ceramic superconductors can produce larger magnetic fields than traditional superconductors, due to their high critical currents. With higher fields experimenters can choose between increasing the accelerator's energy or decreasing its size, because the net energy is proportional to both the radius and magnetic field strength. The technology for making magnet coils out of brittle ceramic materials is now available. By 2003, scientists at the National High Magnetic Field Laboratory in Tallahassee, Florida, had achieved steady magnetic fields of 25 T using a pure superconducting magnet and 45 T using a hybrid magnet that is part superconducting and part normal conductor.

Incidentally, it is not possible to make a large electromagnet simply by winding large coils of superconducting wire. Statistically infrequent events known as *flux jumps* occasionally make the wire a normal conductor in small regions. Unfortunately this small region grows and propagates rapidly due to resistive heating. When one part of the wire undergoes resistive heating, the heat flow from warmer regions soon drives the cooler regions above T_c. For this reason, long superconducting wires must be embedded in a copper matrix. Copper is such a good conductor of heat at low temperatures that it shunts away heat much faster than the superconductor gone normal. This allows the superconductor to recover from those temporary fluctuations.

Magnetic resonance imaging

At least one significant medical application of superconducting magnets is in wide use today. It is called **magnetic resonance imaging (MRI)**. The physical process used in MRI is the well-established quantitative analysis technique called nuclear magnetic resonance (NMR).* NMR was demonstrated in 1946 by Felix Bloch and Edward Purcell, who were awarded the Nobel Prize in Physics for this achievement in 1952.

Beginning in the 1970s, scientists began using NMR techniques in medical imaging. Two of the leaders in the field were Paul Lauterbur and Peter Mansfield, who received the 2003 Nobel Prize in Medicine for their work. In MRI the data is processed by a computer with the goal of making a picture of the sample. Medical workers now use MRI routinely to obtain clear pictures of the body's soft tissues, allowing them to detect tumors and other disorders of the brain, muscles, organs, and connective tissues. Figure 10.42 illustrates that MRI can create a sharp picture of the brain without interference from the skull. The MRI image in

*It has been suggested that the word *nuclear* was dropped so that the general public would not have an irrational fear of the procedure. Of course MRI has nothing to do with gamma radiation or radioactive isotopes!

Figure 10.42 A picture of the soft tissues of the brain produced by magnetic resonance imaging. *Courtesy Tacoma Magnetic Imaging.*

Figure 10.43 is a cross section of a patient's abdomen. Multiple cross sections can be used to scan for tumors and other irregularities.

With high-field (several-tesla) superconducting magnets the MRI diagnosis can be made earlier than with other methods and without surgical intrusion, an important fact. Many lives have already been saved by MRI, and many others have been vastly improved. The major drawback is that the cost of a single MRI image is on the order of $1000. Higher temperature superconductors may eventually reduce the cost and thereby make MRI available to more people.* Just as with Schawlow and the laser, Kamerlingh Onnes could not have imagined this life-saving technology when he first pondered superconductivity.

---

*For a good review see F. W. Wehrli, The origins and future of nuclear magnetic resonance imaging, *Physics Today* **45**(6), 34–42 (June 1992).

Tumor

Figure 10.43 This magnetic resonance image is a horizontal view of the thoracic region of a patient and was used to detect a malignant tumor in the liver. *Courtesy of Haleigh Kurtz.*

Summary

The Coulomb force holds together the atoms of a molecule or solid. In a diatomic molecule, quantum theory can be used to deduce the allowed rotational and vibrational energy levels:

$$E_{rot} = \frac{\hbar^2 \ell(\ell + 1)}{2I} \tag{10.2}$$

and

$$E_{vibr} = \left(n + \tfrac{1}{2}\right)\hbar\omega \tag{10.3}$$

Transitions between these levels are observed in both emission and absorption spectra. In Raman scattering a photon is scattered from a molecule, and the change in frequency of the scattered photon is used to deduce rotational properties of the molecule.

A photon incident upon a molecule in an excited state can cause a stimulated emission, the basic mechanism in the operation of a laser. Three-level and four-level systems can be used to control the output and efficiency of the laser. The monochromatic, intense beam has a number of applications in research and industry, including holography and biomedical uses.

Many solids exhibit regular crystal structures. Solids generally tend to expand when heated. It can be shown that the mean separation between atoms in a solid is very nearly proportional to temperature in the classical limit:

$$\langle x \rangle = \frac{3b}{4a^2\beta} = \frac{3bkT}{4a^2} \tag{10.26}$$

Good electrical conductors tend to be good thermal conductors too. Using the quantum theory of conduction, we find the correct relationship between the electrical and thermal conductivity in a solid (the Lorenz number L) to be

$$L = \frac{K}{\sigma T} = \frac{\pi^2 k^2}{3e^2} \tag{10.36}$$

A few solids, known as *ferromagnets*, have permanent magnetic moments. In most materials a magnetic moment is induced by an applied magnetic field. The magnetic susceptibility (which is positive for paramagnets and negative for diamagnets) relates the induced magnetization to the applied magnetic field:

$$\chi = \frac{\mu_0 M}{B} \tag{10.37}$$

Superconductors exhibit the complete loss of electrical resistance and expulsion of magnetic fields below their transition temperature T_c. A successful theory (the BCS theory) of superconductivity was found in the 1950s, and the BCS theory has been used to explain the behavior of most superconductors. Different elements and compounds have different values of T_c, and the search for a higher T_c has driven a great deal of research. In the late 1980s compounds with $T_c > 100$ K were discovered.

The search for higher T_c values continues today, with the aim of using higher temperature superconductors in a number of applications. Low-loss energy generation and transmission, magnetic levitation systems, superconducting electronic devices, and high-field magnets are some of the applications made possible by the extraordinary properties of superconductors. Today superconductors are already used in research, industry, and medicine, and the extension of superconductors into other areas is an exciting possibility for the future.

Questions

1. Explain in your own words why the sky is blue.
2. Explain why $n > m$ is required in Equation (10.1) in order to produce the potential energy curve shown in Figure 10.1.
3. Are the force constants for diatomic molecules (Table 10.1) at all similar to those for common laboratory springs? (Remember your introductory college or high school lab experience.) Explain.
4. Explain how to use the rotational spectra to determine the equilibrium separation between the two nuclei in a diatomic molecule.
5. Why do the gases He, Ne, and Xe tend to be monatomic rather than diatomic?
6. Do you expect the fundamental vibrational frequency to be higher for HCl or NaCl? Explain.
7. Is it necessary that the substance used in a laser have at least three energy levels? Why?
8. Critique the following statement: "The Pauli exclusion principle is responsible for keeping solids from collapsing to zero volume."
9. The average nearest-neighbor distance between nuclei in solid NaCl is 0.282 nm, but the distance is 0.236 nm

in a free NaCl molecule. How do you account for the difference?

10. What patterns do you notice within the groups of salts in Table 10.2? Explain those patterns.

11. What makes elements good candidates for paramagnetism? For diamagnetism?

12. Explain why the paramagnetic susceptibilities of rare earth elements tend to be higher than those of the transition elements.

13. Considering the electronic configurations of the five ferromagnetic elements, justify why they should be ferromagnetic. Why are some elements in the same columns as these five, with similar electronic configurations, not ferromagnetic?

14. Notice that ferromagnetic elements tend to come from the middle of the rows of rare earth elements and transition elements in the periodic table. Explain.

15. Why should elements and compounds with positive paramagnetic susceptibilities not be good candidates for superconductivity?

16. Explain similarities and differences between the Meissner effect and Lenz's law.

17. Consider the superconducting transition temperatures of the elements as shown in Table 10.5. In cases in which there is more than one superconductor in a column of the periodic table, are the transition temperatures consistent with the spirit of the isotope effect (that is, does the heavier element have a lower T_c)?

18. A superconducting ring can carry a current for an indefinite length of time. Isn't this a perpetual motion machine, which violates the first or second law of thermodynamics?

19. Consider a sample at a temperature initially above its superconducting T_c and in a magnetic field. When the sample is cooled to below T_c, currents are generated to expel the magnetic flux from the interior. What is the source of the energy for these currents?

20. Would you expect that a material with little or no crystal structure (a "glass") could exhibit superconductivity? Why or why not?

21. A superconducting wire and a low-resistance copper wire are connected together in parallel. When a potential difference is applied, is it really true that the copper wire carries no current? Explain.

Problems

Note: The more challenging problems are indicated by a shaded box.

10.1 Molecular Bonding and Spectra

1. Consider again the rotational energy states of the N_2 molecule as described in Example 10.1. Find the energy involved in a transition (a) from the $\ell = 2$ to $\ell = 1$ state, and (b) from the $\ell = 15$ to $\ell = 14$ state.

2. (a) Use the data in Table 10.1 to find the approximate spacing between vibrational energy levels in NO. (b) What temperature would be needed to excite this vibration thermally?

3. Estimate the amplitude of the smallest vibration of the HCl molecule (see Example 10.2).

4. The distance between the centers of the H and Cl atoms in the HCl molecule is approximately 0.128 nm. (a) Find the angular velocity of the molecule about its center of mass when $\ell = 1$ and $\ell = 10$. (b) What is the speed of the H atom in each of the cases in (a)? (c) What value of ℓ is required in order for the H atom to have a speed of $0.1c$? (d) Estimate the classical temperature associated with the ℓ you found in (c).

5. Derive an expression for the allowed rotational levels in a homopolar diatomic molecule using the Bohr quantization rule for angular momentum. Discuss your result in comparison with the correct quantum-mechanical result, Equation (10.2).

6. The wavelength of a microwave absorption line in CO corresponding to a transition from $\ell = 0$ to $\ell = 1$ is 1.30 mm. (a) Calculate the rotational inertia of the CO molecule. (b) Show that it is impossible for this amount of energy (corresponding to a photon of wavelength 1.30 mm) to be absorbed by CO in a *vibrational* transition.

7. If the energy of a vibrational transition from the $n = 0$ state to the $n = 1$ state in CO could be absorbed in a rotational transition that begins in the ground state ($\ell = 0$), what would be the value of ℓ for the final state? Explain why such a rotational transition is impossible.

8. Show that $I = \mu R^2$ for a diatomic molecule, where R is the distance between the two atomic centers and $\mu = m_1 m_2/(m_1 + m_2)$ is the reduced mass. (*Note:* You should assume that the atoms are point particles.)

9. The energy of a transition from the $\ell = 2$ to the $\ell = 3$ state in CO is 1.43×10^{-3} eV. (a) Compute the rotational inertia of the CO molecule. (b) What is the average separation between the centers of the C and O atoms?

10. Consider the model of the H_2O molecule shown on page 384. (a) Find the rotational inertia of H_2O about the dashed line. (b) Estimate the energies of the first two rotational energy levels ($\ell = 1$ and $\ell = 2$). (c) What is the wavelength of a photon required to excite a transition from $\ell = 0$ to $\ell = 1$? (*Note:* This is roughly the wavelength of radiation used in a microwave oven.)

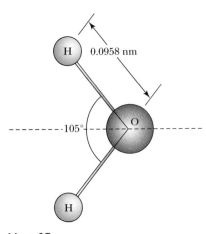

Problem 10

11. Consider the rotational energy of a single helium atom. Assume that the electrons are uniformly distributed throughout a sphere of radius equal to the Bohr radius and that the nucleus is a uniform solid sphere of radius 1.9×10^{-15} m. (a) Estimate the energy of the first (nonzero) rotational energy level in helium. (b) Is this state likely to be observed? Why or why not?

12. Consider the NaCl molecule, for which the rotational inertia is 1.46×10^{-45} kg $\cdot$ m^2. If infrared radiation with wavelength 30 μm is Raman-scattered from a free NaCl molecule, what are the allowed wavelengths of the scattered radiation?

13. This problem deals with the splitting of rotational energy levels of diatomic molecules. If one atom of the molecule has more than one stable isotope, then both isotopes are normally present in a sample. Show that the fractional change $\Delta f/f$ in the observed frequency of a photon emitted in a transition between adjacent rotational states is equal to the fractional difference in the reduced mass $\Delta\mu/\mu$ for molecules containing two different isotopes.

14. (a) At $T = 293$ K what are the relative Maxwell-Boltzmann factors for N$_2$ molecules in the $\ell = 0$, $\ell = 1$, and $\ell = 2$ states? Use $I \approx 10^{-46}$ kg $\cdot$ m^2. (b) Use your answer to (a) along with the fact that there is also a degeneracy factor of $2\ell + 1$ on the ℓth angular momentum level to find the relative populations of the $\ell = 0$, $\ell = 1$, and $\ell = 2$ states of N$_2$ at room temperature. (c) Explain why the $2\ell + 1$ degeneracy factor is more important for lower rotational states, but the Maxwell-Boltzmann factor dominates for higher states.

15. Using the HCl absorption spectrum shown in Figure 10.8, (a) compute the rotational inertia I of the molecule and (b) compute the force constant κ and compare with the value given in Table 10.1.

16. The equilibrium separation between the two ions in the KCl molecule is 0.267 nm. (a) Assuming that the K$^+$ and Cl$^-$ ions are point particles, compute the electric dipole moment of the molecule. (b) Compute the ratio of your result in (a) to the measured electric dipole moment of 5.41×10^{-29} C $\cdot$ m. This ratio is known as the *fractional ionic character* of the molecular bond.

17. Find the energy of the photon required to excite the transition from the ground state to the first excited vibrational state in HF. In what part of the electromagnetic spectrum is this?

10.2 Stimulated Emission and Lasers

18. (a) How many photons are emitted each second from a 5.0-mW helium-neon laser ($\lambda = 632.8$ nm)? (b) If the laser contains 0.02 mole of neon gas, what fraction of the neon atoms in the tube participate in the lasing process during each second of operation? (c) Comment on the relatively low numerical result in (b).

19. A laser emits 5.50×10^{18} photons per second, using a transition from an excited state with energy 1.15 eV to a ground state with energy 0 eV. (a) What is the laser's power output? (b) What is the wavelength?

20. A particular ruby laser emits a short burst of radiation (with $\lambda = 690$ nm) rather than a continuous beam. The burst lasts for 1.0 ns and has a power output of 500 MW. How many atoms made a transition from the excited state to the ground state in order to create this pulse?

21. (a) For the helium-neon laser, estimate the Doppler broadening (see Chapter 9, Problem 3) of the output wavelength 632.8 nm at $T = 293$ K. (b) Estimate the broadening of the same wavelength due to the Heisenberg uncertainty principle, assuming that the metastable state has a lifetime of about 1 ms.

22. Consider the problem of using laser light to measure the distance from the Earth to the moon. (a) What is the maximum uncertainty in timing the round trip for a light pulse in order to determine the distance with an uncertainty of 1 meter? (b) Estimate the effect of the Earth's atmosphere on this experiment, using the fact that the speed of light in air (at sea level) is slower than the speed of light in vacuum by a factor of about 1.0003. Assume an 8-km-high atmosphere of uniform sea level density.

23. What is the minimum fraction of the lasing molecules in a three-level laser that must be in the excited state in order for the laser to operate? Answer the same question for a four-level laser.

24. The 3s state of neon (see Figure 10.15) is 16.6 eV above the ground state. (a) Estimate the relative populations of the ground state and the 3s state at $T = 293$ K. (b) Repeat for $T = 200$ K. (c) Repeat for $T = 500$ K. (d) What implications (if any) do your answers for parts (a)–(c) have for the operation of a He-Ne laser at various temperatures?

10.3 Structural Properties of Solids

25. The density of solid KCl is about 1980 kg/m^3. Compute the nearest-neighbor distance in KCl, that is, the distance between neighboring K$^+$ and Cl$^-$ ions. Note that KCl has the same lattice structure as NaCl.

26. Show that the Madelung constant for a one-dimensional lattice of alternating positive and negative ions is $\alpha = 2 \ln 2$.

27. Write the first five terms of the Madelung constant for a two-dimensional lattice of alternating positive and negative ions.

28. Use Equation (10.19) to evaluate the net force ($F = -dV/dr$) on an atom in a sodium chloride lattice. Then show that the force can be expressed as

$$F = \frac{\alpha e^2}{4\pi\epsilon_0 r_0^2}\left(-\frac{r_0^2}{r^2} + e^{-(r - r_0)/\rho}\right)$$

29. Starting with the result of Problem 28, approximate $r \approx r_0 + \delta r$ (where $\delta r \ll r_0$) and show that

$$F \approx \frac{\alpha e^2}{4\pi\epsilon_0 r_0^2}\left[-\frac{r_0}{(r_0 + \delta r)^2} + e^{-\delta r/\rho}\right]$$

Perform a Taylor series expansion about $r = r_0$, keeping terms up to and including order $(\delta r)^2$. Show that

$$F \approx K_1\delta r + K_2(\delta r)^2$$

where

$$K_1 = \frac{\alpha e^2}{4\pi\epsilon_0 r_0^2}(2/r_0 - 1/\rho)$$

and

$$K_2 = \frac{\alpha e^2}{4\pi\epsilon_0 r_0^2}[-3/r_0^2 + 1/(2\rho^2)]$$

30. If we consider again the linear term (δr) in the result of Problem 29, we have a harmonic oscillator. (a) Find an expression for the frequency of oscillation and evaluate using the correct values of α, r_0, and ρ for NaCl. (b) Find the photon wavelength corresponding to the frequency you computed in (a) and compare with the observed absorption wavelength for NaCl of about 61 μm.

31. Refer again to the result of Problem 29. (a) Use the fact that the average net force on an ion must be zero (if there is no overall translation of the crystal) to show that the average values of δr and $(\delta r)^2$ are related by $\overline{\delta r} = -(K_2/K_1)\overline{(\delta r)^2}$. (b) According to the equipartition theorem the mean potential energy of an oscillator is $\frac{1}{2}K_1\overline{(\delta r)^2} = \frac{1}{2}kT$. Use this to show that $\overline{\delta r} = (K_2/K_1^2)\,kT$, and thereby show that the coefficient of thermal expansion is approximately $(1/r_0) \times (kK_2/K_1^2)$. (c) Use the result of (b) to evaluate the coefficient of thermal expansion for NaCl and compare with the experimental value of $4 \times 10^{-6}\,\text{K}^{-1}$.

10.4 Thermal and Magnetic Properties of Solids

32. Copper has an electrical conductivity of 6.0×10^7 $\Omega^{-1} \cdot \text{m}^{-1}$ at 293 K. Use the Wiedemann-Franz law without quantum corrections to compute the thermal conductivity of copper at the same temperature, and compare with the experimental result of 400 $\text{W} \cdot \text{K}^{-1} \cdot \text{m}^{-1}$.

33. Use the data in Figure 10.23 to estimate the numerical value of the constant b/a^2 in the thermal expansion formula, Equation (10.26). Make sure you express the correct units.

34. (a) Derive Equation (10.26). (b) Evaluate the constant $3bk/4a^2$ in Equation (10.26) for copper, given that the coefficient of linear expansion [defined as $\alpha = \Delta x/(x\,\Delta T)$] for copper is found experimentally to be $1.67 \times 10^{-5}\,\text{K}^{-1}$ at $T = 293$ K.

35. (a) Explain why the parameter a in Equation (10.22) is essentially half the effective force constant of a spring connecting adjacent atoms. (b) Then use this result along with the result of Problem 34b to estimate a value for the parameter b in Equation (10.22), the coefficient of the x^3 term in the potential energy.

36. (a) Show that the ideal gas law can be written as

$$PV = \frac{2N\overline{E}}{3}$$

where N is the number of particles in the sample and $\overline{E}$ is the mean energy. (b) Use the result of (a) to estimate the pressure of the conduction electrons in copper, assuming an ideal Fermi electron gas. Comment on the numerical result, noting that 1 atm = 1.01×10^5 Pa.

37. Show that the *bulk modulus*, defined as

$$B = -V\frac{\partial P}{\partial V}$$

(where P is pressure and V is volume) can be written as

$$B = \frac{5P}{3} = \frac{2NE_\text{F}}{3V}$$

for a Fermi electron gas with Fermi energy E_F. (*Hint:* Use the relationship between P and V given in Problem 36a.)

38. (a) From the result of Problem 37, compute the bulk modulus of pure silver. (b) Compare your result with the experimental value of 1.01×10^{11} N/m^2.

39. Retrace the derivation of the induced diamagnetic moment in Equation (10.38), assuming that (a) the electron orbits clockwise and the magnetic field points out of the page, and (b) the electron orbits counterclockwise and the magnetic field points into the page.

40. Using the result of the preceding problem, what is the induced magnetic moment of a diamagnetic atom with an outer shell having three electrons in a p shell with $m_\ell = 0$, $m_\ell = 1$, and $m_\ell = -1$?

41. Starting with Equation (10.41) derive an expression for $\overline{\mu}$ valid in the *low-temperature* limit $kT \ll \mu B$.

42. (a) Plot $\overline{\mu}$ versus $\mu B/kT$ over the range $\mu B/kT = 0$ to $\mu B/kT = 4$. (b) Compute $\overline{\mu}$ at $\mu B/kT = 5$ and compare your results with the approximation used for $\overline{\mu}$ in Problem 41. (c) Compute $\overline{\mu}$ at $\mu B/kT = 0.10$ and compare your results with the approximate value given in Section 10.4, $\tanh(\beta\mu B) \approx \beta\mu B$.

43. Prove that magnetic susceptibility χ is a dimensionless quantity. Note that the definition in Equation (10.37) presumes SI units.

44. (a) Compute the maximum magnetization of a bulk sample of iron, assuming perfect alignment of the spins and one unpaired spin per atom. (b) Compare with the observed maximum magnetization of about 1.6×10^6 A/m. (c) Based on your results in (a) and (b), what can you say about the actual number of unpaired spins per atom of iron?

10.5 Superconductivity

45. At what temperature (expressed as a fraction of T_c) is $B_c = 0.1B_c(0)$, according to the BCS theory? Repeat for $B_c = 0.5B_c(0)$ and $B_c = 0.9B_c(0)$.

46. It is found that for a given pure metal superconductor, photons of wavelength 0.568 mm are sufficient to break the Cooper pairs at $T = 2.0$ K. Identify the superconductor.

47. Compute T_c of the mercury isotopes ^{201}Hg and ^{204}Hg.

48. Estimate the BCS prediction for the change in T_c if all of the ^{16}O atoms in $YBa_2Cu_3O_7$ (assume $T_c = 93.0$ K) could be replaced by ^{18}O. This change is *not* observed experimentally!

49. From the data given in the text, estimate T_c for $Tl_2Ba_2Ca_{n-1}O_{2n+4}$ with $n = 4$.

50. A solenoid made of superconducting wire has exactly 25 turns/cm of length, and a diameter of 2.8 cm. If the wire carries a current of 5.0 A, what is the magnetic field strength near the center of the solenoid? What is the magnetic flux through a cross section of the solenoid taken near the center? To how many flux quanta does this correspond? Comment on the number of flux quanta.

51. Recall that the magnetic field at the surface of a uniform cylindrical wire of radius R carrying a current I is

$$B = \frac{\mu_0 I}{2\pi R}$$

Find the minimum possible diameter for a wire of pure niobium so that the $T = 0$ critical field would not be exceeded if the wire carried a current of 4.5 A.

52. In a normal conductor heat is generated at a rate $I^2 R$. Therefore a current-carrying conductor must dissipate heat effectively or it can melt or overheat the device in which it is used. Consider a long cylindrical copper wire (resistivity 1.72×10^{-8} $\Omega \cdot$m) of diameter 0.75 mm. If the wire can dissipate 100 W/m^2 along its surface, what is the maximum current this wire can carry?

53. (a) Compute the maximum current that a niobium wire 0.75 mm in diameter can carry at $T = 4.2$ K. (b) Compare your result in (a) with the copper wire of the same diameter described in Problem 52.

10.6 Applications of Superconductivity

54. What is the maximum uncertainty in the measurement of the oscillation frequency in a Josephson junction if the voltage standard of 1 mV is to be maintained within 1 part in 10^{10}? Assume a reference frequency of 483.6 GHz.

55. Find the minimum acceleration needed for a maglev train to reach a speed of 430 km/h in 3.0 km, one tenth of the length of the 30-km track in Shanghai. Express your answer as a fraction of g, the free-fall acceleration near the Earth. Would this acceleration be noticeable?

56. (a) Compute the escape speed of a particle from the Earth's surface. Earth's radius is 6378 km and its mass is 5.98×10^{24} kg. (b) Find the mean speed $\bar{v}$ for a helium atom at a temperature of 293 K. (c) Comment on the fact that your answer to (b) is less than the answer to (a). Why then does helium not remain in the atmosphere in significant quantities?

57. A superconducting Nb_3Sn magnet can achieve a peak magnetic field of 13.5 T in a magnet designed for use in the Large Hadron Collider. Find the maximum energy that a singly charged particle (for example, a proton or electron) can have if that field is maintained around a circular ring of circumference 27 km. (*Note:* In reality, particle energies are about 35% less because the peak field is not maintained throughout the ring.)

General Problems

58. In the persistent current experiment described in Section 10.5, let us assume that the current persisted without detectable reduction for exactly 2.5 years. Given that the inductance of the ring was approximately 3.14×10^{-8} H and the sensitivity of the current measurement was 1 part in 10^9, (a) estimate an upper bound for the resistance of the ring, and (b) estimate how long at least 90% of the current would be certain to remain.

59. From thermodynamics the entropy difference per unit volume between the normal and superconducting states is

$$\frac{\Delta S}{V} = -\frac{\partial}{\partial T}\left(\frac{B^2}{2\mu_0}\right)$$

where $B^2/2\mu_0$ is the magnetic energy density needed to return a superconductor to the normal state. Use this fact to compute the entropy difference between the normal and superconducting states in 1 mole of niobium at a temperature of 6.0 K.

60. A 1.0-m length of copper wire with a resistance of 1.50 Ω is placed in series with a 1.0-m length of superconducting wire. When a 12.0-V battery is placed across the series combination, find (a) the current in the circuit, and (b) the potential difference across the two ends of the copper wire.

Semiconductor Theory and Devices

It is evident that many years of research by a great many people, both before and after the discovery of the transistor effect, has been required to bring our knowledge of semiconductors to its present development. We were fortunate to be involved at a particularly opportune time and to add another small step in the control of Nature for the benefit of mankind.

John Bardeen, 1956 Nobel lecture

In the last half of the twentieth century, a revolution occurred in electronics. The resulting development and widespread use of integrated circuits have changed the way we live, even more than the lasers and superconductors you learned about in Chapter 10. Automobiles, televisions, audio equipment, microwave ovens, and other home appliances now contain microcomputers to enhance their efficiency. The electronics revolution has had an even more profound effect on the work of scientists and engineers. In the 1950s, when the space program was in its early years, the slide rule, pencil, and paper were still the standard tools of the physicist and the aeronautical engineer. Computers were available for the most difficult calculations, but they filled an entire room and lacked the speed and computational power of desktop personal computers available today—not to mention that their cost was orders of magnitude higher than what we pay for personal computers today.

The remarkable properties of semiconductor materials have made possible the advances just described. In this chapter we examine those properties and see how they are used in various devices. You will see how it is possible to understand the behavior of semiconductors by using the quantum theory of solids. We intend to present just enough of the theory in descriptive fashion to allow you to appreciate the beauty and utility of semiconductor materials.

11.1 Band Theory of Solids

In Chapter 10 you learned about structural, thermal, and magnetic properties of solids. Here we concentrate on electrical conduction. There are three categories of solids, based on their conducting properties: conductors, semiconductors, and insulators. As seen in Table 11.1, the electrical conductivity at room temperature

Table 11.1 Electrical Resistivity and Conductivity of Selected Materials at 293 K		
Material	Resistivity $(\Omega \cdot m)$	Conductivity $(\Omega^{-1} \cdot m^{-1})$
Metals		
Silver	1.59×10^{-8}	6.29×10^{7}
Copper	1.72×10^{-8}	5.81×10^{7}
Gold	2.44×10^{-8}	4.10×10^{7}
Aluminum	2.82×10^{-8}	3.55×10^{7}
Tungsten	5.6×10^{-8}	1.8×10^{7}
Platinum	1.1×10^{-7}	9.1×10^{6}
Lead	2.2×10^{-7}	4.5×10^{6}
Alloys		
Constantan	4.9×10^{-7}	2.0×10^{6}
Nichrome	1.5×10^{-6}	6.7×10^{5}
Semiconductors		
Carbon	3.5×10^{-5}	2.9×10^{4}
Germanium	0.46	2.2
Silicon	640	1.6×10^{-3}
Insulators		
Wood	10^{8}–10^{11}	10^{-8}–10^{-11}
Rubber	10^{13}	10^{-13}
Amber	5×10^{14}	2×10^{-15}
Glass	10^{10}–10^{14}	10^{-10}–10^{-14}
Quartz (fused)	7.5×10^{17}	1.3×10^{-18}

From R. Serway and J. Jewett, *Physics for Scientists and Engineers*, 6th ed., Belmont, CA: Brooks/Cole-Thomson (2004), p. 837.

is quite different for each of these three kinds of solids. Metals and alloys have the highest conductivities, followed by semiconductors, and then insulators. We have already modeled the electrical conductivity of ordinary metals in Section 9.6.

The free-electron model used in Chapter 9 does not apply to semiconductors and insulators. These materials simply lack enough free electrons to conduct in a free-electron mode. It may be tempting to apply the free-electron Fermi-Dirac model to semiconductors and perhaps explain their relatively low conductivity at room temperature by a relative lack of free electrons. If the electrical conductivity of semiconductors is about 10 orders of magnitude lower, could it be that the charge carrier density is lower by that factor? It turns out that carrier density is only part of the story. In fact there is a different conduction mechanism for semiconductors than for normal conductors. Striking evidence of this fact is seen in the resistivity-versus-temperature graphs presented in Figure 11.1. While the free-electron theory correctly predicts a linear increase in resistivity with

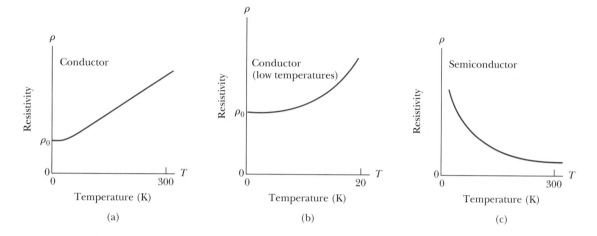

Figure 11.1 (a) Resistivity versus temperature for a typical conductor. Notice the linear rise in resistivity with increasing temperature at all but very low temperatures. (b) Resistivity versus temperature for a typical conductor at very low temperatures. Notice that the curve flattens and approaches a nonzero resistance as $T \to 0$. (c) Resistivity versus temperature for a typical semiconductor. The resistivity increases dramatically as $T \to 0$.

temperature, semiconductors generally exhibit *decreasing* resistivity with increasing temperature.

We need a new theory, known as the **band theory**, in order to account for this and other properties of semiconductors. The essential feature of the band theory is that the allowed energy states for electrons are nearly continuous over certain ranges, called **energy bands**, with forbidden **energy gaps** between the bands. William Shockley, one of the co-inventors of the transistor (Section 11.3), suggested a simple and straightforward justification of the existence of electronic energy bands in solids. We first consider what happens when two atoms of hydrogen are brought together. (The argument is good for any kind of atom; we choose hydrogen because it has known wave functions.) When the two atoms are far apart, the electronic wave functions can be thought of as noninteracting. As the atoms are brought closer together, the wave functions begin to overlap. But because any linear combination of wave functions is possible, there can be either a symmetric or an antisymmetric overlap (see Figure 11.2). These two situations correspond to two slightly different energies. An electron in the symmetric state has a nonzero probability of being halfway between the two atoms (Figure 11.2b), while an electron in the antisymmetric state ($\Psi_A - \Psi_B$) shown in Figure 11.2c has a zero

Band theory

Energy bands

Energy gaps

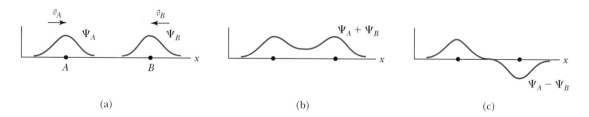

Figure 11.2 A rough representation of the wave functions of two approaching hydrogen atoms. When they are far apart (a) there is negligible overlap of their wave functions. In (b) and (c) the atoms are closer, and the wave functions begin to overlap. In (b) they combine with the same sign (symmetric state), and in (c) with opposite signs (antisymmetric state).

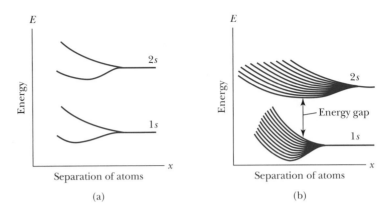

Figure 11.3 The 1s and 2s energy-level splittings of approaching hydrogen atoms for (a) 2 atoms, and (b) 11 atoms. Notice the splitting of each energy level into a nearly continuous band.

probability of being at that location. This causes the binding to be slightly stronger in the symmetric case, and hence the energy of that state is lower. As a result, there is a splitting of all possible energy levels ($1s$, $2s$, and so on), as is seen in Figure 11.3a. In each case the symmetric state ($\Psi_A + \Psi_B$) has the lower energy.

When more atoms are added (as in a real solid), there is a further splitting of energy levels. With a large number of atoms the levels are split into nearly continuous energy bands, with each band consisting of a number of closely spaced energy levels, as illustrated in Figure 11.3b. An energy gap may or may not exist between bands, depending on a number of factors, including the type of atom or atoms in the solid, lattice spacing, lattice structure, and temperature. We shall consider the effects of the existence and size of the energy gap later in this chapter.

Kronig-Penney Model

An effective way to understand the energy gap in semiconductors is to model the interaction between the electrons and the lattice of atoms. This interaction is more important in semiconductors than in good conductors, because the much higher resistivity implies tighter binding and/or more interaction. R. de L. Kronig and W. G. Penney developed a useful one-dimensional model of the electron-lattice interaction in 1931.* They assumed that an electron experiences the potential shown in Figure 11.4, an infinite one-dimensional array of finite potential wells. Each potential well models attraction to an atom in the lattice, so the size of the wells must correspond roughly to the lattice spacing.

The Kronig-Penney method for finding the allowed energy levels for the electron is the same as we used in Chapter 6. The electrons are not free; therefore we assume that the total energy E of an electron is less than the height V_0 of each

*R. de L. Kronig and W. G. Penney, *Proceedings of the Royal Society of London* **A130**, 499 (1931).

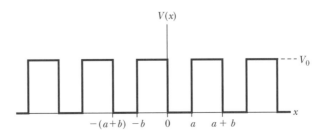

Figure 11.4 The Kronig-Penney square-well potential.

barrier/well in the Kronig-Penney potential shown in Figure 11.4. The electron is essentially free in the gap $0 < x < a$, where it has a wave function of the form

$$\psi = Ae^{ikx} + Be^{-ikx} \tag{11.1}$$

where the wave number k is given by the usual relation $k^2 = 2mE/\hbar^2$. In the barrier region $a < x < a + b$, however, the electron can tunnel through. As we saw in Chapter 6, this means that the wave function loses its sinusoidal character and becomes

$$\psi = Ce^{\kappa x} + De^{-\kappa x} \tag{11.2}$$

with $\kappa^2 = 2m(V_0 - E)/\hbar^2$.

Next, Kronig and Penney used the procedure of matching wave functions and first derivatives thereof at the various boundaries. The fact that these are finite potential wells makes the solution a lengthy one, and we shall not present it here. Application of the appropriate boundary conditions yields the important relation

$$\frac{\kappa^2 b}{2k} \sin(ka) + \cos(ka) = \cos(Ka) \tag{11.3}$$

where K is another wave number. When the left side of Equation (11.3) is plotted against the argument ka (Figure 11.5a), the relation in Equation (11.3) cannot be satisfied for all values of κ and k, because the sine and cosine functions are restricted to the range -1 to $+1$. There is no solution whenever the left side of Equation (11.3) has a value outside the range -1 to $+1$, because the right side of the equation *must* have a value in the range -1 to $+1$. This leads to **forbidden zones** for the wave numbers, and hence there are gaps in the corresponding energies. Figure 11.5b shows how the energy gaps and allowed energy bands alternate with each other as one sweeps over increasing energies. The gaps occur regularly at $ka = n\pi$, for integer values of n. Then with $k = n\pi/a = 2\pi/\lambda$, we see that $\lambda = 2a/n$. Thus, twice the lattice spacing ($2a$) corresponds to an integer multiple of the free-particle wavelength ($n\lambda$), and a free particle with this wavelength would be reflected by the lattice.

Before proceeding, we note some important differences between this simplified Kronig-Penney model and the single potential well studied in Chapter 6. First, for an infinite lattice the allowed energies within each band are continuous rather than discrete. In a real crystal the lattice is not infinite, but even if chains are only thousands of atoms long, the allowed energies are nearly continuous. Second, note that in a real three-dimensional crystal it is appropriate to speak

Forbidden zones

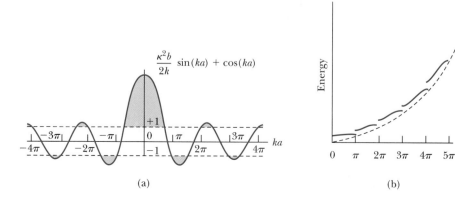

(a)   (b)

Figure 11.5 (a) Plot of the left side of Equation (11.3) versus ka for $\kappa^2 ba/2 = 3\pi/2$. Allowed energy values must correspond to the values of $k = \sqrt{2mE/\hbar^2}$ for which the plotted function lies between -1 and $+1$. Forbidden values are shaded in light blue. (b) The corresponding plot of energy versus ka for $\kappa^2 ba/2 = 3\pi/2$, showing the forbidden energy zones (gaps).

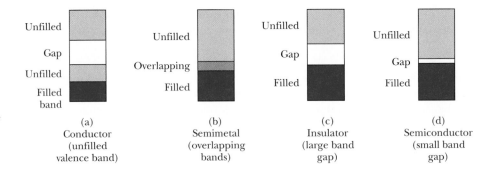

Figure 11.6 Possible band structures: (a) a conductor with an unfilled valence band, (b) a conductor with overlapping valence and conduction bands (a semimetal), (c) an insulator due to its large band gap, and (d) a semiconductor (due to its small band gap).

Wave vector
Brillouin zones

of a **wave vector** $\vec{k}$. The allowed ranges for $\vec{k}$ constitute what are referred to in solid state theory as **Brillouin zones**. Finally, in a real crystal the potential function is somewhat more complicated than the Kronig-Penney squares. As a result, the energy gaps are by no means uniform in size. The gap sizes may be changed by the introduction of impurities or imperfections of the lattice. These facts concerning the energy gaps are of paramount importance in understanding the electronic behavior of semiconductors.

Band Theory and Conductivity

Band theory can help us understand what makes a conductor, insulator, or semiconductor. Good conductors like copper can be understood using the free-electron model, but it is also possible to make a conductor using a material with its highest band filled, in which case no electron in that band can be considered free. If this filled band overlaps with the next higher band, however (so that effectively there is no gap between these two bands, as shown in Figure 11.6b), then an applied electric field can make an electron from the filled band jump to the higher level. This allows conduction to take place, although typically with slightly

Semimetals

higher resistance than in normal metals. Such materials are known as **semimetals**. Several widely used semimetals are arsenic, bismuth, and antimony.

Valence band

Conduction band

The band structures of insulators and semiconductors resemble each other qualitatively. Normally there exists in both insulators and semiconductors a filled energy band (referred to as the **valence band**) separated from the next higher band (referred to as the **conduction band**) by an energy gap. If this gap is at least several electron volts, the material is an insulator, as shown in Figure 11.6c. It is too difficult for an applied field to overcome that large an energy gap, and thermal excitations lack the energy to promote sufficient numbers of electrons to the conduction band. But if the energy gap is smaller—typically on the order of about 1 electron volt—it is possible for enough electrons to be excited thermally into the conduction band, so that an applied electric field can produce a modest current. The result is a semiconductor, with the band structure illustrated in Figure 11.6d. We shall explain this mechanism more fully in the next section.

11.2 Semiconductor Theory

At $T = 0$ we expect all of the atoms in a solid to be in the ground state. The distribution of electrons (fermions) at the various energy levels is governed by the Fermi-Dirac distribution of Equation (9.34):

$$F_{\text{FD}} = \frac{1}{\exp[\beta(E - E_{\text{F}})] + 1} \tag{9.34}$$

where $\beta = (kT)^{-1}$ and E_F is the Fermi energy. As was shown in Figure 9.9, more and more atoms are found in excited states when the temperature is increased from $T = 0$.

The increased number of electrons in excited states explains the temperature dependence of the resistivity of semiconductors. Only those electrons that have jumped from the valence band to the conduction band are available to participate in the conduction process in a semiconductor. More and more electrons are found in the conduction band as the temperature is increased, and the resistivity of the semiconductor therefore decreases.

Although it is not possible to use the Fermi-Dirac factor to derive an exact expression for the resistivity of a semiconductor as a function of temperature, we can make a couple of observations. First, the energy E in the exponential factor makes it clear why the band gap is so crucial. An increase in the band gap by a factor of 10 (say from 1 eV to 10 eV) will, for a given temperature, increase the value of $\exp(\beta E)$ by a factor of $\exp(9\beta E)$. This generally makes the factor F_{FD} so small that the material has to be an insulator. Our second observation is that, based on this analysis, one may expect the resistance of a semiconductor to decrease exponentially with increasing temperature. This is approximately true—although not exactly, because the function F_{FD} is not a simple exponential, and because the band gap does vary somewhat with temperature (Table 11.2).

A useful empirical expression developed by Clement and Quinnell for the temperature variation of standard carbon resistors is

$$\log R + \frac{K}{\log R} = A + \frac{B}{T}$$

(11.4)   **Clement-Quinnell equation**

Table 11.2 Energy Gaps for Selected Semiconductor Materials at $T = 0$ K and $T = 300$ K

Material	E_g (eV)	
	$T = 0$ K	$T = 300$ K
Si	1.17	1.11
Ge	0.74	0.66
InSb	0.23	0.17
InAs	0.43	0.36
InP	1.42	1.27
GaP	2.32	2.25
GaAs	1.52	1.43
GaSb	0.81	0.68
CdSe	1.84	1.74
CdTe	1.61	1.44
ZnO	3.44	3.2
ZnS	3.91	3.6

From C. Kittel, *Introduction to Solid State Physics*, 6th ed., New York: Wiley (1986), p. 185.

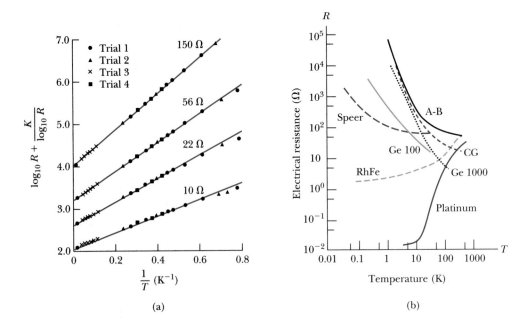

Figure 11.7 (a) An experimental test of the Clement-Quinnell equation, using resistance versus temperature data for four standard carbon resistors. The fit is quite good up to $1/T \approx 0.6$, corresponding to $T \approx 1.6$ K. (b) Resistance versus temperature curves for some thermometers used in research. A-B is an Allen-Bradley carbon resistor of the type used to produce the curves in (a). Speer is a carbon resistor, and CG is a carbon-glass resistor. Ge 100 and 1000 are germanium resistors. *From G. White*, Experimental Techniques in Low Temperature Physics, *Oxford: Oxford University Press (1979).*

where A, B, and K are constants. Figure 11.7a shows the data Clement and Quinnell obtained, plotted in such a way as to test Equation (11.4). The plot of $\log R + K/\log R$ versus $1/T$ is a straight line up to $1/T \approx 0.6$, so we conclude that the Clement-Quinnell equation (Equation 11.4) is good down to $T \approx 1\ \text{K}/0.6 = 1.7$ K. In Figure 11.7b we see a simple plot of resistance versus temperature for a number of different resistance thermometers. For all semiconductor materials shown, the variation of R with T is particularly rapid in the low-temperature range, from $T = 0$ up to about $T = 20$ K. For this reason carbon and other semiconductors are widely used as resistance thermometers in low-temperature physics.

Example **11.1**

Find the relative number of electrons with energies 0.1 eV, 1 eV, and 10 eV above the valence band at room temperature (293 K). *Note:* The Fermi energy is at the top of the valence band.

Strategy Equation (9.34) for the Fermi-Dirac factor may be used for this comparison. Note that 1 eV = 1.6×10^{-19} J. The Fermi energy is at the top of the valence band, so the energy above the valence band is $E - E_F$.

Solution For $E - E_F = 1$ eV, $(E - E_F)/kT = (1.6 \times 10^{-19}\,\text{J})/$ $(1.38 \times 10^{-23}\,\text{J/K})(293\,\text{K}) = 39.61$. Then

$$F_{FD}(0.1\ \text{eV}) = \frac{1}{e^{3.961} + 1} = 0.019$$

$$F_{FD}(1\ \text{eV}) = \frac{1}{e^{39.61} + 1} = 6.27 \times 10^{-18}$$

$$F_{FD}(10\ \text{eV}) = \frac{1}{e^{396.1} + 1} = 1.0 \times 10^{-172}$$

This example illustrates how strongly the Fermi-Dirac factor depends on the size of the band gap. The number of electrons available for conduction drops off sharply as the band gap increases.

When electrons move into the conduction band, they leave behind vacancies in the valence band. These vacancies are called **holes**. Because holes represent the absence of negative charges, it is useful to think of them as positive charges. Whereas the electrons move in a direction opposite to the applied electric field, the holes move in the direction of the electric field. A semiconductor in which there is a balance between the number of electrons in the conduction band and the number of holes in the valence band is called an **intrinsic semiconductor**. Examples of intrinsic semiconductors include pure carbon and germanium.

Holes

Intrinsic semiconductor

It is possible to fine-tune a semiconductor's properties by adding a small amount of another material, called a *dopant*, to the semiconductor. The resulting compound is called an **impurity semiconductor**. As an example, consider what happens when we add a small amount of arsenic to silicon. Each arsenic atom replaces a silicon atom in the lattice. What does this do to the conductive properties of the material? Notice that silicon has four electrons ($3s^2 3p^2$) in its outermost shell (this corresponds to the valence band) and arsenic has five ($4s^2 4p^3$). Therefore, while four of arsenic's outer-shell electrons participate in covalent bonding with its nearest neighbors (just as another silicon atom would), the fifth electron is very weakly bound. In fact, it takes only about 0.05 eV to move this extra electron into the conduction band. The effect is that adding only a small amount of arsenic to silicon greatly increases the electrical conductivity.

Impurity semiconductor

The addition of arsenic to silicon creates what is known as an **n-type** semiconductor (*n* for negative), because it is the electrons close to the conduction band that will eventually carry electrical current. The new arsenic energy levels just below the conduction band are called **donor levels** (see Figure 11.8a), because an electron there is easily donated to the conduction band.

n-type semiconductors

Donor levels

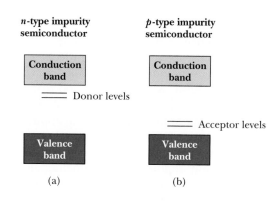

(a)

(b)

Figure 11.8 Energy bands of impurity semiconductors, showing (a) donor levels in an *n*-type impurity semiconductor and (b) acceptor levels in a *p*-type impurity semiconductor.

Special Topic

THE QUANTUM HALL EFFECT

The quantum Hall effect has been the subject of intense investigation by semiconductor physicists since its discovery by the German physicist Klaus von Klitzing in 1980 (Nobel Prize, 1985). Schematically, the apparatus for observing the quantum Hall effect is similar to that used for the normal Hall effect. A thin strip of material is positioned perpendicular to a uniform magnetic field. A current is then made to flow along the length of the strip, and voltage leads detect the potential difference in two directions: along the direction of current flow and perpendicular to the current flow (see Figure 11.9). Both of these voltages are directly proportional to the current flowing through the strip. Therefore in order to characterize the material being studied it is customary to divide both voltages by the current to obtain resistance. The sample's ordinary electrical resistance is the voltage drop in the direction of current flow divided by the current, and the Hall resistance is the voltage in the perpendicular direction divided by the current.

In the normal Hall effect, the Hall resistance increases in direct proportion to the strength of the applied magnetic field. Von Klitzing discovered the quantum Hall effect in semiconductor materials in which the current-carrying electrons were confined to an extremely thin layer (5–10 nm thick), when the sample was placed in a very strong magnetic field (2–10 T) and cooled to about 1 K. Under these conditions von Klitzing found that the Hall resistance, when plotted against the increasing magnetic field, showed well-defined steps at levels of h/ne^2, where h is Planck's constant, e is the electronic charge, and n is an integer (Figure A). Strictly speaking, it is the electrical conductance (reciprocal of the electrical resistance) that is quantized in units of e^2/h. However, we shall stick to the convention of speaking in terms of the Hall resistance. The "base" Hall resistance h/e^2 is about 25,812.8 Ω and has been renamed the von Klitzing constant R_K, with $R_K = h/e^2$. A second striking feature of the quantum Hall effect is that when the Hall resistance is on a plateau, the ordinary electrical resistance is practically zero (again refer to Figure A).

In 1982 another curious phenomenon was found by Daniel Tsui and Horst Störmer. They discovered that some semiconductor materials exhibited a *fractional* quantum Hall effect. Electrical conductances have been found in fractions of 1/2, 1/3, 1/5, 1/7, 2/3, 4/3, 5/3, 4/5, 6/7, and 5/2 of the base value e^2/h. The fractional quantum Hall effect can be observed only in samples made of ultra-pure materials.

The 1998 Nobel Prize for Physics was awarded to Tsui and Störmer, along with Robert Laughlin, who explained the fractional quantum Hall effect theoretically using a quantum fluid model.

Semiconductor theorists have developed models that explain both the quantum Hall effect and the fractional quantum Hall effect. These models are too

---

Now consider what happens when indium ($5s^2 5p^1$) is added to silicon. Indium has one less electron in its outer shell than silicon. The result is one extra hole per indium atom. The existence of these holes creates extra energy levels just above the valence band, because it takes relatively little energy to move another **Acceptor levels** electron into a hole. Those new indium levels are called **acceptor levels** (see Figure 11.8b), because they can easily accept an electron from the valence band. Again, the result is an increased flow of current (or, equivalently, lower electrical resistance) as the electrons move to fill holes under an applied electric field. It is always easier to think in terms of the flow of positive charges (holes) in the di-**p-type semiconductors** rection of the applied field, so we call this a **p-type** semiconductor (p for positive).

In addition to intrinsic and impurity semiconductors, there are many **com-**
Compound **pound semiconductors**, which consist of equal numbers of two kinds of atoms.
semiconductors Examples of compound semiconductors include GaAs, InP, GaN, SiC, and InSb.

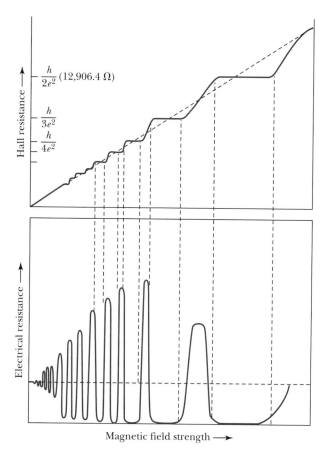

Hall resistance

$\frac{h}{2e^2}$ (12,906.4 Ω)

$\frac{h}{3e^2}$

$\frac{h}{4e^2}$

Electrical resistance

Magnetic field strength →

Figure A The quantized steps in the Hall resistance are evident in the top graph. In the bottom graph we see the corresponding disappearance of electrical resistance. *B. I. Halperin*, Scientific American **254**(4), 52–60 (April 1986).

elaborate to detail here.* We can say, however, that they are based on constructing electron wave functions in *two* dimensions (remember, the electrons are basically confined to a plane). Such a two-dimensional structure is possible in field effect transistors (discussed in Section 11.3) and in devices known as *heterojunctions*, in which electrons are confined to an interface region between two semiconductor materials.

One way that the quantum Hall effect and fractional quantum Hall effect have been applied is in the field of *metrology*, which seeks to develop measurements and standards for industrial and research purposes. The quantum Hall effects can be used to provide very accurate standards for resistance measurements. Because the base resistance $R_K = h/e^2$ depends only on the physical constants h and e, it should be possible to fabricate resistance standards that are relatively insensitive to how they were made, as long as they do exhibit a quantum Hall effect.

*See D. R. Yennie, *Reviews of Modern Physics* **59**, 781–824 (1987) for many of the details.

In a compound semiconductor, a mixed form of covalent and ionic bonding serves to create the same type of tetrahedral bonding as in pure Ge and Si. Compound semiconductors have band gaps and can be doped like intrinsic semiconductors to form p-type and n-type semiconductors.

In introductory physics some students find it bothersome at first that the particles that actually move in a conductor, the electrons, travel opposite to the direction of the applied field and opposite to what we call "positive current."* One may well ask to what extent the holes are real, because in both p-type and n-type semiconductors electrons are moving. It can be shown experimentally (using the

*This accident of history is usually blamed on Benjamin Franklin, who popularized the "single-fluid" model of electricity. Franklin considered electrical forces to be the result of excesses (hence "positive") and deficiencies (hence "negative") of an electrical fluid. Of course his designation is rather arbitrary, and if he had chosen the reverse, electrons would be positive.

normal Hall effect) that *p*-type materials really do behave as if the charge carriers were positive. The magnitude and sign of the Hall voltage allow one to calculate the density and sign of the charge carriers in a conductor or semiconductor and verify that charges in carriers in *p*-type and *n*-type materials have different signs. In fact, hole conduction is not a phenomenon limited to semiconductors. Among metallic conductors, zinc is well known to exhibit positive charge carriers.

Conceptual Example 11.2

What kind (*p*-type or *n*-type) of semiconductor is made if pure silicon is doped with a small amount of (a) gallium; (b) antimony?

Solution (a) Silicon has four electrons in its outer shell $(3s^2 3p^2)$, but gallium has three $(4s^2 4p^1)$. With this electron deficiency, silicon doped with gallium should be a *p*-type semiconductor. (b) Now silicon is being doped with antimony, which has an outer shell $5s^2 5p^3$. There is an extra electron, making this an *n*-type semiconductor.

Example 11.3

A very thin rectangular strip of zinc has been deposited on an insulating substrate. Let the length, width, and thickness of the strip be *x*, *y*, and *z*, respectively. The length and width are measured to be 10.0 cm and 2.0 cm. When a potential difference of 20 mV is applied as shown in Figure 11.9a, the current through the strip is 400 mA.

(a) Use the fact that the resistivity of zinc is (at room temperature) $\rho = 5.92 \times 10^{-8} \ \Omega \cdot \text{m}$ to find the thickness of the strip.

(b) Now a magnetic field of 0.25 T is applied perpendicular to the strip as shown in Figure 11.9b. A Hall voltage $V_H = +0.56 \ \mu\text{V}$ appears when the voltmeter leads (+ and −) are connected as shown. Determine the sign of the charge carriers and their number density.

Strategy (a) The resistance of a wire with uniform cross-sectional area is related to the resistivity by $R = \rho x / A$, where *A* is the cross-sectional area and *x* is the length of the wire. In this case $A = yz$. This allows us to find the thickness of a wire with known resistivity.

(b) In equilibrium the magnetic force on a charge carrier is equal to the electric force due to the Hall voltage. Thus $eE = evB$. From basic electrical conductivity, the electron drift speed is $v = I/neA$. These two relations allow us to solve for the charge carrier density *n*.

Solution (a) The resistance of the sample is $R = \rho x / A$. From Ohm's law, $R = V/I$, so $R = V/I = \rho x / yz$. Therefore

$$z = \frac{\rho x I}{yV} = \frac{(5.92 \times 10^{-8} \ \Omega \cdot \text{m})(0.10 \ \text{m})(0.400 \ \text{A})}{(0.02 \ \text{m})(0.02 \ \text{V})}$$

$$= 5.92 \times 10^{-6} \ \text{m} \qquad (11.5)$$

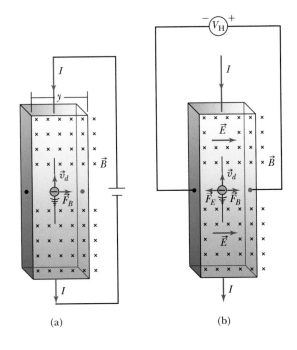

Figure 11.9 A thin strip of metal immersed in a magnetic field is used to test the Hall effect. (a) Here negative charge carriers are forced to the right. (b) In this configuration, the buildup of negative charge on the right side (with a corresponding positive charge on the left) sets up the electric field as shown. This creates an electric force on the charge carriers equal and opposite to the magnetic force. The voltmeter (reading V_H) can detect the magnitude and sign of the potential difference across the strip.

(b) Combining the relations $eE = evB$ and $v = I/neA$,

$$\frac{eV_H}{y} = eB\frac{I}{neA} \tag{11.6}$$

where we have used the fact that for this geometry $E = V_H/y$. Finally, because $A = yz$, Equation (11.6) reduces to

$$n = \frac{IB}{eV_H z} \tag{11.7}$$

$$= \frac{(0.400 \text{ A})(0.25 \text{ T})}{(1.6 \times 10^{-19} \text{ C})(5.6 \times 10^{-7} \text{ V})(5.92 \times 10^{-6} \text{ m})}$$

$$= 1.88 \times 10^{29} \text{ m}^{-3}$$

This is a very high density (slightly higher, in fact, than one finds for copper). It turns out that the higher the density of conductors (electrons or holes), the thinner the strip needs to be in order to obtain a decent Hall voltage. That is why this experiment would be much easier to do with a semiconductor, most of which have a much lower value of n.

Now in order to determine the sign, notice that if the charge carriers were negative, the voltmeter as connected in Figure 11.9b would read *negative*. If instead the majority charge carriers are positive, the magnetic field will cause them to drift to the right on the strip. This is consistent with a positive voltmeter reading, and therefore we conclude that the majority charge carriers for zinc are *positive*. This is a somewhat unusual fact, though zinc is not unique—the majority carriers in Cd and Be are also positive. For most metals the charge carriers are negative.

Thermoelectric Effect

The **thermoelectric effect** is often used to study the properties of semiconductors. When there is a temperature gradient in a thermoelectric material, an electric field appears. It is easier to understand why this happens in a pure metal, in which case we can assume a gas of free electrons. As in an ideal gas, the density of free electrons is greater at the colder end of the wire, and therefore the electrical potential should be higher at the warmer end and lower at the colder end. Of course the free-electron model is not valid for semiconductors; nevertheless, the conducting properties of a semiconductor are temperature dependent, as we have seen, and therefore it is reasonable to believe that semiconductors should exhibit a thermoelectric effect.

In one dimension the induced electric field E in a semiconductor is proportional to the temperature gradient, so that

$$E = Q\frac{dT}{dx} \tag{11.8}$$

where Q is called the **thermoelectric power**. It turns out that the direction of the induced field depends on whether the semiconductor is p-type or n-type (see Figure 11.10), so the thermoelectric effect can be used to determine the extent to which n- or p-type carriers dominate in a complex system. More detailed analysis yields information about carrier concentration and band structures.

Thermoelectric power

The thermoelectric effect we have just described is sometimes called the **Seebeck effect**. It is the most commonly used thermoelectric effect, but there are two others. In a normal conductor, heat is generated at the rate of I^2R. But a temperature gradient across the conductor causes additional heat to be generated. This is known as the **Thomson effect**. Strangely enough, the Thomson effect is entirely reversible. If the direction of the current is toward the higher temperature end of the conductor, heat is generated, but if the current flows toward the lower temperature end, heat is absorbed from the surroundings. The **Peltier effect** occurs when heat is generated at a junction between two conductors as current passes through the junction.

An important application of the Seebeck thermoelectric effect is in thermometry. The thermoelectric power of a given conductor varies as a function of

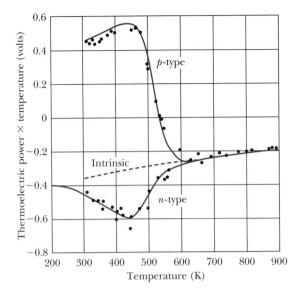

Figure 11.10 Variation in thermoelectric power × temperature with temperature for *p*-type and *n*-type silicon. *From T. H. Geballe and G. W. Hall, Physical Review* **98**, *940 (1955).*

Thermocouple

temperature, and the variation can be quite different for two different conductors. This difference makes possible the operation of a **thermocouple** (Figure 11.11). Two conductors A and B are joined at each end. One end is held at a reference temperature T_0, and the other is placed at some unknown temperature T. The differential thermopowers of A and B cause a voltage V_{AB} to be induced between the two ends of the thermocouple. Knowing the temperature variation of each thermopower allows one to calculate the temperature difference $T - T_0$ (and hence the unknown temperature T) as a function of V_{AB}. In practice it is not necessary

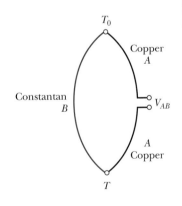

Figure 11.11 Schematic diagram of a thermocouple circuit. Two materials A and B (in this case, copper and constantan) are joined at their ends. One junction is held at a reference temperature T_0, and the other junction is used to measure the temperature T.

Table 11.3 Thermocouple Voltage (in millivolts) versus Temperature (in °C) for an Iron-Constantan Thermocouple

	0	1	2	3	4	5	6	7	8	9
°C					Millivolts					
0	0.00	0.05	0.10	0.15	0.20	0.25	0.30	0.35	0.40	0.45
10	0.50	0.56	0.61	0.66	0.71	0.76	0.81	0.86	0.91	0.97
20	1.02	1.07	1.12	1.17	1.22	1.28	1.33	1.38	1.43	1.48
30	1.54	1.59	1.64	1.69	1.74	1.80	1.85	1.90	1.95	2.00
40	2.06	2.11	2.16	2.22	2.27	2.32	2.37	2.42	2.48	2.53
50	2.58	2.64	2.69	2.74	2.80	2.85	2.90	2.96	3.01	3.06
60	3.11	3.17	3.22	3.27	3.33	3.38	3.43	3.49	3.54	3.60
70	3.65	3.70	3.76	3.81	3.86	3.92	3.97	4.02	4.08	4.13
80	4.19	4.24	4.29	4.35	4.40	4.46	4.51	4.56	4.62	4.67
90	4.73	4.78	4.83	4.89	4.94	5.00	5.05	5.11	5.16	5.21
100	5.27	5.32	5.38	5.43	5.48	5.54	5.59	5.65	5.70	5.76

From *CRC Handbook of Chemistry and Physics*, 55th ed., Cleveland: CRC Press (1974), p. E-109.

to measure thermopowers to use a thermocouple. Well-established tables for voltage versus temperature can be found in many handbooks for different conducting pairs over wide ranges of temperature (see Table 11.3).

11.3 Semiconductor Devices

Their unique electronic properties have made semiconductors useful in a wide range of applications in science and industry. In this section we shall describe some of the most important semiconductor devices and explain how they are used. But we can only scratch the surface; the range of applications is wide, and semiconductor technology is still evolving at a rapid rate. Many of you who are students today will help shape the electronics revolution as it continues in this century.

Diodes

We begin with a simple device known as a **pn-junction diode**, in which p-type and n-type semiconductors are joined together, as shown in Figure 11.12. The principal characteristic of a pn-junction diode is that it allows current to flow easily in one direction but hardly at all in the other direction. We call these situations **forward bias** and **reverse bias**, respectively.

pn-junction diode

Forward and reverse bias

To explain how this happens, we must first consider the situation when no external voltage is applied (the "no bias" case—Figure 11.12a). Free electrons from the n side can drift through random motion to the p side, and their migration leaves a small net positive charge on the n side. This flow of electrons is shown as the electron recombination current I_r. Equilibrium is achieved very quickly, because the potential difference set up by the charge migration (with the n side now at a higher potential than the p side) tends to prohibit further migration. There is typically a small current of electrons from the p side to the n side, due to the fact that at normal temperatures, electrons on the p side can be thermally excited from the valence band to an acceptor level. The thermal electron current is designated I_t in Figure 11.12a. With no external voltage source, the electron currents I_t and I_r exactly cancel each other, thus preserving the equilibrium.

In the reverse bias case, a potential difference is applied across the junction as shown in Figure 11.12b. In a normal conductor, electrons would tend to flow

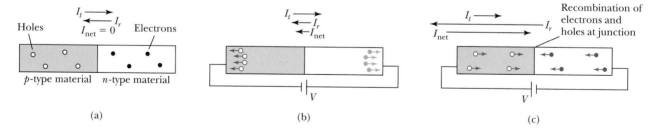

(a)   (b)   (c)

Figure 11.12 The operation of a pn-junction diode. (a) This is the no-bias case. The small thermal electron current (I_t) is offset by the electron recombination current (I_r). The net positive current (I_{net}) is zero. (b) With a DC voltage applied as shown, the diode is in reverse bias. Now I_r is slightly less than I_t. Thus there is a small net flow of electrons from p to n and positive current from n to p. (c) Here the diode is in forward bias. Because current can readily flow from p to n, I_r can be much greater than I_t. [*Note:* In each case, I_t and I_r are electron (negative) currents, but I_{net} indicates positive current.]

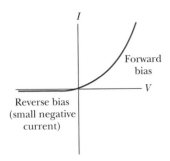

Figure 11.13 A typical *I-V* curve for a *pn*-junction diode.

freely from the negative toward the positive terminal. But in the *pn* junction, there remains the tendency for electrons to drift back from the *n* side toward the *p* side whenever an imbalance is created. This compensates for most of the electron flow. The result is only a small net flow of electrons from the *p* side to the *n* side, or in other words, a small positive current from the high-potential side of the battery to the low-potential side of the battery.

In forward bias (Figure 11.12c) a potential difference is applied with the positive terminal connected to the *p* side and the negative terminal to the *n* side of the junction. Now we are pushing the electrons the way they would tend to move anyway. The only compensating factor, the thermal flow of electrons from *p* to *n* just described, is typically too small to retard the flow of electrons. The result is a steady flow of positive current from higher to lower potential, inhibited only by the natural resistance of the device. Figure 11.13 shows the behavior of the *pn*-junction diode in both forward and reverse bias.

We can use the tools of statistical physics developed in Chapter 9 to model the *I-V* characteristics of the *pn*-junction diode and thereby obtain a quantitative justification of the empirical curve shown in Figure 11.13. Let I_0 and V_0 be the current and voltage in the no-bias situation, and let N be the number of electrons on the *n* side in the conduction band. At room temperature, Maxwell-Boltzmann statistics are sufficient to describe the behavior of the electrons. In the no-bias case the number of electrons able to move from the *n* side to the *p* side is proportional to $Ne^{-eV_0/kT}$, and therefore I_0 is proportional to the same factor. Under the influence of a forward bias voltage *V*, however, the number is proportional to $Ne^{-e(V_0-V)/kT} = Ne^{-eV_0/kT}e^{eV/kT}$. Therefore the electron current under forward bias must be $I = I_0e^{eV/kT}$. Because there is still an additional current $-I_0$ in forward bias due to the motion of holes from the *n* side to the *p* side, the total current in forward bias is

$$I = I_0(e^{eV/kT} - 1) \tag{11.9}$$

Equation (11.9) is a rather good approximation of the *I-V* curve of Figure 11.13.

Example 11.4

Consider a simple *pn*-junction diode. Suppose this diode carries a current of 50 mA with a forward bias voltage of 200 mV at room temperature (293 K). What is the current when a *reverse* bias of 200 mV is applied?

Strategy We could use Equation (11.9) to solve for I_0 in the forward bias case and then use that I_0 to find the current in reverse bias. But it is simpler to use the fact that the forward bias current I_f and the reverse bias current I_r are related by

$$\frac{I_r}{I_f} = \frac{I_0(e^{eV_r/kT} - 1)}{I_0(e^{eV_f/kT} - 1)}$$

Solution Using the current ratio along with the numerical values $I_f = 50$ mA, $V_f = +200$ mV, and $V_r = -200$ mV, we find that

$$\frac{eV_f}{kT} = \frac{(1.602 \times 10^{-19}\ \text{C})(0.200\ \text{V})}{(1.38 \times 10^{-23}\ \text{J/K})(293\ \text{K})} = 7.921$$

Therefore

$$I_r = I_f \frac{e^{eV_r/kT} - 1}{e^{eV_f/kT} - 1} = (50\ \text{mA})\frac{e^{-7.921} - 1}{e^{7.921} - 1} \approx (50\ \text{mA})\frac{-1}{e^{7.921}}$$

$$I_r = \frac{-50\ \text{mA}}{2754} = -18\ \mu\text{A}$$

This is an extremely small current, showing clearly the "one-way" property of the diode.

Bridge Rectifiers

The diode is an important tool in many kinds of electrical circuits. As an example, consider the **bridge rectifier** circuit shown in Figure 11.14. The bridge rectifier is set up so that it allows current to flow in only one direction through the resistor R when an alternating current supply is placed across the bridge. The current through the resistor is then a rectified sine wave of the form

$$I = I_{max} \left| \sin(\omega t) \right| \tag{11.10}$$

This is the first step in changing alternating current to direct current. The design of a power supply can be completed by adding capacitors and resistors in appropriate proportions. This is an important application, because direct current is needed in many devices and the current that we get from our wall sockets is alternating current.

(a)

(b)

Figure 11.14 (a) The circuit diagram symbol for a diode, with p and n sides indicated. (b) Circuit diagram for a diode bridge rectifier.

Zener Diodes

The **Zener diode** is made to operate under reverse bias once a sufficiently high voltage has been reached. The I-V curve of a Zener diode is shown in Figure 11.15. Notice that under reverse bias and low voltage the current assumes a low negative value, just as in a normal pn-junction diode. But when a sufficiently large reverse bias voltage is reached, the current increases at a very high rate.

Depending on which semiconductor materials are used to make the diode, the Zener I-V phenomenon may occur in one of two ways. In the process known as *Zener breakdown*, the applied voltage induces electrons from the valence band on the p side to move to the conduction band on the n side. Once a high reverse bias voltage is reached, large numbers of electrons can be pulled immediately into the conduction band, thus accounting for the sudden breakdown. In another process known as *avalanche multiplication*, the applied voltage is high enough to accelerate electrons to energies sufficient to ionize atoms through collision. The "avalanche" occurs when electrons released in this process are in turn accelerated and ionize other atoms. Again, this happens rather suddenly, accounting for a sharp increase in current at a particular voltage. In heavily doped material, Zener breakdown is likely to occur first.

A common application of Zener diodes is in regulated voltage sources, which are components in many electronic instruments. The idea behind the operation of this regulator circuit is that, as long as we operate on the steep part of the I-V curve, any change (say an increase) in the supply voltage V tends to be compensated for by a sharp increase in the current through the Zener diode (Figure 11.16). Then the voltage across the resistor increases, which in turn tends to keep the output voltage $V_Z = V - IR$ fairly constant.

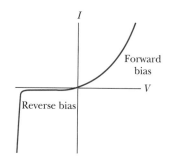

Figure 11.15 A typical I-V curve for a Zener diode.

Figure 11.16 A Zener diode reference circuit.

Light-Emitting Diodes

Another important kind of diode is the **light-emitting diode (LED)**. Whenever an electron makes a transition from the conduction band to the valence band (effectively recombining the electron and hole) there is a release of energy in the form of a photon (Figure 11.17). In some materials the energy levels are spaced so that the photon is in the visible part of the spectrum. In that case, the continuous flow of current through the LED results in a continuous stream of nearly monochromatic light.

Figure 11.17 Schematic of a LED. A photon is released as an electron falls from the conduction band to the valence band. The band gap may be large enough that the photon will be in the visible portion of the spectrum.

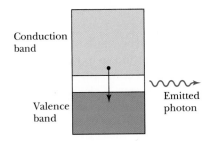

LEDs are found in many places today. The visible displays of many clocks, automobile dashboards, and other instruments consist of combinations of LEDs. Automobile manufacturers have begun using the LED in taillights, and they appear in newer traffic signals. The LED can also be used as a principal component of a laser. Photons released from the LED can be used to stimulate the emission of other photons, as described in Section 10.2.

Photovoltaic Cells

An exciting application closely related to the LED is the **solar cell**, also known as the **photovoltaic cell**. Simply put, a solar cell takes incoming light energy and turns it into electrical energy. A good way to think of the solar cell is to consider the LED in reverse (Figure 11.18). A *pn*-junction diode can absorb a photon of solar radiation by having an electron make a transition from the valence band to the conduction band. In doing so, both a conducting electron and a hole have been created. If a circuit is connected to the *pn* junction, the holes and electrons will move so as to create an electric current, with positive current flowing from the *p* side to the *n* side. Even though the efficiency of most solar cells is low, their widespread use could potentially generate significant amounts of electricity. Remember that the "solar constant" (the energy per unit area of solar radiation reaching the Earth) is over 1400 W/m^2, and more than half of this makes it through the atmosphere to the Earth's surface. There has been tremendous progress in recent years toward making solar cells more efficient.

The photovoltaic effect was first observed in 1839 by Alexandre-Edmond Becquerel. For over 100 years the discovery remained a curiosity for scientists with

(a)   (b)

Figure 11.18 (a) Schematic of a photovoltaic cell. Note the similarity to Figure 11.17. (b) A schematic showing more of the working parts of a real photovoltaic cell. *From H. M. Hubbard, Science* **244**, *297–303 (21 April 1989).*

access to materials that naturally exhibit the photovoltaic effect. In the 1950s, however, researchers interested in using semiconductors in electronics began to develop better ways of manufacturing semiconductor devices, including photovoltaics. In 1954 a silicon-based solar cell with an efficiency of 6% was made at Bell Laboratories (efficiency is defined as the ratio of electrical power output to the power of incident radiation). Since then the advances in design and manufacture of semiconductor devices have been applied to solar cells. As a result, both silicon and gallium arsenide solar cells can now be made with efficiencies of over 20%. Lower manufacturing costs and higher efficiencies have reduced the cost of photovoltaic electricity by more than a factor of 100, from $15 per kilowatt-hour in the 1950s to about $0.10 per kilowatt-hour in the early 2000s. This cost is still almost one order of magnitude higher than the cost of producing electricity using our other principal renewable energy source, hydroelectric power, and it is somewhat higher than the cost of using nonrenewable sources like coal and oil. Continued progress in reducing the cost of solar cells, combined with increased costs of producing energy from other sources, will make the use of solar cells increasingly attractive as the twenty-first century unfolds.

The most widely used and studied photovoltaic materials are silicon-based. The technology exists for making large single crystals of silicon. It is in silicon-based cells that efficiencies of over 20% are reached. Unfortunately, the cost of making good single crystals of silicon is prohibitive. The cost of making cells with polycrystalline and amorphous silicon is lower, but so is the efficiency of the solar cells made with these materials.

There is now hope that other thin-film materials can approach the efficiency of silicon devices. The prime candidate is GaAs and its alloys, including AlGaAs and InGaAs. Recently a device with an efficiency of 25% was made on a reusable GaAs substrate. These reusable single crystals of GaAs are now made fairly inexpensively by growing them on single crystals of Ge.

Another advantage of GaAs is that it makes more efficient use of the solar spectrum than Si. Figure 11.19 shows that the response of GaAs and its alloys is almost entirely within the visible portion of the spectrum, where the output of solar radiation is highest (recall the Wien law from Chapter 3). Recently the use of multiple layers, with each layer sensitive to a different wavelength range, has proved successful. Figure 11.20 shows a two-layer device created by Boeing. The upper layer semiconductor, GaAs, has a band gap of 1.42 eV. That energy corresponds to a photon wavelength of

$$\lambda = hc/E = 875 \text{ nm}$$

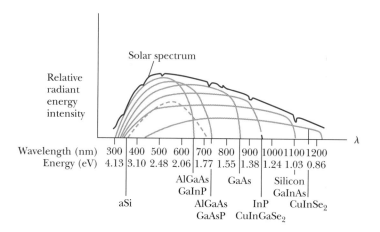

Figure 11.19 The response of various solar cells as a function of wavelength. Note that aSi is amorphous silicon. Due to the variation in band gaps of the different semiconductors, solar cells are more sensitive to some wavelengths than to others. *From H. M. Hubbard,* Science **244**, *297–303 (21 April 1989).*

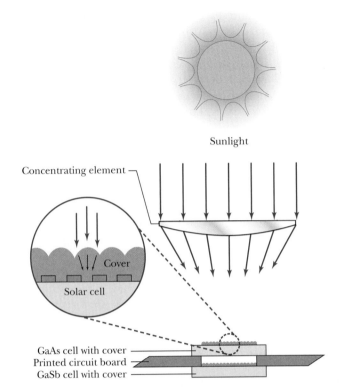

Figure 11.20 Boeing's 31% efficient solar cell, containing a gallium arsenide upper layer and a gallium antimonide lower layer used to capture infrared radiation. *From B. W. Henderson,* Aviation Week & Space Technology *131, 61 (23 October 1989).*

A photon with $\lambda = 875$ nm is in the near infrared portion of the spectrum. The lower layer is made of GaSb, with a band gap of 0.72 eV (corresponding to $\lambda = 1730$ nm). Photons with wavelengths shorter than 875 nm will have sufficient energy to boost an electron from the valence to the conduction band in GaAs, and therefore the top layer is able to use near-IR, visible, and ultraviolet solar radiation to produce energy. Less-energetic IR photons make it through to the GaSb layer, where they can be absorbed and converted to electrical energy if their wavelength is less than 1730 nm. This device has achieved a proven efficiency of 31%.

Other multilayer systems are currently being studied, with the hope of raising the efficiency even further. For example, one promising system multilayer system uses the compound $In_{1-x}Ga_xN$, with the value of x tuned to match the solar spectrum over a wider energy range (0.7 eV to 3.4 eV).* Estimates are that a solar cell based on this system may reach an efficiency of 50%.

The use of solar radiation as an energy source has some intrinsic advantages and disadvantages compared with other sources. One outstanding advantage relative to fossil-fuel burning is that solar cells generate no environmental waste in their use and relatively little in their manufacture. The release of polluting particles, poisonous gases, and greenhouse gases (especially CO_2) in burning fossil fuels may make it desirable to replace fuel-burning generators with solar-cell generators even before the absolute cost per kilowatt-hour of solar cells becomes much lower. Another advantage of solar cells is that they can be made in various sizes. Large arrays (Figure 11.21) can be used to produce electricity for cities and factories, while smaller arrays can be put on rooftops in vacant areas to generate power for homes and small businesses. The principal disadvantage of even very

*W. Wu et al., Unusual properties of the fundamental band gap of InN, *Applied Physics Letters* **80**, 3967–3969 (2002).

Figure 11.21 A photovoltaic center in Carrisa Plains, California. *Harald Sund, The Image Bank/Getty Images.*

efficient solar cells is that it is not always sunny. The use of solar energy at night or during cloudy periods would require large, generally inefficient storage systems. Perhaps this is where the superconducting storage rings discussed in Sections 10.5 and 10.6 can best be put to use. (See also Problems 12 and 26.)

Example 11.5

Suppose an array of 30% efficient solar cells has an effective area of 100 m by 100 m. The cells are tilted so as to receive maximum solar flux, an average of 800 W/m² for a day with 12 hours of daylight. How much energy does this array produce each day? Compare that energy with the output of a 100-MW conventional power plant.

Strategy Energy = power × time. With 1 W = 1 J/s, it is necessary to convert the time to units of s. The solar cell runs for 12 hours, but the power plant runs for 24 hours.

Solution The solar cells have an area of 100 m × 100 m = 10^4 m². The time is

$$t = 12 \text{ h} \times \frac{60 \text{ min}}{\text{h}} \times \frac{60 \text{ s}}{\text{min}} = 43{,}200 \text{ s}$$

Therefore the energy produced in one 12-hour day is

$$E = 800 \text{ W/m}^2 \times 10^4 \text{ m}^2 \times 43{,}200 \text{ s} = 3.5 \times 10^{11} \text{ J}$$

The power plant operates for 24 hours (86,400 s) at a rate of 100 MW = 10^8 W, and produces

$$E = 10^8 \text{ W} \times 86{,}400 \text{ s} = 8.6 \times 10^{12} \text{ J}$$

of energy, which is 25 times more than the solar cell array produces. Producing a comparable amount of energy requires a larger solar array.

Transistors

Another use of semiconductor technology is in the fabrication of **transistors**, devices that amplify voltages or currents in many kinds of circuits. The first transistor was developed in 1948 by John Bardeen, William Shockley, and Walter Brattain (Nobel Prize, 1956).* As an example we consider an ***npn*-junction transistor**, which consists of a thin layer of *p*-type semiconductor sandwiched between two *n*-type semiconductors. The three terminals (one on each semiconducting material) are known as the **collector**, **emitter**, and **base**. A good way of thinking of the operation of the *npn*-junction transistor is to think of two *pn*-junction diodes back

npn-junction transistor

Collector, emitter, base

*For a history of transistor development, see *Physics Today* 34–39 (December 1997).

John Bardeen (1908–1991), pictured left, received his first Nobel Prize in 1956 for work on the semiconductor transistor. See his biographical profile in Chapter 10.

William Shockley (1910–1989), pictured center, earned his undergraduate degree at Caltech and his Ph.D. from MIT in 1936. After obtaining his doctorate, Shockley went to work at Bell Labs, joining the research group of C. J. Davisson (see Chapter 5). In the late 1940s, he teamed with Bardeen and Brattain to develop the semiconductor transistor. Shockley left Bell Labs in 1954 to start Shockley Semiconductor Laboratories, one of the first semiconductor labs in what is now known as Silicon Valley in California.

Walter Brattain (1902–1987), pictured right, grew up in the state of Washington and earned his undergraduate degree from Whitman College. After earning his Ph.D. from the University of Minnesota, Brattain spent the bulk of his career at Bell Labs. In addition to his transistor work with Bardeen and Shockley, Brattain is known for using the photoelectric effect to learn more about the surfaces of semiconductors. *AIP Niels Bohr Library, Brattain Collection.*

to back (Figure 11.22b). Then the emitter side consists of a diode in forward bias, and the collector side consists of a diode in reverse bias. The base is therefore the p side of each diode. Explicitly showing the *npn* construction of each transistor would prove cumbersome in a complicated circuit diagram containing many transistors. A more compact notation for those cases is shown in Figure 11.22c and d.

Consider now the *npn* junction in the circuit shown in Figure 11.23a. If the emitter is more heavily doped than the base, then there is a heavy flow of electrons from left to right into the base. The base is made thin enough so that virtually all of those electrons can pass through the collector and into the output portion of the circuit. As a result the output current is a very high fraction of the input current. The key now is to look at the input and output voltages. Because the base-collector combination is essentially a diode connected in reverse bias, the voltage on the output side can be made higher than the voltage on the input side. Recall that the output and input currents are comparable, so the resulting output power (current × voltage) is much higher than the input power.

In the circuit we have just described, the transistor is used as a voltage amplifier. The circuit in Figure 11.23a can be modified to serve as a current amplifier by moving the input signal to a position between the base and ground, as shown in Figure 11.23b. As we have already shown, a very small current flows in that branch of the circuit. Therefore, in this configuration the output current will be much higher than the input current. It is also possible to make a ***pnp*-junction transistor**, which may be understood using the same model as we used for the *npn* junction, but with hole conduction taking the place of electron conduction.

***pnp*-junction transistor**

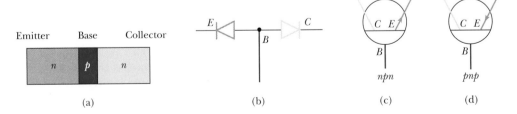

Emitter   Base   Collector

n p n

(a)

E C B

(b)

C E B *npn*

(c)

C E B *pnp*

(d)

Figure 11.22 (a) In the *npn* transistor, the base is a *p*-type material, and the emitter and collector are *n*-type. (b) The two-diode model of the *npn* transistor. (c) The *npn* transistor symbol used in circuit diagrams. (d) The *pnp* transistor symbol used in circuit diagrams.

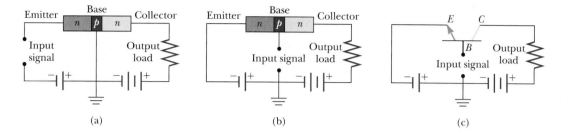

(a)　　　　　　(b)　　　　　　(c)

Figure 11.23 (a) The *npn* transistor in a voltage amplifier circuit. (b) The circuit has been modified to put the input between base and ground, thus making a current amplifier. (c) The same circuit as in (b) using the transistor circuit symbol.

As an example of an amplifier circuit, consider Figure 11.24. The voltages V_{bb} and V_{cc} are fixed. The resistances R_c and R_e may in part be separate from the transistor, but they must include the intrinsic base-collector and base-emitter resistances, respectively. We wish to amplify a signal V_s. Let us assume that there is a current gain β, which means that $I_c = \beta I_b$. In order to calculate the voltage gain, we first apply Kirchhoff's loop rule to the left-hand loop to obtain

$$I_b = \frac{V_s + V_{bb}}{R_b + (1 + \beta)R_e} \qquad (11.11)$$

Then

$$I_c = \beta I_b = \beta \frac{V_s + V_{bb}}{R_b + (1 + \beta)R_e} \qquad (11.12)$$

The output voltage on the load resistor is

$$V_c = I_c R_c = \beta \frac{V_s + V_{bb}}{R_b + (1 + \beta)R_e} R_c \qquad (11.13)$$

Now we can calculate the voltage gain A_v:

$$A_v = \frac{V_c}{V_s} = \beta \frac{1 + V_{bb}/V_s}{R_b + (1 + \beta)R_e} R_c \qquad (11.14)$$

Because the voltages V_{cc} and V_{bb} are present only to establish the proper bias on the transistor, it is possible in practice to make $V_{bb} \ll V_s$, so that

$$A_v \approx \beta \frac{R_c}{R_b + (1 + \beta)R_e} \qquad (11.15)$$

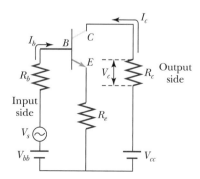

Figure 11.24 A transistor amplifier circuit.

With the appropriate choice of resistors, one can in theory achieve any desired voltage gain. In real circuits, however, there are severe limitations based on the characteristics of the particular transistor and power limitations in various parts of the circuit. Often, capacitors are put into the circuit to limit power surges. It is also not unusual to amplify a small signal through several (or many) stages before reaching the desired output voltage. For example, the initial output from a magnetic or optical reading device is typically on the order of microvolts, and it has to be amplified several times before it is strong enough to drive the speakers of an audio system.

Solid-state transistors are not made by fastening together pieces of *n*-type and *p*-type semiconductors. It is possible to make layers of *n*- and *p*-type materials by diffusing or implanting acceptor and donor atoms at the appropriate thicknesses in a slab of pure germanium or silicon. It is this technology that has made possible the exceptional miniaturization of electronic circuits, which we shall discuss in more detail later.

Field Effect Transistors

The *npn*-junction and *pnp*-junction devices we have just described are referred to as a group by the term **bipolar transistors**. Another type of transistor is the **field effect transistor (FET)**. A schematic diagram of a FET is shown in Figure 11.25. The three terminals of the FET are known as the *drain*, *source*, and *gate*, and these correspond to the collector, emitter, and base, respectively, of a bipolar transistor. Comparing the circuit of Figure 11.25 with the corresponding circuit for the bipolar transistor (Figure 11.23), we see the principal difference is that the *p*-type gate is connected in reverse bias with *both* the *n*-type source and *n*-type drain, while the *p*-type base is connected in reverse bias with the *n*-type collector but a forward bias with the *n*-type emitter. The effect is a severe depletion of charge carriers, both holes and electrons, at the *pn* junction of the FET, and therefore very little current can flow through that junction. For that reason the FET is said to have a high input impedance and is voltage controlled, while the bipolar transistor is current controlled. This turns out to be an advantage in the design of many kinds of circuits, because the FET draws a relatively small amount of current.

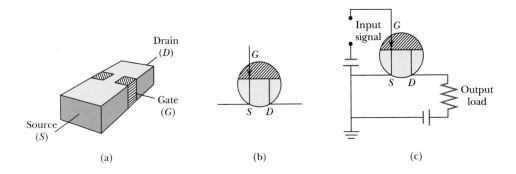

Figure 11.25 (a) A schematic of a FET. The two gate regions are connected internally. (b) The circuit symbol for the FET, assuming the source-to-drain channel is of *n*-type material and the gate is *p*-type. If the channel is *p*-type and the gate *n*-type, then the arrow is reversed. (c) An amplifier circuit containing a FET.

In common use today is the metal oxide semiconductor FET, or **MOSFET**. In a MOSFET the gate is some kind of metal, and it is separated from the channel by a thin layer of oxide (an insulator), usually silicon dioxide. The oxide layer makes the input impedance of the MOSFET much higher than that of the standard FET. It can be made as high as 10^{15} Ω by making the oxide layer thicker. The other principal advantage of MOSFETs is that they can be made extremely small using thin-film deposition methods. This aids in the miniaturization process in the design and manufacture of integrated circuits.

MOSFET

Schottky Barriers

A closely related device is called the **Schottky barrier**, in which direct contact is made between a metal and a semiconductor. If the semiconductor is n-type, electrons from it tend to migrate into the metal, leaving a depleted region within the semiconductor. This will happen as long as the work function of the metal is higher (or lower, in the case of a p-type semiconductor) than that of the semiconductor. The width of the depleted region depends on the properties of the particular metal and semiconductor being used, but it is typically on the order of microns. The I-V characteristics of the Schottky barrier are similar to those of the pn-junction diode, as shown in Equation (11.9) and Figure 11.13. When a p-type semiconductor is used, the behavior is similar but the depletion region has a deficit of holes.

One variation of the Schottky barrier is known as an **ohmic contact**. The ohmic contact also involves contact between a metal and a semiconductor, but now the work function of the semiconductor is higher than that of the metal if the semiconductor is n-type, and lower than that of the metal for a p-type. In this case the interface region will be enriched in majority carriers (whether p or n), and current can flow easily across the boundary.

In Figure 11.26a, a typical Schottky-barrier-gate FET (also known as a metal semiconductor FET, or MESFET) is shown. In this device the doped n-type GaAs serves as the channel. The contacts between the AlGe layers and the doped GaAs layer are ohmic, while the Al gate contact forms a Schottky barrier. When the gate is connected in reverse bias, a higher input impedance is obtained, just as when obtained using the silicon dioxide layer in the MOSFET. The CrPt interface serves only to prevent the Au and Al from forming an alloy.

In Figure 11.26b we see the MESFET amplifier circuit gain plotted as a function of frequency. The superior characteristics of the MESFET are due to the fact

(a)

(b)

Figure 11.26 (a) Schematic drawing of a typical Schottky-barrier FET. (b) Gain versus frequency for two different substrate materials, Si and GaAs. *From D. A. Fraser*, Physics of Semiconductor Devices, *Oxford: Clarendon Press (1979).*

that no diffusion of impurities into any of the layers is needed. The source, gate, and drain are simply evaporated or sputtered directly onto the doped GaAs. Another benefit of this is that the channel can be made extremely narrow (a fraction of a micron), making the device smaller and its operation faster.

Semiconductor Lasers

In recent years semiconductor lasers have been used widely in scientific research and industrial applications. Like the gas lasers described in Section 10.2, semiconductor lasers operate using population inversion—an artificially high number of electrons in excited states. In a semiconductor laser, the band gap determines the energy difference between the excited state and the ground state. Rather than the optical pumping and electrical discharge used to operate gas **Injection pumping** lasers, semiconductor lasers use **injection pumping**, where a large forward current is passed through a diode. This creates electron-hole pairs, with electrons in the conduction band and holes in the valence band. A photon is emitted when an electron falls back to the valence band to recombine with the hole. Thus the photon energy is determined by the band gap in the semiconductor.

The first semiconductor laser was made in 1962 and consisted of a GaAs *pn* junction with a band gap of 1.47 eV. It is straightforward to compute the photon wavelength produced in a transition from the conduction band to the valence band:

$$\lambda = \frac{hc}{\Delta E} = \frac{1240 \text{ eV} \cdot \text{nm}}{1.47 \text{ eV}} = 844 \text{ nm}$$

which is in the near infrared region. Because of the band gaps found in typical semiconductors, outputs of most semiconductor lasers are in the infrared or red portions of the spectrum.

Since their development, semiconductor lasers have been used in a number of applications, most notably in fiber-optics communication. One advantage of using semiconductor lasers in this application is their small size; like other semiconductor devices, semiconductor lasers can be made quite small, with dimensions typically on the order of 10^{-4} m. They are solid-state devices, so semiconductor lasers are more robust than gas-filled tubes.

For many years, the relatively small band gaps (1 eV to 2 eV) in semiconductors limited the operation of semiconductor lasers to longer wavelengths. For example, a 2-eV photon has a wavelength of 620 nm, and lower energies correspond to even longer wavelengths. Over the years scientists worked to develop semiconductor lasers with shorter wavelengths. In 1996 Shuji Nakamura demonstrated a blue-violet laser based on the semiconductor gallium nitride. By doping the gallium nitride with varying amounts of indium, Nakamura produced laser wavelengths ranging from 390 nm to 440 nm. A different scheme involves a "frequency-doubling" device that takes the output from an 830-nm diode and turns it into 415-nm laser light. Blue-violet lasers will replace the infrared lasers now used in compact disk (CD) players and the red 640-nm lasers used in digital video-disk (DVD) players. With shorter laser wavelengths, information can be packed with higher density onto a CD or DVD, with less chance of error in reading the information. The first blue-laser DVD players became commercially available in 2003, and the blue-laser DVD disks can store up to 23 gigabytes (GB) of data (1 byte = 8 bits), almost five times as much as the older 4.7-GB red-laser disks.

Example 11.6

A CD stores 700 MB of data. The region of the CD that stores data has an inner radius of 2.30 cm and outer radius of 5.80 cm, and the width of the data track is 1.60 μm. What is the surface area allowed for each data bit? What are the approximate dimensions of each bit along the track?

Strategy Note that 1 byte (B) = 8 bits. The usable surface area is the difference between the area of a disk of radius 5.80 cm and a disk of radius 2.30 cm—effectively subtracting out the "hole" in the middle of the disk. Then the area per bit is that surface area divided by the number of bits.

Solution The usable area is

$$A = \pi[(0.058 \text{ m})^2 - (0.023 \text{ m})^2] = 8.91 \times 10^{-3} \text{ m}^2$$

The number of bits is 700 MB $\times$ 8 bits/B = 5.60×10^9 bits. The area per bit is

$$\frac{\text{Area}}{\text{Bit}} = \frac{8.91 \times 10^{-3} \text{ m}^2}{5.60 \times 10^9 \text{ bits}} = 1.6 \times 10^{-12} \text{ m}^2/\text{bit}$$

With a track width of 1.6 μm, the length of a bit (area divided by width) is

$$\text{Bit length} = \frac{1.6 \times 10^{-12} \text{ m}^2}{1.6 \times 10^{-6} \text{ m}} = 1.0 \times 10^{-6} \text{ m}$$

The bit length is just 1 micron. Thus the length and width of the bit are comparable to the IR laser wavelength (780 nm) used to scan it.

Integrated Circuits

The most important use of all these semiconductor devices today is not in discrete components, but rather in **integrated circuits**. It is possible to fabricate a silicon-based substrate (commonly called a **chip**) containing a hundred million or more components, including resistors, capacitors, transistors, and logic switches. This extreme miniaturization has done two things for the electronics industry. First, it has made it possible to put sophisticated integrated circuits into even the smallest products, such as wristwatches and cameras. Second, it has enabled those integrated circuits to work much faster. Although there are still limits based on the speeds at which switches can operate, the transit time of an electronic signal has been reduced by packing the components so closely together. Those signals travel at essentially the speed of light—0.3 μm/fs, in units relevant to the distance/time scale of integrated circuits.

Integrated circuits have had a tremendous impact on the work of scientists and engineers by enhancing the computing power available to them. The Manhattan Project (which produced the atomic bomb in 1945) and other scientific work during World War II inspired the development of large mechanical computers that could perform many computations relatively quickly. The ENIAC (Electronic Numerical Integrator and Computer), built in 1945, could multiply 333 ten-digit numbers per second. Unfortunately it weighed 30 tons, contained over 17,000 vacuum tubes, 70,000 resistors, 10,000 capacitors, and 1500 relay switches. It was 2.5 m high and wide and 25 m long (see Figure 11.27), and it used 174 kW of power in normal operation! By 1963 the best computers (for example, the Control Data 6600) still filled a normal-sized room (6 m $\times$ 6 m) but could perform one million operations per second. In 1971 the first true commercial integrated circuit was released: the Intel 4004 (Figure 11.28), which had 2300 transistors and could perform 60,000 operations per second. Today, personal computers, which fit on desktops and are plugged into a standard 110-volt outlet, can perform more than 10^9 operations per second, while the largest supercomputers are thousands of times faster. The record of 10^{12} operations per second was passed in

Figure 11.27 Working on the ENIAC computer. *Courtesy of the School of Engineering and Applied Science, University of Pennsylvania.*

1996, and by 2004 several supercomputers had achieved rates of over 3×10^{13} operations per second.

One indication of the rapid progress in computing is the increase in the number of transistors that can fit on a single microchip. That number was just 2300 for the Intel 4004 chip in 1971. The number of transistors per chip grew to 134,000 with the release of the 80286 chip in 1982; to 275,000 on the 386 chip in 1985; and to 1.2 million on the 486 chip in 1989. The rapid growth in microchip technology continues in the twenty-first century. The Pentium 4 chip, produced by Intel in 2002, contained 55 million transistors on a single chip. In order to achieve such high densities, it is necessary to build extremely small components. In the Pentium 4, for example, circuit elements are about 1.3×10^{-7} m across in each direction. Notice that this is less than the wavelength of visible light. For that reason chip manufacturers must now use ultraviolet lithographic techniques. By 2010 manufacturers hope to use shorter ultraviolet wavelengths to produce circuit elements less than 1×10^{-7} m across, thus producing even smaller and faster microprocessors. The rate of progress is summarized in Moore's law, which states that microchip capacity doubles roughly every 18 to 24 months (see Figure 11.29).

Figure 11.28 The Intel 4004, the first commercial microprocessor (1971). This integrated circuit measured just 3 mm by 4 mm, and with 2300 transistors it could perform 60,000 operations per second. *Courtesy of Intel Corporation.*

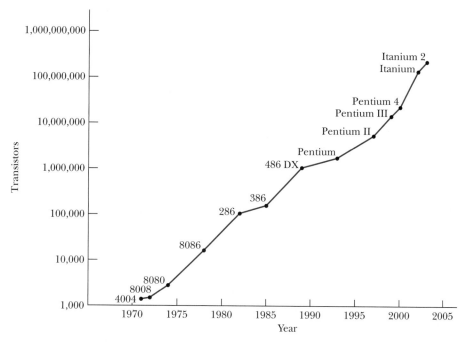

Figure 11.29 Moore's law, showing the progress in computing power over a 30-year span, illustrated here with Intel chip names. The Pentium 4 contains over 50 million transistors. *Courtesy of Intel Corporation. Graph from* http://www.intel.com/research/silicon/mooreslaw.htm.

This "law" was reasonably accurate from the 1970s all the way through the early years of the twenty-first century.

An important factor in technology besides the computing itself is the storage and retrieval of information. By the early twenty-first century there were two standard media: magnetic and optical. It is possible to store and reliably retrieve more than 10^{13} bits of information per m^2 of magnetic disk or tape. This is approaching the theoretical limit, because at this density as few as 100 magnetic particles are used to store each bit of information. At higher densities one begins to have problems with reliability, due to statistical fluctuations in the magnetizations of the particles and overlap of magnetic fields from one small bit to another. You may have both magnetic and optical storage in your computer: the hard disk drive in a computer is magnetic, and a CD drive is an optical reader (and writer in many cases). The standard CD, with a 12-cm diameter, has a 700-MB capacity. Thus, the density of data storage on a CD is somewhat lower than the density on a magnetic device (see Problem 31).

11.4 Nanotechnology

Richard Feynman (1918–1988; Nobel Prize in Physics, 1965) was widely regarded as one of the outstanding theoretical physicists of the twentieth century. He was noted not only for his grasp of many subfields of physics, but also for his vision of the place of physics in the wider world. Feynman achieved widespread acclaim outside the world of physics in 1986 for the role he played as a member of the Presidential commission that examined the crash of the space shuttle *Challenger*. His writings and public lectures helped physics come alive for physicists and non-physicists alike. At an American Physical Society meeting in 1959, Feynman gave a now-famous address entitled: "There's Plenty of Room at the Bottom." The gist of his message lies in the following excerpt:

What I want to talk about is the problem of manipulating and controlling things on a small scale. . . . What I have demonstrated is that there is room—that you can decrease the size of things in a practical way. I now want to show that there is plenty of room. I will not now discuss how we are going to do it, but only what is possible now in principle. . . . We are not doing it simply because we haven't yet gotten around to it.

In the years following Feynman's visionary statement, slow but steady progress was made toward "decreasing the size of things in a practical way." This includes some discoveries and devices you've already studied in this book, such as electronic circuitry (Section 11.3) and the scanning tunneling microscope and atomic force microscope (Section 6.7). In the late 1990s and early twenty-first century, there was a flurry of activity in this field of **nanotechnology**, which is generally defined as the scientific study and manufacture of materials on a submicron scale. Although the prefix *nano-* implies working on a nanometer scale, nanotechnology can involve working on scales ranging from single atoms, on the order of tenths of 1 nm, up to about 1 micron, or 1000 nm.

Nanotechnology defined

Because it refers to work done on specific distance scales without regard to discipline, nanotechnology now includes many fields of research, not only in physics, but also in engineering, chemistry, and the life sciences. By its nature, nanotechnology is interdisciplinary. The excitement generated by this interdisciplinary work, along with the emergence of even more exciting applications, has led to huge increases in funding of nanotechnology research. In this section we'll only be able to discuss a few of the primary research areas and applications, concentrating on those most closely related to the modern physics you've already seen.

Carbon Nanotubes

In 1991, following the discovery of C_{60} buckminsterfullerenes, or "buckyballs" (see Section 10.5), Japanese physicist Sumio Iijima discovered another geometric arrangement of pure carbon into large molecules. In this arrangement, known as a carbon nanotube, hexagonal arrays of carbon atoms lie along a cylindrical tube instead of a spherical ball. One way to envision the nanotube is to think of a flat sheet of carbon in a hexagonal lattice rolled into a tube form (see Figure 11.30). There's virtually no limit to the length of such a tube, so there's no standard chemical symbol (like C_{60}) to describe the nanotube.

The basic structure shown in Figure 11.30 leads to two types of nanotubes. A single-walled nanotube has just the single shell of hexagons as shown. In a multi-walled nanotube, multiple layers are nested like the rings in a tree trunk. Single-walled nanotubes tend to have fewer defects, and they are therefore stronger structurally. However, single-walled nanotubes are also more expensive and difficult to make. The best type of nanotube usually depends on the application.*

For their size, carbon nanotubes are extremely strong. Their measured tensile strength is significantly stronger than that of standard carbon fibers. Also, the tubelike structure tends to bend like a drinking straw rather than break when under compression. Accordingly, carbon nanotubes are used as structural reinforcements in the manufacture of composite materials. For example, the batteries in most cell phones and notebook computers today use nanotubes in this way.

Other nanotube applications depend on their outstanding electronic and thermal properties. Nanotubes have very high electrical and thermal conductivities, along with stability up to very high temperatures. This combination of properties—carrying high current while carrying away the resistive heat generated—

Figure 11.30 Model of a carbon nanotube, illustrating the hexagonal carbon pattern superimposed on a tubelike structure. There is virtually no limit to the length of the tube. © *Chris Ewels/ www.ewels.info.*

*For an overview, see M. S. Dresselhaus, G. Dresselhaus, and P. Avouris, eds., *Carbon Nanotubes: Synthesis, Structure, Properties, and Applications*, Berlin: Springer-Verlag (2001).

(a)   (b)

Figure 11.31 Carbon nanotube used to make a nanoscale transistor. (a) A schematic of the transistor design. (b) Photograph of the nanotube over metal electrodes. *Prof. Cees Dekker/Delft University of Technology.*

leads to extremely high maximum current densities, which rival or even exceed the critical current densities of the high-temperature superconductors described in Section 10.5. The electronic properties of nanotubes have been used to fabricate nanoscale transistors. The first of these was made in 2001 by researchers at the Delft University of Technology in Holland.* A schematic for the transistor design is shown in Figure 11.31a, and a photograph of the nanotube placed over two metal electrodes is shown in Figure 11.31b.

Another exciting nanotube application is in flat-panel displays for televisions and computer monitors. These flat-panel displays are field-emission displays (FED), rather than the old cathode-ray tube (CRT) technology. In a FED, electrons are emitted from sharp tips arranged in pixels, rather than scanning an electron beam across the pixel array. Carbon nanotubes can be fabricated with tips just a few nanometers across, making for an outstanding pointlike source. Several television manufacturers had developed prototypes by early 2005.

Another exotic application involves a combination of nanotubes and buckyballs, known as a "peapod," because analogous to the organic peapod, a row of buckyballs is nested within the nanotube (see Figure 11.32 and the front cover of this book). The electronic properties of the nanotube depend heavily on the placement of the balls within the tube. Hence, this arrangement is described as having tunable electronic properties, which can be adjusted as desired for nanoscale applications.

Nanoscale Electronics

Much of the progress in semiconductor electronics (integrated circuits) that we discussed in Section 11.3 could be described as nanotechnology, because computer chip components have been micron-sized or smaller for many years. You've

---

*Henk W. Ch. Postma et al., Carbon nanotube single-electron transistors at room temperature, *Science* **293**, 76–79 (2001).

Figure 11.32 Nanotube "pea-pod" in which buckyballs lie within a nanotube, shown with electron waves superimposed. Adjusting placement of the balls allows for tuning of the electronic properties of the system. *Courtesy of D. Hornbaker and A. Yazdani, University of Illinois.*

already seen the carbon nanotube as another example of nanoscale electronics, when the nanotube is being used to carry electricity. Many other advances also fall under the heading of nanoscale electronics.

One problem in the development of truly small-scale electronic devices is that the connecting wires in any circuit need to be as small as possible, so that they do not overwhelm the nanoscale components they connect. In addition to the nanotubes we've already described, semiconductor wires (for example, indium phosphide) have been fabricated with diameters as small as 5 nm. These **nanowires** have been shown useful in connecting nanoscale transistors and memory circuits. The transistors themselves, referred to in this context as **nanotransistors**, have now been made several different ways. These include using the nanotubes as mentioned above and even small organic molecules.*

Nanowires

Nanotransistors

One goal of nanotechnology researchers is to reproduce common electronic devices on a nanoscale. For example, nanoscale field effect transistors have been produced that incorporate nanotubes. Nanotubes have also been used as the basis for optoelectronic devices that might replace common LEDs. In these devices, electrons and holes injected from opposite ends of the nanotube can recombine as in a standard LED, with the recombination energy given off as a photon of light.[†] These and other nanoscale electronic devices appear promising but are not yet in widespread use, because the standard semiconductor technology is well developed, with mass-production techniques in place. Researchers are still searching for the best systems for mass-producing and marketing nanoscale electronic devices.

Nanotechnology and the Life Sciences

As we have seen, scientific research in nanotechnology is a fairly recent development. However, nature has been engaged in what might be regarded as nanoscale engineering in living things for most of Earth's history. The complex molecules needed for the variety of life on Earth are themselves examples of nanoscale design. Examples of unusual materials designed for specific purposes include the molecules that make up claws, feathers, and even tooth enamel.

*Charles M. Lieber, The incredible shrinking circuit, *Scientific American* **285**(3), 58–64 (September 2001).

[†]Phaedon Avouris and Joerg Apenzeller, Electronics and optoelectronics with carbon nanotubes, *The Industrial Physicist* **10**(3), 18–21 (June/July 2004).

(a)   (b)

Figure 11.33 (a) Photo of keratin fibers from a gecko's foot. (b) Plastic fibers fashioned with electron-beam lithography, made in an attempt to reproduce the adhesive powers of the gecko's keratin hairs. *Courtesy of the Center for Nanoscience and Nanotechnology, University of Manchester, U.K.*

Scientists and engineers now seek to understand the structure and function of these specially designed molecules, so that they can create similar molecules artificially that they hope will duplicate some of the functions. For example, scientists searching for better adhesives have studied the gecko, which has sticky feet that allow it to hang on a vertical tree trunk or even upside down on a limb. Figure 11.33a shows the 200-nm-wide keratin hairs on the soles of the gecko's feet that allow it to do this. The attractive force between those hairs and another surface come from capillary and van der Waals forces. Using those keratin hairs as a model, researchers have developed flexible plastic fibers, just 2 μm long and 500 nm in diameter (Figure 11.33b), that adhere about 30% as well as the gecko's feet.* This is just one example of small-scale engineering that mimics the functions found in plants and animals.

The novel properties of genetic material might lead to interesting applications. It has been suggested that the DNA molecule could store large amounts of data in its sequence of four bases (adenine, guanine, cytosine, and thymine). In 1994, Leonard Adleman showed that natural enzymes can process this information in a parallel manner, offering an alternative to traditional semiconductor-based computing.† Although technical hurdles exist before a practical DNA computer can be realized, it has been suggested that DNA computing could merge with other nanoscale efforts, such as the nanowires and nanotube transistors already discussed.

In the life sciences proper, nanoscale research is paramount, especially in the field of molecular genetics. In recent years, scientists have succeeded in mapping the human genome, which will help us understand many of the body's functions and disorders. Genetic engineers can now make "designer genes" that can replace faulty natural genes. Researchers also seek to take advantage of human stem cells' ability to grow into different body cell types, to serve as a repair mechanism.

Information Science

In Section 11.3 we described the progress in over a half-century of computing technology. Progress has relied on making components smaller and smaller. However, now that components have reached the submicron scale, Moore's law cannot continue to hold much longer unless new nanoscale technologies are found.

It's possible that current photolithographic techniques for making computer chips could be extended into the hard-UV or soft x-ray range, with wavelengths on the order of 1 nm, to fabricate silicon-based chips on that scale. Some novel

*Eric J. Lerner, Biomimetic nanotechnology, *The Industrial Physicist* **10**(4), 16–19 (August/September 2004).

†Leonard M. Adleman, Computing with DNA, *Scientific American* **279**(2), 54–61 (August 1998).

schemes for storing and retrieving data rely on technology similar to the atomic force microscope, where a cantilever is used to create and erase nanometer-scale pits (representing bits with value 0 or 1) in a polymer substrate.*

Quantum mechanics becomes increasingly important as computing devices reach smaller and smaller dimensions. For example, if you had to rely on the measurement of a single-spin magnetic moment (up/down, corresponding to 1 or 0), errors can be introduced as the magnetic moment interacts with its neighbors, or suffers from random fluctuations, or is affected by the measurement process as a result of the Heisenberg uncertainty principle. However, in the 1990s physicists learned that it is possible to take advantage of quantum effects, specifically the superposition of quantum states, to store and process information more efficiently than a traditional computer. To date, such **quantum computers** have been built in prototype but not mass-produced.

Quantum computers

A separate physical issue, which we won't address at length here, is the thermodynamics of computing. Rolf Landauer, an IBM physicist, stressed the notion that all information is physical and therefore subject to the laws of thermodynamics.[†] Landauer showed that erasing a bit of information resulted in the inevitable generation of at least $k \ln 2$ of entropy. That small amount of entropy is not significant for a traditional computer, but becomes relatively more important as computing reaches the nanometer scale. We haven't reached that scale yet, and there's still room at the bottom!

*Peter Vettiger and Gerd Binnig, The nanodrive project, *Scientific American* **288**(1), 46–53 (January 2003).

[†]See Rolf Landauer, The physical nature of information, *Physics Letters A* **217**, 188–193 (1996).

Summary

The properties of semiconductors are responsible for their widespread use today in computers, electronic instruments, and other applications. In a semiconductor there exists a small band gap (about 1 eV) between the valence band and the conduction band. When a modest electric field is applied, sufficient numbers of electrons can overcome the band gap in order to conduct.

Normal conductors have resistivities that increase with increasing temperature, but in semiconductors the resistivity decreases with increasing temperature, due to the increased statistical probability that the conduction band will be occupied. An empirical formula relating the resistance of many semiconductors to temperature is the Clement-Quinnell equation

$$\log R + \frac{K}{\log R} = A + \frac{B}{T} \qquad (11.4)$$

where A, B, and K are constants.

Some semiconductors are n-type, meaning that the majority carriers are negative electrons. Other semiconductors are p-type, meaning that the majority carriers are positive holes. The Hall effect can be used to determine the sign and magnitude of the charge carriers.

Semiconductors exhibit the thermoelectric effect, the presence of an electric field when a temperature gradient is established. The thermoelectric effect is the basis for the design of thermocouples.

Semiconductors are used to make a variety of electronic devices. Materials of p-type and n-type can be placed end to end to form a pn-junction diode, which permits current to flow easily in one direction (forward bias) but hardly at all in the opposite direction (reverse bias). A Zener diode will operate in reverse bias once a sufficiently high voltage is applied.

Other useful semiconductor devices include light-emitting diodes and photovoltaic (solar) cells. Solar cells are currently the focus of an intense research effort. It is hoped that more efficient and cheaper solar cells can provide a useful source of energy by converting sunlight to electrical energy. This will become increasingly important as nonrenewable sources of energy are depleted.

Semiconducting transistors are important components in computers and other electronic instruments. Transistors can be used to amplify signals (either voltage or current).

Related devices, such as field effect transistors, MOSFETs, and Schottky barriers are also used extensively. The most important use of these electronic devices today is in integrated circuits. Silicon-based chips contain millions of components. Miniaturization has led to significant increases in the speed and efficiency of electronic circuits as well as to greater convenience. Perhaps the most significant application has been in the design of modern computers. Thanks to improvements in semiconductor technology, the large, complex computers of several decades ago have been supplanted by much smaller, cheaper, and easier-to-use microprocessors and personal computers in use today.

The drive for further miniaturization has led to a boom in nanotechnology, both in basic science and its applications. Carbon nanotubes have remarkable structural and electronic properties that make them useful in many applications. Nanotechnology has applications in electronic devices, including information storage and retrieval, as well as the life sciences.

Questions

1. Why is the free-electron model not applicable to semiconductors and insulators?
2. How does a semimetal differ from a normal conductor?
3. Compare the resistivities of conductors, semiconductors, and insulators at room temperature.
4. Why does the addition of impurities to a conductor increase its resistivity, while the addition of impurities to a semiconductor generally decreases its resistivity?
5. Would you expect carbon resistors to be useful as thermometers at room temperature and above? Explain.
6. Why are semiconductors referred to as non-ohmic?
7. Use the size of the energy gap to decide whether the following should tend to be transparent to visible light: conductors, semiconductors, and insulators.
8. Repeat Question 7 considering near infrared light with $\lambda = 1$ micron.

9. A semiconductor has an energy gap E_g. Explain what happens when the semiconductor is bombarded with electromagnetic radiation with wavelength $\lambda < hc/E_g$. Repeat for $\lambda > hc/E_g$.
10. Describe the effect of high temperatures on a semiconductor diode.
11. What role do capacitors play in the conversion of alternating current to direct current?
12. Why is it appropriate to think of a photovoltaic cell as a LED operating in reverse?
13. What are some possible applications of carbon nanotubes?
14. Why are quantum effects important in electronic circuits nanometers in size, but not in those circuits microns in size?

Problems

Note: The more challenging problems are indicated by a shaded box.

11.2 Semiconductor Theory

1. For a certain resistor with a normal (that is, room-temperature) resistance of 150 Ω, the values of the constants in Equation (11.4) are found to be $A = 4.05$, $B = 4.22$, and $K = 4.11$, in units such that R will be in ohms and T in kelvins. What is the resistance of this resistor at (a) $T = 77$ K, (b) $T = 20$ K, and (c) $T = 1$ K?

2. For a nominal 10-Ω resistor (as described in Figure 11.7) the resistance at various temperatures is as follows: $R = 10$ Ω at $T = 293$ K, $R = 40$ Ω at $T = 10$ K, and $R = 5800$ Ω at $T = 1$ K. Determine the constants A, B, and K in Equation (11.4) for this resistor.

3. Consider the experimental arrangement of Figure 11.9, set up to observe the Hall effect. With the power supplies and meters in this configuration, what will be the sign of the voltage on the voltmeter if the sample is a semiconductor in which the majority of charge carriers are holes? Explain.

4. Hall effect data in an actual student laboratory was as follows: for a doped indium arsenide strip of thickness 0.15 mm, with a current of 100 mA flowing through the strip and a magnetic field of 36 mT perpendicular to the strip, the measured Hall voltage was 8.4 mV. (a) Use these data to find the density of charge carriers. (b) Find the density of charge carriers using the complete data set shown in the table on page 422.

I = 100 mA

B (mT)	V_H (mV)
17.5	4.4
27	6.4
36	8.4
31	7.4
25	6.1
47	10.9
50	11.5
59.5	13.6

5. A pure lead bar 10 cm long is maintained with one end at $T = 273$ K and the other at 283 K. The thermoelectric potential difference thus induced across the ends is 12.6 μV. Find the thermoelectric power for lead in this temperature range. (*Note:* Q varies nonlinearly with temperature, but over this narrow temperature range you may use a linear approximation.)

6. The reference junction of an iron-constantan thermocouple is maintained at 0°C, and the other side is at an unknown temperature. Find the unknown temperature if the other side is 2.43 mV higher in potential than the reference junction (see Table 11.3).

7. Assuming that the potential corresponding to *any* temperature T in Table 11.3 could be known (say through a computer model) to at least five significant digits, what maximum uncertainty would be needed in your voltmeter if you were to use an iron-constantan thermocouple to measure temperatures to within 0.01°C?

8. What kind (*p*-type or *n*-type) of semiconductor is made if pure germanium is doped with a small amount of (a) aluminum; (b) selenium?

11.3 Semiconductor Devices

9. Find the wavelength of the photon needed to cause an electron to jump from the valence to the conduction band in (a) germanium, with a band gap of 0.67 eV; (b) silicon, with a band gap of 1.1 eV.

10. When an electron in the compound semiconductor AlAs makes a transition from the conduction band to the valence band, a 574-nm photon is emitted. What is the size of the band gap?

11. Find the fraction of the standard solar flux reaching the Earth (about 1000 W/m²) available to a solar collector lying flat on the Earth's surface at each of the following places at noon on the winter solstice, spring equinox, and summer solstice: (a) San Diego, latitude 33°; (b) Regina, latitude 50°; (c) St. Petersburg, Russia, latitude 60°.

12. Assuming that the *average daily* solar constant at a particular place is 200 W/m², how large an array of 30% efficient solar cells is required to equal the power output of a typical power plant, about 10^9 W?

13. For the diode described in Example 11.4, find the forward bias current with $V = 250$ mV at (a) $T = 250$ K, (b) $T = 300$ K, and (c) $T = 500$ K.

11.4 Nanotechnology

14. A single-walled carbon nanotube has 2.3×10^{19} carbon atoms per m² along its surface. The nanotube diameter is 1.4 nm. (a) Find the mass density of the nanotube in kg/m³. (b) Compare your answer to (a) with the density of steel, about 7800 kg/m³.

15. Using the data in the preceding problem, find the density of a nanotube peapod that also contains one buckyball (C_{60}) for each nm of its length.

General Problems

16. The Fermi-Dirac factor is expressed in Equation (9.34) as

$$F_{FD} = \frac{1}{\exp[\beta(E - E_F)] + 1}$$

In a semiconductor or insulator, with an energy gap E_g between the valence and conduction bands, we can take E_F to be halfway between the bands, so that $E_F = E_g/2$. (a) Show that for a typical semiconductor or insulator at room temperature the Fermi-Dirac factor is approximately equal to $\exp(-E_g/2kT)$. (b) Use the result in (a) to compute the Fermi-Dirac factor for a typical insulator, with $E_g = 6.0$ eV at $T = 293$ K. (c) Repeat for a semiconductor, silicon, with $E_g = 1.1$ eV at $T = 293$ K. (d) Your result in (c) is still small, but sufficiently large to explain why there will be conduction. Explain.

17. Consider what happens when silicon is doped with arsenic. Suppose that the extra, weakly bound electron from arsenic moves in the first Bohr orbit around a silicon atom. The first Bohr orbit has a radius of

$$a_0 = \frac{4\pi\epsilon_0 \hbar^2}{me^2}$$

Due to the effects of screening, it is necessary to replace ϵ_0 with the electric permittivity $\epsilon = \kappa\epsilon_0$, where κ is the dielectric constant ($\kappa = 11.7$ for silicon). (a) Compute the effective Bohr radius for this electron. (b) Compare your result with the lattice spacing in silicon, about 0.235 nm, and comment on the result.

18. Follow the same procedure as in Problem 17 to find the binding energy E_0 for the first Bohr orbit of the extra electron in silicon. Comment on the result.

19. (a) Using the thermocouple data in Table 11.3, use a computer to perform a linear least-squares fit of voltage versus temperature for the range 0°C to 50°C and for the range 50°C to 100°C. Over which range is the fit better? (b) Repeat your analysis of (a), this time using a second-order least-squares fit (voltage will be a function of T and T^2). Compare your results with those you obtained in (a).

20. A certain diode has a reverse bias current of 1.75 μA. Now this diode is connected in forward bias in the circuit shown, in series with a resistor and with a constant voltage source of 6.0 volts. (a) Find the value of resistance R such that a current of 80 mA will flow at room temperature (293 K). (b) Under the condition described in (a), find the voltage drop across the resistor.

6 volts

21. Assuming a temperature of 293 K, for what value V of the bias voltage in Equation (11.9) does (a) $I = 8I_0$; (b) $I = -0.8I_0$?

22. A light-emitting diode made of the semiconductor GaAsP gives off red light ($\lambda = 650$ nm). Determine the energy gap for this semiconductor.

23. How large an energy gap is required for a blue semiconductor laser with $\lambda = 460$ nm?

24. (a) Find the length of each side of a square computer chip with 12 million transistors, if each transistor occupies a square of side 0.13 μm. (b) Find the number of transistors on a chip of the same size you found in (a) if the transistor size can be reduced to 50 nm on a side.

25. Early research on semiconductor materials focused on silicon and germanium, which have band gaps of 1.1 eV and 0.67 eV, respectively. (a) Using the result of Problem 16a compute the Fermi-Dirac factor for silicon and germanium at $T = 0°C$ and $T = 75°C$ (this represents a fair temperature range for semiconductors in use). On the basis of your computed value for F_{FD} for germanium at the higher temperature, explain why silicon is preferred in most applications.

26. Suppose the average solar flux reaching the United States is 200 W/m². This average is taken over a whole year and takes into account seasonal effects and weather (clouds), and assumes fixed solar cells. (a) Find the total energy produced in one year by a 1-m² cell producing energy with an efficiency of 15%. (b) How much area would have to be covered with these solar cells in order to supply the United States with all its electricity, if the yearly electrical energy consumption in the United States is about 2.8×10^{12} kW · h? (c) Real solar arrays require about 2.5 times as much area as you found in (b) in order to keep one array of cells from shading another. What fraction of the land area of the United States (about 9.6×10^6 km²) would have to be covered with solar cells to meet the nation's energy requirements?

27. In this problem you will examine the temperature dependence of the forward/reverse bias modes of a diode. (a) For a pn-junction diode compute the ratio of forward bias current to reverse bias current with an applied voltage of 1.50 volts at each of the following temperatures: 77 K, 273 K, 340 K, 500 K. (b) Comment on the results with respect to possible applications.

28. Pure silicon is used as a photon detector. An incoming photon can strike the surface and excite electrons from the valence band to the conduction band, where they can be counted. (a) Compute the number of electrons you would expect to count if a silicon detector is struck with a 1.04-MeV gamma ray produced in the decay of a ^{136}Cs nucleus. (b) Explain why the counting of electrons should be more precise if the detector is cooled well below room temperature.

29. Suppose you apply the same voltage, 0.10 V, to a pn-junction diode, first in forward bias and then in reverse bias, at room temperature (293 K). What is the ratio of currents in forward and reverse bias?

30. A carbon nanotube has a diameter of 1.6 nm. Young's modulus for the nanotube is 1050 GPa. With one end of the tube fixed, how large a force must be applied to the other end to increase the tube's length by 1%?

31. A DVD has an effective inner radius of 2.3 cm and outer radius of 5.8 cm. The disk's capacity is 4.7 gigabytes, where 1 byte = 8 bits. Find the number of bits stored per square meter, and compare with the density on a magnetic device, 1×10^{13} bits/m².

The Atomic Nucleus

It is said that Cockcroft and Walton were interested in raising the voltage of their equipment, its reliability, and so on, more and more, as so often happens when you are involved with technical problems, and that eventually Rutherford lost patience and said, "If you don't put a scintillation screen in and look for alpha particles by the end of the week, I'll sack the lot of you." And they went and found them [the first nuclear transmutations].

Sir Rudolf Peierls in Nuclear Physics in Retrospect, ed. Roger Stuewer

Ernest Rutherford can rightly be called the "Father of the Nucleus." As discussed in Chapter 4, he proposed a model of atomic structure that placed the heavy, positively charged nucleus at the center and the much lighter electrons at the periphery.

Around 1900, early investigators (including Becquerel, Rutherford, and Marie and Pierre Curie) found that radioactive emissions from atoms comprise three types of radiation, called α (alpha), β (beta), and γ (gamma). Alpha radiation is the least penetrating—it can be stopped by a piece of paper. Beta rays are more penetrating, and they are also common, appearing in the radioactive emissions of many nuclei. Gamma radiation is the most penetrating—it can pass through the human hand, for example. Many early experiments established that α rays were doubly charged positive particles, β rays were probably electrons, and γ rays were electrically neutral. Rutherford proved in a series of experiments that alpha particles were the nuclei of helium atoms.

We begin this chapter by discussing the discovery of the constituents of the nucleus, one of science's most interesting series of experiments. We then study the properties of the nucleus and its constituents, the neutrons and protons. Finally, we discuss nuclear forces and why some nuclei are stable and others are unstable, that is, radioactive. The study of nuclear radioactivity has many important applications, which we shall discuss in Chapter 13.

12.1 Discovery of the Neutron

Although Rutherford proposed the atomic structure with the massive nucleus at the center in 1911, it was not until 1932 that scientists knew which particles compose the nucleus. This study is still ongoing (see Chapter 14), because as physicists strive to find the essence of the fundamental nuclear particles, they continue to find even more particles.

In the early 1900s the nucleus had been erroneously assumed to consist of protons and electrons. However, there are several reasons why electrons cannot exist within the nucleus.

1. *Nuclear size:* We showed previously, in Examples 5.10 and 5.11, that in order to confine an electron in a space as small as a nucleus, the uncertainty principle puts a lower limit on its kinetic energy that is much larger than any kinetic energy observed for an electron emitted from nuclei.
2. *Nuclear spin:* Protons and electrons have spin 1/2. If a deuteron (mass number $A = 2$ and atomic number $Z = 1$) consists of protons and electrons, the deuteron must contain 2 protons and 1 electron in order to have $A = 2$ and $Z = 1$. However, a nucleus composed of 3 fermions must result in a half-integral spin, whereas the nuclear spin of the deuteron has been measured to be 1.
3. *Nuclear magnetic moment:* The magnetic moment of an electron is over 1000 times larger than that of a proton. In a model including the electron in the nucleus, we would expect the nuclear magnetic moment to be on the same order as that of the electron. However, the measured nuclear magnetic moments are on the same order of magnitude as the proton's, so it appears an electron is not a part of the nucleus.

Electrons can't exist within the nucleus

Example 12.1

What is the minimum kinetic energy of a proton in a medium-sized nucleus having a diameter of 8.0×10^{-15} m?

Strategy We will use the uncertainty principle just as we did in Example 5.10, where we found the minimum kinetic energy for an electron in a nucleus. We use the uncertainty principle to find the uncertainty Δp and then use this value to determine the minimum kinetic energy.

Solution We start with the uncertainty principle involving momentum.

$$\Delta p \, \Delta x \geq \frac{\hbar}{2}$$

$$\Delta p \geq \frac{\hbar}{2 \, \Delta x} = \frac{6.58 \times 10^{-16} \text{ eV} \cdot \text{s}}{2(8.0 \times 10^{-15} \text{ m})}$$

$$\Delta p \geq 0.041 \text{ eV} \cdot \text{s/m}$$

The momentum p must be at least as large as Δp. Hence $p_{\min} = \Delta p$, and we have for $p_{\min}c$:

$$p_{\min}c = (0.041 \text{ eV} \cdot \text{s/m})(3.0 \times 10^8 \text{ m/s}) = 12 \text{ MeV}$$

Because 12 MeV is only about 1% of the proton's rest energy (938 MeV, see inside front cover), we can treat the problem nonrelativistically. The kinetic energy of a proton in this nucleus must be at least as large as

$$K = \frac{(p_{\min})^2}{2m} = \frac{(p_{\min}c)^2}{2mc^2} = \frac{(12 \text{ MeV})^2}{2(938 \text{ MeV})} = 0.08 \text{ MeV}$$

which is an entirely reasonable experimental value. The result for a neutron would be similar.

We have seen that there is strong experimental and theoretical evidence that electrons are not bound within the nucleus. Although all this evidence was not available to Rutherford in 1920, he proposed that a neutral particle, called a *neutron*, might exist. A nucleus composed of protons and neutrons would not be ruled out on the basis of the three arguments just presented about the electron.

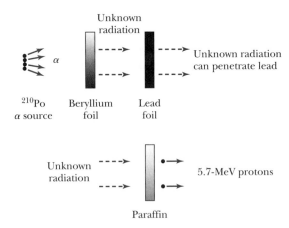

Figure 12.1 Schematic diagram of events leading to neutron discovery. A polonium α-particle source emits α particles that produce unknown radiation when incident on beryllium. This unknown radiation is so penetrating that it can pass through a lead foil, which indicates it may be gamma rays. When the unknown radiation is incident on paraffin, however, 5.7-MeV protons are produced. Only gamma rays with energy above 50 MeV are able to do this, and they are unlikely to be produced by a nucleus. Chadwick suggested the radiation was a neutral particle of about the same mass as a proton.

Walther Bothe (1891–1957) was born near Berlin and studied under Max Planck at the University of Berlin. He worked in Berlin until 1930, and after a brief period in Giessen, worked at the Max Planck Institute in Heidelberg from 1932 until his death. Bothe pioneered the use of coincidences between multiple particle detectors to study the Compton effect, cosmic rays, and many nuclear reactions. © *The Nobel Foundation.*

The discovery of the neutron is a classic experimental investigation. In 1930 the German physicists Walther Bothe and Herbert Becker were using a radioactive polonium source that emitted α particles. Bothe and Becker found that when these α particles bombarded beryllium, a very penetrating radiation was produced. Irène Curie and Frédéric Joliot subsequently showed in 1932 that the radiation could even penetrate several centimeters of lead (see Figure 12.1). This radiation could not be charged particles, because charged particles of the energies available then could not penetrate even a short distance through lead. It was naturally assumed that electromagnetic radiation (photons) was produced in the α bombardment of beryllium. Photons are called *gamma rays* when they have a nuclear origin. Gamma rays produced in the nucleus have energies on the order of MeV (as compared to the order of keV x-ray photons produced by transitions in an atom).

Curie and Joliot performed several measurements to study the effects of this new penetrating radiation (produced by α + Be) on various materials. When the radiation passed through paraffin (which contains hydrogen), they found that protons with energies up to 5.7 MeV were ejected. The simplest assumption was that the radiation—assumed to be gamma rays—scattered by the Compton process from the hydrogen nuclei and knocked the protons out of the paraffin. However, the hypothesis of Compton scattering requires gamma-ray energies of at least 50 MeV to produce 5.7-MeV protons (see Problem 1). Energies as large as 50 MeV were unprecedented at that time. No known reaction could produce such high-energy gamma rays.

In 1932 James Chadwick proposed that the new radiation produced by α + Be consisted of neutrons, the hypothetical electrically neutral particles with the mass of the proton. A neutron passes through material rather easily because it is charge neutral and has only a very small electromagnetic interaction via its magnetic moment. The nuclear force is very short range, and a neutron having MeV energies may have only a 10^{-6} probability of interacting with a nucleus. Chadwick correctly surmised that if neutrons of about 5.7 MeV of kinetic energy were produced in the α + Be reaction, the neutrons could then collide elastically with the protons in paraffin, thereby accounting for the 5.7-MeV protons. In his published results* Chadwick used his experimental data to estimate the neutron's mass as somewhere between 1.005 u and 1.008 u, not far from the modern value of 1.0087 u. For this careful experimental work and theoretical analysis, Chadwick is recognized for having confirmed the neutron's existence.

12.2   Nuclear Properties

The primary constituents of nuclei are the proton and neutron, and the nuclear mass is roughly the sum of its constituent proton and neutron masses (the slight difference being the binding energy of the nucleus). The nuclear charge is $+e$ times the number (Z) of protons ($e = 1.6 \times 10^{-19}$ C). Helium has $Z = 2$, oxygen has $Z = 8$, and uranium has $Z = 92$.

The simplest form of hydrogen has a single proton for a nucleus. However, we know that several forms of hydrogen exist. *Deuterium*—sometimes called "heavy hydrogen"—has a neutron as well as a proton in its nucleus. Another isotope of hydrogen is called *tritium*—it has two neutrons and one proton. The nuclei of the deuterium and tritium atoms are called *deuterons* and *tritons*, respectively.

The atomic (and nuclear) mass number A is the total integral number of protons and neutrons in a nucleus. Atoms with the same Z, but different A, are called **isotopes**. For example, deuterium ($A = 2$) and tritium ($A = 3$) are both isotopes of regular hydrogen ($A = 1$). The atomic mass M is the mass of the entire atom (including electrons), measured, for example, with a *mass spectrograph*.

We will designate an atomic nucleus by the symbol

$$^{A}_{Z}X_{N}$$

where Z = atomic number (number of protons)

N = neutron number (number of neutrons)

A = mass number ($Z + N$)

X = chemical element symbol

Each nuclear species with a given Z and A is called a **nuclide**. As we discussed in Chapter 8, each Z characterizes a chemical element, symbol X, for example Al for aluminum ($Z = 13$) and Ca for calcium ($Z = 20$). Although it is superfluous, we will sometimes include Z to help us remember the number of protons when dealing with elements with which we are not so familiar. When we write the nuclidic symbol for the better known elements, Z is often omitted. The nuclidic symbol A is always shown, but N is often omitted, because $A = N + Z$. Thus, $^{16}_{8}O_{8}$, $^{16}_{8}O$, and ^{16}O all represent the most abundant isotope of oxygen with $Z = 8$, $N = 8$, $A = 16$. Other stable isotopes of oxygen are ^{17}O and ^{18}O, which differ from ^{16}O only in having more neutrons. Nuclides with the same neutron number are called *isotones* (for example, $^{14}_{6}C$, $^{15}_{7}N$, $^{16}_{8}O$, and $^{17}_{9}F$). Nuclides with the same value of A are called *isobars* (for example, $^{16}_{6}C$, $^{16}_{7}N$, $^{16}_{8}O$, and $^{16}_{9}F$).

The chemical properties of an atom are determined by its electron configuration. Because the numbers of electrons and protons are equal in a neutral atom, the chemical properties are essentially determined by Z. The dependence of the chemical properties on N is negligible.

Irène Curie (1897–1956) and her husband Frédéric Joliot (1900–1958) married in 1926 while both were working with Irène's mother Marie Curie at the Radium Institute of Paris. They both took the last name Joliot-Curie. They received the 1935 Nobel Prize in Chemistry "for their synthesis of new radioactive elements." Their contributions to nuclear physics were enormous. Frédéric worked with the Resistance during World War II, and after the war they both helped develop France's first atomic pile reaction. Frédéric founded the French Communist Party, and both were relieved of positions in the early 1950s for political reasons due to their association with communism. Like her mother, Irène died of leukemia, likely due to her exposure to radioactive material. *Bettmann/Corbis.*

*J. Chadwick, The existence of a neutron, *Proceedings of the Royal Society of London, Series A* **136**, 692–708 (1932).

James Chadwick (1891–1974) studied with Ernest Rutherford at the University of Manchester and worked with Hans Geiger in Berlin in 1914 where he was interned as an enemy alien during World War I. After discovering the neutron in 1932, he received the Nobel Prize in Physics in 1935. Chadwick was a professor in Liverpool and Cambridge. He was the leading supporter of the atomic bomb development in Britain during World War II. *Burrell & Hardman, Liverpool, courtesy AIP Emilio Segrè Visual Archives, W. F. Meggers Gallery of Nobel Laureates.*

Table 12.1   Some Nucleon and Electron Properties

Particle	Symbol	Rest Energy (MeV)	Charge	Mass (u)	Spin	Magnetic Moment
Proton	p	938.272	$+e$	1.0072765	1/2	2.79 μ_N
Neutron	n	939.566	0	1.0086649	1/2	$-1.91\ \mu_N$
Electron	e	0.51100	$-e$	5.4858×10^{-4}	1/2	$-1.00116\ \mu_B$

Atomic masses are measured in atomic mass units, which are denoted by the symbol u. Atomic mass units are defined in terms of the mass of the isotope ^{12}C, whose atomic mass is defined to be exactly 12 u. The atomic masses of many nuclides are given in Appendix 8. The reason we present atomic masses rather than nuclear masses will be explained later. As we have seen in Chapter 2, the atomic mass unit works out to be

$$1\ u = 1.66054 \times 10^{-27}\ kg = 931.49\ MeV/c^2 \tag{12.1}$$

The masses of the proton and neutron are given in Table 12.1. The fact that neutrons and protons have almost the same mass is no accident. As we shall see in Chapter 14, both neutrons and protons, collectively called **nucleons**, are constructed of other particles called *quarks*. Neutrons are slightly more massive than protons.

Sizes and Shapes of Nuclei

The size of the nucleus has been determined in a variety of ways. Rutherford concluded from the alpha-particle scattering experiments of his assistants Geiger and Marsden that the range of the nuclear force must be less than about 10^{-14} m, because deviations from Coulomb's law at that distance could be inferred from their data (see Section 4.2).

Let us assume that nuclei are spheres of radius R. Particles, such as electrons, protons, neutrons, and alphas, scatter when projected close to the nucleus. It is not immediately obvious whether the maximum interaction distance measured in such collisions refers to the nuclear size (*matter radius*), or whether the nuclear force extends beyond the nuclear matter (*force radius*). Electrons do not respond to the nuclear force but scatter from the electromagnetic field of the nucleus. Thus, electron scattering measures the nuclear *charge radius*.

The nuclear force between nucleons is the strongest of the three known forces (nuclear, gravitational, and electroweak) at short distances. As such, the nuclear force is often called the **strong** force, and physicists use the terms *nuclear* and *strong* force interchangeably. Because neutrons interact only with the nuclear force, the scattering of neutrons determines the nuclear force radius. Through many measurements using beams of different particles, physicists have found that

Nuclear force radius ≈ mass radius ≈ charge radius

The nuclear radius may be approximated from a spherical charge distribution to be

Nuclear radius

$$R = r_0 A^{1/3} \tag{12.2}$$

ρ/ρ₀ ... Radial distance (fm)

Figure 12.2 The shape of the Fermi distribution for the nuclear charge density given by Equation (12.3), where $\rho_0(r)$ is the central charge density, R is the distance at which the nuclear density has dropped to 50% of its central value, and $t = 4.4a$ is the surface thickness, measured from 90% to 10% of the central density.

where $r_0 \approx 1.2 \times 10^{-15}$ m. Various measurements for r_0 range between 1.0 and 1.5×10^{-15} m. Because the nucleus is so small, we use the **femtometer**, abbreviated fm, with 1 fm = 10^{-15} m. An alternative term for 10^{-15} m, fermi, refers to Enrico Fermi, one of the founders of nuclear physics. The term fermi, which has the same abbreviation, fm, is more commonly used than femtometer; so 2.4 fm is read as 2.4 fermis or 2.4 femtometers.

Robert Hofstadter (Nobel Prize, 1961) and his colleagues at Stanford University in the 1950s performed the first precision electron-scattering measurements of the nuclear charge distribution using electron energies from 100 to 500 MeV. In order to probe the actual shape of most nuclei, we need a particle having a short wavelength. The de Broglie wavelength of a 500-MeV electron is about 2.5 fm, and by now, measurements have been made with much shorter wavelengths using higher energy electrons. These measurements are approximately described for all but the lightest nuclei by the Fermi distribution for the nuclear charge density $\rho(r)$ of the following form (see Figure 12.2):

$$\rho(r) = \frac{\rho_0}{1 + e^{(r-R)/a}} \tag{12.3}$$

The shape of this distribution is shown in Figure 12.2 where ρ_0 is the central nuclear density, R is the distance at which the nuclear density has dropped to 50% of its central value, and $t = 4.4a$ is the surface thickness, measured from 90% to 10% of the central density.

Robert Hofstadter (1915–1990) was educated at the College of the City of New York and Princeton University. Hofstadter used high-energy electrons at Stanford University to measure the charge distribution inside the nucleus, thereby determining the radius of the nucleus. He also studied the structure of the neutron and proton, proving that they were similar particles, differing only in their charge properties. *Courtesy of Stanford University.*

Example 12.2

What is the nuclear radius of ^{40}Ca? What energy electrons and protons are required to probe the size of ^{40}Ca if one wants to "see" at least half the radius?

Strategy We use Equation (12.2) to determine the radius of ^{40}Ca. From the radius we can determine the size of the probe, which subsequently determines the de Broglie wavelength and the momentum p. The total energy and kinetic energy can be found from the momentum.

Solution We determine R from Equation (12.2).

$$R = 1.2 \text{ fm } (40)^{1/3} = 4.1 \text{ fm}$$

In order to distinguish a distance at least half the radius, we need a de Broglie wavelength of 2.0 fm. If we use the relation for the de Broglie wavelength ($\lambda = h/p$), we have $p = h/\lambda$. The total energy of the probing particle is

$$E^2 = (mc^2)^2 + (pc)^2 = (mc^2)^2 + \frac{h^2c^2}{\lambda^2}$$

For a wavelength of 2.0 fm, the last term becomes

$$\frac{h^2 c^2}{\lambda^2} = \left(\frac{1240 \text{ MeV} \cdot \text{fm}}{2.0 \text{ fm}}\right)^2 = 3.8 \times 10^5 \text{ MeV}^2$$

Now if we insert the rest energy mc^2 of the appropriate particle, we can determine the total energy and required kinetic energy of either the electron or proton.

Electron energy:

$$E^2 = (0.511 \text{ MeV})^2 + 3.8 \times 10^5 \text{ MeV}^2$$

$$E = 620 \text{ MeV}$$

$$K = E - mc^2 = 620 \text{ MeV} - 0.5 \text{ MeV} = 620 \text{ MeV}$$

Proton energy:

$$E^2 = (938.3 \text{ MeV})^2 + 3.8 \times 10^5 \text{ MeV}^2$$

$$E = 1120 \text{ MeV}$$

$$K = E - mc^2 = 1120 \text{ MeV} - 938 \text{ MeV} = 180 \text{ MeV}$$

We have quoted the results to two significant figures, because of the uncertainty in the radius R.

Example 12.3

What is the ratio of the radii of ^{238}U and ^{4}He?

Strategy We use Equation (12.2) to determine the ratio of the two radii.

Solution

$$R(^{238}\text{U}) = (1.2 \text{ fm})(238)^{1/3} = 7.4 \text{ fm}$$

$$R(^4\text{He}) = (1.2 \text{ fm})(4)^{1/3} = 1.9 \text{ fm}$$

The ratio is

$$\frac{R(^{238}\text{U})}{R(^4\text{He})} = \frac{7.4 \text{ fm}}{1.9 \text{ fm}} = 3.9$$

Even though ^{238}U has 60 times the number of nucleons of ^{4}He, its radius is only four times greater.

If we approximate the nuclear shape as a sphere, we have $V = 4\pi R^3/3$, or, by using Equation (12.2) for R,

$$V = \tfrac{4}{3}\pi r_0^3 A \tag{12.4}$$

The nuclear mass density (mass/volume) can be determined from $(A \text{ u})/V$ to be $2.3 \times 10^{17} \text{ kg/m}^3$. The nucleus is about 10^{14} times denser than ordinary matter!

Intrinsic Spin

The neutron and proton are fermions with spin quantum numbers $s = 1/2$. The spin quantization rules are those we have already learned for the electron (see Chapter 7).

Intrinsic Magnetic Moment

The proton's intrinsic magnetic moment points in the same direction as its intrinsic spin angular momentum because the proton's charge is positive. This is contrasted with the electron, where the spin and magnetic moment point in opposite directions (see Figure 7.4). Nuclear magnetic moments are measured in units of the nuclear magneton μ_N, which is defined, by analogy to the Bohr magneton for electrons, by the relation

$$\mu_N = \frac{e\hbar}{2m_p} \tag{12.5}$$

Note that the divisor in calculating μ_N is the proton mass m_p (rather than the electron mass), which makes the nuclear magneton some 1800 times smaller than the Bohr magneton.

The proton magnetic moment is measured to be $\mu_p = 2.79\mu_N$. This contrasts strongly with the magnetic moment of the electron, $\mu_e = -1.00116\mu_B$. What is even more surprising is that the neutron, which is electrically neutral, also has a magnetic moment, $\mu_n = -1.91\mu_N$. The negative sign indicates that the magnetic moment points opposite to the neutron spin. The large deviation from unity of the proton's magnetic moment and the fact that the neutron even has a magnetic moment indicate that nucleons are more complicated structurally than electrons. The *nonzero* neutron magnetic moment implies that the neutron has negative and positive internal charge components at different radii, and hence a complex internal *charge distribution*.

12.3   The Deuteron

After the proton, the next simplest nucleus is the deuteron, the nucleus of ^{2}H. A deuteron consists of one proton and one neutron and allows our first look at the nuclear force. First, let us determine how strongly the neutron and proton are bound together in a deuteron. The deuteron mass is 2.013553 u, and the mass of a deuterium atom is 2.014102 u. The difference in masses is $2.014102\,\text{u} - 2.013553\,\text{u} = 0.000549\,\text{u}$, which is just the mass of an electron. The electron binding energy (13.6 eV for hydrogen) is so small that it can be neglected for our purposes. The deuteron nucleus is bound by an energy B_d, which represents mass-energy. The mass of a deuteron is then

$$m_d = m_p + m_n - B_d/c^2 \tag{12.6}$$

The deuteron mass is less than the sum of the masses of a neutron and proton by just the nuclear binding energy B_d. If we add an electron mass to each side of Equation (12.6), we have

$$m_d + m_e = m_p + m_n + m_e - B_d/c^2 \tag{12.7}$$

But $m_d + m_e$ is the atomic deuterium mass $M(^2\text{H})$ and $m_p + m_e$ is the atomic hydrogen mass (if we neglect the small amount of electron binding energy). Thus Equation (12.7) becomes

$$M(^2\text{H}) = m_n + M(^1\text{H}) - B_d/c^2 \tag{12.8}$$

and we can use *atomic masses*. Because the electron masses cancel in almost all nuclear-mass difference calculations like Equation (12.8), we routinely use atomic masses (see Appendix 8) rather than nuclear masses.* Note that we use uppercase M for atomic masses and lowercase m for nuclear and particle masses. The binding energy B_d of the deuteron is now easily determined:

$$m_n = 1.008665\,\text{u} \qquad \text{Neutron mass}$$

$$M(^1\text{H}) = 1.007825\,\text{u} \qquad \text{Atomic hydrogen mass}$$

---

*What is quoted in many reference tables of atomic masses is the mass excess Δ, given by $M - A$. This avoids having to quote the masses to so many significant figures, because M and A are almost equal. We have listed in Appendix 8 the actual atomic mass M rather than Δ in order to make the calculations more transparent to the student.

$$M(^2\text{H}) = 2.014102 \text{ u} \qquad \text{Atomic deuterium mass}$$

$$B_d/c = m_n + M(^1\text{H}) - M(^2\text{H}) = 0.002388 \text{ u}$$

We convert this mass-difference to energy using u = 931.5 MeV/c^2.

$$B_d = 0.002388 \ c^2 \cdot \text{u}\left(\frac{931.5 \text{ MeV}}{c^2 \cdot \text{u}}\right) = 2.224 \text{ MeV} \tag{12.9}$$

Our neglect of the atomic electron binding energy of 13.6 eV is justified, because the nuclear binding energy of 2.2 MeV is almost one million times greater. Even for heavier nuclei we normally neglect the electron binding energies. The electron binding energies cancel to a great extent in an equation like Equation (12.8) for heavy masses anyway.

The binding energy of any nucleus $^A_Z X$ is the energy required to separate the nucleus into free neutrons and protons. It can be determined using the atomic masses $M(^1\text{H})$ and $M(^A_Z X)$:

Nuclear binding energies

$$B(^A_Z X) = [Nm_n + ZM(^1\text{H}) - M(^A_Z X)]c^2 \tag{12.10}$$

Experimental Determination of Nuclear Binding Energies We can check our result for the 2.22-MeV binding energy of the deuteron by using a nuclear reaction. We scatter gamma rays (photons) from deuterium gas and look for the breakup of a deuteron into a neutron and a proton:

$$\gamma + d \rightarrow n + p \tag{12.11}$$

This type of nuclear reaction is called *photodisintegration* or a *photonuclear reaction*, because a photon causes the target nucleus to change form (Figure 12.3). The mass-energy relation corresponding to Equation (12.11) is

$$hf + M(^2\text{H})c^2 = m_n c^2 + M(^1\text{H})c^2 + K_n + K_p \tag{12.12}$$

where hf is the incident photon energy and K_n and K_p are the neutron and proton kinetic energies, respectively. If we want to find the minimum energy required for the photodisintegration, we let $K_n = K_p = 0$. We then find

$$hf_{\min} = m_n c^2 + M(^1\text{H})c^2 - M(^2\text{H})c^2 = B_d \tag{12.13}$$

Equation (12.13) is not quite correct, because momentum must also be conserved in the reaction (K_n and K_p can't both be zero). The precise relation is

$$hf_{\min} = B_d\left[1 + \frac{B_d}{2M(^2\text{H})c^2}\right] \tag{12.14}$$

This value of $hf_{\min}$ is almost exactly B_d, the deuteron binding energy, because the second term is so small. Experiment shows that a photon of energy less than 2.22 MeV cannot dissociate a deuteron.

Gamma ray   Deuteron   Neutron   Proton

Figure 12.3 A gamma ray of energy greater than 2.22 MeV is able to dissociate a deuteron into a neutron and a proton. This photodisintegration effect confirms that the binding energy of the deuteron is 2.22 MeV.

Deuteron Spin and Magnetic Moment Another striking property of a deuteron is its nuclear spin quantum number of 1. This indicates the neutron and proton spins are aligned parallel to each other. The nuclear magnetic moment of a deuteron is $0.86\mu_N$, which is close to the sum of the values for the free proton and neutron: $2.79\mu_N - 1.91\mu_N = 0.88\mu_N$. This supports our hypothesis of parallel spins.

12.4 Nuclear Forces

Many techniques are used to study nuclear forces. The most straightforward are based on scattering experiments. We examine the nuclear force by first studying the simplest systems. We looked at the deuteron in the previous section. In scattering neutrons from protons, a deuteron is sometimes formed in the nuclear reaction:

$$n + p \rightarrow d + \gamma \qquad (12.15)$$

We can also study the angular distribution of neutrons elastically scattered by protons, as shown in Figure 12.4a. Neutron + proton and proton + proton elastic scattering reveals that the nuclear potential is shaped roughly as shown in Figure 12.4b. The internucleon potential has a "hard core" that prevents the nucleons from approaching each other much closer than about 0.4 fm. A proton has a charge radius up to about 1 fm. Physicists believe the neutron is roughly the same size. Two nucleons within about 2 fm of each other feel an attractive nuclear force. Outside about 3 fm the nuclear force is essentially zero. We call the nuclear force *short range* because it falls to zero so abruptly with interparticle separation. Because the nuclear force is short range, nucleons mostly interact with their nearest-neighbor nucleons. The nuclear force is said to be *saturable*, because the interior nucleons are completely surrounded by other nucleons with which they interact. However, nucleons on the nuclear surface are not so completely bound, and their nuclear force is not saturated. Of course, we are speaking classically about phenomena that must be described quantum mechanically.

Nuclear forces are short range

The only difference between the *np* and *pp* potentials shown in Figure 12.4b is the Coulomb potential shown for $r \geq 3$ fm for the *pp* force. Inside 3 fm the nuclear force clearly dominates, but outside 3 fm only the Coulomb force is effective. The depth of the nucleon-nucleon potential is about 40 MeV; the *np* potential is slightly greater because of the absence of the Coulomb force.

The nuclear force is known to be spin dependent because the bound state of the deuteron has the neutron and proton spins aligned, but there is no bound state with the spins antialigned (that is, coupled to total spin 0).

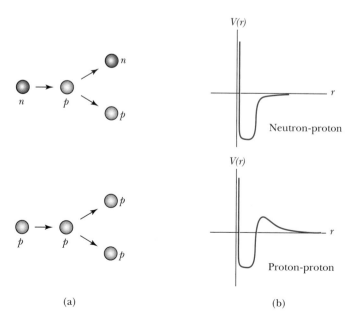

(a)   (b)

Figure 12.4 (a) A detailed study of neutron + proton and proton + proton scattering reveals (b) the shape of the potential describing each interaction. The proton-proton interaction includes the Coulomb effect (not to scale).

The neutron-neutron system is more difficult to study because free neutrons (not bound in a nucleus) are not stable, and we do not know how to construct a target of free neutrons. However, indirect evidence, together with analyses of experiments where moving neutrons scatter from each other (as in simultaneous nuclear bomb explosions), indicates the *nn* potential is similar to the *np* potential. The nuclear potential between two nucleons seems independent of their charges. We call this *charge independence of nuclear forces*. For many purposes, the neutron and proton can be considered different charge states of the same particle. This is why we use the term *nucleon* to refer to either neutrons or protons.

Nuclear forces are charge independent

12.5   Nuclear Stability

In Equation (12.10) we presented a method to determine the binding energies of nuclides in terms of atomic masses. If B is positive, then the nuclide is said to be stable against dissociating into free neutrons and protons. We need to generalize Equation (12.10), however, because a nucleus containing A nucleons is said to be *stable* if its mass is smaller than that of any other possible combination of A nucleons. The binding energy of a nucleus $_Z^A X$ against dissociation into any other possible combination of nucleons, for example nuclei R and S, is

$$B = [M(R) + M(S) - M(_Z^A X)]c^2 \tag{12.16}$$

In particular, the energy required to remove one proton (or neutron) from a nuclide is called the proton (or neutron) *separation energy*, and Equation (12.16) is useful for finding this energy. Even if B is negative for a particular dissociation, there may be other reasons why the nucleus is stable.

Example 12.4

Show that the nuclide ^{8}Be has a positive binding energy but is unstable with respect to decay into two alpha particles.

Strategy We use Equation (12.10) to find the binding energy of ^{8}Be and Equation (12.16) to find the binding energy of ^{8}Be with respect to decay to two alpha particles. The nuclide will be unstable if the binding energy B is negative.

Solution The binding energy of ^{8}Be is

$$B(^8\text{Be}) = [4m_n + 4M(^1\text{H}) - M(^8\text{Be})]c^2$$

We look up the atomic masses of ^{1}H and ^{8}Be in Appendix 8 and calculate the binding energy to be

$$B(^8\text{Be}) = [4(1.008665 \text{ u}) + 4(1.007825 \text{ u}) - 8.005305 \text{ u}]$$
$$\times c^2\left(\frac{931.5 \text{ MeV}}{c^2 \cdot \text{u}}\right) = 56.5 \text{ MeV}$$

So ^{8}Be has a positive binding energy. Now we calculate the binding energy of the decay of ^{8}Be into two α particles,

^{8}Be $\rightarrow 2\alpha$, by using Equation (12.16):

$$B(^8\text{Be} \rightarrow 2\alpha) = [2M(^4\text{He}) - M(^8\text{Be})]c^2$$
$$= [2(4.002603 \text{ u}) - 8.005305 \text{ u}]$$
$$\times c^2\left(\frac{931.5 \text{ MeV}}{c^2 \cdot \text{u}}\right) = -0.093 \text{ MeV}$$

Because the latter B is negative, ^{8}Be is *unstable* against decay to two alpha particles. From the standpoint of energy, there is no reason why a ^{8}Be nucleus will not decay into two alpha particles. Sometimes a nuclide may be stable even if another combination of A nucleons has a lower mass, because some conservation law, such as spin angular momentum, prevents the radioactive decay, but in this case we find experimentally that ^{8}Be does spontaneously decay into two alpha particles. The instability of ^{8}Be is responsible for the fact that stars consist mostly of hydrogen and helium. Because of the instability of ^{8}Be, it is difficult for helium nuclei to join together to make heavier nuclei.

Figure 12.5 A plot of the known nuclides with neutron number N versus proton number Z. The solid points represent stable nuclides, and the shaded area represents unstable nuclei. A smooth line through the solid points would represent the line of stability.

In Figure 12.5 we exhibit all known stable nuclei as well as many known unstable nuclei that are long-lived enough to be observed. Because experimentalists are currently able to measure the half-lives of small quantities of material, Figure 12.5 includes nuclides that decay in a millisecond or less. The line representing the stable nuclides is called the **line of stability** (see Figure 12.5). There are several important facts we can extract from Figure 12.5. First, it appears that for $A \leq 40$, nature prefers the number of protons and neutrons in the nucleus to be about the same, $Z \approx N$. However, for $A \geq 40$, there is a decided preference for $N > Z$. We can understand this difference in the following way. As we noted earlier, the strength of the nuclear force is independent of whether the particles are nn, np, or pp. Equal numbers of neutrons and protons may give the most attractive average internucleon nuclear force, but the Coulomb force must be considered as well. As the number of protons increases, the Coulomb force between all the protons becomes stronger and stronger until it eventually affects the binding significantly.

Line of stability

The electrostatic energy required to contain a charge Ze evenly spread throughout a sphere of radius R can be calculated by determining the work required to bring the charge inside the sphere from infinity (see Problem 55) and is determined to be

$$\Delta E_{\text{Coul}} = \frac{3}{5} \frac{(Ze)^2}{4\pi\epsilon_0 R} \tag{12.17}$$

For a single proton, Equation (12.17) gives for the self-energy

$$\Delta E_{\text{Coul}} = \frac{3}{5}\frac{e^2}{4\pi\epsilon_0 R}$$

This term represents the work done to assemble the proton itself, and we do not want to include it in the electrostatic self-energy of a nucleus composed of Z protons. Therefore we must subtract Z such terms from the total given in Equation (12.17) to give us the total Coulomb repulsion energy in a nucleus:

$$\Delta E_{\text{Coul}} = \frac{3}{5}\frac{Z(Z-1)e^2}{4\pi\epsilon_0 R} \tag{12.18}$$

Example 12.5

Show that Equation (12.18) can be written as

$$\Delta E_{\text{Coul}} = 0.72[Z(Z-1)]A^{-1/3}\,\text{MeV} \tag{12.19}$$

and use this equation to calculate the total Coulomb energy of $^{238}_{92}\text{U}$.

Strategy We use Equation (12.2) for R (with $r_0 = 1.2$ fm) and insert it into Equation (12.18) to find the energy first in joules and then in MeV.

Solution

$$\Delta E_{\text{Coul}} = \frac{3}{5}[Z(Z-1)](1.6\times10^{-19}\,\text{C})^2$$

$$\times (9\times10^9\,\text{N}\cdot\text{m}^2/\text{C}^2)\frac{1}{1.2\times10^{-15}\,\text{m}\cdot A^{1/3}}$$

$$= 1.15\times10^{-13}[Z(Z-1)]A^{-1/3}\,\text{J}\frac{1\,\text{MeV}}{1.6\times10^{-13}\,\text{J}}$$

$$= 0.72[Z(Z-1)]A^{-1/3}\,\text{MeV}$$

Now we insert $Z = 92$ and $A = 238$ into Equation (12.19) to find

$$E_{\text{Coul}} = 0.72(92)(91)(238)^{-1/3} = 970\,\text{MeV}$$

Is this large or small? If we look up the masses in Appendix 8, we calculate that the total binding energy of $^{238}_{92}\text{U}$ with respect to dissociation into its component nucleons is

$$B_{\text{tot}} = [146(1.008665\,\text{u}) + 92(1.007825\,\text{u})$$

$$- 238.050783\,\text{u}]c^2\left(\frac{931.5\,\text{MeV}}{c^2\cdot\text{u}}\right) = 1802\,\text{MeV}$$

The total Coulomb energy is a significant fraction of the binding energy of a large nucleus.

We conclude that for heavy nuclei, the nucleus will have a preference for fewer protons than neutrons because of the large Coulomb repulsion energy. In fact, Figure 12.5 reveals there are no stable nuclei with $Z > 83$ because of the increasingly larger Coulomb force. The heaviest known stable nucleus is $^{209}_{83}\text{Bi}$. All nuclei with $Z > 83$ and $A > 209$ will eventually decay spontaneously into some combination of smaller masses. Adding one proton to a heavy nucleus adds a constant amount of nuclear binding energy, but the repulsive Coulomb energy increases as $\Delta E_{\text{Coul}} \approx (Z+1)^2 - Z^2 \approx 2Z$. Because the Coulomb force is long range, the proton interacts electromagnetically with all the protons already in the nucleus. And because this repulsive energy increases with Z, nuclei with higher Z eventually become unstable. The neutrons dilute the Coulomb repulsion slightly because they intersperse among the protons, causing the protons to be slightly farther apart.

Another interesting fact discernible from Figure 12.5 is that most stable nuclides have *both* even Z and even N (called "even-even" nuclides). Only four stable nuclides, all light nuclei, have odd Z and odd N (called "odd-odd" nuclides). These nuclides are ${}^{2}_{1}$H, ${}^{6}_{3}$Li, ${}^{10}_{5}$B, and ${}^{14}_{7}$N. All the other stable nuclides are odd-even or even-odd, that is, with either an odd number of Z or N. Nature apparently prefers nuclei with even numbers of protons and neutrons.

We can understand this empirical observation in terms of the Pauli exclusion principle. Neutrons and protons are distinguishable fermions; hence they separately obey the exclusion principle. Only two neutrons (or protons) may coexist in each spatial orbital (quantum state), one with spin "up" and the other with spin "down." Each nuclear energy level is thus able to hold two particles whose spins are paired to 0. This configuration of opposite spins is particularly stable because placing the same number of particles in any other arrangement will produce a (less stable) state of higher energy. Therein lies the preference for even N and Z.

Conceptual Example 12.6

Only four stable nuclides have odd numbers for both N and Z. Predict whether adding both a proton and a neutron to each of these nuclides, ${}^{2}_{1}$H, ${}^{6}_{3}$Li, ${}^{10}_{5}$B, and ${}^{14}_{7}$N, will produce a stable nucleus. Look up in Appendix 8 whether they are stable or not.

Solution Adding a proton and a neutron to each nuclide will make them each even-even. There is a good likelihood that each new nuclide is stable, because nucleons prefer to have even numbers.

When we look up the nuclides in Appendix 8, we find that only ${}^{8}_{4}$Be is unstable, which is a special case because it decays to two alpha particles. Alpha particles are particularly stable because both Z and N are equal to 2, which is a "magic number" (see Problem 62).

Niels Bohr, Carl F. von Weizsäcker, and others were able in the 1930s to explain many nuclear phenomena by treating the nucleus as a collection of interacting particles in a liquid drop. This model of the nucleus is known as the **liquid drop model**. In the preceding discussion we understood qualitatively the line of stability curve displayed in Figure 12.5. In much the same way von Weizsäcker proposed in 1935 his semi-empirical mass formula based on the liquid drop model.

Written in terms of the total binding energy, the semi-empirical mass formula is

$$B({}^{A}_{Z}X) = a_V A - a_A A^{2/3} - \frac{3}{5}\frac{Z(Z-1)e^2}{4\pi\epsilon_0 r} - a_S\frac{(N-Z)^2}{A} + \delta \qquad (12.20)$$

Liquid drop model

The von Weizsäcker semi-empirical mass formula

This is actually the binding energy. The mass is found by using Equation (12.10). The volume term (a_V) indicates that the binding energy is approximately the sum of all the interactions between the nucleons. Because the nuclear force is short range and each nucleon interacts only with its nearest neighbors, this interaction is proportional to A, the total number of nucleons.

The second term, called the *surface effect*, is simply a correction to the first term (similar to surface tension), because the nucleons on the nuclear surface are not completely surrounded by other nucleons. The surface nucleons do not have saturated interactions, and a correction should be made proportional to the liquid drop surface, $4\pi R^2$. Because $R \sim A^{1/3}$, the correction is proportional to $A^{2/3}$.

The third term is the Coulomb energy discussed and presented in Equations (12.17) and (12.18). A simple result is Equation (12.19).

The fourth term is due to the symmetry energy, also previously discussed. In the absence of Coulomb forces, the nucleus prefers to have $N \approx Z$. This term has a quantum-mechanical origin, depending on the exclusion principle. Notice that the sign of the fourth term is independent of the sign of $N - Z$.

The last term is due to the pairing energy and reflects the fact that the nucleus is more stable for even-even nuclides. We can determine this term empirically. One set of values given by Enrico Fermi* for the parameters of Equation (12.20) is

$$a_V = 14 \text{ MeV} \qquad \text{Volume}$$

$$a_A = 13 \text{ MeV} \qquad \text{Surface}$$

$$a_S = 19 \text{ MeV} \qquad \text{Symmetry}$$

$$\text{Pairing } \delta = \begin{cases} +\Delta & \text{for even-even nuclei} \\ 0 & \text{for odd-}A \text{ (even-odd, odd-even) nuclei} \\ -\Delta & \text{for odd-odd nuclei} \end{cases}$$

where $\Delta = 33 \text{ MeV} \cdot A^{-3/4}$.

The entire table of stable isotopes can be understood by applying the ideas in the von Weizsäcker semi-empirical mass formula. No nuclide heavier than $^{238}_{92}\text{U}$ has been found in nature. Such nuclides, if they ever existed, must have decayed so quickly that quantities sufficient to measure no longer exist. Many of the nuclides between $^{209}_{83}\text{Bi}$ and $^{238}_{92}\text{U}$ are still found in nature, either because their decay rates are slow enough that they have not sufficiently decayed since their formation in the interior of stars or because they are produced continuously by the radioactive decay of another nuclide.

Binding energy per nucleon

To compare the relative stability of different nuclides, it's important to know the **binding energy per nucleon**. By calculating the binding energy of each known nucleus, and dividing by its mass number, we obtain the plot shown in Figure 12.6. We see that the average binding energy per nucleon peaks near $A = 56$ and slowly decreases for the heaviest nuclei. For the lighter nuclei the curve increases rapidly from hydrogen until all the nucleons are surrounded by other nucleons. This curve demonstrates the saturation effect of nuclear forces. After the very light

*See B. Cohen, *Concepts of Nuclear Physics*, New York: McGraw-Hill (1971).

Figure 12.6 The binding energy per nucleon versus the mass number A. Notice the subpeaks at ^4He, ^{12}C, and ^{16}O.

nuclei $(A < 20)$, the curve is reasonably flat at about 8 MeV/nucleon. There are sharp peaks for the even-even nuclides ^{4}He, ^{12}C, and ^{16}O, which are particularly tightly bound.

Example 12.7

Calculate the binding energy per nucleon for $^{20}_{10}$Ne, $^{56}_{26}$Fe, and $^{238}_{92}$U.

Strategy We first find the binding energy of each of these nuclides using Equation (12.10) and then divide by the mass number to obtain the binding energy per nucleon.

Solution

$$B(^{20}_{10}\text{Ne}) = [10m_n + 10M(^1\text{H}) - M(^{20}_{10}\text{Ne})]c^2$$

$$= [10(1.008665 \text{ u}) + 10(1.007825 \text{ u})$$

$$- 19.992440 \text{ u}]c^2\left(\frac{931.5 \text{ MeV}}{c^2 \cdot \text{u}}\right)$$

$$= 161 \text{ MeV}$$

$$\frac{B(^{20}_{10}\text{Ne})}{20 \text{ nucleons}} = 8.03 \text{ MeV/nucleon}$$

$$B(^{56}_{26}\text{Fe}) = [30m_n + 26M(^1\text{H}) - M(^{56}_{26}\text{Fe})]c^2$$

$$= [30(1.008665 \text{ u}) + 26(1.007825 \text{ u})$$

$$- 55.934942 \text{ u}]c^2\left(\frac{931.5 \text{ MeV}}{c^2 \cdot \text{u}}\right)$$

$$= 492 \text{ MeV}$$

$$\frac{B(^{56}_{26}\text{Fe})}{56 \text{ nucleons}} = 8.79 \text{ MeV/nucleon}$$

$$B(^{238}_{92}\text{U}) = [146m_n + 92M(^1\text{H}) - M(^{238}_{92}\text{U})]c^2$$

$$= [146(1.008665 \text{ u}) + 92(1.007825 \text{ u})$$

$$- 238.050783 \text{ u}]c^2\left(\frac{931.5 \text{ MeV}}{c^2 \cdot \text{u}}\right)$$

$$= 1800 \text{ MeV}$$

$$\frac{B(^{238}_{92}\text{U})}{238 \text{ nucleons}} = 7.57 \text{ MeV/nucleon}$$

All three nuclides have a binding energy per nucleon near 8 MeV, with ^{56}Fe having the largest binding energy per nucleon, as shown in Figure 12.6.

Nuclear Models

Physicists do not fully understand the nuclear force or how nucleons interact inside the nucleus. Current research focuses on the constituent quarks (see Chapter 14) that make up the nucleons. Because the nuclear force is not precisely known, physicists have relied on a multitude of models to explain nuclear behavior. These models have been more or less successful in explaining various nuclear properties.

The models generally fall into two categories:

1. *Independent-particle models*, in which the nucleons move nearly independently in a common nuclear potential. The shell model has been the most successful of these.
2. *Strong-interaction models*, in which the nucleons are strongly coupled together. The liquid drop model already discussed is characteristic of these models and has been quite successful in explaining nuclear masses as well as nuclear fission (see Chapter 13).

Space does not permit us a full discussion of each of the many models. We have already discussed the liquid drop model in this section, so we now present the simplest of the independent-particle models. We show in Figure 12.7 a representation

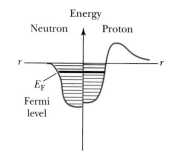

Figure 12.7 Diagram of nuclear potential wells as felt by neutrons and protons. Neutrons are more strongly bound than protons because of the Coulomb potential. All levels below the Fermi energy E_F are filled.

of the nuclear potential felt by the neutron and the proton. Because of the Coulomb interaction, the shape and depth of the proton potential is somewhat different than that of the neutron. For example, typical depths are about 43 MeV for neutrons but only 37 MeV for protons. Energy levels, which represent states that can be filled by the nucleons, are shown inside the potential. Note that nuclei have a Fermi energy level, just as do atoms, which is the highest energy level filled in the nucleus. A typical Fermi energy level has a depth of about 8 MeV. In the ground state of a nucleus, all the energy levels below the Fermi level are filled, but when a nucleus becomes excited, one or more of the nucleons is raised to one of the previously unoccupied levels above the Fermi level.

Nuclei are formed by a collection of nucleons. The nucleons sort themselves into the lowest possible energy levels. In Figure 12.8 we exhibit energy-level diagrams for several possible nuclides between ^{12}C and ^{16}O. These energy-level diagrams assume the zero of the energy scale to be the bottom of the nuclear potential, so we can deal with positive energy values. Both ^{12}C and ^{16}O are particularly stable because they are even-even. Note that we show the neutron energy levels slightly lower than the proton levels because of the additional Coulomb repulsion of the protons. If we add one proton to ^{12}C to make $^{13}_{7}$N, we find it is unstable; whereas if we add a neutron to make ^{13}C, we find it to be stable. Even when we add another neutron to produce ^{14}C, we find it is barely unstable. In this mass region, nature prefers the number of neutrons and protons to be about equal $(N \approx Z)$, but it doesn't want $Z > N$. This helps explain why ^{13}C is stable, but not ^{13}N; ^{14}C has too many more neutrons (8) than protons (6) in this mass region to be stable. When we add a proton to ^{13}C we make ^{14}N, one of the few stable odd-odd nuclides. If we next add a proton to ^{14}N we obtain the unstable ^{15}O. However, if we add a neutron to ^{14}N we find stable ^{15}N, again indicating neutron energy levels to be lower in energy than the corresponding proton ones. Finally, if we add one more proton to ^{15}N, we pair the extra proton and make the extra-stable ^{16}O. However, if we add a neutron to ^{15}N, we have the very unstable nuclide ^{16}N. Only ^{14}N and ^{15}N are stable isotopes of nitrogen.

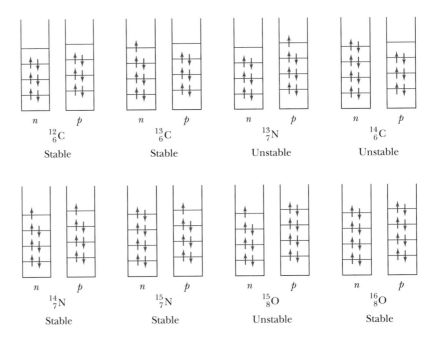

Figure 12.8 Schematic diagram of proton and neutron energy levels for several nuclei between ^{12}C and ^{16}O. The nuclei ^{12}C and ^{16}O are particularly stable, but the effects of $N \approx Z$ and the spin-pairing effects are important in this region.

The shell model of nuclei (see Problems 62, 63) takes advantage of the pairing effect and places only two neutrons or two protons in each shell (or energy level). The ordering of the energy levels is established by angular momentum rules, which couple the nucleon spins in a prescribed manner similar to that already discussed for jj coupling in Chapter 8.

12.6 Radioactive Decay

The discoverers of radioactivity were Wilhelm Röntgen, Henri Becquerel, and Marie Curie in the late 1890s. Marie Curie and her husband Pierre discovered polonium and radium in 1898. We saw in the previous section that many nuclei are unstable and can decay spontaneously to some other combination of A nucleons that has a lower mass. These decays take different forms. The simplest is that of a gamma ray, which represents the nucleus changing from an excited state to a lower energy state (no change in N or Z). Other modes of decay include emission of α particles, β particles, protons, neutrons, and fission. We will discuss α, β, and γ decay in the next section and defer the discussion of fission to the next chapter. In this section we consider the nature of the radioactive decay law.

The general form of the law of radioactivity is the same for all decays because it is a statistical process. Given a sample of radioactive material we measure the disintegrations or decays per unit time, which we define as **activity**. If we have N unstable atoms of a material, the activity R is given by

$$\text{Activity} = -\frac{dN}{dt} = R \tag{12.21}$$

where we insert the minus sign to make R positive (dN/dt is negative because the total number N is decreasing with time). The SI unit of activity is the becquerel (1 Bq = 1 decay/s). More commonly used until recently was the curie (Ci) which is 3.7×10^{10} decays/s. In keeping with the worldwide trend to use SI units, here we will use primarily the becquerel for activity. A typical radioactive source used in a student laboratory experiment (for example, ^{226}Ra or ^{210}Po α emitters), the ^{241}Am α emitter used in a smoke alarm, or the radium used in a luminous watch may contain material having an activity of only about 10^4 Bq (a few μCi), and these are exempt from U.S. federal licensing requirements.

We observe experimentally that the activity of a given sample falls off exponentially with time. We'll now explain theoretically why this occurs. If $N(t)$ is the number of radioactive nuclei in a sample at time t, and λ (called the **decay constant**) is the probability per unit time that any given nucleus will decay, then the activity R is

$$R = \lambda N(t) \tag{12.22}$$

The number dN of nuclei decaying during the time interval dt is

$$dN(t) = -R\, dt = -\lambda N(t)\, dt \tag{12.23}$$

where we have used Equations (12.21) and (12.22). If we rearrange and integrate this equation, we have

$$\int \frac{dN}{N} = -\int \lambda\, dt$$

$$\ln N = -\lambda t + \text{constant}$$

$$N(t) = e^{-\lambda t + \text{constant}}$$

Marie Curie (1867–1934) (on the right) and Irène Joliot-Curie (1897–1956) are the most famous mother-daughter pair in science. Marie won two Nobel Prizes, one in physics with her husband Pierre in 1903 and another in chemistry in 1911, for her work in radiation phenomena and the discovery of the elements radium and polonium. Irène, who began working in her mother's lab as a teenager, won a Nobel Prize in Chemistry in 1935 with her husband Frédéric Joliot-Curie for their production of new radioactive elements. *AIP Emilio Segrè Visual Archives, William G. Myers Collection.*

Decay constant

If we let $N(t = 0) \equiv N_0$, the previous equation becomes

Radioactive decay law

$$N(t) = N_0 e^{-\lambda t} \tag{12.24}$$

This is the **radioactive decay law**, and it applies to all decays. The exponential decay rate is consistent with experimental observation. The activity R is

$$R = \lambda N(t) = \lambda N_0 e^{-\lambda t} = R_0 e^{-\lambda t} \tag{12.25}$$

where R_0 is the initial activity at $t = 0$. The activity of a radioactive sample also falls off exponentially.

It is more common to refer to the half-life $t_{1/2}$ or the mean lifetime τ rather than its decay constant. The half-life is the time it takes one half of the radioactive nuclei to decay.

$$N(t_{1/2}) = \frac{N_0}{2} = N_0 e^{-\lambda t_{1/2}} \tag{12.26}$$

$$\ln\left(\frac{1}{2}\right) = \ln(e^{-\lambda t_{1/2}}) = -\lambda t_{1/2}$$

The half-life is determined to be

Half-life

$$t_{1/2} = \frac{-\ln(1/2)}{\lambda} = \frac{\ln(2)}{\lambda} = \frac{0.693}{\lambda} \tag{12.27}$$

The mean (or average) lifetime τ is calculated to be (see Problem 27)

Mean lifetime

$$\tau = \frac{1}{\lambda} = \frac{t_{1/2}}{\ln(2)} \tag{12.28}$$

The number of radioactive nuclei as a function of time is displayed in Figure 12.9. Because the decay of a radioactive nucleus is a statistical process, it will take a very large sample to give a curve as smooth as that shown in Figure 12.9.

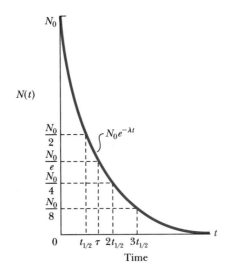

Figure 12.9 The number of remaining nuclei $N(t)$ as a function of time t for a sample of radioactive nuclides. The half-life $t_{1/2}$ and mean lifetime τ are indicated for the exponential radioactive decay law.

Conceptual Example 12.8

A radioactive sample of mass 1.0 mg has a half-life of 1.0 h, which means that about 50% of the 1-mg sample will decay in 1.0 h. What fraction decays in 2.0 h?

Solution At first thought, we might think that another 0.5-mg sample might decay in the next hour, leaving none of the original sample, but that is incorrect. During the

second-hour period the probability is still 50% for each remaining nucleus to decay. Therefore during the second hour, the amount that decays is 50% of 0.5 mg, or 0.25 mg. The total amount decayed in the first two hours is then 0.50 mg + 0.25 mg = 0.75 mg, or 75% of the original sample.

Example 12.9

A sample of ^{210}Po which α decays with $t_{1/2} = 138$ days is observed by a student to have 2000 disintegrations/s (2000 Bq).
(a) What is the activity in μCi for this source?
(b) What is the mass of the ^{210}Po sample?

Strategy (a) We already know the activity in disintegrations/s or Bq. We simply use the conversion to μCi. (b) Because we know the activity (2000 decays/s), we can use Equations (12.22) and (12.28) to find the number of radioactive nuclei. From this we can determine the mass of the ^{210}Po sample.

Solution (a) We multiply the activity of 2000 decays/s by the factor that converts decays/s to Ci.

$$2000 \text{ decays/s}\left(\frac{1 \text{ Ci}}{3.7 \times 10^{10} \text{ decays/s}}\right) = 0.054 \times 10^{-6} \text{ Ci}$$

$$= 0.054 \ \mu\text{Ci}$$

(b) Equations (12.22) and (12.28) give the number of radioactive nuclei.

$$N = \frac{R}{\lambda} = \frac{(R)(t_{1/2})}{\ln(2)} = \frac{2000 \text{ decays/s}}{\ln(2)}(138 \text{ days})\frac{24 \text{ h}}{1 \text{ day}}\frac{3600 \text{ s}}{1 \text{ h}}$$

$$= 3.44 \times 10^{10} \text{ nuclei}$$

We use Avogadro's number to determine the mass from the number of atoms (nuclei).

$$\text{Mass} = 3.44 \times 10^{10} \text{ atoms} \frac{1 \text{ mol}}{6.02 \times 10^{23} \text{ atoms}} \frac{0.21 \text{ kg}}{1 \text{ mol}}$$

$$= 1.2 \times 10^{-14} \text{ kg}$$

This is an extremely small mass!

Example 12.10

A sample of ^{18}F is used internally as a medical diagnostic tool by observing this isotope's positron decay ($t_{1/2} = 110$ min). How much time does it take for 99% of the ^{18}F to decay?

Strategy We use the radioactive decay law, Equation (12.24), to determine the time needed.

$$N = N_0 e^{-\lambda t} = N_0 e^{-\ln(2)t/t_{1/2}}$$

Solution If we want 99% of the initial sample to decay, then only 1% will be left, and $N/N_0 = 0.01$. We then have

$$\frac{N}{N_0} = 0.01 = e^{-\ln(2)t/t_{1/2}}$$

If we take the natural logarithm, we have

$$\ln(0.01) = -\ln(2)\left(\frac{t}{t_{1/2}}\right)$$

$$t = -\left[\frac{\ln(0.01)}{\ln(2)}\right]t_{1/2} = -\left(\frac{-4.61}{0.693}\right)(110 \text{ min})$$

$$= 731 \text{ min} = 12.2 \text{ h}$$

Example 12.11

What is the alpha activity of a 10-kg sample of ^{235}U that is used in a nuclear reactor?

Strategy We find the number of radioactive atoms by using Avogadro's number and the gram-molecular weight. We find in Appendix 8 that ^{235}U has a half-life for emitting α particles of $t_{1/2} = 7.04 \times 10^8$ y. Then we use Equation (12.22) to find the activity.

Solution The number of ^{235}U atoms in a 10-kg sample is

$$N = M\frac{N_A}{M(^{238}U)}$$

$$= (10 \text{ kg})\left(\frac{10^3 \text{ g}}{1 \text{ kg}}\right)\left(\frac{6.02 \times 10^{23} \text{ atoms/mol}}{235 \text{ g/mol}}\right)$$

$$= 2.56 \times 10^{25} \text{ atoms} = 2.56 \times 10^{25} \text{ nuclei}$$

The activity is

$$R = \lambda N = \frac{\ln(2) \cdot N}{t_{1/2}}$$

$$= \frac{\ln(2) \cdot (2.56 \times 10^{25} \text{ nuclei})}{7.04 \times 10^8 \text{ y}}$$

$$= 2.52 \times 10^{16} \text{ decays/y} = 8.0 \times 10^8 \text{ Bq}$$

12.7   Alpha, Beta, and Gamma Decay

The three common decay modes of nuclei (α, β, and γ) were all observed by the early twentieth century. When a nucleus decays, all the conservation laws must be observed: mass-energy, linear momentum, angular momentum, and electric charge. To these laws we add another one for radioactive decay, called the **conservation of nucleons**. It states that *the total number of nucleons (A, the mass number) must be conserved in a low-energy nuclear reaction (say, less than 100 MeV) or decay.* Neutrons may be converted into protons, and vice versa, but the total number of nucleons must remain constant. At higher energies enough rest energy may be available to create nucleons, but other conservation laws to be discussed in Chapter 14 still apply.

Conservation of nucleons

Radioactive decay may occur for a nucleus when some other combination of the A nucleons has a lower mass. Let the radioactive nucleus $^A_Z X$ be called the parent and have the mass $M(^A_Z X)$. Two or more products can be produced in the decay. In the case of two products let the mass of the lighter one be M_y and the mass of the heavier one (normally called the *daughter*) be M_D. The conservation of energy is

$$M(^A_Z X) = M_D + M_y + Q/c^2 \tag{12.29a}$$

where Q is the energy released and is equal to the total kinetic energy of the reaction products. It is given in terms of the masses by

Disintegration energy

$$Q = [M(^A_Z X) - M_D - M_y]c^2 \tag{12.29b}$$

Note that the **disintegration energy** Q is the negative of the binding energy B [see Equation (12.16)]. The binding energy normally refers to stable nuclei, whereas Q is normally used with unstable nuclei. If $B > 0$, a nuclide is bound and stable; if $Q > 0$, a nuclide is unbound, unstable, and may decay. Looking at the naturally abundant radioactive nuclei, we find that decays emitting nucleons do not occur, because the masses are such that $Q < 0$. If nucleon decay were possible for these radioactive nuclei, it would have taken place too quickly for the nuclei to be naturally abundant.

Example 12.12

Show that $^{230}_{92}$U does not decay by emitting a neutron or proton.

Strategy The decays in question are (a) $^{230}_{92}$U $\to n + {}^{229}_{92}$U and (b) $^{230}_{92}$U $\to p + {}^{229}_{91}$Pa. We determine whether the decays occur by looking up the atomic masses in Appendix 8 and use Equation (12.29b) to see if Q is positive or negative.

Solution

(a) $M({}^{230}_{92}$U$) = 230.033927$ u; $m_n = 1.008665$ u; $M({}^{229}_{92}$U$) = 229.033496$ u.

$$Q = [230.033927 \text{ u} - 229.033496 \text{ u} - 1.008665 \text{ u}]c^2$$
$$\times \left(\frac{931.5 \text{ MeV}}{c^2 \cdot \text{u}}\right) = -7.7 \text{MeV}$$

Because $Q < 0$, neutron decay is not allowed.

(b) $m({}^{1}$H$) = 1.007825$ u; $M({}^{229}_{91}$Pa$) = 229.032089$ u.

$$Q = [230.033927 \text{ u} - 229.032089 \text{ u} - 1.007825 \text{ u}]c^2$$
$$\times \left(\frac{931.5 \text{ MeV}}{c^2 \cdot \text{u}}\right) = -5.6 \text{ MeV}$$

Because $Q < 0$, proton decay is not allowed.

In both cases the decay is not allowed, because the mass of the products is greater than that of the parent. The nucleus ^{230}U is stable against nucleon emission.

Alpha Decay

Is it possible for a collection of nucleons inside the nucleus to decay? The nucleus ^{4}He is particularly stable. Its binding energy is 28.3 MeV. The combination of two neutrons and two protons is particularly strong because of the pairing effects previously discussed. If the last two protons and two neutrons in a nucleus are bound by less than 28.3 MeV, then the emission of an alpha particle, called *alpha decay*, is energetically possible. For alpha decay, Equation (12.29) becomes

$$^A_Z X \to {}^{A-4}_{Z-2}D + \alpha \qquad\qquad (12.30) \quad \textbf{Alpha decay}$$

$$Q = [M({}^A_Z X) - M({}^{A-4}_{Z-2}D) - M({}^4\text{He})]c^2 \qquad\qquad (12.31)$$

If $Q > 0$, then alpha decay is possible.

Consider the nucleus $^{230}_{92}$U that we studied in Example 12.12. The alpha-decay reaction is given by

$$^{230}_{92}\text{U} \to \alpha + {}^{226}_{90}\text{Th}$$

We look up the appropriate masses to find

$$M({}^{230}_{92}\text{U}) = 230.033927 \text{ u}; \quad M({}^4\text{He}) = 4.002603 \text{ u}; \quad M({}^{226}_{90}\text{Th}) = 226.024891 \text{ u}$$

If we insert the masses into Equation (12.31), we find

$$Q = [M({}^{230}\text{U}) - M({}^{226}\text{Th}) - M({}^4\text{He})]c^2$$

$$= [230.033927 \text{ u} - 226.024891 \text{ u} - 4.002603 \text{ u}]c^2\left(\frac{931.5 \text{ MeV}}{c^2 \cdot \text{u}}\right)$$

$$= 6.0 \text{ MeV}$$

Alpha decay is allowed, because $Q > 0$. The mass of the products is less than the mass of the decaying nuclide. Many of the nuclei above $A = 150$ in fact are

Figure 12.10 The potential energy barrier for an alpha particle is shown. The Coulomb barrier V_B is much greater than the typical alpha-particle energies produced by radioactive sources. Classically a 5-MeV particle inside the nucleus or scattered from outside cannot penetrate the barrier.

susceptible to alpha decay. These heavy nuclei have increasingly stronger Coulomb repulsion as protons are added. The expulsion of two protons (along with two neutrons) in the form of an alpha particle may decrease this Coulomb energy and make the resulting nucleus more stable.

We might wonder why any nuclei exist with $A > 150$. First, nuclei are not necessarily made up of a collection of alpha particles. In order for alpha decay to occur, two neutrons and two protons group together within the nucleus prior to decay. Second, the alpha particle, even when formed, has great difficulty in overcoming the nuclear attraction from the remaining nucleons to escape. Consider the potential energy diagram shown in Figure 12.10. The barrier height V_B for alpha particles is normally greater than 20 MeV. The kinetic energies of alpha particles emitted from nuclei range from 4 to 10 MeV. It is classically impossible for the alpha particles to escape, because the potential energy barrier is greater than the kinetic energy. If we project 5-MeV α particles onto a heavy nucleus we find that the alpha particle is repelled by the Coulomb force (see Figure 12.10) and doesn't get close enough to feel the attraction of the short-range nuclear force. It is virtually impossible classically for the alpha particle to reach the nucleus. How then can the alpha particle ever surmount the barrier if it is trapped inside the potential barrier? As we discussed in Chapter 6, the alpha particles are able to tunnel through the barrier. This is a pure quantum-mechanical effect, and there is a small, but finite, chance for the alpha particle to appear on the other side of the barrier. The probability depends critically on the barrier height and width. A higher energy alpha particle, E_2 in Figure 12.11, has a much higher probability than does a lower energy alpha particle, E_1, to tunnel through the barrier. The higher tunneling probability corresponds to a shorter lifetime for the radioactive nuclide. In Figure 12.12 we compare the lifetimes of various alpha emitters with the kinetic energies of the alpha particles. We see that there is a strong correlation between lower energies and greater difficulty of escaping (longer lifetimes).

In Example 6.17 we discussed the α decay of ^{238}U. It might be worthwhile to look at that example again. Because of the low probability of tunneling, we

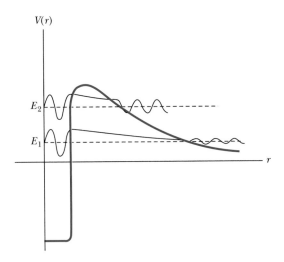

Figure 12.11 Quantum theory allows an alpha particle to tunnel through the barrier. A higher energy alpha particle E_2 has a much higher probability (shorter lifetime) than a lower energy alpha particle E_1. The waves shown are schematics only.

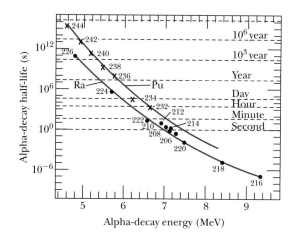

Figure 12.12 The half-lives for several radioactive alpha emitters of radium and plutonium isotopes are plotted versus their alpha energy. The two curves show that the higher energy alpha particles result from nuclei having a much shorter lifetime; these alpha particles have a higher probability of tunneling through the Coulomb barrier.

showed that the α particle must make about 10^{41} traverses back and forth across the nucleus before it can escape. Because only two products occur in alpha decay, we can calculate the kinetic energy of the α particle from the disintegration energy Q. Assume the parent nucleus is initially at rest so that the total momentum is zero. As shown in Figure 12.13, the final momenta of the daughter p_D and alpha particle p_α have the same magnitude and opposite directions. By using the conservation of energy and conservation of linear momentum, we can determine a unique energy for the alpha particle.

$$Q = K_\alpha + K_D$$

$$p_\alpha = p_D$$

$$K_\alpha = Q - K_D = Q - \frac{p_D^2}{2M_D} = Q - \frac{p_\alpha^2}{2M_D}$$

$$K_\alpha = Q - \frac{2M_a K_\alpha}{2M_D} = Q - \frac{M_\alpha}{M_D} K_\alpha$$

$$K_\alpha \left(1 + \frac{M_\alpha}{M_D} \right) = Q$$

$$K_\alpha = \frac{M_D}{M_D + M_\alpha} Q \approx \left(\frac{A-4}{A} \right) Q \qquad (12.32)$$

Because the parent mass A is normally over 150, the alpha particle takes most of the kinetic energy.

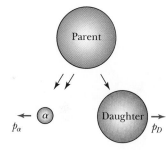

Figure 12.13 When a radioactive parent at rest alpha decays to an alpha particle and a heavy daughter, conservation of mass-energy and momentum still must occur. The momentum of the alpha particle and daughter are equal and opposite.

Beta Decay

Radioactive decay occurs because some nuclides are not stable. In Figure 12.5 we showed a plot of stable nuclei. Note that in alpha decay the parent nucleus reverts to a daughter nucleus that is down 2 units in neutron number N (Figure 12.5) and to the left 2 units in atomic number Z. In many cases alpha decay leaves the daughter nucleus farther from the line of stability than the parent. Unstable nuclei may move closer to the line of stability by undergoing beta decay. The simplest example of beta decay is the decay of a free neutron.

$$n \rightarrow p + \beta^- \qquad (12.33)$$

Figure 12.14 The relative intensity of electrons as a function of kinetic energy is shown for the beta decay of ^{14}C. If there were only two products in beta decay, the electron energy would be monoenergetic and the energy equal to K_{max}.

As we discussed in Section 12.1, electrons cannot exist within the nucleus, so when beta decay occurs for a nuclide, the beta particle, denoted by β^- (we now know it is an electron), is created at the time of the decay. We showed in Figure 12.8 that ^{14}C is unstable—it has an excess of neutrons. The beta decay of ^{14}C to form ^{14}N, a stable nucleus, can be written as

$$^{14}_{6}C \rightarrow {}^{14}_{7}N + \beta^- \qquad (12.34)$$

This decay produces two products, like α decay, and we expect to measure a monoenergetic electron spectrum (see Problem 38). However, the electron energy spectrum from the beta decay of ^{14}C (see Figure 12.14) shows a continuous energy spectrum up to a maximum energy. This experimental result was a major puzzle for many years. In addition to the strange energy spectrum, there was a problem with spin conservation. In neutron decay, the spin 1/2 neutron cannot decay to two spin 1/2 particles, a proton and an electron. Also ^{14}C has spin 0, ^{14}N has spin 1, and the electron has spin 1/2. We cannot combine spin 1/2 and 1 to obtain a spin of 0. Both the electron energy spectrum and the spin angular momentum conservation posed major difficulties with our understanding of beta decay.

The correct explanation was proposed in 1930 by Wolfgang Pauli, who suggested that a third particle—later called a **neutrino**—must also be produced in beta decay. The neutrino, with the symbol ν, has spin quantum number 1/2, charge 0, and carries away the additional energy missing in Figure 12.14. In Figure 12.14 an occasional electron is detected with the kinetic energy K_{max} required to conserve energy, but in the great majority of cases the electron's kinetic energy is less than K_{max}. This means the neutrino has little or no mass, and its energy may be all kinetic, like the photon. The photon cannot be the missing particle because it has spin 1. Pauli's suggestion seemed to explain the difficulties, and all circumstantial evidence supported the neutrino hypothesis. However, the detection of the elusive neutrino was difficult, and its existence was not proven experimentally until 1956 by C. Cowan and F. Reines (see Special Topic, "Neutrino Detection"). Reines received the Nobel Prize in 1995 for this discovery.

Neutrinos interact so weakly with matter that they pass right through the Earth with little chance of being absorbed. They have no charge and do not interact *electromagnetically*. They are not affected by the *strong* force of the nucleus. Recent experimental evidence indicates the mass of the neutrino is very small, as we shall see in Chapter 14. We now believe that beta decay is the reflection of a special kind of force, called simply the *weak* interaction. Neutrinos are created (or absorbed) in weak processes, including β decay. The electromagnetic and weak forces are two manifestations of the *electroweak* force and will be discussed in Chapter 14.

Frederick Reines (1918–1998, left) and Clyde Cowan (1919–1974), while staff members at Los Alamos National Laboratory, discovered the neutrino in 1956, almost three decades after it had been postulated by Pauli. Reines received the Nobel Prize in Physics for this discovery in 1995 in his and Cowan's name. The Nobel Prize is not awarded posthumously. Reines became the founding dean of the School of Physical Sciences of the new University of California, Irvine, in 1966, where he remained. See the Special Topic box on Neutrino Detection. *Courtesy of Frederick Reines.*

β^- Decay We now know there are *anti*neutrinos (symbol $\bar{\nu}$) as well as neutrinos. The beta decay of a free neutron and of ^{14}C is now correctly written as

$$n \rightarrow p + \beta^- + \bar{\nu} \qquad \beta^- \text{ decay} \qquad (12.35)$$

$$^{14}\text{C} \rightarrow {}^{14}\text{N} + \beta^- + \bar{\nu} \qquad \beta^- \text{decay} \qquad (12.36)$$

where it is the antineutrino $\bar{\nu}$ that is actually produced in β^- decay.

In the general beta decay of the parent nuclide $^A_Z X$ to the daughter $^A_{Z+1} D$, the reaction is

$$^A_Z X \rightarrow {}^A_{Z+1} D + \beta^- + \bar{\nu} \qquad \beta^- \text{ decay} \qquad (12.37) \quad \textbf{β^- decay}$$

The disintegration energy Q is given by

$$Q = [M(^A_Z X) - M(^A_{Z+1} D)]c^2 \qquad \beta^- \text{ decay} \qquad (12.38)$$

In order for β^- decay to occur, we must have $Q > 0$. When using Equation (12.38) we must be careful to use atomic masses because the Z of the decaying nuclide changes and the number of electron masses has been accounted for in Equation (12.38). In β^- decay, the nucleus A is constant, but Z changes to $Z + 1$, so that in Figure 12.5, it is the unstable nuclei to the left of the line of stability that are moving closer (one number down and one number right) to the line of stability.

β^+ Decay One might ask what happens for unstable nuclides with too many protons, that is, for nuclei to the right of the line of stability in Figure 12.5. Nature does allow such a transition, and in such cases a positive electron, called a *positron* (e^+), is produced. The positron is the antiparticle of the electron. In beta decay the electron and positron are normally referred to as β^- and β^+, respectively. Using β^- and β^+ helps remind us that the electron and positron are created in the nucleus during beta decay. Current experimental evidence indicates that a free proton does not decay ($t_{1/2} > 10^{32}$ y), but a proton bound within the nucleus may transmutate if energy is taken from the nucleus, and the result is a more stable nucleus. Because ^{14}O is a nucleus with an excess of protons, it is a good candidate for β^+ decay. The nucleus ^{14}O is unstable and decays by emitting a positron to become the stable ^{14}N. The reaction is

$$^{14}\text{O} \rightarrow {}^{14}\text{N} + \beta^+ + \nu \qquad \beta^+ \text{ decay} \qquad (12.39)$$

Nuclei near ^{14}N with an excess neutron (^{14}C) or proton (^{14}O) will decay to ^{14}N by the appropriate beta decay.

The general β^+ decay is written as

$$^A_Z X \rightarrow {}^A_{Z-1} D + \beta^+ + \nu \qquad \beta^+ \text{ decay} \qquad (12.40) \quad \textbf{β^+ decay}$$

A careful analysis of the atomic masses shows that the disintegration energy Q for positron decay is

$$Q = [M(^A_Z X) - M(^A_{Z-1} D) - 2m_e]c^2 \qquad \beta^+ \text{ decay} \qquad (12.41)$$

In this case the electron masses do not cancel when atomic masses are used. The mass of the positron is equal to that of an electron.

Note that in Figure 12.5 the unstable nuclei to the right of the line of stability move one number up and one number to the left when they β^+ decay, again moving closer to the line of stability.

Special Topic

NEUTRINO DETECTION

*I*magine particles that have little probability of being stopped while passing completely through the Earth. They have no charge and very little mass—yet they are everywhere throughout the universe, perhaps as many as 10^9 in every m^3 of "empty" space. A million of them pass through your eyeballs every second, but you have no chance of seeing them. Such is the neutrino, and no wonder it took over a quarter of a century to experimentally prove its existence. Pauli first proposed something like the neutrino in 1930 to solve the problem of nonconservation of energy in beta decay.

Frederick Reines and Clyde Cowan decided in 1951, while working at Los Alamos, to detect the existence of the neutrino by observing the inverse beta process

$$\bar{\nu} + p \rightarrow n + \beta^+ \qquad (B12.1)$$

where $\bar{\nu}$ denotes an antineutrino. At first they planned to detect antineutrinos resulting from an atomic bomb blast, but they eventually decided to use neutrinos coming from a nuclear reactor. They took their initial apparatus to a Hanford, Washington, reactor in 1953, but were not convinced they had detected neutrinos. Their experiment was called *Project Poltergeist* because,

Figure B An antineutrino from the reactor scatters from a proton to create a neutron and β^+ that eventually produces two gamma rays detected in coincidence in the two scintillators S2 and S3.

even though most physicists believed the neutrino existed, some thought it might prove not to be observable. In 1956 they took a larger detector to the Savannah, Georgia, reactor, where the neutrino rate from the beta decay of fission fragments was $10^{13} \, \bar{\nu} \cdot \text{cm}^{-2} \cdot \text{s}^{-1}$. A well-shielded area 12 m below ground helped reduce cosmic ray background. A schematic of their Savannah detector is shown in Figure A. It consists of three liquid scintillator detectors (S1, S2, and S3), each viewed by 110 photomultiplier tubes. Between each scintillator was placed a target (T1 and T2) containing CdCl$_2$ dissolved in water.

The detection scheme is shown in Figure B. An antineutrino from the reactor interacts with a proton in target T2, for example, producing a neutron and a positron. The positron quickly slows down, annihilates with an electron, and produces two 0.5-MeV gamma rays, $e^+ + e^- \rightarrow 2\gamma$. The γ rays pass through the water target and are detected in coincidence in the scintillators S2 and S3. The neutron undergoes several collisions, slows down, and is eventually captured by ^{114}Cd, which has a very large neutron capture cross section. Several gamma rays may be produced by the neutron capture, and they are detected in one or both of the scintillators. The signature of the antineutrino detection is a delayed coincidence between the gamma ray resulting from β^+ and those from the neutron. The 1956 experiment at the Savannah River reactor convincingly showed the neutrino had been detected.

Meanwhile Raymond Davis, Jr., of Brookhaven National Laboratory had designed an experiment

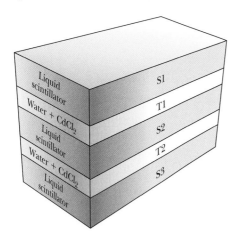

Figure A Schematic diagram of Cowan and Reines's neutrino detector used at the Savannah River reactor. Neutrinos scatter from protons in the water. The height of the detector is 2 m.

using 4000 liters of CCl_4 to detect solar neutrinos. The reaction in this case is

$$\nu + {}^{37}Cl \rightarrow \beta^- + {}^{37}Ar \qquad (B12.2)$$

Helium gas was bubbled through the solution to remove the ${}^{37}Ar$ that was later detected as it decayed back to ${}^{37}Cl$. No reactions were observed when this apparatus was placed next to a reactor. That is because a reactor produces *anti*neutrinos, whereas reaction (B12.2) requires neutrinos. This is further evidence that Cowan and Reines were detecting antineutrinos. Davis detected 2000 neutrinos over a period of 30 years and proved that nuclear fusion provided the energy from the sun. Davis shared in the Nobel Prize in Physics in 2002 for his pioneering contributions in the detection of solar neutrinos.

In the 1960s Davis constructed a large 100,000-gallon detector containing perchloroethylene cleaning fluid (C_2Cl_4), which was placed 1500 m below ground in the Homestake Gold Mine in South Dakota. Such a depth shields against most of the cosmic ray background. This detector was used for 30 years to detect the rate of neutrinos coming from our sun (solar neutrinos). The rate found is only about one half of the rate expected, and the concern over this result has led to several more neutrino detectors being constructed throughout the world (see Chapter 14) as the field of neutrino physics has flourished.

In recent years the number of neutrinos released from the sun has been calculated using theoretical models of the sun. At least five neutrino experiments have found the neutrino flux from the sun to be lower than that expected by about a factor of two. This is known as the "solar neutrino problem" or SNP, and it was not understood until 2002 (see Chapter 14). Experiments include those under the Caucasus Mountains near Baksan in Russia, in the Gran Sasso tunnel in Italy, in the Soudan mine in Minnesota, in a mine in Kamioka, Japan (where it was shown indirectly in 1998 that neutrinos have mass), and in a mine near Sudbury, Ontario (see Figure C). Detectors have included both expensive gallium and ultrapure water with photomultiplier tubes. When absorbed by gallium, neutrinos induce a reaction that produces radioactive germanium. The water-detector operation involves an antineutrino being absorbed by a proton, producing a

Figure C The Sudbury Neutrino Observatory detector is 18 m in diameter and has 10,000 photomultiplier tubes around it. Note the size with respect to the two men at the bottom. The photomultiplier tubes are looking at heavy water (with deuterium replacing the hydrogen atoms in water). Measurements have been under way since 1999. *Courtesy of the Sudbury Neutrino Observatory.*

neutron and β^+; the fast-moving positron produces light, which is detected by photomultiplier tubes.

A unique experiment in the Soudan mine to detect the arrival and type of neutrinos that have traveled 735 km through the Earth from the Fermilab accelerator near Chicago is expected to begin operation in 2005. Similar operations are under consideration around the world. The detection of neutrinos has become a significant pursuit because of the importance of neutrinos in the solar neutrino problem, proton decay, symmetry violations, and mass of the universe. Neutrinos will be discussed further in Chapters 14 and 16, because the fields of neutrino physics and neutrino astronomy are flourishing.

Electron Capture There is one other important possibility for beta decay. Because inner K-shell and L-shell electrons are tightly bound and their (classical) orbits are highly elliptical, these electrons spend a reasonable amount of time passing through the nucleus, thereby increasing the possibility of atomic electron capture (EC). A proton in the nucleus absorbs the e^-, producing a neutron and a neutrino. The reaction for a proton is

$$p + e^- \rightarrow n + \nu \tag{12.42}$$

The general reaction is written as

Electron capture
$$^A_Z X + e^- \rightarrow ^{\ \ A}_{Z-1}D + \nu \qquad \text{Electron capture} \tag{12.43}$$

When one of the inner atomic electrons is captured, another electron will take its place, producing a series of characteristic atomic x-ray spectra. This is a signature of electron capture, because these x rays are produced in the absence of any other kind of radiation. X rays can also be produced in other kinds of nuclear decay, because the decay products (for example, an α particle) may knock out electrons.

Electron capture has the same effect as positron decay; a proton is converted to a neutron. Electron capture occurs more frequently for higher-Z nuclides because the inner atomic electron shells are more tightly bound and there is a greater probability of an electron being absorbed. The disintegration energy Q for electron capture is

$$Q = [M(^A_Z X) - M(^{\ \ A}_{Z-1}D)]c^2 \qquad \text{Electron capture} \tag{12.44}$$

Because $Q > 0$ for β^+ or EC to occur, there will be some cases where EC is possible, but not β^+ decay, because of the difference between Equations (12.41) and (12.44).

Example 12.13

Show that the relations expressed for the disintegration energy Q in Equations (12.38), (12.41), and (12.44) are correct.

Strategy We begin with the actual particle equilibrium definitions for each of the beta decays (β^-, β^+, and EC) and change it to an energy equation. We neglect any neutrino mass and atomic binding energies and eventually use atomic masses.

Solution β^- decay: We begin with β^- decay and write the mass-energy equation for the reaction in Equation (12.37).

$$M_{\text{nucl}}(^A_Z X) = M_{\text{nucl}}(^{\ \ A}_{Z+1}D) + m_e + Q/c^2$$

where we use M_{nucl} to indicate the nuclear mass. In order to change to atomic masses we add Zm_e to each side above.

$$M_{\text{nucl}}(^A_Z X) + Zm_e = M_{\text{nucl}}(^{\ \ A}_{Z+1}D) + (Z+1)m_e + Q/c^2$$

Because we are neglecting the difference in atomic binding energies on the two sides of the equation, we now write this equation in terms of atomic masses.

$$M(^A_Z X) = M(^{\ \ A}_{Z+1}D) + Q/c^2$$

We solve this equation for Q to determine Equation (12.38).

$$Q = [M(^A_Z X) - M(^{\ \ A}_{Z+1}D)]c^2 \qquad \beta^- \text{ decay} \tag{12.38}$$

β^+ decay: We write the mass-energy equation for the reaction in Equation (12.40) and follow a procedure similar to that above.

$$M_{\text{nucl}}(^A_Z X) = M_{\text{nucl}}(^{\ \ A}_{Z-1}D) + m_e + Q/c^2$$

We again add Zm_e to each side and have

$$M_{\text{nucl}}(^A_Z X) + Zm_e = M_{\text{nucl}}(^{\ \ A}_{Z-1}D) + (Z+1)m_e + Q/c^2$$

In this case we only need $(Z-1)m_e$ for the daughter atomic mass, which gives us a remaining mass of $2m_e$.

$$M(^A_Z X) + m_e = M_{\text{nucl}}(^{\ \ A}_{Z-1}D) + 2m_e + Q/c^2$$

We solve this equation for Q to determine Equation (12.41).

$$Q = [M(^A_Z X) - M(^{\ \ A}_{Z-1}D) - 2m_e]c^2 \qquad \beta^+ \text{ decay} \tag{12.41}$$

Electron capture: The mass-energy equation for the electron capture reaction of Equation (12.43) is

$$M_{\text{nucl}}(^A_ZX) + m_e = M_{\text{nucl}}(_{Z-1}^A D) + Q/c^2$$

We add $(Z-1)m_e$ to each side above and obtain

$$M_{\text{nucl}}(^A_ZX) + Zm_e = M_{\text{nucl}}(_{Z-1}^A D) + (Z-1)m_e + Q/c^2$$

In this case we have just the right number of electron masses to change to atomic masses.

$$M(^A_ZX) = M(_{Z-1}^A D) + Q/c^2$$

We solve this equation for Q to find Equation (12.44).

$$Q = [M(^A_ZX) - M(_{Z-1}^A D)]c^2 \quad \text{Electron capture} \quad (12.44)$$

Example 12.14

Show that ^{55}Fe may undergo electron capture, but not β^+ decay.

Strategy The two possible reactions are

$$^{55}_{26}\text{Fe} \rightarrow ^{55}_{25}\text{Mn} + \beta^+ + \nu \quad \beta^+ \text{ decay}$$

$$^{55}_{26}\text{Fe} + e^- \rightarrow ^{55}_{25}\text{Mn} + \nu \quad \text{Electron capture}$$

We determine the disintegration energies Q for each reaction, Equations (12.41) and (12.44). The reactions are possible if Q is positive.

Solution We first find the masses from Appendix 8 needed to determine the disintegration energy Q of Equations (12.41) and (12.44):

$$M(^{55}_{26}\text{Fe}) = 54.938298 \text{ u}, M(^{55}_{25}\text{Mn}) = 54.938050 \text{ u, and}$$

$$m_e = 0.000549 \text{ u}.$$

β^+ **decay:**

$$Q = [54.938298 \text{ u} - 54.938050 \text{ u} - 2(0.000549 \text{ u})]c^2$$

$$\times \left(\frac{931.5 \text{ MeV}}{c^2 \cdot \text{u}}\right)$$

$$= -0.79 \text{ MeV} \quad \beta^+ \text{ decay not allowed}$$

Electron capture:

$$Q = [54.938298 \text{ u} - 54.938050 \text{ u}]c^2\left(\frac{931.5 \text{ MeV}}{c^2 \cdot \text{u}}\right)$$

$$= 0.23 \text{ MeV} \quad \text{Electron capture is allowed}$$

We find experimentally that only electron capture is allowed, with $t_{1/2} = 2.7$ years.

Example 12.15

Find whether alpha decay or any of the beta decays are allowed for $^{226}_{89}$Ac.

Strategy For each of the possible four reactions, we first write the reaction, list the disintegration energy Q equation, look up the appropriate masses, and calculate the disintegration energy Q. If $Q > 0$, the decay is allowed.

Solution We outline the solution, but leave looking up the masses and doing the math to you.

Alpha decay: $^{226}_{89}\text{Ac} \rightarrow ^{222}_{87}\text{Fr} + \alpha$

$$Q = [M(^{226}_{89}\text{Ac}) - M(^{222}_{87}\text{Fr}) - M(^4\text{He})]c^2$$

$$= 5.50 \text{ MeV} \quad \text{Alpha decay is allowed}$$

β^- **decay:** $^{226}_{89}\text{Ac} \rightarrow ^{226}_{90}\text{Th} + \beta^- + \bar{\nu}$

$$Q = [M(^{226}_{89}\text{Ac}) - M(^{226}_{90}\text{Th})]c^2$$

$$= 1.12 \text{ MeV} \quad \beta^- \text{ decay is allowed}$$

β^+ **decay:** $^{226}_{89}\text{Ac} \rightarrow ^{226}_{88}\text{Ra} + \beta^+ + \nu$

$$Q = [M(^{226}_{89}\text{Ac}) - M(^{226}_{88}\text{Ra}) - 2m_e]c^2$$

$$= -0.38 \text{ MeV} \quad \beta^+ \text{ decay is not allowed}$$

Electron capture: $^{226}_{89}\text{Ac} + e^- \rightarrow ^{226}_{88}\text{Ra} + \nu$

$$Q = [M(^{226}_{89}\text{Ac}) - M(^{226}_{88}\text{Ra})]c^2$$

$$= 0.64 \text{ MeV} \quad \text{Electron capture is allowed}$$

We find that α decay, β^- decay, and electron capture are all possible from the same nucleus. Experiment shows that alpha decay occurs only 0.006% of the time for ^{226}Ac, β^- decay 83%, and electron capture 17%.

Figure 12.15 The relevant energy levels for the alpha decay of $^{230}_{92}$U. The alpha-particle energies and their percentage of occurrence are shown for several decays for $^{230}_{92}$U → $^{226}_{90}$Th. The decay percentages may not add up to 100% because less likely decays to higher-lying excited states are not shown. *From R. B. Firestone and V. S. Shirley, eds., Table of Isotopes, 8th ed., New York: Wiley (1996).*

Gamma Decay

The nuclide ^{230}U can alpha decay to the ground state or any of the low-lying excited states of ^{226}Th (see Figure 12.15). Similarly, experimental evidence indicates that ^{226}Ac beta decays primarily to the ground and two excited states of ^{226}Th. The energies calculated for the disintegration energy Q for alpha and beta decay give the appropriate transitions to the ground state. If the decay proceeds to an excited state of energy E_x (for example, from the ground state of ^{230}U to the $E_x = 0.072$ MeV excited state of ^{226}Th), rather than to the ground state, then the disintegration energy Q for the transition to the excited state can be determined with respect to the transition to the ground state. If we label the disintegration energy Q to the ground state as Q_0, the Q for a transition to the excited state E_x is given by

$$Q = Q_0 - E_x \tag{12.45}$$

For example, the disintegration value Q for the α decay transition from ^{230}U to the excited state at $E_x = 0.072$ MeV of ^{226}Th (see Figure 12.15) is given in terms of $Q_0 = 5.992$ MeV and $E_x = 0.072$ MeV by $Q = Q_0 - E_x = 5.992$ MeV − 0.072 MeV = 5.920 MeV.

A nucleus has excited states in much the same way atoms do. The excitation energies, however, tend to be much larger, many keV or even MeV, as a result of the stronger nuclear interaction. One of the possibilities for the nucleus to rid itself of this extra energy is to emit a photon (gamma ray) and undergo a transition to some lower energy state. The gamma-ray energy hf is given by the difference of the higher energy state $E_>$ and the lower one $E_<$.

$$hf = E_> - E_< \tag{12.46}$$

In order to conserve momentum, the nucleus normally must absorb some of this energy difference. However, for a nucleus initially at rest, Equation (12.46) is a very good approximation.

Example 12.16

Consider the γ decay from the 0.072-MeV excited state to the ground state of ^{226}Th at rest shown in Figure 12.15. Find an exact expression for the gamma-ray energy by including both the conservation of momentum and energy. Determine the error obtained by using the approximate value in Equation (12.46).

Strategy We need to account for the conservation of momentum as well as that of energy to find the exact gamma-ray energy. We denote the final momentum of ^{226}Th by p. Because the decaying nucleus is initially at rest, the total linear momentum is zero, and the linear momentum p of the daughter nucleus must have the same magnitude but opposite direction to the momentum of the gamma ray, h/λ.

$$p = \frac{h}{\lambda} = \frac{hf}{c}$$

The conservation of energy gives

$$hf + \frac{p^2}{2M} = E_> - E_<$$

where M is the mass of ^{226}Th. We solve these two equations to find the gamma-ray energy hf.

Solution We substitute the relation for the momentum p into the energy equation and find

$$hf + \frac{(hf)^2}{2Mc^2} = E_> - E_< \tag{12.47}$$

We could solve Equation (12.47) for hf by using the quadratic equation, but that would be tedious. Let us determine whether we can use an approximation. Rewrite Equation (12.47) as

$$hf\left(1 + \frac{hf}{2Mc^2}\right) = E_> - E_<$$

$$hf = \frac{E_> - E_<}{1 + \dfrac{hf}{2Mc^2}}$$

If the value $hf/2Mc^2 = x$ is very small, then we can use the binomial expansion $(1 + x)^{-1} = 1 - x + x^2 - \cdots$. We determine x to be

$$x = \frac{hf}{2Mc^2} = \frac{0.072 \text{ MeV}}{2(226 \text{ u} \cdot c^2)\left(\dfrac{931.5 \text{ MeV}}{c^2 \cdot \text{u}}\right)} = 1.7 \times 10^{-7}$$

So the error in using the approximate Equation (12.46), which amounts to letting $(1 + x)^{-1} \approx 1$, amounts to an error of only about 10^{-7}. We can safely use Equation (12.46) where $hf \approx E_> - E_< = 0.072 \text{ MeV}$.

The decay of an excited state of $^A X^*$ (where * indicates an excited state) to its ground state is denoted by

$$^A X^* \rightarrow {}^A X + \gamma \qquad (12.48) \qquad \textbf{Gamma decay}$$

A transition between two nuclear excited states $E_>$ and $E_<$ is denoted by

$$^A X^*(E_>) \rightarrow {}^A X^*(E_<) + \gamma \qquad (12.49)$$

The excited state energies or *levels* are characteristic of each nuclide, and a careful study of the gamma energies is usually sufficient to identify a particular nuclide. The gamma rays are normally emitted soon after the nucleus is created in an excited state. Coincidence measurements between the β- and/or α-decay products and the subsequent γ rays are a powerful technique in experimental nuclear physics to study properties of nuclear excited states.

We can contrast nuclear spectroscopy with that of the optical spectra of atoms. The solution of the Schrödinger equation for the Coulomb force of the atom is possible, but no such success has yet been obtained for the nuclear force. Although nuclear models are used to predict certain characteristics of excited nuclear levels, no encompassing theory exists.

There are sophisticated, quantum-mechanical selection rules that determine the details of gamma-ray transitions between nuclear states. Sometimes these selection rules prohibit a certain transition, and the excited state may live for a long time—even years! These states are called **isomers** or **isomeric states** and are denoted by a small m for *metastable* next to the mass number A. An example is the spin 9 state of $^{210\text{m}}_{83}\text{Bi}$ at 0.271 MeV excitation energy, which has $t_{1/2} = 3 \times 10^6$ y (alpha decay) and apparently does not gamma decay. The lower energy states have spins 0 and 1, and such a large spin difference transition is prohibitive for gamma decay. The ground state of $^{210}_{83}\text{Bi}$ has a half-life $t_{1/2} = 5$ days (99+% β^- decay, $10^{-4}\%$ α decay). Another isomer is $^{93\text{m}}_{41}\text{Nb}$ ($E_x = 0.03$ MeV, $t_{1/2} = 13.6$ y), which eventually does gamma decay to the ground state. In this case we say the decay is prohibited, but that just means the probability of its occurring is very small. Most isomeric states, of course, have much shorter lifetimes. An extremely useful isomer for clinical work in medicine is $^{99\text{m}}\text{Tc}$ (gamma emitter, $t_{1/2} = 6$ h), which will be discussed in Chapter 13.

Isomers

12.8 Radioactive Nuclides

Most scientists now believe that the universe was created in a tremendous explosion (the *Big Bang*) about 13 billion years ago (13×10^9 y). Neutrons and protons fused together to form deuterons and light nuclei within the first few minutes.

Table 12.2 Some Naturally Occurring Radioactive Nuclides		
Nuclide	$t_{1/2}$ **(y)**	**Natural Abundance**
$^{40}_{19}$K	1.28×10^9	0.01%
$^{87}_{37}$Rb	4.8×10^{10}	27.8%
$^{113}_{48}$Cd	9×10^{15}	12.2%
$^{115}_{49}$In	4.4×10^{14}	95.7%
$^{128}_{52}$Te	7.7×10^{24}	31.7%
$^{130}_{52}$Te	2.7×10^{21}	33.8%
$^{138}_{57}$La	1.1×10^{11}	0.09%
$^{144}_{60}$Nd	2.3×10^{15}	23.8%
$^{147}_{62}$Sm	1.1×10^{11}	15.0%
$^{148}_{62}$Sm	7×10^{15}	11.3%

The heavy elements were formed much later by nuclear reactions within stars. We will discuss these processes in Chapters 13 and 16.

We say that the unstable nuclei found in nature exhibit *natural radioactivity*. Those radioactive nuclides made in the laboratory (for example, with accelerators or reactors) exhibit *artificial radioactivity* and include all known nuclides heavier than ^{238}U. There are many natural radioactive nuclides left on Earth with lifetimes long enough to be observed. Only those with half-lives longer than a few billion years could have existed since primordial times; most of them are heavy elements, but several with $A < 150$ are listed in Table 12.2. Some nuclides with a long half-life have been produced as a result of the decay of another radioactive nucleus.

In addition to nuclear fission, heavy radioactive nuclides can change their mass number only by alpha decay ($^A X \rightarrow {}^{A-4}D$) but can change their charge number Z by either alpha or beta decay. As a result, there are only four paths that the heavy naturally occurring radioactive nuclides may take as they decay to stable end products. The four paths have mass numbers expressed by either $4n$, $4n + 1$, $4n + 2$, or $4n + 3$ (n = integer), because only alpha decay can change the mass number. These four series are listed in Table 12.3. All of these radioactive series

Table 12.3 The Four Radioactive Series				
Mass Numbers	**Series Name**	**Parent**	$t_{1/2}$ **(y)**	**End Product**
$4n$	Thorium	$^{232}_{90}$Th	1.40×10^{10}	$^{208}_{82}$Pb
$4n + 1$	Neptunium	$^{237}_{93}$Np	2.14×10^6	$^{209}_{83}$Bi
$4n + 2$	Uranium	$^{238}_{92}$U	4.47×10^9	$^{206}_{82}$Pb
$4n + 3$	Actinium	$^{235}_{92}$U	7.04×10^8	$^{207}_{82}$Pb

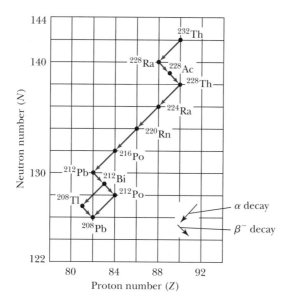

Figure 12.16 The predominant path for the decay chain of $^{232}_{90}$Th. The mass number can only change by alpha decay, but both alpha and beta decay can change the atomic number Z. A branch is shown for ^{212}Bi. Not all branching is shown.

occur in nature except that of neptunium. The member of the neptunium series with the longest half-life, ^{237}Np ($t_{1/2} = 2.14 \times 10^6$ y), has a lifetime so much less than the age of our solar system that virtually all the members have already decayed.

The sequence of one of the radioactive series, ^{232}Th, is shown in Figure 12.16. Note that ^{212}Bi can decay by either alpha or beta decay; this is called *branching*. The subsequent decay is usually a beta or alpha decay, respectively, to eventually reach the same end product. Normally one path of the branch is heavily favored, but both paths are shown for $^{212}_{83}$Bi in Figure 12.16 because it has a 36% probability of alpha decay and a 64% probability of beta decay. As shown in Figure 12.16 the effect of the successive alpha and beta decays is to bring the nuclide closer to the line of stability until a stable nuclide is finally reached.

Time Dating Using Lead Isotopes

Because the isotope ^{204}Pb is not radioactive and no other nuclide decays to it, its abundance is presumably constant. The stable isotopes ^{206}Pb and ^{207}Pb, on the other hand, are at the end of the radioactive chains of ^{238}U and ^{235}U, respectively. However, because ^{235}U has a relatively short half-life (0.70×10^9 y) compared to the age of the Earth, most of the ^{235}U has already decayed to ^{207}Pb, and the ratio of ^{207}Pb/^{204}Pb has been relatively constant over the past 2 billion years. Because the half-life of ^{238}U is so long, the ratio of ^{206}Pb/^{204}Pb is still increasing. A plot of the abundance ratio of ^{206}Pb/^{204}Pb versus ^{207}Pb/^{204}Pb can be a sensitive indicator of the age of lead ores, as shown in Figure 12.17. Such techniques have been used to show that meteorites (Meteor Crater in Arizona and elsewhere), believed to be left over from the formation of the solar system, are 4.55 billion years old. This is in agreement with dating measurements made on moon rocks returned to Earth. Although no 4.5-billion-year-old terrestrial rocks have yet been found on Earth, indirect evidence based on radioactive dating techniques leads us to believe the Earth was formed about 4.55 billion years ago. The oldest terrestrial rocks, dated 3.96 billion years old, were found in 1986 near the Great

Figure 12.17 The growth curve for lead ores from various deposits is shown. The age of the specimens can be obtained from the abundance ratio of $^{206}Pb/^{204}Pb$ versus $^{207}Pb/^{204}Pb$. When extrapolated backward in time, the growth curve gives an age of 4.6 billion years for the specimens from the Meteor Crater in Arizona. *From S. Moorbath,* Scientific American **236**, 92 *(1977).*

Slave Lake in the Northwest Territory of Canada. Rocks as old as 3.8 billion years were found in Greenland. (See Special Topic, "The Formation and Age of the Earth.")

Example 12.17

Assume that all the ^{206}Pb found in a given sample of uranium ore resulted from decay of ^{238}U and that the ratio of $^{206}Pb/^{238}U$ is 0.60. How old is the ore?

Strategy Let N_0 be the original number of ^{238}U nuclei that existed. The ^{238}U nuclei eventually decay to ^{206}Pb, and the longest time in the radioactive decay chain $^{238}U \rightarrow \, ^{206}Pb$ is the half-life of ^{238}U, $t_{1/2} = 4.47 \times 10^9$ y. The numbers of nuclei for ^{238}U and ^{206}Pb are then

$$N(^{238}U) = N_0 e^{-\lambda t}$$

$$N(^{206}Pb) = N_0 - N(^{238}U) = N_0(1 - e^{-\lambda t})$$

The abundance ratio is

$$R' = \frac{N(^{206}Pb)}{N(^{238}U)} = \frac{1 - e^{-\lambda t}}{e^{-\lambda t}} = e^{\lambda t} - 1 \qquad (12.50)$$

We can solve Equation (12.50) for t, because we know experimentally the ratio R' and the decay constant λ for ^{238}U.

Solution The result for t from Equation (12.50) is

$$t = \frac{1}{\lambda}\ln(R' + 1) = \frac{t_{1/2}}{\ln(2)}\ln(R' + 1)$$

$$= \frac{4.47 \times 10^9 \text{ y}}{\ln(2)}\ln(1.60) = 3.0 \times 10^9 \text{ y}$$

The sample of uranium ore is about 3 billion years old.

Radioactive Carbon Dating

Radioactive ^{14}C is produced in our atmosphere by the bombardment of ^{14}N by neutrons produced by cosmic rays.

$$n + {}^{14}N \rightarrow {}^{14}C + p \qquad (12.51)$$

A natural equilibrium of ^{14}C to ^{12}C exists for molecules of CO_2 in the atmosphere. All living organisms use or breathe CO_2 from the atmosphere (see Figure 12.18). However, when the living organisms die, their intake of ^{14}C ceases, and the ratio of $^{14}C/^{12}C$ $(=R)$ decreases as ^{14}C decays. In order to use ^{14}C for dating, corrections must be made for the changes in composition of the Earth's atmosphere and for variations in the flux of cosmic rays due to changes in the Earth's magnetic field. Evidence early in the 1990s indicated that the period just before 9000 years ago had a higher $^{14}C/^{12}C$ ratio by a factor of about 1.5 than it does today. At the time of death, the initial decay rate is 14 decays/min per gram of carbon $(R = 1.2 \times 10^{-12})$. This method has been calibrated with tree-ring counting to ages of about 10,000 years. Because the half-life of ^{14}C is 5730 years, it is convenient to use the $^{14}C/^{12}C$ ratio to determine the age of objects (see Figure 12.19) over a wide range up to perhaps 45,000 years ago. Indeed, quite sophisticated mass spectrometry techniques using accelerators to accelerate carbon ions have enabled scientists to test samples as small as 10^5 atoms (about 10^{-20} kg). Much older dates can now be obtained than by using the previous techniques of actually measuring the ^{14}C decay rate, which required samples of up to 0.01 kg. Willard Libby received the 1960 Nobel Prize in Chemistry for this ingenious technique.

Other particularly useful radioisotopes for dating purposes include ^{10}Be, ^{26}Al, ^{36}Cl, and ^{129}I. The half-life of ^{10}Be is 1.5 million years and may be useful in studying the evolution both of humans and of the ice ages. The isotope ^{36}Cl is particularly well suited for dating and tracing groundwater movement and for determining the suitability of radioactive waste depositories. The dating of ^{10}Be and ^{26}Al in marine sediments has confirmed their extraterrestrial origin, possibly from comets.

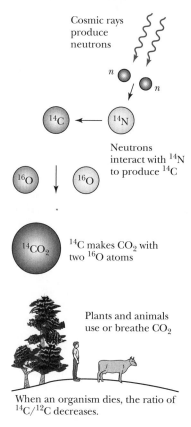

Cosmic rays produce neutrons

n

$^{14}C \leftarrow {}^{14}N$

Neutrons interact with ^{14}N to produce ^{14}C

^{16}O ^{16}O

$^{14}CO_2$ ^{14}C makes CO_2 with two ^{16}O atoms

Plants and animals use or breathe CO_2

When an organism dies, the ratio of $^{14}C/^{12}C$ decreases.

Figure 12.18 Cosmic rays produce neutrons, which react with ^{14}N in the atmosphere to produce ^{14}C. The ^{14}C nuclei enter into living organisms in the form of CO_2. After the living organism dies, the ratio of $^{14}C/^{12}C$ in its remains decreases according to the decay rate of ^{14}C.

Figure 12.19 Although most radiocarbon dating is performed on bone or charcoal, many other organic materials such as those shown here from the British Museum (mummy wrapping, tree rings, ancient Egyptian rope, and reeds) can be dated. © *The Trustees of the British Museum.*

Special Topic

THE FORMATION AND AGE OF THE EARTH

*I*t is believed that the Earth and the solar system were formed 4.55 billion years ago when gravitational forces caused dust and gases to accrete (glob together). This process continues today, albeit at a slower rate, due to meteorites and other particles striking Earth. For the first half billion years, the Earth was relatively cool (~1000°C) and probably solid. The main constituents were most likely nickel, iron, and silicates. As millions of years passed, heat energy was added from collisions and radioactive decay of uranium, thorium, and potassium, among other radioactive elements. As Earth's temperature gradually increased, heavy elements like iron and nickel melted before the silicates, and because the metals are heavier, they sank toward Earth's center. The lighter material floated to the top, and material of intermediate density formed the mantle (see Figure A). The Earth was a

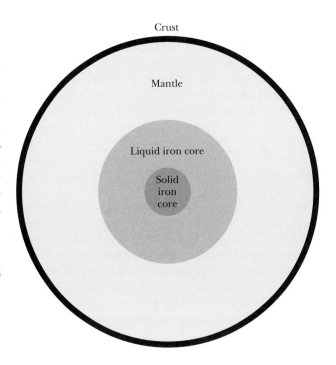

Figure A Schematic diagram of the Earth's composition. The crust is only about 30 km thick and is rich in oxygen, silicon, and aluminum. The mantle extends to a depth of perhaps 2900 km and consists of dense rock and metal oxides. It contains more than 80% of the Earth's volume. The core is composed primarily of iron and nickel and has a temperature as high as 7000°C, greater than the Sun's surface temperature. Radioactivity continues to add most of the heat inside the Earth.

Example 12.18

A bone suspected to have originated during the period of the Roman emperors was found in Great Britain. Accelerator techniques gave its $^{14}C/^{12}C$ ratio as 1.10×10^{-12}. Is the bone old enough to have Roman origins?

Strategy Remember that the initial ratio of $^{14}C/^{12}C$ at the time of death was $R_0 = 1.2 \times 10^{-12}$. We use the radioactive decay law to determine the time t that it will take for the ratio to decrease to 1.10×10^{-12}.

Solution The number of ^{14}C atoms decays as $e^{-\lambda t}$.

$$N(^{14}C) = N_0 e^{-\lambda t}$$

The ratio of ions is given by

$$R = \frac{N(^{14}C)}{N(^{14}C)} = \frac{N_0(^{14}C)e^{-\lambda t}}{N(^{12}C)} = R_0 e^{-\lambda t}$$

where R_0 is the original ratio. We can solve this equation for t.

turbulent place with gigantic heaving and bubbling on the surface. Volcanoes exploded and flowing lava was commonplace.

Eventually the Earth's surface cooled, and the crust of the Earth formed. Earth slowly took on its present appearance, but radioactivity continued to raise the temperature inside the Earth. It is the thermal energy generated by radioactivity that turns iron in the outer core into a convecting dynamo that produces the Earth's magnetic field. This heat also leaks out into the mantle and eventually moves the tectonic plates and produces volcanoes. Temperatures increase toward the center of the Earth and reach more than 6000°C. Despite such high temperatures, the inner core is solid because of the great pressure.

How old is the Earth? According to ancient Hindu philosophy, the Earth is about 2 billion years old. Archbishop James Ussher (1581–1656) of Ireland, who was also vice chancellor of Dublin's Trinity College, wrote a treatise in which he determined that Earth was created on Saturday evening, October 22, 4004 BC. Dr. John Lightfoot, a distinguished Greek and biblical scholar and vice chancellor of Cambridge University, determined in 1642 that Earth was created on September 17, 3928 BC. Charles Darwin, the great evolutionist, made an estimate that Earth is several hundred million years old based on geology. Serious efforts to determine Earth's age began in the mid-eighteenth century. William Thomson (Lord Kelvin), an expert on thermodynamics, calculated that the Earth would have been created 20–40 million years ago based on his determination of the sun's age and how long Earth took to cool from its molten state. Thomson's estimate was low because he was unaware of the highly efficient fusion reaction that powers the sun.

The field changed with the discovery of radioactivity by Henri Becquerel in 1896. As early as 1907 the Yale University radiochemist B. B. Boltwood proposed that the lead/uranium ratio in uranium materials might be useful as a geological dating tool. As stated in Section 12.8 no Earth rocks have been found older than 4 billion years, but radiometric tests of moon rocks and some 70 meteorites have shown that the age of Earth is 4.55 billion years with an uncertainty of 1%. This number assumes that the Earth, moon, and meteorites are part of the same evolving solar system. Clair Patterson (1922–1995), professor of geochemistry at Caltech, obtained the number of 4.55 billion years for the age of the Earth in 1953.

$$e^{-\lambda t} = \frac{R}{R_0}$$

$$t = \frac{-\ln(R/R_0)}{\lambda} = -t_{1/2}\frac{\ln(R/R_0)}{\ln(2)}$$

$$= -(5730 \text{ y})\left(\frac{-0.087}{0.693}\right) = 719 \text{ y}$$

where we have inserted the known values of $t_{1/2}$, R_0, and R to find the age of the bone. The bone does not date from the Roman Empire, but from the medieval period.

Summary

The discovery of the neutron in 1932 solved several outstanding problems, including the understanding of the nuclear constituents and the origin of very penetrating radiation.

A nuclide $^A_Z X$ has mass M and is composed of Z protons and N neutrons. Its mass number is $A = Z + N$. Masses are measured in terms of atomic mass units u. The radius of a nucleus is $R = r_0 A^{1/3}$, where $r_0 \approx 1.2 \times 10^{-15}$ m $= 1.2$ fm. Electron scattering is useful to measure the size and shapes of nuclei. The properties of the nucleons are as follows:

Property	Neutron	Proton
Mass (u)	1.008665	1.007276
Charge (e)	0	+1
Spin ($\hbar$)	1/2	1/2
Magnetic moment ($e\hbar/2m_p$)	−1.91	+2.79

The study of the deuteron and nucleon-nucleon scattering indicates that the nuclear force is attractive and much stronger than the Coulomb force. However, it is effective only over a short range (up to about 3 fm). The nuclear force is charge independent and has a hard core.

A nuclide is stable if its mass is smaller than any other possible combination of the A nucleons. Stable nuclides tend to have $N \approx Z$ for small A and $N > Z$ for medium and large A. The total binding energy for a nuclide is

$$B(^A_Z X) = [Nm_n + ZM(^1H) - M(^A_Z X)]c^2 \qquad (12.10)$$

The von Weizsäcker semi-empirical mass formula is useful in predicting the nuclear binding energy. There are no stable nuclei with $Z > 83$ or $A > 209$. Nuclei tend to be more stable with an even number of protons and/or neutrons. Nuclei near ^{56}Fe have the highest binding energies per nucleon, and the average binding energy per nucleon for most nuclei is about 8 MeV.

The radioactive decay law is $N = N_0 e^{-\lambda t}$, where λ is the decay constant and the half-life $t_{1/2} = 0.693/\lambda$. The activity $R = \lambda N$. A becquerel (Bq) is 1 decay/s. Radioactive decay occurs when the disintegration energy $Q > 0$. The four kinds of alpha and beta decay are

Alpha decay
$$^A_Z X \rightarrow {}^{A-4}_{Z-2}D + \alpha \qquad (12.30)$$
$$Q = [M(^A_Z X) - M(^{A-4}_{Z-2}D) - M(^4He)]c^2 \qquad (12.31)$$

β^- decay
$$^A_Z X \rightarrow {}^A_{Z+1}D + \beta^- + \bar{\nu} \qquad (12.37)$$
$$Q = [M(^A_Z X) - M(^A_{Z-1}D)]c^2 \qquad (12.38)$$

β^+ decay
$$^A_Z X \rightarrow {}^A_{Z-1}D + \beta^+ + \nu \qquad (12.40)$$
$$Q = [M(^A_Z X) - M(^A_{Z-1}D) - 2m_e]c^2 \qquad (12.41)$$

Electron capture
$$^A_Z X + e^- \rightarrow {}^A_{Z-1}D + \nu \qquad (12.43)$$
$$Q = [M(^A_Z X) - M(^A_{Z-1}D)]c^2 \qquad (12.44)$$

There are only four radioactive series. For example, two of them begin with uranium isotopes, ^{235}U and ^{238}U. Radioisotopes are useful to date objects like the age of the Earth and ancient objects. Radiocarbon ^{14}C is one of the most useful.

Questions

1. Explain why neutrons are not prohibited from being in the nucleus for the same three reasons, discussed in the text, that electrons are excluded.
2. Why does the atomic number Z determine the chemical properties of a nuclide? What differences in chemical properties would you expect for ^{16}O, ^{17}O, and ^{18}O, which are all stable? What about ^{15}O, which is unstable?
3. Explain how the nuclear charge radius could be different from the nuclear mass or nuclear force radii.
4. Do you believe it is easier to measure atomic masses or nuclear masses? Explain how you could experimentally measure both for ^{2}H. What about for ^{40}Ca?

5. Do you think it is significant that the nucleus has a hard core? What would happen if it did not?
6. Is the nuclear potential shown in Figure 12.4 consistent with a short-range nuclear force? Explain.
7. How likely is it that there are further, yet undetected, stable nuclides that are not shown in Figure 12.5? Explain.
8. Why does the binding energy curve of Figure 12.6 rise so fast for the light nuclei, but fall off so slowly for the heavy nuclei?
9. Hundreds of nuclides are known to decay by alpha emission. Why is decay by ^{3}He emission never (or rarely) observed?

10. For several decades it was believed that the neutrino was massless. If that were so, explain why it still must be distinct from a photon, because both must travel at the speed of light.

11. Why is electron capture more probable than β^+ decay for the very heavy radioactive elements?

12. Why do unstable nuclei below the line of stability in Figure 12.5 undergo β^+ decay, whereas unstable nuclei above the line of stability undergo β^- decay?

13. Why are radioactive nuclei still producing heat inside the Earth?

Problems

Note: The more challenging problems are indicated by a shaded box.

12.1 Discovery of the Neutron

1. Assume that gamma rays are Compton-scattered by hydrogen. What energy gamma rays are needed to produce 5.7-MeV protons?

2. Are the nuclear spins of the following nuclei integral or half-integral: ^{3}He, ^{6}Li, ^{7}Li, ^{18}F, ^{19}F?

12.2 Nuclear Properties

3. What are the number of protons, number of neutrons, mass number, atomic number, charge, and atomic mass for the following nuclei: $^3_2$He, $^4_2$He, $^{17}_8$O, $^{42}_{20}$Ca, $^{210}_{82}$Pb, and $^{235}_{92}$U?

4. What are the number of neutrons and protons for the following nuclides: ^{6}Li, ^{13}C, ^{40}K, and ^{64}Cu?

5. List from Appendix 8 all the isotopes, isobars, and isotones of ^{40}Ca.

6. Write down the nuclidic symbol and percentage abundances of all the nuclides having atomic number 7, 23, and 37. Use Appendix 8.

7. Write down the nuclidic symbol and half-lives for the unstable nuclei having $Z = 18$ that beta decay and for $Z = 102$ that spontaneous fission. Use Appendix 8.

8. What is the ratio of density of the nucleus to that of water?

9. What is the ratio of the magnetic moment of the proton to that of the electron?

10. Calculate the density and mass of the nuclide ^{56}Fe in SI units.

12.3 The Deuteron

11. What is the ratio of the electron binding energy in deuterium to the rest energy of the deuteron? Is it reasonable to ignore electronic binding energies when doing nuclear calculations?

12. Consider the photodisintegration of a deuteron at rest. Using both the conservation of energy and momentum, what is the minimum photon energy required? What percentage error does neglecting the conservation of momentum make?

12.4 Nuclear Forces

13. Compute the gravitational and Coulomb force between two protons in ^{3}He. Assume the distance between the protons is equal to the nuclear radius. If the average nuclear potential is an attractive 40 MeV effective over a distance of 3.0 fm, compare the nuclear force with both the gravitational and Coulomb forces.

14. Consider two protons in the ^{27}Al nucleus located 2.0 fm apart. How strong does the nuclear force need to be to overcome the Coulomb force?

12.5 Nuclear Stability

15. (a) Show that the equation giving the binding energy of the last neutron in a nucleus $^A_Z X$ is

$$B = [M(^{A-1}_Z X) + m_n - M(^A_Z X)]c^2$$

(b) Calculate the binding energy of the most loosely bound neutron of ^{6}Li, ^{17}O, and ^{207}Pb.

16. (a) Show that the equation giving the binding energy of the last proton in a nucleus $^A_Z X$ is

$$B = [M(^{A-1}_{Z-1} Y) + M(^1 H) - M(^A_Z X)]c^2$$

(b) Calculate the binding energy of the most loosely bound proton of ^{8}Be, ^{15}O, and ^{32}S.

17. What is the energy released when three α particles combine to form ^{12}C?

18. Estimate the nuclear spins of the ^{3}He and ^{4}He nuclei. Explain your reasoning.

19. The energy required to remove an inner K-shell electron from a silver atom is 25.6 keV. Compare this electron binding energy (the most tightly bound electron) with the binding energy of the most loosely bound proton of $^{107}_{47}$Ag.

20. Compare the total Coulomb repulsion between protons for ^{4}He, ^{40}Ca, and ^{208}Pb. Assume the protons in ^{4}He are, on the average, a distance equal to the nuclear radius apart and use Equation (12.18) for the Coulomb repulsion of the larger nuclei.

21. Continue adding neutrons into ^{16}O as in Figure 12.8. We find that ^{17}O and ^{18}O are stable, but not ^{19}O. Explain.

22. Explain why $^{42}_{20}$Ca is stable, but not $^{41}_{20}$Ca. If $^{42}_{20}$Ca is stable, why is $^{42}_{22}$Ti unstable?

23. Use the von Weizsäcker semi-empirical mass formula to determine the mass (in MeV/c^2) of ^{48}Ca. Compare this with the mass determined by the information in Appendix 8.

24. (a) Use the von Weizsäcker semi-empirical mass formula to determine the binding energy per nucleon for ^{18}C, ^{18}N, ^{18}O, and ^{18}Ne. (b) Which of these nuclides is the most stable? Is that consistent with what you expect? Explain. (c) Compare the value for ^{18}O with the same value determined from the information in Appendix 8.

12.6 Radioactive Decay

25. A radioactive sample of ^{60}Co ($t_{1/2}$ = 5.271 y) has a β^- activity of 2.4×10^7 Bq. How many grams of ^{60}Co are present?

26. An unknown radioactive sample is observed to decrease in activity by a factor of five in a one-hour period. What is its half-life?

27. Show that the mean (or average) lifetime of a radioactive sample is $\tau = 1/\lambda = t_{1/2}/\ln(2)$.

28. For the reactor described in Example 12.11, compute the alpha activity of the ^{238}U present in the fuel cell. Assume that the uranium has been enriched to 4% ^{235}U, which is typical for a commercial power reactor fuel cell.

29. Potassium is a useful element in the human body and is present at a level of about 0.3% of body weight. Calculate the $^{40}_{19}$K activity in a 70-kg person. (^{40}K has 0.012% natural abundance.)

30. The nuclide $^{18}_9$F (β^+ emitter, $t_{1/2}$ = 109.8 min) is a useful radioactive tracer for human consumption. An amount of ^{18}F having an activity of 1.0×10^7 Bq is administered to a patient. What is the activity 48 hours later?

31. Tritium ($t_{1/2}$ = 12.33 y) is mostly produced for military purposes. If we had stopped producing tritium in 1960 with a 2000-kg stockpile, how much would be left in the year 2010?

32. If we have the same mass quantities of the following nuclides, rank the activities of the following material: ^{3}H (tritium), ^{222}Rn (radon gas), and ^{239}Pu (alpha source for power generation).

12.7 Alpha, Beta, and Gamma Decay

33. Use atomic masses to show that nucleon decay does not occur for $^{52}_{26}$Fe, although this nuclide is highly unstable. If one could produce $^{40}_{26}$Fe, do you believe it might nucleon decay? Explain.

34. In Example 12.12 we showed that $^{230}_{92}$U does not decay by nucleon emission. What are the neutron and proton separation energies?

35. Show directly using masses that protons do not undergo any of the beta decays.

36. Calculate whether $^{144}_{62}$Sm and $^{147}_{62}$Sm may alpha decay. The natural abundance of ^{144}Sm is 3.1% and that of ^{147}Sm is 15.0%. How can this be explained?

37. How much kinetic energy does the daughter have when $^{241}_{95}$Am undergoes α decay from rest?

38. Show from the conservation of energy and momentum that, if Equation (12.34) correctly describes the β^- decay of ^{14}C initially at rest, the electron energy spectrum must be monoenergetic and not like that shown in Figure 12.14.

39. Find which of the α and β decays are allowed for $^{80}_{35}$Br.

40. Find which of the α and β decays are allowed for $^{227}_{89}$Ac.

41. Show that α, β^-, β^+, and EC decay are possible for the nucleus $^{230}_{91}$Pa.

42. List all the possible energies of γ decay for ^{230}U based on Figure 12.15.

43. Calculate the partial pressure of helium gas for a volume of 1.0×10^{-6} m^3 of $^{222}_{86}$Rn gas after 6 days. The radon gas was originally placed in an evacuated container at 1.0 atm, and the temperature remains constant at 0°C. What is the partial pressure of the radon gas after 6 days? ($t_{1/2}$ = 3.82 d for ^{222}Rn)

44. The nuclide ^{60}Co decays by β^-. Yet ^{60}Co is often used, especially in medical applications, as a source of γ rays. Explain how this is possible.

45. Explain why β^- predominates over β^+ decay in the natural radioactive decay chains of the heavy elements.

46. Give reasons why ^{14}O can β^+ decay, but the proton cannot. (*Hint:* Use masses to prove this.)

12.8 Radioactive Nuclides

47. Two rocks are found that have different ratios R' of ^{238}U to ^{206}Pb: R' = 0.76 and 3.1. What are the ages of the two rocks? Did they likely have the same origin?

48. Use Table 12.2 to list some radioactive nuclides that may be useful for dating the age of the Earth.

49. If scientists are only able to determine the ratio of R' in Equation (12.50) to ±0.01, what is the minimum time possible for dating?

50. If the age of the Earth is 4.5 billion years, what should be the ratio of ^{206}Pb/^{238}U in a uranium-bearing rock as old as the Earth?

51. Using only Z and A values, calculate the number of α and β particles produced from the decay of $^{235}_{92}$U to its stable end product $^{207}_{82}$Pb.

52. The Earth is about 4.5 billion years old. If ^{235}U is 0.72% abundant today, how abundant was it when the Earth formed? (*Hint:* Look in Appendix 8 for useful information.)

53. Consider 100 g of ^{252}Fm, which decays in a sequence of five α decays to eventually reach ^{232}Th. (a) How much of the original sample is ^{252}Fm and how much is ^{248}Cf after one day? (b) After one month? (c) Explain why it is mostly curium after 5 years. (d) What isotope is it mostly after 100 years? (e) Approximate how long it will take for the sample to be mostly thorium.

54. Note in Table 12.2 that the half-lives of two abundant isotopes of tellurium are more than 10^{21} years.

(a) What is the decay rate per unit mass of ^{128}Te, which decays by emitting two β^-, in units of s$^{-1}\cdot$kg^{-1}? (b) How much mass of a natural sample of tellurium would it take to measure a decay rate of 10 β^-/s for ^{128}Te?

General Problems

55. Show that the total Coulomb self-energy of a sphere of radius R containing a charge Ze evenly spread throughout the sphere is given by

$$\Delta E_{\text{Coul}} = \frac{3}{5}\frac{(Ze)^2}{4\pi\epsilon_0 R} \qquad (12.17)$$

Hint: Calculate the work done to bring a charge from infinity into a spherical shell of radius r and then integrate the spherical shell from 0 to R.

56. Use the uncertainty principle $\Delta p\,\Delta x \geq \hbar/2$ to calculate the minimum kinetic energy of a nucleon known to be confined in ^{2}H. What is the de Broglie wavelength of this nucleon? Is this reasonable?

57. The nucleus $^{180}_{73}$Ta is unusual because it has both odd Z and odd N, yet it is barely unstable with a half-life of 8 h. It has an isomeric state at excitation energy 0.075 MeV that experimental measurements indicate has a half-life greater than 10^{15} y. For many years it was believed that the long-lived state was the ground state and might be stable. All the stable odd Z and N nuclei are smaller than ^{16}O. Why are all heavy elements with both odd Z and N unstable? The spins of the ^{180}Ta ground state and isomeric states are believed to be 1^+ and 9^-, respectively. Explain how it might be possible that it was believed for so long that ^{180}Ta was stable.

58. The only stable isotope of holmium is $^{165}_{67}$Ho. Explain this. Can ^{165}Ho α decay? Is it likely to α decay? Explain.

59. The nuclide $^{226}_{88}$Ra decays to gaseous $^{222}_{86}$Rn with a $t_{1/2}$ = 1600 y. The nuclide $^{222}_{86}$Rn in turn α decays with a shorter lifetime, $t_{1/2}$ = 3.82 days. If the radium is originally placed in an evacuated closed container, the amount of radon gas builds up and can be measured. It is found that radon gas builds up to a constant value, and that as much ^{222}Rn is being produced as decays. This process is called *secular equilibrium*, and the activities are equal. (a) Show that radon builds up at the rate $dN_2/dt = \lambda_1 N_1 - \lambda_2 N_2$ where λ_1, N_1 and λ_2, N_2 are the decay constants and number of nuclei present for radium and radon, respectively. (b) Because the decay of ^{226}Ra is so slow, assume that N_1 is constant and show that

$$N_2 = \frac{\lambda_1}{\lambda_2}N_1(1 - e^{-\lambda_2 t})$$

(c) Show that after a long time, secular equilibrium is reached.

60. Rudolf Mössbauer discovered in 1957 that transitions from an excited nuclear state occur with negligible nuclear recoil when the nucleus is embedded in a large crystal lattice, because the entire lattice absorbs the recoil. A transition like that in ^{191}Ir from the 129-keV excited state to the ground state has a lifetime of 1.9×10^{-10} s. (a) Determine the energy width of the decay. (b) Similarly, if the photon is absorbed by ^{191}Ir embedded in a crystal, the recoil is negligible. However, even a slight motion of the absorber will lead to a Doppler shift sufficient to destroy the resonance absorption. Calculate the speed necessary to shift the energy absorption by 5Γ, where Γ is the nuclear decay width.

61. Radon gas in the form of ^{222}Rn is a health hazard because it is a gas that occurs as a result of one of the naturally occurring radioactive decay chains. It tends to collect in basements and can be inhaled by humans. (a) Which decay chain produces this isotope of radon? (b) Show that ^{222}Rn produces five more disintegrations before a stable isotope is reached. (c) Choose one of the paths of the decay chain from ^{222}Rn to the stable isotope and sum the half-lives. Approximate the number of days it would take for more than half these decays to occur for a given amount of radon.

62. Just as there are atomic shell features, there are also nuclear shell structures. Certain values of the number of protons Z and neutrons N are called "magic," because certain nuclides with these magic numbers are more tightly bound and have enhanced abundances. These numbers are 2, 8, 20, 28, 50, 82, and 126. (a) Use this fact to explain why ^{4}He and ^{16}O have peaks in the binding energy curve of Figure 12.6. (b) Explain why ^{208}Pb is one of the heaviest stable nuclides. (c) Discuss the abundances of the calcium isotopes with respect to their value of N. The German physicists Maria Goeppert-Mayer and J. Hans D. Jensen won the Nobel Prize in Physics in 1963 for their discoveries concerning the *nuclear shell model*.

63. Use the nuclear shell model of the previous problem to list five stable nuclides that have magic numbers for both Z and N.

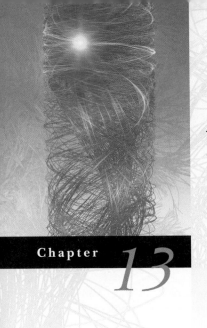

Nuclear Interactions and Applications

Chapter 13

Ernest Lawrence, upon hearing the first self-sustaining chain reaction would be developed at the University of Chicago in 1942 rather than at his University of California, Berkeley lab said, "You'll never get the chain reaction going here. The whole tempo of the University of Chicago is too slow."

Quoted by Arthur Compton in Atomic Quest

We studied the basics of nuclear physics in Chapter 12. In the first decades of the twentieth century, investigators studied the atomic nucleus by using naturally radioactive emitters as sources of high-speed particles to probe the nucleus. The invention of particle accelerators in the 1930s enabled physicists to control the intensity and energy of these high-speed particles. These particle accelerators instigated a new era in nuclear and particle physics that continues today. In this chapter we begin by studying nuclear reactions and mechanisms. Then we discuss nuclear power generation by fission and fusion reactors. Applications of nuclear particle techniques include such varied phenomena as medicine, agriculture, archaeology, art, and crime detection. The search for the discovery of new elements is an exciting and ongoing pursuit of physicists. In Chapter 14 we will discuss particle physics and the development of particle accelerators.

13.1 Nuclear Reactions

Rutherford produced the first nuclear reaction in a laboratory experiment in 1919. He used 7.7-MeV alpha particles from the decay of $^{214}_{84}$Po to bombard a nitrogen target, and he observed protons being emitted. Although he was not certain of the exact nuclear reaction taking place, he convinced himself that a nuclear transformation had taken place. We now know that the reaction was

$$\alpha + {}^{14}_{7}\text{N} \rightarrow p + {}^{17}_{8}\text{O} \tag{13.1}$$

The first nucleus written is the projectile (α or ^{4}He nucleus), and the second is the target (^{14}N), normally at rest. These two nuclei interact and undergo a transmutation to one or more final particles. The detected particle is normally listed

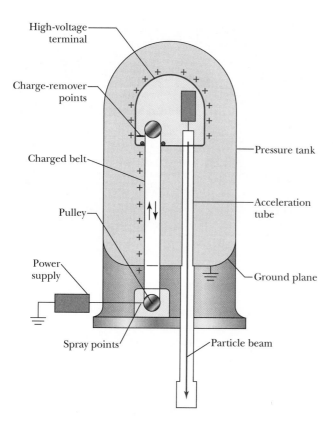

High-voltage terminal

Charge-remover points

Charged belt

Pulley

Power supply

Spray points

Pressure tank

Acceleration tube

Ground plane

Particle beam

Figure 13.1 Schematic diagram of a Van de Graaff electrostatic accelerator. Charge is transferred to an insulated moving belt by spray points and moved to a high-voltage terminal where the charge is removed. The charge builds up on the terminal, creating a large electrostatic potential. Ionized particles are accelerated to ground potential in an acceleration tube, where they become a particle beam that can be steered to a target.

first after the arrow (p or ^{1}H), and the residual nucleus listed last (^{17}O). A shorthand way of writing this reaction is

$$^{14}\text{N}(\alpha, p)^{17}\text{O} \tag{13.2}$$

A general reaction like $x + X \rightarrow y + Y$ is written in shorthand as

$$X(x, y)Y \tag{13.3}$$

Normally p, d, t, α are used as symbols for the nuclei ^{1}H, ^{2}H, ^{3}H, and ^{4}He when they are the projectile or detected particle.

The study of nuclear reactions was helped considerably by three important technological advances in accelerator development:

1. The high-voltage multiplier circuit, developed in 1932 by the British physicists J. D. Cockcroft and E. T. S. Walton (both received the Nobel Prize in Physics, 1951)
2. The first electrostatic generator (Van de Graaff accelerator), developed in 1931 by R. Van de Graaff in the United States (see Figure 13.1)
3. The first cyclotron, built in 1932 at the University of California in Berkeley by E. O. Lawrence (Nobel Prize in Physics, 1939) and M. S. Livingston (see Figure 13.2)

These three accelerators and the ones that followed them allowed nuclear reactions to be studied in the laboratory using accelerated particles of controlled energy and intensity. This allowed detailed study of specific characteristics of the nucleus. Accelerator technology development continues today to allow physicists to delve even deeper into the understanding of nature's forces (see Chapter 14).

(a)

(b)

Figure 13.2 (a) The components of a simple cyclotron. Particles are injected into the middle between the dees and move in circular paths perpendicular to the magnetic field. Each time they pass between the two dees they are accelerated by a radio-frequency (RF) voltage. Eventually the radius of the particle's path becomes large and the particles are extracted as a particle beam. (b) E. O. Lawrence (1901–1958), on the right, working on the 60-inch-diameter cyclotron at the University of California at Berkeley in the 1930s. Lawrence was born in South Dakota and received his college degrees from the University of South Dakota (B.A. in chemistry, 1922), University of Minnesota (M.A. in physics, 1923), and Yale University (Ph.D. in physics, 1925). He went to the University of California, Berkeley, in 1928 at age 27 and within three years he had invented the cyclotron and was made the youngest full professor at Berkeley. He remained there until his death. *AIP Niels Bohr Library.*

We already mentioned several nuclear reactions in the previous chapter. Among them is the production of neutrons in the $^9\mathrm{Be}(\alpha, n)\,^{12}\mathrm{C}$ reaction. **Nuclear photodisintegration** is the initiation of a nuclear reaction by a photon, as in the $^2\mathrm{H}(\gamma, n)\,^1\mathrm{H}$ reaction to measure the binding energy of the deuteron. **Neutron or proton radioactive capture** occurs when the nucleon is absorbed by the target nucleus, with energy and momentum conserved by γ-ray emission. Examples are $^9\mathrm{Be}(p, \gamma)\,^{10}\mathrm{B}$ and $^{16}\mathrm{O}(n, \gamma)\,^{17}\mathrm{O}$.

If we want to draw attention to the fact that one of the residual particles is left in an excited state, we put a superscript asterisk on the nucleus's symbol: $^{12}\mathrm{C} + \,^{12}\mathrm{C} \rightarrow \alpha + \,^{20}\mathrm{Ne}^*$ or $^{12}\mathrm{C}(^{12}\mathrm{C}, \alpha)\,^{20}\mathrm{Ne}^*$. To indicate the explicit excited state that a nucleus is in, we write the energy in parentheses after the nucleus: $^{12}\mathrm{C} + \,^{12}\mathrm{C} \rightarrow \alpha + \,^{20}\mathrm{Ne}(4.247\ \mathrm{MeV})$. The asterisk would be redundant in this case.

Entrance and exit channels   The projectile and target are said to be in the *entrance channel* of a nuclear reaction; for example, $^{12}\mathrm{C}$ and $^{12}\mathrm{C}$ represent a beam of carbon ions incident on a carbon target. The reaction products are in the *exit channel*, for example, α and $^{20}\mathrm{Ne}$. If the conservation laws allow, there may be many exit channels for a given entrance channel. For example, $^{12}\mathrm{C}(^{12}\mathrm{C}, n)\,^{23}\mathrm{Mg}$, $^{12}\mathrm{C}(^{12}\mathrm{C}, p)\,^{23}\mathrm{Na}$, $^{12}\mathrm{C}(^{12}\mathrm{C}, \,^3\mathrm{He})\,^{21}\mathrm{Ne}$, and $^{12}\mathrm{C}(^{12}\mathrm{C}, \,^6\mathrm{Li})\,^{18}\mathrm{F}$ all have different exit channels for the same entrance channel.

Elastic and inelastic scattering   In *elastic scattering* the entrance and exit channels are identical, and the particles in the exit channels are not in excited states. In *inelastic scattering* the entrance and exit channels are also identical, but one or more of the reaction products in the exit channel are left in an excited state as in the $\alpha + \,^{20}\mathrm{Ne} \rightarrow \alpha + \,^{20}\mathrm{Ne}^*$

reaction. We have listed the preceding reactions as if two particles are always in the exit channel. That is certainly not the case. Reactions like

$$\alpha + {}^{20}_{10}\text{Ne} \rightarrow n + p + {}^{22}_{11}\text{Na} \qquad (13.4)$$

can also occur.

Cross Sections

The properties of the nucleus have mostly been studied by detecting one or more particles in the exit channel in a nuclear reaction. The probability of a particular nuclear reaction occurring is determined by measuring the *cross section* σ. The cross section is determined by measuring the number of particles produced in a given nuclear reaction, which in turn allows physicists to learn about the nuclear force that caused the nuclear reaction. We introduced the concept of the cross section in Section 4.2, when we discussed how Rutherford and his colleagues "discovered" the nucleus in their early experiments. We follow the same procedure as in Section 4.2 and define n = number of target atoms/volume, t = target thickness, A = area of the target, ρ = density, N_A = Avogadro's number, N_M = atoms/molecule, and M_g = gram-molecular weight. The number of target nuclei N_s is

$$N_s = ntA = \frac{\rho N_A N_M t A}{M_g} \qquad (4.10)$$

where the parameters are as defined in Section 4.2. The probability of the particle being scattered is proportional to the product of the cross section times the total number of target nuclei N_s. We normalize this product to obtain the probability of scattering by dividing by the total target area A.

$$\text{Probability of scattering} = \frac{N_s \sigma}{A} = \frac{ntA\sigma}{A} = nt\sigma \qquad (13.5)$$

Probability of scattering

The product nt is the number of target nuclei exposed per unit area, and it is related to the probability of scattering by multiplying by the cross section.

We measure the cross section by counting the number of detected particles as a function of the number of incoming particles. If the cross sections are measured as a function of the scattering angle θ (angle between the incoming beam of particles and the detected particle), we call them **differential cross sections** $\sigma(\theta)$. The geometry is shown in Figure 13.3. Differential cross sections are determined by the number of particles scattered into a small solid angle $d\Omega$, measured in units of steradian (abbreviated sr), surrounding the scattering angle θ. In spherical coordinates, with θ being the angle measured from the incident beam direction, the solid angle $d\Omega = \sin\theta \, d\theta \, d\phi$. The differential cross section can also be written as $d\sigma/d\Omega$, which expresses more clearly the number of particles scattered into the

Differential cross section

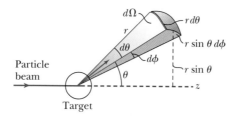

$$d\Omega = \frac{(r\sin\theta)(r\,d\theta)(d\phi)}{r^2} = \sin\theta \, d\theta \, d\phi$$

Figure 13.3 A particle from a beam interacts with a target, sending a reaction particle into a differential solid angle $d\Omega$. The solid angle is defined by the angles θ and ϕ in spherical coordinates: $d\Omega = \sin\theta \, d\theta \, d\phi$.

differential solid angle $d\Omega$: $\sigma(\theta) = d\sigma/d\Omega$. If the cross sections are integrated over the entire range of scattering angles, we call them **total cross sections** σ_T.

Total cross section

$$\sigma_T = \int \sigma(\theta)\, d\Omega = \int_0^{2\pi} d\phi \int_0^{\pi} \frac{d\sigma}{d\Omega} \sin\theta\, d\theta$$

In general, the cross sections depend on the incident kinetic energy and other properties including the spins of the particles. Total cross sections are traditionally measured in units of barns (b) with 1 barn $\equiv 10^{-28}$ m^2 = 100 fm^2, so that a barn is about the cross-sectional area of an $A = 100$ nucleus. The units of differential cross section are barns/steradian, or b/sr.

Example 13.1

J. L. Black and colleagues* measured the total cross section for the ^{12}C$(\alpha, n)^{15}$O reaction at an incident energy of $E_\alpha = 14.6$ MeV to be $\sigma_T = 25$ mb. If a 1.0-μA α-particle beam (^{4}He^{++}) is incident on a 4.0-mm^2 carbon target of thickness 1.0 μm (density = 1.9 g/cm^3) for one hour, how many neutrons are produced?

Strategy In order to find the number N_n of neutrons produced, we find the probability of scattering and multiply it by the number N_I of incident α particles: $N_n = N_I P$. To find the probability of scattering, we first need to determine n, the number of nuclei/volume, which we established in Equation (4.8):

$$n = \frac{\rho N_A N_M}{M_g} \frac{\text{atoms}}{\text{cm}^3} \qquad (4.8)$$

We can calculate the probability of scattering $P = nt\sigma$ [Equation (13.5)], because we can determine n and thickness t, and we are given the cross section σ. Lastly, we determine the number N_I of incident α particles from the beam current and time the beam is on the target.

Solution Some of the values needed are

$\rho = 1.9$ g/cm^3 $N_A = 6.02 \times 10^{23}$ molecules/mol

$N_M = 1$ atom/molecule $M_g = 12$ g/mol

If we substitute these values in the equation for n, we have

$$n = \left(\frac{1.9\ \text{g}}{\text{cm}^3}\right)\left(\frac{6.02 \times 10^{23}\ \text{molecules}}{\text{mol}}\right)$$

$$\times \left(\frac{1\ \text{atom}}{\text{molecule}}\right)\left(\frac{\text{mol}}{12\ \text{g}}\right)$$

$$= 9.53 \times 10^{22}\ \frac{\text{atoms}}{\text{cm}^3}$$

The probability of scattering can now be determined from Equation (13.5).

$$P = nt\sigma = (9.53 \times 10^{22}\ \text{nuclei/cm}^3)(1.0 \times 10^{-6}\ \text{m})$$

$$\times (25 \times 10^{-31}\ \text{m}^2)(10^6\ \text{cm}^3/\text{m}^3)$$

$$= 2.4 \times 10^{-7}$$

The number of incident α particles N_I on the target can be determined by the beam current and length of time the beam is on the target.

$$N_I = (1.0\ \mu\text{A})\left(\frac{10^{-6}\ \text{C/s}}{\mu\text{A}}\right)(1.0\ \text{h})\left(3600\ \frac{\text{s}}{\text{h}}\right)$$

$$\times \left[\frac{1\ \text{alpha}}{2(1.6 \times 10^{-19}\ \text{C})}\right]$$

$$= 1.1 \times 10^{16}\ \text{alphas}$$

Note that we have taken the charge of the incident α particles to be $+2e$.

The ratio of detected neutrons to incident alpha particles (N_n/N_I) is the probability P of scattering. We therefore have

$$N_n = N_I P = N_I n t \sigma$$

$$= (1.1 \times 10^{16}\ \text{alphas})(2.4 \times 10^{-7}\ \text{neutrons/alpha})$$

$$= 2.6 \times 10^9\ \text{neutrons}$$

*J. L. Black et al., *Nuclear Physics* **115**, 683 (1968).

Example 13.2

E. M. Bernstein and colleagues* measured the differential cross section $\sigma(\theta)$ at the same energy $E_\alpha = 14.6$ MeV and at the same scattering angle for the $\alpha + {}^{12}C \rightarrow n + {}^{15}O$ and $\alpha + {}^{12}C \rightarrow p + {}^{15}N$ reactions. They found differential cross sections of 3 mb/sr and 0.2 mb/sr, respectively, for the neutron and proton production at the same scattering angle θ. How much more likely is it that a neutron is produced than a proton?

Solution The probability of scattering is simply $nt\sigma$, so let P_n and P_p be the neutron and the proton probability, respectively. We also denote the neutron and proton differential

cross section at scattering angle θ by $\sigma_n(\theta)$ and $\sigma_p(\theta)$, respectively.

$$\frac{P_n}{P_p} = \frac{nt\sigma_n(\theta)}{nt\sigma_p(\theta)} = \frac{\sigma_n(\theta)}{\sigma_p(\theta)} = \frac{3\text{ mb/sr}}{0.2\text{ mb/sr}} = 15$$

For the particular scattering angle θ, this ratio is larger than might be expected from the fact that neutrons and protons have similar nuclear interactions. Other factors, such as overcoming the Coulomb barrier and the existence of resonances, are important and will be discussed in Section 13.3.

*E. M. Bernstein et al., *Physical Review C* **3**, 427 (1971).

13.2   Reaction Kinematics

In this chapter we are discussing low-energy nuclear reactions, in which the kinetic energies are typically much lower than the rest energies. Therefore we can use nonrelativistic kinematics in most cases. Consider the reaction $x + X \rightarrow y + Y$ depicted in Figure 13.4a, where the momentum and kinetic energy of the projectile x are $\vec{p}_x$ and K_x, respectively. The target X is assumed to be at rest, so $\vec{p}_X = 0$ and $K_X = 0$.

The conservation of energy for the reaction is

$$M_x c^2 + K_x + M_X c^2 = M_y c^2 + K_y + M_Y c^2 + K_Y \qquad (13.6)$$

If we rearrange this equation and put all the masses on one side and all the kinetic energies on the other side, we find a quantity similar to the disintegration energy of Chapter 12:

$$Q = M_x c^2 + M_X c^2 - (M_y c^2 + M_Y c^2) = K_y + K_Y - K_x \qquad (13.7) \quad \textbf{Q value}$$

Before

After

(a)

Before

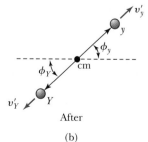

After

(b)

Figure 13.4 (a) Two-body nuclear reaction, $X(x, y)Y$, as observed in the laboratory system showing the bodies before and after. (b) The same two-body nuclear reaction, now observed in the center-of-mass system in which the center of mass (cm) is at rest.

The difference between the final and initial kinetic energies is precisely the difference between the initial and final mass energies. We call this difference the **Q value** or energy release. Nuclear masses are listed in Equation (13.7), but as we learned in Chapter 12, we may use atomic masses. The electron masses cancel, and the electron binding energies are so small that we neglect their differences when using masses to determine the Q value. When using the tabulated atomic masses from Appendix 8 in Equation (13.7), we determine the ground state Q value, that is, all the nuclei are in their lowest energy state. If one or more of the nuclei are left in an excited state E^*, that amount of energy must be added to the mass energy listed in Equation (13.7) in order to calculate the reaction Q value. We don't need to worry about this particular point if we use the kinetic energies in Equation (13.7) to determine the Q value.

Energy is released in a nuclear reaction when $Q > 0$, and we call this an **exoergic** (or *exothermic*) reaction. When $Q < 0$, kinetic energy is converted to mass energy and we call the reaction **endoergic** (or *endothermic*). In an elastic collision, $x + X \rightarrow x + X$, we must have $Q = 0$. In an inelastic collision, $x + X \rightarrow x + X^*$, we must have $Q < 0$.

Endoergic and exoergic reactions

Example 13.3

Calculate the ground state Q value for the reaction $^{14}\text{N}(\alpha, p)^{17}\text{O}$ in which Rutherford first observed a nuclear reaction. The kinetic energy of the α particles was 7.7 MeV. What was the sum of the kinetic energies of the exit channel?

Strategy We can determine the Q value with Equation (13.7) by using the atomic masses listed in Appendix 8. They are

$M(^4\text{He}) = 4.002603$ u $M(^1\text{H}) = 1.007825$ u

$M(^{14}\text{N}) = 14.003074$ u $M(^{17}\text{O}) = 16.999132$ u

Notice that we must use the atomic masses for ^1H and ^4He in order to have the complete cancellation of the electron masses. We can also determine the kinetic energies of the exit channel from Equation (13.7).

Solution The ground state Q value is calculated from Equation (13.7).

$$\frac{Q}{c^2} = M(^4\text{He}) + M(^{14}\text{N}) - [M(^1\text{H}) + M(^{17}\text{O})]$$

$$= 4.002603 \text{ u} + 14.003074 \text{ u}$$

$$- (1.007825 \text{ u} + 16.999132 \text{ u})$$

$$= -0.001280 \text{ u}$$

$$Q = (-0.001280 \ c^2 \cdot \text{u})\left(\frac{931.5 \text{ MeV}}{c^2 \cdot \text{u}}\right) = -1.192 \text{ MeV}$$

The reaction is endoergic. Equation (13.7) gives the sum of the kinetic energies of the products.

$$Q = K(p) + K(^{17}\text{O}) - K(\alpha)$$

$$K(p) + K(^{17}\text{O}) = Q + K(\alpha)$$

$$= -1.192 \text{ MeV} + 7.7 \text{ MeV} = 6.5 \text{ MeV}$$

The final reaction products share 6.5 MeV of energy.

An endoergic ($Q < 0$) reaction will not occur unless there is enough kinetic energy K_x to supply the needed nuclear rearrangement energy. Because linear momentum must also be conserved, the energy must be available in the center-of-mass (cm) system where v_{cm} is zero. The total momentum is zero in the cm system (see Figure 13.4b). The minimum kinetic energy needed to initiate a nuclear reaction is called the **threshold energy** K_{th}, and for this energy the particles y and Y will be at rest in the cm system. Of course, particle y will still be moving in the laboratory system.

Let us examine conservation of energy in the center-of-mass system. The speed of the center of mass is given by

$$v_{cm} = v_x \left(\frac{M_x}{M_x + M_X} \right) \tag{13.8}$$

As shown in Figure 13.4b, the speeds of x and X in the center-of-mass system are $v'_x = v_x - v_{cm}$ (to the right) and $v'_X = v_{cm}$ (to the left), respectively, where the primes indicate the quantities in the cm system. At threshold we must have $v'_y = v'_Y = 0$ in the center-of-mass system. The conservation of energy in the center-of-mass system is

$$\frac{1}{2} M_x (v_x - v_{cm})^2 + \frac{1}{2} M_X v_{cm}^2 + M_x c^2 + M_X c^2$$

$$= M_y c^2 + M_Y c^2 + \frac{1}{2} M_y v'^2_y + \frac{1}{2} M_Y v'^2_Y \tag{13.9}$$

However, because $v'_y = v'_Y = 0$ at threshold, we have, upon rearranging terms,

$$\frac{1}{2} M_x \left(v_x - \frac{M_x v_x}{M_x + M_X} \right)^2 + \frac{1}{2} M_X \frac{M_x^2 v_x^2}{(M_x + M_X)^2} = -Q$$

This equation reduces to

$$\frac{1}{2} v_x^2 \left(\frac{M_x M_X}{M_x + M_X} \right) = -Q$$

But the threshold energy is defined by $K_{th} = \frac{1}{2} M_x v_x^2$, so we have

$$K_{th} = -Q \left(\frac{M_x + M_X}{M_X} \right) \tag{13.10}$$ **Threshold energy**

Equation (13.10) is the threshold energy calculated nonrelativistically for an endoergic reaction. The Q value determined in Example 13.3 for the $^{14}N(\alpha, p)^{17}O$ reaction was -1.192 MeV. The threshold kinetic energy is 1.191 MeV $\times$ $(4 + 14)/14$, which gives 1.531 MeV. The reaction discussed in Example 13.3 will not take place if the incident α particle has less than 1.531 MeV kinetic energy.

13.3   Reaction Mechanisms

As we discussed in Chapter 12, physicists use models to describe the nucleus because they have been unable to completely understand the nuclear force. The primary technique for studying nuclei and nuclear forces has been to perform scattering reactions. Physicists measure cross sections, which are proportional to the reaction probabilities, which in turn depend on details of nuclear structure and the strengths and ranges of the interaction.

There are also different types of nuclear reactions—each occurring over varying bombarding energies. For example, reaction mechanisms initiated by projectiles like electrons are much different than those initiated by alpha particles. For heavy charged particles (protons and alpha particles) at low energies, $E < 10$ MeV, the scattering is dominated by the Coulomb force, and the *compound nucleus* reaction mechanism (discussed in the next section) is appropriate. *Direct reactions* dominate for bombarding energies in the range 10–100 MeV. At energies above about 140 MeV, a new particle called a *pion* emerges (see Chapter 14); the pion interacts strongly through the nuclear force. The energy region above 200 MeV to a few GeV is the realm of *medium-energy physics*. Experiments with protons, pions, and electrons dominate this region, which has some overlap with *high-energy* or *elementary particle physics* (covered in Chapter 14).

The Compound Nucleus

Niels Bohr proposed in 1936 that nuclear reactions take place through formation and decay of a compound nucleus. The **compound nucleus** is a composite of the projectile and target nuclei, usually in a high state of excitation. For the entrance channel nuclei x and X forming the compound nucleus CN, the excitation energy of the compound nucleus is

$$E(\text{CN}^*) = M_x c^2 + M_X c^2 - M_{CN} c^2 \tag{13.11}$$

In the $^{12}\text{C}(\alpha, n)^{15}\text{O}$ reaction discussed in Examples 13.1 and 13.2, enough energy is available from just the masses to leave $^{16}\text{O}^*$ in an excited state of 7.2 MeV when an α particle and ^{12}C join to produce $^{16}\text{O}^*$. The kinetic energy *available in the center of mass* K'_{cm} (see Problem 18)

$$K'_{\text{cm}} = \frac{M_X}{M_x + M_X} K_{\text{lab}} \tag{13.12}$$

is available to excite the compound nucleus to even higher excitation energies than that from just the masses.

Once formed, the compound nucleus may exist for a relatively long time compared to the time taken by the bombarding particle to cross the nucleus. This latter time is sometimes referred to as the *nuclear time scale* t_N. For a 5-MeV proton ($v \approx 0.1c$) crossing a typical nuclear diameter of about 9 fm, $t_N \approx 3 \times 10^{-22}$ s. Compound nuclei may live as long as 10^{-15} s, or $10^6 t_N$. Bohr's hypothesis was that this is such a long time that the nucleus "forgets" how it was formed, and the excitation energy is shared by all the nucleons in the nucleus. When the compound nucleus finally does decay from its highly excited state, it decays into all the possible exit channels according to statistical rules consistent with the conservation laws. Examples of the formation and decay of ^{16}O are shown in Figure 13.5. In Figure 13.6 we show the excitation energy in $^{16}\text{O}^*$ equivalent to the various entrance channel rest energies.

We mentioned earlier that nuclei have discrete energy levels, much as atoms do, and in Figure 13.6 we exhibit the low-lying states of ^{16}O. The lowest state, called the *ground state*, is the reference; its energy is 0.0 MeV. The first excited state is 6.05 MeV, with the next state close by at 6.13 MeV. The ^{16}O ground state is a particularly strongly bound nuclear state. To excite ^{16}O takes an anomalously large amount of energy (6.05 MeV).

Figure 13.5 Several different entrance-channel two-body reactions may form the nucleus ^{16}O. Similarly, the excited ^{16}O nucleus may decay to one of many exit channels that the conservation laws allow.

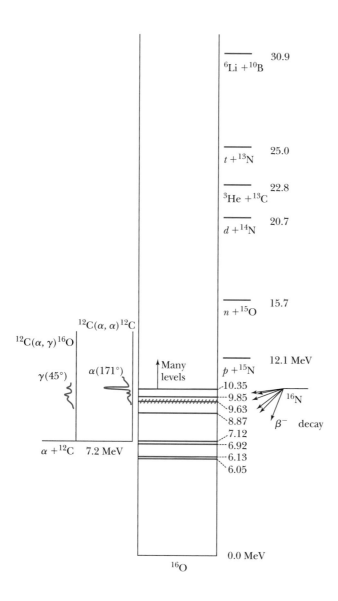

Figure 13.6 Energy-level diagram of the nucleus ^{16}O. The first excited state is at 6.05 MeV. If $d + {}^{14}$N could join together at rest to form ^{16}O, the excitation energy of ^{16}O would be 20.7 MeV. The nucleus ^{16}O is much more tightly bound than any of the two-body systems shown, with $\alpha + {}^{12}$C being the lowest at 7.2 MeV in ^{16}O. On the left are shown excitation functions for $\alpha + {}^{12}$C for the outgoing channels of γ at 45° and α at 171°. Note that resonances are observed in some reactions at bombarding energies appropriate for a particular resonance in ^{16}O; see, for example, 10.35 MeV. Note that ^{16}N can beta decay to levels in ^{16}O with excitation energies below about 10 MeV. The crosshatched level at 9.63 MeV is quite broad.

Example 13.4

In Example 13.2 we noted that the ^{12}C$(\alpha, n)^{15}$O reaction cross section was much larger than that of the ^{12}C$(\alpha, p)^{15}$N reaction at $E_\alpha = 14.6$ MeV. To what final excitation energy is ^{16}O* excited in this reaction?

Strategy We need to find the sum of two energies. The first is the kinetic energy available in the center of mass [Equation (13.12)]. The second is the excitation energy $E(^{16}$O*$)$ due to just the masses when an α particle and ^{12}C form to make ^{16}O [Equation (13.11)].

Solution The available kinetic energy in the center of mass is, from Equation (13.12),

$$K'_{cm} = \frac{12}{4 + 12}(14.6 \text{ MeV}) = 11.0 \text{ MeV}$$

The excitation energy $E(^{16}$O*$)$ due to just the masses when an α particle and ^{12}C form to make ^{16}O is determined from Equation (13.11) to be

$$E(^{16}\text{O*}) = M(^4\text{He})c^2 + M(^{12}\text{C})c^2 - M(^{16}\text{O})c^2$$

The appropriate masses are $M(^4\text{He}) = 4.002603$ u, $M(^{12}\text{C}) = 12.0$ u, and $M(^{16}\text{O}) = 15.994915$ u. The ground state excitation energy of $^{16}\text{O}^*$ becomes

$$E(^{16}\text{O}^*) = (4.002603 \text{ u} + 12.0 \text{ u} - 15.994915 \text{ u})c^2$$

$$= 0.007688\, c^2 \cdot \text{u}$$

$$= (0.007688\, c^2 \cdot \text{u})\left(\frac{931.5 \text{ MeV}}{c^2 \cdot \text{u}}\right) = 7.16 \text{ MeV}$$

This energy, 7.2 MeV, is indicated for $\alpha + {}^{12}\text{C}$ in Figure 13.6. The final excitation energy in $^{16}\text{O}^*$ is the sum of the available K'_{cm} and 7.2 MeV.

$$E_{\text{final}}(^{16}\text{O}^*) = 11.0 \text{ MeV} + 7.2 \text{ MeV} = 18.2 \text{ MeV}$$

Nuclear physicists study nuclear excited states by varying the projectile bombarding energy K_x and measuring the cross section at each energy, generally at fixed angles for the outgoing particles. This is called an *excitation function*. Excitation functions for reactions proceeding to the final ground states of $^{12}\text{C}(\alpha, \alpha)^{12}\text{C}$ and $^{12}\text{C}(\alpha, \gamma)^{16}\text{O}$ are shown on the left in Figure 13.6. Notice that there are peaks and sudden changes in the smooth curves. Such sharp peaks in the excitation function of the reacting particles are called **resonances**, and they represent a quantum state of the compound nucleus being formed; in this case an excited state of ^{16}O is formed when α and ^{12}C interact. This is confirmed in the excitation functions of Figure 13.6 where the bumps coincide with particular energy levels in $^{16}\text{O}^*$.

Resonances

Now we can understand why $^{12}\text{C}(\alpha, n)^{15}\text{O}$ had such a large cross section in Example 13.2. The reaction populated a resonance near $E(^{16}\text{O}^*) = 18.2$ MeV as determined in Example 13.4. The quantum numbers of the $^{12}\text{C}(\alpha, n)^{15}\text{O}$ exit channel select this energy level in $^{16}\text{O}^*$ to be populated, whereas the quantum numbers for the $^{12}\text{C}(\alpha, p)^{15}\text{N}$ exit channel apparently do not.

The uncertainty principle may be used to relate the energy width of a particular nuclear state (called Γ) to its lifetime (called τ). The relationship is

$$\Gamma\tau > \frac{\hbar}{2} \tag{13.13}$$

If the width of a certain nuclear state is measured to be Γ by an excitation function, Equation (13.13) can be used to determine its lifetime. Ground states for stable nuclei, for example ^{16}O, have an infinite lifetime and therefore have zero energy width.

Neutron Activation Because neutrons have zero net charge, they interact more easily with nuclei at low energies than do charged particles, because of the Coulomb barrier. If the nuclide $^{113}_{48}\text{Cd}$ interacts with a neutron, the compound nucleus $^{114}_{48}\text{Cd}^*$ may be formed. We call this process *neutron activation*. One common mode for decay of this compound nucleus is the emission of a γ ray. The reaction $^{113}\text{Cd}(n, \gamma)^{114}\text{Cd}$, an example of *neutron radioactive capture*, produces γ-ray energies that depend on the energy-level structure of ^{114}Cd. The γ-ray energies and intensities are characteristic of the ^{114}Cd nucleus and no other. They are like a unique fingerprint that lets us determine the compound nucleus to be ^{114}Cd and thus indicate the original presence of ^{113}Cd. The general technique, called *neutron activation analysis*, is a powerful practical method for identifying elements without damaging the sample. We will discuss it further in Section 13.7.

The neutron capture reaction (n, γ) often has a large cross section, which can be as large as thousands of barns at a resonance. As neutrons pass through matter, they lose energy by having many elastic and inelastic collisions. They eventually reach a (thermal) kinetic energy $3kT/2$ when they attain thermal equilibrium with their surroundings. The average neutron capture cross section (at energies up to about 100 keV) varies empirically as $1/v$, where v is the neutron's velocity (see the cross section for Ag in Figure 13.7). The $1/v$ dependence can be explained in terms of the time the neutron spends near the nucleus. This time is $2r/v$ ($2r$ = nuclear diameter). The longer the neutron is within the range of the nuclear force, the higher the probability it will be captured, because the average capture probability per unit time is nearly independent of incident energy.

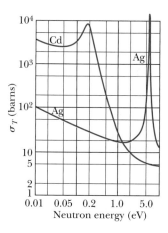

Figure 13.7 Total neutron capture cross section at low energies for silver and cadmium. The peaks indicate the presence of resonances in the compound nucleus. Note the $1/v$ dependence for the cross section of Ag at low energies. Cadmium's high cross section makes it an excellent material to control neutron flux in nuclear reactors.

Direct Reactions

As the energy of the bombarding particle rises, the excitation energy of the compound nucleus becomes high. The compound nuclear excited states become broad (large Γ), and the number of nuclear states becomes very large. As the lifetimes of the states decrease, they approach that of the nuclear time scale. In addition, the compound nuclear states overlap, and the idea of a compound nucleus loses its utility for analyzing reactions. If a compound nucleus state is formed, it is very likely to emit one or more nucleons (particularly neutrons) to rid itself of the extra excitation energy. The nuclear force is much stronger than the Coulomb force, so nucleon decay will almost always occur before γ decay when the conservation laws allow it.

For higher bombarding energies, the bombarding particle spends much less time within the range of the nuclear force. Simply *stripping* one or more nucleons off the projectile or *picking up* one or more nucleons from the target becomes more probable. It is also possible for the projectile to *knock out* energetic nucleons from the target nucleus. These are called **direct reactions**.

The chief advantage of direct reactions is that the final residual nucleus may be left in any one of many low-lying excited states. By using different direct reactions, the nuclear excited states can be studied in a variety of ways to learn more about nuclear structure. Many different reactions may be used, for example, to study ^{16}O energy levels; ^{15}N(^{3}He, d)^{16}O, ^{13}C(α, n)^{16}O, ^{12}C(^{6}Li, d)^{16}O, ^{18}O(p, t)^{16}O, and ^{17}O(d, t)^{16}O are just a few of many. All the states shown for ^{16}O in Figure 13.6 are populated in the ^{14}N(^{3}He, p)^{16}O reaction.

13.4 Fission

We saw in Chapter 12 that nuclei near $A = 56$ have the highest average binding energy per nucleon. Some nuclei with $A > 100$ are able to alpha decay, and many nuclei with $A > 220$ are unstable with respect to fission. In fission a nucleus separates into two *fission fragments*. As we will show, one fragment is typically somewhat larger than the other. We can determine which nuclei are able to fission by using Equation (12.20) for the binding energy.

Fission occurs for heavy nuclei because of the increased Coulomb forces between the protons. We can understand fission by using the semi-empirical mass formula based on the liquid drop model (Chapter 12). For a spherical nucleus with mass number $A \sim 240$, the attractive short-range nuclear forces (volume term) more than offset the Coulomb repulsive term. However, as a nucleus becomes deformed (nonspherical), the surface energy is increased, and the effect of the short-range nuclear interactions is reduced. Nucleons on the surface are not surrounded by other nucleons, and the unsaturated nuclear force reduces the overall nuclear attraction. For a certain deformation, a critical energy is reached, and the *fission barrier* is overcome. This is understood in terms of the liquid drop model where the Coulomb force pushes a deformed drop further apart. A careful

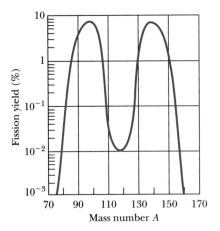

Figure 13.8 The percentage distribution of fission fragments yielded by the thermal neutron induced fission of ^{235}U. Notice that asymmetric fission is much more probable than symmetric fission (equal masses). *From R. D. Evans, The Atomic Nucleus, New York: McGraw-Hill (1955).*

Spontaneous fission   examination of the semi-empirical mass formula reveals that *spontaneous fission* occurs for nuclei with $Z^2/A \geq 49$ ($Z \approx 115, A \approx 270$). The term Z^2/A comes from the surface energy term ($A^{2/3}$) and the Coulomb energy term ($\sim Z^2/A^{1/3}$). Spontaneous fission can also occur for $Z^2/A < 49$ by the process of tunneling through the Coulomb barrier, but the half-lives are much increased. The naturally occurring nuclide with the highest spontaneous fission rate is ^{238}U, which has a half-life for fission of 8.2×10^{15} y. This is to be compared with its alpha decay half-life of 4.5×10^9 y. Thus ^{238}U is over a million times more likely to alpha decay than to fission.

Induced Fission

The previous discussion has concerned nuclei in their naturally occurring ground states. Fission may also be *induced* by a nuclear reaction. A neutron absorbed by a heavy nucleus forms a highly excited compound nucleus that may quickly fission. An example of induced fission is

Induced fission
$$n + {}^{235}_{92}\text{U} \rightarrow {}^{236}_{92}\text{U}^* \rightarrow {}^{99}_{40}\text{Zr} + {}^{134}_{52}\text{Te} + 3n \qquad (13.14)$$

The fission products have a ratio of N/Z much too high to be stable for their A value. Normally two or three neutrons are emitted during fission. There are many possibilities for the Z and A of the fission products, as shown in Figure 13.8. Symmetric fission (products with equal Z) is possible, but the most probable fission is asymmetric (one mass larger than the other), as shown by the two peaks in Figure 13.8.

Example 13.5

Calculate the ground state Q value of the induced fission reaction in Equation (13.14) if the neutron is thermal. A neutron is said to be *thermal* when it is in thermal equilibrium with its environment; it then has an average kinetic energy given by $\frac{3}{2}kT$.

Strategy Because the kinetic energy of a thermal neutron is so small, its kinetic energy can be neglected; even for a temperature of 10^6 K, the thermal energy is only 130 eV. The Q value is given by Equation (13.7).

Solution We look up the atomic masses in Appendix 8 and determine the Q value to be

$$Q = \{M(^{235}\text{U}) + m(n) - [M(^{99}\text{Zr}) + M(^{134}\text{Te}) + 3m(n)]\}c^2$$

$$= [235.0439 \text{ u} - 98.9165 \text{ u} - 133.9115 \text{ u}$$
$$- 2(1.0087 \text{ u})]c^2$$
$$= 0.1985 \, c^2 \cdot \text{u}$$

$$= (0.1985 \, c^2 \cdot \text{u})\left(\frac{931.5 \text{ MeV}}{c^2 \cdot \text{u}}\right) = 185 \text{ MeV}$$

Even if the fission is induced by a thermal neutron of negligible kinetic energy on the nuclear scale, a tremendous amount of energy is released.

Thermal Neutron Fission

The result of Example 13.5 can be understood by examining the binding-energy curve of Figure 12.6. The binding energy per nucleon of ^{236}U is about 7.6 MeV. The binding energy per nucleon of the fission fragments, however, is about 8.4 MeV. Because of this difference in binding energies, the energy released per nucleon is 0.8 MeV. The total energy released is 236 nucleons times 0.8 MeV/nucleon or 190 MeV. This rough calculation of the binding-energy difference agrees with the Q value determined in Example 13.5.

We can use the liquid drop model to understand how fission can be induced so easily in heavy nuclei. When ^{235}U absorbs a thermal neutron to form ^{236}U*, the excitation energy of ^{236}U* is 6.5 MeV. This nucleus, in a highly excited and unstable state, is agitated and becomes deformed as shown in Figure 13.9. Finally, it becomes so deformed that the Coulomb force overcomes the nuclear force (which acts only over very short distances), and the nucleus separates—much like a liquid drop.

Experiment shows that ^{235}U and ^{239}Pu fission easily after absorbing thermal neutrons (very low energy neutrons). The nuclide ^{238}U needs a neutron of at least 1 MeV kinetic energy to easily fission. If lower energy neutrons are absorbed by ^{238}U, the resulting ^{239}U* is more likely to decay by emitting a γ ray. As we discussed previously, the cross section for heavy nuclei to absorb a low-energy neutron varies as $1/v$, so the cross section is larger for lower energy neutrons.

Figure 13.9 After absorbing a neutron, a large nucleus becomes excited and unstable. It becomes deformed, and eventually the Coulomb force causes it to fission and produce two fragments of unequal mass and, in addition, three neutrons.

Example 13.6

Calculate the excitation energy of the compound nuclei produced when ^{235}U and ^{238}U absorb thermal neutrons.

Strategy The two reactions are

$$n + {}^{235}\text{U} \rightarrow {}^{236}\text{U}^*$$
$$n + {}^{238}\text{U} \rightarrow {}^{239}\text{U}^*$$

As we did in Example 13.4, we find the excitation energy from the atomic masses using Equation (13.11). A thermal neutron has a negligible kinetic energy (about 0.03 eV).

Solution We have from Equation (13.11)

$$E(^{236}\text{U}^*) = [m(n) + M(^{235}\text{U}) - M(^{236}\text{U})]c^2$$
$$= [1.0087 \text{ u} + 235.0439 \text{ u} - 236.0456 \text{ u}]c^2$$
$$= (0.0070 \, c^2 \cdot \text{u})\left(\frac{931.5 \text{ MeV}}{c^2 \cdot \text{u}}\right) = 6.5 \text{ MeV}$$

$$E(^{239}\text{U*}) = [m(n) + M(^{238}\text{U}) - M(^{239}\text{U})]c^2$$

$$= [1.0087\ \text{u} + 238.0508\ \text{u} - 239.0543\ \text{u}]c^2$$

$$= (0.0056\ c^2 \cdot \text{u})\left(\frac{931.5\ \text{MeV}}{c^2 \cdot \text{u}}\right) = 4.8\ \text{MeV}$$

Thus ^{236}U* has almost 2 MeV more excitation energy than ^{239}U* when both are produced by thermal neutron absorption. This helps explain why ^{235}U more easily undergoes thermal neutron fission.

Fission fragments are highly unstable because they are so neutron rich. This occurs because heavy nuclei deviate further and further away from the $N = Z$ line of Figure 13.10. After the fission, the resulting fission fragments are relatively

Prompt neutrons   further away from the line of stability, also seen in Figure 13.10. *Prompt neutrons* are emitted simultaneously with the fissioning process. Even after prompt neutrons are released, the fission fragments undergo beta decay, releasing more energy. Most of the $\approx$200 MeV released in fission goes to the kinetic energy of the fission products, but the neutrons, beta particles, neutrinos, and gamma rays typically carry away 30–40 MeV of the kinetic energy.

Chain Reactions

Because several neutrons are produced in fission, these neutrons may subsequently produce other fissions. This is the basis of the *self-sustaining chain reaction*. If slightly more than one neutron, on the average, results in another fission, the

Critical chain reaction   chain reaction becomes *critical*. Because a sufficient amount of mass is required to increase the chances of a neutron being absorbed, a *critical mass* of fissionable material must be present. If less than one neutron, on the average, produces another fission, the reaction is said to be *subcritical*. If more than one neutron, on the average, produces another fission, the reaction is said to be *supercritical*. An atomic bomb is an extreme example of a supercritical fission chain reaction.

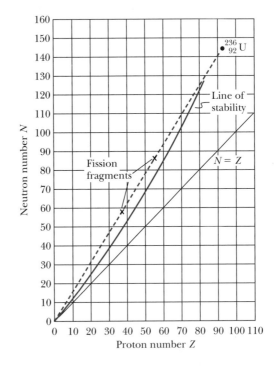

Figure 13.10 When ^{236}U fissions, its fragments are very neutron rich and the nuclides are far away from the line of stability. As a result, neutrons are produced in the fission process. The fission fragments beta decay to return to the line of stability.

Example 13.7

It only takes a few microseconds for a neutron to be absorbed and cause another fission. Assume that 1.01 neutrons are captured on the average within 5 μs of each fission. Determine how many fissions will occur within 50 ms and the total energy produced.

Strategy Every time a fission occurs, 1.01 neutrons will be produced. We determine how many cycles N occur within 50 ms and then determine the total number $(1.01)^N$ of fissions. If 185 MeV is produced for each fission, we can determine the amount of energy produced.

Solution Within 50 ms the number of cycles of fission is 50 ms/5 μs = 10,000. The number of fissions is $(1.01)^{10,000} = 1.6 \times 10^{43}$. The total amount of energy produced is

$$\text{Energy} = (1.6 \times 10^{43} \text{ fissions})\left(\frac{185 \text{ MeV}}{\text{fission}}\right)$$

$$= 3 \times 10^{45} \text{ MeV}$$

This is 5×10^{32} J, but the total world energy use in one year is less than 10^{21} J. Fortunately, the process does not occur so quickly due to delayed neutrons, as we discuss next.

As the previous example shows, we must have some way to control a runaway fission reaction. A critical-mass fission reaction can be controlled by absorbing neutrons. A self-sustaining controlled fission process depends on the fact that not all the neutrons are *prompt*. Some of the neutrons are *delayed* by several seconds and are emitted by daughter nuclides resulting from the (slow) beta decay of the fission fragments. These delayed neutrons allow the control of a nuclear reactor. The application of *control rods* regulates the absorption of neutrons to sustain a controlled reaction.

Delayed neutrons

13.5 Fission Reactors

Because so much energy is released in nuclear fission, it is a useful energy source for commercial power production. The energy content of several fuels is shown in Table 13.1, and the fuel requirements for a 1000-megawatt (MWe, the e indicating electrical power) power plant are shown in Table 13.2.

Several components are important for a controlled nuclear reactor:

Reactor components

1. Fissionable fuel
2. Moderator to slow down neutrons
3. Control rods for safety and to control criticality of reactor
4. Reflector to surround moderator and fuel in order to contain neutrons and thereby improve efficiency
5. Reactor vessel and radiation shield
6. Energy transfer systems if commercial power is desired

We have already learned that ^{235}U fissions with thermal neutrons. Because only $2\frac{1}{2}$ neutrons, on the average, result from each fission, it is important not to lose neutrons. Two main effects can "poison" reactors: (1) neutrons may be absorbed without producing fission [for example, by neutron radiative capture ^{235}U(n, γ) ^{236}U], and (2) neutrons may escape from the fuel zone.

In order to produce a critical mass of ^{235}U, it is necessary to process natural uranium ore to enrich the ^{235}U content (0.7%) from the more abundant ^{238}U (99%). Such enrichment is difficult because chemical techniques cannot be used. A single giant uranium processing plant in Paducah, Kentucky, enriches uranium (in the form of UF_6) by a gaseous diffusion process for use in the United States. The molecules of ^{235}U diffuse slightly more easily than ^{238}U, and after

Table 13.1	Energy Content of Fuels	
Material	**Amount**	**Energy (J)**
Coal	1 kg	3×10^7
Oil	1 barrel (0.16 m^3)	6×10^9
Natural gas	1 ft^3 (0.028 m^3)	10^6
Wood	1 kg	10^7
Gasoline	1 gallon (0.0038 m^3)	10^{10}
Uranium (fission)	1 kg	10^{14}
Uranium (fusion)	1 kg	2×10^{14}

Table 13.2	Daily Fuel Requirements for 1000-MWe Power Plant	
Material	**Amount**	
Coal	8×10^6 kg	(1 trainload/day)
Oil	40,000 barrels (6400 m^3)	(1 tanker/week)
Natural gas	2.5×10^6 ft^3 (7.1×10^4 m^3)	
Uranium	3 kg	

about 1500 stages of diffusion, the more valuable ^{235}U is obtained in enriched form. Only France and the United States still use the diffusion process. The favored enrichment process in the rest of the world is the gas ultracentrifuge process, which is more compact and efficient. Interest in a laser enrichment technique has dampened in recent years due to high development costs.

Fission neutrons typically have 1–2 MeV of kinetic energy, and because the fission cross section increases as $1/v$ at low energies, slowing down the neutrons helps to increase the chance of producing another fission. A *moderator* is used to elastically scatter the high-energy neutrons and thus reduce their energies. A neutron loses the most energy in a single collision with a light stationary particle. Hydrogen (in water), carbon (graphite), and beryllium are all good moderators.

The simplest method to reduce the loss of neutrons escaping from the fissionable fuel is to make the fuel zone larger. The fuel elements are normally placed in regular arrays within the moderator as shown in Figure 13.11. Fission neutrons

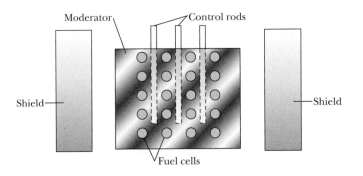

Figure 13.11 Cross-section schematic of an idealized nuclear reactor.

from one fuel cell will be moderated before entering an adjacent fuel cell to produce another fission. The fuel must be clumped in cells because of large $^{238}U(n, \gamma)$ cross sections at 7 and 21 eV, which would cause the neutrons to be absorbed as they slow down. The neutrons should be moderated outside the fuel cells.

Control rods control the reaction rates of nuclear reactors. Cadmium is an excellent control material to absorb neutrons because of its extremely large (n, γ) cross section (see Figure 13.7). The delayed neutrons produced in fission allow the mechanical movement of the rods to control the fission reaction. A "fail-safe" system automatically drops the control rods into the reactor in an emergency shutdown. If the fuel and moderator are surrounded by a material with a very low neutron capture cross section, there is a reasonable chance that after one or even many scatterings, the neutron will be backscattered or "reflected" back into the fuel area. Water is often used both as moderator and reflector. One of the earliest nuclear reactors used graphite as a moderator (Figure 13.12). Finally, the reactor must be contained within a secure vessel with adequate radiation shielding to protect personnel.

The world's first controlled nuclear chain reaction was constructed at the University of Chicago by Enrico Fermi. It first operated in December 2, 1942, and ran for $4\frac{1}{2}$ minutes under the stands of the university's football stadium, although there was concern that the city of Chicago would be blown up. Graphite blocks were used as the moderator, and cadmium sheets wrapped around wooden rods stuck inside holes served to control the chain reaction. Fermi called it an "atomic pile." (See photo and biography of Enrico Fermi in Chapter 9.)

If commercial power is desired, reactor designers must add a method to transfer energy. The most common method is to pass hot water heated by the reactor through some form of heat exchanger. In *boiling water reactors* (BWRs) the moderating water turns into steam, which drives a turbine producing electricity as shown in Figure 13.13. In *pressurized water reactors* (PWRs) the moderating water is under high pressure and circulates from the reactor to an external heat exchanger where it produces steam, which drives a turbine. This two-step process is shown in Figure 13.14. Boiling water reactors are inherently simpler than pressurized water reactors. However, the possibility that the steam driving the turbine may become radioactive is greater with the BWR. The two-step process of the PWR helps to isolate the power generation system from possible radioactive contamination.

Reactors are designed and operated for different purposes. We have been describing primarily *power* reactors used for commercial electricity production.

Figure 13.12 The graphite reactor, built in only 11 months in 1943 at Oak Ridge National Laboratory, operated until 1963, producing for many years most of the world's supply of radioisotopes for medicine, agriculture, and industry. The facility is now a National Historic Landmark, and the control room and reactor face are accessible to the public. Note that the uranium elements were loaded inside the reactor through the horizontal holes in the reactor side. *Courtesy of Oak Ridge National Laboratory, Oak Ridge, Tennessee.*

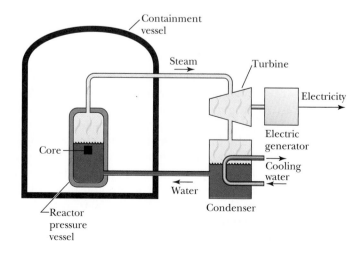

Figure 13.13 Schematic diagram of a boiling water nuclear reactor. Note that steam directly heated by the reactor core drives the turbine.

Figure 13.14 Schematic diagram of a pressurized water nuclear reactor. Note that the steam driving the turbine is one step removed from the water in the reactor core.

More than 400 power reactors are in operation worldwide today and produce 16% of the world's electricity. The United States has more than 100 of these reactors, and over 50 are in France, where power reactors produce 77% of the electricity. Among many countries heavily dependent on nuclear power for electricity are Lithuania (82%), Belgium (60%), Sweden (46%), Switzerland (41%), Japan (35%), and South Korea (34%), whereas in the United States about 20% of electricity comes from reactors. Although the percentage of the world's electricity from power reactors has been slowly decreasing since 1998, the amount of electricity from power reactors has been increasing for several years and is expected to continue doing so for several years before decreasing. This decrease will occur, especially in the United States, because existing nuclear reactors are not being replaced as they age.

Research reactors are operated to produce high neutron fluxes for research such as neutron-scattering experiments. Smaller reactors, on the order of 100 MW, are operating in some colder countries like Russia for *heat production* to warm both homes and businesses. Other reactors are designed to produce *radioisotopes* for industrial and medical purposes. Several *training* reactors are located on college campuses.

Nuclear Reactor Problems

Problems certainly exist with nuclear power plants. The danger of a serious accident in which radioactive elements are released into the atmosphere or groundwater is of great concern to the general public. Another drawback is that thermal pollution both in the atmosphere and in lakes and rivers used for cooling may be a significant ecological problem; this is presently under close scrutiny.

A more serious problem is the safe disposal of the radioactive wastes produced in the fissioning process, because some fission fragments have a half-life of thousands of years. The safe storage of nuclear wastes in proposed locations like Yucca Mountain in Nevada has been under debate for 50 years, and the problem is still unresolved. The proliferation of fissionable fuel, even from used fuel elements, to countries capable of atomic weapon production is of great concern.

A radiological weapon or "dirty bomb" could be made with a conventional explosive (TNT or a fertilizer–fuel oil mixture) mixed together with radioactive materials. The explosion would vaporize the radioactive material and spread it over a wide area. The economic chaos and widespread fear generated by a dirty bomb would perhaps be more devastating than the death and radiation illness it would cause.

Serious reactor accidents

Two widely publicized accidents at nuclear reactor facilities—one at Three Mile Island in Pennsylvania in 1979, the other at Chernobyl in Ukraine in 1986—have significantly dampened the general public's support for nuclear reactors. The number of deaths as a result of the Chernobyl accident has been officially reported to be 31, but it has been estimated that the toll could reach tens of thousands from cancer and other radiation-related diseases over the widely affected land area. No new nuclear power plants have been ordered since 1978 in the United States, and no plant has been completed since 1995.

Though it is important to consider problems with nuclear power, these problems must be evaluated in comparison with problems posed by other methods of power generation. Despite the two major nuclear reactor accidents, the safety record of nuclear power is generally accepted as better than most other methods of power generation except natural gas, of which there is a limited supply. Hydroelectric power is also reasonably safe, but its supply is also limited. Though nuclear energy production does have environmental impact, this impact may not be as severe as that of other methods of energy production. For example, burning of fossil fuels causes air pollution and greenhouse-effect gases. Over a hundred thousand lives have been lost in the United States due to coal mining accidents. A million Chinese are believed to suffer currently from black lung disease, which is caused by prolonged inhaling of coal dust. Hydroelectric dams cause environmental changes for fish and other wildlife.

A faculty group at the Massachusetts Institute of Technology released an interdisciplinary study in 2003 (see http://web.mit.edu/nuclearpower/) called "The Future of Nuclear Power" in which it analyzed "what would be required to retain nuclear power as a significant option for reducing greenhouse gas emissions and meeting growing needs for electricity supply." They concluded that a large expansion of nuclear power can succeed only if four critical problems are overcome: lower costs, improved safety, better nuclear waste management, and lower proliferation risk. They suggest how these might be accomplished.

Breeder Reactors

A promising choice for a more advanced kind of reactor is the *breeder* reactor, which produces more fissionable fuel than it consumes. When ^{238}U undergoes the (n, γ) reaction, the resulting ^{239}U beta decays to ^{239}Np and subsequently to ^{239}Pu in the chain

$$n + {}^{238}_{92}\text{U} \rightarrow {}^{239}_{92}\text{U}^* \rightarrow \gamma + {}^{239}_{92}\text{U}$$
$$\hookrightarrow \beta^- + {}^{239}_{93}\text{Np} + \bar{\nu}$$
$$\hookrightarrow \beta^- + {}^{239}_{94}\text{Pu} + \bar{\nu} \quad (13.15)$$

The resulting ^{239}Pu nucleus has a half-life of 24,100 years and is an α emitter. The intermediate beta decays take place in a matter of days. The plutonium is easily separated from uranium by chemical means, and ^{239}Pu fissions easily with both thermal and fast neutrons. Thermal neutrons are those in equilibrium with their environment and have generally much less than 1 keV of energy. Fast neutrons

Special Topic

EARLY FISSION REACTORS

*I*n the late 1930s European scientists were busy study-ing the properties of the heaviest nuclei. Several groups, including Enrico Fermi and his collaborators in Italy, the Curies and Savitch in France, and O. Hahn and F. Strassman in Germany, were bombarding ura-nium with neutrons. In 1938 Hahn and Strassman ob-served that barium ($Z = 56$) had been formed from their neutron bombardment of uranium ($Z = 92$). The Austrian-born Lise Meitner was a well-known phys-icist who had spent almost her entire career in Germany before fleeing to the Nobel Institute in Stockholm in 1938 where she joined her nephew O. R. Frisch. Meitner received word from Hahn of his experi-mental results during the turmoil leading up to World War II. Meitner and Frisch were the first to report in 1939 the correct analysis of the Hahn and Strassman experiment as being due to the fission of uranium (see Figure A).

By 1939 Fermi had fled Italy, again due to the per-secution of some scientists, and was at Columbia Uni-versity in New York. Bohr brought word of the Hahn and Strassman experiment, and the resulting excite-ment caused a flurry of experiments in the United States as well as at nuclear laboratories throughout the world. Fermi was one of the best nuclear experimental physicists of the period, and he also had a strong grasp of theoretical physics. The Hungarian-born Leo Szi-lard, working with Fermi, encouraged Einstein to write President Franklin Roosevelt and encourage the effort that was to lead to the Manhattan Project of World War II.

Bohr and John Wheeler showed theoretically that the nuclide ^{235}U undergoes fission more readily than the more abundant ^{238}U. The fission of ^{235}U was known to occur with slow (low energy) neutrons, so a process

Figure A The German Otto Hahn (1879–1968) and Austrian Lise Meitner (1878–1968) worked together in Berlin beginning in 1907 for almost 30 years doing research on radioactive substances. He was a chemist, and she was a physicist. In the 1930s they did impor-tant work identifying the products of neutron bombardment of uranium. Meitner was forced to leave Berlin in 1938 because of her Jewish ancestry, and she settled in the Nobel Institute in Stock-holm. At Meitner's suggestion at a secret meeting with Hahn in Copenhagen in November 1938, Hahn and Fritz Strassman quickly found that a product of the neutron bombardment of uranium was barium. During the Christmas season 1938, Meitner and her nephew Otto Frisch, who was working at the Bohr Institute in Copenhagen, figured out that nuclear fission was taking place. Hahn received the Nobel Prize in Chemistry in 1944, but the Nobel Prize committee overlooked Meitner's contributions. Both Hahn and Meitner received many awards in their long lives. The element 109 was named meitnerium after Meitner. *AIP Emilio Segrè Visual Archives, Brittle Books Collection.*

to slow down the neutrons produced in nuclear fission was needed in order to produce the chain reaction.

The effort toward producing the first controlled chain reaction moved to the University of Chicago, where in a former squash court under the stands of the football stadium (Stagg Field), the team led by Fermi succeeded on December 2, 1942. The race

have energies of up to a few MeV. Because ^{239}Pu fission produces an average of 2.7 neutrons (compared with 2.3 for ^{235}U), it is a highly desirable fuel for both re-actors and bombs. The critical mass can be made smaller for ^{239}Pu because 1.7 of the neutrons may be lost on the average, rather than 1.3 for fissioning of ^{235}U.

Fast breeder reactors have been built that convert ^{238}U to ^{239}Pu. The reactors are designed to use fast neutrons. If the loss of neutrons is kept small, we occasionally have a fission event that not only produces another fission event to keep the

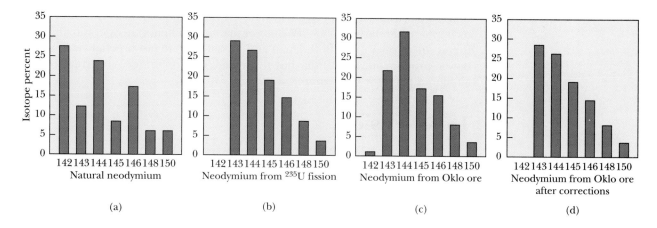

Figure B The isotopic analysis of neodymium from various sources. The composition of (a) natural neodymium is characteristically different from that of (b) neodymium produced from $n + ^{235}$U fission. (c) The composition of neodymium from the Oklo mines, which requires corrections due to the combination of natural neodymium and changes due to neutron absorption. (d) The resulting analysis is very similar to (b); this shows a fission reaction may have taken place in the ore. *From George A. Cowan, A natural fission reactor, Scientific American 235, 36 (July 1976).*

toward producing the atomic bomb developed quickly, with the first atomic bombs detonated in 1945. Eventually, in 1955, the first electricity for public use was generated by a nuclear reactor in Idaho. The *Nautilus*, the world's first nuclear submarine, was launched in 1954, and the world's first large-scale nuclear reactor began producing electricity in England in 1956. The nuclear age was under way.

There is some evidence that a natural fission reactor occurred in the Republic of Gabon on the west coast of Africa almost 2 billion years ago. The present natural abundance of ^{235}U is 0.7% and of ^{238}U 99.3%. It is believed that the relative abundances were about equal when the Earth was formed, but more of the ^{235}U has decayed because of its shorter half-life. However, 2 billion years ago the natural abundance of ^{235}U was 3.7%, and calculations indicate a critical mass of ^{235}U could have been present in natural uranium ore.

The groundwater or possibly water flooding served as a moderator to slow down the fission neutrons. When the natural reactor became too hot, the water probably boiled and the neutrons were not slowed, thus reducing the probability of causing another fission. This served as an automatic control much as the water moderator might do in a reactor today. The isotopic abundances of the fission products found in the uranium ore in Gabon (see Figure B) lead to the conclusion that for several hundred thousand years a natural fission reactor may have existed.

Scientists believe that at least six reactor zones exist in the Oklo mine of Gabon and that conditions probably existed elsewhere for similar natural fission reactors. Such an event is not possible today because of the low percentage of ^{235}U in natural uranium ores.

reaction sustained but also converts ^{238}U to ^{239}Pu. The amount of ^{239}Pu will then slowly build up (referred to as "breeding" ^{239}Pu). This ^{239}Pu can eventually be used as fuel for another power reactor. There is 99.3% natural ^{238}U and only 0.7% natural ^{235}U, so breeder reactors hold the promise of providing an almost unlimited supply of fissionable material. One of the downsides of such reactors is that plutonium is highly toxic, and there is concern about its use in unauthorized weapons production.

The United States had experimental breeder reactors operational as early as 1951, but in 1984 the government decided to postpone indefinitely the construction of a prototype power plant because of safety concerns and new projections of electricity usage. France has moved much more quickly in fast breeder technology and started a 233-MWe power plant (the Phénix) in operation in 1974 that is still in operation. Both Japan and Russia operate fast breeder reactors for power production, and India has a breeder reactor under construction and ambitious plans for three more.

13.6   Fusion

Except for nuclear fission and geothermal power, all known terrestrial sources of energy are derived from sunlight. This includes combustion of wood, coal, gas, oil, and both water and wind power. The only primary source in widespread use that is *not* derived from the sun is nuclear power (which incidentally is also the sun's energy source). Energy emitted by stars arises from *nuclear fusion* reactions, in which the enormous heat and pressure of the star's core cause light nuclei to fuse together. This process contrasts with nuclear fission, in which large nuclei divide. The origin of fusion can be understood by referring back to Figure 12.6. Nuclei near $A = 56$ have the highest binding energy per nucleon. When ^{236}U* fissions, it divides into nuclei having a larger binding energy per nucleon, thereby releasing energy. Similarly, if two light nuclei fuse together, they also form a nucleus with a larger binding energy per nucleon and energy is released. The most energy is released if two isotopes of hydrogen, ^{2}H and ^{3}H, fuse together in the reaction

$$^2\text{H} + {}^3\text{H} \rightarrow n + {}^4\text{He} \qquad Q = 17.6 \text{ MeV} \qquad (13.16)$$

About 3.5 MeV per nucleon is released because of the strong binding of ^{4}He. Less than 1 MeV per nucleon is released in fission (200 MeV/236 nucleons = 0.85 MeV/nucleon). The lower-mass side of Figure 12.6 is much steeper than the higher-mass side, explaining why the nuclear fusion process, smaller masses fusing to form larger masses, is a prolific source of energy.

Formation of Elements

When the primordial Big Bang occurred 13.7 billion years ago, the light elements of hydrogen and helium were formed in the first few minutes. It was millions of years later before the heavier elements were formed in stars through nuclear fusion. We examine these fascinating phenomena further in Chapter 16, but now we want to study two of the main cycles for producing energy in stars.

The first is the **proton-proton chain**, which includes a series of reactions that eventually converts four protons into an alpha particle. As stars form due to gravitational attraction of interstellar matter, the heat produced by the attraction is enough to cause protons to overcome their Coulomb repulsion and fuse by the following reaction:

$$^1\text{H} + {}^1\text{H} \rightarrow {}^2\text{H} + \beta^+ + \nu \qquad Q = 0.42 \text{ MeV} \qquad (13.17)$$

This reaction produces ^{2}H and is a special kind of weak-interaction beta decay process. It is extremely slow, because only 1 collision in about 10^{26} produces a reaction. This is good, because otherwise the sun would explode! The deuterons are then able to combine with ^{1}H to produce ^{3}He:

$$^2\text{H} + {}^1\text{H} \rightarrow {}^3\text{He} + \gamma \qquad Q = 5.49 \text{ MeV} \qquad (13.18)$$

The ^{3}He atoms can then combine to produce ^{4}He:

$$^3\text{He} + {}^3\text{He} \rightarrow {}^4\text{He} + {}^1\text{H} + {}^1\text{H} \qquad Q = 12.86 \text{ MeV} \qquad (13.19)$$

Note that two each of reactions (13.17) and (13.18) must occur to produce (13.19). A total of six ^{1}H are required to produce ^{4}He and two ^{1}H. This process consumes four protons. The total Q value, or kinetic energy produced, when six ^{1}H produce ^{4}He is 24.7 MeV. An additional 2 MeV is derived from the annihilation of an electron and a positron for a total of 26.7 MeV.

The proton-proton chain is particularly slow because reaction (13.17) limits the entire process. As the reaction proceeds, however, the star's temperature increases, and eventually ^{12}C nuclei are formed by a process that converts three ^{4}He into ^{12}C.

Another cycle due to carbon is also able to produce ^{4}He. The series of reactions responsible for the **carbon** or **CNO cycle** are

$$^1\text{H} + {}^{12}\text{C} \rightarrow {}^{13}\text{N} + \gamma$$
$$\qquad\qquad \hookrightarrow {}^{13}\text{C} + \beta^+ + \nu \qquad t_{1/2} = 9.96 \text{ min}$$
$$^1\text{H} + {}^{13}\text{C} \rightarrow {}^{14}\text{N} + \gamma$$
$$^1\text{H} + {}^{14}\text{N} \rightarrow {}^{15}\text{O} + \gamma$$
$$\qquad\qquad \hookrightarrow {}^{15}\text{N} + \beta^+ + \nu \qquad t_{1/2} = 2.04 \text{ min}$$
$$^1\text{H} + {}^{15}\text{N} \rightarrow {}^{12}\text{C} + {}^4\text{He} \qquad\qquad\qquad (13.20)$$

Note that four ^{1}H and one ^{12}C nuclei are required to produce ^{4}He and ^{12}C; the ^{12}C nucleus merely serves as a catalyst. We believe the proton-proton chain is probably responsible for most of our sun's energy, but the carbon cycle is a much more rapid fusion reaction. It requires higher temperatures (perhaps 20×10^6 K) than are present in the sun, because of the higher Coulomb barrier of ^{12}C relative to ^{1}H for the protons. Hans Bethe (Nobel Prize in Physics, 1967) was instrumental in showing that the proton-proton and CNO cycles are the only nuclear reactions that can supply the energy in stars.

A hydrostatic equilibrium exists in the sun between the gravitational attraction tending to contract a star and a gas pressure pushing out due to all the particles. As the lighter nuclides are "burned up" to produce the heavier nuclides, the gravitational attraction succeeds in contracting the star's mass into a smaller volume and the temperature increases. A higher temperature allows the nuclides with higher Z to fuse. This process continues in a star until a large part of the star's mass is converted to iron. The star then collapses under its own gravitational attraction to become, depending on its mass, a white dwarf star, neutron star, or black hole. It may even undergo a supernova explosion (see Chapter 16).

Hans Bethe (1906–) was born in Strasbourg, Germany (now France) and was educated in Frankfurt and Munich, where he received his Ph.D. in 1928. After serving at several institutions in Germany, he emigrated in 1933 first to England and then in 1935 to Cornell University where he has spent the remainder of his career, except for several leaves of absence. For example, during World War II he worked first at MIT on developing microwave radar, and in 1943–1946 he served as director of the theoretical physics division at Los Alamos National Laboratory. Bethe's illustrious career was mostly in nuclear physics, but also included astrophysics, atomic and molecular physics, and solid state physics. © *The Nobel Foundation.*

Example **13.8**

One of the processes by which *supergiant* stars produce heavier masses is the reaction $^4\text{He} + {}^{12}\text{C} \rightarrow {}^{16}\text{O} + \gamma$. (a) How much energy is expended in this reaction? (b) Explain why such a fusion reaction cannot occur in low-mass stars.

Strategy (a) To find the energy expended we find the Q value from Equation (13.7). (b) We consider the progression of a star's life to determine what fusion reactions can occur.

Solution (a) We look up the masses in Appendix 8 to find the Q value:

$$Q = [M(^4\text{He}) + M(^{12}\text{C}) - M(^{16}\text{O})]c^2$$

$$Q = (4.002603 \text{ u} + 12.0 \text{ u} - 15.994915 \text{ u})c^2$$

$$= 0.007688 \text{ u} \cdot c^2$$

$$Q = 0.007688 \text{ u} \cdot c^2 \left(931.5 \frac{\text{MeV}}{\text{u} \cdot c^2} \right) = 7.2 \text{ MeV}$$

(b) Stars begin by burning hydrogen to form helium. As the helium is exhausted, the stars contract to higher densities and temperatures, which then allows helium to burn. Because there are no mass-5 and mass-8 stable nuclides, there is a hurdle to get beyond mass-4 nuclides. The hurdle is cleared only if the star is large enough so that as it contracts, helium can burn to create heavier masses. Our sun can produce carbon, but the burning of alpha particles and carbon requires temperatures as high as 300 million K in order to overcome the Coulomb barrier and undergo nuclear fusion. Our sun does not reach this temperature.

Nuclear Fusion on Earth

Many scientists believe that controlled nuclear fusion ultimately represents our best source of terrestrial energy. Among the several possible fusion reactions, three of the simplest involve the three isotopes of hydrogen.

$$^2\text{H} + {}^2\text{H} \rightarrow n + {}^3\text{He} \qquad Q = 3.3 \text{ MeV} \qquad (13.21)$$

$$^2\text{H} + {}^2\text{H} \rightarrow p + {}^3\text{H} \qquad Q = 4.0 \text{ MeV} \qquad (13.22)$$

$$^2\text{H} + {}^3\text{H} \rightarrow n + {}^4\text{He} \qquad Q = 17.6 \text{ MeV} \qquad (13.23)$$

Deuterium exists in vast quantities in seawater. If we estimate there are 10^{21} liters of water on Earth, the natural abundance of deuterium (0.015%) gives 10^{43} deuterons. These deuterons, when fused together in reaction (13.21), would produce over 10^{30} J of energy, enough to support present world energy consumption for a few billion years.

Three main conditions are necessary for controlled nuclear fusion:

1. The temperature must be hot enough to allow the ions, for example, deuterium and tritium, to overcome the Coulomb barrier and fuse their nuclei together. This requires a temperature of 100–200 million K.
2. The ions have to be confined together in close proximity to allow the ions to fuse. A suitable ion density is 2–3×10^{20} ions/m^3.
3. The ions must be held together in close proximity at high temperature long enough to avoid plasma cooling. A suitable time is 1–2 s.

The suitable values given above assume magnetic confinement, which we will discuss soon. The product of the plasma density n and the containment time τ must have a minimum value at a sufficiently high temperature in order to initiate fusion and produce as much energy as it consumes. The minimum value is

Lawson criterion
$$n\tau \geq 3 \times 10^{20} \text{ s/m}^3 \qquad (13.24)$$

This relation is called the *Lawson criterion* after the British physicist J. D. Lawson who first derived it in 1957. A triple product of $n\tau T$ called the *fusion product* is sometimes used (where T is the ion temperature).

Fusion product
$$n\tau T \geq 6 \times 10^{28} \text{ s} \cdot \text{K/m}^3 \quad \text{or} \quad 6 \times 10^{21} \text{ s} \cdot \text{keV/m}^3 \qquad (13.25)$$

The factor Q is used to represent the ratio of the power produced in the fusion reaction to the power required to produce the fusion (heat). This Q factor is not to be confused with the Q value. The breakeven point is $Q = 1$, and ignition occurs for $Q \gg 1$. For controlled fusion produced in the laboratory, temperatures on the order of 20 keV are satisfactory. For uncontrolled fusion ($Q = \infty$), as in the hydrogen bomb or "H-bomb," high temperatures and densities are achieved over a very brief time by using an atomic bomb as a trigger. Fusion bombs do not produce radiation effects nearly as severe as those of fission bombs, because the primary products (n, p, and ^{4}He) are not dangerously radioactive.

Unfortunately, there are enough thermonuclear warheads (H-bombs) to destroy most of life on Earth. Some scientists have predicted that even a limited nuclear war could produce so much dust in the Earth's atmosphere that sunlight would be partially blocked. This might lead to a "nuclear autumn" (or even "winter"), with a drop in temperature of up to 30°C that could last for several months.

Example **13.9**

Calculate the ignition temperature needed for the reaction (13.23).

Strategy We need to calculate how much thermal energy ($\frac{3}{2}kT$) is needed to overcome the Coulomb barrier. We will use 3 fm as the distance where the nuclear force first becomes effective. The particles must approach each other to at least this distance. We will assume that the minimum kinetic energy we need is the Coulomb potential energy at 3 fm.

Solution The charges of the participants in reaction (13.23) are both $+e$. The Coulomb potential energy that must be overcome is

$$V = \frac{q_1 q_2}{4\pi\epsilon_0 r}$$

$$= \frac{(9 \times 10^9 \text{ N} \cdot \text{m}^2/\text{C}^2)(1.6 \times 10^{-19} \text{ C})^2}{3 \times 10^{-15} \text{ m}} = 7.7 \times 10^{-14} \text{ J}$$

The thermal energy required is equal to $\frac{3}{2}kT$, so we have for the ignition temperature

$$T = \frac{2V}{3k} = \frac{2(7.7 \times 10^{-14} \text{ J})}{3(1.38 \times 10^{-23} \text{ J/K})} = 3.7 \times 10^9 \text{ K}$$

We have calculated that a temperature of almost 4 billion K (or °C) is needed to ignite the deuteron-triton (D + T) reaction. This is an overestimate for several reasons. First, a deuteron and a triton are extended objects, and their nucleons will probably feel an attraction before the centers of the nuclei are 3 fm apart. Second, the protons in D + T will tend to repel each other as the two nuclides approach, compared to the behavior of the neutrons. A more appropriate distance to use in the Coulomb potential energy could be a distance as great as 5 fm, which would result in a lower temperature. Third (and most important), the distribution of energies for plasma (ionized particles) in thermal equilibrium at temperature T follows a statistical process. We have used the mean energy, $\frac{3}{2}kT$. However, far out on the tail of the distribution (see Figure 9.7), there are many particles with energies several times greater than $\frac{3}{2}kT$. It only takes a few particles out of the total of 10^{20}–10^{22} particles/m^3 to initiate the reaction. Fourth, we have assumed that each particle needs $\frac{3}{2}kT$ energy. If the collision were head-on, then each of the D and T ions would need only half this energy. More accurate ignition temperature estimates for the D + T fusion reaction are in the range of 100–200 million K, which seems a reasonable correction to our original estimate.

Controlled Thermonuclear Reactions

A controlled thermonuclear reaction of nuclear fusion in the laboratory is one of the primary goals of science and engineering, because of the large potential of energy production. Scientists do not expect to reach this goal for several more decades. Because of the large amount of energy produced and the relatively small Coulomb barrier, the first fusion reaction will most likely be the D + T reaction, Equation (13.23). The tritium will be derived from two possible reactions:

$$n + {}^6\text{Li} \rightarrow {}^3\text{H} + {}^4\text{He} \tag{13.26}$$

$$n + {}^7\text{Li} \rightarrow {}^3\text{H} + {}^4\text{He} + n \tag{13.27}$$

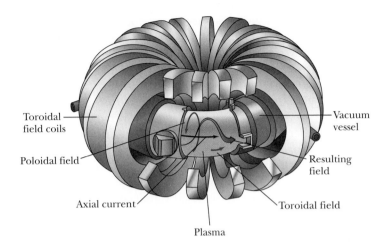

Figure 13.15 Diagram of a toroidal fusion device, the tokamak. The toroidal and poloidal fields are the most important magnetic fields used to contain the plasma inside the vacuum vessel.

The lithium is required to generate the tritium and is also used as the heat transfer medium and a neutron radiation shield.

For $Q = 1$ (breakeven), a product of $n\tau T$ of a few times 10^{21} keV·s·m^{-3} will be required for a commercial reactor using D + T. The problem of controlled fusion involves significant scientific and engineering difficulties. The two major schemes to control thermonuclear reactions are magnetic confinement fusion (MCF) and inertial confinement fusion (ICF). We will describe the main ideas of each.

Tokamak

Magnetic Confinement of Plasma The primary effort of research laboratories around the world for several years has been a device called the **tokamak**, first developed in the USSR in the 1960s. Several other MCF schemes have been tried. A schematic diagram of a tokamak is shown in Figure 13.15. As many as six separate magnetic fields may be used to contain and heat the plasma.

A schematic cross section of a typical magnetic containment vessel for a tokamak that might eventually be used as a commercial device is shown in Figure 13.16. The center is the location of the hot plasma where the D + T reaction takes place. The plasma is surrounded by vacuum to keep out impurities that would poison the reaction. A wall, which is subjected to intense radiation, surrounds the plasma and vacuum region. The plasma must be kept from touching the enclosure that contains the plasma, which is possibly the most hostile environment of any device yet designed. The next layer is a lithium blanket, which absorbs neutrons to breed more tritium. Next comes the radiation shield to prevent

Figure 13.16 Schematic cross section of a typical magnetic confinement vessel (a tokamak) that might be used for commercial power production. Heat absorbed in the blanket is used to produce electricity.

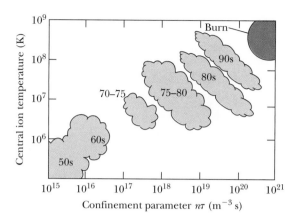

Figure 13.17 Plot of plasma ion temperature versus the Lawson confinement parameter $n\tau$. Numbers inside the plot indicate years during the twentieth century when the achievements occurred; note the gradual progress made since the 1950s.

radiation from reaching the magnets, which may be superconducting in a commercial reactor. The lithium blanket and radiation shield are not present in existing test machines.

The heating of the plasma to sufficiently high temperatures begins with the resistive heating from the electric current flowing in the plasma. Because this is insufficient to attain the high ignition temperature, there are two other schemes to add additional heat: (1) injection of high-energy (40–120 keV) neutral (so they pass through the magnetic field) fuel atoms that interact with the plasma, and (2) radio-frequency (RF) induction heating of the plasma (similar to a microwave oven).

The best results to date have come from the Princeton University Tokamak Fusion Test Reactor (TFTR) in the United States (shut down in 1997) and the Joint European Torus (JET) in England. Both fusion reactors have used tritium as fuel. In 1997 JET reached a power output of 16 MJ for 1 s and had a Q value of 0.65, the highest to date. We show in Figure 13.17 a plot of plasma ion temperature versus the confinement parameter $n\tau$. During the past few decades significant progress has been made in reaching the elusive values $Q = 1$ and higher needed for power production.

The most significant fusion project under consideration now is a large toka-mak fusion reactor called the International Thermonuclear Experimental Reactor (ITER). It is a partnership of the European Union, Japan, Russia, South Korea, China, and the United States. If it is actually ever built, the ITER is expected to generate self-sustained fusion power of 500 megawatts for as long as 500 s. A commercial power station is not envisioned until after the middle of this century. It might have the form shown in Figure 13.18.

ITER

Inertial Confinement The concept of inertial confinement fusion is to use an intense high-powered beam of heavy ions or light (laser) called a *driver* to implode a pea-sized target (a few mm in diameter) composed of D + T to a density and temperature high enough to cause fusion ignition. Some inertial confinement results are included in Figure 13.17. Several institutions in the United States are doing research and development in laser fusion including Lawrence Livermore National Laboratory and the University of Rochester. The National Ignition Facility at Livermore will use 192 lasers to create a thermonuclear burn for research purposes. It is currently under construction and is expected to be in operation in 2008. Sandia National Laboratories has used a device called a *Z-pinch* that uses a huge jolt of current to create a powerful magnetic field that squeezes ions into

Z-pinch

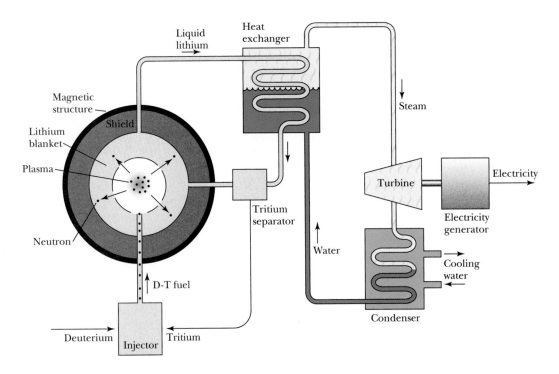

Figure 13.18 Diagram of a possible fusion reactor configuration using magnetic confinement. The tritium is obtained from neutron reactions with lithium, and the tritium must be separated before going back to the fusion reactor as fuel. As in fission reactors, the heat produced by fusion is used to run a turbine, which in turn generates electrical power.

implosion and heats the plasma. Sandia has proposed an upgrade that may be a serious contender in the fusion race.

13.7   Special Applications

In this chapter and the previous one we have mentioned several applications of nuclear science. Particle beams, radioactive nuclides, and nuclear effects have many applications. Most depend on some specific isotope of a radioactive element, called a **radioisotope**. The usefulness of a given radioisotope may depend on the specific decay particle it produces, for example, an α particle, β ray, γ ray, or fission fragment, and even on the half-life of the radioisotope. Radioisotopes are produced for useful purposes by different methods:

Radioisotopes

1. By particle accelerators as reaction products
2. In nuclear reactors as fission fragments or decay products
3. In nuclear reactors using neutron activation

An important accelerator-produced radioisotope is ^{67}Ga, which is useful in diagnostic medicine. An important fission fragment is ^{99}Mo, which beta decays to ^{99m}Tc, probably the most useful radioisotope in medicine. Neutron activation is used in a whole host of applications from crime detection to the production of nuclides heavier than ^{238}U.

One other important area of applications is the search for a very small concentration of a particular element, called a *trace* element. Irradiating an object suspected of containing a trace element may produce a radioactive nucleus that can be inspected for its particular decay, and even for the γ rays produced by excited states of the trace element. Such techniques have been invaluable in detecting minute quantities of trace elements for forensic science and environmental purposes.

Medicine

Radioisotopes are useful in medical research, diagnostics, and treatment. They have been used to study virtually every organ and tissue in the body. The β^- emitters ^{3}H, ^{14}C, and ^{32}P are all widely used in medical research. Over 1100 radioisotopes are available for clinical use. By far the most widely used is ^{99m}Tc, mentioned at the end of Section 12.7. An isomer of technetium with a half-life of 6 h, it produces a 140-keV gamma ray when it decays to the ground state of ^{99}Tc. A patient is given an intravenous injection of ^{99m}Tc, for example, TcO_4. The ^{99m}Tc is trapped by cancerous cells, for example in the thyroid or salivary glands, the choroid plexus of the brain, or the gastric mucosa. A few minutes after the injection, the patient is scanned by an array of NaI γ-ray detectors that are able to pinpoint the ^{99m}Tc activity. This is only one of several applications for ^{99m}Tc.

The nuclide ^{99}Mo is produced as a fission product in a nuclear reactor or as a *by-product material* in a nuclear reactor, formed by neutron activation of the stable isotope ^{98}Mo in the reaction ^{98}Mo$(n, \gamma)^{99}$Mo. The latter method is particularly easy because the ^{98}Mo can be inserted encapsulated into the reactor. After retrieval, some of the material has been converted to ^{99}Mo. After the radioisotope ^{99}Mo is separated out, it is shipped typically once a week by commercial companies producing radioisotopes to hospitals all over the country as a "^{99m}Tc generator" of about 10^{11} Bq. The ^{99}Mo beta decays to form the useful ^{99m}Tc.

$$^{99}_{42}\text{Mo} \rightarrow \beta^- + {}^{99m}_{43}\text{Tc} \qquad\qquad t_{1/2} = 66 \text{ h} \qquad\qquad (13.28)$$
$$\quad\qquad\qquad \hookrightarrow \gamma + {}^{99}_{43}\text{Tc} \qquad t_{1/2} = 6 \text{ h}, E_\gamma = 140 \text{ keV}$$

The radionuclide ^{123}I ($t_{1/2} = 13$ h) is useful for a thyroid function test. ^{123}I is produced by an accelerator in a reaction such as

$$\alpha + {}^{121}_{51}\text{Sb} \rightarrow 2n + {}^{123}_{53}\text{I} \qquad\qquad (13.29)$$

After being administered to the patient, ^{123}I decays by electron capture to excited states of $^{123}_{52}$Te.

$$^{123}_{53}\text{I} + e^- \rightarrow {}^{123}_{52}\text{Te}^* \qquad\qquad (13.30)$$
$$\qquad\qquad \hookrightarrow \gamma + {}^{123}_{52}\text{Te}$$

About 98% of the time a 159-keV γ ray is produced, which is detected by an array of NaI detectors.

The radionuclide $^{67}_{31}$Ga is useful as a tumor-localizing agent for Hodgkin's disease. The nuclide ^{67}Ga electron captures $^{67}_{30}$Zn*, which then gamma decays with energies of 93 keV, 185 keV, 300 keV, and 394 keV, among others.

Other useful radioisotopes are ^{153}Gd for detection of bone mineral loss or osteoporosis in elderly people; ^{192}Ir for a large number of cancer treatments and also for industrial radiography of welds in steel, oil well rigs, and pipelines; ^{241}Am for oil exploration and smoke alarms; and ^{85}Kr for leak testing of sealed electrical components such as transistors. Over 10,000 men in the United States are treated each year with ^{125}I seeds implanted into the prostate to inhibit cancer. In addition to ^{125}I, which has a half-life of 60 days, ^{103}Pd (17-day half-life) is also used. Both nuclides emit useful low-energy gamma-ray energies.

Figure 13.19 Photo of brain scan using PET and MRI combined. *Courtesy of C. Pelizzari, University of Chicago Radiation Oncology.*

Radioisotopes are also used in **tomography**, a technique for displaying images of practically any part of the body to look for abnormal physical shapes or for testing functional characteristics of organs. By using detectors (either surrounding the body or rotating around the body) together with computers, three-dimensional images of the body can be obtained. Tomography now includes various techniques—for example, single-photon emission computed tomography (SPECT), positron emission tomography (PET, see Section 3.9), and magnetic resonance imaging (MRI, see Section 10.6), among others. In the first two techniques the radiation (normally γ rays) is detected externally to determine the location of the radionuclide inside the body. The use of very short-lived *radiopharmaceuticals*, compounds produced with the radioisotopes, has resulted in about 50 accelerators placed in medical centers around the country to produce the necessary radioisotopes. A PET scan used in conjunction with a noninvasive MRI of the brain is shown in Figure 13.19.

Archaeology

The use of ^{14}C in radioactive dating has already been discussed in Chapter 12. Investigators can now measure a large number of trace elements in many ancient specimens and then compare the results with the concentrations of components having the same origin. Materials may include glass, metal, pottery, stone, minerals, paper, and fabric. For example, this technique has been used to examine pottery fragments excavated from the ancient Agora in Athens and from pyramids in Egypt.

Probably the greatest controversy in New World archaeology is the question of who the first settlers were and when they came to the Americas. Many experts believe they crossed a land bridge over the Bering Strait from Siberia to Alaska. Archaeologists found distinctive fluted spear points at a site near Clovis, New Mexico, in 1932. Subsequent ^{14}C radioactive dating indicates that humans had a settlement there 12,000 years ago. Several claims have surfaced in the past few years, especially from South America, that dispute this earliest finding, but no conclusive proof has been confirmed. The controversy rages on.

The Chauvet Cave, discovered in France in 1995, is one of the most important archaeological finds in decades. More than 300 paintings and engravings and many traces of human activity, including hearths, flintstones, and footprints, were found. These works are believed, from ^{14}C radioactive dating, to be from the Paleolithic era, some 32,000 years ago.

Art

Neutron activation

Neutron activation is a nondestructive technique that is becoming more widely used to examine oil paintings. A thermal neutron beam from a nuclear reactor is spread broadly and evenly over the painting. Several elements within the painting become radioactive. X-ray films sensitive to beta emissions from the radioactive nuclei are subsequently placed next to the painting for varying lengths of time. This method, called an *autoradiograph*, has been used by art historians to identify modern repairs of the painting as well as to see the underdrawings of the original figures in the painting. For example, it was used to examine Van Dyck's *Saint Rosalie Interceding for the Plague-Stricken of Palermo*, from the New York Metropolitan Museum of Art collection and revealed an over-painted self-portrait of Van Dyck himself.

Crime Detection

Neutron activation analysis is also useful to search for particular elements indicative of crime. Because this sophisticated technique requires access to a nuclear reactor, few police departments other than national agencies have this capability on a day-to-day basis.

The examination of gunshots by measuring trace amounts of barium and antimony from the gunpowder has proven to be 100 to 1000 times more sensitive than looking for the residue itself. Amounts as small as 0.005 μg of barium and 0.001 μg of antimony may be detected by (n, γ) techniques. Specialists are able to ascertain firing distances up to 2 m and whether a hole or slit in some material like cloth, flesh, wood, leather, and so on, was caused by a bullet. Sensitive trace elements include barium, antimony, lead (from the primer and the bullet), and copper (from jacketed bullets, the cartridge case, and the primer case).

Scientists are also able to detect toxic elements in hair by neutron activation analysis. Human head hair grows about 10 cm/y. Small amounts of arsenic and mercury may be detected, and the time of poisoning may even be determined. A famous study by several Scottish scientists examined samples of Napoleon's hair, taken during the last five years of his life and just after his death, for arsenic content. The beta emitter ^{76}As ($t_{1/2} = 26.4$ h) was studied after being activated in a nuclear reactor. Analysis of strands of hair as short as 1 mm (about 3 days' growth) showed that Napoleon had arsenic concentrations as high as 40 times normal. Some speculate that Napoleon was poisoned, but a more reasonable explanation is that his physicians were using arsenic to treat his many illnesses.

Was Napoleon poisoned?

Mining and Oil

Radioactive sources have long been useful in the petroleum industry. For example, when inserted in the oil in pipelines, they can signal a change in the product being shipped, or simply log the arrival time. Geologists and petroleum engineers use radioactive sources routinely to search for oil and gas. A source and detector are inserted down an exploratory drill hole to examine the material at different depths. Neutron sources called PuBe (plutonium and beryllium) or AmBe (americium and beryllium) are particularly useful, as shown in Figure 13.20. Small Van de Graaff accelerators producing 14-MeV neutrons have been inserted into a casing that has already been placed inside the borehole. Casing diameters are typically between 14 and 25 cm, leaving enough room for the spectroscopic tool. The neutrons activate nuclei in the material surrounding the borehole, and these nuclei produce gamma decays characteristic of the particular element. NaI and, more recently, cooled germanium detectors pick up characteristic elemental

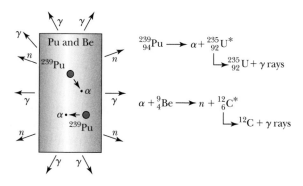

Figure 13.20 A plutonium-beryllium (PuBe) source consists of a can of beryllium powder mixed with ^{239}Pu, which α decays. The alpha interacts with the ^{9}Be and produces a neutron. Copious numbers of gamma rays also result, mostly from the decay of ^{235}U*.

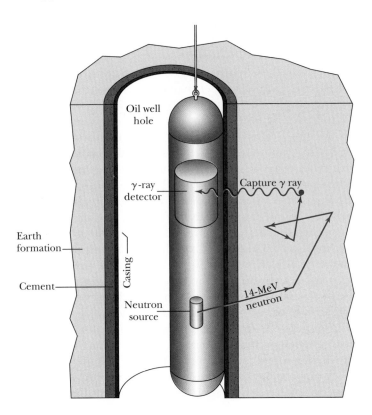

Figure 13.21 Example of a system used to search for oil and gas in a well boring. The (n, γ) reaction is used to examine for particular elements indicative of oil and gas.

decays when they pass through different geological formations. Especially of interest are the oil- and gas-bearing regions (see Figure 13.21).

Materials

The problem of radiation damage of an electronic device on a large single chip of silicon has long been recognized. For example, alpha-particle decay from a contaminant uranium or thorium nucleus could cause "soft" computer errors as it ionizes the silicon. Cosmic ray particles can create similar problems, especially in satellites and space probes. Scientists have discovered, however, that fast-neutron irradiation of bulk computer memory components can decrease the soft-error rate by a factor of 10. Apparently the neutrons reduce the intrinsic resistivity in the silicon substrate so that the extraneous ionization caused later is much less likely to reset a bit.

Phosphorus-doped silicon may also be produced with fast-neutron irradiation. Natural silicon consists of 3.1% of the isotope ^{30}Si, which undergoes the reaction

$$n + {}^{30}_{14}\text{Si} \rightarrow {}^{31}\text{Si} \tag{13.31}$$
$$\hookrightarrow \beta^- + {}^{31}_{15}\text{P} \qquad t_{1/2} = 2.62 \text{ h}$$

Because the ^{31}Si will be evenly distributed throughout the silicon, the phosphorus doping will also be much more uniformly distributed than could have been achieved by diffusive doping techniques. Silicon treated in this manner (called *neutron-transmutation-doped silicon*) is better able to handle high power levels in rectifiers, among several other uses.

Neutrons have long been used to study properties of materials, primarily because the neutron's wave nature and neutral charge allow it to easily probe atomic dimensions. Such effects include crystal structures, magnetic properties, inter-atomic and lattice forces, alloys, structures and dynamics of liquids, superconductors, phase transitions, and voids in materials and oil shale. Neutrons are particularly useful because they have no charge and do not ionize the material, as do charged particles and photons. They penetrate matter easily and introduce uniform lattice distortions or impurities. Because they have a magnetic dipole moment, neutrons can probe bulk magnetization and spin phenomena. Thermal neutrons have just the right momenta and energies to probe vibrational states, their acoustic modes, and the underlying interatomic forces in solid lattices.

Example 13.10

Neutrons are used to study structures of solids and their properties. What energy (and temperature) neutrons are needed if the atomic structures are of the size 0.06 nm?

Strategy We use the de Broglie wavelength relation to match the wavelength needed with an appropriate momentum. From the momentum we determine the kinetic energy and temperature from $K = \frac{3}{2}kT$.

Solution We determine the neutron momentum from the de Broglie wavelength relation with $\lambda = 0.06$ nm.

$$p = \frac{h}{\lambda} = \frac{6.63 \times 10^{-34}\,\text{J}\cdot\text{s}}{0.06 \times 10^{-9}\,\text{m}} = 1.1 \times 10^{-23}\,\text{kg}\cdot\text{m/s}$$

Because we expect the kinetic energy to be low, we use a nonrelativistic relation to determine the kinetic energy.

$$K = \frac{p^2}{2m} = \frac{(1.1 \times 10^{-23}\,\text{kg}\cdot\text{m/s})^2}{2(1.67 \times 10^{-27}\,\text{kg})} = 3.6 \times 10^{-20}\,\text{J}$$

$$= 0.23\,\text{eV}$$

Thus we see that the nonrelativistic relation is certainly adequate here. In thermal equilibrium, the temperature is found from $K = \frac{3}{2}kT$.

$$T = \frac{2K}{3k} = \frac{2(3.6 \times 10^{-20}\,\text{J})}{3(1.38 \times 10^{-23}\,\text{J/K})} = 1740\,\text{K}$$

Such energy is easily obtained by thermalizing neutrons from a nuclear reactor.

Small Power Systems

Alpha-emitting radioactive sources like ^{241}Am or ^{238}Pu have been used as power sources in heart pacemakers. Smoke detectors use ^{241}Am sources of alpha particles as current generators. The scattering of the alpha particles by the smoke particles reduces the current flowing to a sensitive solid-state device, which results in an alarm.

Spacecraft have been powered by radioisotope generators (RTGs) since the early 1960s. These devices use the heat produced by the α decay of ^{238}Pu to produce electricity in a thermocouple circuit. About two dozen U.S. spacecraft have used RTGs including Apollo, Pioneer 10 and 11, Viking 1 and 2, Voyager 1 and 2, Galileo, Ulysses, and more recently Cassini, which was launched in 1997, arrived at Saturn in 2004, and produced some fantastic photos. The Cassini mission is expected to end in 2008. Current RTGs supply almost 300 watts each, and usually more than one unit is placed on a spacecraft (Cassini has three). All past and present operational RTGs have exceeded their original design requirements both in power output and longevity. Voyagers 1 and 2 were both launched in 1977 and their RTGs are expected to be operational through at least 2020, a period of over 40 years! Voyager 1 is the most remote human emissary in space; at the beginning of 2005 it was 1.4×10^{10} km from Earth, and its signals took 14 hours to travel to Earth.

Special Topic

THE SEARCH FOR NEW ELEMENTS

Francium ($Z = 87$) was the last element to be discovered in nature (1939). It was found in the decay products of uranium by Marguerite Perey working at the Curie Laboratory in 1939. Even before this, the first synthesized element, **technetium** ($Z = 43$), was artificially created in a nuclear reaction in 1937 at the University of California, Berkeley, by Carlo Perrier and Emilio Segrè. After bombarding molybdenum with deuterons, the discoverers took the products back to Italy where technetium was identified. This discovery was made possible by the invention of the cyclotron at Berkeley by E. O. Lawrence, for whom the Lawrence Berkeley National Laboratory (LBNL) is named.

During the next three decades the discovery of new elements was dominated by researchers in Berkeley except during their relocation to the University of Chicago during World War II. **Astatine** ($Z = 85$) was discovered by Dale Corson and colleagues in 1940 at Berkeley by bombarding bismuth with α particles. Also in 1940 at Berkeley, Edwin McMillan and Philip Abelson discovered the first *transuranium* element, which is an element with more protons than uranium ($Z > 92$). They found **neptunium** ($Z = 93$) by bombarding uranium with neutrons. Similarly, in 1940 Glenn Seaborg, Arthur Wahl, and Joseph Kennedy discovered **plutonium** ($Z = 94$) by bombarding uranium with deuterons, which produced a heavy isotope of neptunium that beta decayed to plutonium. McMillan and Seaborg received the 1951 Nobel Prize in Chemistry. Plutonium was discovered in secret in 1940, and it was not announced publicly until 1946 because of its use in atomic bomb production in World War II. Plutonium has often been used as an energy source for deep space probes to produce electricity.

Americium ($Z = 95$) was discovered at Berkeley in 1944 by Seaborg and others by successive neutron capture reactions by plutonium in a nuclear reactor. Americium is used in smoke detectors. Using a different reaction, the next element, **curium** ($Z = 96$), was discovered by Seaborg, Ralph James, and Albert Ghiorso at Berkeley in 1944 by bombarding a tiny amount of plutonium with α particles.

Three elements (promethium, einsteinium, and fermium) were discovered during the development of nuclear weapons. **Promethium** ($Z = 61$) was the last of the "lighter elements" to be discovered. It was found at Oak Ridge, Tennessee, in 1947 by J. A. Marinsky, Lawrence Glendenin, and Charles Coryell by chemical identification of residues from a nuclear reactor. Earlier claims of the discovery of promethium in 1924 and 1941 were not substantiated. Both **einsteinium** and **fermium** (elements 99 and 100, respectively) were discovered in 1952 by a group of scientists led by Ghiorso from Argonne National Lab, Los Alamos National Lab, and Berkeley. They isolated the new elements from the radioactive debris from the first large hydrogen bomb test in the Pacific Ocean. The fermium isotope produced in the blast had been produced by 17 successive neutron captures in uranium, followed by beta decay.

After World War II several more new elements were discovered at Berkeley by various groups of scientists led by Seaborg and Ghiorso. These include **berkelium** (97) in 1949, **californium** (98) in 1950, **mendelevium** (101) in 1955, **lawrencium** (103) in 1961, **rutherfordium** (104) in 1969, **dubnium** (105) in 1967, and **seaborgium** (106) in 1974. All these new elements were synthesized by bombardment of transuranic elements by particles from cyclotrons. In the 1960s a new group of scientists doing particle accelerator experi-

Considerable research and development have been done in the United States on a small nuclear reactor to be used in space. The latest program was terminated in 1994 because there was no clear mission that would justify construction. Possible uses of a space reactor include outer planetary exploration, manned science outposts on the moon, and astronaut visits to Mars. Space nuclear power has always been highly controversial. The United States has put only one reactor in space (1965), but the former Soviet Union is reported to have ments at the Joint Institute for Nuclear Research in Dubna, Russia, came on the scene. They also claimed the discovery of elements 103, 104, 105, and 106. Some controversy about the discovery precedence led to a long delay in the naming of the new elements. The International Union of Pure and Applied Chemistry (IUPAC), together with the International Union of Pure and Applied Physics (IUPAP), is the international body generally recognized with the responsibility and authority to decide on the priority of element discovery and the naming of new elements. The discovery team (or teams) is normally requested by IUPAC to suggest a name for the element that IUPAC ultimately approves. The naming of elements 101 to 106 was not settled until 1997 by acknowledging the efforts of both labs.

An international team working at the Nobel Institute for Physics in Stockholm claimed the discovery of **nobelium** (102) in 1957 by bombarding curium with carbon ions. The Berkeley group announced in 1958 that they could not reproduce the results, and the Dubna group soon agreed with the Berkeley findings. Both Berkeley and Dubna were later able to make authenticated discoveries of element 102.

Since 1980 a team led by Peter Armbruster working at the Institute for Heavy Ion Research at Darmstadt, Germany, has dominated the search for discovery of new elements. This group bombarded targets with accelerated heavy ions, and in fairly quick succession discovered **bohrium** (107) in 1981, **hassium** (108) in 1984, and **meitnerium** (109) in 1982. This last element resulted from the bombardment of an iron target by high-energy bismuth ions. There was no controversy over which team first discovered elements 107, 108, and 109.

It was a decade later when the next element was discovered in 1994, and all three labs (Berkeley, Dubna, and Darmstadt) claimed the discovery of element 110 in accelerator experiments. Eventually a Joint Working Party (JWP) of IUPAC-IUPAP confirmed the German discovery of element 110 in 2001 and approved its name, **darmstadtium**.

The German group at Darmstadt discovered element 111 in 1994 and element 112 in 1996. However, the discovery of element 111 was not fully recognized by the JWP until 2003, and the JWP requires more confirmation before it accepts the discovery of element 112. Element 111 was named **roentgenium** in 2004. Scientists are now moving into the region where physicists predict the island of "magic numbers" of the shell model may result in some particularly stable nuclides around $Z = 114$, 120, or 126 and $N = 184$. A joint group between Lawrence Livermore National Lab and the Joint Institute for Nuclear Research in Dubna, Russia, announced in 2004 that they had discovered elements 113 and 115. A team from the same two laboratories claimed discovery of element 114 in 1999, but it has not been confirmed (only one atom was produced). In 1999 researchers at Berkeley reported the discovery of elements 116 and 118, but in 2002 the discovery was retracted as "a result of fabricated research data and scientific misconduct by one individual," according to the Berkeley lab director.

A collaboration of several international researchers working at Dubna in 2000 reported the discovery of element 116, but the discovery has not yet been confirmed. Experimenters have not yet been able to reach the magic number for neutrons, $N = 184$, and the next few years will be exciting as they strive to discover the hoped-for island of stability. See Figure 13.22 for the current status of superheavy nuclei as the high Z and N island of stability is approached. The future is bright for superheavy element experimenters.

---

placed almost three dozen, with some spectacular mishaps. The most famous one spread its radioactive debris across northwestern Canada in 1978.

New Elements

No **transuranic** elements—those with atomic number greater than $Z = 92$ (uranium)—are found in nature because of their short half-lives. However, with the use of reactors and especially accelerators, scientists have been able to produce

Figure 13.22 The element number Z (number of protons) is plotted versus the number of neutrons N for elements as heavy as lead and higher. The shades of blue indicate the level of the known or predicted stability. Note the island of stability that has been predicted around $N = 184$ and $Z = 114$. Physicists have been steadily discovering elements that approach the predicted island of stability during the last few decades (see dots for N in the mid-170s). *Adapted from* Science **303**, 740 (2004).

Glenn T. Seaborg (1912–1999) is shown here in 1951 at the University of California at Berkeley with the apparatus that he used to chemically separate out the newly created transuranic elements. Seaborg, a nuclear chemist, had a distinguished career in both research and public service, having served as chairman of the U.S. Atomic Energy Commission. *Courtesy of Lawrence Berkeley National Laboratory.*

22 of these new elements up to $Z = 116$ (depending on whether several of the discoveries past $Z = 111$ are confirmed). Over 150 new isotopes heavier than uranium have been discovered. Neptunium ($Z = 93$) and plutonium ($Z = 94$) were discovered in Berkeley in 1940. Glenn Seaborg (Nobel Prize in Chemistry, 1951) was able to reclassify the periodic table and place the actinide series under the lanthanide series. Seaborg, a nuclear chemist, used his knowledge of chemistry and physics to predict the chemical properties of all the elements with $Z = 97$ to 103.

Physicists have reasons to suspect from shell model calculations that superheavy elements with atomic numbers of 110–120 and 184 neutrons may be particularly long-lived. In particular, the element with $Z = 114$ and $N = 184$ might be "doubly magic." Figure 13.22 shows a plot of the known nuclei in the region of lead and heavier. Physicists are slowly approaching the predicted "island of stability" where it is hoped that superheavy elements can be produced with lifetimes suitable for extended study. These elements should have particularly interesting chemical and physical properties. Several scientific laboratories around the world, especially in the United States, Germany, and Russia, are involved in the search. See Special Topic, "The Search for New Elements" in this chapter for more information on these interesting but difficult experiments.

Certain isotopes including ^{252}Cf (californium) have been produced in sufficient quantities in high-flux reactors to be generally available for special applications. Discovered in 1953, ^{252}Cf has $t_{1/2} = 2.6$ y with both alpha emission (97%) and spontaneous fission (3%). Its high spontaneous fission rate makes it particularly useful as a source of fission fragments as well as neutrons. It is extremely useful in treating malignant tumors. A tiny amount of ^{252}Cf may be placed adjacent to the tumor, so that the body tissues surrounding the malignant tumor are not nearly as damaged as they can be in normal γ-ray radiotherapy projected from outside the body. The potential use of ^{252}Cf as a neutron source includes inspection of welds in aluminum aircraft, detection of corrosion effects in operating aircraft, oil-well logging of rock formation characteristics (see Figure 13.21), fuel-rod scans, fission waste monitors, copper and nickel ore analyzers, determination of sulfur content in coal or oil, moisture monitors, and cement analyzers.

Summary

The construction of accelerators in the 1930s heralded a new era for physicists, allowing them to more easily study the nuclear force by performing nuclear reactions. The energy released in a nuclear reaction $x + X \rightarrow y + Y$ is called the Q value and can be determined from the atomic masses.

$$Q = (M_x + M_X - M_y - M_Y)c^2 \qquad (13.7)$$

Different nuclear reaction mechanisms include compound nucleus, Coulomb excitation, and direct reactions, among others. Nuclei have excited states, which may appear as resonances in a compound nucleus reaction. The lifetimes τ of nuclear states are related to their widths Γ by the uncertainty principle, $\Gamma \tau \approx \hbar/2$.

Heavy nuclei may fission into two nearly equal fission fragments because of the increasingly large Coulomb force. Spontaneous fission occurs for nuclei with $Z^2/A \geq 49$, and fission can be induced with a nuclear reaction. The nuclide ^{235}U fissions with the absorption of a slow neutron. A self-sustaining chain reaction is possible, because fission produces neutrons that can cause another fission.

Nuclear reactors may be built in different ways for special purposes. A breeder reactor produces more fissionable fuel than it consumes. Nuclear fusion is an efficient energy source, and isotopes of hydrogen, ^{2}H and ^{3}H, appear to be the most useful. Although fusion is the source of our sun's energy, it has not yet been controlled on Earth.

Applications of nuclear science are plentiful and include medicine, archaeology, art, crime detection, mining, oil, material studies and production, and small power systems. The search for new elements continues to discover new elements in the quest for the island of stability for super-heavy elements.

Questions

1. Rutherford was able to initiate nuclear reactions with α particles before 1920. Why wasn't he able to initiate nuclear reactions with protons?

2. In Example 13.2 we learned that the ^{12}C(α, n)^{15}O cross section is much larger than the ^{12}C(α, p)^{15}N reaction for $E_\alpha = 14.6$ MeV. We believe this is evidence of a resonance in ^{16}O. If it is a resonance, why aren't both neutron and proton exit channels strongly populated? Why do we conclude the difference must be due to quantum numbers in the exit channel? Can the Coulomb barrier in the exit channels make a difference?

3. Why do the lifetimes of nuclear excited states decrease for higher excitation energies?

4. Why is the density of nuclear excited states larger for higher excitation energies?

5. Both deuterons and alpha particles can cause direct reactions by stripping. Which are more effective? Explain.

6. Discuss the changes in the cross section for neutron-induced and proton-induced reactions as the initial kinetic energy is decreased from 50 MeV. Ignore resonances.

7. Think about how a chain reaction could be controlled without delayed neutrons. Is it possible? What would be the difficulties?

8. Think carefully about the fissioning process. Does it seem peculiar that symmetric fission is not the most probable? Does the distribution shown in Figure 13.8 seem reasonable? Explain.

9. Why is it useful to slow down neutrons produced by fission in a nuclear reactor?

10. All the moderators mentioned in this chapter to slow down neutrons are light nuclei. Why are light nuclei used for moderators instead of heavy nuclei?

11. Why is fission fuel placed in 4-m-long rods placed parallel but separated, rather than in one lump of mass?

12. Discuss how each of the following sources of energy is ultimately derived from the sun: wood, coal, gas, oil, water, and wind.

13. Why does a star's temperature increase as fusion proceeds? Why does it require higher temperatures for the carbon cycle than for the proton-proton chain?

14. The fusion process continues until a star consists of nuclei near ^{56}Fe. Explain why this occurs.

15. The first wall of a magnetic fusion containment vessel has been said to contain the most hostile environment yet designed by man. Justify this statement.

16. Neutron-activation analysis is much more widely used than charged-particle activation. Why do you suppose that is true?

17. Explain in your own words the origin of the names of elements 97 through 102; that is, who or what the elements were named after and the reasons for doing so.

18. Explain in your own words the origin of the names of elements 103 through 108; that is, who or what the elements were named after and the reasons for doing so.

19. Explain in your own words the origin of the names of elements 109 through 114; that is, who or what the elements were named after and the reasons for doing so. You can skip those elements for which IUPAC has not yet officially assigned a name.

20. How many new elements have been discovered that are not mentioned in this textbook? Discuss them.

21. Small research nuclear reactors, like those mostly used in universities, are often submerged in concrete structures that look like swimming pools. The water serves as a moderator of the neutrons. They often have a blue glow in the swimming pool around the reactor. What is the origin of the blue color? *Hint:* Look up Cerenkov radiation.

Problems

Note: The more challenging problems are indicated by a shaded box.

13.1 Nuclear Reactions

1. Write down the precise nuclide identification for the missing element x for the following reactions.
(a) $^{16}O(d, x)^{14}N$,   (b) $^7Li(x, n)^7Be$,   (c) $^{13}C(\alpha, n)x$,
(d) $x(d, p)^{74}Ge$,   (e) $^{107}Ag(^3He, d)x$,   and
(f) $^{162}Dy(x, d)^{163}Ho$.

2. For each of the reactions listed in Problem 1, write down one other possible exit channel.

3. The cross section for a 2-MeV neutron being absorbed by a ^{238}U nucleus and producing fission is 0.6 barn. For a pure ^{238}U sample of thickness 3 cm, what is the probability of a neutron producing fission? ($\rho = 19$ g/cm^3 for uranium)

4. List at least three entrance channels using stable nuclei that can produce the exit channel $d + ^{20}Ne$.

5. The cross section for neutrons of energy 10 eV being captured by silver ($\rho = 10.5$ g/cm^3) is 17 barns. What is the probability of a neutron being captured as it passes through a layer of silver 2 mm thick?

6. In order to measure the cross section of the $^{12}C(\alpha, p)$ ^{15}N reaction of Example 13.2, a detector subtending a solid angle of 3×10^3 sr is used at the scattering angle θ. A 0.2-μA beam of 14.6-MeV α particles is incident on a ^{12}C target of thickness 100 μg/cm^2 for one hour. If the differential cross section is 0.2 mb/sr, how many protons are detected at the angle θ?

7. Write down the complete reaction for an ^{16}O target for the following reactions. List which products are stable. (a) (n, α), (b) (d, n), (c) (γ, p), (d) (α, p), (e) $(d, ^3He)$, (f) $(^7Li, p)$.

13.2 Reaction Kinematics

8. Calculate the ground state Q values for the following reactions. Are the reactions endothermic or exothermic? (a) $^{16}O(d, \alpha)^{14}N$, (b) $^{12}C(^{12}C, d)^{22}Na$, (c) $^{23}Na(p, ^{12}C)^{12}C$.

9. For the endothermic reactions of the previous problem, calculate the threshold kinetic energy.

10. A state in $^{16}O^*$ at an excitation energy of 9.63 MeV has a broad width $\Gamma = 510$ keV. It is indicated in Figure 13.6 by hatch marks. In the excitation functions for $^{12}C(\alpha, \alpha)^{12}C$ and $^{12}C(\alpha, \gamma)^{16}O$ shown in Figure 13.6, broad peaks reflect this state. (a) At what laboratory bombarding energy K_α will the resonance be observed? (The actual peak may be shifted slightly due to interference effects.) (b) What is the approximate lifetime of this excited state?

11. In a certain nuclear reaction initiated by 5.5-MeV α particles, the outgoing particles are measured to have kinetic energies of 1.1 MeV and 6.4 MeV. (a) What is the Q value of the reaction? (b) If exactly the same reaction were initiated by 10-MeV α particles, what is the Q value? (The outgoing energies will change.)

12. Calculate the Q value and threshold energy for the $^{20}Ne(\alpha, ^{12}C)^{12}C$ reaction. What will be the kinetic energies of the ^{12}C nuclides if the alpha particle initially has 45 MeV of kinetic energy in the lab?

13. The threshold kinetic energy is calculated nonrelativistically in Equation (13.10). For the reaction $A(a, b)$ B show that the threshold kinetic energy calculated relativistically is

$$K_{th} = -\frac{Q(m_a + m_A + m_b + m_B)}{2m_A}$$

14. A slow neutron is absorbed by ^{10}B in the reaction $^{10}B(n, \gamma)^{11}B$. What is the energy of the γ ray?

15. In a PuBe source, plutonium produces α particles of about 5 MeV. These α particles interact with beryllium by the $^9Be(\alpha, n)^{12}C$ reaction. How much kinetic energy do the reaction products have?

16. Calculate the ground state Q value and the threshold kinetic energy for the reactions (a) $^{16}O(\alpha, p)^{19}F$ and (b) $^{12}C(d, ^3He)^{11}B$.

17. ^{60}Co is produced by neutron activation of ^{59}Co placed in a nuclear reactor where the neutron flux is 10^{18} neutrons/m$^2\cdot$s. The cross section is 20 b, and the sample of ^{59}Co has mass 40 mg. (a) If the ^{59}Co is left in the reactor for one week, how many ^{60}Co nuclei are produced? (b) What would be the activity of the ^{60}Co? (density of cobalt = 8.9 g/cm^3) (c) Describe a procedure for producing 10^{14} Bq of ^{60}Co for medical use.

13.3 Reaction Mechanisms

18. Consider the reaction $X(x, y)Y$ depicted in Figure 13.4, where the target X is at rest. The energy of the center of mass is given by

$$K_{cm} = \frac{1}{2}(M_x + M_X)v_{cm}^2$$

where v_{cm} is the speed of the center of mass given by Equation (13.8). Show that the energy available in the center-of-mass system K'_{cm} is given by Equation (13.12):

$$K'_{cm} = K_{lab} - K_{cm} = \frac{M_X}{M_x + M_X}K_{lab}$$

where $K_{lab} = M_x v_x^2/2$.

19. A 7.7-MeV alpha particle initiates the $^{14}N(\alpha, p)^{17}O$ reaction in air. What is the excitation energy of the compound nucleus ^{18}F?

20. A 14-MeV neutron is captured by a ^{208}Pb nucleus. (a) At what excitation energy is the resulting ^{209}Pb? (b) What decay mechanism would you expect for this highly excited ^{209}Pb nucleus?

21. (a) Make an estimate for the Coulomb barrier that the alpha particle must overcome for the reaction $^{14}N(\alpha, p)^{17}O$ in Example 13.3. (b) Also make an estimate for the proton kinetic energy at a forward scattering angle. (c) Will the proton have enough energy to tunnel out of the nucleus?

22. The ground state of ^{17}Ne is unstable. Its half-life has been measured to be 109 ms. (a) What is the energy width of the state? (b) List two possible decay mechanisms.

23. The first excited state of ^{17}F is at 0.495 MeV. Can the $p + {}^{16}O$ reaction populate this state? Give your reasons.

24. ^{239}Pu absorbs a thermal neutron and the resulting nucleus gamma decays to the ground state. (a) What is the energy of the gamma ray? (b) What would be the energy of the gamma ray if a 1-MeV neutron is absorbed by ^{239}Pu at rest?

25. List as many nuclear reactions as you can that use deuterons and alpha particles for projectiles with stable targets that will populate ^{22}Ne as the final state in direct reactions.

13.4 Fission

26. Calculate how much energy is released when ^{239}Pu absorbs a thermal neutron and fissions in the reaction

$$n + {}^{239}_{94}Pu \rightarrow {}^{240}_{94}Pu^* \rightarrow {}^{95}_{40}Zr + {}^{142}_{54}Xe + 3n$$

27. A sample of shale contains 0.04% ^{238}U by weight. Calculate the number of spontaneous fissions in one day in a 10^6-kg pile of the shale by determining (a) the mass of ^{238}U present, (b) the number of ^{238}U atoms, (c) the fission activity, and finally (d) the number of fissions. The density of uranium is 19 g/cm^3. The spontaneous fission activity rate is 0.69 fissions/kg · s.

28. Calculate the percentage abundance of ^{235}U and ^{238}U 2 billion years ago if the abundance today is 0.7% and 99.3%, respectively. The higher percentage of ^{235}U probably allowed natural nuclear reactors to occur. Explain why such a reaction could not occur today.

29. Use the information in Figure 13.8 to write at least three common sets of fission fragments for the fission products of ^{236}U.

13.5 Fission Reactors

30. A fission reactor operates at the 1000-MWe level. Assume all this energy comes from the 200 MeV released by fission caused by thermal neutron absorption by ^{235}U. At what daily rate is the mass of ^{235}U used? (In practice, of course, the energy conversion is not 100% efficient, nor is all the ^{235}U in a fuel cell used.)

31. Calculate the energy released in kilowatt hours from the fission of 1 kg of ^{235}U. Compare this with the energy released from the combustion of 1 kg of coal. The heat of combustion of coal is about 29,000 Btu/kg.

13.6 Fusion

32. Neutrons in equilibrium with their surroundings at temperature T are called thermal neutrons and have an average kinetic energy $\frac{3}{2}kT$. Calculate the thermal neutron energy for (a) room temperature (300 K), and (b) the sun (15×10^6 K).

33. Determine the ground state Q values for each of the reactions in the carbon cycle and show that the overall energy released is the same as for the proton-proton chain (26.7 MeV).

34. There is a bottleneck in producing masses higher than ^4He, because there are no mass-5 or mass-8 stable nuclides. For older stars with high densities and high temperatures ($T > 100$ million K), three alpha particles can form ^{12}C. This occurs by two alpha particles first forming ^8Be, and ^8Be reacting with another alpha particle to form ^{12}C before ^8Be can decay back to two alpha particles. (a) Explain why this has to happen for very hot stars and high density. (b) Calculate how much energy is given up when three alpha particles form ^{12}C.

35. Following the triple-alpha process to form ^{12}C (see the previous problem), a variety of nuclear reactions can form heavier nuclide masses. In one of them, $^{16}O + {}^{16}O \rightarrow {}^{32}S + \gamma$, the temperature must be greater than about 3 billion K. (a) Why does the temperature have to be so high? (b) Calculate how much energy is released in the reaction. It is reactions like this that allow nuclei in the iron region to be formed.

36. One of the fusion reactions that goes on in massive stars is silicon burning, $^{28}Si + {}^{28}Si \rightarrow {}^{56}Ni + \gamma$. This reaction is how fusion reactions eventually reach the most stable iron/nickel region. It is also a precursor to the end of a star's life and may lead to a supernova. (a) Calculate the ignition temperature required for

this reaction. (b) How much energy is expended in this reaction?

37. One of the fusion reactions that goes on in massive stars is carbon burning, $^{12}C + {}^{12}C \rightarrow {}^{24}Mg + \gamma$. (a) Calculate the ignition temperature required for this reaction. (b) How much energy is expended in this reaction?

38. Assume that two-thirds of the Earth's surface is covered with water to an average depth of 3 km. Calculate how many nuclei of deuterium exist (^2H is 0.015% abundant). Estimate using reaction (13.22) how many joules of energy this represents through fusion.

39. The ignition temperature of fusion reactions is referred to both in temperature and in kinetic energy. (a) Explain why this is done. (b) What is the relation between the two? (c) At what temperature is the energy 6 keV?

40. The following reactions may be useful in producing energy for fusion reactions. Find their Q values. $^4He(^3He, \gamma)^7Be$, $^2H(d, p)^3H$, $^2H(p, \gamma)^3He$, $^{12}C(p, \gamma)$ ^{13}N, $^3He(^3He, pp)^4He$, $^7Li(p, \alpha)^4He$, $^3H(d, n)^4He$, $^3He(d, p)^4He$.

41. Determine how hot the environment must be in order for the first reaction of the CNO cycle to occur. (*Hint:* First find the threshold kinetic energy for the proton and the Coulomb barrier. After determining the kinetic energy, determine the temperature.)

42. One of the possibilities for producing energy in a star after the hydrogen has burned to helium is $3\alpha \rightarrow {}^{12}C$ (that is, three alpha particles react to form ^{12}C). How much energy is released in this process?

43. For a thermal neutron (300 K), what is its (a) energy, (b) speed, and (c) de Broglie wavelength?

13.7 Special Applications

44. In order to determine the wear of an automobile engine, a steel compression ring is placed in a nuclear reactor where it becomes neutron activated due to the formation of $^{59}Fe(t_{1/2} = 44.5$ days, $\beta^-)$. The activity of the ring when placed in the engine is 4×10^5 Bq. Over the next 60 days, the car is driven 100,000 km on a test track. The engine oil is extracted and the activity rate of the oil is measured to be 512 β^-/min. What fraction of the ring was worn off during the test?

45. (a) Why does a ^{99m}Tc generator need to be shipped once a week to hospitals? (b) What is the activity of a 10^{11}-Bq ^{99m}Tc generator source 9 days after it was produced? (c) If the activity is 0.9×10^{11} Bq on Monday morning when it arrives, what will be the activity at the same time on Friday morning, the last day of the working week?

46. The Los Angeles County police want to use neutron activation analysis to look for a tiny residue of barium in gunpowder. The suspected residue is placed in a nuclear reactor, where it is activated by the neutron flux. Natural barium contains 71.7% ^{138}Ba. The β^- emitter ^{139}Ba is produced in the $^{138}Ba(n, \gamma)^{139}Ba$ reaction. The half-life of ^{139}Ba is 83.1 min. ^{139}Ba beta decays to ^{139}La, 72% going to the ground state and 27% going to the first excited state at 0.166 MeV. Scientists think they need a count rate for the 166-keV γ ray (decay to the ground state) of at least 1000 Bq 30 min after the residue is removed from the reactor in order to make a positive identification of barium. (a) How many ^{139}Ba nuclei must be present at the end of the activation? (Remember the decay and fraction going to the first excited state.) (b) How many grams of ^{139}Ba must be produced? If the original amount of barium was 0.01 μg, what fraction of the ^{138}Ba was activated?

47. A 4.0×10^5 Bq ^{241}Am alpha source is used in a smoke alarm. The device is arranged so that 15% of the decay alphas are detected. (a) What current is detected? (b) If the introduction of smoke causes a 10% change in the intensity of the alpha particles, what sensitivity must the electronic circuit have to cause an alarm?

48. Consider a spacecraft's power source consisting of ^{210}Po, which emits a 5.3-MeV alpha particle, $t_{1/2} = 138$ days. (a) How many kg of ^{210}Po are needed to initially produce a power source of 5.0 kW? (b) If the power source must produce 7.0 kW after 2.0 years in space, how much ^{210}Po is needed?

49. A hospital has a 3.0×10^{14} Bq ^{60}Co source for cancer therapy. What is the rate of γ rays incident on a patient of area 0.30 m^2 located 4.0 m from the source? ^{60}Co emits a 1.1- and a 1.3-MeV γ ray for each disintegration.

50. Rework Example 13.10 if the neutron is to probe the diameter of a ^{238}U nucleus. Could neutrons from a nuclear reactor be used? Explain.

51. Assume that a 10.0-kg sample of ^{239}Pu is used to produce electrical power from its α decay. If your device is 60% efficient in producing electrical power, how much power can be produced?

52. We mentioned several superheavy elements that had been observed, but not yet confirmed or officially approved by the International Union of Pure and Applied Chemistry. (a) List those elements. (b) Research and discuss their status: Have they been confirmed? Has IUPAC approved them?

53. An inflated catheter is used in balloon angioplasty to open up arteries that are occluded with plaque formation. Stents are placed in the arteries to support the arterial wall. Radioisotopes have been incorporated into the stents to inhibit the reclosing of the artery (called *restenosis*). Almost a half million patients in the United States receive intravascular therapy each year. (a) Research the current status of using radioisotopes in this process. How many patients are treated in the United States each year using it? (b) Which radioisotopes are primarily used? Are they beta or gamma emitters? Why would one be favored over the other?

General Problems

54. Compare the following: (a) total atomic binding energy of 1.0 kg of hydrogen atoms, (b) nuclear binding energy of 1.0 kg of deuterons, (c) annihilation energy of 0.50 kg of protons with 0.50 kg of antiprotons.

55. One method used to determine unknown atomic masses consists of precisely measuring the kinetic energies of the particles involved in a nuclear reaction and using known atomic masses. The mass of ^{34}Si is determined by the ^{30}Si(^{18}O, ^{14}O)^{34}Si reaction initiated by 100-MeV ^{18}O particles. The outgoing particles have 86.63 MeV of energy, which can be determined only by measuring the ^{14}O energy and using the conservation of momentum and energy. (a) What is the Q value of the reaction? (b) What is the mass of ^{34}Si assuming the other three masses involved are known (see Appendix 8)?

56. ^{90}Sr is one of the most deadly products of nuclear fission. Assume that 4% of the fission fragment yield from a ^{235}U atomic bomb is ^{90}Sr. In a nuclear interchange on the planet Inhospitable, 1000 atomic bombs, each corresponding to the fission of 100 kg of ^{235}U, are detonated. (a) How many atoms of ^{90}Sr are released? (b) Assuming the ^{90}Sr is spread evenly over the planet of diameter 12,000 km, what is the resulting activity for each m^2? The half-life of ^{90}Sr is 28.8 y.

57. Assume a temperature of 2.0×10^8 K in a controlled thermonuclear reactor. (a) Calculate the most probable energy of deuterons at this temperature. (b) Use the Maxwell-Boltzmann distribution from Chapter 9 to determine the fraction of deuterons having an energy that is 2, 5, and 10 times the most probable energy.

58. A PuBe source has a neutron activity of 10^5 Bq. The neutrons are produced by the ^{9}Be(α, n)^{12}C reaction with an effective cross section of 90 mb and thickness of 3 cm. (a) What is the probability of an incident alpha particle interacting with a ^{9}Be nucleus? (b) What must be the rate of alpha particles incident on ^{9}Be? (c) What must be the amount of mass of the ^{239}Pu producing the alpha particles?

59. A typical person of mass 70 kg contains 0.35% potassium, by weight. Of the potassium, 0.012% is ^{40}K, an unstable nucleus which decays through β^- (89.3%) and electron capture (10.7%) with a $t_{1/2} = 1.28 \times 10^9$ y. What is the ^{40}K activity due to β^- decay in a typical person's body?

60. The yields of nuclear fission bomb weapons are measured in terms of the equivalent amount of energy produced by 1 kiloton of TNT (1 kiloton TNT = 4.2×10^{12} J). The bomb dropped on Hiroshima, Japan, on August 6, 1945, was believed to yield 15–20 kilotons. Assume that the bomb yield was 15 kilotons of TNT and that each fission yields 200 MeV. What is the minimum mass of ^{235}U that this bomb (called "Little Boy") could have contained?

Elementary Particles

If I could remember the names of all these particles, I'd be a botanist.

Enrico Fermi

I have done a terrible thing: I have postulated a particle that cannot be detected.

Wolfgang Pauli (after postulating the existence of the neutrino)

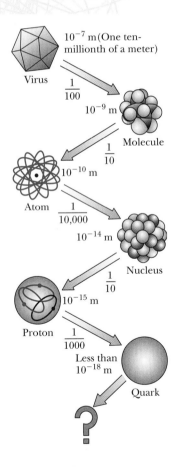

Figure 14.1 Starting from a virus, the structure of matter can be divided into smaller and smaller entities down to the quark and to whatever lies beyond. *Courtesy of Universities Research Association.*

10^{-7} m (One ten-millionth of a meter)

Virus

$\frac{1}{100}$

10^{-9} m

Molecule

$\frac{1}{10}$

10^{-10} m

Atom

$\frac{1}{10,000}$

10^{-14} m

Nucleus

$\frac{1}{10}$

10^{-15} m

Proton

$\frac{1}{1000}$

Less than 10^{-18} m

Quark

?

We began our study of subatomic physics in Chapter 12. We investigated the nucleus in Chapters 12 and 13. Now we want to delve deeper, because finding answers to some of the basic questions about nature is a foremost goal of science: What are the basic building blocks of matter? What is inside the nucleus? What are the forces that hold matter together? How did the universe begin? Will the universe end, and if so, how and when?

We try to use the ideas, concepts, and laws of physics to answer these questions. The ancient Greek philosophers, among them Aristotle, supposed things were made of earth, fire, wind, and water. Democritus derived the word *atom* from the Greek word *atomos*, which refers to an object that cannot be further cut into pieces. So the atom was meant to be the smallest indivisible unit of matter. In this book we have shown that matter is made of molecules and atoms, and electromagnetic forces are responsible for holding atoms together. At a smaller level, an atom is made of electrons and a nucleus. The electromagnetic force is responsible for attracting the electrons to the nucleus, while the strong (nuclear) force is responsible for keeping neutrons and protons together in the nucleus. Previously we have thought of electrons, neutrons, and protons as **elementary particles**, because we believe they are basic building blocks of matter. However, in this chapter we learn that the term elementary particle is used loosely to refer to hundreds of particles, most of which are unstable. Adding energy to neutrons and protons can produce new elementary particles. Some of these particles are indeed very crucial to our understanding of matter and of the forces that hold matter together. We see in this chapter that neutrons and protons are made up of even more fundamental particles called *quarks*. Although quarks cannot be observed outside the nucleus, we believe they must exist in order to explain experimental data. How far can this division into smaller and smaller units of matter continue? (See Figure 14.1.)

We will find in this chapter, and especially in Chapter 16 on cosmology, that there is an intertwined connection between the oldest science, astronomy with its grand dimensions, and particle physics, in which matter is studied in the tiniest dimensions. The question of how the universe started and how it may end is in fact buried in the mysteries of elementary particle physics. For this reason many physicists believe that particle physics is at the forefront of physics, and indeed of all science. The mysteries are deep and profound, but finding the answers gives us great satisfaction. Although the subject is complex and requires the meshing of relativity, quantum mechanics, electrodynamics, and gravitation, we hope to convey some of the excitement and flavor of particle physics. This excitement is even evident in the tools that experimental physicists use. In the last section of this chapter we discuss some of the accelerators used by physicists, especially by nuclear and particle physicists.

14.1   Early Discoveries

In 1930 the known elementary particles were the proton, electron, and photon. Thomson had identified the electron in 1897, and Einstein's work on the photoelectric effect can be said to have defined the photon (originally called a *quantum*) in 1905. The proton is the nucleus of the hydrogen atom. Despite the rapid progress of physics in the first couple of decades of the twentieth century, no more elementary particles were discovered until 1932, when Chadwick proved the existence of the neutron, and Carl Anderson identified the positron in cosmic rays.

The Positron

Paul A. M. Dirac (1902–1984), a British theoretical physicist, received his bachelor's degree in electrical engineering and doctoral degree in mathematics (see Dirac's photo and biography in Chapter 9). His training and brilliant insight into nature allowed him to make many contributions to physics, including his own general form of quantum mechanics, as well as showing that Heisenberg's matrix mechanics and Schrödinger's wave mechanics were special cases of his own general theory. Dirac also generalized the concept of "action-at-a-distance." When discussing gravitation and electromagnetism, the concept of fields is useful to understand the forces on objects in the external fields. Dirac developed the early form of quantum electrodynamics in which the absorption and emission of photons is a quantum process of the radiation field itself. His quantum electrodynamics (QED) theory was later generalized by Richard Feynman, Julian Schwinger, and Sin-itiro Tomonaga to produce the most accurately tested theory of physics today. For their work Feynman, Schwinger, and Tomonaga were awarded the 1965 Nobel Prize for Physics.

Perhaps Dirac's greatest success was his 1928 relativistic theory of the electron, for which he received the Nobel Prize for Physics in 1933. When Dirac used his considerable mathematical skills to combine quantum mechanics with relativity, he found that his wave equation had negative, as well as positive, energy solutions. His theory can be interpreted as a vacuum being filled with an infinite sea of electrons with negative energies. If enough energy is transferred to the sea, an electron can be ejected with positive energy. This leaves behind a hole that is the *positron*, denoted by e^+. Notice the analogy with semiconductors where holes appear instead of positrons. Dirac's theory, along with refinements made by other

Carl D. Anderson (1905–1991) was born in New York City and educated at the California Institute of Technology (B.S., 1927; Ph.D., 1930). He began cosmic ray studies in 1930 with his graduate advisor Professor Robert Millikan, which led to the discovery of the positron in 1932 for which he received the Nobel Prize for Physics in 1936. He also discovered the muon in 1936. He spent his entire career at Caltech. *Bettmann/Corbis.*

Quantum electrodynamics

Figure 14.2 Example of a Feynman spacetime diagram. Electrons interact through mediation of a photon. The axes are normally omitted.

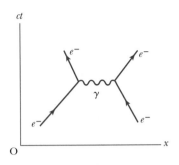

Hideki Yukawa (1907–1981) was born in Tokyo and raised in Kyoto. His early career was spent at Kyoto University and Osaka University. In 1935 he published his famous paper that proposed a new field theory of nuclear forces and predicted the existence of the meson, for which he received the Nobel Prize for Physics in 1949. He spent time in the United States in the 1940s and 1950s at Princeton and Columbia Universities. © *The Nobel Foundation.*

scientists, opened the possibility of **antiparticles**, which have the same mass and lifetime as their associated particles, and the same magnitude, but the opposite sign for such physical quantities as electric charge and various quantum numbers. All particles, even neutral ones (with some notable exceptions like the neutral pion), have antiparticles (as discussed further in Section 14.3).

Cosmic rays are highly energetic particles, mostly protons, that cross interstellar space and enter the Earth's atmosphere, where their interaction with particles creates cosmic "showers" of many distinct particles. Cosmic rays contain the highest particle energies ever observed (up to 10^{21} eV), although they normally are in the GeV (10^9 eV) range. The Austrian physicist V. F. Hess (1883–1964; Nobel Prize for Physics, 1936) discovered cosmic rays in 1912 by detecting them at high altitudes during balloon flights. In 1932 the American physicist Carl D. Anderson discovered positrons by observing the paths of cosmic ray showers passing through a cloud chamber placed in a magnetic field.

The ultimate fate of positrons (antielectrons) is annihilation with electrons. After a positron slows down by passing through matter, it is attracted by the Coulomb force to an electron, where it annihilates through the reaction

$$e^+ + e^- \rightarrow 2\gamma \tag{14.1}$$

We have already discussed this reaction in Chapter 3. The characteristic of positron annihilation is the emission of two oppositely directed 0.511-MeV gamma rays with the kinetic energy coming from the rest energies of the electron and positron. The kinetic energies of the charged particles are usually very small when they meet and are neglected.

Feynman presented a particularly simple graphical technique to describe interactions. For example, when two electrons approach each other, according to the quantum theory of fields, they exchange a series of photons. These photons are called *virtual*, because they cannot be directly observed. The action of the electromagnetic field (for example, the Coulomb force) can be interpreted as the exchange of photons. In this case we say that the photons are the *carriers* or *mediators* of the electromagnetic force. The procedure is represented schematically in a spacetime diagram like that shown in Figure 14.2, called a **Feynman diagram**, in which a photon is exchanged between two electrons. Feynman diagrams have been generalized to represent quite complicated interactions. The rules are quite simple. Straight lines represent trajectories of particles with mass; wavy lines represent the force carriers.

Yukawa's Meson

The idea of the photon being the mediator of the electromagnetic force had been discussed by several physicists besides Dirac, including H. Bethe, E. Fermi, C. Møller, and G. Breit. The Japanese physicist Hideki Yukawa had the idea of developing a quantum field theory that would describe the force between nucleons analogous to the electromagnetic force. In order to do this, he had to determine the carrier or mediator of the nuclear strong force analogous to the photon in the electromagnetic force. What particle is exchanged between nucleons to keep the nucleons strongly attracted?

In this case the mediating particle is called a meson, derived from the Greek word *meso*, which means "middle." The attraction of nucleons depends on the exchange of a virtual meson, much like the exchange of a virtual photon in the Coulomb attraction of an electron and proton. The energy ΔE required to create the meson is $\Delta E = m_\pi c^2$ where m_π is the meson mass. The Feynman diagram indicating the meson exchange between a neutron and a proton is displayed in Figure 14.3. Quantum theory shows that if the range of the force is small (10^{-15} m), the mass of the mediating particle must be large.

Example 14.1

Use the Heisenberg uncertainty principle to estimate the mass of the meson if the range of the nuclear force R_N is about 1 fm.

Strategy The uncertainty principle allows energy conservation to be violated during a short time Δt. We will let energy conservation be violated to the extent of the excess energy ΔE required to create the meson. The fastest speed possible for the meson is c, so the distance the meson travels in time Δt is $c\,\Delta t$. This distance $c\,\Delta t$ must be about the range of the nuclear force R_N.

Solution The energy ΔE and the time Δt are related by the uncertainty principle, $\Delta E\,\Delta t \approx \hbar/2$, so Δt is given by $\Delta t \approx \hbar/(2\,\Delta E)$. The value of ΔE must be at least as large as is needed to create the mass particle, the meson in this case, so $\Delta E = m_\pi c^2$. We combine these results to give

$$\Delta t \approx \frac{\hbar}{2\,\Delta E} = \frac{\hbar}{2 m_\pi c^2} \qquad (14.2)$$

We solve Equation (14.2) for the meson mass and use $R_N = c\,\Delta t$:

$$m_\pi c^2 = \frac{\hbar}{2\,\Delta t} = \frac{\hbar c}{2 R_N} \qquad (14.3)$$

If the mean range of the nuclear force is about 1 fm, we have

$$m_\pi c^2 = \frac{1.973 \times 10^2 \text{ eV}\cdot\text{nm}}{2 \times 10^{-15} \text{ m}} \approx 100 \text{ MeV}$$

Equation (14.3) is useful to relate the effective length of any force R and the mass m that mediates the force. We write R as

$$R = \frac{\hbar}{2mc} = \frac{\hbar c}{2mc^2} \qquad (14.4)$$

In 1935 Yukawa predicted a particle mass of about 200 electron masses for the virtual meson or about 100 MeV. No accelerators at the time could create such a high-energy particle, but some physicists believed that mesons could be observed in cosmic rays.

In 1938 Carl Anderson and his collaborators observed a particle in cosmic radiation that at first was believed to be Yukawa's particle. Initially called a mu-meson, it proved not to be a member of the meson family at all and is now called a **muon**. It has a mass of 106 MeV/c^2, but subsequent experiments showed that it did not interact strongly with the nucleus. This new particle could not be the propagator of the strong force; it interacts only through the weak and electromagnetic interactions.

Yukawa's meson, called a **pion** (or *pi-meson* or π-meson), was finally identified in 1947 by C. F. Powell (1903–1969) and G. P. Occhialini (1907–1993). Charged pions $\pi^\pm$ have masses of 140 MeV/c^2, and a neutral pion π^0 was later discovered that has a mass of 135 MeV/c^2. Yukawa said the spin of the meson must be zero (or integral), which the pion has, but the muon has spin 1/2. We will see later that neutrons, protons, and pions are made of quarks, and that the nuclear interaction actually takes place between the quarks. It is preferable to speak of the "strong" force rather than the "nuclear" force, because the interaction is at a level more basic than the nucleus.

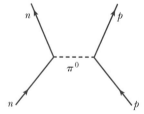

Figure 14.3 A Feynman diagram indicating the exchange of a pion (Yukawa's meson) between a neutron and a proton.

14.2 The Fundamental Interactions

In Chapter 1 we discussed the fundamental forces in nature responsible for all interactions. Those forces include the gravitational, electroweak, and strong forces. For all practical purposes, there are really four fundamental forces, because we often treat the electromagnetic and weak interactions as separate despite the fact that the electroweak interaction is a unification of the electromagnetic and weak interactions, just as the electromagnetic interaction is a unification of electric and magnetic interactions. The electromagnetic and weak forces act independently except at particle energies available only in cosmic rays, produced by accelerators, or in the early stages of the creation of the universe. We will discuss this further in Chapter 16.

We have learned that the fundamental forces act through the exchange or mediation of particles according to the quantum theory of fields. The exchanged particle in the electromagnetic interaction is the photon. All particles having either electric charge or a magnetic moment (and also the photon) interact with the electromagnetic interaction. The electromagnetic interaction has very long range.

In the 1960s Sheldon Glashow, Steven Weinberg, and Abdus Salam (Nobel Prize for Physics, 1979) predicted that particles, which they called W (for weak) and Z, should exist that are responsible for the weak interaction. This theory, called the *electroweak* theory, unified the electromagnetic and weak interactions much as Maxwell had unified electricity and magnetism into the electromagnetic theory a hundred years earlier. The details of the electroweak theory are too complex to be discussed here, but we know the W particle must be massive, because the weak interaction is very short range (remember the uncertainty principle argument in Example 14.1, $R = \hbar/2mc$). The W is a boson (integral spin). There are three versions of it: $W^{\pm}$ (mass 80.4 GeV/c^2) and Z (mass 91.2 GeV/c^2), all with spin 1. In 1983 these massive bosons were first observed at CERN (the European Organization for Nuclear Research) by a group led by C. Rubbia (Nobel Prize for Physics, 1984; see Special Topic, "Experimental Ingenuity," page 534), confirming the theoretical predictions and the unification of the interactions. A Feynman diagram of the neutron beta decay is displayed in Figure 14.4, showing the W^- as the carrier of the interaction.

Figure 14.4 A Feynman diagram indicating the beta decay of a neutron. Note that the W^- mediates this beta decay.

We previously mentioned (Section 14.1) that Yukawa's pion is responsible for the nuclear force. Now we know there are other mesons that interact with the strong force. Later, in Section 14.5, we will see that the nucleons and mesons are part of a general group of particles formed from even more fundamental particles called **quarks**. The particle that mediates the strong interaction between quarks is called a **gluon** (for the "glue" that holds the quarks together); it is massless and has spin 1, just like the photon. Thus, we believe that at a more fundamental level it is the gluons that are responsible for the strong force. No one has observed an isolated quark or a gluon, because they stay hidden within particles, confined by the strong force. Experimental evidence (to be presented in Section 14.5) convinces us that quarks and gluons exist. Particles that interact by the strong interaction are called **hadrons**; examples include the neutron, proton, and mesons.

Gluons

Hadrons

Gravitons

It has been suggested that the particle responsible for the gravitational interaction be called a **graviton**. The graviton is the mediator of gravity in quantum field theory and has been postulated because of the success of the photon in quantum electrodynamics theory. It must be massless, travel at the speed of light, have spin 2, and interact with all particles that have mass-energy. The graviton has never been observed because of its extremely weak interaction with objects.

The four interactions are shown schematically in Figure 14.5, and the mediators are listed in Table 14.1. Note that all the mediating particles are bosons with integral spin.

The quest to unify all the fundamental forces of nature into one theory has been an ultimate goal of some physicists for over a century. Maxwell unified the electric and magnetic forces into the electromagnetic theory in the 1860s. Glashow, Weinberg, and Salam unified the electromagnetic and weak interactions into the electroweak interaction in the 1960s. But the unsuccessful attempts are legend. Dirac badly wanted the proton to be the particle in the negative energy states of the electron in his relativistic electrodynamics theory of 1928. His theory would then encompass all the particles then known (proton, electron, and photon). Heisenberg was purported to have discovered a unified theory in the 1950s, but the details were never forthcoming. This rumor prompted Pauli's famous work of art shown in Figure 14.6 and his quote that "the details remain to be sketched in."

The Standard Model The most widely accepted theory of elementary particle physics at present is the **Standard Model**. It is a simple, comprehensive theory that explains hundreds of particles and complex interactions with six quarks,

Abdus Salam (1926–1996), Sheldon Glashow (1932–), and Steven Weinberg (1933–) received the Nobel Prize for Physics in 1979 for formulating the electroweak theory, which unifies the electromagnetic and weak interactions. Salam was born and educated in Pakistan before going to Cambridge University, where he received his Ph.D. in 1952. After returning to Pakistan for a few years, he worked at Cambridge and then the Imperial College, London, in 1957. In 1964 he helped found and became the director for the International Center for Theoretical Physics in Trieste, Italy. Glashow and Weinberg were both born in New York City, attended the Bronx High School of Science, and graduated from Cornell University in 1954. After Glashow received his Ph.D. from Harvard in 1959, he served on the faculties of the University of California at Berkeley and Harvard. After graduating from Cornell, Weinberg spent a year at the Niels Bohr Institute in Copenhagen. He then received his Ph.D. from Princeton in 1957 and spent time on the faculties of Columbia, UC-Berkeley, MIT, and Harvard before moving to the University of Texas, Austin, in 1982. *Left: © The Nobel Foundation; center: AIP Emilio Segrè Visual Archives, Physics Today Collection; right: AIP Emilio Segrè Visual Archives, Weber Collection.*

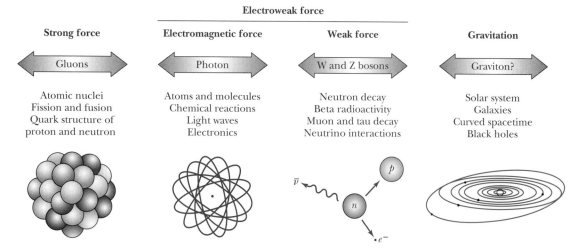

Figure 14.5 Some manifestations of the fundamental forces of nature. The mediating particles are shown as well as the areas in which the forces are effective. *Courtesy Universities Research Association.*

Table 14.1 **The Fundamental Interactions**			
Interaction	Relative Strength	Range	Mediating Particle
Strong	1	10^{-15} m	Gluons
Electroweak:			
Electromagnetic	10^{-2}	∞	Photons
Weak	10^{-6}	10^{-18} m	$W^{\pm}$, Z bosons
Gravitation	10^{-43}	∞	Graviton

Figure 14.6 Wolfgang Pauli was a skeptic of elementary particle theories that purported to explain everything. Once, in the 1950s when a rumor of such a theory by Werner Heisenberg was circulating, Pauli noted that the details of Heisenberg's theory remained to be sketched in. He drew the figure shown and announced, "Below is the proof that I am as great an artist as Rembrandt; the details remain to be sketched in." *From O. W. Greenberg,* American Scientist *76, 361 (1988).*

six leptons, and three force-mediating particles. It is a combination of the electroweak theory and quantum chromodynamics (QCD), but does not include gravity. The Standard Model was developed in the 1960s and 1970s and verified by experimental studies in the 1980s. It predicted the existence of the W and Z bosons, the gluon, charmed quark, and top quark, all of which were subsequently found with their expected properties.

The details of the Standard Model are too complicated to present here, but much of the next few sections is based on it. The only missing parameter of the Standard Model is the mass of the Higgs boson (see next section), which is under extensive investigation. The model has been so successful in elementary particle physics that disagreements with its predictions now make news, not its successes. There are now enough disagreements with the Standard Model to indicate that it is an approximation of a yet more fundamental theory (see Section 14.7). The particle classifications of Section 14.3, the conservation laws and symmetries described in Section 14.4, the discussion of quarks in Section 14.5, and the families of matter in Section 14.6 are all based on and consistent with the Standard Model unless otherwise stated.

Example 14.2

Use the mass of the W^{-} particle to estimate the range of the weak interaction responsible for the neutron beta decay.

Strategy This calculation is much like the one we did in Example 14.1, when we discussed Yukawa's meson. In that case we knew the nuclear force range, but in this case, the force range is unknown. We use Equation (14.4) to find the force range R_{W}.

Solution The range R_{W} becomes

$$R_{\mathrm{W}} = \frac{\hbar c}{2m_{\mathrm{W}}c^2}$$

$$= \frac{1.973 \times 10^2 \text{ eV} \cdot \text{nm}}{2(80.4 \text{ GeV}/c^2)(c^2)} = 1.2 \times 10^{-18} \text{ m}$$

In this case, it may not be true that the W^{-} travels near the speed of light, because it is such a massive particle. The calculation here is therefore an upper limit. Note that in this case the violation of the conservation of energy is massive, so the violation must occur over a very short period of time. We calculate Δt from Equation (14.2) to be

$$\Delta t = \frac{\hbar}{2m_{\mathrm{W}}c^2}$$

$$\Delta t = \frac{1.055 \times 10^{-34} \text{ J} \cdot \text{s}}{2(80.4 \text{ GeV})} \left(\frac{1 \text{ GeV}}{1.6 \times 10^{-10} \text{ J}} \right) \approx 4 \times 10^{-27} \text{ s}$$

The lifetime of the neutron is much longer than the time it takes for the decay process itself.

14.3 Classification of Elementary Particles

As the number of elementary particles discovered continued to increase in the 1950s and 1960s, physicists proposed various schemes to make sense of what some referred to as the elementary particle "zoo." Since the 1960s experimentalists have needed accelerators with ever-increasing energies in order to test elementary particle theories. Eventually, an understanding of the zoo developed due to many contributions. In this section we present a compact picture of the organization of elementary particles that represents some of the most important advances in particle physics during the past 40 years.

We discussed in Chapter 9 that particles with half-integral spin are called *fermions* and those with integral spin are called *bosons*. This is a particularly useful way to classify elementary particles because *all stable matter in the universe appears to be composed, at some level, of constituent fermions*. We have already discussed some bosons in the previous section. Photons, gluons, $W^\pm$, and the Z are called **gauge bosons** and are responsible for the strong and electroweak interactions. Gravitons are also bosons, having spin 2. Fermions exert attractive or repulsive forces on each other by exchanging gauge bosons, which are the force carriers.

Fermions and bosons

Gauge bosons

One other boson that has been predicted, but not yet detected, is necessary in quantum field theory to explain why the $W^\pm$ and Z have such large masses, yet the photon has no mass. This missing boson is called the **Higgs particle** (or **Higgs boson**) after Peter Higgs, who first proposed it. We don't know whether the Higgs particle is an elementary boson or a composite particle. The Standard Model proposes that there is a field called the *Higgs field* that permeates space. By interacting with this field, particles acquire mass. Particles that interact strongly with the Higgs field have heavy mass; particles that interact weakly have small mass. The Higgs field has at least one particle associated with it, and that is the Higgs particle (or Higgs boson). The properties of the gauge and Higgs bosons, as well as the graviton, are given in Table 14.2.

Higgs boson

The search for the Higgs boson is of the highest priority in elementary particle physics. The last accelerator facility currently in operation that had any chance of observing it is the Tevatron at Fermilab where both the D-zero and CDF (Collider Detector at Fermilab) collaborations have experiments under way. These collaborations use huge experimental apparatus (Figure 14.7) and have large numbers of scientists, many of whom have specific duties. In 2004 the lower

---

Table 14.2   Boson Properties: Gauge, Higgs, and Graviton

Boson	Mass	Spin	Electric Charge	Comments
Gauge:				
Photon	0	1	0	Stable
W^+, W^-	$80.42 \text{ GeV}/c^2$	1	1, −1	$\Gamma = 2.12$ GeV, decays observed
Z	$91.19 \text{ GeV}/c^2$	1	0	$\Gamma = 2.50$ GeV, decays observed
Gluon	0	1	0	Bound in hadrons, not free
Higgs H^0	$>117 \text{ GeV}/c^2$	0	0	Not observed
Graviton	0	2	0	Stable, not observed

Review of particle physics, Particle Data Group, *Physics Letters B* **592**, 1 (2004).

Figure 14.7 The detector systems used in particle physics are usually gigantic. They consist of several magnets, multiwire counters, scintillation counters, and so on. This photo shows the CDF (Collider Detector at Fermilab) apparatus on the Tevatron at Fermilab. Note the size of the detector compared to the persons standing in the foreground. *Fermilab Visual Media Services.*

limit of the mass of the Higgs boson was raised to $117 \text{ GeV}/c^2$, which probably puts its detection out of the range of the Tevatron. Physicists may have to wait until the new Large Hadron Collider (LHC) accelerator at CERN is operational, expected in 2007 or 2008, before the Higgs boson is found—if ever.

We shall spend the remainder of this section discussing fermions, which come in two varieties (leptons and quarks), and the mesons, which are bosons also made from quarks. Leptons include charged particles (electrons and muons) as well as uncharged particles (neutrinos). Quarks make up hadrons. We discuss leptons and hadrons in turn.

Leptons

The leptons are perhaps the simplest of the elementary particles. They appear to be pointlike, that is, with no apparent internal structure, and seem to be truly elementary. Thus far there has been no plausible suggestion they are formed from some more fundamental particles. There are only six leptons (listed in Table 14.3), plus their six antiparticles. We have already discussed the electron and muon. Each of the charged particles has an associated neutrino, named after its charged partner (for example, muon neutrino ν_μ). The muon decays into an electron, and the tau can decay into an electron, a muon, or even hadrons (which is most probable). The muon decay (by the weak interaction) is

$$\mu \rightarrow e + \nu_\mu + \bar{\nu}_e \tag{14.5}$$

We are already familiar with the electron neutrino that occurs in the beta decay of the neutron (Chapter 12). Neutrinos have zero charge. Their masses are known to be very small. The precise mass of neutrinos may have a bearing on current cosmological theories of the universe because of the gravitational attraction of mass. All leptons have spin 1/2, and all three neutrinos have been identified experimentally. Neutrinos are particularly difficult to detect because they have no charge and little mass, and they interact so weakly with matter that their

Table 14.3			The Leptons		
Particle Name	Symbol	Anti-particle	Mass (MeV/c^2)	Mean Lifetime (s)	Main Decay Modes
Electron	e^-	e^+	0.511	Stable	
e-Neutrino	ν_e	$\bar{\nu}_e$	$<3 \times 10^{-6}$		
Muon	μ^-	μ^+	105.7	2.2×10^{-6}	$e^- \bar{\nu}_e \nu_\mu$
μ-Neutrino	ν_μ	$\bar{\nu}_\mu$	<0.19		
Tau	τ^-	τ^+	1777	2.9×10^{-13}	$\mu^- \bar{\nu}_\mu \nu_\tau, e^- \bar{\nu}_e \nu_\tau$
τ-Neutrino	ν_τ	$\bar{\nu}_\tau$	<18		

Review of particle physics, Particle Data Group, *Physics Letters B* **592**, 1 (2004).

probability of being stopped while going through our planet Earth is exceedingly small. Leptons do not experience the strong force.

Hadrons

As mentioned previously, hadrons are particles that interact through the strong force. There are two classes of hadrons: **mesons** and **baryons**. Mesons are particles with integral spin having masses greater than that of the muon (106 MeV/c^2; note that the muon is a lepton and not a meson). All baryons have masses at least as large as the proton and have half-integral spins.

Mesons and baryons

Mesons We have already discussed the pion in Section 14.1; the pion is a meson that can either have charge or be neutral. Mesons are also bosons because of their integral spin. The meson family is rather large and consists of many variations, distinguished according to their composition of quarks. We list in Table 14.4 only a few of the mesons. In addition to the pion there is also a K meson, which exists in both charged ($K^\pm$) and neutral forms (K^0). The K^- meson is the antiparticle of the K^+, and their common decay mode is into muons or pions. The K^0 meson is particularly interesting because it has two decay lifetimes: K_S^0 has a shorter mean lifetime of 9×10^{-11} s and decays to $\pi^+ \pi^-$ or $2\pi^0$, whereas K_L^0 has a longer mean lifetime of 5×10^{-8} s and has many decay modes, including the two-pion mode (only 0.3% of all K_L^0 decays), which violates the combined conservation laws of charge and parity (see Section 14.4). All mesons are unstable and not abundant in nature. They are routinely produced in cosmic radiation and in nuclear and particle physics experiments. The π^0 is its own antiparticle.

Baryons The neutron and proton are the best-known baryons, and the baryons are a prolific group. Baryons having nonzero strangeness numbers (a new quantum number to be discussed in the next section) are called hyperons. The proton is the only stable baryon, but some theories predict that it is also unstable with a lifetime greater than 10^{30} years. Ongoing experiments have set a lower limit for the proton lifetime to be almost 10^{33} years (see Section 14.4). The longest-lived baryons are listed in Table 14.4 along with the mesons. They include the lambda (Λ), sigma ($\Sigma^\pm$, Σ^0), xi (Ξ^0, Ξ^-), and omega (Ω^-). All baryons, except the proton, eventually decay into protons.

Table 14.4   The Hadrons

Particle Name	Symbol	Anti-particle	Mass (MeV/c^2)	Mean Lifetime (s)	Main Decay Modes	Spin	Baryon Number B	Strangeness Number S	Charm Number C
Mesons									
Pion	π^-	π^+	140	2.6×10^{-8}	$\mu^+\nu_\mu$	0	0	0	0
	π^0	Self	135	8.4×10^{-17}	2γ	0	0	0	0
Kaon	K^+	K^-	494	1.2×10^{-8}	$\mu^+\nu_\mu, \pi^+\pi^0$	0	0	1	0
	K_S^0	$\overline{K}_S^0$	498	8.9×10^{-11}	$\pi^+\pi^-, 2\pi^0$	0	0	1	0
	K_L^0	$\overline{K}_L^0$	498	5.2×10^{-8}	$\pi^\pm e^\mp \nu_e, 3\pi^0,$ $\pi^\pm \mu^\mp \nu_\mu,$ $\pi^+\pi^-\pi^0$	0	0	1	0
Eta	η^0	Self	547	5×10^{-19}	$2\gamma, 3\pi^0,$ $\pi^+\pi^-\pi^0$	0	0	0	0
Charmed D's	D^+	D^-	1868	1.0×10^{-12}	$e^+, K^\pm, K^0,$ $\overline{K}^0 +$ anything	0	0	0	1
	D^0	$\overline{D}^0$	1864	4.1×10^{-13}	Same as D^+	0	0	0	1
	D_S^+	$\overline{D}_S^-$	1969	4.9×10^{-13}	Various	0	0	1	1
Bottom B's	B^+	B^-	5279	1.7×10^{-12}	Various	0	0	0	0
	B^0	$\overline{B}^0$	5279	1.5×10^{-12}	Various	0	0	0	0
J/Psi	J/ψ	Self	3097	10^{-20}	Various	0	0	0	0
Upsilon	$\Upsilon(1S)$	Self	9460	10^{-20}	Various	0	0	0	0
Baryons									
Proton	p	$\overline{p}$	938.3	Stable (?)		$\frac{1}{2}$	1	0	0
Neutron	n	$\overline{n}$	939.6	886	$pe^-\overline{\nu}_e$	$\frac{1}{2}$	1	0	0
Lambda	Λ	$\overline{\Lambda}$	1116	2.6×10^{-10}	$p\pi^-, n\pi^0$	$\frac{1}{2}$	1	-1	0
Sigmas	Σ^+	$\overline{\Sigma}^-$	1189	8.0×10^{-11}	$p\pi^0, n\pi^+$	$\frac{1}{2}$	1	-1	0
	Σ^0	$\overline{\Sigma}^0$	1193	7.4×10^{-20}	$\Lambda\gamma$	$\frac{1}{2}$	1	-1	0
	Σ^-	$\overline{\Sigma}^+$	1197	1.5×10^{-10}	$n\pi^-$	$\frac{1}{2}$	1	-1	0
Xi	Ξ^0	$\overline{\Xi}^0$	1315	2.9×10^{-10}	$\Lambda\pi^0$	$\frac{1}{2}$	1	-2	0
	Ξ^-	Ξ^+	1321	1.6×10^{-10}	$\Lambda\pi^-$	$\frac{1}{2}$	1	-2	0
Omega	Ω^-	Ω^+	1672	0.82×10^{-10}	$\Lambda K^-, \Xi^0\pi^-$	$\frac{1}{2}$	1	-3	0
Charmed lambda	Λ_C^+	$\overline{\Lambda}_C^-$	2285	2.0×10^{-13}	Various	$\frac{1}{2}$	1	0	1

Review of particle physics, Particle Data Group, *Physics Letters B* **592**, 1 (2004).

Particles and Lifetimes

The lifetimes of particles are also indications of their force interactions. Particles that decay through the strong interaction are usually the shortest-lived, normally decaying in less than 10^{-20} s. The decays caused by the electromagnetic interaction generally have lifetimes on the order of 10^{-16} s, and the weak interaction decays are even slower, longer than 10^{-10} s. There are several important exceptions to these general statements; for example, some nuclear beta decays take a long time, including the beta decay of the free neutron, which has a lifetime of about 15 min.

The length of a particle's lifetime has sometimes been used to define what we mean by a *particle*. The argument is that an object cannot rightfully be called a particle if its short lifetime prevents a direct observation. There is no single definition of a particle. An experimental physicist might say that a particle is an object having a well-defined charge and mass that behaves like a point (particle) while being accelerated or being detected.

A theoretical physicist might define a particle as an object having a complete set of numbers for charge, spin, mass, lifetime, and various other quantum numbers like charm, strangeness, and isospin (some of which we have not yet discussed). Some particles are observed only as a resonance, and although there is a precise definition for a resonance, its complexity prevents us from pursuing it here. We previously discussed compound nucleus resonances in Chapter 13; the resonances we are discussing here are wavelike phenomena and occur in particle scattering. The first excited state of the nucleon at 1232 MeV is a good example of a resonance called the delta [$\Delta(1232)$]; its lifetime is about 10^{-23} s and is too short for the delta to be directly observed. It decays to a nucleon and a pion and is easily observed in pion and electron scattering from a proton (see Figure 14.8). Objects having lifetimes as long as 10^{-10} to 10^{-14} s are normally regarded as particles because these times are long enough for the particles to be detected and

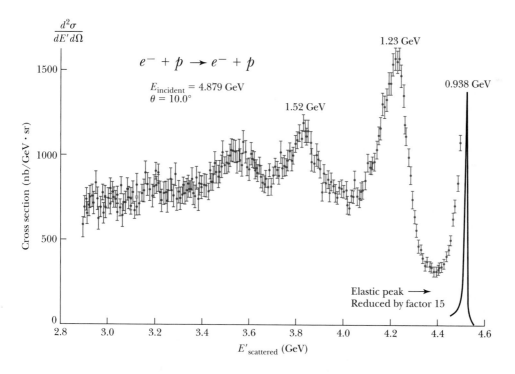

Figure 14.8 The spectrum of scattered electrons from 4.879-GeV electrons scattered inelastically from protons taken at DESY (Hamburg) showing the presence of the first excited state of the proton, called the $\Delta(1232)$, at an invariant mass energy of 1.23 GeV (the ground state has 0.938 GeV). The particle is short-lived and quickly decays. Peaks due to other higher-lying resonances are also seen at 1.52 GeV and multiple resonances near 1.68 GeV. *After W. Bartel et al.*, Physics Letters B *28, 148 (1968).*

induce other reactions. The lambda, sigma, xi, and omega baryons are examples of particles.

We call certain particles *fundamental*; this means that they are not composed of other, smaller particles. We believe leptons, quarks, and gauge bosons are fundamental particles. Although the Z and W bosons have very short lifetimes, they are regarded as particles, so a definition of particles dependent only on lifetimes is too restrictive. Other particles are *composites*, made from the fundamental particles.

We previously related the mean lifetime of a particle τ to an uncertainty Δt of the system. In discussing Yukawa's estimation of the meson mass, we used the uncertainty principle $\Delta E \, \Delta t \geq \hbar/2$ to estimate the time Δt:

$$\Delta t \approx \frac{\hbar}{2 \, \Delta E} \tag{14.2}$$

The full width at half maximum (FWHM) is used to describe the characteristic of a resonance like that in the $\Delta(1232)$. The FWHM is $2 \, \Delta E$ and is called the width Γ. So we have

$$\Delta t = \tau \approx \frac{\hbar}{2 \, \Delta E} = \frac{\hbar}{\Gamma} \quad \text{or} \quad \tau \approx \frac{\hbar}{\Gamma} \tag{14.6}$$

Conceptual Example 14.3

The J/ψ particle was discovered at both the Stanford Linear Accelerator Center (SLAC) and Brookhaven National Laboratory in 1974. The SLAC data from the head-on collision of e^+ and e^- are shown in Figure 14.9. The width of the state is $\Gamma = 0.063$ MeV. Discuss the evidence that this resonance represents an elementary particle and determine the particle's lifetime.

Solution The data shown in Figure 14.9 represent a clear experimental example of a resonance. The peak near 3.1 GeV was also observed in other outgoing reaction channels, indicating that the resonant peak occurs in the "compound" system and not in the outgoing reaction channel.

The width of the resonance observed in Figure 14.9 is on the order of 5 MeV and is limited by experimental resolution, namely, the beam momentum. With better experimental energy resolution, the resonance would become much sharper. As stated, the actual energy width of the resonant state has been established to be 0.063 MeV and was determined by other means. We use Equation (14.6) to determine the lifetime of the J/ψ particle.

$$\tau = \frac{\hbar}{\Gamma} = \frac{1.055 \times 10^{-34} \, \text{J} \cdot \text{s}}{0.063 \, \text{MeV}} \left(\frac{1 \, \text{MeV}}{1.6 \times 10^{-13} \, \text{J}} \right) \approx 10^{-20} \, \text{s}$$

The lifetime of 10^{-20} s is very long compared to the characteristic time taken by the e^+ and e^-, traveling practically at the speed of light, to interact over the distance of a few fermi.

Figure 14.9 The experimental results of Burton Richter and his group at SLAC showing the observation of the J/ψ resonance near a mass 3.1 GeV/c^2. The Nobel Prize–winning experiment was done with e^+ and e^- colliding head-on in the SPEAR storage ring at SLAC. The energies are in the center-of-mass system (CMS). *Data from Augustin et al.*, Physical Review Letters *33, 1406 (1974). Figure after D. H. Perkins*, Introduction to High Energy Physics, *Reading, MA: Addison Wesley (1982), p. 205.*

14.4 Conservation Laws and Symmetries

As the number of known elementary particles continued to increase, physicists were often perplexed by the occurrence of some reactions and decays. Physicists like to have clear rules or laws that determine whether a certain process can occur or not. It seems that everything occurs in nature that is not forbidden. Certain conservation laws are already familiar from our study of classical physics. These include mass-energy, charge, linear momentum, and angular momentum. These are absolute conservation laws: they are always obeyed. In this section we introduce additional conservation laws that are helpful in understanding the many possibilities of elementary particle interactions. As we shall see, some of these laws are absolute, but others may be valid for only one or two of the fundamental interactions.

Baryon Conservation

In low-energy nuclear reactions, the number of nucleons is always conserved. We now know empirically that this is part of a more general conservation law for baryon number. We assign a new quantum number called *baryon number* that has the value $B = +1$ for baryons, -1 for antibaryons, and 0 for all other particles. *The conservation of baryon number requires the same total baryon number before and after the reaction.* Although there are no known violations of baryon conservation, there are theoretical indications that it was violated sometime in the beginning of the universe when temperatures were quite high. This is thought to account for the preponderance of matter over antimatter in the universe today.

Conceptual Example 14.4

Would the discovery of proton decay be a violation of baryon conservation?

Solution There are no known baryons with lighter mass than the proton. Therefore, if proton decay occurred, it would be a violation of baryon conservation unless a new, lighter baryon was also discovered. The lifetime of the proton is believed to be greater than 10^{33} years, but active searches for its decay continue.

Let us examine this conservation law in a few examples. In the neutron decay, $n \rightarrow p + e + \bar{\nu}_e$, the baryon number is $+1$ on the left and the right sides, so baryon number is conserved. The antiproton was discovered in the reaction $p + p \rightarrow p + p + p + \bar{p}$. Note that as many as four particles must be produced in the reaction in order to create one antiproton; the conservation of baryons requires it: $B = 1 + 1 = 2$ on the left and $B = 1 + 1 + 1 - 1 = 2$ on the right. No fewer than three protons must be on the right side of the reaction in order to create the one antiproton.

Conceptual Example **14.5**

Examine the conservation of baryon number in the reaction producing the Ω^- particle shown in the bubble chamber photograph of Figure 14.10.

Solution This should be a straightforward application of the baryon conservation law. The reaction is

$$K^- + p \rightarrow K^0 + K^+ + \Omega^- \tag{14.7}$$

The kaons all have $B = 0$, so on the left side we have $B = 0 + 1 = 1$. The Ω^- has $B = +1$, so on the right side we have $B = 0 + 0 + 1 = 1$. This was the experiment that first identified the Ω^- particle, some three years after Gell-Mann predicted its existence.

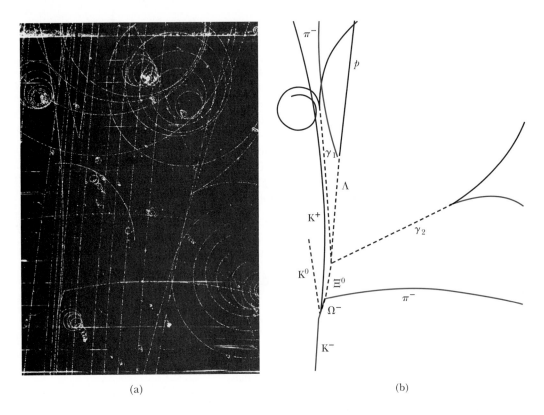

(a)   (b)

Figure 14.10 A photograph and the schematic diagram of a reaction from a liquid hydrogen bubble chamber at Brookhaven National Lab showing the production of Ω^- by the interaction of a K^- meson with a proton (a hydrogen nucleus in the bubble chamber). The neutral particles (dashed lines) leave no tracks in the bubble chamber. The Ω^- decays, producing the Ξ^0 and a π^-. Notice the variety of particles produced in the reaction and the eventual decays of short-lived particles. *Courtesy of Brookhaven National Laboratory.*

Lepton Conservation

The leptons are all fundamental particles, and *there is a conservation of leptons for each of the three kinds (families) of leptons; the number of leptons from each family is the same both before and after a reaction.* We let $L_e = +1$ for the electron and the electron neutrino, $L_e = -1$ for their antiparticles, and $L_e = 0$ for all other particles. We assign the quantum numbers L_μ for the muon and its neutrino and L_τ for the tau and its neutrino in a similar way. We now have three additional conservation laws, one each for L_e, L_μ, and L_τ, that are obeyed in most reactions and decays. We say *most* reaction and decays because the neutrino oscillations (see Section 14.7), which occur for neutrinos in flight, violate the conservation of leptons within individual families. The total lepton number $L = L_e + L_\mu + L_\tau$ is still conserved, except again at very high temperatures at the beginning of the universe, as for baryon conservation.

Let us examine a few reactions to study the effects of these conservation laws. First, we look again at the neutron beta decay, $n \rightarrow p + e + \bar{\nu}_e$. The values of L_μ and L_τ are zero on both sides. On the left side we have $L_e = 0$, and on the right side we have $L_e = 0 + 1 - 1 = 0$, so L_e is also conserved. It now becomes clear why the antineutrino, rather than the neutrino, is produced in beta decay; the antiparticles play an important role in the weak interactions.

Conceptual Example 14.6

Examine the lepton conservation laws in the decay of the π^- and μ^-.

Solution The decays are

$$\pi^- \rightarrow \mu^- + \bar{\nu}_\mu$$

$$\mu^- \rightarrow e^- + \bar{\nu}_e + \nu_\mu$$

The π^- first decays to μ^-, which in turn decays into e^-. However, we must also have neutrinos in order to conserve lepton number. The first reaction has $L_\mu = 0$ on the left and $L_\mu = +1 - 1 = 0$ on the right. For the second reaction, we must conserve both L_e and L_μ. On the left side we have $L_e = 0$ and $L_\mu = 1$, and on the right side we have $L_e = 1 - 1 + 0 = 0$ and $L_\mu = 0 + 0 + 1 = 1$, so both L_e and L_μ are conserved.

Strangeness

In the early 1950s physicists had considerable difficulty understanding the myriad of observed reactions and decays. For example, the behavior of the K mesons seemed very odd. There is no conservation law for the production of mesons, but it appeared that K mesons, as well as the Λ and Σ baryons, were always produced in pairs in the proton reaction studied most often, namely the $p + p$ reaction. In addition, the very fast decay of the π^0 meson into two photons (10^{-16} s) is the preferred mode of decay. One would expect the K^0 meson to also decay into two photons very quickly, but it does not. The long and short decay lifetimes of the K^0 are 10^{-8} and 10^{-10} s, respectively.

This strange behavior was understood by assigning a new quantum number called *strangeness* to certain particles. Strangeness is conserved in the strong and electromagnetic interactions, but not in the weak interaction. The values of the strangeness quantum number S are listed in Table 14.4. The kaons have $S = +1$, lambda and sigmas have $S = -1$, the xi has $S = -2$, and the omega has $S = -3$. Their antiparticles all have the opposite sign for the S quantum number. When the strange particles are produced by the $p + p$ strong interaction, they must be produced in pairs to conserve strangeness. Although π^0 can decay into two photons by the strong interaction, it is not possible for K^0 to decay at all by the strong interaction. The K^0 is the lightest $S = 0$ particle, and there is no other strange particle to which it can decay. It can decay only by the weak interaction, which violates strangeness conservation. Because the typical decay times of the weak interaction are on the order of 10^{-10} s, this explains the longer decay time for K^0. Only $\Delta S = \pm 1$ violations are allowed by the weak interaction.

Example 14.7

Explain the unusually short lifetime of the Σ^0 relative to the other hyperons (baryons with nonzero strangeness numbers) shown in Table 14.4.

Strategy We should be able to explain the lifetimes by examining the conservation laws. All the lifetimes for the hyperons are on the order of 10^{-10} s except for the Σ^0, which has a lifetime of 7×10^{-20} s. We suspect that something is different about the Σ^0 because of its short lifetime.

Solution Note that the decay is $\Sigma^0 \rightarrow \Lambda + \gamma$, which has $S = -1$ on the left side and $S = -1 + 0 = -1$ on the right side. It is an allowed transition. The Σ^0 is able to decay by the strong interaction to another strange particle with the same value of strangeness.

All the other hyperons in Table 14.4, however, have a decay that violates strangeness. Both Σ^+ and Σ^- decay to nucleons, which violates strangeness ($|\Delta S| = 1$). The Ξ^0 and Ξ^- both decay to Λ, which violates strangeness because the left side has $S = -2$ whereas the right side has $S = -1$. Similarly, the Ω^- decays to either Ξ^0 or Λ; both violate strangeness because $S = -3$ for Ω^-. These decay times are all on the order of 10^{-10} s, which is characteristic of the weak interaction.

The decay of the Ω^- into the Λ is particularly interesting. At first glance it appears to violate strangeness by $|\Delta S| = 2$, but note that the reaction is actually $\Omega^- \rightarrow \Lambda + K^-$, so $S = -3$ on the left side, and $S = -1 - 1 = -2$ on the right side; thus, $\Delta S = 1$, because both Λ and K^- have $S = -1$.

One more quantity, called **hypercharge**, has also become widely used as a quantum number. The hypercharge quantum number Y is defined by $Y = S + B$. Hypercharge is the sum of the strangeness and baryon quantum numbers and is conserved in strong interactions. Because hypercharge is not independent of strangeness, the conservation laws of hypercharge and strangeness are also related. The hypercharge and strangeness conservation laws hold for the strong and electromagnetic interactions, but are violated for the weak interaction.

Symmetries

Symmetry of equations describing a system under some operation is a useful aid to physicists in understanding particle reactions. Familiar symmetry operations, for example, are translation or rotation of a system in space. Because symmetries lead directly to conservation laws, it is important that we discuss three symmetry operators called *parity, charge conjugation*, and *time reversal*.

The conservation of parity P describes the inversion symmetry of space, that is, $x \rightarrow -x$, $y \rightarrow -y$, and $z \rightarrow -z$. Inversion, if valid, does not change the laws of physics. The conservation of parity is valid for the strong and electromagnetic interactions, but T. D. Lee and C. N. Yang (Nobel Prize for Physics, 1957) pointed out in 1956 that there was no experimental evidence for parity conservation in the weak interaction. They suggested, as a test, an experiment involving nuclear beta decay, and in May 1956 Lee talked to C. S. Wu, who was a well-known beta-decay physicist. She and her colleagues found evidence late in 1956 for the nonconservation of parity in the beta decay of ^{60}Co, which verified the suspicions of Lee and Yang.

Charge conjugation C reverses the sign of the particle's charge and magnetic moment. It has the effect of interchanging every particle with its antiparticle. Charge conjugation is also not conserved in the weak interactions, but it is valid for the strong and electromagnetic interactions. Even though both C and P are violated for the weak interaction, for several years after Lee and Yang's work, it was believed that when both charge conjugation and parity operations are performed (called CP), conservation was still valid. We can also understand this from

Tsung-Dao Lee (1926–) and Chen Ning Yang (1922–) were born in China where they were educated before coming to the United States to earn doctorates at the University of Chicago (Lee in 1950 and Yang in 1948). Lee worked at the University of California, Berkeley, and the Institute of Advanced Study at Princeton before becoming a professor at Columbia in 1953. Yang was at the Institute of Advanced Study from 1949 to 1966, and at the State University of New York at Stony Brook from 1966 to 1999. He returned to China in the 1990s and is now at Tsinghua University in Beijing. They received the Nobel Prize for Physics in 1957 for their investigation of parity. *Alan W. Richards, courtesy AIP Emilio Segrè Visual Archives.*

the general theoretical result that when all three operations are performed (*CPT*), where *T* is the time reversal symmetry, conservation holds. We believe that nature will proceed in the same manner both forward and backward in time for microscopic systems. We speak of the *invariance* of the symmetry operators, such as *T*, *CP*, and *CPT*.

This was the situation in 1964 when V. Fitch and J. Cronin (Nobel Prize, 1980) with their colleagues found that the K_L^0 meson decayed 0.3% of the time into two pions, rather than into three pions. Even though the decay into two pions is rare, it violates *CP* conservation and has tremendous ramifications for time reversal symmetry because of the strong belief in *CPT* invariance. The complete understanding and consequences of this result are still in some doubt.

The symmetries discussed here are just a small part of those used in particle physics, which relies heavily on group theory, the branch of mathematics that utilizes symmetry. We will return to symmetry in Section 14.7.

14.5 Quarks

We are now prepared to discuss quarks and how they form the many baryons and mesons that have been discovered experimentally. In 1961 Murray Gell-Mann and Yuval Ne'eman independently proposed a classification system called the **eightfold way** that separated the known particles into multiplets based on charge, hypercharge, and another quantum number called *isospin*, which we have not previously discussed. Isospin is a characteristic that can be used to classify different charged particles that have similar mass and interaction properties. The neutron and proton are members of an isospin multiplet we call the *nucleon*. In this case the isospin quantum number (*I*) has the value 1/2, with the proton having the substate value +1/2 ("spin up") and the neutron having −1/2 ("spin down"). Isospin is conserved in strong interactions, but not in electromagnetic interactions.

After the eightfold way was developed, it was noticed that some members of the multiplets were missing. Because of physicists' strong belief in symmetry, experimentalists set to work to find them, a task made easier because many of the particles' properties were predicted by the theoretical model. The Ω^- was detected in 1964 at Brookhaven National Laboratory (see Figure 14.10 and Example 14.5) in this manner, a discovery that confirmed the usefulness of the eightfold way.

However, as other particles were discovered it soon became clear that the eightfold way was not the final answer. In 1963 Gell-Mann and, independently, George Zweig proposed that hadrons were formed from fractionally charged particles called *quarks*.* The quark theory was unusually successful in describing properties of the particles and in understanding particle reactions and decay. Three quarks were proposed, named the *up* (*u*), *down* (*d*), and *strange* (*s*), with the charges $+2e/3$, $-e/3$, and $-e/3$, respectively. The strange quark has the strangeness value of -1, whereas the other two quarks have $S = 0$. Quarks are believed to be essentially pointlike, just like leptons.

With these three quarks, all the known hadrons could be specified by some combination of quarks and antiquarks. But there was still a problem. We discussed

Val Fitch (1923–), together with James Cronin, discovered a decay of the K meson that violated a widely held belief in the symmetry of charge conjugation and parity. Fitch and Cronin, both born in the United States, were professors at Princeton University at the time of their surprising experimental discovery, for which they won the 1980 Nobel Prize for Physics. *Photo by Clem Fiori. Courtesy AIP Emilio Segrè Visual Archives.*

---

*The term was coined by Gell-Mann from a line in James Joyce's book *Finnegans Wake* that says "three quarks for Muster Mark"; Gell-Mann is well known for his many other interests besides producing innovative theories.

Murray Gell-Mann (1929–) was born in New York City and entered Yale University at age 15, where he received his B.S. in physics. He obtained his Ph.D. from the Massachusetts Institute of Technology in 1951 and worked at the Institute for Advanced Study in Princeton and the University of Illinois before going to Caltech in 1955, where he remained until 1993. While still in his 20s he proposed the new quantum number strangeness, which explained some existing results and led to others. In 1962 he proposed the eightfold way, which organized elementary particles into families and led to the discovery of the Ω^- particle. He later proposed the idea of quarks, and still later, he contributed to the idea of the color force. He has been at the Santa Fe Institute in New Mexico since 1993. He received the Nobel Prize for Physics in 1969 for his work on the theory of elementary particles. He has many other interests including natural history, archaeology, depth psychology, biological evolution, and cultural evolution. © The Nobel Foundation.

in the last section how the strangeness quantum number was able to explain the lifetimes of some of the known particles. For example, the differences in lifetimes of the hyperons were discussed in Example 14.7 and were explained in terms of the violation of strangeness conservation. Similarly, it was thought that another new quantum number might be able to explain some additional discrepancies in the lifetimes of some of the known particles, and a fourth quark called the *charmed* quark (*c*) was proposed in 1970. A new quantum number called **charm C** was introduced so that the new quark would have $C = +1$ while its antiquark would have -1. Particles without the charmed quark have $C = 0$. Charm is similar to strangeness in that it is conserved in the strong and electromagnetic interactions, but not in the weak interactions. This behavior was sufficient to explain the particle lifetime difficulties.

Experimentalists continued in their never-ending search for new elementary particles. In 1974 two groups working independently, one led by Burton Richter at SLAC and another by Samuel Ting from MIT at Brookhaven National Laboratory, found evidence in particle-scattering experiments for a heavy meson that required the existence of the charmed quark and its antiquark (see Example 14.3). Both Richter and Ting received the Nobel Prize for Physics in 1976 for the discovery of this new particle, named the *J/ψ particle* (the teams assigned different names). Other particles soon followed that had even higher masses and also needed the properties of charm.

In 1977 physicists led by Leon Lederman at Fermi National Laboratory discovered the massive Υ (upsilon) meson, which contains the **bottom** quark (*b*). It was also soon verified by scanning resonances from e^+e^- scattering, where three narrow resonances near 10 GeV of the upsilon meson require the existence of the *b* quark. The Standard Model then predicted another quark called the **top** quark (*t*), and after a long search the top quark was discovered at the Fermi National Laboratory in 1995. These last two quarks are sometimes called *truth* and *beauty*. Quantum numbers called **bottomness B** and **topness T** are assigned to these quarks.

Quark Description of Particles

We can now present the given quark properties and see how they are used to make up the hadrons. In Table 14.5 we give the name, symbol, mass, charge, and the quantum numbers for strangeness, charm, bottomness, and topness. The spin of all quarks (and antiquarks) is 1/2.

A meson consists of a quark-antiquark pair, which gives the required baryon number of 0. Baryons normally consist of three quarks. We present the quark content of several mesons and baryons in Table 14.6. The structure is quite simple. For example, a π^- consists of $\overline{u}d$, which gives a charge of $(-2e/3) + (-e/3) = -e$, and the two spins couple to give 0 $(-\frac{1}{2} + \frac{1}{2} = 0)$. A proton is uud, which gives a charge of $(2e/3) + (2e/3) + (-e/3) = +e$; its baryon number is $\frac{1}{3} + \frac{1}{3} + \frac{1}{3} = 1$; and two of the quarks' spins couple to zero, leaving a spin $\frac{1}{2}$ for the proton $(\frac{1}{2} + \frac{1}{2} - \frac{1}{2} = \frac{1}{2})$.

What about the quark composition of the Ω^-, which has a strangeness of $S = -3$? We look in Table 14.6 and find that its quark composition is sss. According to the properties in Table 14.5 its charge must be $3(-e/3) = -e$, and its spin is due to three quark spins aligned, $3(1/2) = 3/2$. Both of these values are correct. There is no other possibility for a stable omega (lifetime $\sim 10^{-10}$ s) in agreement with Table 14.4.

Table 14.5 **Quark Properties**

Quark Name	Symbol	Mass* (GeV/c^2)	Charge	Baryon Number	Strangeness S	Charm C	Bottomness B	Topness T
Up	u	0.0015 to 0.004	$2e/3$	$\frac{1}{3}$	0	0	0	0
Down	d	0.004 to 0.008	$-e/3$	$\frac{1}{3}$	0	0	0	0
Strange	s	0.08 to 0.130	$-e/3$	$\frac{1}{3}$	-1	0	0	0
Charmed	c	1.15 to 1.35	$2e/3$	$\frac{1}{3}$	0	1	0	0
Bottom	b	4.1 to 4.3	$-e/3$	$\frac{1}{3}$	0	0	-1	0
Top	t	178	$2e/3$	$\frac{1}{3}$	0	0	0	1

Antiquarks, $\bar{u}, \bar{d}, \bar{s}, \bar{c}, \bar{b}$, and $\bar{t}$, have opposite signs for charge, baryon number, S, C, B, and T.

Review of particle physics, Particle Data Group, *Physics Letters B* **592**, 1 (2004). Mass of top quark is new 2004 value from Fermilab.

*The u, d, and s quark masses are estimates of so-called current-quark masses.

Conceptual Example 14.8

Check the neutron beta decay reaction for the quark composition of each particle and check that the charge and baryon numbers are correct.

Solution The neutron beta decay is given by

$$n \rightarrow p + e + \bar{\nu}_e$$

The corresponding quark composition for each of the particles is given by

$$\text{Neutron beta decay:} \quad udd \rightarrow uud$$

where we don't have to worry about the quark composition of the electron and its antineutrino because they are leptons, not hadrons. We have already checked that the quark composition of the proton gives a charge $+e$, so let us check that udd gives a zero charge for the neutron: $(2e/3) + (-e/3) + (-e/3) = 0$. The baryon number on both sides is 1, because there are three quarks. In neutron beta decay a *down* quark becomes an *up* quark.

Example 14.9

The primary decay of the Ξ^- is $\Xi^- \rightarrow \Lambda^0 + \pi^-$. The two decay products in turn decay. Look up the first decay listed in Table 14.4 for both the Λ^0 and π^- and write the original decay and two subsequent decays in terms of their quark composition.

Strategy We look up the two subsequent decays and find $\Lambda^0 \rightarrow p + \pi^-$ and $\pi^- \rightarrow \mu^+ + \nu_\mu$. We use Table 14.6 to find the quark composition. The quark structures of the antiparticles are just the corresponding antiquarks of the particles.

Solution We list first the reaction and then the quark composition. Keep in mind that these decays all occur by the weak interaction in which quark transformations can occur.

We have not discussed this, but it is due to the fact that these decays occur by exchange of the $W^\pm$ gauge bosons.

$$\Xi^- \rightarrow \Lambda^0 + \pi^- \qquad dss \rightarrow uds + \bar{u}d$$

A s quark is changed to a d quark, and an $\bar{u}u$ pair is created.

$$\Lambda^0 \rightarrow p + \pi^- \qquad uds \rightarrow uud + \bar{u}d$$

Again a s quark is changed to a d quark, and an $\bar{u}u$ pair is created.

$$\pi^- \rightarrow \mu^+ + \nu_\mu \qquad \bar{u}d \rightarrow \text{no quarks}$$

The d quark can change to an u quark and annihilate the existing $\bar{u}$ quark.

Table 14.6 **Quark Composition of Selected Hadrons**	
Particle	Quark Composition
Mesons	
π^+	$u\bar{d}$
π^-	$\bar{u}d$
K^+	$u\bar{s}$
K^0	$d\bar{s}$
D^+	$c\bar{d}$
D^0	$c\bar{u}$
Baryons	
p	uud
n	udd
Λ	uds
Σ^+	uus
Σ^0	uds
Ξ^0	uss
Ξ^-	dss
Ω^-	sss
Λ_C^+	udc

Color

There is one difficulty that perhaps you have noticed. Because the quarks have spin 1/2, they are all fermions. According to the Pauli exclusion principle, no two fermions can exist in the same state. Yet we have three strange quarks in the Ω^-. This is not possible unless some other quantum number distinguishes each of these quarks in one particle. Establishing a new quantum number called color circumvents this problem. A theory named **quantum chromodynamics (QCD)** is based on this concept. There are three colors for quarks, which for simplicity we shall call red (R), green (G), and blue (B). This color designation has absolutely nothing to do with the visual colors that we see. It is merely an attempt to distinguish this new property, which in some ways is analogous to the behavior of colored light. We can then call the corresponding colors for antiquarks antired ($\bar{R}$), antigreen ($\bar{G}$), and antiblue ($\bar{B}$). *Color* is the "charge" of the strong nuclear force, analogous to *electric charge* for electromagnetism.

The two theories, quantum electrodynamics and quantum chromodynamics, are similar in structure; color is often called *color charge* and the force between quarks is sometimes referred to as *color force*. Back in Section 14.2 we mentioned that *gluons* are the particles that hold the quarks together. We show a Feynman diagram of two quarks interacting in Figure 14.11. A red quark comes in from the left and interacts with a blue quark coming in from the right. They exchange a gluon, changing the blue quark into a red one and the red quark into a blue one.

A color and its anticolor cancel out. We call this *colorless* (or white). All hadrons are colorless. In Figure 14.11 the gluon itself must have the color $B\bar{R}$ in order for the diagram to work. Quarks change color when they emit or absorb a gluon, and quarks of the same color repel, whereas quarks of different color attract.

To finish the story we should mention that the six different kinds of quarks are referred to as *flavors*. There are six flavors of quarks (u, d, s, c, b, t). Each flavor has three colors. Finally, how many different gluons are possible? Using the three colors red, blue, and green, there are nine possible combinations for a gluon. They are $B\bar{B}$, $B\bar{R}$, $B\bar{G}$, $R\bar{B}$, $R\bar{R}$, $R\bar{G}$, $G\bar{B}$, $G\bar{R}$, and $G\bar{G}$. Note in Figure 14.11 that the gluon is $B\bar{R}$ and not BR. The combination $B\bar{B} + R\bar{R} + G\bar{G}$ does not have any net color change and cannot be independent. Therefore, there are only eight independent gluons, and that is what quantum chromodynamics predicts. Gluons can interact with each other, because each gluon carries a color charge. Note that in this case gluons, as the mediator of the strong force, are much different from photons, the mediator of the electromagnetic force.

To date no one has ever clearly observed a free quark. However, in 1967, Jerome Friedman, Henry Kendall, and Richard Taylor (Nobel Prize, 1990) performed experiments at the Stanford Linear Accelerator (SLAC) by scattering 20-GeV electrons deep into protons. They found a larger number of scattered electrons at backward angles than would be expected if protons were uniform spheres of matter. Their experiment was interpreted as evidence for pointlike quarks inside the proton. Notice the similarity between this experiment and that of Rutherford almost 50 years earlier. Both concluded there was something hard inside the object (proton in the former and atom in the latter). Yet it was several years after 1967 before the idea of quarks was widely accepted by the general physics community.

Figure 14.11 A Feynman diagram showing the exchange of a gluon ($B\bar{R}$) between a quark having color R and a quark having color B. The colors of the quarks are changed in the interaction.

Confinement

Physicists now believe that free quarks cannot be observed; they can only exist within hadrons. This is called **confinement**. David Gross, H. David Politzer, and

Frank Wilczek received the 2004 Nobel Prize for Physics for their explanation of asymptotic freedom that explains quark behavior in the strong interaction. We can see what happens with a simple diagram like that in Figure 14.12, which shows three quarks confined within a neutron. At low energies the three quarks are easily contained and are free to move around. If an incident photon scatters from the neutron, one of the quarks may become so energetic that it tries to escape. Quarks have the property, however, that the color force transmitted by the exchanged gluons increases as the quarks get further apart. As one of the quarks moves away, the restoring force increases. If there is so much energy that the color force can't confine the quarks within the neutron, then this extra energy will be able to create a quark-antiquark pair, and a meson is created. In the case shown in Figure 14.12, a proton and π^- are the final result. Apparently a single quark will not escape when a photon or electron interacts at high energy with a baryon like a neutron or proton. With enough energy several mesons may be produced, as long as all the conservation laws and quark rules are observed. The production of the delta resonance in Section 14.3 is an example of this. High-energy electrons scattered from the proton produce the delta resonance at 1232 MeV, which subsequently decays to a $p + \pi^0$ or $n + \pi^+$. In both cases a quark-antiquark pair is produced.

Consider a hadron in which one of the quarks is being pulled away. The color-force field "stretches" between the quarks as shown in Figure 14.13. As the quark is pulled away, more and more energy is added to the field to keep it from stretching, much like the force on a spring that is stretched. In the quark case, however, the color-force field eventually "snaps" (like a spring breaking) into two new quarks, rather than pulling the quark out of the hadron. The energy is converted into the mass of new quarks, and the system relaxes back to the unstretched state (see Figure 14.13). Quark masses cannot be measured directly, because they are confined within hadrons and are not observed as physical particles.

$$\gamma + n \rightarrow p + \pi^-$$

Figure 14.12 When a high-energy γ ray is scattered from a neutron, a free quark cannot escape because of confinement. For high enough energies, an antiquark-quark pair is created (for example, $u\bar{u}$), and a pion and proton are the final particles.

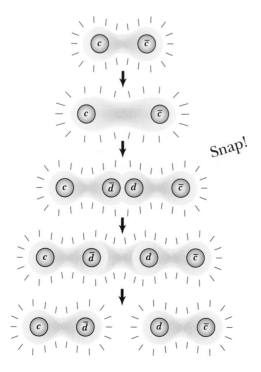

Snap!

Figure 14.13 Quarks cannot be pulled away from a hadron. As the color-force field increases to pull out the quark, the increasing energy eventually goes into creating a quark pair. An analogy might be a spring that is pulled apart. More and more energy goes into the potential energy of the spring that eventually snaps. The energy in this case creates the new quark pair, and the system relaxes.

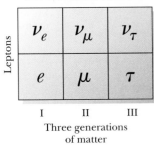

I   II   III

Three generations
of matter

Figure 14.14 The three generations (or families) of matter. Note that both quarks and leptons exist in three distinct sets. One of each charge type of quark and lepton make up a generation. All visible matter in the universe is made from the first generation; second- and third-generation particles are unstable and decay into first-generation particles.

14.6   The Families of Matter

Now that we have had a brief review of the particle classifications and have learned how the hadrons are made from the quarks, we should summarize. We presently believe that the two varieties of fermions, called *leptons* and *quarks*, are fundamental particles. These fundamental particles can be divided into three simple families or generations as shown in Figure 14.14. Each generation consists of two leptons and two quarks. The two leptons are a charged lepton and its associated neutrino. The quarks are combined by twos or threes to make up the hadrons.

Most of the mass in the universe is made from the components in the first generation (electrons and u and d quarks). The second generation consists of the muon, its neutrino, and the charmed and strange quarks. The members of this generation are found in certain astrophysical objects of high energy and in cosmic rays, and are produced in high-energy accelerators. The third generation consists of the tau and its neutrino and two more quarks, the bottom (or beauty) and top (or truth). The members of this third generation existed in the early moments of the creation of the universe and can be created with very high energy accelerators. Note that in each group, the mass increases with the generation number, with the third generation having the greatest mass.

Leptons are essentially pointlike, because they have no internal structure. There are three leptons with mass and three others with little mass (the *neutrinos*). Quarks and antiquarks make up the *hadrons* (*mesons* and *baryons*). Quarks may also be pointlike ($<10^{-18}$ m) and are confined together, never being in a free state. There are six flavors of quarks (up, down, strange, charmed, bottom, and top) and there are three colors (green, red, and blue) for each flavor. Rules for combining the colored quarks allow us to represent all known hadrons.

Bosons mediate the four fundamental forces of nature: gluons are responsible for the strong interaction, photons for the electromagnetic interaction, $W^{\pm}$ and Z for the weak interaction, and the as yet unobserved graviton for the gravitational interaction. In our study of nuclear physics we discussed the pion as the mediator of the strong force. At a more fundamental level, we can now say that the gluon is responsible. The gluon is responsible for the attraction between the antiquark and quark that make up the pion, and the gluon is responsible for the attraction between the quarks that make up the nucleons.

14.7   Beyond the Standard Model

Although the Standard Model has been successful in particle physics, it doesn't answer all the questions. For example, it is not by itself able to predict the particle masses. It is believed that the Higgs boson may be the key to unlocking this "black box" mystery of particle masses. Why are there only three generations or families of fundamental particles? Do quarks and/or leptons actually consist of more fundamental particles? All of our past experience would lead us to suspect this to be true. The understanding of the nuclear components was completed in 1932 with Chadwick's discovery of the neutron. The next level of understanding, using the quark model, did not come for another 35 years. If there is another level, we may take even longer to reach it, because experiments at such a level become more difficult and expensive to perform.

Neutrino Oscillations

One of the most perplexing problems over the last three decades has been the solar neutrino problem (see Section 12.7 and its Special Topic, "Neutrino Detection"). The number of solar neutrinos reaching Earth from the sun is a factor of 2–3 too small if our understanding of the energy-producing (nuclear fusion) processes in the sun is correct. Only electron neutrinos are produced in nuclear fusion in the sun. After several laboratories reached the same conclusion about the solar neutrino measurement, suggestions were made that other processes were going on. Neutrinos come in three varieties or flavors: electron, muon, and tau. In 1998 the Super-Kamiokande Collaboration based in Japan announced that researchers had seen neutrinos generated in the Earth's atmosphere (from cosmic rays) changing or "oscillating" into another flavor. This could only happen if neutrinos have mass.

Solar neutrino problem

Neutrino mass

By now much data have been gathered, particularly by the Super-Kamiokande group and by the Sudbury Neutrino Observatory in Canada. Physicists have seen various oscillations between the three flavors of neutrinos. The probability that a neutrino will change flavor is dependent upon the distance the neutrino travels divided by its energy. Thus the distance from the sun to the Earth offers a measurable probability of neutrino oscillation, which does account for the low number of electron neutrinos reaching Earth from the sun.

A whole new field of neutrino physics is flourishing. For example, several accelerators are producing neutrinos and sending them hundreds of kilometers to detectors located deep underground in mines or tunnels. Each of the three neutrino flavors is a mixing of three elements: ν_1, ν_2, and ν_3. Much remains to be done before we fully understand this mixing, including measuring the neutrinos' masses and the rate at which each neutrino flavor oscillates with the others. Future experiments will determine if the neutrino has a CP violation component, which would also affect the problem of matter over antimatter in the universe. The Standard Model allows mixing patterns in quarks and charged leptons, but it was not expected for neutrinos. Theorists have already proposed variations of the Standard Model to explain the neutrino oscillation observations, but future experimental data will be more challenging to explain as the parameters are narrowed.

Matter-Antimatter

According to existing theories of the creation of our universe in the Big Bang, matter and antimatter should have been created in exactly equal quantities. It appears that matter dominates over antimatter now in our universe, and the reason for this has concerned physicists and cosmologists for years. Fortunately, we have an opportunity to understand this dilemma. The tiny violation of CP symmetry in the kaon decay tilts the scales in terms of matter over antimatter. However, Standard Model estimates indicate that this CP violation is too small by at least 10 orders of magnitude to explain the predominance of matter.

The Standard Model predicts that experiments with B mesons might reveal considerably more about CP violation than kaons possibly could. B mesons are similar to kaons, but have a strange quark replaced by the much more massive bottom quark. Remarkable experiments done in the United States and Japan have developed "B factories" that have produced copious numbers of B meson pairs (B and $\bar{B}$). The amount of CP violation in the B meson decays may be greater than that allowed by the Standard Model, and physicists are now exploring theories that go beyond the Standard Model. This result promises the prospect of new physics as research in this field continues.

B factories   The results discussed above are produced in B factories housed in electron accelerators at Stanford in the United States and KEK (High Energy Accelerator Research Organization) in Japan. The upsilon (Υ) particle is produced in head-on collisions of electrons and positrons at the resonance energy of 10.58 GeV. The upsilon in turn decays into a B meson and anti-B meson. Experiments at other accelerator installations are either under way or soon will be. The field of B meson physics promises to yield a rich harvest of new physics results.

Grand Unifying Theories

Einstein is known to have worked during the last 30 years of his life on a *unified field theory* that would encompass some of the four fundamental forces. Neither he nor anyone else has completely succeeded. There have been several attempts toward a grand unified theory (GUT) to combine the weak, electromagnetic, and strong interactions. The theory of Howard Georgi and Sheldon Glashow of the 1970s has been one of the most successful, but the simplest version of their model predicted a proton lifetime too short by orders of magnitude. The various GUT theories make several predictions, including the following:

1. The proton is unstable with a lifetime of 10^{29} to 10^{31} years. Current experimental measurements have shown the lifetime to be greater than 10^{32} years.
2. Neutrinos may have a small, but finite, mass. This has been confirmed.
3. Massive magnetic monopoles may exist. There is presently no confirmed experimental evidence for magnetic monopoles.
4. The proton and electron electric charges should have the same magnitude.

The unification of the strong and electroweak interactions is an important part of cosmological attempts to understand the origin of the universe. This will be examined more thoroughly in Chapter 16.

String Theory Although string theory can be traced back to the late 1960s, a revolution occurred in 1984 when physicists showed that certain mathematical inconsistencies in the theory could be resolved. For the last two decades there has been a tremendous amount of effort by theorists in string theory, which has had several variations. The addition of supersymmetry resulted in the name *theory of*

Superstring theory   *superstrings.* In this revolutionary theory the universe is described by as many as ten dimensions rather than the four (x, y, z, t) that we have discussed. In superstring theory elementary particles do not exist as points, but rather as tiny, wiggling loops that are only 10^{-35} m in length. Further work has revealed that they describe not just strings, but other objects including membranes and higher-dimensional objects. The addition of membranes has resulted in "brane" theories (see M-theory below). Particles of 10^{19} GeV (the Planck energy) would be required to probe a size as small as 10^{-35} m. Superstring theory seems so far removed from the real world that it would appear to be difficult to bridge the wide gap between theory and experiment. However, at the present time, superstring theory is a promising approach to unify the four fundamental forces, including gravity. Gravity is not something that is added to superstring theory; gravity has to be in superstring theory. The theory has a mode that includes the existence of the graviton.

Supersymmetry   **Supersymmetry** is a necessary ingredient in many of the theories trying to unify the forces of nature. The symmetry relates fermions and bosons. All fermions will

Superpartners   have a *superpartner* that is a boson of equal mass, and vice versa. The superpartner spins will differ by $\hbar/2$. Supersymmetry (called SUSY) is contained in modern superstring theories like the ones discussed above, in M-theory (discussed below), and the Minimal Supersymmetric Standard Model. Presently, none of the known leptons, quarks, or gauge bosons can be identified with a superpartner of any other particle type. If superpartners exist, the extrapolation in Figure 14.15 for the strong, weak, and electromagnetic forces all coincide at a single energy. The details of supersymmetry are beyond the level of this book.

At some fundamental level, matter is believed to be composed of spin 1/2 particles (quarks and leptons). The gauge bosons mediate the color and electroweak interactions. At some very high energy, supersymmetry is expected to be valid. However, if supersymmetry is relevant at all to nature, it must be broken spontaneously, and we refer to *spontaneous symmetry breaking*. An example is shown in Figure 14.15, where we show the coupling strengths of the four interactions on a logarithmic scale as a function of energy (or mass). At some high energy there is unification that is spontaneously broken at lower energies. As the universe evolves, there is a richness of physics as various symmetries are broken. The violation of *CP* in kaon decay is a good example in which symmetry breaking led to new physics results.

M-Theory More recently theorists have proposed a successor to superstring theory called M-theory. M-theory has 11 dimensions and predicts that strings coexist with membranes, called "branes" for short. The M in M-theory is variously said to stand for *m*embrane, *m*other of all strings, *m*agic, *m*atrix, or *m*ystery. Most theorists think a theory that produces the Standard Model from first principles is perhaps still decades away.

The number of particles that have been predicted from a variety of different theories include the fancifully named sleptons, squarks, axions, winos, photinos, zinos, gluinos, and preons. Only through experiments will scientists be able to wade through the vast number of unifying theories to weed out the good from the bad. The next few years and decades will be exciting in this field of physics.

14.8 Accelerators

Although cosmic rays contain high-energy particles, the small intensity of those particles only allows physicists to perform limited experiments. Particle physics was not able to develop fully until particle accelerators were constructed with high enough energies to create particles with a mass of about 1 GeV/c^2 or greater. Some early accelerators were briefly mentioned in Chapter 13.

The precursor of most modern accelerators is the cyclotron shown in Figure 13.2. The pioneering work of E. O. Lawrence at the University of California, Berkeley, led to the development of the cyclotron in the 1930s. Cyclotrons rely on charged particles moving in a circular orbit perpendicular to a magnetic field. The radius of curvature is given by $R = p/(qB)$ where p is the particle's momentum and q is its charge. The orbital frequency (cyclotron frequency) is given by

$$f = \frac{\omega}{2\pi} = \frac{v}{2\pi R} = \frac{mv}{2\pi mR} = \frac{p}{R}\frac{1}{2\pi m} = \frac{qB}{2\pi m} \qquad (14.8)$$

Cyclotron frequency

where m is the mass of the particle. In a cyclotron, the gap between the two "dees" (in the shape of semicircles) contains a radio-frequency (RF) voltage that accelerates the particle every time it passes through the gap. The charged particle travels thousands of orbits, gaining two bursts of energy during each orbit.

Figure 14.15 The coupling strength of the four fundamental interactions is plotted versus energy. Both are log scales. Unification may occur at some very high energy, but at lower energies symmetry breaking separates the four interactions.

Branes

EXPERIMENTAL INGENUITY

Some people think of physicists as long-haired old men holed up in offices with indecipherable scribbling on a blackboard. Other people think of laboratories crammed with wires, pipes, and weird-looking apparatus on which people are crawling, peering, adjusting, and taking measurements. Although these stereotypes may be exaggerated, physicists do tend to be either theorists or experimentalists. The best-known scientists are perhaps the ones who propose radical new breakthroughs in thought that propel us forward in great jumps in our understanding of nature. Certainly Newton, Maxwell, and Einstein are in this group. But these great leaps in intellect generally require much experimental measurement to understand nature and to test the theories that have been proposed.

The field of experimental particle physics began around 1950, but we might argue that it started with the advent of particle accelerators around 1930. Experiments did not actually begin until many years later when energies reached about 1 GeV. Since 1955, six Nobel Prizes have been awarded for theoretical work in elementary particle physics (1957, 1965, 1969, and 1979, 1999, and 2004, see Appendix 9), but eleven have been awarded for experimental work (1959, 1960, 1961, 1968, 1976, 1980, 1984, 1988, 1990, 1992, and 1995). Several of these experiments resulted in an increased understanding of physics, but some of them were for pure experimental development, which eventually led to experiments performed by others. It is this latter category that we want to highlight here because of the experimenters' special ingenuity.

We begin with the invention of the bubble chamber by physicist Donald Glaser (Figure A) at the University of California at Berkeley in the early 1950s. Bubble chambers, having much greater densities than cloud chambers, allowed much higher energy particles to be detected. Glaser tried several liquids that could be superheated, just short of boiling, and eventually settled on liquid hydrogen and xenon. He developed an ingenious method of photographing the tracks of the particles passing through the liquid. His efforts revolutionized experimental particle physics, and

bubble chambers were the mainstay of high-energy particle detectors for many years.

Though Glaser was the inventor of the bubble chamber, his colleague at Berkeley, Luis Alvarez (Figure B), was its developer, and although nominated to receive the Nobel Prize with Glaser, Alvarez did not receive the Prize until 1968, eight years after Glaser. Alvarez and his colleagues used a large bubble chamber to discover the first of the new "resonance" particles. Although his Nobel Prize citation mentioned his development of the bubble chamber and measurements with it, Alvarez undoubtedly was rewarded for his many contributions to science, which include 40 patents comprising a wide range of projects including radar, color television, and a golf-training device. He was particularly well known for his efforts in looking for unknown tombs beneath the Egyptian pyramids.

Figure A Donald Glaser (1926–) is shown examining a detector at Lawrence Berkeley Laboratory around 1960. The bubble chamber was mostly developed while Glaser was a professor at the University of Michigan during the period 1950–1959. *Courtesy of the Lawrence Radiation Laboratory, University of California, Berkeley.*

In contrast to the flamboyant Rubbia, the shy and unassuming van der Meer was responsible for inventing the stochastic cooling of the beam particles that allowed the accumulation of sufficient numbers of antiprotons to make Rubbia's experiment feasible. Although a surprise, the selection of the Dutch engineer van der Meer turned physicist to share the Nobel Prize for Physics in 1984 with Rubbia was highly applauded by the physics community because of his tremendous ingenuity in the very complex method of stochastic cooling.

Similarly, Georges Charpak of CERN received the Nobel Prize for Physics in 1992 for his invention of the multiwire proportional counter, which is still today a mainstay in several areas of physics, particularly in nuclear and particle physics.

The research labs of the world are full of scientists and engineers developing clever new ways of building and operating equipment that will lead to important advances in the future.

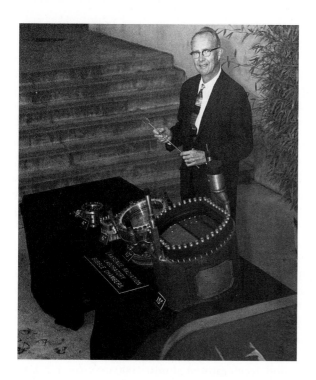

Figure B Luis Alvarez (1911–1988), shown here with his bubble chamber, was born in California and spent almost his entire career at the University of California, Berkeley. Alvarez was also well known for his work on analyzing the film tape of John Kennedy's assassination, his searches for burial chambers in pyramids, and research with his son Walter on the demise of dinosaurs due to the debris caused by an asteroid colliding with Earth. *Courtesy of Orlando Ernest Lawrence Berkeley Laboratory.*

Figure C Carlo Rubbia (wearing the tie) and his colleagues await the start-up of the LEP accelerator at CERN in 1989. *Courtesy of CERN.*

Carlo Rubbia (Figure C) and Simon van der Meer represent an unusual pair of Nobel Prize winners whose success depended on each other. Rubbia was an aggressive leader who made convincing arguments in the late 1970s to reconfigure an existing CERN accelerator to scatter protons and antiprotons head-on. He conducted the successful experiment to discover the W and Z particles in the early 1980s. Rubbia managed a huge team of hundreds of technicians, engineers, and scientists who built and operated the detector.

There are three main types of accelerators used presently in particle physics experiments: synchrotrons, linear accelerators, and colliders. We discuss each of these in turn.

Synchrotrons

Eventually at higher energies relativistic effects limit the cyclotron technique. The equation for the radius of curvature doesn't change as long as the relativistic momentum is used, but the orbital frequency becomes

Synchrotron frequency

$$f = \frac{qB}{2\pi m}\sqrt{1 - \frac{v^2}{c^2}} \tag{14.9}$$

In *synchrocyclotrons* the frequency of the RF voltage between the dees is adjusted to match this changing frequency. In *synchrotrons* the magnetic field is changed to keep the radius ($R = p/qB$) constant and match the frequency. In both cases the particles are accelerated in pulses or bunches, because the variation of RF voltage or magnetic field can only match one particular particle momentum. The Bevatron at Berkeley was an early (1954) high-energy proton synchrotron. Its energy was 6.4 GeV, just enough for E. G. Segrè (1905–1989) and O. Chamberlain (1920–) to produce antiprotons in 1955, for which they received the Nobel Prize for Physics in 1959.

Most modern proton accelerators are synchrotrons. The magnetic fields do not span the entire circle but rather encircle a closed pipe at a fixed radius. The proton synchrotron (called the Tevatron) at the Fermi National Lab is currently the world's highest-energy proton accelerator at almost 1000 GeV (or 1 TeV).

Synchrotron Radiation One difficulty with cyclic accelerators is that when charged particles are accelerated, they radiate electromagnetic energy called *synchrotron radiation*. This problem is particularly severe when electrons, moving very close to the speed of light, move in curved paths. If the radius of curvature is small, electrons can radiate as much energy as they gain. The Large Electron-Positron (LEP) ring (now also housing the Large Hadron Collider) at the European Center for Nuclear Research (CERN) has a radius of 4.3 km in order to limit synchrotron radiation losses.

Light sources

Physicists have learned to take advantage of these synchrotron radiation losses and now build special electron accelerators (called **light sources**) that produce copious amounts of photon radiation used for both basic and applied research in physics, chemistry, materials science, metallurgy, biology, and medicine. For example, the radiation is used in x-ray lithography to produce miniaturized computer chips with higher speed and greater capacity. Synchrotron radiation accelerators are being constructed at an increasing rate, even by industry for applied purposes. There are dozens of accelerators throughout the world used by thousands of physicists, representing universities, government laboratories, and industries.

Linear Accelerators

Linacs

Linear accelerators or **linacs** typically have straight electric-field-free regions between gaps of RF voltage boosts (Figure 14.16). Because the particles gain speed with each boost, and the voltage boost is on for a fixed period of time, the distance between gaps becomes increasingly larger as the particles accelerate. Linacs are sometimes used as preacceleration devices for large circular accelerators. The

Figure 14.16 A schematic diagram of one type of linear accelerator. Each succeeding drift tube has to be longer because of the increasing speed of the particle. An RF voltage accelerates the charged particle in the region between the drift tubes. *Adapted from A. Arya,* Elementary Modern Physics, *Reading, MA: Addison-Wesley, 1974.*

longest linear accelerator is the 3-km-long Stanford Linear Accelerator at SLAC, which accelerates electrons up to 50 GeV. The SLAC linear accelerator gains its energy from a more complicated traveling wave, which accelerates the electron continuously along its 3-km path. It operated originally at 20 GeV, but a very successful upgrade allowed it to operate at 50 GeV.

Fixed-Target Accelerators

Most of the accelerators discussed so far have fixed targets. The accelerated particles are directed at a fixed target, where the reaction takes place. Because of conservation of momentum, the energy of the beam's particle is not fully available to create reactions and produce new particles. If we consider the reaction $m_1 + m_2 \rightarrow anything$, with the bombarding particle of mass m_1 having kinetic energy K on the fixed target of mass m_2, the amount of energy available in the center-of-mass system is

$$E_{cm} = \sqrt{(m_1 c^2 + m_2 c^2)^2 + 2m_2 c^2 K} \qquad (14.10)$$

In Chapter 13 we found the threshold kinetic energy required to initiate a nuclear reaction with a fixed target. If we use relativistic relations, the threshold kinetic energy is

$$K_{th} = (-Q)\frac{\text{total masses involved in reaction}}{2m_2} \qquad (14.11)$$

See the following example for an indication of how little energy is available for reaction studies with fixed-target accelerators.

Example 14.10

(a) How much energy was available in the center of mass for the experiment of Segrè and Chamberlain, who used 6.4-GeV protons on a fixed proton target to produce antiprotons in the reaction given here?

$$p + p \rightarrow p + p + p + \bar{p}$$

(b) How much beam energy was necessary to produce the antiprotons? (c) How much energy is available for a similar reaction with 1-TeV protons from the Tevatron on a fixed proton target?

Strategy (a) We can use Equation (14.10) to calculate the energy available in the center of mass. (b) The threshold kinetic energy can be found by using Equation (14.11). We use the Q value defined in Equation (13.7). (c) We use Equation (14.10) to determine the center-of-mass energy available for the reaction for the Tevatron.

Solution (a) If we use Equation (14.10) and the rest energy of the proton (938 MeV), we have

$$E_{cm} = \sqrt{[2(0.938 \text{ GeV})]^2 + 2(0.938 \text{ GeV})(6.4 \text{ GeV})}$$

$$= 3.94 \text{ GeV}$$

The total mass of the reaction products is $4(0.938 \text{ GeV}/c^2) =$ 3.75 GeV/c^2, so the Bevatron was constructed with just enough energy to create the four particles in the final state.

(b) Equation (13.7) gives the Q value:

Q = (Initial mass energies) − (final mass energies)

$$= 2m_pc^2 - 4m_pc^2 = -2m_pc^2$$

$$= -2(0.938 \text{ GeV}) = -1.88 \text{ GeV}$$

If we insert this value into Equation (14.11), we obtain

$$K_{\text{th}} = (-Q)\frac{6m_p}{2m_p} = (1.88 \text{ GeV})(3) = 5.6 \text{ GeV}$$

The kinetic energy of 6.4 GeV was clearly enough to initiate the reaction, which is consistent with our result in (a).

(c) Now we insert the kinetic energy of 1 TeV into Equation (14.10) to determine

$$E_{\text{cm}} = \sqrt{[2(0.938 \text{ GeV})]^2 + 2(0.938 \text{ GeV})(1000 \text{ GeV})}$$

$$= 43 \text{ GeV}$$

Because of the conservation of momentum requirement, there is a tremendous reduction in the available energy for reactions for the 1000-GeV proton beam from the Tevatron.

Colliders

Because of the limited energy available for reactions like that found for the Tevatron in Example 14.10c, physicists decided they had to resort to colliding beam experiments, in which the particles meet head-on. If the colliding particles have equal masses and kinetic energies, the total momentum is zero and all the energy is available for the reaction and the creation of new particles. By the 1960s physicists had gained enough experience in accelerator technology to build **colliders**, a new concept for accelerators. Colliding-beam accelerators usually have to utilize storage rings in order to collect enough particles to ensure sufficient intensity for the expected reaction. Two accelerators that have had electrons and positrons colliding are the CERN Large Electron-Positron (LEP) storage ring and the SLAC Stanford Linear Collider (SLC). The CERN Super Proton Synchrotron (SPS) used an ingenious technique shown in Figure 14.17 to cause 270-GeV protons and antiprotons to collide head-on. The Tevatron has 1-TeV protons and antiprotons colliding. These machines require fantastic precision in order to steer particles moving at nearly the speed of light to interact head-on.

The LEP collider at CERN ceased operation in 2000, but its tunnel is 27 km in circumference located 100 m beneath ground straddling the border of France and Switzerland. Presently under development in the LEP tunnel at CERN is what will be the world's highest energy accelerator, the Large Hadron Collider

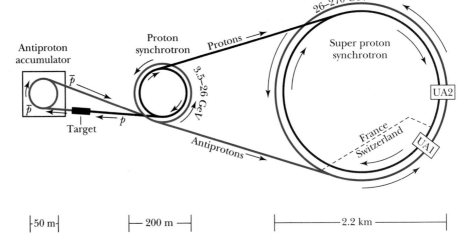

Figure 14.17 A schematic diagram of the Super Proton Synchrotron at CERN, which straddles France and Switzerland, indicating how the protons and antiprotons are created, are accelerated, and finally interact in two detector regions at UA1 and UA2. The protons originate from the synchrotron accelerator system and produce antiprotons in the target in front of the antiproton accumulator. *After A. Kernan*, American Scientist **74**, 21 (1986).

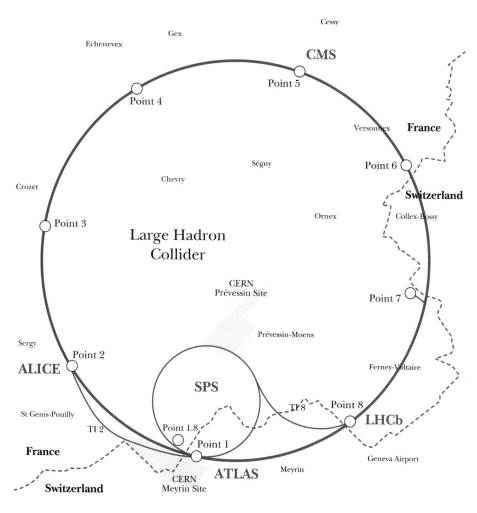

Figure 14.18 The LHC (Large Hadron Collider) accelerator facility at CERN straddles the border of France and Switzerland. It is 8.6 km in diameter and lies 100 m underground. Notice that the accelerator is located under communities and the countryside. The locations of the four detectors (the large CMS and ALICE and the smaller ATLAS and LHCb detectors) are noted. The detectors are placed at several of the eight labeled access points. The main site of CERN is at Meyrin (see shaded blue area). Although most of CERN is in France, the main entrance gate is in Switzerland. *Adapted from CERN.*

(LHC), in which 7-TeV protons will collide head-on with 7-TeV protons for a total energy of 14 TeV (see Figure 14.18). The SPS accelerator described earlier will be a preaccelerator for the LHC. In addition, beams of lead particles will collide at the LHC with a collision energy of 1150 TeV. Many countries, including the United States, are partners in this project, which should be operational by 2007 or 2008. The LHC will be the premier elementary particle physics accelerator facility for at least the next decade. The next proposed accelerator, which may be decades away, is a large linear collider that will smash electrons and positrons together head-on.

Example 14.11

How much energy would a fixed-target proton accelerator require to match the energy available in the Tevatron at Fermilab for a $p + \bar{p}$ reaction?

Strategy The energy available in the Tevatron is the sum of the colliding beams, or 2 TeV. We use Equation (14.10) to determine the kinetic energy K needed for a fixed-target experiment.

Solution If we insert 2 TeV (2000 GeV) into Equation (14.10), we can solve for K, the kinetic energy.

$$2000 \text{ GeV} = \sqrt{[2(0.938 \text{ GeV})]^2 + 2(0.938 \text{ GeV})K}$$

The second term on the right-hand side will have to be much larger than the first term, so we can neglect the first term. We then solve for K.

$$K = \frac{(2000 \text{ GeV})^2}{2(0.938 \text{ GeV})} = 2.1 \times 10^6 \text{ GeV} = 2.1 \times 10^{15} \text{ eV}$$

Such a fixed-target accelerator cannot currently be constructed.

The highest energies currently available for particle physics research are in cosmic rays, which have up to 10^{21} GeV energy. We discussed cosmic rays in Section 14.1 because of their importance in the early discoveries of elementary particles. Although the intensity of these high-energy cosmic rays is exceedingly small, the field of cosmic ray physics continues to flourish because of the uniqueness of such high energies, with ongoing experiments on mountains, in balloons, and in space vehicles.

Summary

The beginnings of elementary particle physics occurred in the 1930s with the discoveries of the positron, the neutron, and the muon. The development of accelerators and particle detectors during the 1930s and the following decades has given physicists the tools needed to delve deeper into the structure of the nucleus and finally the elementary particles. Accelerators have included cyclotrons, synchrotrons, linear accelerators, and variations including storage rings and colliders.

The fundamental interactions are the gravitational, electroweak, and strong. For all practical purposes, except at very high energies, the electroweak is two interactions: electromagnetic and weak. Gluons are the mediators of the strong force, photons mediate the electromagnetic interaction, the $W^{\pm}$ and Z bosons mediate the weak interaction, and an as yet unobserved particle, the graviton, mediates the gravitational interaction.

Elementary particles are classified in various ways. For example, fermions and bosons have half-integral and integral spins, respectively. Stable particles appear to be composed, at some level, of constituent fermions. We believe leptons are truly elementary, pointlike particles: electrons, muons, taus, and their respective neutrinos. Hadrons are particles that interact through the strong force and consist of two classes: mesons (integral spin) and baryons (half-integral spin). Mesons are unstable. The only stable free baryon is the proton, but some theories predict it is also unstable. Neutrons and protons are stable when bound in nuclei. Baryons heavier than the nucleons are called *hyperons*.

The building blocks of matter can be described by three simple families shown in Figure 14.14. Each family consists of two leptons and two quarks. The two leptons are a charged lepton and its associated neutrino. Quarks are combined by twos (mesons) and threes (baryons) to make the hadrons.

The conservation of baryon number and three separate laws for the conservation of leptons appear to be universally valid. Symmetry breaking seems to be spontaneous and widespread. For example, some other conservation laws appear to be valid for one or more of the interactions, but not for all of them. These include strangeness, charge conjugation, and parity.

A breakthrough in the understanding of elementary particle structure began in the 1960s with the introduction of quarks. Quarks have fractional electric charges and only exist as constituents of hadrons. There are six quarks: up, down, strange, charmed, bottom, and top. Quantum chromodynamics (QCD) theory establishes how quarks are combined to form elementary particles. A property called *color* is required to understand how quarks and antiquarks combine.

The Standard Model combines the electroweak and QCD theories and has been quite successful in explaining elementary particle physics. For example, another particle (or group of particles) called the Higgs boson is expected, but has not yet been identified. The neutrino oscillations have solved a problem with the solar neutrino rate. A whole new field of neutrino physics is flourishing. The preponderance of matter over antimatter is under active research, as is the search for the Higgs boson. Grand unifying theories (GUTs) have taken another step in unifying the fundamental interactions, but some of their predictions have already been negated. Research on GUTs and new theories called *superstrings* continues.

The new LHC accelerator facility at CERN is scheduled to become operational in 2007 or 2008 and promises an exciting future for particle physics.

Questions

1. What are the characteristics of the following conservation laws: mass-energy, electric charge, linear momentum, angular momentum, baryon number, lepton number? Explain how they are related to fundamental laws of nature.
2. Why are storage rings useful for high-energy accelerators?
3. Why are colliding beams useful for particle physics experiments?
4. Can a baryon be produced when an antibaryon interacts with a meson? Explain.
5. What kinds of neutrinos are produced in the reaction $\mu^+ + e^- \rightarrow 2\nu$? Explain.
6. What mediating particles are exchanged between two positrons? Between two quarks?
7. Which families of particles seem to be truly elementary?

8. How you can determine whether particles are mesons or baryons by looking at their quark structure?
9. Does it appear that the total baryon number of the universe is zero? What would that mean?
10. Explain how a magnet may be used to distinguish a range of energies for protons. How can a monoenergetic beam of protons be obtained?
11. Why is it a problem when a particle gets out of phase with the frequency of a pulsed accelerator?
12. Explain why electron accelerators that are not linear must have large radii. The largest such accelerator is the LHC/LEP at CERN, which has a radius of 4.3 km.
13. The next proposed accelerator is an electron-positron collider that may be many kilometers long. Why must it be so long? Why can't it be circular to save space?

Problems

Note: The more challenging problems are indicated by a shaded box.

14.1 Early Discoveries

1. What are the frequencies of two photons produced when a proton and antiproton annihilate each other at rest?
2. The mass of the charged pion is 140 MeV/c^2. Determine the range of the nuclear force if the pion is the mediator.

14.2 The Fundamental Interactions

3. The strong interaction must interact within the time it takes a high-energy nucleon to cross the nucleus. Use an appropriate speed and distance to estimate the time for the strong interaction to occur.
4. In order to probe another particle with linear dimensions D, the wavelength of the probing particle must be $\lambda \leq D$. In order to learn details, the wavelength should be substantially less, perhaps as small as $0.1D$. Calculate the kinetic energies of an electron and a proton needed to probe the details of a neutron. Assume the diameter of the neutron is 1.5 fm.
5. Some details of elementary particles may need to be probed on distance scales as small as 10^{-18} m. Calculate the kinetic energies of an electron and a proton needed to probe details this small. (*Hint:* See the previous problem, but we don't need to divide by the factor of 10 to determine the wavelength.)

6. The omega meson ω (mass = 782 MeV/c^2) is believed to be the mediator for short-range repulsive force. Estimate the range of this force.
7. Suppose that the Higgs boson is discovered and that it has a mass of 150 GeV/c^2. Physicists believe it may be the mediator of a new force. Use its mass to determine the range of the force.

14.3 Classification of Elementary Particles

8. Supply the missing neutrinos in the following reactions or decays.
   (a) $\mu^+ \rightarrow e^+ + ?$ (d) $K^- \rightarrow \mu^- + ?$
   (b) $? + p \rightarrow n + e^+$ (e) $? + n \rightarrow p + \mu^-$
   (c) $\pi^- \rightarrow \mu^- + ?$
9. Estimate the approximate baryon number of the Earth.
10. The mass of the Ω^- baryon is 1672.45 ± 0.29 MeV/c^2, where the latter number represents the experimental uncertainty. The lifetime is quoted as 0.821×10^{-10} s. (a) What is the intrinsic value of its resonance width Γ based on its lifetime? (b) How does this compare with the experimental uncertainty in the mass-energy?
11. The values of the mean lifetime are not listed in the 2004 *Review of Particle Physics* (see http://pdg.lbl.gov) for the J/ψ and upsilon [Y(1S)] mesons. We have listed approximate values in Table 14.4. The 2004 *Review* gives the full-width Γ values of 91.0 keV and 53.0 keV, respectively, for the two mesons. Determine their lifetimes.

14.4 Conservation Laws and Symmetries

12. Consider the two following reactions:

$$K^0 + p \rightarrow \Sigma^+ + \pi^{''}$$

$$\overline{K}^0 + p \rightarrow \Sigma^+ + \pi^0$$

Do both of these reactions obey the conservation rules? Does this explain why K^0 is not its own anti-particle?

13. Explain why each of the following reactions is forbidden.
    (a) $p + p \rightarrow p + p + n$
    (b) $p + p \rightarrow p + \pi^+ + \gamma$

14. Explain why each of the following reactions is forbidden.
    (a) $p + \overline{p} \rightarrow \mu^+ + e^-$
    (b) $\gamma + p \rightarrow n + \pi^0$
    (c) $\gamma \rightarrow \pi^+ + \pi^-$

15. Determine the energy of a γ ray produced in the decay of Σ^0 at rest into $\Lambda + \gamma$.

16. A π^0 of kinetic energy 600 MeV decays in flight into 2 γ rays of equal energies. Determine the angles of the γ rays from the incident π^0 direction.

17. Complete the following reactions.
    (a) $\mu^- + p \rightarrow n + ?$
    (b) $n + p \rightarrow \Sigma^0 + n + ?$

14.5 Quarks

18. Show that electric charge, baryon number, and strangeness for the neutron, Σ^+, and Λ_C^+ are all equal to the sum of their quark configurations.

19. Show that electric charge, baryon number, and strangeness for the π^+, K^+, and D^0 are all equal to the sum of their quark configurations.

20. Determine the quark composition of the D^0 and D^+ charmed mesons.

21. Determine the quark composition of the B^+, B^-, and B^0 mesons.

22. What kind of particle would you expect to be made of the $\overline{c}u$ configuration?

23. The Ω^- baryon decays primarily through the following reaction: $\Omega^- \rightarrow \Lambda^0 + K^-$. Subsequently the Λ^0 and K^- also decay by the first reaction listed for each in the Table 14.4 column Main Decay Modes. (a) Are these strong or weak decay interactions? (b) Write out each reaction using its particle symbols and its quark composition.

24. Consider particles with the following quark composition. Determine their quantum numbers and the particle names. (a) $c\overline{d}$ (b) uds (c) $\overline{sss}$ (d) $\overline{c}d$.

14.7 Beyond the Standard Model

25. Assume the half-life of the proton is 10^{33} y. How many decays per year would you expect in a tank of water containing 100,000 gallons of water? Assume the bound protons can decay.

26. Some GUT theories allow the proton to be unstable. What conservation laws are broken in the following proton decays?
    (a) $p \rightarrow \pi^+ + \overline{\nu}_e$
    (b) $p \rightarrow \mu^+ + \pi^0$
    (c) $p \rightarrow e^+ + K^0 + \nu_e$

27. Assume that half of the mass of a 50-kg person consists of protons. If the half-life of the proton is 10^{32} years, calculate the number of proton decays per day from the body.

28. A Σ^+ particle of kinetic energy 3 GeV is produced inside a bubble chamber. Ignoring energy losses, what is the mean distance the Σ^+ travels in the detector?

14.8 Accelerators

29. In modern collider experiments the beam energy K has much greater energy than its rest energy ($K \gg mc^2$). If two particles each of mass m collide head-on with each having beam energy K, the energy in the center of mass $E_{cm} = 2K$. Show that two particles of the same mass m colliding head-on with each having kinetic energy $K(K \gg mc^2)$ have the same center-of-mass energy as the same particle of energy K_{lab} in a fixed-target accelerator colliding with the same target particle, where

$$K_{lab} \approx \frac{2K^2}{mc^2}$$

30. CERN constructed the first proton-proton collider, called the Intersecting Storage Rings (ISR), in 1971. Each proton beam had a kinetic energy of 31 GeV. Calculate how much energy a fixed-target accelerator would need in order to have the same energy available in the center-of-mass energy.

31. Calculate the speed of the 7-TeV protons that will be accelerated in the LHC at CERN.

32. A cyclotron is used to accelerate protons to 60 MeV. If these protons are elastically scattered from deuterons (^{2}H) and tritons (^{3}H), what are the maximum energies of the ^{2}H and ^{3}H?

33. Calculate the minimum kinetic energy of a proton to be scattered from a fixed proton target in order to produce an antiproton.

34. A magnetic field of 1 T is used to accelerate 10-MeV protons in a cyclotron. (a) What is the radius of the magnet? (b) What is the cyclotron frequency?

35. Show that for higher particle energies the simple cyclotron frequency in Equation (14.8) becomes limited by relativistic effects. Show that Equation (14.9) is the correct orbital frequency:

$$f = \frac{qB}{2\pi m} \sqrt{1 - \frac{v^2}{c^2}}$$

36. Show that Equation (14.10) is the correct relativistic result for the amount of energy available in the center-of-mass system when the reaction $m_1 + m_2 \rightarrow anything$

occurs with a bombarding particle of mass m_1 and kinetic energy K on a fixed target of mass m_2. Show that it reduces in the nonrelativistic limit to Equation (13.12).

37. In a fixed-target experiment with $m_1 = m_2 = m$, derive approximate expressions for E_{cm} in the following limiting cases: (a) $K \ll mc^2$ and (b) $K \gg mc^2$. Discuss your results in both cases.

38. Calculate the electron velocity in the 50-GeV beam at SLAC.

39. What is the ratio of cyclotron frequencies calculated relativistically and nonrelativistically for (a) a 10-MeV proton, (b) a 100-MeV proton, and (c) a 1-GeV proton?

40. A 33-GeV proton is said to take about half a second to make some 160,000 revolutions around the 0.8-km circumference of the Alternating Gradient Synchrotron at Brookhaven. Check this statement.

General Problems

41. Consider the reaction $e^- + e^+ \rightarrow \bar{\nu}_e + ?$. (a) Determine the quantum numbers for the missing particle or particles (baryon, lepton, charge, spin, and so on). (b) Consider whether one or two particles is needed to complete the reaction. What are possible particles?

42. Determine which of the following decays is not allowed and explain why.
(a) $\Xi^0 \rightarrow \pi^- + \Sigma^+$
(b) $\Xi^0 \rightarrow \pi^- + p$
(c) $\Xi^0 \rightarrow K^- + p$

43. Consider the following reaction: $p + p \rightarrow p + \Lambda + K^+$. How much bombarding energy is required for this reaction (a) if the second proton is a stationary target, and (b) if the protons collide head-on?

44. Determine which of the following decays is not allowed and explain why.
(a) $\pi^+ \rightarrow \mu^+ + n$ (c) $\Lambda \rightarrow p + \pi^-$
(b) $\mu^- \rightarrow e^- + \gamma$ (d) $p \rightarrow \pi^+ + \pi^0$

45. Determine which of the following decays or reactions is not allowed and explain why.
(a) $\bar{p} + p \rightarrow \bar{\Lambda} + \Lambda$ (d) $\pi^- + p \rightarrow \pi^- + \Sigma^+$
(b) $n \rightarrow p + e^- + \nu_e$ (e) $\pi^+ \rightarrow \mu^+ + \nu_\mu$
(c) $\Xi^0 \rightarrow n + \pi^0$

46. Determine which of the following decays or reactions is not allowed and explain why.
(a) $p \rightarrow e^+ + \pi^0$ (c) $p + p \rightarrow p + p + \pi^0$
(b) $\pi^+ + p \rightarrow \Lambda + K^0$ (d) $n \rightarrow p + e^- + \bar{\nu}_e$

47. Determine which of the following decays or reactions is not allowed and explain why.
(a) $\Xi^- \rightarrow \Lambda + \pi^-$ (c) $p + p \rightarrow p + \pi^+ + \bar{\Lambda} + \bar{K}^0$
(b) $\Lambda \rightarrow p + \pi^0$ (d) $\Omega^+ \rightarrow \Xi^+ + \pi^0$

48. What is the relation between the de Broglie wavelength λ, mass m, and kinetic energy K for (a) a low-energy proton, and (b) a very high energy proton?

49. Consider the 7-TeV protons that will be accelerated in the LHC collider at CERN. (a) Use the results of Problems 29, 36, or 37 to find the available center-of-mass energy if these protons collide with other protons in a fixed-target experiment. (b) Compare your results in (a) with the center-of-mass energy available in a colliding-beam experiment.

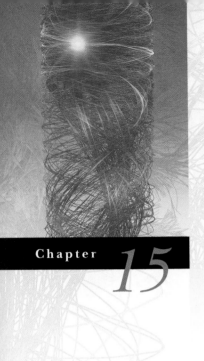

Chapter 15

General Relativity

There is nothing in the world except empty, curved space. Matter, charge, electromagnetism, and other fields are only manifestations of the curvature.

John Archibald Wheeler

Time and space and gravitation have no separate existence from matter.

Albert Einstein

Although Albert Einstein's *special* theory of relativity was tremendously useful in the field of physics, it was not until the introduction of his *general* theory of relativity in 1916 that Einstein became a celebrity. Perhaps because the general theory is so much more complicated than the special theory, Einstein proposed several experiments to test his general theory. A solar eclipse in 1919 afforded scientists an opportunity to test Einstein's theory by measuring the gravitational bending of light. When their results confirmed the predictions of the general theory of relativity, Einstein's fame was sealed. In this chapter we discuss this and other remarkable experiments that support Einstein's general theory. We also mention some other general relativistic effects that are the subject of present-day research.

General relativity is really the story of gravity. Gravity is the governing force of the universe because it holds the planets in our solar system to the sun, binds the stars into galaxies, and determines the fate of our universe. Many of the subjects discussed in this chapter on the general theory of relativity will appear again in the next chapter on cosmology. Gravity and cosmology are bound together throughout the cosmos.

15.1 Tenets of General Relativity

In the first few decades after Einstein presented his general theory of relativity in 1916, it was sometimes said that only a few scientists truly understood it. That probably was because the mathematics is so complex. We will not delve into the mathematics of metrics, matrices, and tensors, but rather approach general relativity from two concepts: (1) the principle of equivalence, which is an extension of Einstein's first postulate of special relativity (see Section 2.3) and (2) the curvature of spacetime due to gravity. We discuss these tenets in this section.

Principle of Equivalence

The special theory of relativity encompasses inertial frames of reference moving at uniform relative velocities. The special theory can be used to explain the physics in any of the relative inertial systems. But what about systems moving in nonuniform motion with respect to one another? Linearly accelerating and rotating systems are examples of such motion. Is this kind of nonuniform motion also relative? Einstein believed so, and after several years of thinking about the problem, he presented his general theory of relativity.

We imagine another *gedanken* experiment to help us understand nonuniform motion. Consider an astronaut sitting in a confined space on a rocket placed on Earth ready to blast off for a mission to Mars (see Figure 15.1a). The astronaut is strapped into a chair that is mounted on a weighing scale that indicates a mass M. The astronaut drops a safety manual that falls to the floor. Now contrast this situation with the one shown in Figure 15.1b. The rocket is now on its way to Mars and is far enough away from Earth that the gravitational attraction is negligible. However, the rocket is still accelerating in order to gain speed for the trip to Mars. If the acceleration at this point has exactly the same magnitude as g on Earth, then the weighing scale indicates precisely the same mass M that it did on Earth, and the safety manual, when dropped, still falls with the same acceleration as measured by the astronaut. The question is: how can the astronaut tell whether

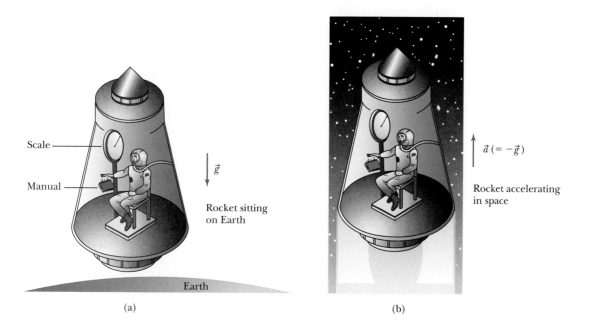

Scale

Manual

$\vec{g}$

Rocket sitting
on Earth

Earth

(a)

$\vec{a}\,(=-\vec{g})$

Rocket accelerating
in space

(b)

Figure 15.1 An astronaut sits in a confined space inside a rocket with no windows. The astronaut sits on a chair mounted on an ordinary weighing scale and drops a safety manual to the floor. In (a) the rocket is sitting at rest on Earth, and in (b) the rocket is accelerating upward with acceleration equal to $-\vec{g}$. The astronaut can do no experiment to distinguish between gravity and acceleration (forget the noise!).

the rocket is sitting on Earth or accelerating in space? Einstein provided the answer in the **principle of equivalence**:

Principle of equivalence

There is no experiment that can be done in a small confined space that can detect the difference between a uniform gravitational field and an equivalent uniform acceleration.

Astronauts experience this equivalence to a certain degree when they are in free fall, rotating around the Earth in the space shuttle orbiter. Does the fact they feel weightless mean they are not attracted to the Earth by gravity? No. In fact, they are "falling around the Earth." The inside of the orbiter is almost an inertial reference system, and the astronauts can do experiments inside the orbiter on a small particle that obeys Newton's laws. However, according to an inertial reference system fixed in the distant stars, the shuttle is in an accelerated reference system. If the astronauts cannot see outside their capsule, can they tell whether they are somewhere in outer space or "falling around the Earth"? Can they understand precisely why they are weightless?

With considerable experimentation, the astronauts can figure out they are in a gravitational field because weightlessness actually occurs only at the center of mass of the orbiter. If they put a drop of water in a corner of the orbiter, it does not have a precisely spherical shape because the gravitational field of the Earth is not uniform. The drop becomes very slightly bulged on both sides closest and farthest from the Earth because of the change in the Earth's gravitational field with distance. In this manner the astronauts can tell they are in a gravitational field. The forces are called *tidal*, because the same forces are responsible for the ocean tides.

Let us look briefly at gravitational and inertial forces. According to Newton's second law, a body accelerates in reaction to a force according to its *inertial mass m_I*:

$$\vec{F} = m_I \vec{a}$$

Inertial mass m_I measures how strongly an object resists a change in its motion.

According to the law of gravitation, a *gravitational mass m_G* reacts to a field $\vec{g}$ like that due to Earth. The field $\vec{g}$ depends on the location of all the masses other than m_G. Gravitational force measures how strongly an object m_G is attracted to other masses.

$$\vec{F} = m_G \vec{g}$$

For the same force we set the two preceding equations equal and find

$$\vec{a} = \left(\frac{m_G}{m_I} \right) \vec{g} \tag{15.1}$$

According to the principle of equivalence, the inertial and gravitational masses are equal.

The principle of equivalence has been proposed in various forms for hundreds of years—including versions by Galileo, Newton, and Einstein. You encountered it in introductory physics when the equivalence of inertial and gravitational masses were discussed. Galileo and Newton both performed tests of inertial and gravitational mass equivalence using a pendulum and found them equivalent to 1 part in 10^3. Pendulums of equal lengths would have periods proportional to the ratio $\sqrt{m_I/m_G}$ (see Problem 1). During a couple of decades around 1900, Lorand Eötvös performed a remarkable series of experiments showing the two masses are equivalent to 1 part in 10^8. More recent experiments show they are equivalent to 1 part in 10^{11}. The equivalence of inertial and gravitational masses helps in understanding Einstein's assertion that inertial and gravitational forces are equivalent. Einstein believed that if no experiment can distinguish inertial and gravitational forces, then they must be the same thing.

The principle of equivalence is the key to Einstein's general theory of relativity. The advanced mathematics needed prevents us from delving too far into the theory. We will only present some results and several predictions of the theory. For a few years after Einstein presented his general theory, there was a flurry of activity to test several predictions that Einstein had made. While none of these early experiments seemed to contradict Einstein's general theory, none of them were able to prove the theory conclusively. And there have been alternate theories to Einstein's that also explained the data available. It was not until the 1960s, aided by advances in technology like space probes, that experiments were finally conducted that confirmed Einstein's general theory. Few scientists now doubt it, and the general theory has opened new fields of study in astrophysics and cosmology that include research in mathematics, particle physics, nuclear physics, and astronomy. We will return to these matters again in the next chapter when we discuss cosmology and the origin of the universe.

There is one simple, but perhaps surprising, prediction of the equivalence principle. Consider a rocket accelerating through a region of space where the gravitational force is negligible. A small hole allows a burst of light from a distant star to enter the spacecraft. Consider in Figure 15.2 what the astronaut inside sees. By the time the horizontally moving light pulse hits the opposite wall, the rocket has moved up considerably (Figure 15.2a). In Figure 15.2b we show what happens to the light pulse *according to the astronaut inside*: the light pulse *curves downward* as it travels through the spacecraft. This effect occurs because the rocket is accelerating vertically while the light pulse traverses the spacecraft. Now consider the same thing happening on the rocket placed on Earth in a gravitational field. According to the equivalence principle, exactly the same thing happens: *the light pulse is attracted by the gravitational field and curves downward* as shown in Figure 15.2c. We have greatly exaggerated the effect in Figure 15.2.

If we think about it a little, this result should not really come as a great surprise. We showed in Section 2.12 that energy and mass are equivalent. A light pulse has energy, and it therefore can act as if it had a mass. We can think of the bending of the light pulse as simply the gravitational attraction of light. The bending of light on Earth due to gravitational attraction is tiny. A beam of light sent across a distance as wide as the United States and under a gravitational attraction of g would be deflected less than 1 mm. However, gravitational effects on light have been observed experimentally, even on Earth, as discussed in the next section.

Spacetime Curvature

We already learned in our discussion of the special theory of relativity in Chapter 2 that space and time are interrelated. Einstein thought of gravity not as a force but as a curvature of spacetime. Spacetime is flat in empty space, but spacetime becomes highly changeable near matter. It is like a flexible material that accommodates matter by curving in the vicinity of the object. We can visualize this in two dimensions by imagining what a rubber sheet would do if a bowling ball is placed on it. The rubber sheet would stretch to accommodate the bowling ball, as shown in Figure 15.3a. Spacetime is curved by massive bodies, and we can imagine a

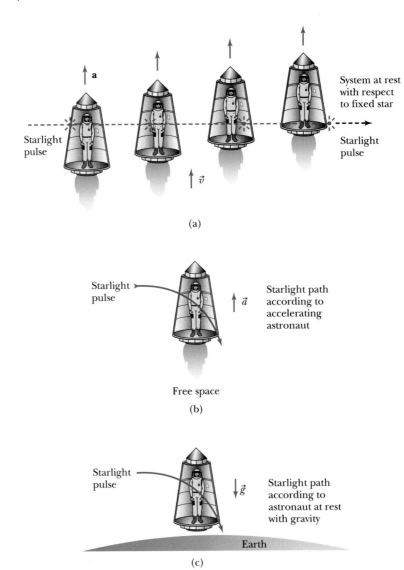

Figure 15.2 Starlight enters a small hole in a spacecraft while the rocket is accelerating. (a) The burst of starlight will hit a spot on the opposite wall at a point lower than where the light came in (greatly exaggerated here). (b) According to the astronaut inside, the light pulse curved downward and must have been affected by the acceleration. (c) According to the equivalence principle, the same thing must happen on the Earth because of gravity.

small mass moving toward the massive body, not because of the gravitational force, but because it rolls in on the spacetime curvature toward the massive body (see Figure 15.3b).

The flow of time is determined by the magnitude of the gravitational field nearby. For example, light travels a longer path in strong gravitational fields,

Figure 15.3 (a) A suggestion of spacetime curvature in two dimensions. Imagine a bowling ball being placed on a rubber sheet. The sheet deforms to accommodate the heavy mass. The lines represent spacetime. (b) We can imagine a smaller mass being attracted to the larger mass by rolling along the spacetime curvature.

because the spacetime geometry has changed and expanded. Simply put, mass-energy curves space and time. We can summarize as follows:

Mass-energy tells spacetime how to curve.

Spacetime curvature tells matter how to move.

15.2   Tests of General Relativity

We review several of the many tests of general relativity in this section. Because of extreme experimental difficulties, only a few experimental tests were done before the 1960s. A constant stream of experiments has been done in recent years.

Bending of Light

According to Einstein's predictions, light should be bent while passing through a strong gravitational field. During a solar eclipse of the sun by the moon, most of the sun's light is blocked on Earth, which affords the opportunity to view starlight passing close to the sun. Light from a distant star passing close to the sun will be bent away from its normal direction while passing close to the sun, as shown in Figure 15.4. A total eclipse of the sun was to occur in May 1919, and following the end of World War I preparations were quickly made by Arthur Eddington and colleagues to mount observations in South America and Africa, both good places to observe that particular eclipse. Einstein's general theory predicted a deflection of 1.75 seconds of arc, and the two measurements found 1.98 ± 0.16 and 1.61 ± 0.40 seconds. Einstein became an international celebrity because of the experiment and the publicity surrounding it. Since the eclipse of 1919, many experiments, using both starlight and radio waves from quasars, have confirmed Einstein's predictions about the bending of light with increasingly good accuracy.

Spectacular evidence has now been found for the gravitational bending of light. The effect is called **gravitational lensing**. If light from a distant object like a quasar (see Chapter 16) passes by a nearby galaxy on its way to us on Earth, the light can be bent multiple times as it passes in different directions around the galaxy. The result can be like that shown in Figure 15.5.

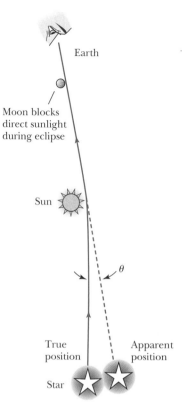

Figure 15.4 Starlight passing close to the sun is bent due to gravitational attraction. The effect is that the apparent position of a star is not always its true position.

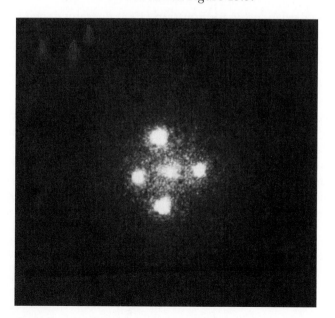

Figure 15.5 A Hubble Space Telescope image of a gravitational lens effect. This image is called the "Einstein Cross." The light of a distant quasar has been bent around a nearby galaxy on its way to Earth, forming the four outside images. *NASA.*

Gravitational Redshift

The second test of general relativity also relies on the gravitational attraction on light. Imagine a light pulse being emitted from the surface of the Earth to travel vertically upward. The gravitational attraction of the Earth cannot slow down light, but it can do work on the light pulse to lower its energy. This is similar to a rock being thrown straight up. As it goes up, its gravitational potential energy increases while its kinetic energy decreases. A similar thing happens to a light pulse. A light pulse's energy depends on its frequency f through Planck's constant, $E = hf$. As the light pulse travels up vertically, it loses kinetic energy and its frequency decreases. Its wavelength increases, so the wavelengths of visible light are shifted toward the red end of the visible spectrum. We say that the light is *redshifted*, and the general phenomenon is referred to as the **gravitational redshift**.

Let's assume we do the experiment close to the surface of the Earth where to a good approximation g is constant. The energy lost when traveling upward a distance H is mgH. If f is the frequency of the light at the bottom, and f' the frequency at the top, energy conservation gives

$$hf = hf' + mgH \qquad (15.2)$$

We can substitute an effective mass of light by letting $m = E/c^2 = hf/c^2$ to obtain

$$hf = hf' + hf \frac{gH}{c^2}$$

If we cancel Planck's constant h and let $\Delta f = f - f'$, we have

$$\frac{\Delta f}{f} = \frac{gH}{c^2} \qquad (15.3)$$

Measurements comparing frequency differences are very sensitive, and such an experiment was performed by Pound and Rebka in 1960 in a tower at Harvard University using gamma rays from radioactive ^{57}Co ($^{57}Co + e^- \rightarrow {}^{57}Fe^* \rightarrow {}^{57}Fe + \gamma$ ray). They sent γ rays down the tower, so the γ rays gained energy, increasing their frequency. In this case, a *blueshift* occurs. Pound and Rebka used the Mössbauer effect* to obtain the needed sensitivity of $\Delta f/f \approx 10^{-15}$.

The general relationship for Equation (15.3) does not depend on the gravitational potential energy mgH but rather on $GMm(1/r_1 - 1/r_2)$, where G is the universal gravitational constant. The ratio of the difference in frequency is

$$\frac{\Delta f}{f} = -\frac{GM}{c^2}\left(\frac{1}{r_1} - \frac{1}{r_2}\right) \qquad (15.4)$$

where clocks measuring the frequency are placed at distances r_1 and r_2 from the center of the gravitational field. When $r_1 < r_2$, Δf is negative. A signal leaving r_1 has a lower frequency when arriving at r_2. This means that an atom emitting light signals in a strong gravitational field (r_1 is small near the mass center) has its wavelength redshifted when the light arrives at r_2 far away from the mass center.

Gravitational redshifts from stars are difficult to measure, because of the larger *Doppler redshift* due to the star moving away from us at high speed; a few

---

*The energy of photons emitted or absorbed by nuclei is shifted due to the recoil of the nucleus. The Doppler effect can offset somewhat the change in frequency of the photon. In 1958 R. L. Mössbauer (Nobel Prize, 1961) performed experiments in which the nucleus was embedded in a solid crystal, thereby producing essentially a *recoil-free* emission or absorption of photons. This causes a tremendous increase in sensitivity in the resonance emission or absorption of photons, a situation that allows the measurement of extremely small differences in energies (or photon frequencies).

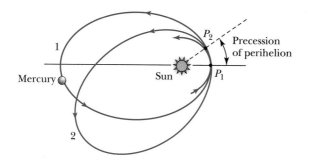

stars are observed to be moving toward us. Several experiments using light from the sun and from white dwarfs have confirmed the existence of the gravitational redshift. A very accurate experiment was done by comparing the frequency of an atomic clock flown on a Scout D rocket to an altitude of 10,000 km with the frequency of a similar clock on the ground. The measurement agreed with Einstein's general relativity theory to within 0.02%.

Perihelion Shift of Mercury

The orbits of the planets are ellipses, and the point closest to the sun in a planetary orbit is called the *perihelion*. It has been known for hundreds of years that Mercury's orbit precesses about the sun as shown in Figure 15.6. The point of perihelion precesses very slowly, but the rate of precession has been accurately measured. Most of the effect we observe is due to the Earth's rotation, but after that is subtracted there is still a perihelion shift of 575 seconds of arc *per century*. Most of this shift is due to the gravitational perturbation caused by the other planets, but in 1859 Urbain Jean Joseph Le Verrier (who had earlier discovered the planet Neptune) announced that after even these corrections were accounted for, there was still a remaining perihelion shift amounting to 43 seconds of arc per century. This was a great mystery to scientists at the time, and the answer had to wait some 60 years, until Einstein worked out his general theory of relativity. Einstein showed that general relativity also predicted a perihelion shift, and his calculation showed that the shift was just the missing 43 seconds needed! Einstein was overjoyed by this calculation; together with the observed deflection of starlight by the sun in 1919, it gave great credence to his theory.

Example 15.1

In the test of Hafele and Keating of flying atomic clocks around the Earth (see Chapter 2), the gravitational redshift had to be considered. Calculate the effect and compare it with the special relativity time dilation effect. Assume the jet airplane travels 300 m/s and the circumference of the Earth is about 4×10^7 m.

Strategy The ratio $\Delta f / f$ will be equal to the time difference ratio $\Delta T / T$ measured by the clocks on Earth and on the jet airplanes. Because the airplane's altitude is negligible in comparison to the Earth's radius, we use the simpler of the

two equations for the gravitational redshift, Equation (15.3). We use the time dilation Equation (2.19) from Chapter 2 to determine the special relativity effect.

Solution Equation (15.3) gives us

$$\frac{\Delta T}{T} = \frac{gH}{c^2}$$

The value of H for the clock on the Earth's surface is r_e, and the value for the flying airplane is $r_e + A$, where A is the altitude of the airplane, about 33,000 feet (10,000 m). The value of H is the difference in the height of the two clocks in the Earth's gravitational field g, and $H = A$. We neglect the change of g over this altitude, and $\Delta T/T$ becomes

$$\frac{\Delta T}{T} = \frac{(9.8 \text{ m/s}^2)(10{,}000 \text{ m})}{(3 \times 10^8 \text{ m/s})^2} = 1.09 \times 10^{-12}$$

The eastward and westward airplane trips took about $T = 45$ hours flying time. The difference in the two clocks due to the gravitational redshift is

$$\Delta T = (1.09 \times 10^{-12})(45 \text{ h})\left(\frac{3600 \text{ s}}{1 \text{ h}}\right)$$

$$= 177 \times 10^{-9} \text{ s} = 177 \text{ ns}$$

A clock fixed on Earth will measure a flight time T_0 of

$$T_0 = \frac{4 \times 10^7 \text{ m}}{300 \text{ m/s}} = 1.33 \times 10^5 \text{ s}$$

Because a clock in the airplane will run slowly, an observer on Earth will say the time measured on the airplane is $T = T_0\sqrt{1 - \beta^2}$ where $\beta = v/c = (300 \text{ m/s})/(3 \times 10^8 \text{ m/s}) = 10^{-6}$. The time difference is

$$\Delta T = T_0 - T = T_0(1 - \sqrt{1 - \beta^2})$$

Because β is so small, we can use a power series expansion of the square root and ignore all terms smaller than β^2. The special relativity effect is

$$\Delta T = T_0[1 - (1 - \beta^2/2 + \cdots)] = \frac{\beta^2 T_0}{2}$$

$$= \frac{1}{2}(10^{-6})^2(1.33 \times 10^5 \text{ s}) = 6.65 \times 10^{-8} \text{ s} = 66.5 \text{ ns}$$

The *gravitational time dilation* effect of 177 ns is larger than the approximate 66.5-ns *special relativity time dilation* effect. There is also a third correction due to the rotation of the Earth.

Light Retardation

Light retardation is similar to the deflection of light (or any electromagnetic radiation) by gravitational fields. As light passes by a massive object, the path taken by the light is longer because of the spacetime curvature, as we show in Figure 15.7 with a radio wave. The longer path causes a time delay for a light pulse traveling close to the sun. The time difference between similar paths, with one being close to the sun, can be compared with the general relativity prediction.

Irwin Shapiro showed that such an effect could be measured by sending a radar wave to Venus, where it was reflected back to Earth. The position of Venus has to be in the "superior conjunction" position on the other side of the sun from the Earth as shown in Figure 15.8. In this path, the light signal passed close to the sun and experienced a time delay. Shapiro's measurement on January 25, 1970, of a time delay of about 200 μs was in excellent agreement with the general theory (Figure 15.9).

Several experiments have since been done with spacecraft to measure this effect. The experiment to land part of the Viking spacecraft on Mars in 1976 produced agreement with the general theory to within an experimental uncertainty of about 0.1%. Physicists reported in 2003 on a measurement done with the NASA *Cassini* spacecraft. The spacecraft, on its way to its highly successful rendezvous with Saturn in 2004, passed on the opposite side of the sun from Earth in 2002. Researchers measured the frequency change of radio waves that passed

Figure 15.7 Light or radio waves from a satellite on the opposite side of the sun from Earth pass close to the sun. As the light passes close to the sun, the light is gravitationally attracted by the sun and encounters a longer path due to spacetime curvature on its way to Earth. The effect, called light retardation, is that the light takes longer to reach Earth.

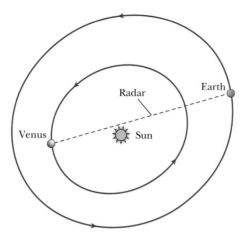

Figure 15.8 The superior conjunction position of Venus when it is on the other side of the sun from the Earth. Radar waves sent between the two planets will be delayed slightly due to the gravitational attraction of the sun.

Figure 15.9 Shapiro's 1970 data of the time-delay measurements of the radar waves between Earth and Venus. The solid line is the general relativity prediction. *From Shapiro et al.*, Physical Review Letters *26, 1132 (1971).*

close to the sun while they were traveling to and from the spacecraft from Earth. The radio waves passing close to the sun traveled an extra distance due to the spacetime curvature. The results were the best to date for this kind of measurement and agreed with the general theory to 2 parts in 10^5.

15.3   Gravitational Waves

The four types of experiments discussed in the previous section represent tests that have confirmed the validity of general relativity. Other phenomena predicted by general relativity are currently under extensive investigation. One of these is gravitational waves, which is the subject of this section. Gravitational waves are not too hard to imagine. We know that electromagnetic waves are produced by oscillating charges. When a charge accelerates, the electric field surrounding the charge redistributes itself. This change in the electric field produces an electromagnetic wave, which is easily detected. In much the same way, an accelerated mass should also produce gravitational waves. It is a classical effect,

Russell Hulse (1950–) was born in New York City where he attended the Bronx High School of Science before attending Cooper Union College. While Hulse was a graduate student at the University of Massachusetts (Ph.D., 1975) working with Joseph Taylor, they discovered the first binary pulsar in 1974 using the Arecibo Observatory in Puerto Rico. They received the Nobel Prize for Physics in 1993, which was the first prize given for work in general relativity. *AIP Emilio Segrè Visual Archives, W. F. Meggers Gallery of Nobel Laureates.*

not a quantum one, but the effect is very small. Gravitational waves carry energy and momentum, travel at the speed of light, and are characterized by frequency and wavelength. As gravitational waves pass through spacetime, they cause small ripples. The stretching and shrinking is on the order of 1 part in 10^{21} even due to a strong gravitational wave source.

Einstein's general theory of relativity is also a modern theory of gravitation, but does not include quantum theory. Einstein showed that his general theory had wave solutions, but because the expected wave amplitudes are so small, they were not taken seriously until 1968 when Joseph Weber announced that he had detected gravitational radiation (waves) from space. Weber's announcement spurred a new field of gravitational astronomy, even though subsequent investigations by other experimenters have not confirmed his results. It seems unlikely that gravitational waves produced in the laboratory could ever be detected, but possibly waves from an astronomical phenomenon such as the collapse of two neutron stars rotating around each other, a neutron star falling into a black hole, or the gravitational collapse of a star to form a black hole (see Chapter 16) may have enough accelerating mass to produce detectable radiation. Neutron stars are collapsed stars of extremely high density and will be discussed in Section 15.4. It might even be possible to detect gravitational radiation coming from the first fraction of a second after the origin of the universe (the Big Bang). Physicists conjecture that gravitational waves cause the space between objects to expand and contract. The difficulty of ever proving this theory is that the predicted effects of gravitational waves would alter distances by less than 1 part in 10^{21} by the time the waves reach Earth. This effect has been likened to noticing a single grain of sand added to all the beaches of Long Island, New York.

Astronomers and physicists believe with some conviction that they have identified a system exhibiting gravitational radiation, although no direct detection of gravitational waves has occurred. Russell Hulse and Joseph Taylor (Nobel Prize for Physics, 1993) discovered in 1974 a binary system consisting of a pulsar (a rapidly spinning neutron star) and another star rotating around each other with a period of about 8 h. The unidentified second star is also believed to be a neutron star. If the system (called PSR 1913 + 16) is radiating gravitational waves, it loses energy, and the two stars come closer together, spiraling faster and faster around each other. As the gravitational radiation increases, the stars will finally lose enough energy to crash into one another. The predicted and observed decrease in the orbital period is in good agreement with the production of gravitational waves.

One possible method of detecting gravitational waves (shown in Figure 15.10) takes advantage of optical interference. Four test masses, two on each arm of the apparatus, react to a gravitational wave passing through, changing the path lengths L_1 and L_2 slightly. This basic Michelson interferometer device takes advantage of the Fabry-Perot method by placing mirrors on the masses, so that the light bounces back and forth between masses (2 km or 4 km) many times before interfering. This greatly increases the sensitivity of the device.

Detectors placed in the states of Washington and Louisiana (over 3000 km apart) and in three foreign countries aid in looking for correlations in the gravitational waves. In the United States this ambitious project, called LIGO for Laser Interferometer Gravitational-Wave Observatory, is led by groups from Caltech and MIT; it has two detectors in Louisiana and one in Washington state. Three others are located in Germany, Japan, and Italy. Even though the six detectors vary somewhat in size, sensitivity, and operation, they all work in approximately the same manner. As a gravitational wave passes through Earth, it arrives at each detector at a slightly different time. This allows the researchers to determine the

Joseph Taylor (1941–) was born in Philadelphia and educated at Haverford College and Harvard (Ph.D., 1968). He has been on the faculties of the University of Massachusetts and Princeton University. *AIP Emilio Segrè Visual Archives, Physics Today Collection.*

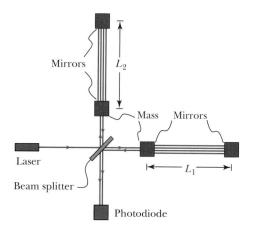

Figure 15.10 A schematic diagram of the LIGO device. A gravitational wave passing through large masses suspended on thin wires causes the distance $L_1 - L_2$ to change. The interference pattern observed in the photodiode should be capable of recognizing the gravitational wave.

source of the gravitational wave and to eliminate noise sources. The LIGO detectors are designed to have a peak sensitivity of 3 parts in 10^{23} at 180 Hz. Five of the six detectors were operational in 2005, but were having difficulty reaching their design sensitivities.

NASA and the European Space Agency (ESA) are jointly developing a space-based gravitational wave observatory named Laser Interferometer Space Antenna (LISA) that will use laser interferometry between three spacecraft located 5 million km apart in space in orbit around the sun. Each spacecraft will carry two telescopes with associated lasers and optical systems. The spacecraft will approximate a triangle in the plane of the Earth's orbit around the sun. This array is being designed to be sensitive to gravitational waves with frequencies as low as 10 millihertz. Even though the spacecraft will not be susceptible to the same kind of noise as the ground-based observatories, they will still have difficulties accounting for the solar wind and solar radiation pressure to obtain the sensitivities needed. This ambitious project is not expected to be launched before 2012.

15.4 Black Holes

While a star is burning, the heat and radiation pressure produced by the thermonuclear reactions push out the star's matter and balance the force of gravity. When the star's fuel is used up, no heat or radiation pressure is left to counteract the tremendous force of gravity, which becomes dominant. The star's remaining mass collapses into an incredibly dense ball, much smaller than the burning star. The amount of mass left in the star determines what it becomes. A star the size of our sun will become a *white dwarf*. Stars somewhat larger than the sun, but having a mass less than about 3 solar masses, will become *neutron stars*. We will discuss white dwarfs and neutron stars further in Section 16.3. When stars having more than 3 solar masses collapse, an astounding phenomenon occurs. The local gravitational force is so strong that nothing can ever leave the collapsed star, not even light! Robert Oppenheimer and Hartland Snyder predicted in 1939 that the gravitational collapse of a star could produce such a body, now called a **black hole**. Because nothing can escape a black hole, it is very hard to detect.

Let's discuss the case of a spherical object like a star as the star's fuel is used up, and the gravitational attraction starts to contract the mass. The star becomes smaller and smaller until the point is reached when the gravitational force is so strong that nothing inside the star can escape. Karl Schwarzschild determined the

Karl Schwarzschild (1873–1916) was born in Germany and studied at Strasbourg and Munich (Ph.D., 1896). He was a professor at Göttingen and Potsdam before volunteering for the German army in World War I. While serving in Russia in 1915 he did important work on both relativity and quantum theory, which resulted in his pioneering work on black holes. He was the most celebrated astronomer of his day in Germany. He contracted a disease while serving in Russia and died at the young age of 42. *AIP Emilio Segrè Visual Archives.*

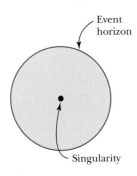

Figure 15.11 At the center of a black hole is a *singularity*. Within a certain distance of the singularity, called the *event horizon*, the gravitational pull is so strong that nothing, including light, can escape. The event horizon is not a physical boundary, but rather a limit beyond which nothing can leave. The event horizon describes the size of the black hole.

radius of the collapsed star in 1916 using Einstein's general theory of relativity. He showed that this radius, $r = r_S$, now called the *Schwarzschild radius*, is given by

$$r_S = \frac{2GM}{c^2} \qquad \text{Schwarzschild radius} \qquad (15.5)$$

This expression is for a nonrotating spherical mass. The body continues contracting past the Schwarzschild radius, but nothing happening inside the collapsing body can affect the spacetime region outside r_S. Any light ray or particle emitted inside the black hole is kept from going outside by the gravitational force.

An observer outside the Schwarzschild radius cannot even tell when the radius passes through r_S. The light leaving the body right before the body crossed over to $r < r_S$ is so strongly attracted to the body that it takes progressively longer and longer to reach the observer. This is merely an optical effect, because physically the apparent luminosity of the body decreases rapidly with time, and the light is strongly redshifted.

The boundary region or surface of a black hole is called the **event horizon** (see Figure 15.11); no information from inside the event horizon can escape to the outside, not even light. The point at the center of a black hole is called the *singularity*. For a spherical black hole, the event horizon is located at r_S. When people talk about the size of a black hole, they are referring to the size of the event horizon. If we insert the gravitational constant and the speed of light into the Schwarzschild radius, we obtain

Schwarzschild radius

$$r_S = \frac{2GM}{c^2} = (1.5 \times 10^{-27}\,\text{m/kg})M \qquad (15.6)$$

This allows us to calculate the radius of a black hole for a given mass, as seen in the following example.

Example 15.2

Calculate the Schwarzschild radius for the sun and Earth.

Strategy We use the masses of the sun and Earth in Equation (15.6) to find the Schwarzschild radius for each.

Solution If we substitute the mass of the sun and Earth into Equation (15.6), we obtain

$$r_S(\text{sun}) = (1.5 \times 10^{-27}\,\text{m/kg})(2 \times 10^{30}\,\text{kg}) = 3\,\text{km}$$

$$r_S(\text{Earth}) = (1.5 \times 10^{-27}\,\text{m/kg})(6 \times 10^{24}\,\text{kg}) = 9\,\text{mm}$$

If the Earth were a black hole, the dark spot would be very tiny indeed! Of course neither the Earth nor the sun have enough mass to become a black hole.

The problem of detecting a black hole is particularly difficult. If light cannot leave a black hole, how can it be detected? There are several indirect means. As the star collapses during the formation of a black hole, the gravitational redshifts should be large. The signals abruptly cease when the black hole is formed. Observation of such signals would be pure luck. Stephen Hawking has applied quantum theory to the problem of a black hole and has outlined a method by which energy, called *Hawking radiation*, evaporates from the black hole much like a thermal energy spectrum (see Problem 12). The energy is very difficult to detect, and the process is so slow that it would take many ages of the universe for the black hole to dissipate its energy. We can understand this process by imagining a particle being attracted from outside the black hole to the event horizon. The particle gains energy from the gravitational field of the black hole. As it approaches the event horizon, a virtual particle-antiparticle pair is created by utilizing the large energy gain. One member of the newly created pair occasionally goes into the black hole, while the other one escapes. In essence the gravitational energy of the black hole has been used to create the particle-antiparticle pair.

Hawking calculated that the temperature of the blackbody radiating the energy is

$$T = \frac{\hbar c^3}{8\pi kGM} \tag{15.7}$$

In Equation (3.16) we found that the power radiated per unit area at temperature T is given by the Stefan-Boltzmann law, $R(T) = \epsilon\sigma T^4$, where we let the emissivity $\epsilon = 1$ for a blackbody and σ is the Stefan-Boltzmann constant. The power $P(T)$ radiated by a blackbody at temperature T becomes

$$P(T) = (4\pi r_S^2)\sigma T^4 \tag{15.8}$$

where r_S is the blackbody's Schwarzschild radius. If we insert the blackbody temperature from Equation (15.7) into Equation (15.8), we have

$$P(T) = 4\pi\sigma r_S^2\left(\frac{\hbar c^3}{8\pi kGM}\right)^4 \tag{15.9}$$

This result can be used to determine the rate of Hawking radiation emanating from a black hole.

Example 15.3

Determine how much time a black hole having 3 solar masses will take to radiate its energy.

Strategy We use Equation (15.9) to determine how long it takes a blackbody of mass-energy Mc^2 to radiate all its energy.

Solution We use the mass-energy Mc^2 to find that it loses its energy at the rate $P(T) = -d(Mc^2)/dt = -c^2(dM/dt)$, where we have taken the time derivative. We set this result equal to Equation (15.9) and find

$$-c^2\frac{dM}{dt} = 4\pi\sigma r_S^2\left(\frac{\hbar c^3}{8\pi kGM}\right)^4$$

$$= 4\pi\sigma\left(\frac{2GM}{c^2}\right)^2\left(\frac{\hbar c^3}{8\pi kGM}\right)^4 = \frac{2\sigma\hbar^4 c^8}{(8\pi)^3 k^4 G^2}\frac{1}{M^2}$$

and the rate of mass loss becomes

$$\frac{dM}{dt} = -\frac{2\sigma\hbar^4 c^6}{(8\pi)^3 k^4 G^2}\frac{1}{M^2} = -\alpha\frac{1}{M^2} \tag{15.10}$$

where we have collected the constants into α. Evaluation of the constant $\alpha = 3.96 \times 10^{15}$ kg^3/s is left to Problem 16.

We rewrite Equation (15.10) and integrate to find

$$M^2\,dM = -\alpha\,dt$$

$$\int M^2\,dM = -\alpha\int dt$$

$$\frac{M^3}{3} = -\alpha t + C'$$

$$M^3 = -3\alpha t + C = -3\alpha t + M_0^3$$

where the new constant $C = 3C' = M_0^3$ is related to the mass M_0 at $t = 0$. The time for all the mass to radiate can be found by letting $M = 0$, and we find

$$M_0^3 = 3\alpha t$$

$$t = \frac{M_0^3}{3\alpha}$$

Now we find the time for the 3-solar-mass black hole to evaporate.

$$t = \frac{[3(1.99 \times 10^{30} \text{ kg})]^3}{3(3.96 \times 10^{15})} = 1.8 \times 10^{76} \text{ s} = 5.7 \times 10^{68} \text{ y}$$

This is obviously many times the lifetime of our universe, which is 13.7 billion years.

Black holes may be detected indirectly by their gravitational influence on their surroundings. Consider a binary system of a black hole and a companion star. The strong gravitational force of the black hole pulls gaseous matter off the star, but the gas does not go directly into the black hole. Because the companion star has some rotational motion, the gas moves into an orbit around the black hole (see Figure 15.12). This rotating mass collects into an **accretion disk** surrounding the black hole. As the gas moves toward the inside of the disk, it revolves at speeds approaching the speed of light, and the internal friction between layers of gas heats up the gas to high temperatures. This superheated gas emits x rays that can be observed, because the accretion disk is not inside the black hole's event horizon. The strongest galactic x-ray sources are binary systems, and only black holes and neutron stars are believed capable of producing the immense x-ray emissions observed. Experiments indicate that many of the x-ray sources are due to neutron stars, but remember that neutron stars with masses greater than about 3 solar masses will collapse into a black hole. Neutron stars will have 1.5 to 3 solar masses.

A good candidate for a black hole would be an unseen mass in a binary system producing copious amounts of x rays. For larger black holes much of the accretion disk's energy is emitted as electromagnetic radiation before the matter is absorbed by the black hole. Another process called *advection* is also possible. If the black hole accumulates mass at a slower rate, there is not as much friction in the accretion disk, and the energy is absorbed by the black hole. Thus it is possible for black holes, even in a binary system, to only dimly emit radiation.

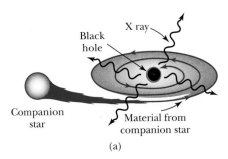

(a)

Figure 15.12 Matter from a companion star is attracted by the strong gravitational pull of (a) a black hole or (b) a neutron star. The matter spirals inward, forming a disk before being pulled into the black hole or neutron star. Friction within the disk creates tremendous temperatures and copious numbers of x rays. *Adapted from drawing by Ramesh Narayan and Michael Garcia, Harvard-Smithsonian Center for Astrophysics.*

(b)

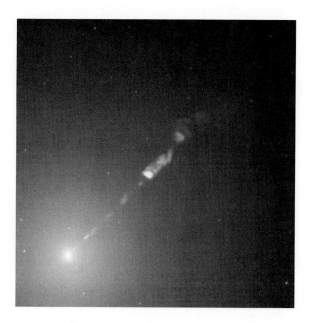

Figure 15.13 There is convincing evidence of a black hole in the center of the giant galaxy M87 seen in the lower left of this Hubble Space Telescope photograph. The highly energetic jet emanating upward and to the right is composed of fast-moving charged particles and has broken into knots as small as 10 lightyears across. *Courtesy of H. Ford (JHU, STSci), R. Harms, Z. Tsvantanov, and NASA.*

It is likely that astronomers and astrophysicists have discovered a number of black holes. The earliest and best-known candidate is Cygnus X-1, which was found in 1965 to be an x-ray source. The X-1 designation indicates it was the first x-ray emitter found in the constellation Cygnus. In the 1970s it became known that Cygnus X-1 and an observable star were a binary system. Many observations and analyses have been made on this system in the intervening decades. It appears to be a classic system having an accretion disk whose gaseous molecules get hot and emit x rays. Data indicate that the mass of the unobserved object must be about 7 solar masses, which is convincing evidence to most scientists that Cygnus X-1 is indeed a black hole.

Black holes come in at least two sizes. Black holes like Cygnus X-1 that have between 5 and 20 masses of the sun (solar masses) are called *stellar black holes*. The other variety is *supermassive* black holes, which are thought to have masses of millions or billions of times that of our sun. There is some controversy about black holes having intermediate sizes.

Supermassive black holes

A convincing case can be made that the unusual giant galaxy M87 has a black hole near its center. This black hole is of the supermassive variety and is thought to have a mass of at least 3 billion solar masses. A disk of hot gas can be seen in the lower left of Figure 15.13, taken with the Hubble Space Telescope. The observations of the rotating gas indicate the mass of the object, while the size of the disk indicates the approximate volume of the central object. These results yield an object with a central density so high that only a black hole is possible. The photo also dramatically indicates a plasma jet shooting mass up and away from the central object.

Figure 15.14 shows a composite photo of ground-based optical and radio telescope images of the NGC 4261 galaxy beside a high-resolution Hubble Space Telescope image of the galaxy core. The radius, speed, and mass of the object can be determined from its rotation. The ring is believed to be the accretion disk feeding the black hole and is thought to have a mass of a billion suns. The black hole, thought to be inside the bright white spot at the center, is too small to be observed in this image. The disk is believed to be cold gas and dust. The gas and dust are compressed and heated as they rush toward the black hole. The jets are

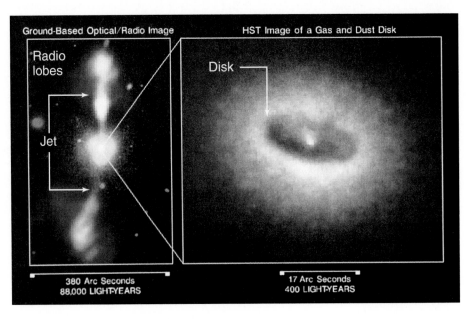

Figure 15.14 The core of the elliptical galaxy NGC 4261 is shown with a Hubble Space Telescope image on the right and a composite photo of an expanded region on the left. A black hole is believed to be inside the bright white spot at the center of the disk on the right side. Hot gas is escaping in jets from the vicinity of the black hole on the left. *Walter Jaffe/Leiden Observatory, Holland, Ford/JHU/STSci, and NASA.*

due to hot gas escaping from the vicinity of the black hole. The images provide a classic case of the phenomena expected near a black hole and provide strong circumstantial evidence of a black hole.

With the tremendous production of the Hubble Space Telescope and the output of new telescopes coming on line, and with better detection systems and computer analysis, more and more candidates for black holes appear each year. The scientific community generally accepts the existence of these strange phenomena that can only be inferred indirectly, because we can't actually see them.

15.5   Frame Dragging

Soon after Einstein presented his general theory of relativity in 1915, the Austrian physicists Josef Lense and Hans Thirring used it to propose in 1918 that a rotating body's gravitational force can literally drag space and time around with it as the body rotates. This effect, sometimes called the Lense-Thirring effect, is normally referred to as **frame dragging**. All celestial bodies that rotate can modify the spacetime curvature, and the larger the gravitational force, the greater the effect. It was one of the last of the general theory's predictions to be confirmed.

Because of the strong gravitational force near black holes, researchers were first able to confirm frame dragging in 1997 by noticing that x-ray emissions from several black holes varied in intensity. This variation was repetitious and could be explained if the object's orbit were precessing. As the matter orbits the black hole, the spacetime that is being dragged around the black hole drags the matter around with it as shown in Figure 15.15. If we could observe the black hole binary system, we might see the accretion disk wobble like a top out of balance. Astrophysicists were first able to observe the spacetime distortion of black holes by observing x rays with NASA's Rossi X-ray Timing Explorer spacecraft.

In 1998 Italian and American researchers were able to use two Earth-orbiting Laser Geodynamics Satellites, LAGEOS I and LAGEOS II, to detect the effects of

GRAVITY PROBE B

The Gravity Probe B (GP-B) was finally launched in 2004 to prove two predictions of the general theory: the effects of spacetime curvature and frame dragging. Originally proposed in 1961, the GP-B has four gyroscopes bathed in liquid helium that measure the Earth's spin with unprecedented accuracy. Each gyroscope has an incredibly smooth quartz sphere the size of a Ping-Pong ball that spins 10,000 times per minute. The spacecraft is in polar orbit, passing over each of the poles, about 650 km above Earth. The spacecraft's telescope is locked onto the guide star IM Pegasus, about 300 lightyears away.

Gravity Probe B is designed to measure two effects: frame dragging and the geodetic effect. Frame dragging is the effect due to a massive body, Earth in this case, dragging spacetime with it as it moves through space. Because Earth is rotating around its own axis, it will drag spacetime with it, and the effect on the gyroscope is that the gyroscope will tilt slightly away from the plane of its orbit because Earth is dragging it. GP-B is supposed to measure the effect to 1%.

The geodetic effect is due to spacetime curvature. This distortion of spacetime has the effect on the gyroscope of tilting the gyroscope's spin axis in the orbit plane. General relativity predicts an effect 150 times larger than the frame dragging, so GP-B will have to measure the effect to 1 part in 10^4. The scientists expect to measure a deviation from the gyroscope's orientation with respect to a guide star. The difficulty is that the change expected is only 0.041 seconds of arc per year for the frame dragging effect—about the same arc made by the diameter of a penny in Los Angeles when observed by someone in New York City.

Some scientists claim the GP-B has been in development for so long that the scientific goals it was designed to address are no longer a priority; the focus of research in the field has changed based on experimental results obtained since the project was originally planned. But other scientists say the GP-B will be the first direct experiment to test its general relativity objectives. After several months of calibration the spacecraft locked onto the guide star IM Pegasi, and data collection started in 2004. The experiment is expected to be completed in late 2005.

frame dragging near Earth. The LAGEOS satellites are passive and dedicated to laser ranging, in which Earth-based lasers are reflected off the satellites. The researchers were able to detect that the plane of the satellites shifted about 2 meters per year in the direction of the Earth's rotation, in agreement with the prediction of general relativity. This relativistic effect is about 10 million times smaller than classical Newtonian disturbances and required painstaking analysis of data that required four years to acquire. The Gravity Probe B satellite is expected to reduce the uncertainty of this effect by a factor of more than 10.

It may seem that general theory has no applications in our modern life, but the use of Global Positioning System (GPS) receivers to know where we are at work and play must utilize general relativity corrections. The receivers operate by passively receiving signals from multiple satellites and then determining parameters like position, altitude, and velocity. The determination of these parameters depends on very accurate atomic clocks in the satellites. Special relativity corrections are applied because the satellites are moving with respect to Earth. General relativity corrections are also made because the satellites are in orbit more than 20,000 km above the Earth. When the GPS system was implemented, it was found that general relativistic corrections needed to be routinely made.

Figure 15.15 An artist's conception of a spinning black hole (invisible at the center) twisting spacetime as it turns. The lines curving to the outside represent spacetime. Excess energy shoots off in jets of hot gas along the directions of the rotation axes. The white ball in the center represents the event horizon. Note the similarities to Figure 15.14. *Artwork used by permission of artist Joe Bergeron.*

What is perhaps so remarkable about the general theory of relativity is that Einstein did not develop it as a way to explain existing experimental data. Einstein's imagination was perhaps the greatest of any scientist's. By imagining gravity as warping spacetime, and not as a force at a distance, Einstein made an intellectual leap that is still astounding. The theory ranks as one of the greatest achievements of the twentieth century along with quantum theory. It has been said that if Einstein had not developed his general theory, it might have been decades before another physicist had the creative genius to produce it.

Summary

Einstein's general theory of relativity is a theory of gravity that replaces Newton's force laws by a system based on curvature of spacetime. Einstein's principle of equivalence states that there is no experiment that can be done in a small confined space that can tell the difference between a uniform gravitational field and an equivalent uniform acceleration. The principle of equivalence leads to the bending of light in a gravitational field. Spacetime curvature occurs near masses.

There are several tests of general relativity. The bending of light was first measured in 1919 and has been observed in several experiments since. Light can gain and lose energy when passing through gravitational fields. When light loses energy, the frequency of visible light is decreased, and we say it is *redshifted*. To first order, in small regions where g is approximately constant, the gravitational redshift is given by

$$\frac{\Delta f}{f} = \frac{gH}{c^2} \tag{15.3}$$

Many experiments confirm the existence of gravitational redshifts. The general theory also predicts a noticeable effect on the perihelion of Mercury's orbit. Such a calculation

accounted for an anomaly in astronomical measurements known since 1859. The deflection of light by gravitational fields has been detected by light passing close to the sun between Venus and the Earth. Other, more precise experiments have also been done.

One prediction made by general relativity that has not yet been directly confirmed is the existence of gravitational waves, which is currently of great interest. Black holes are objects that have collapsed under their own gravitational attraction. Nothing can escape a black hole, not even light. The size of a nonrotating spherical black hole is given by the Schwarzschild radius.

$$r_{\mathrm{S}} = \frac{2GM}{c^2} \tag{15.5}$$

Scientists believe that the existence of black holes has been confirmed, despite the difficulty of their detection.

More recently the existence of frame dragging has been confirmed for black holes and perhaps for the Earth as well. It is expected that the Gravity Probe B satellite will confirm the effects of frame dragging and the geodetic effect.

Questions

1. Laser light from Earth is received for an experiment by an Earth satellite. Is the light redshifted or blueshifted? What happens to the light if it is reflected back to Earth?

2. Explain why true weightlessness occurs only at the center of mass of the space shuttle orbiter as it rotates around the Earth.

3. Why does a drop of water become bulged in the space shuttle due to the Earth's gravitational field? Draw the water drop showing the direction to the Earth's center.

4. Devise a way for the occupants of a spaceship to know whether they are being pulled into a black hole. What can they do if they determine they are within the Schwarzschild radius?

5. Astronauts riding in the space shuttle are said to be in "zero-free" or "micro" gravity. Explain why this is not really so. Is the net force on them zero?

6. In the experiment discussed in Chapter 2 of the atomic clocks flown around Earth, what was the effect regarding general relativity?

7. In 1919 when the gravitational deflection of light was measured, why did the scientists travel to Africa and South America?

8. How likely is it for a black hole to collide with Earth? Would we have much warning?

9. Why is it difficult to experimentally test models of general relativity?

10. We mention in the text that gravitational redshifts can be observed and measured during the collapse of a star into a black hole. When might the redshifts cease?

11. We mentioned that astronauts can tell whether they are in outer space or "falling around the Earth" by observing a water drop in the corner of their spacecraft. What are the *tidal* forces that were mentioned and where do they come from? Why are they called "tidal" on Earth?

12. Why can we conclude from Equation (15.1) that the inertial and gravitational masses are equal?

Problems

Note: The more challenging problems are indicated by a shaded box.

15.1 Tenets of General Relativity

1. Devise an experiment like the one Newton performed to test the equivalence of inertial and gravitational masses. Using different masses on pendula of equal length, show that the period depends on the ratio of $\sqrt{m_I/m_G}$.

15.2 Tests of General Relativity

2. Controllers want to communicate with a satellite in orbit 400 km above the Earth. If they use a signal of frequency 100 MHz, what is the gravitational redshift? Assume g is constant.

3. For clocks near the surface of the Earth, show that Equation (15.4) reduces to Equation (15.3) for $\Delta f/f$.

4. Repeat Example 15.1 using the more accurate Equation (15.4) for the gravitational redshift. Compare with the result of Example 15.1.

5. In Shapiro's experiment on the time delay during the superior conjunction of Venus and Earth, how much time did it take for the radar signals to travel the round trip to Venus? What percentage change was he looking for?

6. Calculate the gravitational redshift of radiation of wavelength 550 nm (the middle of the visible range) that is emitted from a neutron star having a mass of about 5.0×10^{30} kg and a radius of about 10 km.

7. Radiation is emitted from the sun over a wide range of wavelengths. Calculate the gravitational redshift of light of wavelength 550 nm (the middle of the visible range) that is emitted from the sun.

8. Assume the experiment of Pound and Rebka is performed on the top of the Empire State Building (height = 381 m). What are the change in frequency and the percentage change in frequency due to the gravitational redshift?

9. In the experiment of Pound and Rebka, a 14.4-keV gamma ray fell through a distance of 22.5 m near the Earth's surface. What are the change in frequency and the percentage change in frequency due to the gravitational redshift?

15.4 Black Holes

10. What is the value of the Schwarzschild radius for the moon? ($m_{moon} = 7.35 \times 10^{22}$ kg)

11. Calculate the Schwarzschild radius for Jupiter. ($m_{Jupiter} = 1.9 \times 10^{27}$ kg)

12. Stephen Hawking has predicted the temperature of a black hole to be $T = \hbar c^3/8\pi kGM$, where k is Boltzmann's constant. Calculate the temperature of a black hole with the mass of the sun.

13. Calculate the mass and Schwarzschild radius of a black hole at room temperature. (See Problem 12.) How many solar masses is this?

14. The black hole at the center of the NGC 4261 galaxy is thought to have a mass of 1 billion suns. (a) Calculate its Schwarzschild radius and compare it with the size of our solar system. (b) How long would this black hole take to evaporate by Hawking radiation?

15. (a) Use the known lifetime of the universe to determine the mass of a black hole that would evaporate all its mass during that time. (b) How likely is it that a black hole could exist of this mass?

16. Determine the constant α in Equation (15.10).

General Problems

17. The Global Positioning System satellites operate at an altitude of 20,200 km and use communication frequencies of 1575.42 MHz. Find the gravitational frequency change with respect to Earth.

18. Derive Equation (15.4).

19. One of the communication frequencies that the space shuttle orbiter uses is 294 MHz. Find the gravitational frequency change with respect to Earth when the orbiter is at an altitude of 300 km. Assume g is constant.

20. Weightlessness occurs only at the center of mass of the space shuttle orbiter as it rotates 300 km above the Earth. Calculate the effective g that an astronaut

in the orbiter would feel who is 3 m closer to the Earth than the center of mass. You may choose to ignore relativistic effects.

21. Find the mass of a particle whose Compton wavelength is πr_S where r_S is the Schwarzschild radius. This mass is called the *Planck mass* m_P, and the energy required to create the mass is called the *Planck energy*, $E_P = m_P c^2$. Determine the values of both the Planck mass and energy.

22. The length scale on which the quantized nature of gravity should first become evident is called the *Planck length*. (a) Determine it using dimensional analysis using the fundamental constants G, h, and c. (b) Determine it by finding the de Broglie wavelength of the Planck mass of Problem 21. Are the values close to the value of 10^{-35} m?

23. Use the fundamental constants G, h, and c and dimensional analysis to determine a time constant called the *Planck time*. How much time would it take light to travel the *Planck length* discovered in the previous problem? Are these two times consistent?

24. A communications satellite is at a geosynchronous orbit position (35,870 km above the Earth's surface) and communicates with the Earth at a frequency of 2×10^9 Hz. What is the frequency change due to gravity?

Cosmology—The Beginning and the End

I too can see the stars on a desert night, and feel them. But do I see less or more? The vastness of the heavens stretched my imagination—stuck on this carousel my little eye can catch one-million-year-old light. A vast pattern—of which I am a part—perhaps my stuff was belched from some forgotten star, as one is belching there.

Richard Feynman

Most physicists and astronomers believe our universe evolved from a primordial event called the *Big Bang*. In this chapter we present some of the experimental evidence supporting this belief. The modern science of cosmology is based on the general theory of relativity and has strong ties to elementary particle physics; that is why we have waited until now to present cosmology. Our knowledge of cosmology is increasing, and especially so in the last two decades. The field is rapidly changing, but the foundation is growing brick by brick. We consider in this chapter some of the unexplained evidence and conflicting theories.

What students perhaps find most surprising during their first exposure to cosmology and astrophysics is how much these disciplines depend on other fields of physics. The formation of stars depends on quantum physics and gravitation. The understanding of nucleosynthesis of elements depends on exact measurements of nuclear physics cross sections to determine how certain elements were formed in the early stages of the universe. As we will see in Section 16.2 on the Big Bang, elementary particle physics and cosmology have continued to merge so much that the fields now overlap considerably. The fundamental forces are subjects of extreme interest to both cosmologists and elementary particle physicists. Strongly held theories of today can be disproved by the experimental observations of tomorrow. New Earth- and space-based telescopes are providing answers to our most perplexing questions, and new accelerator experiments may help increase our knowledge about the origin of the universe.

Not only do we attempt in this chapter to understand the origins of our universe, we examine what the demise of life on Earth will be like. Our sun will undoubtedly burn out eventually, but current evidence seems to indicate that our universe will expand forever.

16.1   Evidence of the Big Bang

Over the last half century, scientists have proposed various theories about the origin of the universe. In the 1950s and early 1960s there were two rival theories of cosmology, but data were insufficient to prove either of them. Both theories accounted for the known expansion of the universe, which was demonstrated conclusively by the redshift data of Edwin Hubble in 1929. One theory, known as the **steady-state theory**, held that matter is being continuously created; as the universe expands, the density of the universe remains constant. The other theory, the Big Bang model, proposes that the universe was created in a primeval fireball of incredible density and high temperature. Accumulating evidence in favor of the Big Bang model, which will be presented in Section 16.2, meant that it finally won out over the steady-state theory.

Before presenting the Big Bang model, we will present three important pieces of evidence that have led most scientists to accept it as the most likely model for the origin of the universe. The pieces of evidence include the following:

1. The observations between 1929 and 1952 by Edwin Hubble, using the giant telescopes of Mount Wilson and Mount Palomar, that the galaxies of the universe are moving away from each other at high speeds. The universe is apparently expanding outward from some primordial event.
2. The observation in 1964 by Arno Penzias and Robert Wilson, two Bell Laboratory scientists, that a cosmic microwave background radiation permeates the universe. This background radiation has been attributed to the Big Bang.
3. There is good agreement between predictions of the primordial nucleosynthesis of the elements and the known abundance of elements in the universe. This applies to the light elements that were produced in the early stages of the Big Bang.

Although there are still controversies concerning some of the predictions and results of the Big Bang model, its main ideas are widely accepted. It has been modified somewhat since it was first proposed, as we shall see in Section 16.5, where we will discuss some of the difficulties. We examine here the three pieces of evidence in some detail: the expansion of the universe as determined by Hubble, the cosmic microwave background radiation, and the relative abundance of the light elements.

Edwin Hubble (1889–1953) was born in Missouri and was better known in his youth for athletics than for academics. He graduated from the University of Chicago and won a Rhodes scholarship to Oxford University where he studied law. He decided to pursue astronomy, however, and was appointed to the staff of Mount Wilson Observatory in 1919, which had the two largest telescopes in the world. Hubble's career in astronomy is exceptional, and between 1922 and 1936 he solved four of the central problems in astronomy. These include a classification system for nebulae, the Cepheids, distribution of galaxies, and the linear velocity-distance relation, which led to the theory of the expanding universe. *New York Times/Getty Images.*

Hubble's Measurements

Edwin Hubble's career spanned four decades of brilliant observation and understanding of our universe. He began by showing in the mid-1920s that there were indeed galaxies other than our own. Even though this pioneering work conflicted with other published data of the time, Hubble was able to win support for his ideas by carefully and painstakingly presenting a persuasive case.

The recessional velocity of astronomical objects is inferred from the shift toward lower frequencies (redshift) of certain spectral lines emitted by very distant objects. We derived the redshift relation in Chapter 2 [Equation (2.33)]. Using data like those shown in Figure 16.1, Hubble was able to determine the recessional velocities of many objects. Other astronomers, particularly V. M. Slipher of Lowell Observatory in 1912, had reported that certain nebulae appeared to be receding at high radial velocities from us, but it was Hubble who put this experimental fact

Cluster nebula in	Distance in lightyears	Redshift

Virgo — 7.8×10^7 — H + K — 1,200 km/s

Ursa Major — 1.0×10^9 — 15,000 km/s

Corona Borealis — 1.4×10^9 — 22,000 km/s

Boötes — 2.5×10^9 — 39,000 km/s

Hydra — 4×10^9 — 61,000 km/s

Figure 16.1 Redshift data for various galaxies are shown with their distance in lightyears. The spectrum of each galaxy is the wide, hazy band placed in the middle between the laboratory comparison spectra. The K and H lines of calcium are shown redshifted by the arrow and move to the right for higher velocities. These early data convincingly showed that the universe was expanding. *Courtesy of The Observatories of the Carnegie Institution of Washington.*

on firm footing. Galaxies receding from us with speed v are related to the distance R from Earth by the relation known as **Hubble's law**:

$$v = HR \qquad (16.1)$$

Hubble's law

The parameter H is called the **Hubble parameter**, and it is related to a scale factor a that is proportional to the distance between galaxies by

Hubble parameter

$$H = \frac{1}{a}\frac{da}{dt} \qquad (16.2)$$

Because the universe has been expanding, Hubble's parameter is not constant, but changes slowly over long periods of time. Its value today is sometimes denoted by H_0 and called the *Hubble constant*.

In order to determine whether Hubble's law is valid, it is necessary to know the distance R to objects for which the redshift has been measured. Hubble developed sophisticated techniques that used the brightness of stars and galaxies to determine the distances R. Hubble was able to do this with some certainty for stars out to distances of 10 million lightyears and, with some additional assumptions, for galaxies out to distances of 500 million lightyears.

Together with his gifted colleague Milton Humason, Hubble examined, over a period of many years, hundreds of stellar objects to determine their redshifts. By 1929 he was able to show that 49 galaxies (which he called *extragalactic nebulae*) fit the velocity-distance relationship. Hubble's results required painstaking mea-

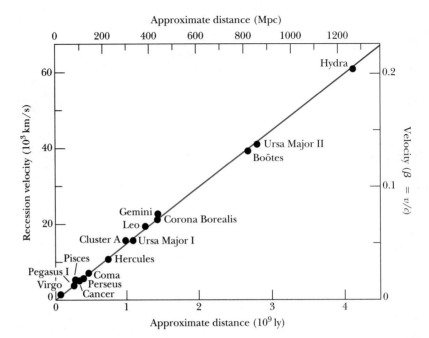

Figure 16.2 An analysis of 15 clusters of galaxies for the recession velocity as a function of distance. The solid line represents a Hubble constant of 15 km/s per million lightyears (or 49 km · s^{-1} · Mpc^{-1}). Various analyses of the distances give different values of H. *From G. Siegfried Kutter,* The Universe and Life, *Boston: Jones and Bartlett (1987), p. 50.*

surements of brightness and redshifts. By 1935 Hubble and Humason had catalogued the redshifts of 100 additional galaxies; these data unequivocally showed that the galaxies farthest away from us were moving at the highest speeds.

Hubble showed that Equation (16.1) is valid. The linearity between v and R remains valid today (Figure 16.2), although the distance measurements have been corrected over the intervening years as more observational data were collected. Today we believe that Hubble's constant H_0 is about 22 km/s per million lightyears.

It is not necessary for Earth to be at the center of the universe in order to observe the expansion. We show in Figure 16.3 a balloon with dots. Notice that as the balloon is inflated, *all* the dots move further apart from each other. The surface of the balloon is two-dimensional; a three-dimensional example often quoted is raisins in bread dough. As the bread bakes, it rises and expands in three dimensions, and the raisins separate. The raisins all move further apart, with the ones on the outside moving faster. Something similar happens as the universe expands and the galaxies separate. We will discuss in Section 16.6 the possibility of using Hubble's constant to determine the age of the universe.

Figure 16.3 A representation of how galaxies are receding with respect to each other. As the balloon is inflated, each dot is further away from the other dots. As the universe expands it seems to remain homogeneous.

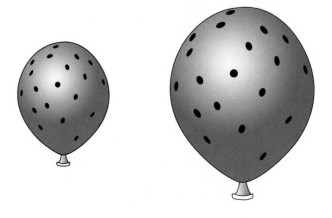

Example 16.1

Astronomers use a measure of distance called the *parsec*, abbreviated pc, more often than the lightyear, another measure of distance. A star that is one parsec away has a parallax of one second of arc relative to the Earth's orbit about the sun. (Parallax is the apparent displacement of a body due to its observation from two different positions.) One parsec is about 3.26 lightyears. Determine Hubble's constant in km/s per Mpc.

Strategy We need to use conversions and the value of the Hubble constant. One Mpc is a megaparsec or 10^6 pc.

Solution We use the value just given for Hubble's constant, 22 km/s per million lightyears, and make the conversion to Mpc.

$$\frac{22 \text{ km/s}}{\text{million ly}} \frac{3.26 \text{ ly}}{\text{pc}} = 72 \frac{\text{km/s}}{\text{Mpc}}$$

Hubble's constant is quoted using either unit. We are using the value 71 km $\cdot$ s^{-1} $\cdot$ Mpc^{-1} in this text.

Cosmic Microwave Background Radiation

In 1964 Arno Penzias and Robert Wilson were studying radio emissions in the microwave wavelength region. They kept picking up an annoying constant signal in their specially designed low-noise antenna. When repeated attempts to eliminate the source of the noise were not successful, Penzias and Wilson were at a loss to understand their result. In a conversation with another colleague they learned of calculations by P. James E. Peebles and his group at Princeton that predicted a remnant background of radiation from the Big Bang. Another Princeton physicist, Robert Dicke, was mounting an experiment to measure it. It was this remnant background radiation that Penzias and Wilson were observing.

Because of the rapid expansion and cooling of the universe, there came a point when protons and electrons could form atoms that the photon radiation could no longer ionize. Matter had decoupled from radiation, and this happened at a temperature of 3000 K. That blackbody radiation characteristic of 3000 K several billion years ago has Doppler-shifted to a peak near 3 K today. The calculated redshift based on the extremely high velocity of that part of the universe with respect to Earth today is about a factor of 1000. The blackbody radiation spectrum is shown in Figure 16.4. Subsequent satellite measurements have mapped out the complete blackbody spectrum to amazing accuracies and shown it to be nearly, but not completely, isotropic. We will return to this subject later. The observation of the cosmic microwave background radiation by Penzias and Wilson (Nobel Prize for Physics, 1978) was a triumph of the Big Bang model, and the result was very difficult to explain with the steady-state model. After the initial observation it became widely known that George Gamow, Ralph Alpher, and Robert Herman had performed calculations in the late 1940s and early 1950s that had predicted the Big Bang remnant radiation would appear in the range of 5–7 K.

Nucleosynthesis

A few minutes after the Big Bang, the universe had cooled enough so that neutrons and protons could undergo thermonuclear fusion and form light elements. The cosmic microwave background radiation provides a view back to about 400,000 years after the creation of the universe, but the formation of elements began after only a few minutes. Therefore, the nucleosynthesis of elements provides a stringent test of the Big Bang model. By measuring the present relative

Arno Penzias (1933–, on right) and Robert Wilson (1936–) were radio astronomers working at Bell Labs in 1965 when they discovered the cosmic microwave background radiation, for which they received the 1978 Nobel Prize for Physics. Bell Labs had a giant antenna that was no longer used when Penzias and Wilson began using it for a telescope. However, they first had to solve an annoying background noise problem. Despite all their efforts, even shoveling out the pigeon droppings from inside the antenna, they could not get rid of the noise, which of course turned out to be the cosmic microwave radiation. *Ted Thai/Getty Images.*

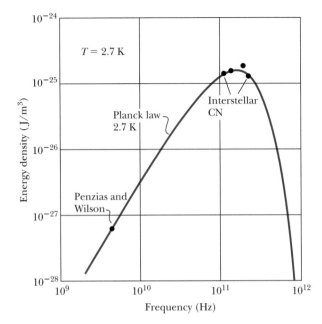

Figure 16.4 Calculated black-body radiation distribution is shown as a function of frequency; the datum point that Penzias and Wilson measured is noted. Radiation due to the cyanogen (CN) radical in interstellar space was observed in 1937, but it was incorrectly thought at the time that the radiation was due to collisions of electrons with CN.

abundances of the elements, physicists are able to work backward and test the conditions of the universe that may have existed as early as a fraction of a second, when neutrons and protons were first joined to produce nuclei.

We examined the proton-proton nucleosynthesis cycle in Chapter 13. In reactions like the proton-proton cycle, the light elements ^{2}H, ^{3}He, ^{4}He, and ^{7}Li are formed, and predictions of the Big Bang model can be compared with current observations. Heavier elements are formed in stars, but the vast majority of the presently known mass in the universe is composed of hydrogen and helium. The other elements exist in only minute quantities. The standard model of elementary particle physics, together with nuclear physics, has been used in complex and difficult calculations to compare with current experimental observations of the elemental abundances. The observations and Big Bang predictions are in remarkable agreement, as shown in Figure 16.5, for a density of about 3×10^{-31} g/cm^3.

Figure 16.5 The fractional mass abundances of several light elements are displayed against possible mass densities. The boxes represent experimental observations, and the solid curves are calculations of the standard model of the Big Bang. The best agreement between observation and calculation occurs for a density of 3×10^{-31} g/cm^3. These data strongly support the standard Big Bang model, but do not agree with the critical density, $\sim 9 \times 10^{-30}$ g/cm^3 (see Example 16.8). *From R. A. Malaney and W. A. Fowler,* American Scientist *76, 472 (1988).*

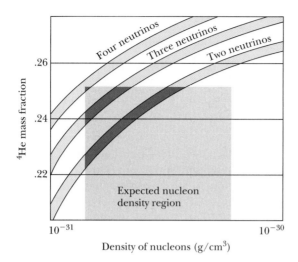

Figure 16.6 The predicted ^{4}He mass fraction is shown as a function of the density of nucleons for three different possible family groups of particles. These results indicate a maximum of four different families having neutrinos. Evidence from elementary particle physics experiments indicates that there are only three families (three neutrinos), which is consistent with this figure. *From D. N. Schramm and G. Steigman,* Scientific American *(June 1988), p. 66.*

We call this the *ordinary matter density* of the universe, which is due primarily to baryons. We will see in Section 16.5 that present evidence indicates that much of the matter in the universe has not been observed. One variation of the Big Bang model that we will discuss in Section 16.5 proposes an inflationary universe, and that model suggests a density of about 9×10^{-30} g/cm^3.

Approximately 25% of the known (or ordinary) mass of the universe is composed of ^{4}He, the remainder being almost all free protons. The synthesis of ^{4}He depended critically on the ratio of neutrons to protons and the density of matter during early stages of the universe. The ^{4}He abundance has been used to suggest that at most four families of elementary particles are possible. The suggestion and data are in good agreement with the current belief from elementary particle physics data that only three families exist (see Figure 16.6).

Example 16.2

If the current mass of the universe consists of 75% protons and 25% ^{4}He, what is the current ratio of protons to neutrons?

Strategy We count the number of protons and neutrons in ^{4}He and add those protons to those of single protons, from which we can find the ratio.

Solution The mass of a proton is about 1 u, and the mass of ^{4}He is about 4 u. If the ratio of mass of free protons to ^{4}He is 3:1 (75% protons, 25% ^{4}He), there must be 12 free protons for every ^{4}He. Therefore out of a total of 16 nucleons (12 free protons and 4 nucleons in ^{4}He), only 2 are neutrons, and the remaining 14 are protons (two are in ^{4}He). The ratio of protons to neutrons is about 7. This is in reasonable agreement with the fraction calculated from temperature considerations made for the time when protons and neutrons formed deuterons. When the temperature dropped below 10^9 K, photons no longer had enough energy to dissociate the deuteron (2.2-MeV binding energy). After that time, free neutrons decayed (they are unstable with a half-life of 10.4 minutes), and deuterons combined to form helium.

16.2   The Big Bang

The Big Bang model rests on two theoretical foundations:

1. The general theory of relativity, which we discussed in the previous chapter.

Cosmological principle

2. The **cosmological principle**, which assumes the universe looks roughly the same everywhere and in every direction. The universe is both isotropic and homogeneous.

The Russian mathematician Alexander Friedmann used these two foundations in 1922 to determine the evolution of the universe represented by his *Friedmann equations*. The solutions to his equations suggest that the universe originated in a "Big Bang" explosion of a very hot, dense state and has been expanding in size ever since.

Georges Lemaître independently arrived at results similar to Friedmann's in 1927. Then in 1935, A. G. Walker and H. P. Robertson independently proved that the Robertson-Walker metric* is the simplest spacetime geometry consistent with an isotropic, homogeneous universe. A single parameter a that changes with time according to the Friedmann equation [see Equation (16.3) below] describes the time evolution of spacetime. These universes go by various names including Friedmann, Lemaître, Robertson, and Walker (for example, FRW or FLRW universes). The idea is consistent with the model shown in Figure 16.3 of the expanding balloon dotted with galaxies.

One of the Friedmann cosmological equations can be written

$$\left(\frac{1}{a}\frac{da}{dt}\right)^2 = \frac{8\pi G \rho_m}{3} - \frac{k}{a^2} + \frac{\Lambda}{3} \tag{16.3}$$

where G is the universal gravitational constant, a is the scale parameter introduced in Equation (16.2), ρ_m is the average mass density of the universe, k is the curvature parameter of the universe, and Λ is the cosmological constant. The constant Λ was introduced by Einstein to form a static universe and a homogeneous and isotropic universe, because astronomers assured him that the universe was not in motion. The cosmological constant term accounts for the energy of perfect vacuum in order to have the homogeneous and isotropic universe. After Hubble presented his evidence for the universe's expansion, Einstein referred to the cosmological constant as his biggest blunder. Until a few years ago, cosmologists normally set $\Lambda = 0$, but now the cosmological constant is under active discussion for reasons we will discuss in Section 16.5.

We rewrite Equation (16.3) using the Hubble parameter H of Equation (16.2).

$$H^2 = \frac{8\pi G \rho_m}{3} - \frac{k}{a^2} + \frac{\Lambda}{3} \tag{16.4}$$

This is sometimes called the *Friedmann equation* and is extremely important in cosmology. It may allow us to determine both the age and size of the universe if we can ascertain the unknown parameters in the equation. We divide both sides of the equation by H^2 to have

$$1 = \frac{8\pi G \rho_m}{3H^2} - \frac{k}{a^2 H^2} + \frac{\Lambda}{3H^2} \tag{16.5}$$

Alexander Friedmann (1888–1925) was born in St. Petersburg, Russia, where he spent most of his life. A gifted mathematician, Friedmann was educated at the University of St. Petersburg. He spent time doing meteorology, and during World War I he flew airplanes on bombing raids. He became an expert on aeronautics before returning to St. Petersburg in 1920 where he became a professor of physics and mathematics at the Polytechnic Institute. He became interested in general relativity and corresponded with Einstein. In 1925 he made a record-breaking balloon flight in which he performed meteorological and medical observations. The world lost a great intellect when he died of typhoid at age 37.

*A metric is a mathematical structure that provides information in spacetime about lengths and times. It is formally a symmetric nondegenerate tensor.

The terms on the right have special meaning in cosmology. The first term is called the mass density Ω_m, the second is the curvature Ω_k, and the last is the vacuum energy density Ω_Λ. Notice that all three terms are dimensionless. We rewrite them as

$$\Omega_m = \frac{8\pi G \rho_m}{3H^2} \qquad \text{Mass density} \qquad (16.6)$$

$$\Omega_k = -\frac{k}{a^2 H^2} \qquad \text{Curvature} \qquad (16.7)$$

$$\Omega_\Lambda = \frac{\Lambda}{3H^2} \qquad \text{Vacuum energy density} \qquad (16.8)$$

We will return to these three terms in Sections 16.6 and 16.7 after we present evidence that indicates the curvature term is zero. The evolution of the universe is then determined by the two terms Ω_m and Ω_Λ, a remarkable situation.

Since the 1960s the Big Bang has generally been accepted as the event that began our universe. We can use our knowledge of nuclear and elementary particle physics (Chapters 12–14) and general relativity (Chapter 15) to understand the formation of our universe since the Big Bang. In this section we define the time $t = 0$ as the beginning of the universe, or the Big Bang, and describe the major steps in the intervening 13.7 billion years until today. The temperature of the universe is shown as a function of time in Figure 16.7.

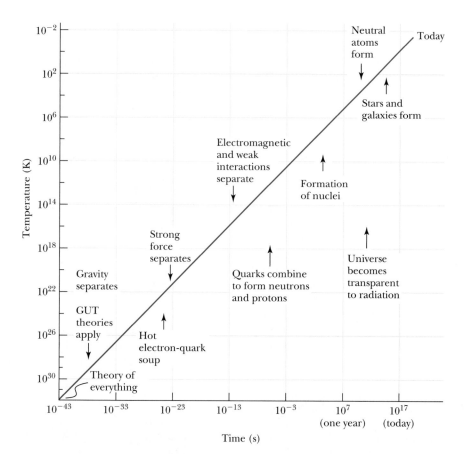

Figure 16.7 The temperature of the universe is displayed as a function of time since the Big Bang. Note that both scales are logarithmic. The comments are placed vertically near the times at which the phenomena occurred, except for the Theory of Everything which might be applicable for the time $<10^{-43}$ s.

The unknown

$t = 0 \rightarrow 10^{-43}$ **s** We have no theories that can tell us what happened in this era, because the known laws of physics do not apply. In the beginning, the universe most likely had infinite mass density and zero spacetime curvature. This condition is known as a *cosmological singularity*. The size of the visible universe by the time 10^{-43} s was probably less than 10^{-52} m. The four forces of strong, electromagnetic, weak, and gravity were all unified into one force, and our present theories of gravity and quantum physics would not apply. Only a theory of gravitation including quantum physics, or quantum gravity, has any hope of explaining what happened. There are no viable candidates for this theory at the present time. The temperature of matter was probably greater than 10^{30} K. We don't know if any of the particles known today could have even existed. The applicable theory has been dubbed the *Theory of Everything*.

Gravity separates

$t = 10^{-43}$ **s** $\rightarrow 10^{-35}$ **s** As strange as it seems, we believe that we have some understanding of what happened in this era. By 10^{-35} s the universe had expanded somewhat to a size of something like 10^{-30} m, and the temperature was slightly reduced, maybe to 10^{28} K. Gravity was established as the first separate force. The grand unified theories (GUTs) try to explain this era, but there is no widely accepted theory as yet.

Quark-electron soup

$t = 10^{-35}$ **s** $\rightarrow 10^{-13}$ **s** The strong force had now broken off and become a separate force. The fundamental particles (quarks and leptons) had formed as well as their antiparticles. The small universe can be described as a hot, quark-electron soup. The universe continued to cool, perhaps to 10^{16} K by 10^{-13} s. No accelerator today has enough particle energy to explore what happened at these high temperatures. At present only some cosmic rays have comparable energies. The size of the universe during this period was on the order of 10^{-1} m.

Neutrons and protons form

Electromagnetic and weak forces separate

$t = 10^{-13}$ **s** $\rightarrow 10^{-3}$ **s** During this period the quarks bound together to form neutrons and protons. Only the two lightest quarks, the up and down, were very effective in the thermodynamics of the universe, because the more massive quarks had such low velocities at this time that they could not transport much energy. There were more protons than neutrons, because protons are slightly less massive. At energies above 100 GeV (corresponding to temperatures of about 10^{15} K) the electromagnetic and weak interactions were unified into the electroweak interaction. However, below 100 GeV, the $W^{\pm}$ and Z bosons behave like massive particles and the photon is massless, so the electromagnetic and weak interactions had broken their symmetry and acted separately. We say that "symmetry is broken" below 100 GeV. At the end of this period the temperature had dropped to about 10^{11} K. The four forces of today had become distinct. By the end of this period the universe mostly consisted of a soup of photons, electrons, neutrinos, protons, and neutrons as well as various antiparticles (for example, positrons and antiprotons). The size of the universe was about 1000 m.

Deuterons form

$t = 10^{-3}$ **s** $\rightarrow 3$ **min** By the end of this period the universe had cooled enough (to 10^9 K) that deuterons did not quickly fly apart when formed, because the cooler photons did not have enough energy to dissociate the deuterons. This was the beginning of nucleosynthesis—the formation of nuclei. The universe had expanded to about 10^{10} m.

Light nuclei form

$t = 3$ **min** $\rightarrow 300,000$ **y** Helium and other light atomic nuclei formed by nucleosynthesis (for example, the proton-proton cycle). Neutron decay occurred, so there were many more protons than neutrons. The universe continued to cool, and the temperature was as low as 10^4 K at the end of this era. The universe consisted primarily of photons, protons, helium nuclei, and electrons. Atoms were not able to form because the intense electromagnetic radiation ionized them almost as soon as they were formed. Photons interacted freely with charged particles through the electromagnetic interaction and were absorbed, emitted, and scattered by matter. The size of the universe was about 10^{21} m, and it continued to expand.

t = **300,000 y → Present** During this period the universe had finally cooled enough that electromagnetic radiation (photons) decoupled from matter. Until about 300,000–700,000 years, the universe was "radiation dominated," meaning that most of the energy was in the form of photons, which were continually being absorbed and emitted by ions. At about 3000 K the temperature was low enough that protons could combine with electrons to form neutral hydrogen atoms. At this point the scattering of photons from neutral hydrogen (as opposed to free protons and electrons) dropped dramatically, and electromagnetic radiation was free to pass throughout the universe. From this time on, the universe is "matter dominated," with more energy in the form of matter than radiation. Because photons are now free to pass throughout the universe, a blackbody radiation of temperature 3000 K should persist forever. Remember from Section 16.1 that this radiation characteristic of 3000 K is redshifted with respect to us, and we consider the vast majority of photons in the universe today are due to the 3 K background. This is the cosmic microwave background radiation discussed in the previous section, and it can be detected on our television sets. It is part of the snowy picture we obtain with an antenna when no strong channel is received. Atoms were able to form, and matter began to clump together to form molecules, gas clouds, stars, and eventually galaxies. The rest is history! From this time on, the universe evolved into the form we recognize today.

Matter-dominated universe

Example 16.3

Nucleosynthesis began around the time 10^{-3} s, when protons and neutrons could finally remain together in the deuteron without flying apart due to the interaction of radiation. Calculate the ratio of protons to neutrons when the temperature of the universe was about 10^{10} K.

Strategy The ratio of protons to neutrons is purely a statistical distribution based on the available energy and the masses of the proton and neutron. The ratio is determined by the Maxwell-Boltzmann distribution from thermodynamics and the difference in masses. Because the proton has lower mass, we expect more protons to exist.

Solution Let $\Delta m = m_n - m_p = 939.566$ MeV/c^2 − 938.272 MeV/c^2 = 1.294 MeV/c^2. The ratio of protons to neutrons is calculated to be

$$\frac{\text{Number of protons}}{\text{Number of neutrons}} = \frac{e^{-m_p c^2 / kT}}{e^{-m_n c^2 / kT}}$$

$$= e^{\Delta m \, c^2 / kT} = \exp\left(\frac{\Delta m \, c^2}{kT}\right) \qquad (16.9)$$

$$= \exp\left(\frac{1.294 \times 10^6 \text{ eV}}{(8.6 \times 10^{-5} \text{ eV/K})(10^{10} \text{ K})}\right)$$

$$= 4.5$$

As the temperature continued to decrease, the ratio of protons to neutrons continued to increase due both to the decay of the neutrons and to the temperature factor T in the exponential. However, the ratio of protons to neutrons bound in nuclei eventually stabilized due to the production of helium.

16.3   Stellar Evolution

Some 400,000 years after the Big Bang, matter in the form of electrons, protons, and ^{4}He drifted throughout the universe much like gas particles in a large room. Eventually, as the temperature continued to decrease, gravitational forces managed to bring some of the matter together into massive gaseous clouds, which formed the basis for stars. As the protons were attracted together by their gravitational interaction, their kinetic energy rose. This process continued as the interior temperature of these infant stars kept increasing. The interior of a gaseous cloud had a higher density and temperature than the outside. The cloud continued to contract until finally the temperature reached about 10^7 K, and the nuclear fusion process began. Nuclear fusion is a characteristic of a star. It may have taken a million or more years for the contraction of the cloud to be able to produce fusion for the star to be born, although some recent results indicate it could have happened as soon as 200,000 years (see Section 16.6). The schematic formation of a star is shown in Figure 16.8.

We previously discussed (Chapter 13) the nuclear fusion process as the energy source of stars. The proton-proton chain releases energy, which is observed as radiation. The eventual result of this fusion process is ^{4}He, which collects at the center of the star. Other processes form ^{12}C and heavier masses if the temperature in the star is high enough.

Of course, we know most about our own star, the sun. We can measure the surface temperature of stars by measuring the color of the radiated light, but it is difficult to know the interior temperature. We believe the surface temperature of the sun to be about 5500 K and the interior temperature to be as high as 6×10^6 K. A star the size of our sun may burn for 10^{10} y, but a larger star may use up its fusion fuel much faster. The light presently received by Earth was most likely produced in the interior of the sun more than 10^5 years ago and underwent many scatterings until it was emitted from the surface of the sun.

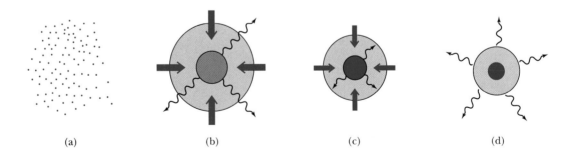

(a)   (b)   (c)   (d)

Figure 16.8 (a) Stars form when interstellar gas and clouds condense by gravitational attraction. (b) As the matter contracts, a core forms that heats up and radiates energy. (c) Eventually the outer region becomes so dense that the radiation from the hot inner core can no longer escape. The collapse slows, but the matter continues to heat and eventually a protostar of high density and temperature forms. For a star of about 1 solar mass (the mass of our sun), the contraction eventually heats up enough to sustain nuclear fusion. (d) The radiation pressure produced by nuclear fusion balances the gravitational contraction, and the star stabilizes into a main sequence star. The star will burn for 10 billion years, converting its hydrogen into helium.

Example 16.4

Estimate the mean temperature of the sun by assuming its protons behave as a gas.

Strategy As gas clouds collected together through the gravitational interaction, the decrease in the gravitational potential energy was accompanied by an increase in the kinetic energy of the particles. We can estimate the mean temperature of the sun by setting the change in the kinetic energy equal to the negative of the change in potential energy, $\Delta(\text{K.E.}) = -\Delta(\text{P.E.})$. In other words, the total energy, K.E. + P.E., is constant. We use the kinetic theory of gases to relate the mean velocity to temperature.

Solution We assume the sun is a uniform sphere of mass M and radius R. Its self-potential energy can be calculated as (see Problem 26)

$$\text{P.E.} = -\frac{3}{5}\frac{GM^2}{R}$$

$$= -\frac{3}{5}\frac{(6.67 \times 10^{-11}\,\text{N}\cdot\text{m}^2/\text{kg}^2)(1.99 \times 10^{30}\,\text{kg})^2}{6.96 \times 10^8\,\text{m}}$$

$$= -2.28 \times 10^{41}\,\text{J}$$

The kinetic energy of the particles within the sun is then 2.28×10^{41} J. If we assume the sun is made entirely of protons, the number of protons in the sun is $N = M/m_p$, where M is the mass of the sun. We write the kinetic energy of the sun as N protons each having speed v.

$$\frac{1}{2}Nm_p v^2 = \frac{1}{2}\frac{M}{m_p}m_p v^2 = 2.28 \times 10^{41}\,\text{J}$$

From this relation we can determine the proton velocity.

$$v^2 = \frac{2(2.28 \times 10^{41}\,\text{J})}{M} = \frac{2(2.28 \times 10^{41}\,\text{J})}{1.99 \times 10^{30}\,\text{kg}}$$

$$= 2.3 \times 10^{11}\,\text{m}^2/\text{s}^2$$

$$v = 4.8 \times 10^5\,\text{m/s}$$

Using the kinetic theory of gases relation for the velocity of a gas particle at temperature T, we can determine the sun's mean temperature from Equation (9.20).

$$v = \sqrt{\frac{3kT}{m}} \qquad (9.20)$$

We determine T to be

$$T = \frac{mv^2}{3k}$$

For a proton, we determine the temperature to be

$$T = \frac{(1.67 \times 10^{-27}\,\text{kg})(2.3 \times 10^{11}\,\text{m}^2/\text{s}^2)}{3(1.38 \times 10^{-23}\,\text{J/K})} = 9 \times 10^6\,\text{K}$$

This is a reasonable estimate for the mean temperature of the sun.

In one of the cycles of the proton-proton chain in the sun, $^3\text{He} + {}^4\text{He} \rightarrow {}^7\text{Be} + \gamma$, the ^7Be undergoes electron capture to produce ^7Li and a neutrino of 0.862 MeV. It was these neutrinos and others that were measured in a long running experiment at the Homestake Gold Mine in South Dakota. Over a 20-year period of the experiment, the measured rate of neutrinos was only about one third of the expected rate. This discrepancy, called the *solar neutrino problem*, was solved just a few years ago by the observation that the neutrinos oscillate into other flavors while in flight (see Chapters 12 and 14).

Solar neutrino problem

The Ultimate Fate of Stars

The final stages of stellar evolution begin when the hydrogen fuel is exhausted. At this point the gravitational attraction continues; the density and temperature increase. The temperature becomes hot enough that the helium nuclei begin to fuse. Heavier elements are subsequently created in nuclear fusion processes that are well understood. The fusion process continues until nuclei near the iron region are produced, where elements have the highest binding energy per nucleon (see Chapter 12). The nuclear fusion process can no longer continue, and the reactions stop.

The star's ultimate fate depends on its mass. For stars somewhat more massive than the sun, the gravitational attraction of the mass continues until the density of the star is incredibly high. Let us look at this process in some detail. Let there be N nucleons, each of mass m, in the star. The gravitational self-potential energy of a uniform sphere of mass Nm and radius R is (see Problem 26)

$$\text{P.E.} = U_{\text{grav}} = -\frac{3}{5}\frac{G(Nm)^2}{R} \tag{16.10}$$

where we have used U for potential energy rather than V to avoid confusion with the volume V. We determine the gravitational pressure by determining the force per unit area.

$$P_{\text{grav}} = \frac{F}{A} = -\frac{1}{4\pi R^2}\frac{dU_{\text{grav}}}{dR} = \frac{3G}{5}\frac{(Nm)^2}{4\pi R^4} \tag{16.11}$$

We write the gravitational pressure in terms of the volume by using $V = \frac{4}{3}\pi R^3$ and obtain

$$P_{\text{grav}} = 0.3G\frac{(Nm)^2}{V^{4/3}} \tag{16.12}$$

Matter is kept from total collapse by the outward electron pressure. This occurs because the Pauli exclusion principle effectively keeps two electrons from occupying the same state. However, for a sufficiently massive star, gravity will eventually force the electrons to interact with the protons through the reaction

$$e^- + p \rightarrow n + \nu_e \tag{16.13}$$

Neutron star   The result, called a **neutron star**, is composed mostly of neutrons.

However, because neutrons also obey the exclusion principle, an outward pressure similar to that of the electrons will also result from neutrons. Using the techniques developed in Chapter 9, the pressure of an electron gas is shown to be

$$P_e = \frac{2f\pi^2}{3}\frac{\hbar^2}{2m_e}\left(\frac{N_e}{V}\right)^{5/3}$$

where f is the fraction of the average energy between 0 and E_F. For a Fermi gas, the average energy is about $\frac{3}{5}E_F$ (see Chapter 9), so we can take $f \approx 3/5$. We can use the same relation for neutrons by replacing N_e by N and m_e by m. The neutron pressure becomes

$$P_n = \frac{3.9\hbar^2(N/V)^{5/3}}{2m} \tag{16.14}$$

In equilibrium, the outward pressure of the neutrons due to the exclusion principle will balance the gravitational pressure. We set Equations (16.12) and (16.14) equal.

$$0.3G\frac{(Nm)^2}{V^{4/3}} = \frac{3.9\hbar^2(N/V)^{5/3}}{2m} \tag{16.15}$$

We solve this equation for the cube root of the volume (see Problem 10).

$$V^{1/3} = \frac{6.5\hbar^2}{N^{1/3}m^3G} \tag{16.16}$$

We can use this result to find the size of neutron stars as in the following example.

Special Topic

PLANCK'S TIME, LENGTH, AND MASS

We have recognized several fundamental constants in our study of physics. They include the gravitational constant G, Planck's constant h, and the speed of light c. The gravitational force depends on G, and several properties of a particle depend on h and c. These include the energy, $E = mc^2$, the wavelength as a function of a particle's momentum, $\lambda = h/p$, the Compton wavelength of a particle, $\lambda_C = h/mc$, and the energy of a photon, $E = hf$.

It is interesting to use dimensional analysis to determine some characteristic values of time, length, and mass using the fundamental constants G, c, and h. Of course, the numerical values depend on the system of units used, in our case SI, but what physical significance might these characteristic values have? The values of the fundamental constants and their dimensions in terms of length (L), time (T), and mass (M) are

Gravitational constant $G = 6.6726 \times 10^{-11} \text{N} \cdot \text{m}^2/\text{kg}^2$
  Dimensions $M^{-1}L^3T^{-2}$

Speed of light $c = 2.9979 \times 10^8 \text{ m/s}$
  Dimensions LT^{-1}

Planck's constant $h = 6.6261 \times 10^{-34} \text{J} \cdot \text{s}$
  Dimensions ML^2T^{-1}

If we use dimensional analysis, we find that we obtain mass by using the combination of $\sqrt{hc/G}$. We call this **Planck's mass** m_P and determine its value to be

$$m_P = \sqrt{\frac{hc}{G}} = 5.46 \times 10^{-8} \text{ kg}$$

The physical significance of Planck's mass is that general relativity does not allow a black hole to be created from a mass smaller than Planck's mass.

We determine a characteristic length, called **Planck's length** λ_P, similarly by using dimensional analysis to determine

$$\lambda_P = \sqrt{\frac{Gh}{c^3}} = 4.05 \times 10^{-35} \text{ m}$$

The time that it takes for light to travel across Planck's length is called **Planck's time**, t_P, but it could also be determined from dimensional analysis. Its value is given by

$$t_P = \frac{\lambda_P}{c} = \sqrt{\frac{Gh}{c^5}} = 1.35 \times 10^{-43} \text{ s}$$

In particular, physicists do not understand what happened before Planck's time in the creation of the universe. In essence, we can consider Planck's time to be the moment of creation of the universe, because the laws of physics as we know them now are not valid before Planck's time of 10^{-43} s.

Example 16.5

Determine the radius of a neutron star with a mass of 2 solar masses.

Strategy We use Equation (16.16) to determine the volume, and from that we can determine the radius. One solar mass = 1.99×10^{30} kg.

Solution First, we find the number of neutrons N.

$$N = \frac{2M_{\text{sun}}}{m_{\text{neutron}}} = \frac{2(1.99 \times 10^{30} \text{ kg})}{1.675 \times 10^{-27} \text{ kg}} = 2.4 \times 10^{57} \text{ neutrons}$$

We now determine the cube root of the volume to be

$V^{1/3}$

$$= \frac{6.5(1.06 \times 10^{-34} \text{J} \cdot \text{s})^2}{(2.4 \times 10^{57})^{1/3}(1.675 \times 10^{-27} \text{ kg})^3(6.67 \times 10^{-11} \text{J} \cdot \text{m/kg}^2)}$$

$$= 1.7 \times 10^4 \text{ m}$$

The radius R is calculated to be, from $V = \frac{4}{3}\pi R^3$,

$$R = \left(\frac{3}{4\pi}\right)^{1/3} V^{1/3} = \left(\frac{3}{4\pi}\right)^{1/3} (1.7 \times 10^4 \text{ m}) = 11 \text{ km}$$

The radius of a neutron star twice as massive as our sun is only 11 km! It is interesting to compare the density of a neutron star with that of a typical nucleus and with a nucleon (see Problem 11).

Neutron stars have densities as high as nuclear densities. They were first predicted in the 1930s but were not clearly observed until 1967, when a **pulsar**, a rapidly rotating neutron star (see Figure 16.9), was discovered. If the collapsing star has a mass greater than 3 solar masses, it can collapse through the neutron star stage and form a *black hole*. These fascinating objects were discussed in Chapter 15.

If the mass of the star is less than about 1.4 solar masses (the *Chandrasekhar limit*), the star can support itself against gravitational collapse, because the free electrons exert a considerable outward pressure. Such hot stars are called **white dwarfs**, and for a solar mass, their radius is about 1% of the solar radius. Such is the future of our sun.

Our sun will become a white dwarf

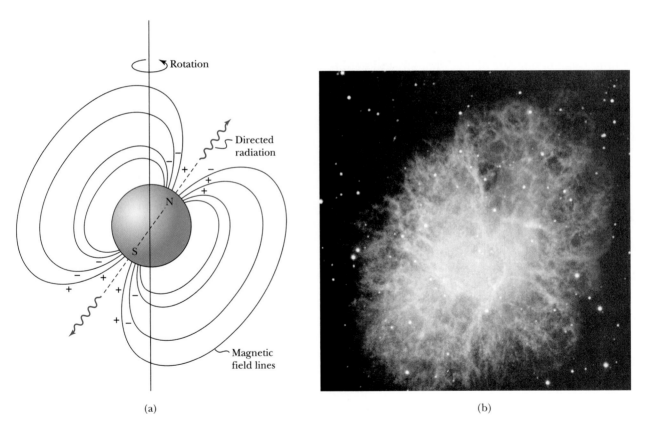

(a)   (b)

Figure 16.9 (a) A pulsar is a rapidly rotating neutron star. Large magnetic fields most likely accelerate charged particles near the pole regions, producing copious amounts of electromagnetic radiation along the magnetic polar regions as shown in the figure. As the star rotates in space, this highly directional radiation appears as a pulsed radiation source to an observer located along one of the polar directions. Pulsars have periods ranging mostly from 1/30 second to 3 minutes. (b) Photo of the Crab Nebula, which contains a pulsar whose supernova explosion in 1054 could be observed in daylight for three weeks. *Courtesy of Max Planck Institute for Astronomy and Calar Alto Observatory; K. Meisenheimer and A. Quetz.*

In some ways, it is quite amazing that quantum physics, through the Pauli exclusion principle, plays such a major role in stellar evolution. Until now, we have tended to think of quantum physics as playing a role only inside the tiny confines of the atom, yet now we are talking about objects as massive as our sun!

16.4   Astronomical Objects

Collections of stars are called **galaxies**; the gravitational attraction between stars keeps a galaxy together. Our own galaxy is the **Milky Way**, and it is believed to be composed of 200 billion stars. The number of galaxies is extremely large, at least 100 billion (10^{11}). In the early part of the twentieth century there was considerable controversy about whether the Milky Way constituted the entire universe. *Nebula* is the name astronomers gave to objects that had a cloudy unresolved appearance. In 1924 Hubble showed that the Andromeda Nebula is quite far away from us ($\sim 10^6$ lightyears) and is in fact a separate galaxy. It is rightfully called the *Andromeda Galaxy*. Although other, smaller galaxies are closer to our Milky Way, the Andromeda Galaxy is the largest nearby one, and on a clear night it can be observed with the naked eye.

Galaxies

Astronomers keep finding galaxies farther and farther away. For example, early in 2004 scientists reported finding an ancient galaxy using the Very Large Telescope in Chile (although the discovery has yet to be confirmed). They utilized the gravitational lensing effect of the Abell 1835 cluster of galaxies to concentrate light. The detected redshift was 10. We are seeing the light from this galaxy when it was only 460 million years old, practically a primeval galaxy in its infancy. Its recorded redshift of 10 means the galaxy is moving away from us at great speed, and its light traveled more than 13 billion years to reach us. The formation of galaxies in the early universe is of intense interest today to cosmologists.

Active Galactic Nuclei and Quasars

In the early 1960s astronomers discovered objects with tremendously strong radio signals that had optical spectra that could not be understood. In 1963 Maarten Schmidt showed that these objects had to be at least 3 billion lightyears away from Earth because of their large redshifts. He was able to decipher the shifted spectral lines of hydrogen. These objects were dubbed **quasars**, short for *quasi-stars* or *quasi-stellar objects*, and can outshine galaxies having hundreds of billions of stars.

Thousands of quasars have now been discovered, and today we know that most of their energy is in the infrared region of the electromagnetic spectrum, but with considerable optical and x-ray components. Only about 10% or fewer of the quasars emit powerful radio signals. Scientists quickly surmised that only a supermassive black hole residing in the center of a host galaxy could produce such an intense energy. Quasars are among the most distant objects in the universe and therefore are among the earliest objects formed in the universe, some more than 12 billion years ago. Part of the observational difficulty is that the enormous distance of the galaxy means that the host galaxy is faint against the much brighter light of the quasar. The fact that quasars can vary in brightness in just a few hours or days suggests their size is only a few light hours or light days across, not much larger than our solar system.

Long ago when young galaxies were forming, the stars in their cores were closely packed. Collisions and mergers of these stars gave rise to the supermassive black holes. These black holes required huge amounts of mass fuel, which was provided by gas from the galaxy's interstellar medium, from a companion galaxy, or from a star that came too close. Because quasars are observed at great distances, they were formed long ago. We do not observe quasars close to our galaxy, so the conditions for the formation of quasars are not favorable today. Therefore, quasars must evolve into objects that are common today, most likely normal, quiescent galaxies. If this is the case, then some normal galaxies should have massive black holes in their centers that are no longer quasars, because they are starved for fuel. Of course we don't know what fraction of normal galaxies passed through the quasar phase.

Unraveling the mystery of quasars was considered so important in the 1970s that it was a major justification for the Hubble Space Telescope (HST). Spectacular images from the HST have indeed shown that quasars reside in galaxies. However, the images have revealed other surprising results. The variety of these galaxies is extensive; some are normal, while others are merging and colliding with surrounding galaxies, sucking their mass from them. We now believe that the supermassive black holes at the center of quasars turn themselves on and off as they consume mass distributions from their galaxies or when galaxies collide. Although quasars remain mysterious, they provide us with a window to study the conditions that prevailed early in the history of the universe.

An amazing array of exotic extragalactic objects is lumped into the category called **active galactic nuclei (AGN)**. These include the extraordinarily luminous quasars, Seyfert galaxies, and blazars. We believe their core contains a supermassive black hole surrounded by an accretion disk providing the mass. As matter spirals into contact with the black hole, huge amounts of electromagnetic radiation and jets of plasma are spewed out into space, often in directions perpendicular to the accretion disk. We show an artist's conception of this in Figure 16.10. When we look directly along the spewing jets, we see remarkable intensities. *Blazars* are active galactic nuclei of high energy with a relativistic jet spewing energy directly toward Earth. They are associated with strong radio emissions and were discovered in 1991 using the Compton Gamma Ray Observatory.

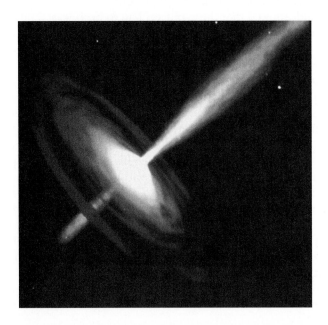

Figure 16.10 An artist's rendering of an active galactic nuclei. The core is particularly bright and is believed to have a supermassive black hole in its center. Gas from the galaxy's interstellar medium, from a star that strays too close, or from another galaxy falls into the massive black hole. An accretion disk forms and just before the rotating mass falls into the black hole, huge amounts of energy are spewed out in jets along the rotation axis. The energy contains radiation across the electromagnetic spectrum including the infrared, radio, ultraviolet, and x-ray wavelengths. *Dana Berry/ STSci/NASA.*

Gamma Ray Astrophysics

Gamma-ray bursts (GRBs) are short flashes of electromagnetic radiation that are observed about once a day at unpredictable times from random directions. In the late 1960s the United States launched the Vela satellites to detect gamma-ray flashes from nuclear detonations for monitoring the Soviet Union's compliance with the Nuclear Test Ban Treaty. In a surprising discovery, Vela did detect gamma flashes, but they were from outer space, not from Earth. After declassification of these data in 1973, the scientific community set out to discover the source of these gamma-ray bursts, which last from a few milliseconds to several minutes. Gamma rays are mostly absorbed in the Earth's atmosphere, so most of what has been learned has been from space-based observatories. As recently as the early 1990s, astronomers were unsure as to where GRBs originated. Did they come from our solar system, our Milky Way Galaxy, or from far away in the cosmos? Only recently have we learned from a combination of satellite and ground-based observations, together with theoretical calculations, that GRBs come from supernovae in distant galaxies.

Gamma-ray bursts

One interesting feature of GRBs is the "afterglow" of lower energy photons, including x rays, light and radio waves, that last for days or even weeks. The radiation from a GRB lies along the blast direction, which can only be detected if the blast direction is pointing at Earth. The observation of the afterglow was an important part of understanding the GRB phenomena. Although the Compton Gamma Ray Observatory was not equipped to detect the afterglow, the Italian-Dutch BeppoSAT spacecraft had both x-ray and gamma-ray detection capability and was able to discover and observe the afterglows during its 1996–2002 research period.

Astronomers used information from the BeppoSAT to alert other astronomers where to direct their optical telescopes in hopes of finding spectral data. Using the powerful Keck and Hubble telescopes, astronomers were able in 1997 to measure redshifts of the optical spectra to prove the bursts came from far away. The directional measurements also proved the GRBs came from random directions.

A gamma-ray burst observed in 1999 had a luminosity believed to be equal to 10^{43} W, which is a factor of 10^{17} times brighter than our sun. Observations on this GRB convinced researchers that the radiation was emanating from a narrow jet of an opening of only a few degrees. By 1998 researchers suspected GRBs were associated with supernova. In 2003 it was shown that the afterglow's optical spectrum was nearly identical to that of a supernova, which seemed to be convincing proof of the supernova link. The most widely accepted idea now is that a supernova explosion occurs at the center of a massive star. A blast wave or fireball moving close to the speed of light collides with stellar material still inside the star and produces the gamma rays, which burst out of the star just ahead of the blast wave. The afterglow is due to the collision of the blast wave or jet with gas and dust around the star.

Questions remain about GRBs. There are actually long- and short-duration GRBs. We have been discussing the long-lived ones. Short-duration GRBs last no longer than 2 s and are probably created by a different process. Newer spacecraft satellites, including NASA's High Energy Transient Explorer 2 (HETE-2, managed by the Massachusetts Institute of Technology), INTEGRAL (European Space Agency), the Swift Gamma-Ray Burst Mission, and the Gamma-ray Large Area Space Telescope (GLAST, scheduled for launch after 2007), are (or will be) able to locate gamma-ray bursts more precisely in real time and detect higher energy gammas.

Ground-based facilities are also expanding. Astrophysicists are taking advantage of the fact that gamma rays normally do not pass through our atmosphere. They are planning to study gamma rays as energetic as 1 TeV by using Cerenkov telescopes to look for the visible light that the gamma rays create with their encounter with the atmosphere. Although this technique has been used for at least 20 years, the new detectors are much larger and will be clustered together to increase the area. Such detectors are already operational in Namibia, Australia, and the Canary Islands. A U.S.-based system should be in operation in 2006.

Novae and Supernovae

Other interesting astronomical objects are **novae** and **supernovae**. Novae are simply stars that suddenly brighten, become visible (thus the word *nova* for "new"), and then fade over some period. There are believed to be two types of supernovae, Type I showing no hydrogen spectral lines and Type II that do. This is no longer a convenient classification scheme, and Type I has been further divided into Ia, Ib, and Ic, according to other criteria. Types Ib, Ic, and II have more in common with each other than they do with Type Ia. Type Ia supernovae are the brightest and are thought to occur in binary systems where a white dwarf acquires material from its companion until the white dwarf can no longer support itself and collapses. Supernovae are crucial, because they are the primary source of heavy elements. Lighter elements like carbon and oxygen are produced by nucleosynthesis within stars and are thrown out in the explosion, where they can be assimilated by other stellar formations. But elements heavier than iron cannot be produced by nuclear fusion inside stars. Cataclysmic explosions in supernovae provide the temperature and pressure to fuse together medium-mass elements to produce heavy elements like uranium. Supernovae are also crucial as light sources to determine distances in the universe.

The Chinese may have spotted several supernovae during the past few thousand years. Good evidence exists that the explosion of the Crab supernova (Type I) occurred in the year 1054. The Crab exploded only 6000 lightyears away, and it must have been an awesome sight. It was as bright as the planet Venus and could even be seen in the daytime! Both the Chinese and Japanese, and possibly American Indians, observed and noted the occurrence. The Crab Nebula, shown in Figure 16.9, is now a beautiful cloud of glowing gas easily seen with a telescope today. Other supernovae were observed by the Danish astronomer Tycho Brahe in 1572 and by Kepler in 1604. Curiously enough, there doesn't seem to have been a supernova visible to the naked eye after 1604 until 1987. The interest in supernovae increased dramatically with the unexpected observance of a supernova in 1987 (called the SN 1987A; see Figure 16.11).

Supernova 1987A

Supernova Explosion Astrophysicists now believe they have a good understanding of how the SN 1987A could have radiated more energy in 10 seconds than 100 of our suns could radiate in 10 billion years! We can describe a Type II supernova in very broad terms. The details are interesting, and since SN 1987A, much has been learned because of the availability of a closely observed and measured occurrence. As hydrogen became exhausted by thermonuclear fusion and the helium content increased, gravity contracted the star to another phase. The helium nuclei began to fuse, and the central core of the star became denser and hotter. Eventually the helium, which had produced carbon and oxygen, was exhausted. Carbon began to fuse when the core temperature reached 8×10^8 K and the

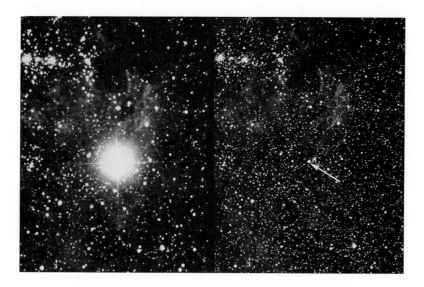

Figure 16.11 Photos after (on left) and before (on right) the supernova 1987A explosion taken by the Anglo-Australian Telescope in Australia. This is the first time that a progenitor star has been identified before the star actually exploded. © *Anglo-Australian Observatory/David Malin.*

nuclei of neon, magnesium, and silicon were produced. Eventually most of the nuclei fused to form nuclei near iron.

Before the explosion, the SN 1987A star had a mass of about 18 solar masses. At the center was a core of about 1.5 solar masses. The iron nuclei became so hot that they spewed out helium nuclei. The temperature and density skyrocketed, and neutrinos were radiated at an incredible rate. The gravitational force became so strong that a neutron star was formed. As the neutrons came closer together, the nuclear strong force eventually became effective and caused the neutrons to attract even more closely together. This attraction stopped only when the neutrons came so close together that the nuclear force became repulsive to keep the nucleons from overlapping. The strongly repulsive nuclear force caused the material falling inward in the gravitational collapse to come to a screeching halt and to rebound. The reaction of the inner core stopping and the outside layers still collapsing under gravitation resulted in a shock wave moving outward. However, the shock wave was not quite able to blast its way back out through the mass of approximately 18 solar masses. The large flux of neutrinos, which normally just pass right through matter, was slowed down by the dense shock wave. The energy absorbed from the neutrinos was sufficient to give the shock wave the boost it needed to continue blasting its way through the outer layers of the star.

Because the explosion requires a high neutrino rate, neutrino production is a key indicator of a supernova explosion. These neutrinos were detected in Japan and the United States *three hours* before the first visible light reached Earth. The neutrinos are able to pass more quickly through the star than visible light, which has a more circuitous path. The weak interaction has just enough strength to allow neutrinos to escape from the core. The travel time to Earth* and the energy spectrum of the neutrinos were consistent with predictions made from gravitational collapse and neutron-star formation. Our understanding of stellar evolution took a giant step forward, and the field of *neutrino astronomy* became well established with the singular event of supernova 1987A. Important information, including an upper mass limit of 16 eV/c^2 for the electron neutrino, was gleaned from the observation and associated calculations of SN 1987A. Computer simu-

---

*Note that SN 1987A occurred 160,000 years ago, and the light and neutrinos took this long to reach Earth.

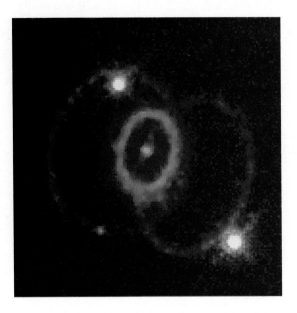

Figure 16.12 Supernova 1987A is the closest supernova since the telescope was invented. Its explosion sent a tremendous amount of gas, light, and neutrinos into interstellar space. In this Hubble Space Telescope image taken in 1994, large strange rings were discovered whose origins are still mysterious and unknown. It is thought that the rings may have been expelled before the main explosion. *Courtesy of Dr. Christopher Burrows, ESA/STSci and NASA.*

lations have been quite successful in understanding the ultraviolet burst some two hours after the explosion, the increase in luminosity for a week, and then the gradual decrease in brightness. The study of SN 1987A gave us confidence that observations of supernovae can be used as an indicator of cosmological distances. SN 1987A occurred in the Large Magellanic Cloud, our nearest neighbor galaxy, and estimates of the distance from this galaxy to Earth based solely on SN 1987A agreed with our best previous estimates (160,000 ly) to within 10%.

Supernova 1987A has been a veritable testing ground for theories and models on supernovae. The ionizing radiation has lit up a complex system of surrounding rings (see Figure 16.12). Astrophysicists are reconstructing how these rings were formed. One model is that the brightest inner ring represents gas thrown off when the progenitor star devoured a small binary companion some 20,000 years previously. The inner ring was lit by x rays and ultraviolet light reaching the ring a few months after the supernova explosion, but it has been getting dimmer ever since. The Hubble Space Telescope first observed the two outer rings in 1994. All three rings were becoming dimmer in the mid-1990s, but a gaseous shock wave triggered by the supernova explosion is working its way out to the brightest inner ring that is 160 billion km in diameter. By 1997 certain spots on the ring were already getting brighter, and by 2007 the ring will be ablaze with a fiery glow. It is hoped that a study of the inner ring will allow astrophysicists to learn more about the supernova's history, especially the origin of the outer rings.

Example 16.6

Because galaxies have been observed with redshifts as great as 10 and quasars with redshifts more than 6, it is useful to have a plot of redshift versus recession velocity. Use Equation (2.33) for the relativistic Doppler shift to calculate such a plot.

Strategy The term *redshift* used by astronomers and physicists refers to $\Delta\lambda/\lambda$. We can use Equation (2.33) to find a relationship between $\Delta\lambda/\lambda$ and β, where $\beta = v/c$.

Solution Equation (2.33) allows us to write

$$\frac{\lambda}{\lambda_0} = \frac{f_0}{f} = \sqrt{\frac{1+\beta}{1-\beta}} \qquad (16.17)$$

We determine $\Delta\lambda/\lambda_0$ from this equation.

$$\frac{\Delta\lambda}{\lambda_0} = \frac{\lambda - \lambda_0}{\lambda_0} = \sqrt{\frac{1+\beta}{1-\beta}} - 1 \qquad (16.18)$$

We show the relationship between the redshift and recession velocity (β) in Figure 16.13. Note that as the redshift increases toward 4, the velocity increases dramatically and approaches the speed of light.

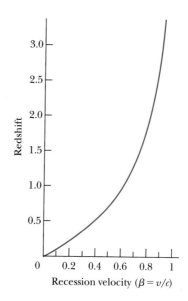

Figure 16.13 The relationship between redshift ($\Delta\lambda/\lambda_0$) and recession velocity ($\beta = v/c$) according to Equation (16.18).

Example 16.7

Astrophysicists were able to deduce a limit on the mass of electron neutrinos based on the spread of arrival times of neutrinos from supernova 1987A using detectors in both Japan and the United States. Find the difference in travel times between a neutrino with very low mass and a neutrino having a mass of 16 eV/c^2. Assume the energy of the neutrino is 20 MeV.

Strategy Let t be the travel time for the low-mass neutrino, t' be the travel time for the neutrino having mass, and Δt be the time difference. We use relativistic relations and relate the mass-energy E with the speed v, which allows us to calculate Δt. The distance of SN 1987A from Earth is $d = 160{,}000$ ly.

Solution We assume the speed of the neutrino with small mass is c. If β is the value of v/c for the neutrino having greater mass, the time relations are

$$t = \frac{d}{c} \qquad t' = \frac{d}{\beta c}$$

$$\Delta t = t' - t = \frac{d}{c}\left(\frac{1}{\beta} - 1\right) = \frac{d}{c}\left(\frac{1-\beta}{\beta}\right)$$

From our earlier study of relativity, we have the total energy $E = \gamma mc^2$, $\gamma^2 = 1/(1-\beta^2)$, and $1 - \beta^2 = 1/\gamma^2$. The last equation can be written

$$(1-\beta)(1+\beta) = \frac{1}{\gamma^2}$$

or

$$(1-\beta) = \frac{1}{1+\beta}\frac{1}{\gamma^2} \approx \frac{1}{2\gamma^2}$$

because $\beta \approx 1$. The equation for Δt becomes, with $\beta \approx 1$,

$$\Delta t = \frac{d}{c}\frac{1}{2\gamma^2} = \frac{d}{2c}\left(\frac{mc^2}{E}\right)^2$$

We put in the appropriate numbers to obtain

$$\Delta t = \frac{1.6 \times 10^5 \text{ ly}}{2c}\left(\frac{16 \text{ eV}}{20 \text{ MeV}}\right)^2 = 5.1 \times 10^{-8}\text{ y} = 1.6 \text{ s}$$

This result is consistent with the actual spread in arrival times of a few seconds. Some assumptions had to be made in the actual calculation. For example, what if the slower neutrinos were emitted first and the faster neutrinos emitted last due to some effect within the supernova? Then the neutrinos might tend to arrive bunched together. A Monte Carlo simulation considering a wide class of possible neutrino emission models was used to assign an upper mass limit of 16 eV/c^2 for the electron neutrinos.

16.5   Problems with the Big Bang

By the early 1980s the Big Bang model was the preferred theory for the origin of the universe. Nevertheless, there were at least three unexplained difficulties that the simple Big Bang model was not adequate to explain. They were

Flat universe

1. Why is the universe flat? Einstein's theory of relativity states that space must bend due to the gravitational attraction. Depending on the amount of matter per unit volume in the universe (its mass density), space curves in on itself such that parallel lines converge above a certain density called the *critical density*. For a mass density less than the critical density, parallel lines diverge, and the universe expands forever. This is called an *open universe*. For a mass density greater than the critical density, the expansion of the universe will eventually be halted and then collapse. This is called a *closed universe*. In the 1980s it appeared that the universe may be just at that critical density where we have a *flat universe*, or very nearly so. Such an occurrence would be quite an extraordinary circumstance.

Horizon problem

2. Why does the universe appear to be so homogeneous no matter in what direction we look? This is called the *horizon* problem. If the universe is 13.7 billion years old, then opposite sides of the universe are 27 billion lightyears apart. How can these regions have microwave radiation that is so similar? The temperature of the universe reflects the 3 K microwave background no matter in what direction we look. How can two regions of the universe have temperatures so similar and be 10^{19} m apart and not able to communicate?

3. Why have we not yet detected magnetic monopoles? The occurrence of magnetic monopoles brings symmetry to Maxwell's equations of electromagnetism and also satisfies other physics theories.

The Inflationary Universe

These difficulties with the Big Bang model were relieved considerably in 1981 by a suggestion of Alan Guth, who proposed a variation of the standard Big Bang model. Guth proposed that at some time between roughly 10^{-35} s and 10^{-31} s after the Big Bang, the size of the universe suddenly expanded by a factor of 10^{50}. This period is called the *inflationary epoch* and is due to the separation of the nuclear and electroweak forces. We have to remember that the size of the universe was incredibly small at the time, so the magnitude of such an expansion is remarkable. It is as if the electron of a hydrogen atom, which is normally only 10^{-10} m from a proton, suddenly found itself 10^{24} lightyears away! After the inflationary period, the universe resumed its evolution according to the standard Big Bang model.

No matter how hard this may be to believe, inflation does mostly solve the three problems just listed. The inflationary theory *requires* the mass density to be very close to the critical density. This solves the flatness problem. Guth argued that the universe was so tiny that it had already reached equilibrium before the inflation occurred; this explains the homogeneous universe. Subsequent to Guth's work, cosmologists have contributed many suggestions to his inflationary idea. Magnetic monopoles would have to occur along the boundaries or walls of different domains. These domains might be likened to different universes. We cannot observe the magnetic monopoles, if they exist and if they are at the edge of the universe. It is a neat, tidy package and much more complicated than we have described here. Inflationary theory is crucially connected to elementary particle theories.

Alan Guth (1947–) was born in New Jersey and educated at the Massachusetts Institute of Technology (Ph.D., 1971). He held postdoctoral positions at Princeton, Columbia, and Cornell Universities in addition to the Stanford Linear Accelerator Center. During this period he created in 1979 the theory of cosmic inflation as a modification of the Big Bang theory. He returned to MIT in 1980 where he is now a professor of physics. He authored the popular book *The Inflationary Universe: The Quest for a New Theory of Cosmic Origins* in 1998. *Courtesy of Alan Guth.*

The Lingering Problems

Despite the significant progress that has been made, there are lingering problems with our understanding of the universe. One of them, the age of the universe, will be discussed in the next section. We discuss some others here.

Formation of Stars and Galaxies Cosmologists had difficulty explaining how the clumping of matter into galaxies, clusters of galaxies, and other strange objects like quasars and emitters of gamma-ray bursts has occurred. This is difficult to do in light of the homogeneous universe as expressed by the cosmological principle. The primary purpose of the Cosmic Background Explorer satellite (COBE) launched in 1989 and the Wilkinson Microwave Anisotropy Probe (WMAP) launched in 2001 by NASA was to investigate the cosmic microwave background radiation (CMB) believed to exist from the remnants of the Big Bang. Today the CMB has cooled to near 3 K and it permeates the universe. We discussed this CMB earlier in Section 16.1, and its existence is strong evidence for the Big Bang model.

By studying the minute details of the CMB, physicists hope to learn about the universe on very large scales, because the radiation has traveled for eons over large distances. First, COBE showed that the homogeneity occurs over a much wider frequency range (see Figure 16.14) than could have been detected by ground-based observers. Both spacecraft probes searched for small fluctuations in the CMB temperature that would be indicative of the tiny *seeds* of inhomogeneity around which the first stars could have formed. COBE (in 1992) and WMAP (in 2002 with 15 times better sensitivity) have observed very tiny differences in the temperature of radiation coming from matter at the edge of the universe (Figure 16.15). These tiny ripples in the temperature represent the universe's condition at only 400,000 years after the Big Bang and are believed to account for the formation of stars and galaxies. Inflationary theory allows small density fluctuations that can later serve as the seeds for gravity to cause matter to clump together to form stars, galaxies, and other structures. WMAP (Figure 16.16) shows that the CMB temperature is precisely 2.725 K and is consistent with the inflationary model of the Big Bang.

COBE, WMAP

How Can Stars Be Older Than the Universe? During much of the 1990s observations of the age of the universe indicated it was 14 billion years old or younger.

Figure 16.14 The spectrum of the intensity of the cosmic microwave background as measured with the COBE satellite. The data are smaller than the points shown, and the solid line is the blackbody radiation calculation for 2.73 K. The agreement is spectacular. The data are plotted as a function of wavenumber (inversely proportional to wavelength). *Courtesy of Nancy Burgess, NASA COBE Science Team.*

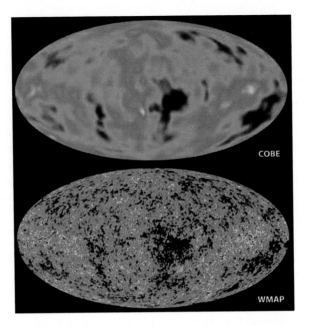

Figure 16.15 In 1995 NASA's COBE satellite detected temperature fluctuations in the cosmos, which was a landmark discovery. Based on the success of the COBE mission, the Wilkinson Microwave Anisotropy Probe (WMAP) was proposed in 1995 and launched in 2001. This WMAP satellite image was released in 2003 and showed a much more detailed image of the tiny fluctuations that became the seeds of galaxy formation. *Adaptation from the NASA/WMAP Science Team.*

At the same time other measurements on stars found that some were 15 billion years old or older. Most astronomers thought the problem was with the age of the universe determinations. Theorists believed the expansion rate measurements were wrong, while experimentalists thought the basic cosmological model was incorrect. Only a few astronomers questioned the age measurements of the stars, because the theoretical model predictions had not changed for 30 years. Finally, in the late 1990s it was found that globular clusters, containing many of the older stars, were further away than previously believed. The methods of determining these distances are too complex to be discussed here, but the resolution is the

Figure 16.16 An artist's depiction of the WMAP in orbit at the second Lagrange Point or "L2." WMAP is a million miles from Earth, four times further than the moon. The L2 orbit is a semistable region of gravity, and WMAP is the first spacecraft to be positioned permanently as an observing station in this orbit. The sun, Earth, and moon are all depicted in this figure. Six months is required for WMAP to make a full sky observation. WMAP's mission was scheduled for 2001–2005. *Adaptation from the NASA/WMAP Science Team.*

realization that the expansion of the universe is accelerating rather than slowing down as had been previously believed. There must be some kind of repulsive force that acts opposite to gravity. This force is associated with a phenomenon known as *dark energy*.

Dark Matter Another problem concerns the discrepancy between the mass of the universe required for the critical density (9×10^{-30} g/cm^3) and the apparent mass density (3×10^{-31} g/cm^3). This is known as the **missing mass problem**. The ratio of the actual mass density of the universe to the critical density is called omega (Ω). Much of the mass of the universe may be in some form, called **dark matter**, which we have not yet observed.

Missing mass problem

Dark matter

Astronomers and astrophysicists have inferred from the motion of stars, galaxies, and other astronomical objects that there doesn't seem to be enough mass in the universe for gravity to keep the objects from flying apart. The amount of matter in the universe is determined from the radiation that we detect. Therefore, it seems that much of the universe consists of this dark matter that does not emit, scatter, or absorb light. The dark matter is also required to enable gravity to amplify the fluctuations of the CMB discussed previously. Scientists thought that dark matter might be due to neutrinos, WIMPs (weakly interacting massive particles), or primordial black holes. Astronomers and physicists have mounted many unsuccessful searches for the missing matter. The searches for dark matter continue. One theory proposes that the problem is with Newton's laws of motion at large distances, not with missing mass. This theory, called MOND, stands for *modified Newtonian dynamics* and proposes that the force is smaller at large distances, thus eliminating the need for additional mass.

The Accelerating Universe In 1998 two independent teams of astronomers using supernovae data were investigating the rate at which the expansion of the universe was slowing down when they reached a surprising decision. Both teams decided their data could only be understood if the expansion was speeding up. The results imply that some mysterious repulsive force is acting against gravity, causing galaxies to move apart at increased speeds. With more data and careful checking, the results still stand. The source of this mysterious force has been called **dark energy**, and preliminary analysis indicated it might represent 75% of the mass-energy of the universe. Other results now seem to confirm the cosmic acceleration concept.

Dark energy

This is not the first time dark energy has been proposed. It has the same effect on the universe as Einstein's cosmological constant. It is a vacuum energy density that produces a negative pressure that drives the expansion of the universe. It was suggested in the 1960s by particle physicists to understand certain phenomena and has occasionally been used by astronomers to explain astronomical phenomena. The accelerating universe is certainly a part of the inflationary period, but that was right after the Big Bang.

Theorists have suggested several explanations, and the cosmological constant is at the top of the list. Theorists have also suggested **quintessence**, a new form of energy with negative pressure. The term "quintessence" is derived from the Greeks' mysterious fifth element (after earth, air, fire, and water) that supposedly held the moon, planets, and stars in position. Although the cosmological constant is indeed constant, observational evidence doesn't indicate that dark energy has always been constant. Dark energy seems to have become effective 5–10 billion years ago, but it had to be created no later than 10^{-35} s after the Big Bang. Quintessence, on the other hand, is a dynamic, time-evolving, spatially changing form

Quintessence

of energy with negative pressure, its supporters argue. A third possibility to explain dark energy is a cosmic field associated with inflation. A fourth possibility could be a problem with general relativity itself. At the present time, no one knows what dark energy actually is or how it originated.

The Cosmological Constant   The cosmological constant can be represented by

$$\Lambda = \frac{8\pi G}{c^4} \rho_V \tag{16.19}$$

where ρ_V is the vacuum energy density. A positive or negative cosmological constant has dramatic results for the future of the universe: if it is negative, the expansion of the universe will slow down, and if positive, the expansion of the universe will continue forever.

Example 16.8

Determine the critical density of the universe.

Strategy   One way to calculate the critical density ρ_c is by assuming the mass and radius of the universe are M and R, respectively. A galaxy of mass m and velocity v will be able to escape the universe with zero velocity if its total energy is precisely zero. We therefore determine the critical density by finding the condition for zero total energy.

Solution   The sum of the kinetic and potential energy of the galaxy with respect to the universe is

$$\text{K.E.} + \text{P.E.} = \frac{1}{2}mv^2 - G\frac{mM}{R} = 0 \tag{16.20}$$

The mass M of the universe is related to the density ρ_c by $M = \frac{4}{3}\pi R^3 \rho_c$. Equation (16.20) becomes

$$\frac{1}{2}mv^2 = G\frac{m}{R}\frac{4}{3}\pi R^3 \rho_c$$

If we use $v = HR$ for the galaxy velocity, we rearrange the previous equation to give

$$\rho_c = \frac{3}{8\pi}\frac{H^2}{G} \tag{16.21}$$

This is also the equation obtained from the general relativistic cosmological model of Einstein-de Sitter. If we insert the value of G and a value of $H = 22$ km/s per million lightyears, then we find $\rho_c = 9 \times 10^{-30}$ g/cm³. This is the value reported in Section 16.1.

16.6   The Age of the Universe

The age of the universe has been perhaps the most sought after and the most controversial value in science for the past few decades. Until the mid-1990s, theory was well ahead of experiment, but with the breakthrough results of the Hubble Space Telescope, other productive satellite telescopes, and the advances in ground-based telescopes together with adaptive optics systems, we have been in a time rich with new data. Results since the turn of the twenty-first century have narrowed on a value near 13.7 ± 0.2 billion years.

The age of the universe can be determined by several methods. In the past, cosmological models have been preferred, but more recently traditional methods have given more precise values that are coming into general agreement. We discuss those results here.

Age of Chemical Elements

We have already discussed in Chapter 12 how radioactive decay can be used to determine the age of material. If we assume that the elements were made not long after the Big Bang, then we can use the ratio of certain elements like $^{87}Sr/^{86}Sr$, because ^{87}Sr is produced by radioactive decay (from ^{87}Rb), whereas ^{86}Sr is stable and should remain constant over time. By comparing the ratios of $^{87}Sr/^{86}Sr$ and $^{87}Rb/^{86}Sr$, the age of a sample can be determined. Such measurements have shown that meteorites hitting Earth are 4.5 billion years old. It is difficult to extrapolate these measurements to determine the age of our galaxy and the universe. Using various techniques results in an age of the universe between 8 and 17.5 billion years. These results should become better with more research.

Radioactive Dating of Stars WMAP showed, much to the surprise of cosmologists, that stars were formed as soon as 200,000 years after the Big Bang. Even though we don't have samples of stellar material, we can examine the relative intensities of elemental spectral lines from old stars. Several measurements have been made in the last decade. The ratios of thorium/europium and uranium/thorium isotopes have led to reported star age values of 15.2 ± 3.5 Gyr, 15.6 ± 4.6 Gyr, 12.5 ± 3 Gyr, and 14.1 ± 2.5 Gyr, where a Gyr is 10^9 or a billion years. Notice that the average of these values is near 14 billion years.

Age of Astronomical Objects

The techniques used by astronomers to determine the ages of various stars and clusters are remarkable.

Old Stars in Globular Clusters Globular clusters are aggregations containing up to millions of stars that are gravitationally bound; however, they are much smaller than galaxies. Thousands of stars in each globular cluster are about the same age. The H-R diagram, named after Ejnar Hertzsprung and Henry Russell who invented it in 1910, compares the temperature and luminosity of stars. When stars are burning their hydrogen to form helium in their cores, their temperature and luminosity fall on a single curve in the H-R diagram. The age of the star is inversely proportional to the luminosity, so an upper limit for the age of the cluster can be determined from the most luminous star. It is the age of these globular clusters that seemed in the mid-1990s to be older than the universe itself, according to other determinations. Better measurements have since shown that the globular clusters were further away than originally believed. Once this correction had been made, their ages of 11 to 13 billion years were within other estimates of the age of the universe.

Age of White Dwarf Stars Stars the size of our sun become white dwarfs after burning all their fuel. Their mass is about that of our sun, but their size is only about that of Earth. Their densities are a million times that of water, but they lack enough mass to become a neutron star or black hole. White dwarf stars produce residual heat radiation. The oldest ones will be the faintest, because they have been cooling the longest. Because they are so faint, they are difficult to see. It is easiest to see them in a binary system, and the Hubble Space Telescope observed in 1995 more than 75 white dwarfs in the globular cluster M4. Conceptually, the determination of the age of a white dwarf is similar to determining how long ago a campfire was burning by measuring the temperature of the remaining coals. Ancient white dwarf stars were found in 2004 to be 12 to 13 billion years old. Other observations indicate these stars were created in a period less than one billion years after the Big Bang. These age determinations are consistent with other values.

Cosmological Determinations

The question of the age of the universe is intertwined with the ultimate fate of the universe. Cosmologists cannot determine the age of the universe without being able to understand how the universe was formed, especially in the early stages. This is because we must interpret the observed electromagnetic radiation (light, ultraviolet, infrared, radio, x ray, and gamma) that comes from billions of years ago. As new data appear that are not consistent with current theories, new theories are required. These theories also predict the fate of the universe. The cosmic microwave background radiation is strong proof of the Big Bang, and inflation theory seems secure. However, the simple inflation theory does not explain all the data.

Hubble constant   The present value of the Hubble parameter H is called the **Hubble constant** and is denoted by H_0. The Hubble constant H_0 determines an upper limit for the age of the universe given by $\tau = 1/H_0$ if the Hubble parameter had always been constant at H_0.

Example 16.9

The inverse of the Hubble parameter has the units of time. Use the value of the Hubble constant 71 (km/s)/Mpc to find the upper limit τ for the age of the universe.

Strategy We will need to understand the relationship between the various astronomical parameters to do this. We begin with the units given by H_0^{-1} and convert the result to years.

Solution We begin with

$$\tau = \frac{1}{H_0} = \frac{1}{71 \frac{km}{s \cdot Mpc}}$$

$$= \frac{1 \, s \cdot Mpc}{71 \, km} \frac{1 \, y}{3.16 \times 10^7 \, s} \frac{10^6 \, pc}{Mpc} \frac{3.26 \, ly}{1 \, pc} \frac{9.46 \times 10^{12} \, km}{ly}$$

$$= 13.7 \times 10^9 \, y$$

The upper limit of the age of the universe is 13.7 Gly.

In order to use cosmology to determine the age of the universe we return to Equation (16.5).

$$1 = \frac{8\pi G \rho_m}{3H^2} - \frac{k}{a^2 H^2} + \frac{\Lambda}{3H^2} \tag{16.5}$$

We rewrite this equation using the terms in Equations (16.6) through (16.8).

$$1 = \Omega_m + \Omega_k + \Omega_\Lambda \tag{16.22}$$

The second term on the right depends on the curvature of the universe, which depends on the geometry of spacetime. We are familiar with curvature, because we live on the surface of a sphere (Earth) and cannot see over the horizon. There are three qualitatively different classes of curvature (see Figure 16.17), each dependent on the value of the curvature parameter k, which in turn determines Ω_k as denoted in Equation (16.7). The value of Ω_k is greater than 1 for the closed geometry (Figure 16.17a), which has the spherical shape, and $\Omega_k < 1$ for the

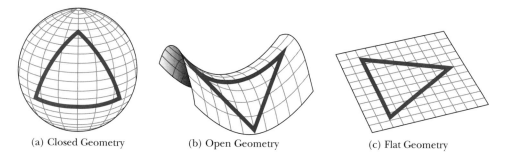

(a) Closed Geometry   (b) Open Geometry   (c) Flat Geometry

Figure 16.17 The three possible shapes of the universe are shown. (a) The spherical or closed geometry has a positive curvature (large mass). (b) The hyperbolic or open universe has a negative curvature (low mass). We saw this before in Chapter 15 in the spacetime geometry near large masses. (c) The Euclidean or flat geometry has no curvature. The shape of the triangle is shown in each geometry.

hyperbolic or open geometry (Figure 16.17b). Inflationary theory indicates the universe should have the flat geometry or zero curvature, $\Omega_k = 0$ (Figure 16.17c).

The geometry of the universe is also dependent on the mass density, in particular the critical density. For a mass density greater than the critical density, the geometry of space is closed. If it is less than the critical density, the geometry is open. If the density is close to the critical density, then the flat geometry should prevail. The WMAP (Wilkinson Microwave Anisotropy Probe) science team was able to determine various cosmological parameters by analyzing the detailed structure of the background fluctuations in the cosmic microwave background radiation. The WMAP results indicate the universe is flat to within a 2% margin of error. Therefore the term Ω_k is 0 in Equation (16.22). Astronomers also found that the Hubble constant is $H_0 = 71 \pm 4$ (km/s)/Mpc, and found the age of the universe to be 13.7 billion years (Figure 16.18).

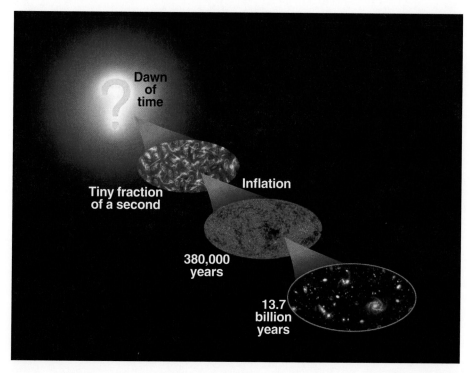

Figure 16.18 Inflation occurs a tiny fraction of a second after the Big Bang. After another 380,000 years, the universe cooled enough (3000 K) that the radiation no longer interacted with matter. It is that radiation, which has cooled to 3 K today, that is the cosmic microwave background radiation. By observing the minute fluctuations throughout the universe, scientists are able to determine what happened after the inflationary period. *Adaptation from the NASA/WMAP Science Team.*

Sloan Digital Sky Survey The Sloan Digital Sky Survey (SDSS) is an ambitious project to map in detail one quarter of the entire sky and to determine the positions and brightness of more than 100 million astronomical objects. The SDSS is also measuring distances to more than a million galaxies and quasars. Scientists have used observations of 3000 known quasars to date the cosmic clustering of diffuse hydrogen gas. Combining these data with the WMAP measurements, they conclude the universe is 13.6 billion years old.

Universe Scale Determination Another method of determining the future of the universe has to do with the scale factor a (the approximate separation distance between galaxies) in Equation (16.2). In Equation (16.2) we defined the Hubble parameter to be $H = (1/a)(da/dt)$. We find the Hubble time τ to be

$$\tau = \frac{1}{H} = \frac{a}{\dfrac{da}{dt}} \tag{16.23}$$

If we draw a tangent in Figure 16.19 (Problem 27) to the a curve at the present time t_0, the intercept of this tangent with $a = 0$ determines the Hubble time τ. Figure 16.19 would give a value of about $\tau = 15$ Gyr, but these graphs are a little arbitrary.

In the case of the flat universe and mass density equal to the critical density, the relationship between a and t is $a^3 = Ct^2$, where C is a constant. Then if we take the derivative, we have

$$3a^2 \, da = 2Ct \, dt$$

$$\frac{da}{dt} = \frac{2C}{3a^2} t$$

Equation (16.23) then gives

$$\tau = a \frac{3a^2}{2C} \frac{1}{t} = \frac{3a^3}{2C} \frac{1}{t} = \frac{3}{2} t \tag{16.24}$$

In the case of the flat universe the age of the universe is $2\tau/3$ ($2/3H$), and if we use a value of 71 km/s per Mpc, we determine $\tau = (H_0)^{-1} = 13.7$ billion years, and the age of the universe would be only 9 billion years. Because there is not

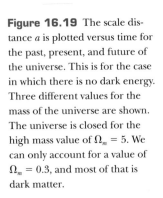

Figure 16.19 The scale distance a is plotted versus time for the past, present, and future of the universe. This is for the case in which there is no dark energy. Three different values for the mass of the universe are shown. The universe is closed for the high mass value of $\Omega_m = 5$. We can only account for a value of $\Omega_m = 0.3$, and most of that is dark matter.

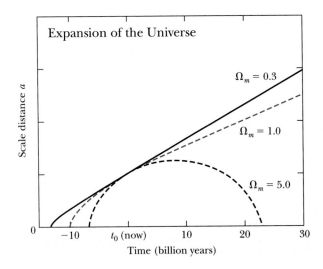

Expansion of the Universe

$\Omega_m = 0.3$

$\Omega_m = 1.0$

$\Omega_m = 5.0$

Scale distance a

0 -10 t_0 (now) 10 20 30

Time (billion years)

so much mass in the universe, a more refined calculation with $\Omega_m = 0.3$ shows $t = \tau = (H_0)^{-1} = 13.7$ Gyr.

Universe Age Conclusion

In this section we have discussed several determinations of the age of the universe. There is little question that the results are coalescing around a value near 14 billion years. Some results indicate a more precise value of 13.7 billion years. Much work will continue to be done to determine the age of the universe.

16.7 The Future

Long before we have to worry about what will happen to the universe, the sun will be a cold dark mass. In this section we first discuss the demise of the sun, of which we are fairly certain. We then summarize what we know about the future of the universe and the possibility of other Earth-like planets.

The Demise of the Sun

The sun is about halfway through its life as a productive star. It became a star about 4.5 billion years ago, and it will be about another 5 billion years before it runs out of its thermonuclear fuel. As the hydrogen fuel is exhausted, the sun will contract under its own gravitational attraction. The sun will heat up even more, with helium at its core and a layer of hydrogen gas outside. The heat will cause the outside layers of the sun to expand. The sun's surface will expand to such a size that it will engulf the planet Mercury, perhaps reaching as far as the Earth. We call such a star a **red giant**. The sun's surface temperature will cool from the present 5500 K to maybe 4000 K. However, the temperature of the Earth will dramatically increase because of the proximity to the extended sun's surface, and no life will then be able to exist on Earth.

Eventually, the hot core of a star uses up its thermonuclear fuel, producing heavier and heavier nuclear masses so that mostly iron remains. The mass of the sun is not large enough for the gravitational attraction to form a neutron star or supernova. The lighter elements in the outer layers of the sun will boil off due to the reduced gravitational attraction at the surface, and the final mass will be about 0.5 to 0.6 times the mass of today's sun. The remaining sun will contract to about the size of the Earth. Slowly the sun will cool down, becoming a white dwarf, and then finally a cold black dwarf. The sun and its planets will be doomed to an eternally frozen death.

Where Is the Missing Mass?

We return to our earlier result, Equation (16.22), determined from the Friedmann equations. The WMAP science team showed from their analysis of the CMB that the curvature term Ω_k is zero, because the universe is flat. This agrees with inflationary theory. This leaves two terms in Equation (16.22); Ω_m is the mass density and Ω_Λ is the dark energy or cosmological constant term. Remember that the term Ω is the ratio of the actual mass density of the universe to the critical density, and the data indicate that it should be 1. However, there is not enough ordinary mass in the universe. Even if we include the dark matter, the term Ω_m seems

Special Topic

FUTURE OF SPACE TELESCOPES

After decades of astrophysics and cosmology being dominated by theoretical models with little data to differentiate between sometimes wildly different projections, observational astronomy finally caught up in the 1990s. Many new space telescopes and ground-based telescopes fitted with adaptive optics systems have all helped bring a myriad of new experimental results. Many cosmologists are simply rubbing their hands with glee at the prospect of delving into all these new data.

With a dozen additional 8-m-diameter ground-based telescopes coming into operation by the year 2010, why do we need telescopes in space? The answer is that the Earth's atmosphere absorbs the majority of the electromagnetic radiation coming from space (see Figure A). Only visible light, some radio waves, and a limited amount of infrared and ultraviolet light can pass through the atmosphere without being absorbed

or distorted by the atmosphere. In order to see x rays, gammas, and most of the ultraviolet and infrared spectrum, we need to have telescopes above the Earth's atmosphere. The first observation of an x-ray source from space was made in 1962 on a rocket that stayed above the atmosphere for 6 minutes. Both the U.S. National Aeronautics and Space Administration (NASA) and the European Space Agency (ESA) have been active in deploying telescopes in space.

In the 1970s NASA scientists conceived of the *Great Observatories* program for space to cover most of the electromagnetic spectrum. The launch of the Hubble Space Telescope in 1990 (and its subsequent repairs and improvements in 1993, 1997, 1999, and 2002), the Compton Gamma Ray Observatory in 1991, the Chandra X-ray Observatory in 1999, and the Spitzer Space Telescope (formerly the Space Infrared Telescope Facility) launched in 2003 have produced spectacular results. The Hubble Space Telescope has been one of the greatest scientific instruments produced, and we read about its discoveries in the daily newspapers.

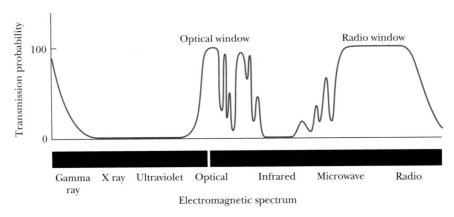

Figure A The Earth's atmosphere is not transparent to much of the electromagnetic spectrum, as shown by the transmission probabilities of the upper line. It is mostly optical and radio frequencies that are able to permeate the atmosphere.

---

to be only about 0.3. WMAP shows that the baryonic matter and atoms represent only 4% of the universe. Dark matter accounts for 23%, and the remainder, 73%, is dark energy. We show in Figure 16.20 what happens to the size of the universe under the various mass values. What we now believe is the correct value ($\Omega_m = 0.3$ and $\Omega_\Lambda = 0.7$) is indicated by the solid blue curve. The size of the universe is expanding and even accelerating its expansion. The universe is a strange place.

These results can be made more clear by using the cosmic triangle where

The Compton Gamma Ray Observatory focused on the huge energy coming from gamma-ray bursts. Chandra is examining x rays from throughout the universe, like colliding galaxies, supernovae, and black holes. Its goal is to increase our understanding of the evolution of the universe, including its origin and ultimate fate. Spitzer is detecting infrared radiation, like heat, given off by objects in space. It will be able to detect cosmic features never seen before. Many objects like planets and unignited stars do not produce visible or ultraviolet light and so are not visible through conventional telescopes. Some of the most fascinating objects and processes in the universe may be hidden behind vast clouds of dust and gas. These objects include black holes, quasars, and regions where stars are forming in galaxies and planets are forming around stars. Spitzer will be able to sense the heat of dark, faint, or hidden objects at the edge of the universe.

Although the Compton Gamma Ray Observatory ceased operation in 2000, other spacecraft with similar missions are in operation or planned. This includes the NASA-supported High Energy Transient Explorer Mission (HETE-2), launched in 2000, to detect x rays and gamma rays to provide precise localization of gamma-ray bursts. The ESA launched the International Gamma-Ray Astrophysics Laboratory (INTEGRAL) in 2002 with detectors for x rays, gamma rays, and visible light to study the formation of new chemical elements and gamma-ray bursts. NASA's Swift spacecraft was launched in 2004 and is investigating gamma-ray bursts and their afterglows with detection capability for gamma-ray, x-ray, ultraviolet, and visible wavelengths. Its goal is to determine the origin of gamma-ray bursts and how the blast wave evolves, and to use this information to study the infant universe. The Italian Space Agency plans to launch the AGILE in 2005 for gamma-ray and x-ray studies. The Gamma-ray Large Area Space Telescope (GLAST) is under study by the United States, Japan, and Europe for possible flight before 2010. It will detect gamma rays from 20 MeV to more than 300 GeV energy that are emitted from active galactic nuclei, pulsars, stellar-mass black holes, supernova remnants, and gamma-ray bursts.

The James Webb Space Telescope (JWST) is presently under development by NASA and international partners such as the ESA for possible launch after 2010. It is intended to be the replacement for the Hubble Space Telescope. Its main feature will be a large 6.5-m-diameter mirror. It will primarily observe infrared light to study high-redshift galaxies and protogalaxies. It would be placed in an L2 orbit 1.5 million miles from Earth.

Many other missions are in the development or planning stages. We can mention only a few. NASA's Space Interferometry Mission will determine the locations of stars with an accuracy several hundred times better than any previous telescope. It will also probe nearby stars for Earth-like planets. Planned for launch around 2009, it will utilize optical interferometry for its measurements. NASA's Constellation-X will deploy four spacecraft, each having 1.6-m-diameter telescopes for measuring the x-ray spectra of active galactic nuclei, quasars, and anything near a black hole. It is planned for launch after 2010. ESA plans to launch the LISA Pathfinder perhaps in 2007 to test the technology needed for the ambitious LISA mission, which will search for gravitational waves (see Chapter 15). ESA's Planck spacecraft, planned for launch in 2007, is a mission similar to the successful COBE and WMAP spacecrafts. It will also be in the L2 orbit and will investigate the cosmic microwave background radiation. The Gaia is a ESA mission, planned for launch around 2010, that will conduct a census of one billion stars in the Milky Way Galaxy in hopes of discovering extrasolar planets and brown dwarfs (failed stars).

the three sides of the triangle are the values of Ω_m, Ω_Λ, and Ω_k (Figure 16.21). Constraints from three different sets of experimental data are included. The Type Ia supernovae data are consistent with an accelerating universe. The WMAP cosmic microwave background radiation data are consistent with a flat universe, and the clusters of galaxies data are consistent with a low-density universe. The intersection of the three sets of data constrains the universe mass parameters to the values we have been discussing: $\Omega_k = 0$, $\Omega_m = 0.3$, and $\Omega_\Lambda = 0.7$.

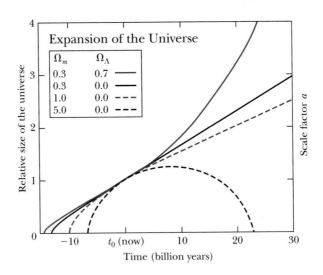

Figure 16.20 Four possible scenarios for the expansion of the universe are shown for various values of Ω_m and Ω_Λ. The black dashed curve represents a high-mass-density universe that eventually contracts and collapses under its own gravitational attraction. The blue dashed curve shows a flat, critical-density universe that continues to expand, but more slowly. The black solid curve represents an open, low-density universe. The solid blue curve represents scientists' best idea of the true universe. The dark energy makes up the missing matter, and its negative attraction causes the universe to expand at an accelerating rate.

Figure 16.21 Three sets of data (supernovae, clusters of galaxies, and WMAP cosmic background radiation) are located inside the cosmic triangle with the mass parameters Ω_m, Ω_Λ, and Ω_k. The intersections of the three data sets constrain the mass parameters and indicate the universe is expanding at an accelerated rate with a flat spacetime geometry.
Adapted from N. A. Bahcall, J. P. Ostriker, S. Perlmutter, and P. J. Steinhardt, Science **284**, 1481 (1999).

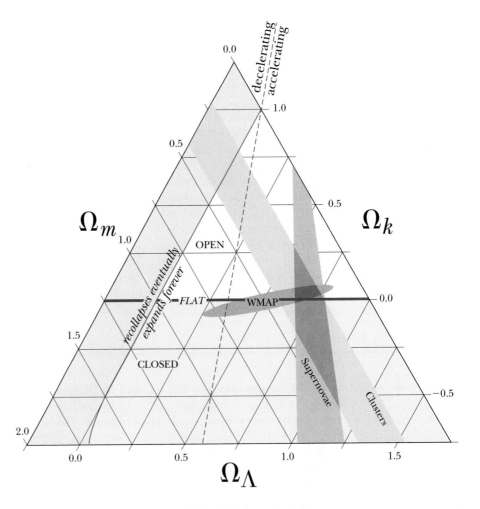

The Future of the Universe

Of this there no longer seems to be much doubt: the universe is flat, but it is expanding, and the expansion is even accelerating. Eventually all the stars in our Milky Way Galaxy, as well as in all other galaxies, will die. Black holes will not be able to find any more mass to consume. The laws of thermodynamics indicate the universe will be a dark and cold place.

Are Other Earths Out There?

We are now in a position to question seriously whether there are other Earth-like planets in the universe. In the last few years, astronomers have identified many candidates for extrasolar planets, that is, planets revolving around stars other than our own sun. Initially, these candidates were identified through a wobble of the star that was induced by an orbiting planet. The wobble's period and magnitude is used to deduce the planet's orbit and minimum mass. Such observations use the Doppler effect and require extensive measurements.

One difficulty is distinguishing between planets and **brown dwarfs**, which are starlike objects with too little mass (<80 Jupiter masses) to create nuclear fusion. A brown dwarf forms the same way a star does, by a gas cloud collapsing in on itself by gravitational attraction. Planets grow from dust and gas accreting in a circumstellar disk. Several observations of swirling dust around a star indicate a planet is forming. The first such observation of a dust disk was found in 1994 around Beta Pictoris, a young star only 59 lightyears away. Extrasolar planets are now being found at the rate of about one per week.

The first observation of an extrasolar planet transiting its star was made in 1999. By observing a transiting planet (Figure 16.22) astronomers are able to measure its diameter and even perform measurements to determine the chemical makeup of its atmosphere. A transit has to occur every time the planet orbits the star. The repeated occurrence of the transit is evidence that the observed transit is really due to a planet. There is now an army of amateur astronomers looking for extrasolar planets using complex computer analysis of their observations. In 2004 such an experiment found evidence for a planet that was subsequently examined by a more powerful telescope and shown to exist.

Figure 16.22 An artist's conception of an extrasolar planet transiting a star other than our sun. Astronomers cannot yet directly observe these planets. They infer their existence from the changing level of light coming from the star. Spectroscopic analysis has been made of a planet's atmosphere by the changing spectral line intensities of radiation from the star passing through the atmosphere. *Lynette Cook.*

Summary

Scientists believe the universe started with a primordial event called the Big Bang. The universe started as an extremely dense, hot fireball. Our current theories have been fairly successful in understanding the events after the initial 10^{-43} s. For example, by the time 10^{-13} s, quarks and leptons were formed, and by 10^{-3} s, neutrons and protons were also formed. Nuclei formed by 3 min, but atoms couldn't form for 400,000 years.

The three pieces of evidence that confirm the Big Bang include the redshift measurements of Edwin Hubble that indicate the universe is expanding, the observation of the cosmic microwave background radiation by Penzias and Wilson, and the agreement between the predictions of the primordial nucleosynthesis of the elements and known abundances. Hubble's law relates the recessional velocity and the distance from Earth R:

$$v = HR \tag{16.1}$$

where H, Hubble's parameter, is related to a scale factor a that is the approximate separation distance between galaxies.

$$H = \frac{1}{a}\frac{da}{dt} \tag{16.2}$$

The Big Bang model depends on two major theoretical foundations: the general theory of relativity and the cosmological principle, which states the universe is isotropic and homogeneous. The basic equation is due to Alexander Friedmann and can be written

$$H^2 = \frac{8\pi G \rho_m}{3} - \frac{k}{a^2} + \frac{\Lambda}{3} \qquad (16.4)$$

Stars were formed when matter collected together due to the gravitational force. Eventually nuclear fusion provides the energy radiated from stars. Stars emit neutrinos due to the nuclear fusion process, and the solar neutrino problem has finally been understood by the oscillation of neutrinos coming from the sun. After the thermonuclear processes are exhausted, massive stars undergo a gravitational contraction so strong that eventually the star consists mostly of neutrons. These neutron stars are incredibly dense. Stars of mass less than about 1.4 solar masses are able to exert enough pressure due to electrons to prevent a gravitational collapse. They eventually become white dwarfs.

There are many astronomical objects in the universe. Several kinds of galaxies, which are collections of many stars, have been observed. Quasars are tremendous sources of energy, and several have been found quite far away that must have been formed early in the evolution of the universe.

Increased interest in supernovae occurred because of the SN 1987A event, which could be seen with the naked eye. This supernova occurred when a massive star underwent gravitational collapse after its thermonuclear fuel was exhausted. A tremendous explosion resulted that sent radiation streaming outward. On Earth we were able to see visible light and detect neutrinos, among other types of radiation.

There has been increased interest in gamma-ray bursts, which originate from throughout the universe. Researchers believe they come from supernovae, and additional facilities are being planned to investigate them.

Some problems with the Big Bang model were mostly solved in the early 1980s by the idea of an inflationary universe. Inflation solved the flatness and homogeneity problems, but other problems remained. There is a discrepancy between the critical density and observed mass density of the universe. It appears that there is not enough mass in the universe to have a flat universe. The Wilkinson Microwave Anisotropy Probe (WMAP) has had spectacular results. Cosmologists now conclude, by putting together considerable data, that the universe is 13.7 billion years old and that the Hubble constant has the value 71 km $\cdot$ s^{-1} $\cdot$ Mpc^{-1}. The universe is flat and will expand forever. The mass density consists of 4% ordinary baryonic matter, 23% dark matter, and 73% dark energy. Dark energy consists of a repulse gravitational energy that is causing the universe expansion to accelerate. We do not know what the dark matter and dark energy actually is.

Many new ground-based and space telescopes are now available, under construction, or being planned that will give new data and results that undoubtedly will be in conflict with present beliefs. This is the golden age of astrophysics and cosmology.

Questions

1. Explain why Hubble's parameter, with its value today called Hubble's constant, is not actually a constant.
2. According to thermodynamic equilibrium, which should be the most abundant and least abundant quarks during the period from 10^{-13} s to 10^{-3} s?
3. If the gravitational attraction is important in a neutron star where the neutrons are close together, then why isn't the gravitational interaction important in a nucleus with many neutrons?
4. If all the distant galaxies are moving away from us, explain why we are not at the center of the universe.
5. How can you explain the fact that the Andromeda Galaxy appears to be approaching us rather than receding?
6. Explain why the universe cannot be older than the Hubble time.
7. Explain why elements heavier than iron are not found in stars.
8. Why isn't it possible to know what is happening to our nearest neighbor stars today (in the next 24 hours)?
9. During which stage of the beginning of the universe would you expect deuterons to be formed? Explain.
10. What happened to the neutrons produced in the early stages of the universe that were not synthesized to deuterons or ^{4}He nuclei?
11. During what time period do free neutrons disappear? Explain.
12. Explain how it might be possible to confuse the redshifts from recession velocities with the gravitational redshifts. How can we distinguish the two?
13. Quasars are known to vary in brightness by just a few hours or days. What can we say about the size of these quasars?
14. Observations from the Compton Gamma Ray Observatory indicate that the gamma-ray bursts have an even distribution throughout the sky. How can we be sure that these bright phenomena are not coming from our own galaxy, the Milky Way?
15. Sometimes dark matter is called "cold dark matter." Why do you think this is done?
16. As mentioned in Section 16.4, the fact that quasars can vary in brightness in just a few hours or days suggests that their size is only a few light hours or light days across. Explain how we can make this statement.

Problems

Note: The more challenging problems are indicated by a shaded box.

16.1 Evidence of the Big Bang

1. Derive the conversion from parsecs to lightyears given the information in Example 16.1.

16.2 The Big Bang

2. Calculate the temperature for which the ratio of free protons to neutrons in the early stages of the universe was 7.

3. At what temperature was the universe hot enough for photons to produce π^0? Approximately what time was this? Let $kT = mc^2$ and use Figure 16.7. Use the mean value of the distribution.

4. Use the thermodynamic equilibrium factor $\exp(\Delta mc^2/kT)$ to determine the relative abundances of the quarks during the time period from 10^{-13} s to 10^{-3} s. Assume the temperature is 10^{14} K and use midrange quark masses from Table 14.5.

5. At what threshold temperature were the electron and muon first formed?

6. If the mass of the electron neutrino is 5 eV/c^2, at what time was it first formed in the universe? What if its mass is 10^{-4} eV/c^2?

7. Was π^0 or π^+ formed first in the early stages of the universe? What is the difference in mean temperature between when they were formed?

8. Calculate the temperature of the universe when photons can no longer disassociate deuterons. Use the mean value of the distribution.

9. Determine the temperature of the universe when it has cooled enough that photons no longer disassociate the hydrogen atom. Use the mean value of the distribution.

16.3 Stellar Evolution

10. Show that the result given in Equation (16.16) for the volume of a neutron star follows from the equation preceding it.

11. Calculate the density of a neutron star from the results given in Example 16.5 and compare that with the density of a nucleon and a nucleus.

12. Show that the radius of a neutron star decreases as the number of neutrons increases. Does this make sense? Shouldn't the radius increase with more neutrons?

13. Calculate the gravitational pressure for (a) the sun, and (b) the neutron star of Example 16.5.

16.4 Astronomical Objects

Note: Use the value $H = 71$ km·s^{-1}·Mpc^{-1} for problems in this section.

14. An object in Hydra is 4 Gly from us. What would we expect its recessional velocity to be?

15. An object in Ursa Major is determined to be receding from us with a velocity of 15,000 km/s. How far from us is it?

16. Use the redshift of 3.8 for 4C41.17, a powerful radio galaxy, to determine the distance of the galaxy from us in (a) Mpc and in (b) lightyears.

17. A galaxy has been reported receding from us with a redshift of 10. How fast is the galaxy moving with respect to us? How far away is it?

18. (a) Research the different types of supernova and explain why Types Ia, Ib, and Ic are labeled differently. (b) Why do Types Ib, Ic, and II have more in common with each other than with Type Ia?

16.5 Problems with the Big Bang

19. (a) Use the observed ordinary mass density of the universe to determine the average number of nucleons per cubic meter throughout the universe. (b) There are 60 stars within 16.6 ly of the sun. If each star averages 1 solar mass, what is the mass density of nucleons in the neighborhood of the sun?

16.7 The Future

20. Examine carefully the size of the universe shown in Figure 16.20. (a) Explain what is happening for each of the four curves. (b) Do any of the curves represent a closed universe? If so, explain.

21. In Example 16.8 show that the critical density ρ_c is 9×10^{-30} g/cm^3.

General Problems

22. Use the blackbody spectrum to determine the peak wavelength for 2.725 K.

23. Calculate the critical density necessary for a closed universe for two extremes of the Hubble constant: 55 km·s^{-1}·Mpc^{-1} and 85 km·s^{-1}·Mpc^{-1}.

24. The time before which we don't know what happened in the universe (10^{-43} s) is called the *Planck time*. The theory needed is a quantum theory of gravity and concerns the three fundamental constants h, G, and c. (a) Use dimensional analysis to determine the exponents m, n, l if the Planck time $t_P = h^m G^n c^l$. (b) Calculate the Planck time using the expression you found in (a).

25. Let the wavelength of a photon produced during the early stages of the universe be λ, and λ_D the Doppler-shifted wavelength we measure today. Show that

$$\lambda_D = \frac{\sqrt{1 + \beta}}{\sqrt{1 - \beta}}\lambda$$

where $\beta = v/c$.

26. On two occasions we have used the gravitational self-energy of a uniform sphere of mass M and radius R. Use integral calculus and start with a mass dm in the sphere. Calculate the work done to bring the remainder of the mass in from infinity. By this technique show that the self-potential energy of the mass is

$$\text{P.E.} = -\frac{3}{5}\frac{GM^2}{R}$$

27. Draw tangents on all the curves in Figure 16.19 and determine the relationship between the Hubble time τ and the age of the universe.

28. Show that the extra time Δt that a neutrino with finite mass takes to reach the Earth from a supernova explosion compared to that taken for a zero mass particle is

$$\Delta t = (2.57\text{ s})\left(\frac{\text{distance}}{50\text{ kpc}}\right)\left(\frac{m_v c^2}{10\text{ eV}}\right)^2\left(\frac{10\text{ MeV}}{E}\right)^2$$

where $m_v c^2$ is the rest energy in eV and E is the energy in MeV of the neutrino.

29. Show that the mass density of radiation ρ_{rad} is given by

$$\rho_{\text{rad}} = \frac{4\sigma T^4}{c^3}$$

where σ is the Stefan-Boltzmann constant and T is the temperature. You might find the Stefan-Boltzmann law and $E = mc^2$ to be useful.

30. Use the mass density of radiation in the previous problem to determine the mass density of radiation when $T = 2.725$ K. How does this compare with the average density of matter in the universe? Does this mean we are in a radiation-dominated or matter-dominated universe?

31. Use the mass density of radiation from Problem 29 to calculate the density for several temperatures between 10^{-2} K and 10^{30} K. Make a plot of ρ_{rad} versus time using Figure 16.7. If the universe changed from being radiation dominated to matter dominated at 400,000 years, at what density for ρ_{rad} and ρ_{matter} did this occur?

32. The exponential falloff in the brightness of supernova 1987A was due to the decay of ^{56}Ni ($t_{1/2} = 6.1$ days) $\rightarrow$ ^{56}Co ($t_{1/2} = 77.1$ days) $\rightarrow$ ^{56}Fe. If the energy were primarily due to the decay of ^{56}Ni, what falloff in brightness by the end of 300 days would we expect? What if it were due to the energy in the decay of ^{56}Co? The actual data showed a decrease in brightness by a factor of about 100 after 300 days.

33. The Lyman alpha line (K_α) of hydrogen is measured in the laboratory to have a wavelength of 121.6 nm. In the quasar PKS 2000-330 the same line is determined to have a wavelength of 582.5 nm. What is its redshift and recession velocity?

34. The redshift parameter z is defined by $\Delta\lambda/\lambda$. Show that the Doppler redshift parameter is related to relative speed β by

$$1 + z = \sqrt{\frac{1 + \beta}{1 - \beta}}$$

35. In cases where the speed is small ($\beta \ll 1$), show that the Doppler redshift parameter is related to β by $z \approx \beta$.

36. In 1998 a galaxy named RD1 was discovered with a redshift of 5.34. (a) What is the speed of this galaxy with respect to us? (b) Use Hubble's law to determine how far away the galaxy is.

37. The first reaction in the proton-proton chain is $p + p \rightarrow d + \beta^+ + \nu$. Calculate the Q value of the reaction and determine the maximum neutrino energy.

38. Inflationary theory indicates the density of the universe should be equal to the critical density. Show that the critical density can be written in the form

$$\rho_c = \left(\frac{H_0}{100}\right)^2(1.9 \times 10^{-29}\text{ g/cm}^3)$$

where H_0 is entered in units of km $\cdot$ s^{-1} $\cdot$ Mpc^{-1}.

39. Assume a power law for the scale factor $a = Ct^n$, where C is a constant. (a) For what values of n are the universe accelerating and decelerating? (b) For deceleration, what is the dependence of H on time?

40. Let the total number of neutrons be N_n, the number of protons be N_p, and $N = N_n + N_p$. Let the fractions be $X_i = N_i/N$. (a) If the probability of a particle having energy E is proportional to the Boltzmann factor, $\exp(-E/kT)$, show that $X_n/X_p = \exp(-1.3\text{ MeV}/kT)$. (b) If the known ratio of neutrons to protons in the universe is 6.7, at what temperature did this occur? (c) What is the kinetic energy associated with this temperature? Is there anything noteworthy about this temperature?

Fundamental Constants

Quantity	Symbol	Value
Speed of light in vacuum	c	299 792 458 m/s (exact)
Permeability of vacuum	μ_0	$4\pi \times 10^{-7}$ N/A^2 (exact)
Permittivity of vacuum	ϵ_0	$8.854\ 187\ 817\ldots \times 10^{-12}$ F/m (exact)
Newtonian gravitational constant	G	$6.6742(10) \times 10^{-11}$ m$^3 \cdot$ kg$^{-1} \cdot$ s^{-2}
Fine structure constant	$\alpha = \dfrac{e^2}{4\pi\epsilon_0 \hbar c}$	$0.007297352568(24) \approx \dfrac{1}{137}$
Planck constant	h	$6.626\ 0693(11) \times 10^{-34}$ J $\cdot$ s
Planck constant $h/2\pi$	$\hbar$	$1.054\ 571\ 68(18) \times 10^{-34}$ J $\cdot$ s
Elementary charge	e	$1.602\ 176\ 53(14) \times 10^{-19}$ C
Magnetic flux quantum	$\Phi_0 = \dfrac{h}{2e}$	$2.067\ 833\ 72(18) \times 10^{-15}$ T $\cdot$ m^2
Josephson constant	$K_J = 2e/h$	$4.835\ 97879(41) \times 10^{14}$ Hz/V
Von Klitzing constant	$R_K = h/e^2$	$25\ 812.807\ 449(86)\ \Omega$
Rydberg constant	$R_\infty = \dfrac{\alpha^2 m_e c}{2h}$	$1.097\ 373\ 1568525(73) \times 10^7$ m^{-1}
Bohr radius	$a_0 = \dfrac{4\pi\epsilon_0 \hbar^2}{m_e e^2}$	$0.529\ 177\ 2108(18) \times 10^{-10}$ m
Compton wavelength	$\lambda_c = \dfrac{h}{m_e c}$	$2.426\ 310\ 238(16) \times 10^{-12}$ m
Classical electron radius	$r_e = \alpha^2 a_0$	$2.817\ 940\ 325(28) \times 10^{-15}$ m
Bohr magneton	$\mu_B = \dfrac{e\hbar}{2m_e}$	$9.274\ 00949(80) \times 10^{-24}$ J/T
Nuclear magneton	$\mu_N = \dfrac{e\hbar}{2m_p}$	$5.050\ 78343(43) \times 10^{-27}$ J/T
Avogadro constant	N_A	$6.022\ 1415(10) \times 10^{23}$ mol^{-1}
Molar gas constant	R	$8.314\ 472(15)$ J $\cdot$ mol$^{-1} \cdot$ K^{-1}
Boltzmann constant	k	$1.380\ 6505(24) \times 10^{-23}$ J/K
Stefan-Boltzmann constant	σ	$5.670\ 400(40) \times 10^{-8}$ W $\cdot$ m$^{-2} \cdot$ K^{-4}
Wien displacement law constant	$\lambda_{max} T$	$2.897\ 7685(51) \times 10^{-3}$ m $\cdot$ K
Atomic mass unit	u	$1.660\ 53886(28) \times 10^{-27}$ kg
Electron mass	m_e	$9.109\ 3826(16) \times 10^{-31}$ kg
Muon mass	m_μ	$1.883\ 53140(33) \times 10^{-28}$ kg
Proton mass	m_p	$1.672\ 62171(29) \times 10^{-27}$ kg
Neutron mass	m_n	$1.674\ 92728(29) \times 10^{-27}$ kg
Deuteron mass	m_d	$3.343\ 58335(57) \times 10^{-27}$ kg

Note: Digits in parentheses represent one standard deviation uncertainty in the final digits of the given value. P. J. Mohr and B. N. Taylor, "The 2002 CODATA Recommended Values of the Fundamental Physical Constants, Web Version 4.0," available at http://physics.nist .gov/constants (National Institute of Standards and Technology, Gaithersburg, MD 20899, 9 December 2003).

Conversion Factors

Length

	m	cm	km	ft	ly	pc
1 meter	1	10^2	10^{-3}	3.281	1.057×10^{-16}	3.241×10^{-17}
1 centimeter	10^{-2}	1	10^{-5}	3.281×10^{-2}	1.057×10^{-18}	3.241×10^{-19}
1 kilometer	10^3	10^5	1	3.281×10^3	1.057×10^{-13}	3.241×10^{-14}
1 foot	0.3048	30.48	3.048×10^{-4}	1	3.222×10^{-17}	9.878×10^{-18}
1 lightyear	9.461×10^{15}	9.461×10^{17}	9.461×10^{12}	3.104×10^{16}	1	0.3066
1 parsec	3.086×10^{16}	3.086×10^{18}	3.086×10^{13}	1.012×10^{17}	3.262	1

Mass

	kg	g	slug	u
1 kilogram	1	10^3	6.852×10^{-2}	6.022×10^{26}
1 gram	10^{-3}	1	6.852×10^{-5}	6.022×10^{23}
1 slug	14.59	1.459×10^4	1	8.789×10^{27}
1 atomic mass unit	1.661×10^{-27}	1.661×10^{-24}	1.138×10^{-28}	1

Time

	s	min	h	d	y
1 second	1	1.667×10^{-2}	2.778×10^{-4}	1.157×10^{-5}	3.169×10^{-8}
1 minute	60	1	1.667×10^{-2}	6.944×10^{-4}	1.901×10^{-6}
1 hour	3600	60	1	4.167×10^{-2}	1.141×10^{-4}
1 day	8.640×10^4	1440	24	1	2.738×10^{-3}
1 year	3.156×10^7	5.259×10^5	8.766×10^3	365.24	1

Speed

	m/s	cm/s	ft/s	mi/h	km/h
1 meter/second	1	10^2	3.281	2.237	3.600
1 centimeter/second	10^{-2}	1	3.281×10^{-2}	2.237×10^{-2}	0.036
1 foot/second	0.3048	30.48	1	0.6818	1.097
1 mile/hour	0.4470	44.70	1.467	1	1.609
1 kilometer/hour	0.2778	27.78	0.9113	0.6214	1

Note: 1 mi/min = 60 mi/h = 88 ft/s

Force

	N	dyn	lb
1 newton	1	10^5	0.2248
1 dyne	10^{-5}	1	2.248×10^{-6}
1 pound	4.448	4.448×10^5	1

Work, Energy, Heat

	J	erg	ft · lb
1 joule	1	10^7	0.7376
1 erg	10^{-7}	1	7.376×10^{-8}
1 ft · lb	1.356	1.356×10^7	1
1 eV	1.602×10^{-19}	1.602×10^{-12}	1.182×10^{-19}
1 cal	4.186	4.186×10^7	3.087
1 Btu	1.055×10^3	1.055×10^{10}	7.779×10^2
1 kWh	3.600×10^6	3.600×10^{13}	2.655×10^6

	eV	cal	Btu	kWh
1 joule	6.242×10^{18}	0.2388	9.478×10^{-4}	2.778×10^{-7}
1 erg	6.242×10^{11}	2.388×10^{-8}	9.478×10^{-11}	2.778×10^{-14}
1 ft · lb	8.464×10^{18}	0.3239	1.285×10^{-3}	3.766×10^{-7}
1 eV	1	3.827×10^{-20}	1.519×10^{-22}	4.450×10^{-26}
1 cal	2.613×10^{19}	1	3.968×10^{-3}	1.163×10^{-6}
1 Btu	6.585×10^{21}	2.520×10^2	1	2.930×10^{-4}
1 kWh	2.247×10^{25}	8.598×10^5	3.412×10^3	1

Mathematical Relations

Expansions

$$(1 \pm x)^n = 1 \pm nx + \frac{n(n-1)}{2!}x^2$$

$$\pm \frac{n(n-1)(n-2)}{3!}x^3 + \cdots \quad |x| < 1$$

$$(1 \pm x)^{1/2} = 1 \pm \tfrac{1}{2}x - \tfrac{1}{8}x^2 \pm \tfrac{1}{16}x^3 - \cdots \quad |x| < 1$$

$$(1 \pm x)^{-1/2} = 1 \mp \tfrac{1}{2}x + \tfrac{3}{8}x^2 \mp \tfrac{5}{16}x^3 + \cdots \quad |x| < 1$$

$$(1 \pm x)^{-1} = 1 \mp x + x^2 \mp x^3 + \cdots \quad |x| < 1$$

$$\sin x = x - \frac{x^3}{3!} + \frac{x^5}{5!} - \frac{x^7}{7!} + \cdots$$

$$\cos x = 1 - \frac{x^2}{2!} + \frac{x^4}{4!} - \frac{x^6}{6!} + \cdots$$

$$\tan x = x + \frac{x^3}{3!} + \frac{2}{15}x^5 + \cdots \quad |x| < \pi/2$$

$$e^x = 1 + x + \frac{x^2}{2!} + \frac{x^3}{3!} + \cdots$$

Functions and Relations

$$\sin(A \pm B) = \sin A \cos B \pm \cos A \sin B$$

$$\cos(A \pm B) = \cos A \cos B \mp \sin A \sin B$$

$$\sinh x = \frac{e^x + e^{-x}}{2}$$

$$\cosh x = \frac{e^x + e^{-x}}{2}$$

Cartesian form: $z = x + iy$

Complex conjugate: $z^* = x - iy, \; i = \sqrt{-1}$

Polar form:
$$z = |z|e^{i\theta}$$
$$z^* = |z|e^{-i\theta}$$
$$zz^* = |z|^2 = x^2 + y^2$$

Real part of z: $\operatorname{Re} z = \tfrac{1}{2}(z + z^*) = x$

Imaginary part of z: $\operatorname{Im} z = -\tfrac{1}{2}(z - z^*) = y$

$$\sin x = \frac{e^{ix} - e^{-ix}}{2i}$$

$$\cos x = \frac{e^{ix} + e^{-ix}}{2}$$

$$e^{ix} = \cos x + i \sin x$$

Integrals

$$\int \sin^2 x \, dx = -\frac{1}{2}\cos x \sin x + \frac{1}{2}x = \frac{1}{2}x - \frac{1}{4}\sin 2x$$

$$\int \sin^3 x \, dx = -\frac{1}{3}\cos x(\sin^2 x + 2)$$

$$\int x \sin^2 x \, dx = \frac{x^2}{4} - \frac{x \sin 2x}{4} - \frac{\cos 2x}{8}$$

$$\int x^2 \sin^2 x \, dx = \frac{x^3}{6} - \left(\frac{x^2}{4} - \frac{1}{8}\right)\sin 2x - \frac{x \cos 2x}{4}$$

$$\int_0^\infty e^{-x/\alpha} \, dx = \alpha$$

$$\int_0^\infty x e^{-x/\alpha} \, dx = \alpha^2$$

$$\int_0^\infty x^2 e^{-x/\alpha} \, dx = 2\alpha^3$$

$$\int_0^\infty x^n e^{-x/a} \, dx = n!\alpha^{n+1}$$

Periodic Table of the Elements

Legend: Metals / Metalloids / Nonmetals

Main table (element name, atomic number, symbol, atomic mass — by group 1–18):

1	2	3	4	5	6	7	8	9	10	11	12	13	14	15	16	17	18
Hydrogen 1 **H** 1.0079																	Helium 2 **He** 4.0026
Lithium 3 **Li** 6.941	Beryllium 4 **Be** 9.0122											Boron 5 **B** 10.811	Carbon 6 **C** 12.0107	Nitrogen 7 **N** 14.0067	Oxygen 8 **O** 15.9994	Fluorine 9 **F** 18.9984	Neon 10 **Ne** 20.1797
Sodium 11 **Na** 22.9898	Magnesium 12 **Mg** 24.3050											Aluminum 13 **Al** 26.9815	Silicon 14 **Si** 28.0855	Phosphorus 15 **P** 30.9738	Sulfur 16 **S** 32.066	Chlorine 17 **Cl** 35.4527	Argon 18 **Ar** 39.948
Potassium 19 **K** 39.0983	Calcium 20 **Ca** 40.078	Scandium 21 **Sc** 44.9559	Titanium 22 **Ti** 47.867	Vanadium 23 **V** 50.9415	Chromium 24 **Cr** 51.9961	Manganese 25 **Mn** 54.9380	Iron 26 **Fe** 55.845	Cobalt 27 **Co** 58.9332	Nickel 28 **Ni** 58.6934	Copper 29 **Cu** 63.546	Zinc 30 **Zn** 65.39	Gallium 31 **Ga** 69.723	Germanium 32 **Ge** 72.64	Arsenic 33 **As** 74.9216	Selenium 34 **Se** 78.96	Bromine 35 **Br** 79.904	Krypton 36 **Kr** 83.80
Rubidium 37 **Rb** 85.4678	Strontium 38 **Sr** 87.62	Yttrium 39 **Y** 88.9059	Zirconium 40 **Zr** 91.224	Niobium 41 **Nb** 92.9064	Molybdenum 42 **Mo** 95.94	Technetium 43 **Tc** (98)	Ruthenium 44 **Ru** 101.07	Rhodium 45 **Rh** 102.9055	Palladium 46 **Pd** 106.42	Silver 47 **Ag** 107.8682	Cadmium 48 **Cd** 112.411	Indium 49 **In** 114.88	Tin 50 **Sn** 118.710	Antimony 51 **Sb** 121.76	Tellurium 52 **Te** 127.60	Iodine 53 **I** 126.9045	Xenon 54 **Xe** 131.29
Cesium 55 **Cs** 132.9054	Barium 56 **Ba** 137.327	Lanthanum 57 **La** 138.9055	Hafnium 72 **Hf** 178.49	Tantalum 73 **Ta** 180.9479	Tungsten 74 **W** 183.84	Rhenium 75 **Re** 186.207	Osmium 76 **Os** 190.23	Iridium 77 **Ir** 192.217	Platinum 78 **Pt** 195.078	Gold 79 **Au** 196.9666	Mercury 80 **Hg** 200.59	Thallium 81 **Tl** 204.3833	Lead 82 **Pb** 207.2	Bismuth 83 **Bi** 208.9804	Polonium 84 **Po** (209)	Astatine 85 **At** (210)	Radon 86 **Rn** (222)
Francium 87 **Fr** (223)	Radium 88 **Ra** (226)	Actinium 89 **Ac** (227)	Rutherfordium 104 **Rf** (261)	Dubnium 105 **Db** (262)	Seaborgium 106 **Sg** (266)	Bohrium 107 **Bh** (264)	Hassium 108 **Hs** (269)	Meitnerium 109 **Mt** (268)	Darmstadtium 110 **Ds** (269)	Roentgenium 111 **Rg** (272)	Ununbium 112 **Uub*** (277)						

Lanthanides:

Cerium 58 **Ce** 140.115	Praseodymium 59 **Pr** 140.9076	Neodymium 60 **Nd** 144.24	Promethium 61 **Pm** (145)	Samarium 62 **Sm** 150.36	Europium 63 **Eu** 151.965	Gadolinium 64 **Gd** 157.25	Terbium 65 **Tb** 158.9253	Dysprosium 66 **Dy** 162.50	Holmium 67 **Ho** 164.9303	Erbium 68 **Er** 167.26	Thulium 69 **Tm** 168.9342	Ytterbium 70 **Yb** 173.04	Lutetium 71 **Lu** 174.967

Actinides:

Thorium 90 **Th** 232.0381	Protactinium 91 **Pa** 231.0359	Uranium 92 **U** 238.0289	Neptunium 93 **Np** (237)	Plutonium 94 **Pu** (244)	Americium 95 **Am** (243)	Curium 96 **Cm** (247)	Berkelium 97 **Bk** (247)	Californium 98 **Cf** (251)	Einsteinium 99 **Es** (252)	Fermium 100 **Fm** (257)	Mendelevium 101 **Md** (258)	Nobelium 102 **No** (259)	Lawrencium 103 **Lr** (260)

Note: Averaged atomic masses are 1997 IUPAC values (up to four decimal places). An entry in parentheses indicates the mass number of the longest-lived isotope of an element that has no stable isotopes.

*The official name has not been agreed upon for this element.

Mean Values and Distributions

Average values are encountered in several forms. A simple example is an arithmetic mean. If N students take an exam and receive scores S_1, S_2, S_3, and so on, the class average is

$$\text{Average score} = \frac{S_1 + S_2 + S_3 + \cdots + S_N}{N}$$

This calculation is more compact if we group the students receiving a given score S_i. If N_i students receive the score S_i, then the class average becomes

$$\text{Average score} = \frac{N_1 S_1 + N_2 S_2 + N_3 S_3 + \cdots}{N}$$

An example from physics of a *weighted* average brings us closer to the idea of mean value. In mechanics the center of mass of N particles of different mass (m_i) placed at various locations (x_i) along the x axis is

$$x_{cm} = \frac{m_1 x_1 + m_2 x_2 + \cdots}{m_1 + m_2 + \cdots} = \frac{\sum_i m_i x_i}{\sum_i m_i}$$

We define the mean value just like the weighted average, with the probabilities of the various possible events serving as the weights. Suppose some physical quantity x can have N possible values, which we shall call x_i, with $i = 1, 2, 3, \ldots, N$. Let P_i be the probability that x has the value x_i. Then the mean value of x is

Mean value

$$\bar{x} = \frac{P_1 x_1 + P_2 x_2 + \cdots}{P_1 + P_2 + \cdots} = \frac{\sum_i P_i x_i}{\sum_i P_i}$$

As usual, it is required that $\sum_i P_i = 1$ (the sum of all probabilities is one), so

$$\bar{x} = \sum P_i x_i \qquad (A5.1)$$

The same definition of mean value holds if we are considering a function $f(x)$:

Mean value of a function

$$\overline{f(x)} = \sum P_i f(x_i) \qquad (A5.2)$$

For any two functions $f(x)$ and $g(x)$ we have

$$\overline{f(x) + g(x)} = \overline{f(x)} + \overline{g(x)}$$

and

$$\overline{c[f(x) + g(x)]} = c\overline{[f(x) + g(x)]} = c\overline{f(x)} + c\overline{g(x)}$$

where c is a constant.

   In physics we are often interested in how much a particular value differs from the mean value. The difference between a particular value and the mean value is known as the deviation (Δx_i):

$$\Delta x_i = x_i - \bar{x} \qquad\qquad \textbf{Deviation}$$

The mean value of all deviations is easily computed:

$$\overline{\Delta x_i} = \overline{x_i - \bar{x}} = \bar{x}_i - \bar{x} = 0$$

This is just as one should expect because the positive and negative deviations will exactly cancel each other. However, we would like to get some idea of how large a typical deviation will be. To this end we define the **standard deviation**:

$$\sigma_x = \left[\overline{(\Delta x_i)^2} \right]^{1/2} = \left[\sum_i P_i (x_i - \bar{x})^2 \right]^{1/2} \qquad (A5.3) \quad \textbf{Standard deviation}$$

Notice that in Equation (A5.3) the negative deviations have been eliminated by squaring. The standard deviation is an indication of the sharpness of the peak of the probability distribution curve. For a purely random (Gaussian) distribution it can be shown that the width of the probability curve at half of the peak value is slightly more than two standard deviations.

   Equation (A5.3) can be rewritten

$$\sigma_x = \left[\sum_i P_i (x_i^2 - 2x_i\bar{x} + \bar{x}^2) \right]^{1/2}$$

$$= \left[\sum_i P_i x_i^2 - 2\bar{x} \sum_i (P_i x_i + \bar{x}^2) \right]^{1/2}$$

The first sum is simply $\overline{x^2}$, and the second is $\bar{x}$. Thus

$$\sigma_x = \left[\overline{x^2} - \bar{x}^2 \right]^{1/2} \qquad (A5.4)$$

   An important feature of many systems is that as N increases, the probability distribution begins to approach a smooth curve. Indeed, it will be to our advantage to think of such a curve as a function, which we shall call a **distribution function**. This will enable us to replace the sums used in this section by integrals, which is clearly a wise thing to do if we want to consider problems in which $N \approx 10^{23}$. The distribution function is not a probability per se, but rather a probability density. For example, if $f(x)$ is the distribution function describing the positions of a large number of particles lying along the x axis, then

$$f(x)\, dx = \text{the probability of finding a particle between } x \text{ and } x + dx$$

It is necessary to define the distribution function in this way so that we can say [by analogy with Equation (A5.1)]

$$\bar{x} = \int x f(x)\, dx$$

$$\overline{x^2} = \int x^2 f(x)\, dx$$

and so on.

Probability Integrals $I_n = \displaystyle\int_0^\infty x^n \exp(-ax^2)\, dx$

Even and odd functions

Before proceeding to the calculation of these integrals, we introduce the concept of even and odd functions. An even function is one for which $f(x) = f(-x)$; similarly, for an odd function $f(x) = -f(-x)$. An example of an even function is the integrand of Equation (9.4):

$$\int_{-\infty}^{\infty} g(v_x)\, dv_x = \int_{-\infty}^{\infty} C' \exp\left(-\frac{1}{2}\beta m v_x^2\right) dv_x$$

This even integrand makes it possible to replace $\displaystyle\int_{-\infty}^{\infty}$ with $2\displaystyle\int_0^{\infty}$ because the curve is perfectly symmetric about $x = 0$ (see Figure 9.2).

On the other hand, consider the odd integrand of Equation (9.6):

$$\bar{v}_x = C' \int_{-\infty}^{\infty} v_x \exp\left(-\frac{1}{2}\beta m v_x^2\right) dv_x$$

The odd function is antisymmetric with respect to $x = 0$, and if we divide the integral into two parts ($-\infty$ to 0 and 0 to ∞), those two parts just cancel. The integral of an odd function from $-\infty$ to ∞ (or indeed over any symmetric limits) is therefore zero.

Based on these properties of even and odd functions, it makes sense to define

$$I_n = \int_0^{\infty} x^n \exp(-ax^2)\, dx \qquad (A6.1)$$

and say that

$$\int_{-\infty}^{\infty} x^n \exp(-ax^2)\, dx = \begin{cases} 2I_n & \text{for even } n \\ 0 & \text{for odd } n \end{cases} \qquad (A6.2)$$

Let us proceed to calculate the I_n, beginning with $n = 0$ [as in Equation (9.4)]. Now

$$I_0 = \int_0^{\infty} \exp(-ax^2)\, dx \qquad (A6.3)$$

It is not possible to do this integral by straightforward substitution. Because this is a definite integral, x is a dummy variable. We would obtain the same result with any other variable, say y:

$$I_0 = \int_0^{\infty} \exp(-ay^2)\, dy \qquad (A6.4)$$

Multiplying Equations (A6.3) and (A6.4) yields

$$I_0^2 = \int_0^{\infty}\int_0^{\infty} \exp[-a(x^2 + y^2)]\, dx\, dy \qquad (A6.5)$$

We now switch to polar coordinates, with $r^2 = x^2 + y^2$ and $dx\,dy = r\,dr\,d\theta$. The limits of integration are 0 to ∞ for r but only 0 to $\pi/2$ for θ, because the limits on x and y restrict us to the first quadrant. Thus

$$I_0^2 = \int_0^\infty \exp(-ar^2)\,r\,dr \int_0^{\pi/2} d\theta \qquad (A6.6)$$

The θ integral yields $\pi/2$, and the r integral is easily performed using a standard substitution (say $u = ar^2$):

$$I_0^2 = \left(\frac{\pi}{2}\right)\left(\frac{1}{2}a\right) = \frac{\pi}{4a}$$

$$I_0 = \sqrt{\frac{\pi}{4a}} = \frac{\sqrt{\pi}}{2}\,a^{-1/2} \qquad (A6.7)$$

Notice that Equation (A6.6) actually contains I_1, so we see that

$$I_1 = \int_0^\infty r\exp(-ar^2)\,dr = \frac{1}{2a} \qquad (A6.8)$$

To calculate I_2 we return for a moment to I_0. Differentiating Equation (A6.3) with respect to a produces

$$\frac{dI_0}{da} = -\int_0^\infty x^2 \exp(-ax^2)\,dx = -I_2 \qquad (A6.9)$$

But by Equation (A6.7)

$$\frac{dI_0}{da} = -\frac{\sqrt{\pi}}{4}a^{-3/2}$$

and so

$$I_2 = \frac{\sqrt{\pi}}{4}a^{-3/2} \qquad (A6.10)$$

In a similar way one may use $\dfrac{dI_1}{da}$ to calculate I_3, $\dfrac{dI_2}{da}$ to calculate I_4, and so on, indefinitely. Some useful results are

$$I_3 = \frac{1}{2}a^{-2}$$

$$I_4 = \frac{3\sqrt{\pi}}{8}a^{-5/2}$$

$$I_5 = a^{-3}$$

It can be shown that in general

$$I_n = \frac{[(n-1)/2]!}{2a^{(n+1)/2}} \qquad\qquad \text{for } odd\ n$$

$$I_n = \frac{1\cdot 3\cdot 5\cdots(n-1)}{2^{(n/2)+1}a^{(n/2)}}\sqrt{\frac{\pi}{a}} \qquad \text{for } even\ n$$

Integrals of the Type $\int_0^\infty \dfrac{x^{n-1}\,dx}{e^x - 1}$

Evaluation of these integrals introduces us to two special mathematical functions: the gamma function and the Riemann zeta function. A good handbook of integrals (see for example *Table of Integrals, Series, and Products* by Gradshteyn and Ryzhik) will give the result

$$\int_0^\infty \frac{x^{n-1}\,dx}{e^x - 1} = \Gamma(n)\,\zeta(n) \tag{A7.1}$$

The gamma function is in turn defined by the integral

The gamma function

$$\Gamma(n) = \int_0^\infty e^{-x} x^{n-1}\,dx \tag{A7.2}$$

In many cases the gamma functions we encounter in physics have integer arguments. Then they are easily evaluated using the identity

$$\Gamma(n) = (n-1)! \tag{A7.3}$$

When n is not an integer one may use the recursion formula

$$\Gamma(n+1) = n\,\Gamma(n) \tag{A7.4}$$

with tabulated values of the gamma function [those tables generally range from $\Gamma(1)$ to $\Gamma(2)$] to find the desired value. Further, the fact that $\Gamma(\frac{3}{2}) = \sqrt{\pi}/2$ is useful, because half-integer powers occur in a number of applications. For other values one may resort to tables or computer packages such as Mathematica.

One may also use the appropriate tables or computer packages to obtain values for the Riemann zeta function. It is helpful to know that the zeta function can sometimes be expressed in closed form. For example,

The Riemann zeta function

$$\zeta(2) = \frac{\pi^2}{6} \quad \text{and} \quad \zeta(4) = \frac{\pi^4}{90} \tag{A7.5}$$

One integral of this form is

$$\int_0^\infty \frac{x^3}{e^x - 1}\,dx$$

which is found in Example 3.7. According to Equation (A7.1) the result is

$$\int_0^\infty \frac{x^3}{e^x - 1}\,dx = \Gamma(4)\,\zeta(4) = 3!\left(\frac{\pi^4}{90}\right) = \frac{\pi^4}{15}$$

This allows us to complete Example 3.7 and obtain the closed form expression for the total radiation emitted by a blackbody at a temperature T:

$$R(T) = \frac{2\pi^5 k^4}{15 h^3 c^2}\,T^4$$

In Section 9.7 we encounter an integral with a fractional value of n ($n = \frac{3}{2}$), with the result $\Gamma(\frac{3}{2}) \, \zeta(\frac{3}{2})$. As noted previously, $\Gamma(\frac{3}{2}) = \sqrt{\pi}/2$. A numerical evaluation yields $\zeta(\frac{3}{2}) \approx 2.61238$, so that $\Gamma(\frac{3}{2}) \, \zeta(\frac{3}{2}) \approx 2.315$, as used in Equation (9.64).

The common thread running through these applications is that they concern collections of bosons. The Bose-Einstein distribution contains the factor $[B_2 \exp(\beta E) - 1]^{-1}$, which can often be put into the form $e^x - 1$ for the purpose of doing this integral.

A P P E N D I X **8**

Atomic Mass Table

Atomic masses are used in both atomic and nuclear physics calculations. Because of space limitations, we have not included all the known isotopes in this compilation, but we have listed the atomic masses of all known stable isotopes and all those with half-lives greater than 10 s. In addition, for the light and heavy elements, we have included some additional isotopes.

We have listed the chemical symbol, atomic number Z, mass number A, atomic masses in atomic mass units (u), either the half-life $t_{1/2}$ or natural abundance on Earth, and the decay modes. A few nuclides, for example, ^{40}K and ^{238}U, are unstable but have such a long half-life that they are still found naturally on Earth. In these few cases, we list both the half-life and natural abundance. The most important decay modes are listed with the most likely ones given first. We present below the legend for the various decay modes.

The m listed after a few of the mass numbers means the data are for a metastable state of the particular nuclide, but we have omitted most metastable nuclides for brevity. An s listed at the end of the atomic mass value means that the mass is obtained from systematics, not measured. The time units for the half-life are s (seconds), m (minutes), h (hours), d (days), and y (years). In some cases, for very short-lived isotopes of light masses, we have listed the width of the state Γ instead of the half-life. The half-life can be determined from the time τ after using the Heisenberg uncertainty principle.

The data for this appendix were obtained from the website (www.nndc .bnl.gov) of the National Nuclear Data Center at Brookhaven National Laboratory. Look at the website for the latest values. The abundances, half-lives, and decay modes are from J. K. Tuli, *Nuclear Wallet Cards*, 5th ed. (1995). The atomic mass values are from G. Audi and A. H. Wapstra, "The 1995 Update to the Atomic Mass Evaluations," *Nuclear Physics* **A595**, pp. 409–480 (1995). We have rounded off the mass values to the nearest 0.000001 u.

The legend for the decay modes is clear except for a few cases. We list them all here.

Listed	Decay
β^-	β^-
$2\beta^-$	Double β^-
ϵ	Electron capture and/or β^+
IT	Isometric transition
$\alpha, 2\alpha$	α or 2α
SF	Spontaneous fission
$\beta^- n, \beta^- p, \beta^- \alpha, \beta^- 2\alpha,$ $\beta^- n\alpha, \beta^- 3\alpha$	β^- decay followed by subsequent particles decaying
$\epsilon n, \epsilon p, \epsilon\alpha, \epsilon 2\alpha, \epsilon 3\alpha, \epsilon$SF	Electron capture and/or β^+ followed by other decays
$n, p, p2\alpha$	Neutron, proton, etc.
^{12}C, ^{14}C	^{12}C, ^{14}C

Z	A	Atomic Mass (u)	$t_{1/2}$ or Abundance	Decay Mode
Neutron (n)				
0	1	1.008665	10.4 m	β^-
Hydrogen (H)				
1	1	1.007825	99.985%	
1	2	2.014102	0.015%	
1	3	3.016049	12.33 y	β^-
Helium (He)				
2	3	3.016029	0.000137%	
2	4	4.002603	99.999863%	
2	5	5.012220	0.60 MeV	α, n
2	6	6.018888	806.7 ms	β^-
2	7	7.028030	160 keV	n
2	8	8.033922	119.0 ms	$\beta^-, \beta^- n$
Lithium (Li)				
3	5	5.012540	1.5 MeV	α, p
3	6	6.015122	7.5%	
3	7	7.016004	92.5%	
3	8	8.022487	838 ms	$\beta^-, \beta^- 2\alpha$
3	9	9.026789	178.3 ms	$\beta^-, \beta^- n$
3	10	10.035481	1.2 MeV	n
3	11	11.043796	8.5 ms	$\beta^-, \beta^- n\alpha$
Beryllium (Be)				
4	6	6.019726	92 keV	$2p$
4	7	7.016929	53.29 d	ϵ
4	8	8.005305	6.8 eV	2α
4	9	9.012182	100%	
4	10	10.013534	1.51×10^6 y	β^-
4	11	11.021658	13.81 s	$\beta^-, \beta^- \alpha$
4	12	12.026921	23.6 ms	$\beta^-, \beta^- n$
Boron (B)				
5	8	8.024607	770 ms	$\epsilon\alpha, \epsilon, \epsilon 2\alpha$
5	9	9.013329	0.54 keV	$2\alpha, p$
5	10	10.012937	19.9%	
5	11	11.009305	80.1%	
5	12	12.014352	20.20 ms	$\beta^-, \beta^- 3\alpha$
5	13	13.017780	17.36 ms	β^-
5	14	14.025404	13.8 ms	β^-
5	15	15.031097	10.5 ms	β^-
Carbon (C)				
6	9	9.031040	126.5 ms	$\epsilon, \epsilon p, \epsilon 2\alpha$
6	10	10.016853	19.255 s	ϵ

Z	A	Atomic Mass (u)	$t_{1/2}$ or Abundance	Decay Mode
Carbon (C)				
6	11	11.011434	20.39 m	ϵ
6	12	12.000000	98.89%	
6	13	13.003355	1.11%	
6	14	14.003242	5730 y	β^-
6	15	15.010599	2.449 s	β^-
6	16	16.014701	0.747 s	β^-
6	17	17.022584	193 ms	$\beta^-, \beta^- n$
6	18	18.026760	88 ms	β^-
Nitrogen (N)				
7	12	12.018613	11.000 ms	$\epsilon, \epsilon 3\alpha$
7	13	13.005739	9.965 m	ϵ
7	14	14.003074	99.63%	
7	15	15.000109	0.37%	
7	16	16.006101	7.13 s	β^-
7	17	17.008450	4.173 s	$\beta^-, \beta^- n$
7	18	18.014082	624 ms	$\beta^-, \beta^- \alpha, \beta^- n$
7	19	19.017027	304 ms	$\beta^-, \beta^- n$
7	20	20.023370	100 ms	$\beta^-, \beta^- n$
7	21	21.027090	95 ms	$\beta^- n, \beta^-$
7	22	22.034440	24 ms	$\beta^- n, \beta^-$
Oxygen (O)				
8	13	13.024810	8.58 ms	ϵ
8	14	14.008595	70.606 s	ϵ
8	15	15.003065	122.24 s	ϵ
8	16	15.994915	99.76%	
8	17	16.999132	0.038%	
8	18	17.999160	0.20%	
8	19	19.003579	26.91 s	β^-
8	20	20.004076	13.51 s	β^-
8	21	21.008655	3.42 s	β^-
8	22	22.009970	2.25 s	β^-
Fluorine (F)				
9	16	16.011466	40 keV	p
9	17	17.002095	64.49 s	ϵ
9	18	18.000938	109.77 m	ϵ
9	19	18.998403	100%	
9	20	19.999981	11.00 s	β^-
9	21	20.999949	4.158 s	β^-
9	22	22.002999	4.23 s	β^-
9	23	23.003570	2.23 s	β^-
9	24	24.008100	340 ms	β^-

Z	A	Atomic Mass (u)	$t_{1/2}$ or Abundance	Decay Mode
Neon (Ne)				
10	17	17.017700	109.2 ms	$\epsilon, \epsilon p$
10	18	18.005697	1.672 s	ϵ
10	19	19.001880	17.22 s	ϵ
10	20	19.992440	90.48%	
10	21	20.993847	0.27%	
10	22	21.991386	9.25%	
10	23	22.994467	37.24 s	β^-
10	24	23.993615	3.38 m	β^-
10	25	24.997790	602 ms	β^-
10	26	26.000460	230 ms	β^-
Sodium (Na)				
11	20	20.007348	447.9 ms	ϵ
11	21	20.997655	22.49 s	ϵ
11	22	21.994437	2.6019 y	ϵ
11	23	22.989770	100%	
11	24	23.990963	14.9590 h	β^-
11	25	24.989954	59.1 s	β^-
11	26	25.992590	1.072 s	β^-
11	27	26.994010	301 ms	$\beta^-, \beta^- n$
Magnesium (Mg)				
12	20	20.018863	95 ms	$\epsilon, \epsilon p$
12	21	21.011714	122 ms	$\epsilon, \epsilon p$
12	22	21.999574	3.857 s	ϵ
12	23	22.994125	11.317 s	ϵ
12	24	23.985042	78.99%	
12	25	24.985837	10.00%	
12	26	25.982593	11.01%	
12	27	26.984341	9.458 m	β^-
12	28	27.983877	20.91 h	β^-
12	29	28.988550	1.30 s	β^-
12	30	29.990460	335 ms	β^-
12	31	30.996550	230 ms	$\beta^-, \beta^- n$
12	32	31.999150	120 ms	$\beta^-, \beta^- n$
Aluminum (Al)				
13	23	23.007265	0.47 s	$\epsilon, \epsilon p$
13	24	23.999941	2.053 s	$\epsilon, \epsilon \alpha$
13	25	24.990429	7.183 s	ϵ
13	26	25.986892	7.4×10^5 y	ϵ
13	27	26.981538	100%	
13	28	27.981910	2.2414 m	β^-
13	29	28.980445	6.56 m	β^-
13	30	29.982960	3.60 s	β^-
13	31	30.983946	644 ms	β^-

Z	A	Atomic Mass (u)	$t_{1/2}$ or Abundance	Decay Mode
Silicon (Si)				
14	24	24.011546	102 ms	$\epsilon, \epsilon p$
14	25	25.004107	220 ms	$\epsilon, \epsilon p$
14	26	25.992330	2.234 s	ϵ
14	27	26.986705	4.16 s	ϵ
14	28	27.976927	92.23%	
14	29	28.976495	4.67%	
14	30	29.973770	3.10%	
14	31	30.975363	157.3 m	β^-
14	32	31.974148	172 y	β^-
14	33	32.978001	6.18 s	β^-
14	34	33.978576	2.77 s	β^-
14	35	34.984580	0.78 s	β^-
14	36	35.986690	0.45 s	$\beta^-, \beta^- n$
Phosphorus (P)				
15	27	26.999190	260 ms	$\epsilon, \epsilon p$
15	28	27.992312	270.3 ms	ϵ
15	29	28.981801	4.140 s	ϵ
15	30	29.978314	2.498 m	ϵ
15	31	30.973762	100%	
15	32	31.973907	14.262 d	β^-
15	33	32.971725	25.34 d	β^-
15	34	33.973636	12.43 s	β^-
15	35	34.973314	47.3 s	β^-
15	36	35.978260	5.6 s	β^-
15	37	36.979610	2.31 s	β^-
15	38	37.984470	0.64 s	$\beta^-, \beta^- n$
Sulfur (S)				
16	30	29.984903	1.178 s	ϵ
16	31	30.979554	2.572 s	ϵ
16	32	31.972071	95.02%	
16	33	32.971459	0.75%	
16	34	33.967867	4.21%	
16	35	34.969032	87.51 d	β^-
16	36	35.967081	0.02%	
16	37	36.971126	5.05 m	β^-
16	38	37.971163	170.3 m	β^-
16	39	38.975140	11.5 s	β^-
16	40	39.975470	8.8 s	β^-
Chlorine (Cl)				
17	33	32.977452	2.511 s	ϵ
17	34	33.973762	1.5264 s	ϵ
17	35	34.968853	75.77%	
17	36	35.968307	3.01×10^5 y	β^-, ϵ
17	37	36.965903	24.23%	
17	38	37.968011	37.24 m	β^-
17	39	38.968008	55.6 m	β^-

Z	A	Atomic Mass (u)	$t_{1/2}$ or Abundance	Decay Mode
Chlorine (Cl) (*continued*)				
17	40	39.970420	1.35 m	β^-
17	41	40.970650	38.4 s	β^-
17	42	41.973170	6.8 s	β^-
17	43	42.974200	3.3 s	β^-
Argon (Ar)				
18	35	34.975257	1.775 s	ϵ
18	36	35.967546	0.337%	
18	37	36.966776	35.04 d	ϵ
18	38	37.962732	0.063%	
18	39	38.964313	269 y	β^-
18	40	39.962383	99.600%	
18	41	40.964501	1.822 h	β^-
18	42	41.963050	32.9 y	β^-
18	43	42.965670	5.37 m	β^-
18	44	43.965365	11.87 m	β^-
18	45	44.968090	21.48 s	β^-
18	46	45.968090	8.4 s	β^-
Potassium (K)				
19	37	36.973377	1.226 s	ϵ
19	38	37.969080	7.636 m	ϵ
19	39	38.963707	93.2581%	
19	40	39.963999	1.277×10^9 y 0.0117%	β^-, ϵ
19	41	40.961826	6.7302%	
19	42	41.962403	12.360 h	β^-
19	43	42.960716	22.3 h	β^-
19	44	43.961560	22.13 m	β^-
19	45	44.960700	17.3 m	β^-
19	46	45.961976	105 s	β^-
19	47	46.961678	17.5 s	β^-
19	48	47.965513	6.8 s	β^-
19	49	48.967450	1.26 s	$\beta^-, \beta^- n$
Calcium (Ca)				
20	38	37.976319	440 ms	ϵ
20	39	38.970718	859.6 ms	ϵ
20	40	39.962591	96.941%	
20	41	40.962278	1.03×10^5 y	ϵ
20	42	41.958618	0.647%	
20	43	42.958767	0.135%	
20	44	43.955481	2.086%	
20	45	44.956186	162.6 d	β^-
20	46	45.953693	0.004%	
20	47	46.954546	4.536 d	β^-
20	48	47.952534	0.187%	
20	49	48.955673	8.718 m	β^-
20	50	49.957518	13.9 s	β^-

Z	A	Atomic Mass (u)	$t_{1/2}$ or Abundance	Decay Mode
Calcium (Ca)				
20	51	50.961470	10.0 s	$\beta^-, \beta^- n$
20	52	51.965100	4.6 s	β^-
Scandium (Sc)				
21	43	42.961151	3.891 h	ϵ
21	44	43.959403	3.927 h	ϵ
21	45	44.955910	100%	
21	46	45.955170	83.79 d	β^-
21	47	46.952408	3.349 d	β^-
21	48	47.952235	43.7 h	β^-
21	49	48.950024	57.2 m	β^-
21	50	49.952187	102.5 s	β^-
21	51	50.953603	12.4 s	β^-
Titanium (Ti)				
22	44	43.959690	49 y	ϵ
22	45	44.958124	184.8 m	ϵ
22	46	45.952630	8.25%	
22	47	46.951764	7.44%	
22	48	47.947947	73.72%	
22	49	48.947871	5.41%	
22	50	49.944792	5.18%	
22	51	50.946616	5.76 m	β^-
22	52	51.946898	1.7 m	β^-
22	53	52.949730	32.7 s	β^-
Vanadium (V)				
23	47	46.954907	32.6 m	ϵ
23	48	47.952255	15.974 d	ϵ
23	49	48.948517	330 d	ϵ
23	50	49.947163	1.4×10^{17} y 0.250%	ϵ, β^-
23	51	50.943964	99.750%	
23	52	51.944780	3.74 m	β^-
23	53	52.944343	1.61 m	β^-
23	54	53.946444	49.8 s	β^-
Chromium (Cr)				
24	48	47.954036	21.56 h	ϵ
24	49	48.951341	42.3 m	ϵ
24	50	49.946050	4.345%	
24	51	50.944772	27.702 d	ϵ
24	52	51.940512	83.79%	
24	53	52.940654	9.50%	
24	54	53.938885	2.365%	
24	55	54.940844	3.497 m	β^-
24	56	55.940645	5.94 m	β^-
24	57	56.943750	21.1 s	β^-

Z	A	Atomic Mass (u)	$t_{1/2}$ or Abundance	Decay Mode
Manganese (Mn)				
25	51	50.948216	46.2 m	ϵ
25	52	51.945570	5.591 d	ϵ
25	53	52.941295	3.74×10^6 y	ϵ
25	54	53.940363	312.12 d	ϵ, β^-
25	55	54.938050	100%	
25	56	55.938909	2.5785 h	β^-
25	57	56.938287	85.4 s	β^-
25	58	57.939990	65.3 s	β^-
25	59	58.940450	4.6 s	β^-
25	60	59.943190	51 s	β^-
Iron (Fe)				
26	51	50.956825	305 ms	ϵ
26	52	51.948117	8.275 h	ϵ
26	53	52.945312	8.51 m	ϵ
26	54	53.939615	5.845%	
26	55	54.938298	2.73 y	ϵ
26	56	55.934942	91.754%	
26	57	56.935399	2.119%	
26	58	57.933280	0.282%	
26	59	58.934880	44.503 d	β^-
26	60	59.934077	1.5×10^6 y	β^-
26	61	60.936749	5.98 m	β^-
26	62	61.936770	68 s	β^-
Cobalt (Co)				
27	55	54.942003	17.53 h	ϵ
27	56	55.939844	77.27 d	ϵ
27	57	56.936296	271.79 d	ϵ
27	58	57.935758	70.82 d	ϵ
27	59	58.933200	100%	
27	60	59.933822	5.2714 y	β^-
27	61	60.932479	1.650 h	β^-
27	62	61.934054	1.50 m	β^-
27	63	62.933615	27.4 s	β^-
Nickel (Ni)				
28	56	55.942136	6.08 d	ϵ
28	57	56.939800	35.60 h	ϵ
28	58	57.935348	68.077%	
28	59	58.934352	7.6×10^4 y	ϵ
28	60	59.930791	26.223%	
28	61	60.931060	1.140%	
28	62	61.928349	3.634%	
28	63	62.929673	100.1 y	β^-
28	64	63.927970	0.926%	
28	65	64.930088	2.517 h	β^-
28	66	65.929115	54.6 h	β^-
28	67	66.931570	21 s	β^-

Z	A	Atomic Mass (u)	$t_{1/2}$ or Abundance	Decay Mode
Nickel (Ni)				
28	68	67.931845	19 s	β^-
28	69	68.935180	11.4 s	β^-
Copper (Cu)				
29	59	58.939504	81.5 s	ϵ
29	60	59.937368	23.7 s	ϵ
29	61	60.933462	3.333 h	ϵ
29	62	61.932587	9.74 m	ϵ
29	63	62.929601	69.17%	
29	64	63.929768	12.700 h	ϵ, β^-
29	65	64.927794	30.83%	
29	66	65.928873	5.088 m	β^-
29	67	66.927750	61.83 h	β^-
29	68	67.929640	31.1 s	β^-
29	69	68.929425	2.85 m	β^-
29	70	69.932409	4.5 s	β^-
29	71	70.932620	19.5 s	β^-
Zinc (Zn)				
30	60	59.941832	2.38 m	ϵ
30	61	60.939514	89.1 s	ϵ
30	62	61.934334	9.186 h	ϵ
30	63	62.933216	38.47 m	ϵ
30	64	63.929147	48.6%	
30	65	64.929245	244.26 d	ϵ
30	66	65.926037	27.9%	
30	67	66.927131	4.1%	
30	68	67.924848	18.8%	
30	69	68.926554	56.4 m	β^-
30	70	69.925325	5×10^{14} y 0.6%	$2\beta^-$
30	71	70.927727	2.45 m	β^-
30	72	71.926861	46.5 h	β^-
30	73	72.929780	23.5 s	β^-
30	74	73.929460	95.6 s	β^-
30	75	74.932940	10.2 s	β^-
Gallium (Ga)				
31	63	62.939140	32.4 s	ϵ
31	64	63.936838	2.630 m	ϵ
31	65	64.932739	15.2 m	ϵ
31	66	65.931592	9.49 h	ϵ
31	67	66.928205	3.2612 d	ϵ
31	68	67.927984	67.629 m	ϵ
31	69	68.925581	60.108%	
31	70	69.926028	21.14 m	β^-, ϵ
31	71	70.924705	39.892%	
31	72	71.926369	14.10 h	β^-
31	73	72.925170	4.86 h	β^-

Z	A	Atomic Mass (u)	$t_{1/2}$ or Abundance	Decay Mode	Z	A	Atomic Mass (u)	$t_{1/2}$ or Abundance	Decay Mode
Gallium (Ga) (*continued*)					**Selenium (Se)**				
31	74	73.926940	8.12 m	β^-	34	73	72.926767	7.15 h	ϵ
31	75	74.926501	126 s	β^-	34	74	73.922477	0.89%	
31	76	75.928930	32.6 s	β^-	34	75	74.922524	119.779 d	ϵ
31	77	76.929280	13.2 s	β^-	34	76	75.919214	9.36%	
					34	77	76.919915	7.63%	
Germanium (Ge)					34	78	77.917310	23.78%	
					34	79	78.918500	6.5×10^5 y	β^-
32	64	63.941570	63.7 s	ϵ	34	80	79.916522	49.61%	
32	65	64.939440	30.9 s	ϵ	34	81	80.917993	18.45 m	β^-
32	66	65.933850	2.26 h	ϵ	34	82	81.916700	1.1×10^{20} y	$2\beta^-$
32	67	66.932738	18.9 m	ϵ				8.73%	
32	68	67.928097	270.82 d	ϵ	34	83	82.919119	22.3 m	β^-
32	69	68.927972	39.05 h	ϵ	34	84	83.918465	3.1 m	β^-
32	70	69.924250	21.23%		34	85	84.922240	31.7 s	β^-
32	71	70.924954	11.43 d	ϵ	34	86	85.924271	15.3 s	β^-
32	72	71.922076	27.66%						
32	73	72.923459	7.73%		**Bromine (Br)**				
32	74	73.921178	35.94%						
32	75	74.922859	82.78 m	β^-	35	71	70.939250s	21.4 s	ϵ
32	76	75.921403	7.44%		35	72	71.936500	78.6 s	ϵ
32	77	76.923548	11.30 h	β^-	35	73	72.931790	3.4 m	ϵ
32	78	77.922853	88.0 m	β^-	35	74	73.929891	25.4 m	ϵ
32	79	78.925400	18.98 s	β^-	35	75	74.925776	96.7 m	ϵ
32	80	79.925445	29.5 s	β^-	35	76	75.924542	16.2 h	ϵ
					35	77	76.921380	57.036 h	ϵ
Arsenic (As)					35	78	77.921146	6.46 m	ϵ, β^-
					35	79	78.918338	50.69%	
33	67	66.939190	42.5 s	ϵ	35	80	79.918530	17.68 m	β^-, ϵ
33	68	67.936790	151.6 s	ϵ	35	81	80.916291	49.31%	
33	69	68.932280	15.2 m	ϵ	35	82	81.916805	35.30 h	β^-
33	70	69.930930	52.6 m	ϵ	35	83	82.915180	2.40 h	β^-
33	71	70.927115	65.28 h	ϵ	35	84	83.916504	31.80 m	β^-
33	72	71.926753	26.0 h	ϵ	35	85	84.915608	2.90 m	β^-
33	73	72.923825	80.30 d	ϵ	35	86	85.918797	55.1 s	β^-
33	74	73.923929	17.77 d	ϵ, β^-	35	87	86.920711	55.60 s	$\beta^-, \beta^- n$
33	75	74.921596	100%		35	88	87.924070	16.34 s	$\beta^-, \beta^- n$
33	76	75.922394	26.32 h	β^-					
33	77	76.920648	38.83 h	β^-	**Krypton (Kr)**				
33	78	77.921829	90.7 m	β^-					
33	79	78.920948	9.01 m	β^-	36	72	71.941910	17.2 s	ϵ
33	80	79.922578	15.2 s	β^-	36	73	72.938930	27.0 s	$\epsilon, \epsilon p$
33	81	80.922133	33.3 s	β^-	36	74	73.933260	11.50 m	ϵ
33	82	81.924500	19.1 s	β^-	36	75	74.931034	4.3 m	ϵ
33	83	82.924980	13.4 s	β^-	36	76	75.925948	14.8 h	ϵ
					36	77	76.924668	74.4 m	ϵ
Selenium (Se)					36	78	77.920386	0.35%	
					36	79	78.920083	35.04 h	ϵ
34	68	67.941870s	35.5 s	ϵ	36	80	79.916378	2.25%	
34	69	68.939560	27.4 s	$\epsilon, \epsilon p$	36	81	80.916592	2.29×10^5 y	ϵ
34	70	69.933500s	41.1 m	ϵ	36	82	81.913485	11.6%	
34	71	70.932270s	4.74 m	ϵ	36	83	82.914136	11.5%	
34	72	71.927112	8.40 d	ϵ	36	84	83.911507	57.0%	

Z	A	Atomic Mass (u)	$t_{1/2}$ or Abundance	Decay Mode
Krypton (Kr) (*continued*)				
36	85	84.912527	10.756 y	β^-
36	86	85.910610	17.3%	
36	87	86.913354	76.3 m	β^-
36	88	87.914447	2.84 h	β^-
36	89	88.917630	3.15 m	β^-
36	90	89.919524	32.32 s	β^-
Rubidium (Rb)				
37	75	74.938569	19.0 s	ϵ
37	76	75.935071	36.5 s	ϵ
37	77	76.930407	3.78 m	ϵ
37	78	77.928141	17.66 m	ϵ
37	79	78.923997	22.9 m	ϵ
37	80	79.922519	33.4 s	ϵ
37	81	80.918994	4.576 h	ϵ
37	82	81.918208	1.273 m	ϵ
37	83	82.915112	86.2 d	ϵ
37	84	83.914385	32.77 d	ϵ, β^-
37	85	84.911789	72.17%	
37	86	85.911167	18.631 d	β^-, ϵ
37	87	86.909184	4.75×10^{10} y 27.83%	β^-
37	88	87.911319	17.78 m	β^-
37	89	88.912280	15.15 m	β^-
37	90	89.914809	158 s	β^-
37	91	90.916534	58.4 s	β^-
Strontium (Sr)				
38	78	77.932179	2.5 m	ϵ
38	79	78.929707	2.25 m	ϵ
38	80	79.924525	106.3 m	ϵ
38	81	80.923213	22.3 m	ϵ
38	82	81.918401	25.55 d	ϵ
38	83	82.917555	32.41 h	ϵ
38	84	83.913425	0.56%	
38	85	84.912933	64.84 d	ϵ
38	86	85.909262	9.86%	
38	87	86.908879	7.00%	
38	88	87.905614	82.58%	
38	89	88.907453	50.53 d	β^-
38	90	89.907738	28.78 y	β^-
38	91	90.910210	9.63 h	β^-
38	92	92.911030	2.71 h	β^-
38	93	92.914022	7.423 m	β^-
38	94	93.915360	75.3 s	β^-
38	95	94.919358	23.90 s	β^-

Z	A	Atomic Mass (u)	$t_{1/2}$ or Abundance	Decay Mode
Yttrium (Y)				
39	80	79.934340s	35 s	ϵ
39	81	80.929130	72.4 s	ϵ
39	83	82.922350	7.08 m	ϵ
39	85	84.916427	2.68 h	ϵ
39	86	85.914888	14.74 h	ϵ
39	87	86.910878	79.8 h	ϵ
39	88	87.909503	106.65 d	ϵ
39	89	88.905848	100%	
39	90	89.907151	64.1 h	β^-
39	91	90.907303	58.51 d	β^-
39	92	91.908947	3.54 h	β^-
39	93	92.909582	10.18 h	β^-
39	94	93.911594	18.7 m	β^-
39	95	94.912824	10.3 m	β^-
Zirconium (Zr)				
40	81	80.936820	15 s	$\epsilon, \epsilon p$
40	82	81.931090	32 s	ϵ
40	83	82.928650	44 s	$\epsilon, \epsilon p$
40	84	83.923250s	25.9 m	ϵ
40	85	84.921470	7.86 m	ϵ
40	86	85.916470	16.5 h	ϵ
40	87	86.914817	1.68 h	ϵ
40	88	87.910226	83.4 d	ϵ
40	89	88.908889	78.41 h	ϵ
40	90	89.904704	51.45%	
40	91	90.905645	11.22%	
40	92	91.905040	17.15%	
40	93	92.906476	1.53×10^6 y	β^-
40	94	93.906316	17.38%	
40	95	94.908043	64.02 d	β^-
40	96	95.908276	3.9×10^{19} y 2.80%	$2\beta^-$
40	97	96.910951	16.90 h	β^-
40	98	97.912746	30.7 s	β^-
40	99	98.916511	2.1 s	β^-
Niobium (Nb)				
41	84	83.933570s	12 s	$\epsilon, \epsilon p$
41	85	84.927910	20.9 s	ϵ
41	86	85.925040	88 s	ϵ
41	87	86.920360	2.6 m	ϵ
41	88	87.917960s	14.5 m	ϵ
41	89	88.913500	1.18 h	ϵ
41	90	89.911264	14.60 h	ϵ
41	91	90.906991	6.8×10^2 y	ϵ
41	92	91.907193	3.47×10^7 y	ϵ, β^-
41	93	92.906378	100%	
41	93m	92.906412	16.13 y	IT

Z	A	Atomic Mass (u)	$t_{1/2}$ or Abundance	Decay Mode
Niobium (Nb) (*continued*)				
41	94	93.907284	2.03×10^4 y	β^-
41	95	94.906835	34.975 d	β^-
41	96	95.908100	23.35 h	β^-
41	97	96.908097	1.227 h	β^-
41	99	98.911618	15.0 s	β^-
Molybdenum (Mo)				
42	86	85.930700	19.6 s	ϵ
42	87	86.927330	14.5 s	ϵ, ϵp
42	88	87.921953	8.0 m	ϵ
42	89	88.919481	2.04 m	ϵ
42	90	89.913936	5.67 h	ϵ
42	91	90.911751	15.49 m	ϵ
42	92	91.906810	14.84%	
42	93	92.906812	4.0×10^3 y	ϵ
42	94	93.905088	9.25%	
42	95	94.905842	15.92%	
42	96	95.904679	16.68%	
42	97	96.906021	9.55%	
42	98	97.905408	24.13%	
42	99	98.907712	65.94 h	β^-
42	100	99.907477	1.2×10^{19} y 9.63%	$2\beta^-$
42	101	100.910347	14.61 m	β^-
42	102	101.910297	11.3 m	β^-
42	103	102.913200	67.5 s	β^-
42	104	103.913760	60 s	β^-
42	105	104.916970	35.6 s	β^-
Technetium (Tc)				
43	90	89.923560	8.7 s	ϵ
43	91	90.918430	3.14 m	ϵ
43	92	91.915260	4.23 m	ϵ
43	93	92.910248	2.75 h	ϵ
43	94	93.909656	293 m	ϵ
43	95	94.907656	20.0 h	ϵ
43	96	95.907871	4.28 d	ϵ
43	97	96.906365	2.6×10^6 y	ϵ
43	98	97.907216	4.2×10^6 y	β^-
43	99	98.906255	2.111×10^5 y	β^-
43	99m	98.906408	6.01 h	IT, β^-
43	100	99.907658	15.8 s	β^-
43	101	100.907314	14.22 m	β^-
43	103	102.909179	54.2 s	β^-
43	104	103.911440	18.3 m	β^-
43	105	104.911660	7.6 m	β^-
43	106	105.914355	35.6 s	β^-
43	107	106.915080	21.2 s	β^-

Z	A	Atomic Mass (u)	$t_{1/2}$ or Abundance	Decay Mode
Ruthenium (Ru)				
44	92	91.920120s	3.65 m	ϵ
44	93	92.917050	59.7 s	ϵ
44	94	93.911360	51.8 m	ϵ
44	95	94.910413	1.643 h	ϵ
44	96	95.907598	5.52%	
44	97	96.907555	2.9 d	ϵ
44	98	97.905287	1.88%	
44	99	98.905939	12.7%	
44	100	99.904220	12.6%	
44	101	100.905582	17.0%	
44	102	101.904350	31.6%	
44	103	102.906324	39.26 d	β^-
44	104	103.905430	18.7%	
44	105	104.907750	4.44 h	β^-
44	106	105.907327	373.59 d	β^-
44	107	106.909910	3.75 m	β^-
44	108	107.910190	4.55 m	β^-
44	109	108.913200	34.5 s	β^-
44	110	109.913970	14.6 s	β^-
Rhodium (Rh)				
45	95	94.915900	5.02 m	ϵ
45	96	95.914518	9.90 m	ϵ
45	97	96.911340	30.7 m	ϵ
45	98	97.910716	8.7 m	ϵ
45	99	98.908132	16.1 d	ϵ
45	100	99.908117	20.8 h	ϵ
45	101	100.906164	3.3 y	ϵ
45	102	101.906843	207 d	ϵ, β^-
45	102m	101.906994	2.9 y	ϵ, IT
45	103	102.905504	100%	
45	104	103.906655	42.3 s	β^-, ϵ
45	105	104.905692	35.36 h	β^-
45	106	105.907285	29.80 s	β^-
45	107	106.906751	21.7 m	β^-
45	109	108.908736	80 s	β^-
45	110	109.910950	3.2 s	β^-
45	111	110.911660s	11 s	β^-
Palladium (Pd)				
46	96	95.918220	2.03 m	ϵ
46	97	96.916480	3.1 m	ϵ
46	98	97.912721	17.7 m	ϵ
46	99	98.911768	21.4 m	ϵ
46	100	99.908505	3.63 d	ϵ
46	101	100.908289	8.47 h	ϵ
46	102	101.905608	1.02%	
46	103	102.906087	16.991 d	ϵ
46	104	103.904035	11.14%	

Z	A	Atomic Mass (u)	$t_{1/2}$ or Abundance	Decay Mode	Z	A	Atomic Mass (u)	$t_{1/2}$ or Abundance	Decay Mode

Palladium (Pd) (*continued*)

Z	A	Atomic Mass (u)	$t_{1/2}$ or Abundance	Decay Mode
46	105	104.905084	22.33%	
46	106	105.903483	27.33%	
46	107	106.905128	6.5×10^6 y	β^-
46	108	107.903894	26.46%	
46	109	108.905954	13.70 h	β^-
46	110	109.905152	11.72%	
46	111	110.907640	23.4 m	β^-
46	112	111.907313	21.03 h	β^-
46	113	112.910150	93 s	β^-
46	114	113.910365	2.42 m	β^-
46	115	114.913680	25 s	β^-
46	116	115.914160	11.8 s	β^-

Silver (Ag)

Z	A	Atomic Mass (u)	$t_{1/2}$ or Abundance	Decay Mode
47	97	96.924000s	19 s	ϵ
47	98	97.921760	46.7 s	ϵ, ϵp
47	99	98.917600	124 s	ϵ
47	100	99.916070	2.01 m	ϵ
47	101	100.912800	11.1 m	ϵ
47	102	101.912000	12.9 m	ϵ
47	103	102.908972	65.7 m	ϵ
47	104	103.908628	69.2 m	ϵ
47	105	104.906528	41.29 d	ϵ
47	106	105.906666	23.96 m	ϵ, β^-
47	107	106.905093	51.839%	
47	108	107.905954	2.37 m	β^-, ϵ
47	108m	107.906071	418 y	ϵ, IT
47	109	108.904756	48.161%	
47	110	109.906110	24.6 s	β^-, ϵ
47	110m	109.906237	249.79 d	β^-, IT
47	111	110.905295	7.45 d	β^-
47	112	111.907004	3.130 h	β^-
47	113	112.906566	5.37 h	β^-
47	115	114.908760	20.0 m	β^-
47	116	115.911360	2.68 m	β^-
47	117	116.911680	72.8 s	β^-

Cadmium (Cd)

Z	A	Atomic Mass (u)	$t_{1/2}$ or Abundance	Decay Mode
48	99	98.925010s	16 s	ϵ, ϵp, $\epsilon \alpha$
48	100	99.920230	49.1 s	ϵ
48	101	100.918680	1.2 m	ϵ
48	102	101.914780	5.5 m	ϵ
48	103	102.913419	7.3 m	ϵ
48	104	103.909848	57.7 m	ϵ
48	105	104.909468	55.5 m	ϵ
48	106	105.906458	1.25%	
48	107	106.906614	6.50 h	ϵ
48	108	107.904183	0.89%	
48	109	108.904986	462.6 d	ϵ
48	110	109.903006	12.49%	

Cadmium (Cd)

Z	A	Atomic Mass (u)	$t_{1/2}$ or Abundance	Decay Mode
48	111	110.904182	12.80%	
48	112	111.902757	24.13%	
48	113	112.904401	9.3×10^{15} y 12.22%	β^-
48	113m	112.904684	14.1 y	β^-, IT
48	114	113.903358	28.73%	
48	115	114.905431	53.46 h	β^-
48	116	115.904755	7.49%	
48	117	116.907218	2.49 h	β^-
48	118	117.906914	50.3 m	β^-
48	119	118.909920	2.69 m	β^-
48	120	119.909851	50.80 s	β^-
48	121	120.912980	13.5 s	β^-

Indium (In)

Z	A	Atomic Mass (u)	$t_{1/2}$ or Abundance	Decay Mode
49	101	100.926560s	16 s	ϵ, ϵp
49	102	101.924710	24 s	ϵ
49	103	102.919914	65 s	ϵ
49	104	103.918340	1.8 m	ϵ
49	105	104.914673	5.07 m	ϵ
49	106	105.913461	6.2 m	ϵ
49	107	106.910292	32.4 m	ϵ
49	108	107.909720	58.0 m	ϵ
49	109	108.907154	4.2 h	ϵ
49	110	109.907169	69.1 m	ϵ
49	110	109.907169	4.9 h	ϵ
49	111	110.905111	2.8047 d	ϵ
49	112	111.905533	14.97 m	ϵ, β^-
49	113	112.904061	4.29%	
49	114	113.904917	71.9 s	β^-, ϵ
49	115	114.903878	4.41×10^{14} y 95.71%	β^-
49	116	115.905260	14.10 s	β^-, ϵ
49	117	116.904516	43.2 m	β^-
49	119	118.905846	2.4 m	β^-
49	120	119.907960	3.08 s, 46.2 s, 47.3 s	β^-
49	121	120.907849	23.1 s	β^-

Tin (Sn)

Z	A	Atomic Mass (u)	$t_{1/2}$ or Abundance	Decay Mode
50	104	103.923190	20.8 s	ϵ
50	105	104.921390	31 s	ϵ, ϵp
50	106	105.916880	115 s	ϵ
50	107	106.915670	2.90 m	ϵ
50	108	107.911970	10.30 m	ϵ
50	109	108.911287	18.0 m	ϵ
50	110	109.907853	4.11 h	ϵ
50	111	110.907735	35.3 m	ϵ
50	112	111.904821	0.97%	

Z	A	Atomic Mass (u)	$t_{1/2}$ or Abundance	Decay Mode
Tin (Sn)(*continued*)				
50	113	112.905173	115.09 d	ϵ
50	114	113.902782	0.65%	
50	115	114.903346	0.34%	
50	116	115.901744	14.54%	
50	117	116.902954	7.68%	
50	118	117.901606	24.22%	
50	119	118.903309	8.58%	
50	119*m*	118.903406	293.1 d	IT
50	120	119.902197	32.59%	
50	121	120.904237	27.06 h	β^-
50	121*m*	120.904243	55 y	IT, β^-
50	122	121.903440	4.63%	
50	123	122.905722	129.2 d	β^-
50	124	123.905275	5.79%	
50	125	124.907785	9.64 d	β^-
50	126	125.907654	1.0×10^5 y	β^-
50	127	126.910351	2.10 h	β^-
50	128	127.910535	59.07 m	β^-
50	129	128.913440	2.23 m	β^-
50	130	129.913850	3.72 m	β^-
50	131	130.916920	56.0 s	β^-
50	132	131.917744	39.7 s	β^-
Antimony (Sb)				
51	109	108.918136	17.0 s	ϵ
51	110	109.916760*s*	23.0 s	ϵ
51	111	110.913210*s*	75 s	ϵ
51	112	111.912395	51.4 s	ϵ
51	113	112.909378	6.67 m	ϵ
51	114	113.909100	3.49 m	ϵ
51	115	114.906599	32.1 m	ϵ
51	116	115.906797	15.8 m	ϵ
51	117	116.904840	2.80 h	ϵ
51	118	117.905532	3.6 m	ϵ
51	119	118.903946	38.19 h	ϵ
51	120	119.905074	15.89 m	ϵ
51	121	120.903818	57.21%	
51	122	121.905175	2.7238 d	β^-, ϵ
51	123	122.904216	42.79%	
51	124	123.905938	60.20 d	β^-
51	125	124.905248	2.7582 y	β^-
51	126	125.907250	12.46 d	β^-
51	127	126.906915	3.85 d	β^-
51	128	127.909167	9.01 h	β^-
51	129	128.909150	4.40 h	β^-
51	130	129.911546	39.5 m	β^-
51	131	130.911950	23.03 m	β^-
51	132	131.914413	2.79 m	β^-
51	133	132.915240	2.5 m	β^-
51	134	133.920550	10.22 s	β^-, $\beta^- n$

Z	A	Atomic Mass (u)	$t_{1/2}$ or Abundance	Decay Mode
Tellurium (Te)				
52	110	109.922410	18.6 s	ϵ, α
52	111	110.921120	19.3 s	ϵ, ϵp
52	112	111.917060	2.0 m	ϵ
52	113	112.915930*s*	1.7 m	ϵ
52	114	113.912060*s*	15.2 m	ϵ
52	115	114.911580	5.8 m	ϵ
52	116	115.908420	2.49 h	ϵ
52	117	116.908634	62 m	ϵ
52	118	117.905825	6.00 d	ϵ
52	119	118.906408	16.03 h	ϵ
52	120	119.904020	0.096%	
52	121	120.904930	16.78 d	ϵ
52	121*m*	120.905245	154 d	IT, ϵ
52	122	121.903047	2.603%	
52	123	122.904273	1.3×10^{13} y 0.908%	ϵ
52	123*m*	122.904538	119.7 d	IT
52	124	123.902820	4.816%	
52	125	124.904425	7.139%	
52	126	125.903306	18.952%	
52	127	126.905217	9.35 h	β^-
52	127*m*	126.905312	109 d	IT, β^-
52	128	127.904461	7.7×10^{24} y 31.687%	$2\beta^-$
52	129	128.906596	69.6 m	β^-
52	130	129.906223	2.7×10^{21} y 33.799%	$2\beta^-$
52	131	130.908522	25.0 m	β^-
52	132	131.908524	3.204 d	β^-
52	133	132.910940	12.5 m	β^-
52	134	133.911540	41.8 m	β^-
52	135	134.916450	19.0 s	β^-
52	136	135.920100	17.5 s	β^-, $\beta^- n$
Iodine (I)				
53	115	114.917920*s*	1.3 m	ϵ
53	117	116.913650	2.22 m	ϵ
53	118	117.913380	13.7 m	ϵ
53	119	118.910180	19.1 m	ϵ
53	120	119.910048	81.0 m	ϵ
53	121	120.907366	2.12 h	ϵ
53	122	121.907592	3.63 m	ϵ
53	123	122.905598	13.27 h	ϵ
53	124	123.906211	4.1760 d	ϵ
53	125	124.904624	59.408 d	ϵ
53	126	125.905619	13.11 d	ϵ, β^-
53	127	126.904468	100%	
53	128	127.905805	24.99 m	β^-, ϵ
53	129	128.904987	1.57×10^7 y	β^-
53	130	129.906674	12.36 h	β^-
53	131	130.906124	8.02070 d	β^-

Z	A	Atomic Mass (u)	$t_{1/2}$ or Abundance	Decay Mode
Iodine (I) (*continued*)				
53	132	131.907995	2.295 h	β^-
53	133	132.907806	20.8 h	β^-
53	134	133.909877	52.5 m	β^-
53	135	134.910050	6.57 h	β^-
53	136	135.914660	83.4 s	β^-
53	137	136.917873	24.5 s	$\beta^-, \beta^- n$
Xenon (Xe)				
54	114	113.928150s	10.0 s	ϵ
54	115	114.926540s	18 s	$\epsilon\alpha, \epsilon, \epsilon p$
54	116	115.921740s	59 s	ϵ
54	117	116.920560	61 s	$\epsilon, \epsilon p$
54	118	117.916570	3.8 m	ϵ
54	119	118.915550	5.8 m	ϵ
54	120	119.912150	40 m	ϵ
54	121	120.911386	40.1 m	ϵ
54	122	121.908550	20.1 h	ϵ
54	123	122.908471	2.08 h	ϵ
54	124	123.905896	0.10%	
54	125	124.906398	16.9 h	ϵ
54	126	125.904269	0.09%	
54	127	126.905180	36.3446 d	ϵ
54	128	127.903530	1.91%	
54	129	128.904780	26.4%	
54	130	129.903508	4.1%	
54	131	130.905082	21.2%	
54	132	131.904154	26.9%	
54	133	132.905906	5.2475 d	β^-
54	134	133.905394	10.4%	
54	135	134.907207	9.14 h	β^-
54	136	135.907220	9.3×10^{19} y 8.9%	$2\beta^-$
54	137	136.911563	3.818 m	β^-
54	138	137.913990	14.08 m	$\beta^-, \beta^- n$
54	139	138.918787	39.68 s	β^-
54	140	139.921640	13.60 s	β^-
54	142	141.929710	1.22 s	β^-
Cesium (Cs)				
55	118	117.926555	14 s	$\epsilon, \epsilon p$
55	119	118.922371	43.0 s	ϵ
55	120	119.920678	57 s	$\epsilon, \epsilon p$
55	120	119.920678	64 s	ϵ
55	121	120.917184	128 s	ϵ
55	122	121.916122	21.2 s	ϵ
55	123	122.912990	5.87 m	ϵ
55	124	123.912246	30.9 s	ϵ
55	125	124.909725	45 m	ϵ
55	126	125.909448	1.63 m	ϵ
55	127	126.907418	6.25 h	ϵ

Z	A	Atomic Mass (u)	$t_{1/2}$ or Abundance	Decay Mode
Cesium (Cs)				
55	128	127.907748	3.66 m	ϵ
55	129	128.906063	32.06 h	ϵ
55	130	129.906706	29.21 m	ϵ, β^-
55	131	130.905460	9.69 d	ϵ
55	132	131.906430	6.479 d	ϵ, β^-
55	133	132.905447	100%	
55	134	133.906713	2.0648 y	β^-, ϵ
55	135	134.905972	2.3×10^6 y	β^-
55	136	135.907306	19 s, 13.16 d	β^-
55	137	136.907084	30.1 y	β^-
55	138	137.911011	32.41 m	β^-
55	139	138.913358	9.27 m	β^-
55	140	139.917277	63.7 s	β^-
55	141	140.920044	24.94 s	$\beta^-, \beta^- n$
Barium (Ba)				
56	120	119.926050	32 s	ϵ
56	121	120.924490	29.5 s	$\epsilon, \epsilon p$
56	122	121.920260s	1.95 m	ϵ
56	123	122.918850s	2.7 m	ϵ
56	124	123.915088	11.9 m	ϵ
56	125	124.914620	3.5 m	ϵ
56	126	125.911244	100 m	ϵ
56	127	126.911120	12.7 m	ϵ
56	128	127.908309	2.43 d	ϵ
56	129	128.908674	2.23 h	ϵ
56	130	129.906310	0.106%	
56	131	130.906931	11.50 d	ϵ
56	132	131.905056	0.101%	
56	133	132.906002	10.52 y	ϵ
56	134	133.904503	2.417%	
56	135	134.905683	6.592%	
56	136	135.904570	7.854%	
56	137	136.905821	11.23%	
56	138	137.905241	71.70%	
56	139	138.908835	83.06 m	β^-
56	140	139.910599	12.752 d	β^-
56	141	140.914406	18.27 m	β^-
56	142	141.916448	10.6 m	β^-
56	143	142.920617	14.33 s	β^-
56	144	143.922940	11.5 s	$\beta^-, \beta^- n$
Lanthanum (La)				
57	123	122.926240s	17 s	ϵ
57	124	123.924530s	29 s	ϵ
57	125	124.920670s	76 s	ϵ
57	126	125.919370s	54 s	ϵ
57	127	126.916160s	3.8 m	ϵ
57	128	127.915450	5.0 m	ϵ
57	129	128.912670	11.6 m	ϵ

Z	A	Atomic Mass (u)	$t_{1/2}$ or Abundance	Decay Mode	Z	A	Atomic Mass (u)	$t_{1/2}$ or Abundance	Decay Mode
Lanthanum (La) *(continued)*					**Praseodymium (Pr)**				
57	130	129.912320s	8.7 m	ϵ	59	131	130.920060	1.53 m	ϵ
57	131	130.910110	59 m	ϵ	59	132	131.919120s	1.6 m	ϵ
57	132	131.910110	4.8 h	ϵ	59	133	132.916200s	6.5 m	ϵ
57	133	132.908400	3.912 h	ϵ	59	134	133.915670s	17 m	ϵ
57	134	133.908490	6.45 m	ϵ	59	135	134.913140	24 m	ϵ
57	135	134.906971	19.5 h	ϵ	59	136	135.912650	13.1 m	ϵ
57	136	135.907650	9.87 m	ϵ	59	137	136.910680	1.28 h	ϵ
57	137	136.906470	6×10^4 y	ϵ	59	138	137.910749	1.45 m	ϵ
57	138	137.907107	1.05×10^{11} y 0.0902%	ϵ, β^-	59	139	138.908932	4.41 h	ϵ
57	139	138.906348	99.9098%		59	140	139.909071	3.39 m	ϵ
57	140	139.909473	1.6781 d	β^-	59	141	140.907648	100%	
57	141	140.910957	3.92 h	β^-	59	142	141.910040	19.12 h	β^-, ϵ
57	142	141.914074	91.1 m	β^-	59	143	142.910812	13.57 d	β^-
57	143	142.916059	14.2 m	β^-	59	144	143.913301	17.28 m	β^-
57	144	143.919590	40.8 s	β^-	59	145	144.914507	5.984 h	β^-
57	145	144.921640	24.8 s	β^-	59	146	145.917590	24.15 m	β^-
					59	147	146.918980	13.4 m	β^-
					59	148	147.922180	2.27 m	β^-
Cerium (Ce)					59	149	148.923791	2.26 m	β^-
					59	151	150.928230	18.90 s	β^-
58	125	124.928540s	9.0 s	$\epsilon, \epsilon p$					
58	126	125.924100s	50 s	ϵ					
58	127	126.922750s	32 s	ϵ	**Neodymium (Nd)**				
58	128	127.918870s	3 m	ϵ					
58	129	128.918090s	3.5 m	ϵ	60	130	129.928780s	28 s	ϵ
58	130	129.914690s	25 m	ϵ	60	131	130.927100	27 s	$\epsilon, \epsilon p$
58	131	130.914420	10.2 m	ϵ	60	132	131.923120s	1.75 m	ϵ
58	131	130.914420	5.0 m	ϵ	60	133	132.922210s	70 s	ϵ
58	132	131.911490s	3.51 h	ϵ	60	134	133.918650s	8.5 m	ϵ
58	133	132.911550s	4.9 h	ϵ	60	135	134.918240s	12.4 m	ϵ
58	134	133.909030	75.9 h	ϵ	60	136	135.915020	50.65 m	ϵ
58	135	134.909146	17.7 h	ϵ	60	137	136.914640	38.5 m	ϵ
58	136	135.907140	0.19%		60	138	137.911930s	5.04 h	ϵ
58	137	136.907780	9.0 h	ϵ	60	139	138.911920	29.7 m	ϵ
58	138	137.905986	0.25%		60	140	139.909310	3.37 d	ϵ
58	139	138.906647	137.640 d	ϵ	60	141	140.909605	2.49 h	ϵ
58	140	139.905434	88.48%		60	142	141.907719	27.13%	
58	141	140.908271	32.501 d	β^-	60	143	142.909810	12.18%	
58	142	141.909240	5×10^{16} y 11.08%	$2\beta^-$	60	144	143.910083	2.29×10^{15} y 23.80%	α
58	143	142.912381	33.039 h	β^-	60	145	144.912569	8.30%	
58	144	143.913643	284.893 d	β^-	60	146	145.913112	17.19%	
58	146	144.917230	3.01 m	β^-	60	147	146.916096	10.98 d	β^-
58	146	145.918690	13.52 m	β^-	60	148	147.916889	5.76%	
58	147	146.922510	56.4 s	β^-	60	149	148.920144	1.728 h	β^-
58	148	147.924390	56 s	β^-	60	150	149.920887	1.1×10^{19} y 5.64%	$2\beta^-$
					60	151	150.923825	12.44 m	β^-
Praseodymium (Pr)					60	152	151.924680	11.4 m	β^-
					60	153	152.927695	28.9 s	β^-
59	129	128.924860s	24 s	ϵ	60	154	153.929480	25.9 s	β^-
59	130	129.923380s	40.0 s	ϵ					

Z	A	Atomic Mass (u)	$t_{1/2}$ or Abundance	Decay Mode
Promethium (Pm)				
61	133	132.929720s	12 s	ϵ
61	134	133.928490s	5 s	ϵ
61	135	134.924620s	40 s	ϵ
61	136	135.923450	47 s, 107 s	ϵ
61	137	136.920710s	2.4 m	ϵ
61	138	137.919450s	10 s	ϵ
61	139	138.916760	4.15 m	ϵ
61	141	140.913607	20.90 m	ϵ
61	142	141.912950	40.5 s	ϵ
61	143	142.910928	265 d	ϵ
61	144	143.912586	363 d	ϵ
61	145	144.912744	17.7 y	ϵ, α
61	146	145.914692	5.53 y	ϵ, β^-
61	147	146.915134	2.6234 y	β^-
61	148	147.917468	5.370 d	β^-
61	149	148.918329	53.08 h	β^-
61	150	149.920979	2.68 h	β^-
61	151	150.921203	28.40 h	β^-
61	152	151.923490	4.12 m	β^-
61	153	152.924113	5.4 m	β^-
61	154	153.926550	1.73 m	β^-
61	155	154.928100	41.5 s	β^-
61	156	155.931060	26.70 s	β^-
61	157	156.933200s	10.56 s	β^-
Samarium (Sm)				
62	134	133.934020s	10 s	ϵ
62	135	134.932350s	10 s	$\epsilon, \epsilon p$
62	136	135.928300s	47 s	ϵ
62	137	136.927050	45 s	ϵ
62	138	137.923540s	3.1 m	ϵ
62	139	138.922302	2.57 m	ϵ
62	140	139.918991	14.82 m	ϵ
62	141	140.918469	10.2 m	ϵ
62	142	141.915193	72.49 m	ϵ
62	143	142.914624	8.83 m	ϵ
62	144	143.911995	3.1%	
62	145	144.913406	340 d	ϵ
62	146	145.913037	10.3×10^7 y	α
62	147	146.914893	1.06×10^{11} y 15.0%	α
62	148	147.914818	7×10^{15} y 11.3%	α
62	149	148.917180	2×10^{15} y 13.8%	α
62	150	149.917271	7.4%	
62	151	150.919928	90 y	β^-
62	152	151.919728	26.7%	
62	153	152.922094	46.27 h	β^-
62	154	153.922205	22.7%	
62	155	154.924636	22.3 m	β^-

Z	A	Atomic Mass (u)	$t_{1/2}$ or Abundance	Decay Mode
Samarium (Sm)				
62	156	155.925526	9.4 h	β^-
62	157	156.928350	482 s	β^-
62	158	157.929990	5.30 m	β^-
62	159	158.933200s	11.37 s	β^-
Europium (Eu)				
63	137	136.935210s	11 s	ϵ
63	138	137.933450s	12.1 s	ϵ
63	139	138.929840s	17.9 s	ϵ
63	141	140.924890	41.4 s	ϵ
63	143	142.920287	2.57 m	ϵ
63	144	143.918774	10.2 s	ϵ
63	145	144.916261	5.93 d	ϵ
63	146	145.917200	4.59 d	ϵ
63	147	146.916741	24.1 d	ϵ, α
63	148	147.918154	54.5 d	ϵ, α
63	149	148.917926	93.1 d	ϵ
63	150	149.919698	36.9 y	ϵ
63	151	150.919846	47.8%	
63	152	151.921740	13.537 y	ϵ, β^-
63	153	152.921226	52.2%	
63	154	153.922975	8.593 y	β^-, ϵ
63	155	154.922889	4.7611 y	β^-
63	156	155.924751	15.19 d	β^-
63	157	156.925419	15.18 h	β^-
63	158	157.927840	45.9 m	β^-
63	159	158.929084	18.1 m	β^-
63	160	159.931970s	38 s	β^-
63	161	160.933680s	26 s	β^-
63	162	161.937040s	10.6 s	β^-
Gadolinium (Gd)				
64	140	139.933950s	15.8 s	ϵ
64	141	140.932210s	14 s	$\epsilon, \epsilon p$
64	142	141.928230s	70.2 s	ϵ
64	143	142.926740	39 s	ϵ
64	144	143.922790s	4.5 m	ϵ
64	145	144.921690	23.0 m	ϵ
64	146	145.918305	48.27 d	ϵ
64	147	146.919089	38.06 h	ϵ
64	148	147.918110	74.6 y	α
64	149	148.919336	9.28 d	ϵ, α
64	150	149.918655	1.79×10^6 y	α
64	151	150.920344	124 d	ϵ, α
64	152	151.919788	1.08×10^{14} y 0.20%	α
64	153	152.921746	241.6 d	ϵ
64	154	153.920862	2.18%	
64	155	154.922619	14.80%	
64	156	155.922120	20.47%	

Z	A	Atomic Mass (u)	$t_{1/2}$ or Abundance	Decay Mode
Gadolinium (Gd) (*continued*)				
64	157	156.923957	15.65%	
64	158	157.924101	24.84%	
64	159	158.926385	18.479 h	β^-
64	160	159.927051	21.86%	
64	161	160.929666	3.66 m	β^-
64	162	161.930981	8.4 m	β^-
64	163	162.933990s	68 s	β^-
64	164	163.935860s	45 s	β^-
Terbium (Tb)				
65	143	142.934750s	12 s	ϵ
65	145	144.928880s	31.6 s	ϵ
65	147	146.924037	1.7 h	ϵ
65	148	147.924300	60 m	ϵ
65	149	148.923242	4.118 h	ϵ, α
65	150	149.923654	3.48 h	ϵ, α
65	151	150.923098	17.609 h	ϵ, α
65	152	151.924070	17.5 h	ϵ, α
65	153	152.923431	2.34 d	ϵ
65	154	153.924690	21.5 h	ϵ, β^-
65	155	154.923500	5.32 d	ϵ
65	156	155.924744	5.35 d	ϵ, β^-
65	157	156.924021	99 y	ϵ
65	158	157.925410	180 y	ϵ, β^-
65	159	158.925343	100%	
65	160	159.927164	72.3 d	β^-
65	161	160.927566	6.88 d	β^-
65	162	161.929480	7.60 m	β^-
65	163	162.930644	19.5 m	β^-
65	164	163.933350	3.0 m	β^-
65	165	164.934880s	2.11 m	β^-
Dysprosium (Dy)				
66	145	144.936950s	10.5 s	ϵ
66	146	145.932720	33.2 s	ϵ
66	147	146.930880	40 s	$\epsilon, \epsilon p$
66	148	147.927180	3.1 m	ϵ
66	149	148.927334	4.20 m	ϵ
66	150	149.925580	7.17 m	ϵ, α
66	151	150.926180	17.9 m	ϵ, α
66	152	151.924714	2.38 h	ϵ, α
66	153	152.925761	6.4 h	ϵ, α
66	154	153.924422	3.0×10^6 y	α
66	155	154.925749	9.9 h	ϵ
66	156	155.924278	0.06%	
66	157	156.925461	8.14 h	ϵ
66	158	157.924405	0.10%	
66	159	158.925736	144.4 d	ϵ
66	160	159.925194	2.34%	
66	161	160.926930	18.9%	

Z	A	Atomic Mass (u)	$t_{1/2}$ or Abundance	Decay Mode
Dysprosium (Dy)				
66	162	161.926795	25.5%	
66	163	162.928728	24.9%	
66	164	163.929171	28.2%	
66	165	164.931700	2.334 h	β^-
66	166	165.932803	81.6 h	β^-
66	167	166.935650	6.20 m	β^-
66	168	167.937230s	8.7 m	β^-
66	169	168.940300	39 s	β^-
Holmium (Ho)				
67	149	148.933790	21.1 s	ϵ
67	150	149.933350s	72 s	ϵ
67	151	150.931681	35.2 s	ϵ, α
67	152	151.931740	161.8 s	ϵ, α
67	153	152.930195	2.02 m	ϵ, α
67	154	153.930596	11.76 m	ϵ, α
67	155	154.929079	48 m	α, ϵ
67	156	155.929710s	56 m	ϵ
67	157	156.928190	12.6 m	ϵ
67	158	157.928950	11.3 m	ϵ
67	159	158.927709	33.05 m	ϵ
67	160	159.928726	25.6 m	ϵ
67	161	160.927852	2.48 h	ϵ
67	162	161.929092	150.0 m	ϵ
67	163	162.928730	4570 y	ϵ
67	164	163.930231	29 m	ϵ, β^-
67	165	164.930319	100%	
67	166	165.932281	26.763 h	β^-
67	166m	165.932287	1.20×10^3 y	β^-
67	167	166.933126	3.1 h	β^-
67	168	167.935500	2.99 m	β^-
67	169	168.936868	4.7 m	β^-
67	170	169.939610	2.76 m	β^-
67	171	170.941460	53 s	β^-
67	172	171.944820s	25 s	β^-
Erbium (Er)				
68	149	148.942170s	4 s	$\epsilon, \epsilon p$
68	150	149.937760s	18.5 s	ϵ
68	151	150.937460s	23.5 s	ϵ
68	152	151.935080	10.3 s	α, ϵ
68	153	152.935093	37.1 s	α, ϵ
68	154	153.932777	3.73 m	ϵ, α
68	155	154.933200	5.3 m	ϵ, α
68	156	155.931020	19.5 m	ϵ
68	157	156.931950	18.65 m	ϵ, α
68	158	157.929910s	2.29 h	ϵ
68	159	158.930681	36 m	ϵ
68	160	159.929080	28.58 h	ϵ

Z	A	Atomic Mass (u)	$t_{1/2}$ or Abundance	Decay Mode		Z	A	Atomic Mass (u)	$t_{1/2}$ or Abundance	Decay Mode

Erbium (Er) (*continued*)

Ytterbium (Yb)

Z	A	Atomic Mass (u)	$t_{1/2}$ or Abundance	Decay Mode
68	161	160.930001	3.21 h	ϵ
68	162	161.928775	0.14%	
68	163	162.930029	75.0 m	ϵ
68	164	163.929197	1.61%	
68	165	164.930723	10.36 h	ϵ
68	166	165.930290	33.6%	
68	167	166.932045	22.95%	
68	168	167.932368	26.8%	
68	169	168.934588	9.40 d	β^-
68	170	169.935460	14.9%	
68	171	170.938026	7.516 h	β^-
68	172	171.939352	49.3 h	β^-
68	173	172.942400s	1.4 m	β^-
68	174	173.944340s	3.3 m	β^-

Thulium (Tm)

Z	A	Atomic Mass (u)	$t_{1/2}$ or Abundance	Decay Mode
69	155	154.939192	21.6 s	ϵ, α
69	156	155.939010	83.8 s	ϵ, α
69	157	156.936760	3.63 m	ϵ
69	158	157.937000s	3.98 m	ϵ
69	159	158.934810	9.13 m	ϵ
69	160	159.935090	9.4 m	ϵ
69	161	160.933400	30.2 m	ϵ
69	162	161.933970	21.70 m	ϵ
69	163	162.932648	1.810 h	ϵ
69	164	163.933451	2.0 m	ϵ
69	164	163.933451	5.1 m	IT, ϵ
69	165	164.932432	30.06 h	ϵ
69	166	165.933553	7.70 h	ϵ
69	167	166.932849	9.25 d	ϵ
69	168	167.934170	93.1 d	ϵ, β^-
69	169	168.934211	100%	
69	170	169.935798	128.6 d	β^-, ϵ
69	171	170.936426	1.92 y	β^-
69	172	171.938396	63.6 h	β^-
69	173	172.939600	8.24 h	β^-
69	174	173.942160	5.4 m	β^-
69	175	174.943830	15.2 m	β^-
69	176	175.946990	1.9 m	β^-
69	177	176.949040s	85 s	β^-

Ytterbium (Yb)

Z	A	Atomic Mass (u)	$t_{1/2}$ or Abundance	Decay Mode
70	156	155.942850	26.1 s	ϵ, α
70	157	156.942660	38.6 s	ϵ, α
70	158	157.939858	1.49 m	ϵ, α
70	159	158.940150	1.58 m	ϵ
70	160	159.937560s	4.8 m	ϵ
70	161	160.937850s	4.2 m	ϵ
70	162	161.935750s	18.87 m	ϵ

Ytterbium (Yb)

Z	A	Atomic Mass (u)	$t_{1/2}$ or Abundance	Decay Mode
70	163	162.936270	11.05 m	ϵ
70	164	163.934520s	75.8 m	ϵ
70	165	164.935398	9.9 m	ϵ
70	166	165.933880	56.7 h	ϵ
70	167	166.934947	17.5 m	ϵ
70	168	167.933894	0.13%	
70	169	168.935187	32.026 d	ϵ
70	170	169.934759	3.05%	
70	171	170.936322	14.3%	
70	172	171.936378	21.9%	
70	173	172.938207	16.12%	
70	174	173.938858	31.8%	
70	175	174.941272	4.185 d	β^-
70	176	175.942568	12.7%	
70	177	176.945257	1.911 h	β^-
70	178	177.946643	74 m	β^-
70	179	178.950170s	8.0 m	β^-
70	180	179.952330s	2.4 m	β^-

Lutetium (Lu)

Z	A	Atomic Mass (u)	$t_{1/2}$ or Abundance	Decay Mode
71	158	157.949170s	10.4 s	ϵ, α
71	159	158.946620	12.1 s	ϵ, α
71	160	159.946020s	36.1 s	ϵ
71	161	160.943540s	72 s	ϵ
71	162	161.943220s	1.37 m	ϵ
71	163	162.941200	238 s	ϵ
71	164	163.941220s	3.14 m	ϵ
71	165	164.939610	12 m	ϵ
71	166	165.939760	2.65 m	ϵ
71	167	166.938310	51.5 m	ϵ
71	168	167.938700	5.5 m	ϵ
71	169	168.937649	34.06 h	ϵ
71	170	169.938472	2.00 d	ϵ
71	171	170.937910	8.24 d	ϵ
71	172	171.939082	6.70 d	ϵ
71	173	172.938927	1.37 y	ϵ
71	174	173.940334	3.31 y	ϵ
71	174m	173.940517	142 d	IT, ϵ
71	175	174.940768	97.41%	
71	176	175.942682	3.73×10^{10} y 2.59%	β^-
71	177	176.943755	6.734 d	β^-
71	177m	176.944796	160.4 d	β^-, IT
71	178	177.945951	28.4 m	β^-
71	179	178.947324	4.59 h	β^-
71	180	179.949880	5.7 m	β^-
71	181	180.951970s	3.5 m	β^-
71	182	181.955210s	2.0 m	β^-
71	183	182.957570s	58 s	β^-
71	184	183.961170s	20 s	β^-

Z	A	Atomic Mass (u)	$t_{1/2}$ or Abundance	Decay Mode
Hafnium (Hf)				
72	160	159.950710	13.0 s	ϵ, α
72	161	160.950330	16.8 s	α, ϵ
72	162	161.947203	37.6 s	ϵ, α
72	163	162.947060s	40.0 s	ϵ
72	164	163.944420s	111 s	ϵ
72	165	164.944540s	76 s	ϵ
72	166	165.942250s	6.77 m	ϵ
72	167	166.942600s	2.05 m	ϵ
72	168	167.940630s	25.95 m	ϵ
72	169	168.941160	3.24 m	ϵ
72	170	169.939650s	16.01 h	ϵ
72	171	170.940490s	12.1 h	ϵ
72	172	171.939460	1.87 y	ϵ
72	173	172.940650s	23.6 h	ϵ
72	174	173.940040	2.0×10^{15} y 0.162%	α
72	175	174.941503	70 d	ϵ
72	176	175.941402	5.206%	
72	177	176.943220	18.606%	
72	178	177.943698	27.297%	
72	178m	177.946324	31 y	IT
72	179	178.945815	13.629%	
72	180	179.946549	35.100%	
72	181	180.949099	42.39 d	β^-
72	182	181.950553	9×10^6 y	β^-
72	183	182.953530	1.067 h	β^-
72	184	183.955450	4.12 h	β^-
72	185	184.958780s	3.5 m	β^-
Tantalum (Ta)				
73	163	162.954320	11.0 s	ϵ, α
73	164	163.953570s	14.2 s	ϵ
73	165	164.950820s	31.0 s	ϵ
73	166	165.950470s	31.5 s	ϵ
73	167	166.947970s	1.33 m	ϵ
73	168	167.947790s	2.0 m	ϵ
73	169	168.945920s	4.9 m	ϵ
73	170	169.946090s	6.76 m	ϵ
73	171	170.944460s	23.3 m	ϵ
73	172	171.944740	36.8 m	ϵ
73	173	172.943540s	3.14 h	ϵ
73	174	173.944170	1.05 h	ϵ
73	175	174.943650s	10.5 h	ϵ
73	176	175.944740	8.09 h	ϵ
73	177	176.944472	56.6 h	ϵ
73	178	177.945750	9.31 m	ϵ
73	178	177.945750	2.36 h	ϵ
73	179	178.945934	1.82 y	ϵ
73	180	179.947466	8.152 h	ϵ, β^-
73	180m	179.947546	1.2×10^{15} y 0.012%	β^-, ϵ

Z	A	Atomic Mass (u)	$t_{1/2}$ or Abundance	Decay Mode
Tantalum (Ta)				
73	181	180.947996	99.988%	
73	182	181.950152	114.43 d	β^-
73	183	182.951373	5.1 d	β^-
73	184	183.954009	8.7 h	β^-
73	185	184.955559	49.4 m	β^-
73	186	185.958550	10.5 m	β^-
Tungsten (W)				
74	166	165.955020	18.8 s	ϵ, α
74	167	166.954670s	19.9 s	α, ϵ
74	168	167.951860s	51 s	ϵ, α
74	169	168.951760s	80 s	ϵ
74	170	169.949290s	2.42 m	ϵ
74	171	170.949460s	2.38 m	ϵ
74	172	171.947420s	6.6 m	ϵ
74	173	172.947830s	7.6 m	ϵ
74	174	173.946160s	31 m	ϵ
74	175	174.946770s	35.2 m	ϵ
74	176	175.945590s	2.5 h	ϵ
74	177	176.946620s	135 m	ϵ
74	178	177.945850	21.6 d	ϵ
74	179	178.947072	37.05 m	ϵ
74	180	179.946706	0.12%	
74	181	180.948198	121.2 d	ϵ
74	182	181.948206	26.498%	
74	183	182.950224	1.1×10^{17} y 14.314%	
74	184	183.950933	3×10^{17} y 30.642%	α?
74	185	184.953421	75.1 d	β^-
74	186	185.954362	28.426%	
74	187	186.957158	23.72 h	β^-
74	188	187.958487	69.4 d	β^-
74	189	188.961910	11.5 m	β^-
74	190	189.963180	30.0 m	β^-
Rhenium (Re)				
75	171	170.955550s	15.2 s	ϵ
75	172m	171.955290s	15 s	ϵ
75	173	172.953060s	1.98 m	ϵ
75	174	173.953110s	2.40 m	ϵ
75	175	174.951390s	5.89 m	ϵ
75	176	175.951570s	5.3 m	ϵ
75	177	176.950270s	14.0 m	ϵ
75	178	177.950850	13.2 m	ϵ
75	179	178.949980	19.5 m	ϵ
75	180	179.950790	2.44 m	ϵ
75	181	180.950065	19.9 h	ϵ
75	182	181.951210	64.0 h	ϵ
75	183	182.950821	70.0 d	ϵ

Z	A	Atomic Mass (u)	$t_{1/2}$ or Abundance	Decay Mode	Z	A	Atomic Mass (u)	$t_{1/2}$ or Abundance	Decay Mode
Rhenium (Re) (*continued*)					**Iridium (Ir)**				
75	184	183.952524	38.0 d	ϵ	77	184	183.957390	3.09 h	ϵ
75	184m	183.952726	169 d	IT, ϵ	77	185	184.956590s	14.4 h	ϵ
75	185	184.952956	37.40%		77	186	185.957951	16.64 h	ϵ
75	186	185.954987	90 h	β^-, ϵ	77	187	186.957361	10.5 h	ϵ
75	186m	185.955147	2.0×10^5 y	IT, β^-	77	188	187.958852	41.5 h	ϵ
75	187	186.955751	4.35×10^{10} y	β^-, α	77	189	188.958716	13.2 d	ϵ
			62.60%		77	190	189.960590	11.78 d	ϵ
75	188	187.958112	17.021 h	β^-	77	191	190.960591	37.3%	
75	189	188.959228	24.3 h	β^-	77	192	191.962602	73.830 d	β^-, ϵ
75	190	189.961820	3.1 m	β^-	77	192m	191.962768	241 y	IT
75	191	190.963124	9.8 m	β^-	77	193	192.962924	62.7%	
75	192	191.965960s	16 s	β^-	77	194	193.965076	19.15 h	β^-
					77	194m	193.965280	171 d	β^-
					77	195	194.965977	2.5 h	β^-
Osmium (Os)					77	196	195.968380	52 s	β^-
76	172	171.960080s	19.2 s	ϵ, α	77	197	196.969636	5.8 m	β^-
76	173	172.959790s	16 s	ϵ, α					
76	174	173.957120s	44 s	ϵ, α					
76	175	174.957080s	1.4 m	ϵ	**Platinum (Pt)**				
76	176	175.954950s	3.6 m	ϵ	78	177	176.968450s	11 s	ϵ, α
76	177	176.955050s	2.8 m	ϵ	78	178	177.965710s	21.1 s	ϵ, α
76	178	177.953350	5.0 m	ϵ	78	179	178.965480s	21.2 s	ϵ, α
76	179	178.953950s	6.5 m	ϵ	78	180	179.963220s	52 s	ϵ, α
76	180	179.952350s	21.5 m	ϵ	78	181	180.963180s	51 s	ϵ, α
76	181	180.953270	105 m	ϵ	78	182	181.961270	2.2 m	ϵ, α
76	182	181.952186	22.10 h	ϵ	78	183	182.961730s	6.5 m	ϵ, α
76	183	182.953110s	13.0 h	ϵ	78	184	183.959900s	17.3 m	ϵ, α
76	184	183.952491	5.6×10^{13} y	α	78	185	184.960750	70.9 m	ϵ
			0.020%		78	186	185.959430	2.0 h	ϵ, α
76	185	184.954043	93.6 d	ϵ	78	187	186.960560s	2.35 h	ϵ
76	186	185.953838	2.0×10^{15} y	α	78	188	187.959396	10.2 d	ϵ, α
			1.58%		78	189	188.960832	10.87 h	ϵ
76	187	186.955748	1.6%		78	190	189.959930	6.5×10^{11} y	α
76	188	187.955836	13.3%					0.01%	
76	189	188.958145	16.1%		78	191	190.961685	2.96 d	ϵ
76	190	189.958445	26.4%		78	192	191.961035	0.79%	
76	191	190.960928	15.4 d	β^-	78	193	192.962985	50 y	ϵ
76	192	191.961479	41.0%		78	194	193.962664	32.9%	
76	193	192.964148	30.11 h	β^-	78	195	194.964774	33.8%	
76	194	193.965179	6.0 y	β^-	78	196	195.964935	25.3%	
76	195	194.968120	6.5 m	β^-	78	197	196.967323	19.8915 h	β^-
76	196	195.969620	34.9 m	β^-	78	198	197.967876	7.2%	
					78	199	198.970576	30.8 m	β^-
Iridium (Ir)					78	200	199.971424	12.5 h	β^-
					78	201	200.974500	2.5 m	β^-
77	177	176.961170s	30 s	ϵ, α	78	202	201.975740s	44 h	β^-
77	178	177.961080s	12 s	ϵ					
77	179	178.959150s	79 s	ϵ					
77	180	179.959250s	1.5 m	ϵ	**Gold (Au)**				
77	181	180.957640	4.90 m	ϵ	79	181	180.969950s	11.4 s	ϵ, α
77	182	181.958130	15 m	ϵ	79	182	181.969620s	15.6 s	ϵ, α
77	183	182.956810s	57 m	ϵ					

Z	A	Atomic Mass (u)	$t_{1/2}$ or Abundance	Decay Mode	Z	A	Atomic Mass (u)	$t_{1/2}$ or Abundance	Decay Mode
Gold (Au) (*continued*)					**Thallium (Tl)**				
79	183	182.967620s	42.0 s	ϵ, α	81	184	183.981760s	11 s	ϵ, α
79	184	183.967470s	12.0 s	ϵ	81	185	184.979100s	19.5 s	ϵ
79	185	184.965810	4.25 m	ϵ, α	81	186	185.978550s	27.5 s	ϵ, α
79	186	185.966000	10.7 m	ϵ	81	187	186.976170s	51 s	ϵ, α
79	187	186.964560s	8.4 m	ϵ	81	189	188.973690s	2.3 m	ϵ
79	188	187.965090s	8.84 m	ϵ	81	193	192.970550s	21.6 m	ϵ
79	189	188.963890s	28.7 m	ϵ, α	81	194	193.971050s	33.0 m	ϵ, α
79	190	189.964699	42.8 m	ϵ, α	81	195	194.969650s	1.16 h	ϵ
79	191	190.963650	3.18 h	ϵ	81	196	195.970520s	1.84 h	ϵ
79	192	191.964810	4.94 h	ϵ	81	197	196.969540	2.84 h	ϵ
79	193	192.964132	17.65 h	ϵ	81	198	197.970470	5.3 h	ϵ
79	194	193.965339	38.02 h	ϵ	81	199	198.969810	7.42 h	ϵ
79	195	194.965018	186.10 d	ϵ	81	200	199.970945	26.1 h	ϵ
79	196	195.966551	6.183 d	ϵ, β^-	81	201	200.970804	72.912 h	ϵ
79	197	196.966552	100%		81	202	201.972091	12.23 d	ϵ
79	198	197.968225	2.6952 d	β^-	81	203	202.972329	29.524%	
79	199	198.9687848	3.139 d	β^-	81	204	203.973849	3.78 y	β^-, ϵ
79	200	199.970720	48.4 m	β^-	81	205	204.974412	70.476%	
79	201	200.971641	26 m	β^-	81	206	205.976095	4.199 m	β^-
79	202	201.973790	28.8 s	β^-	81	207	206.977408	4.77 m	β^-
79	203	202.975137	60 s	β^-	81	208	207.982005	3.053 m	β^-
79	204	203.977710s	39.8 s	β^-	81	209	208.985349	2.20 m	β^-
79	205	204.979610s	31 s	β^-	81	210	209.990066	1.30 m	$\beta^-, \beta^- n$
Mercury (Hg)					**Lead (Pb)**				
80	182	181.974750s	10.83 s	ϵ, α	82	183	182.991930s	300 ms	α, ϵ
80	184	183.971900s	30.6 s	ϵ, α	82	184	183.988200s	0.55 s	α
80	185	184.971980s	49.1 s	ϵ, α	82	185	184.987580s	4.1 s	α
80	186	185.969460	1.38 m	ϵ, α	82	186	185.984300s	4.7 s	α
80	187	186.969790s	2.4 m	ϵ, α	82	188	187.981060s	25.5 s	ϵ, α
80	188	187.967560s	3.25 m	ϵ, α	82	189	188.980880s	51 s	ϵ, α
80	189	188.968130s	7.6 m	ϵ, α	82	190	189.978180	1.2 m	ϵ, α
80	190	189.966280s	20.0 m	ϵ, α	82	191	190.978200s	1.33 m	ϵ, α
80	191	190.967060	49 m	ϵ	82	192	191.975760s	3.5 m	ϵ, α
80	192	191.965570s	4.85 h	ϵ	82	194	193.973970s	12.0 m	ϵ, α
80	193	192.966644	3.80 h	ϵ	82	195	194.974470s	15 m	ϵ
80	194	193.965382	520 y	ϵ	82	196	195.972710s	37 m	ϵ, α
80	195	194.966640	9.9 h	ϵ	82	197	196.973380s	8 m	ϵ
80	196	195.965815	0.15%		82	198	197.971980s	2.40 h	ϵ
80	197	196.967195	64.14 h	ϵ	82	199	198.972910	90 m	ϵ
80	198	197.966752	9.97%		82	200	199.971816	21.5 h	ϵ
80	199	198.968262	16.87%		82	201	200.972850	9.33 h	ϵ
80	200	199.968309	23.10%		82	202	201.972144	52.5×10^3 y	ϵ, α
80	201	200.970285	13.18%		82	203	202.973375	51.873 h	ϵ
80	202	201.970626	29.86%		82	204	203.973029	1.4×10^{17} y	α
80	203	202.972857	46.612 d	β^-				1.4%	
80	204	203.973476	6.87%		82	205	204.974467	1.52×10^7 y	ϵ
80	205	204.976056	5.2 m	β^-	82	206	205.974449	24.1%	
80	206	205.977499	8.15 m	β^-	82	207	206.975881	22.1%	
80	207	206.982580	2.9 m	β^-	82	208	207.976636	52.4%	
80	208	207.985940s	42 m	β^-	82	209	208.981075	3.253 h	β^-

Z	A	Atomic Mass (u)	$t_{1/2}$ or Abundance	Decay Mode
Lead (Pb) (*continued*)				
82	210	209.984173	22.3 y	β^-, α
82	211	210.988731	36.1 m	β^-
82	212	211.991888	10.64 h	β^-
82	213	212.996500s	10.2 m	β^-
82	214	213.999798	26.8 m	β^-
Bismuth (Bi)				
83	191	190.986050s	12 s	α, ϵ
83	192	191.985370s	34.6 s	ϵ, α
83	193	192.983060s	67 s	ϵ, α
83	194	193.982750s	95 s	ϵ
83	195	194.980750s	183 s	ϵ, α
83	196	195.980610s	308 s	ϵ, α
83	197	196.978930	9.33 m	ϵ, α
83	198	197.979020	10.3 m	ϵ
83	199	198.977580	27 m	ϵ
83	200	199.978140	36.4 m	ϵ
83	201	200.976970	108 m	ϵ, α
83	202	201.977670	1.72 h	ϵ, α
83	203	202.976868	11.76 h	ϵ, α
83	204	203.977805	11.22 h	ϵ
83	205	204.977375	15.31 d	ϵ
83	206	205.978483	6.243 d	ϵ
83	207	206.978455	31.55 y	ϵ
83	208	207.979727	3.68×10^5 y	ϵ
83	209	208.980383	100%	
83	210	209.984105	5.013 d	β^-, α
83	210m	209.984396	3.04×10^6 y	α
83	211	210.987258	2.14 m	α, β^-
83	212	211.991272	60.55 m	$\beta^-, \alpha, \beta^-\alpha$
83	213	212.994375	45.59 m	β^-, α
83	214	213.998699	19.9 m	β^-, α
83	215	215.001830	7.6 m	β^-
83	216	216.006200s	3.6 m	β^-
Polonium (Po)				
84	194	193.988280	0.392 s	α
84	195	194.988050s	4.64 s	α, ϵ
84	196	195.985510s	5.8 s	α, ϵ
84	197	196.985570s	53.6 s	ϵ, α
84	198	197.983340s	1.76 m	α, ϵ
84	199	198.983600s	5.48 m	ϵ, α
84	200	199.981740s	11.5 m	ϵ, α
84	201	200.98210s	15.3 m	ϵ, α
84	202	201.980700s	44.7 m	ϵ, α
84	203	202.981410	36.7 m	ϵ, α
84	204	203.980307	3.53 h	ϵ, α
84	205	204.981170	1.66 h	ϵ, α
84	206	205.980465	8.8 d	ϵ, α
84	207	206.981578	5.80 h	ϵ, α

Z	A	Atomic Mass (u)	$t_{1/2}$ or Abundance	Decay Mode
Polonium (Po)				
84	208	207.981231	2.898 y	α, ϵ
84	209	208.982416	102 y	α, ϵ
84	210	209.982857	138.376 d	α
84	211	210.986637	0.516 s	α
84	212	211.988852	0.299 μs	α
84	213	212.992843	4.2 μs	α
84	214	213.995186	164.3 μs	α
84	215	214.999415	1.781 ms	α, β^-
84	216	216.001905	0.145 s	α
84	217	217.006250s	10 s	α, β^-
84	218	218.008966	3.10 m	α, β^-
Astatine (At)				
85	196	195.995700s	0.3 s	α
85	197	196.993290s	0.35 s	α, ϵ
85	198	197.992750s	4.2 s	α, ϵ
85	199	198.990630s	7.2 s	α, ϵ
85	200	199.990290s	43 s	α, ϵ
85	201	200.988490	89 s	α, ϵ
85	202	201.988450	184 s	ϵ, α
85	203	202.986850	7.4 m	ϵ, α
85	204	203.987260	9.2 m	ϵ, α
85	205	204.986040	26.2 m	ϵ, α
85	206	205.986600	30.0 m	ϵ, α
85	207	206.985776	1.80 h	ϵ, α
85	208	207.986583	1.63 h	ϵ, α
85	209	208.986159	5.41 h	ϵ, α
85	210	209.987131	8.1 h	ϵ, α
85	211	210.987481	7.214 h	ϵ, α
85	212	211.990735	0.314 s	$\alpha, \epsilon, \beta^-$
85	213	212.992921	125 ns	α
85	214	213.996356	558 ns	α
85	215	214.998641	0.10 ms	α
85	216	216.002409	0.30 ms	$\alpha, \epsilon, \beta^-$
85	217	217.004710	32.3 ms	α, β^-
85	218	218.008681	1.6 s	α, β^-
85	219	219.011300	56 s	α, β^-
85	220	220.015300s	3.71 m	β^-
85	221	221.018140s	2.3 m	β^-
85	222	222.022330s	54 s	β^-
85	223	223.025340s	50 s	β^-
Radon (Rn)				
86	199	198.998310s	0.62 s	α, ϵ
86	200	199.995680s	1.06 s	α, ϵ
86	201	200.995540s	7.0 s	α, ϵ
86	202	201.993220s	9.85 s	ϵ, α
86	203	202.993320s	45 s	α, ϵ
86	204	203.991370s	1.24 m	α, ϵ
86	205	204.991670s	170 s	ϵ, α

Z	A	Atomic Mass (u)	$t_{1/2}$ or Abundance	Decay Mode		Z	A	Atomic Mass (u)	$t_{1/2}$ or Abundance	Decay Mode
Radon (Rn) (*continued*)						**Francium (Fr)**				
86	206	205.990160s	5.67 m	α, ϵ		87	227	227.031830	2.47 m	β^-
86	207	206.990730	9.25 m	ϵ, α		87	228	228.035720s	39 s	β^-
86	208	207.989631	24.35 m	α, ϵ		87	229	229.038430s	50 s	β^-
86	209	208.990380	28.5 m	ϵ, α		87	230	230.042510s	19.1 s	β^-
86	210	209.989680	2.4 h	α, ϵ		87	231	231.045410s	17.5 s	β^-
86	211	210.990585	14.6 h	ϵ, α						
86	212	211.990689	23.9 m	α		**Radium (Ra)**				
86	213	212.993868	25.0 ms	α						
86	214	213.995346	0.27 μs	α		88	206	206.003780s	0.24 s	α
86	215	214.998729	2.30 μs	α		88	207	207.003730s	1.3 s	α, ϵ
86	216	216.000258	45 μs	α		88	208	208.001780s	1.7 s	α, ϵ
86	217	217.003915	0.54 ms	α		88	209	209.001940s	4.6 s	α, ϵ
86	218	218.005586	35 ms	α		88	210	210.000450s	3.7 s	α, ϵ
86	219	219.009475	3.96 s	α		88	211	211.000890	13 s	α, ϵ
86	220	220.011384	55.6 s	α		88	212	211.999783	13.0 s	α, ϵ
86	221	221.015460s	25 m	β^-, α		88	213	213.000350	2.74 m	α, ϵ
86	222	222.017570	3.8235 d	α		88	214	214.000091	2.46 s	α, ϵ
86	223	223.021790s	23.2 m	β^-		88	215	215.002704	1.59 ms	α
86	224	224.024090s	107 m	β^-		88	216	216.003518	182 ns	α, ϵ
86	225	225.028440s	4.5 m	β^-		88	217	217.006306	1.7 μs	α
86	226	226.030890s	6.0 m	β^-		88	218	218.007124	15.6 μs	α
86	227	227.035410s	22.5 s	β^-		88	219	219.010069	10 ms	α
86	228	228.038080s	65 s	β^-		88	220	220.011015	17 ms	α
						88	221	221.013908	28 s	$\alpha, ^{14}C$
						88	222	222.015362	38.0 s	$\alpha, ^{14}C$
						88	223	223.018497	11.435 d	$\alpha, ^{14}C$
Francium (Fr)						88	224	224.020202	3.66 d	$\alpha, ^{12}C$
						88	225	225.023604	14.9 d	β^-
87	201	201.003990s	48 ms	α, ϵ		88	226	226.025403	1600 y	$\alpha, ^{14}C$
87	202	202.003290s	0.34 s	α, ϵ		88	227	227.029171	42.2 m	β^-
87	203	203.001050s	0.55 s	α, ϵ		88	228	228.031064	5.75 y	β^-
87	204	204.000590s	1.7 s	α, ϵ		88	229	229.034820	4.0 m	β^-
87	205	204.998660	3.85 s	α, ϵ		88	230	230.037080	93 m	β^-
87	206	205.998490	15.9 s	α, ϵ		88	231	231.041220s	1.72 m	β^-
87	207	206.996860	14.8 s	α, ϵ		88	232	232.043690s	250 s	β^-
87	208	207.997130	59.1 s	α, ϵ		88	233	233.048000s	30 s	β^-
87	209	208.995920	50.0 s	α, ϵ		88	234	234.050550s	30 s	β^-
87	210	209.996398	3.18 m	α, ϵ						
87	211	210.995529	3.10 m	α, ϵ						
87	212	211.996195	20.0 m	ϵ, α		**Actinium (Ac)**				
87	213	212.996175	34.6 s	α, ϵ						
87	214	213.998955	5.0 ms	α		89	209	209.009570	0.10 s	α, ϵ
87	215	215.000326	86 ns	α		89	210	210.009260	0.35 s	α, ϵ
87	216	216.003188	0.70 μs	α, ϵ		89	211	211.007650	0.25 s	α
87	217	217.004616	16 μs	α		89	212	212.007810	0.93 s	α, ϵ
87	218	218.007563	1.0 ms	α		89	213	213.006570	0.80 s	α
87	219	219.009241	20 ms	α		89	214	214.006890	8.2 s	α, ϵ
87	220	220.012313	27.4 s	α, β^-		89	215	215.006450	0.17 s	α, ϵ
87	221	221.014246	4.9 m	α, β^-		89	216	216.008721	0.33 ms	α
87	222	222.017544	14.2 m	β^-		89	217	217.009333	69 ns	α, ϵ
87	223	223.019731	22.00 m	β^-, α		89	218	218.011630	1.06 μs	α
87	224	224.023240	3.30 m	β^-		89	219	219.012400	11.8 μs	α
87	225	225.025607	4.0 m	β^-		89	220	220.014750	26.1 ms	α, ϵ
87	226	226.029340	48 s	β^-						

Z	A	Atomic Mass (u)	$t_{1/2}$ or Abundance	Decay Mode
Actinium (Ac) (*continued*)				
89	221	221.015580	52 ms	α
89	222	222.017829	5.0 s	α, ϵ
89	223	223.019126	2.10 m	α, ϵ
89	224	224.021708	2.9 h	$\epsilon, \alpha, \beta^-$
89	225	225.023221	10.0 d	α
89	226	226.026090	29.4 h	$\beta^-, \epsilon, \alpha$
89	227	227.027747	21.773 y	β^-, α
89	228	228.031015	6.15 h	β^-, α
89	229	229.032930	62.7 m	β^-
89	230	230.036030	122 s	β^-
89	231	231.038550	7.5 m	β^-
89	232	232.042020	119 s	β^-
89	233	233.044550s	145 s	β^-
89	234	234.048420s	44 s	β^-
Thorium (Th)				
90	212	212.012920s	30 ms	α, ϵ
90	213	213.012960s	140 ms	α
90	214	214.011450s	100 ms	α
90	215	215.011730	1.2 s	α
90	216	216.011051	0.028 s	α, ϵ
90	217	217.013070	0.252 ms	α
90	218	218.013268	109 ns	α
90	219	219.015520	1.05 μs	α
90	220	220.015733	9.7 μs	α, ϵ
90	221	221.018171	1.68 ms	α
90	222	222.018454	2.2 ms	α
90	223	223.020795	0.60 s	α
90	224	224.021459	1.05 s	α
90	225	225.023941	8.72 m	α, ϵ
90	226	226.024891	30.6 m	α
90	227	227.027699	18.72 d	α
90	228	228.028731	1.9131 y	α
90	229	229.031755	7880 y	α
90	230	230.033127	7.538×10^4 y	α, SF
90	231	231.036297	25.52 h	β^-, α
90	232	232.038050	1.405×10^{10} y 100%	α, SF
90	233	233.041577	22.3 m	β^-
90	234	234.043595	24.10 d	β^-
90	235	235.047500	7.1 m	β^-
90	236	236.049710s	37.5 m	β^-
90	237	237.053890s	5.0 m	β^-
Protactinium (Pa)				
91	215	215.019100	15 ms	α
91	216	216.019110	105 ms	α, ϵ
91	217	217.018290	3.4 ms	α
91	218	218.020010	0.11 ms	α

Z	A	Atomic Mass (u)	$t_{1/2}$ or Abundance	Decay Mode
Protactinium (Pa)				
91	219	219.019880	53 ns	α
91	220	220.021880	0.78 μs	α
91	221	221.021860	5.9 μs	α
91	222	222.023730s	3.3 ms	α
91	223	223.023960	5 ms	α
91	224	224.025610	0.95 s	α, ϵ
91	225	225.026120	1.7 s	α
91	226	226.027933	1.8 m	α, ϵ
91	227	227.028793	38.3 m	α, ϵ
91	228	228.031037	22 h	ϵ, α
91	229	229.032089	1.50 d	ϵ, α
91	230	230.034533	17.4 d	$\epsilon, \beta^-, \alpha$
91	231	231.035879	3.276×10^4 y	α, SF
91	232	232.038582	1.31 d	β^-, ϵ
91	233	233.040240	26.967 d	β^-
91	234	234.043302	6.70 h	β^-
91	235	235.045440	24.5 m	β^-
91	236	236.048680	9.1 m	β^-
91	237	237.051140	8.7 m	β^-
91	238	238.054500	2.3 m	β^-
Uranium (U)				
92	225	225.029380	95 ms	α
92	226	226.029340	0.20 s	α
92	227	227.031140	1.1 m	α
92	228	228.031366	9.1 m	α, ϵ
92	229	229.033496	58 m	ϵ, α
92	230	230.033927	20.8 d	α
92	231	231.036289	4.2 d	ϵ, α
92	232	232.037146	68.9 y	α
92	233	233.039628	1.592×10^5 y	α, SF
92	234	234.040946	2.455×10^5 y 0.0055%	α, SF
92	235	235.043923	703.8×10^6 y 0.720%	α, SF
92	236	236.045562	2.342×10^7 y	α, SF
92	237	237.048724	6.75 d	β^-
92	238	238.050783	4.468×10^9 y 99.2745%	α, SF
92	239	239.054288	23.45 m	β^-
92	240	240.056586	14.1 h	β^-, α
92	242	242.062930s	16.8 m	β^-
Neptunium (Np)				
93	227	227.034960	0.51 s	α
93	228	228.036180s	1.07 m	ϵ, ϵSF
93	229	229.036250	3.85 m	α, ϵ
93	230	230.037810	4.6 m	ϵ, α
93	231	231.038230	48.8 m	ϵ, α
93	232	232.040100s	14.7 m	ϵ

Z	A	Atomic Mass (u)	$t_{1/2}$ or Abundance	Decay Mode
Neptunium (Np) *(continued)*				
93	233	233.040730	36.2 m	ϵ, α
93	234	234.042889	4.4 d	ϵ
93	235	235.044056	396.1 d	ϵ, α
93	236	236.046560	154×10^3 y	$\epsilon, \beta^-, \alpha$
93	237	237.048167	2.144×10^6 y	α, SF
93	238	238.050940	2.117 d	β^-
93	239	239.052931	2.3565 d	β^-
93	240	240.056169	61.9 m	β^-
93	240m	240.056169	7.22 m	β^-, IT
93	241	241.058250	13.9 m	β^-
93	242	242.061640s	2.2 m	β^-
93	242	242.061640s	5.5 m	β^-
93	243	243.064270s	1.8 m	β^-
93	244	244.067850s	2.29 m	
Plutonium (Pu)				
94	230	230.039646	200 s	α
94	232	232.041179	34.1 m	ϵ, α
94	233	233.042990	20.9 m	ϵ, α
94	234	234.043305	8.8 h	ϵ, α
94	235	235.045282	25.3 m	ϵ, α
94	236	236.046048	2.858 y	α, SF
94	237	237.048404	45.2 d	ϵ, α
94	238	238.049553	87.7 y	α, SF
94	239	239.052156	24110 y	α, SF
94	240	240.053808	6564 y	α, SF
94	241	241.056845	14.35 y	β^-, α, SF
94	242	242.058737	3.733×10^5 y	α, SF
94	243	243.061997	4.956 h	β^-
94	244	244.064198	8.08×10^7 y	α, SF
94	245	245.067739	10.5 h	β^-
94	246	246.070198	10.84 d	β^-
94	247	247.074070s	2.27 d	β^-
Americium (Am)				
95	232	232.046590s	79 s	ϵ, α
95	234	234.047790s	2.32 m	α, ϵ
95	237	237.049970	73.0 m	ϵ, α
95	238	238.051980	98 m	ϵ, α
95	239	239.053018	11.9 h	ϵ, α
95	240	240.055288	50.8 h	ϵ, α
95	241	241.056823	432.7 y	α, SF
95	242	242.059543	16.02 h	β^-, ϵ
95	242m	242.059596	141 y	IT, α, SF
95	243	243.061373	7370 y	α, SF
95	244	244.064279	10.1 h	β^-
95	245	245.066445	2.05 h	β^-
95	246	246.069768	39 m	β^-
95	247	247.072090s	23.0 m	β^-

Z	A	Atomic Mass (u)	$t_{1/2}$ or Abundance	Decay Mode
Curium (Cm)				
96	238	238.053020	2.4 h	ϵ, α
96	239	239.054950s	2.9 h	ϵ, α
96	240	240.055519	27 d	α, ϵ, SF
96	241	241.057647	32.8 d	ϵ, α
96	242	242.058829	162.79 d	α, SF
96	243	243.061382	29.1 y	α, ϵ, SF
96	244	244.062746	18.10 y	α, SF
96	245	245.065486	8500 y	α, SF
96	246	246.067218	4730 y	α, SF
96	247	247.070347	1.56×10^7 y	α
96	248	248.072342	3.40×10^5 y	α, SF
96	249	249.075947	64.15 m	β^-
96	250	250.078351	9700 y	SF, α, β^-
96	251	251.082278	16.8 m	β^-
96	252	252.084870s	2 d	β^-
Berkelium (Bk)				
97	240	240.059750s	4.8 m	ϵ, ϵSF
97	242	242.062050s	7.0 m	ϵ
97	243	243.063002	4.5 h	ϵ, α
97	244	244.065168	4.35 h	ϵ, α
97	245	245.066355	4.94 d	ϵ, α
97	246	246.068670	1.80 d	ϵ, α
97	247	247.070299	1380 y	α
97	248	248.073080s	23.7 h	$\beta^-, \epsilon, \alpha$
97	248	248.073080s	9 y	α
97	249	249.074980	320 d	β^-, α, SF
97	250	250.078311	3.217 h	β^-
97	251	251.080753	55.6 m	β^-, α
Californium (Cf)				
98	240	240.062300s	1.06 m	α
98	241	241.063720s	3.78 m	ϵ, α
98	242	242.063690	3.49 m	α
98	243	243.065420s	10.7 m	ϵ, α
98	244	244.065990	19.4 m	α
98	245	245.068040s	45.0 m	ϵ, α
98	246	246.068799	35.7 h	α, ϵ, SF
98	247	247.070992	3.11 h	ϵ, α
98	248	248.072178	333.5 d	α, SF
98	249	249.074847	351 y	α, SF
98	250	250.076400	13.08 y	α, SF
98	251	251.079580	898 y	α
98	252	252.081620	2.645 y	α, SF
98	253	253.085127	17.81 d	β^-, α
98	254	254.087316	60.5 d	SF, α
98	255	255.091040s	85 m	β^-
98	256	256.093440s	12.3 m	SF, β^-, α

Z	A	Atomic Mass (u)	$t_{1/2}$ or Abundance	Decay Mode
Einsteinium (Es)				
99	243	243.069630s	21 s	ϵ, α
99	244	244.070970s	37 s	ϵ, α
99	245	245.071320s	1.1 m	ϵ, α
99	246	246.072970s	7.7 m	ϵ, α
99	247	247.073650s	4.55 m	ϵ, α
99	248	248.075460s	27 m	ϵ, α
99	249	249.076410s	102.2 m	ϵ, α
99	250	250.078650s	8.6 h	ϵ, α
99	251	251.079984	33 h	ϵ, α
99	252	252.082970	471.7 d	$\alpha, \epsilon, \beta^-$
99	253	253.084818	20.47 d	α, SF
99	254	254.088016	275.7 d	α, ϵ, SF, β^-
99	255	255.090266	39.8 d	β^-, α, SF
99	256	256.093590s	25.4 m	β^-
Fermium (Fm)				
100	242	242.073430s	0.8 ms	SF
100	243	243.074510s	0.18 s	α
100	244	244.074080s	3.3 ms	SF
100	245	245.075380s	4.2 s	α, SF
100	246	246.075280	1.1 s	α, SF, ϵ
100	248	248.077184	36 s	α, ϵ, SF
100	249	249.079020s	2.6 m	ϵ, α
100	250	250.079515	30 m	α, ϵ, SF
100	251	251.081566	5.30 h	ϵ, α
100	252	252.082460	25.39 h	α, SF
100	253	253.085176	3.00 d	ϵ, α
100	254	254.086848	3.240 h	α, SF
100	255	255.089955	20.07 h	α, SF
100	256	256.091767	157.6 m	SF, α
100	257	257.095099	100.5 d	α, SF
100	258	258.097070s	370 μs	SF
100	259	259.100590s	1.5 s	SF
Mendelevium (Md)				
101	247	247.081800s	2.9 s	α
101	248	248.082910s	7 s	ϵ, α, SF
101	249	249.083000s	24 s	α, ϵ
101	250	250.084490s	52 s	ϵ, α
101	251	251.084920s	4.0 m	ϵ, α
101	252	252.086630s	4.8 m	ϵ
101	253	253.087280s	6 m	ϵ
101	254	254.089730s	10 m	ϵ
101	254	254.089730s	28 m	ϵ
101	255	255.091075	27 m	ϵ, α, SF
101	256	256.094050	78.1 m	ϵ, α, SF
101	257	257.095535	5.52 h	ϵ, α, SF
101	258	258.098425	60 m	ϵ
101	258	258.098425	51.5 d	α, SF

Z	A	Atomic Mass (u)	$t_{1/2}$ or Abundance	Decay Mode
Mendelevium (Md)				
101	259	259.100500s	1.60 h	SF, α
101	260	260.103650s	27.8 d	SF, $\alpha, \epsilon, \beta^-$
Nobelium (No)				
102	250	250.087490s	0.25 ms	SF, α
102	251	251.088960s	0.8 s	α, ϵ, SF
102	252	252.088966	2.30 s	α, SF
102	253	253.090650s	1.7 m	α, ϵ
102	254	254.090949	55 s	α, ϵ, SF
102	255	255.093232	3.1 m	α, ϵ
102	256	256.094276	2.91 s	α, SF
102	257	257.096850	25 s	α
102	258	258.098200s	1.2 ms	SF, α
102	259	259.101020s	58 m	α, ϵ, SF
102	260	260.102640s	106 ms	SF
Lawrencium (Lr)				
103	252	252.095330s	1 s	α, ϵ, SF
103	253	253.095260s	1.3 s	α, SF, ϵ
103	254	254.096590s	13 s	α, ϵ, SF
103	255	255.096770s	22 s	α, ϵ
103	256	256.098760s	28 s	α, ϵ, SF
103	257	257.099610s	0.646 s	α, SF
103	258	258.101880s	3.9 s	α, ϵ, SF
103	259	259.102990s	6.1 s	α, SF, ϵ
103	260	260.105570s	180 s	α, ϵ, SF
103	261	261.106940s	39 m	SF
103	262	262.109690s	3.6 h	ϵ, SF
Rutherfordium (Rf)				
104	253	253.100680s	1.8 s	α, SF
104	254	254.100170s	0.5 ms	SF, α
104	255	255.101490s	1.5 s	SF, α
104	256	256.101180	6.7 ms	SF, α
104	257	257.103070s	4.7 s	α, ϵ, SF
104	258	258.103570s	12 ms	SF, α
104	259	259.105630s	3.1 s	α, SF, ϵ
104	260	260.106430s	20.1 ms	SF, α
104	261	261.108750s	65 s	α, ϵ, SF
104	262	262.109920s	1.2 s	SF
Dubnium (Db)				
105	255	255.107400s	1.6 s	α, SF
105	256	256.108110s	2.6 s	α, SF, ϵ
105	257	257.107860s	1.3 s	α, SF, ϵ
105	258	258.109440s	4.4 s	α, ϵ, SF
105	258	258.109440s	20 s	ϵ
105	260	260.111430s	1.52 s	α, SF, ϵ

Z	A	Atomic Mass (u)	$t_{1/2}$ or Abundance	Decay Mode
Dubnium (Db)) (*continued***)**				
105	261	261.112110s	1.8 s	α, SF
105	262	262.114150s	34 s	α, SF, ϵ
105	263	263.115080s	27 s	SF, α
Seaborgium (Sg)				
106	259	259.114650s	0.9 s	α, SF
106	260	260.114440	3.6 ms	α, SF
106	261	261.116200s	0.23 s	α, SF
106	263	263.118310s	0.8 s	SF, α
106	265	265.121070s	16 s	α, SF
106	266	266.121930s	20 s	α, SF
Bohrium (Bh)				
107	261	261.121800s	11.8 ms	α, SF
107	262	262.123010s	102 ms	α, SF

Z	A	Atomic Mass (u)	$t_{1/2}$ or Abundance	Decay Mode
Hassium (Hs)				
108	264	264.128410	0.08 ms	α
108	265	265.130000s	1.8 ms	α
108	267	267.131770s	60 ms	α
Meitnerium (Mt)				
109	266	266.137940s	3.4 ms	α, SF
109	268	268.138820s	70 ms	α
Darmstadtium (Ds)				
110	269	269.145140s	0.17 ms	α
110	271	271.146080s	1.1 ms	α
110	272	272.146310s	8.6 ms	SF
Roentgenium (Rg)				
111	272	272.153480s	1.5 ms	α

Nobel Laureates in Physics

The following list gives the names and short descriptions of award citations* for all the physics laureates and a few chemistry laureates whose work was related to physics (denoted by C in front of their name).

Year	Nobel Laureate		Citation for
1901	Wilhelm Konrad Röntgen	1845–1923	Discovery of x rays
1902	Hendrik Antoon Lorentz	1853–1928	Their researches into the influence of magnetism upon radiation phenomena
	Pieter Zeeman	1865–1943	
1903	Antoine Henri Becquerel	1852–1908	His discovery of spontaneous radioactivity
	Pierre Curie	1859–1906	Their joint researches on the radiation phenomena discovered by Prof. Henri Becquerel
	Marie Sklowdowska-Curie	1867–1934	
1904	John William Strutt (Lord Rayleigh)	1842–1919	Investigations of the densities of the most important gases and his discovery of argon
	C Sir William Ramsay	1851–1939	His discovery of the inert gaseous elements in air and his determination of their place in the periodic system
1905	Philipp Eduard Anton von Lenard	1862–1947	His work on cathode rays
1906	Joseph John Thomson	1856–1940	His theoretical and experimental investigations on the conduction of electricity by gases
1907	Albert Abraham Michelson	1852–1931	His optical precision instruments and the spectroscopic and metrological investigations carried out with their aid
1908	Gabriel Lippman	1845–1921	His method of reproducing colors photographically based on the phenomena of interference
	C Ernest Rutherford	1871–1937	His investigations into the disintegration of the elements and the chemistry of radioactive substances
1909	Guglielmo Marconi	1874–1937	Their contributions to the development of wireless telegraphy
	Carl Ferdinand Braun	1850–1918	
1910	Johannes Diderik van der Waals	1837–1923	His work on the state of equations of gases and liquids
1911	Wilhelm Wien	1864–1928	His discoveries regarding the laws governing the radiation of heat
	C Marie Curie	1867–1934	Her services to the advancement of chemistry by the discovery of the elements radium and polonium, and by the isolation of radium and the study of its nature and compounds

*From the Nobel Foundation website: http://nobelprize.org/index.html

Year	Nobel Laureate		Citation for
1912	Nils Gustaf Dalén	1869–1937	His invention of automatic regulators for use in conjunction with gas accumulators for illuminating lighthouses and buoys
1913	Heike Kamerlingh Onnes	1853–1926	His investigations of the properties of matter at low temperatures, which led, *inter alia*, to the production of liquid helium
1914	Max von Laue	1879–1960	His discovery of the diffraction of x rays by crystals
1915	William Henry Bragg William Lawrence Bragg	1862–1942 1890–1971	Their analysis of crystal structure by means of x rays
1917	Charles Glover Barkla	1877–1944	His discovery of the characteristic x rays of the elements
1918	Max Planck	1858–1947	His discovery of energy quanta
1919	Johannes Stark	1874–1957	His discovery of the Doppler effect in canal rays and of the splitting of spectral lines in electric fields
1920	Charles-Édouard Guillaume	1861–1938	The service he has rendered to precise measurement in physics by his discovery of anomalies in nickel steel alloys
1921	Albert Einstein	1879–1955	His services to Theoretical Physics, and especially for his discovery of the law of the photoelectric effect
	C Frederick Soddy	1877–1956	His contributions to our knowledge of the chemistry of radioactive substances, and his investigations into the origin and nature of isotopes
1922	Niels Bohr	1885–1962	His investigation of the structure of atoms and the radiation emanating from them
	C Francis W. Aston	1877–1945	His discovery, by means of his mass spectrograph, of isotopes in a large number of nonradioactive elements, and for his enunciation of the whole-number rule
1923	Robert Andrews Millikan	1868–1953	His work on the elementary charge of electricity and on the photoelectric effect
1924	Karl Manne Georg Siegbahn	1886–1978	His discoveries and researches in the field of x-ray spectroscopy
1925	James Franck Gustav Hertz	1882–1964 1887–1975	Their discovery of the laws governing the impact of an electron upon an atom
1926	Jean-Baptiste Perrin	1870–1942	His work on the discontinuous structure of matter, and especially for his discovery of sedimentation equilibrium
1927	Arthur Holly Compton	1892–1962	His discovery of the effect named after him
	Charles Thomson Rees Wilson	1869–1959	His method of making the paths of electrically charged particles visible by condensation of vapor
1928	Owen Willans Richardson	1879–1959	His work on the thermionic phenomenon, and especially for the discovery of the law named after him
1929	Prince Louis-Victor de Broglie	1892–1987	His discovery of the wave nature of electrons

(continued)

Year	Nobel Laureate		Citation for
1930	Sir Chandrasekhara Venkata Raman	1888–1970	His work on the scattering of light and for the discovery of the effect named after him
1932	Werner Heisenberg	1901–1976	The creation of quantum mechanics, the application of which has, *inter alia*, led to the discovery of the allotropic forms of hydrogen
1933	Erwin Schrödinger	1887–1961	Their discovery of new productive forms
	Paul Adrien Maurice Dirac	1902–1984	of atomic theory
1934	C Harold C. Urey	1893–1981	His discovery of heavy hydrogen
1935	James Chadwick	1891–1974	His discovery of the neutron
	C Frédéric Joliot	1900–1958	In recognition of their synthesis of new
	C Irène Joliot-Curie	1897–1956	radioactive elements
1936	Victor Franz Hess	1883–1964	His discovery of cosmic radiation
	Carl David Anderson	1905–1991	His discovery of the positron
	C Peter Debye	1884–1966	His contributions to our knowledge of molecular structure through his investigations on dipole moments and on the diffraction of x rays and electrons in gases
1937	Clinton Joseph Davisson	1881–1958	Their experimental discovery of the diffraction
	George Paget Thomson	1892–1975	of electrons by crystals
1938	Enrico Fermi	1901–1954	His demonstrations of the existence of new radioactive elements produced by neutron irradiation, and for his related discovery of nuclear reactions brought about by slow neutrons
1939	Ernest Orlando Lawrence	1901–1958	The invention and development of the cyclotron and for results obtained with it, especially with regard to artificial radioactive elements
1943	Otto Stern	1888–1969	His contributions to the development of the molecular ray method and his discovery of the magnetic moment of the proton
1944	Isidor Isaac Rabi	1898–1988	His resonance method for recording the magnetic properties of atomic nuclei
	C Otto Hahn	1879–1968	His discovery of the fission of heavy nuclei
1945	Wolfgang Pauli	1900–1958	His discovery of the Exclusion Principle, also called the Pauli Principle
1946	Percy Williams Bridgman	1882–1961	The invention of an apparatus to produce extremely high pressures and for the discoveries he made in the field of high-pressure physics
1947	Sir Edward Victor Appleton	1892–1965	His investigations of the physics of the upper atmosphere, especially for the discovery of the Appleton layer
1948	Patrick Maynard Stuart Blackett	1897–1974	His development of the Wilson cloud chamber method and his discoveries therewith in nuclear physics and cosmic radiation
1949	Hideki Yukawa	1907–1981	His prediction of the existence of mesons on the basis of theoretical work on nuclear forces
1950	Cecil Frank Powell	1903–1969	His development of the photographic method of studying nuclear processes and his discoveries regarding mesons made with this method

Year	Nobel Laureate		Citation for
1951	Sir John Douglas Cockcroft Ernest Thomas Sinton Walton	1897–1967 1903–1995	Their pioneer work on the transmutation of atomic nuclei by artificially accelerated particles
	C Edwin M. McMillan C Glenn T. Seaborg	1907–1991 1912–1999	Their discoveries in the chemistry of the transuranium elements
1952	Felix Bloch Edward Mills Purcell	1905–1983 1912–1997	The development of new methods for nuclear magnetic precision measurements and discoveries in connection therewith
1953	Frits Zernike	1888–1966	His demonstration of the phase contrast method, especially for his invention of the phase contrast microscope
1954	Max Born	1882–1970	His fundamental research in quantum mechanics, especially his statistical interpretation of the wave function
	Walter Bothe	1891–1957	The coincidence method and his discoveries made therewith
1955	Willis Eugene Lamb, Jr.	1913–	His discoveries concerning the fine structure of the hydrogen spectrum
	Polykarp Kusch	1911–1993	His precision determination of the magnetic moment of the electron
1956	William Shockley John Bardeen Walter Houser Brattain	1910–1989 1908–1991 1902–1987	Their investigations on semiconductors and their discovery of the transistor effect
1957	Chen Ning Yang Tsung Dao Lee	1922– 1926–	Their penetrating investigation of the parity laws, which led to important discoveries regarding elementary particles
1958	Pavel Alekseyevich Cherenkov Ilya Mikhaylovich Frank Igor Yevgenyevich Tamm	1904–1990 1908–1990 1895–1971	Their discovery and interpretation of the Cherenkov effect
1959	Emilio Gino Segrè Owen Chamberlain	1905–1989 1920–	Their discovery of the antiproton
1960	Donald Arthur Glaser	1926–	The invention of the bubble chamber
	C Willard F. Libby	1908–1980	His method to use ^{14}C for age determination in several branches of science
1961	Robert Hofstadter	1915–1990	His pioneering studies of electron scattering in atomic nuclei and for his discoveries concerning the structure of the nucleons achieved thereby
	Rudolf Ludwig Mössbauer	1929–	His researches concerning the resonance absorption of γ rays and his discovery in this connection of the effect that bears his name
1962	Lev Davidovich Landau	1908–1968	His pioneering theories of condensed matter, especially liquid helium
1963	Eugene Paul Wigner	1902–1995	His contributions to the theory of the atomic nucleus and the elementary particles, particularly through the discovery and application of fundamental symmetry principles
	Maria Goeppert Mayer J. Hans D. Jensen	1906–1972 1907–1973	Their discoveries concerning nuclear shell structure

(continued)

Year	Nobel Laureate		Citation for
1964	Charles H. Townes	1915–	Fundamental work in the field of quantum electronics, which has led to the construction of oscillators and amplifiers based on the maser-laser principle
	Nikolai G. Basov	1922–2001	
	Alexander M. Prokhorov	1916–2002	
1965	Sin-Itiro Tomonaga	1906–1979	Their fundamental work in quantum electrodynamics, with profound consequences for the physics of elementary particles
	Julian Schwinger	1918–1994	
	Richard P. Feynman	1918–1988	
1966	Alfred Kastler	1902–1984	The discovery and development of optical methods for studying Hertzian resonance in atoms
1967	Hans Albrecht Bethe	1906–	His contributions to the theory of nuclear reactions, especially his discoveries concerning the energy production in stars
1968	Luis W. Alvarez	1911–1988	His decisive contribution to elementary particle physics, in particular the discovery of a large number of resonance states, made possible through his development of the technique of using the hydrogen bubble chamber and data analysis
1969	Murray Gell-Mann	1929–	His contributions and discoveries concerning the classification of elementary particles and their interactions
1970	Hannes Alfvén	1908–1995	Fundamental work and discoveries in magnetohydrodynamics with fruitful applications in different parts of plasma physics
	Louis-Eugène-Félix Néel	1904–2000	Fundamental work and discoveries concerning antiferromagnetism and ferrimagnetism, which have led to important applications in solid state physics
1971	Dennis Gabor	1900–1979	His invention and development of the holographic method
1972	John Bardeen	1908–1991	Their theory of superconductivity, usually called the BCS theory
	Leon N. Cooper	1930–	
	J. Robert Schrieffer	1931–	
1973	Leo Esaki	1925–	His discovery of tunneling in semiconductors
	Ivar Giaever	1929–	His discovery of tunneling in superconductors
	Brian D. Josephson	1940–	His theoretical predictions of the properties of a supercurrent through a tunnel barrier
1974	Antony Hewish	1924–	The discovery of pulsars
	Sir Martin Ryle	1918–1984	His observations and inventions in radio astronomy
1975	Aage Bohr	1922–	The discovery of the connection between collective motion and particle motion in atomic nuclei and for the theory of the structure of the atomic nucleus based on this connection
	Ben R. Mottelson	1926–	
	L. James Rainwater	1917–1986	
1976	Burton Richter	1931–	Their pioneering work in the discovery of a heavy elementary particle of a new kind
	Samuel Chao Chung Ting	1936–	
1977	Philip Warren Anderson	1923–	Their fundamental theoretical investigations of the electronic structure of magnetic and disordered systems
	Nevill Francis Mott	1905–1996	
	John Hasbrouck Van Vleck	1899–1980	

Year	Nobel Laureate		Citation for
1978	Pyotr L. Kapitza	1894–1984	His basic inventions and discoveries in the area of low-temperature physics
	Arno A. Penzias	1933–	Their discovery of cosmic microwave background radiation
	Robert Woodrow Wilson	1936–	
1979	Sheldon Lee Glashow	1932–	Their contributions to the theory of the unified weak and electromagnetic interaction between elementary particles, including, *inter alia*, the prediction of the weak neutral current
	Abdus Salam	1926–1996	
	Steven Weinberg	1933–	
1980	James W. Cronin	1931–	The discovery of violations of fundamental symmetry principles in the decay of neutral K-mesons
	Val L. Fitch	1923–	
1981	Nicolaas Bloembergen	1920–	Their contributions to the development of laser spectroscopy
	Arthur L. Schawlow	1921–1999	
	Kai M. Siegbahn	1918–	His contribution to the development of high-resolution electron spectroscopy
1982	Kenneth G. Wilson	1936–	His theory for critical phenomena in connection with phase transitions
1983	Subrahmanyan Chandrasekhar	1910–1995	His theoretical studies of the physical processes of importance to the structure and evolution of the stars
	William A. Fowler	1911–1995	His theoretical and experimental studies of the nuclear reactions of importance in the formation of the chemical elements in the universe
1984	Carlo Rubbia	1934–	Their decisive contributions to the large project, which led to the discovery of the field particles W and Z, communicators of the weak interaction
	Simon van der Meer	1925–	
1985	Klaus von Klitzing	1943–	The discovery of the quantized Hall effect
1986	Ernst Ruska	1906–1988	His fundamental work in electron optics and for the design of the first electron microscope
	Gerd Binnig	1947–	Their design of the scanning tunneling microscope
	Heinrich Rohrer	1933–	
1987	J. Georg Bednorz	1950–	Their important breakthrough in the discovery of superconductivity in ceramic materials
	Karl Alex Müller	1927–	
1988	Leon M. Lederman	1922–	The neutrino beam method and the demonstration of the doublet structure of the leptons through the discovery of the muon neutrino
	Melvin Schwartz	1932–	
	Jack Steinberger	1921–	
1989	Hans G. Dehmelt	1922–	Their development of the ion-trap technique
	Wolfgang Paul	1913–1993	
	Norman F. Ramsey	1915–	The invention of the separated oscillatory fields method and its use in the hydrogen maser and other atomic clocks
1990	Jerome I. Friedman	1930–	Their pioneering investigations concerning deep inelastic scattering of electrons on protons and bound neutrons, which have been of essential importance for the development of the quark model in particle physics
	Henry W. Kendall	1926–1999	
	Richard E. Taylor	1929–	

(continued)

Year	Nobel Laureate		Citation for
1991	Pierre-Gilles de Gennes	1932–	His discovering that methods developed for studying order phenomena in simple systems can be generalized to more complex forms of matter, in particular to liquid crystals and polymers
	C Richard R. Ernst	1933–	His contributions to the development of the methodology of high-resolution nuclear magnetic resonance (NMR) spectroscopy
1992	Georges Charpak	1924–	His invention and development of particle detectors, particularly multi-wire proportional counters
1993	Russell A. Hulse	1950–	The discovery of a new type of pulsar
	Joseph H. Taylor Jr.	1941–	
1994	Bertram N. Brockhouse	1918–2003	The development of neutron scattering
	Clifford G. Shull	1915–2001	The development of the neutron diffraction technique
1995	Martin L. Perl	1927–	The discovery of the tau lepton
	Frederick Reines	1918–1998	The detection of the neutrino
1996	David M. Lee	1931–	Their discovery of superfluidity in He-3
	Douglas D. Osheroff	1945–	
	Robert C. Richardson	1937–	
1997	Steven Chu	1948–	Their development of methods to cool and trap atoms with laser light
	Claude Cohen-Tannoudji	1933–	
	William D. Phillips	1948–	
1998	Robert B. Laughlin	1950–	Their discovery of a new form of quantum fluid with fractionally charged excitations
	Horst L. Störmer	1949–	
	Daniel C. Tsui	1939–	
1999	Gerardus 't Hooft	1946–	Their elucidating the quantum structure of electroweak interactions in physics
	Martinus J. G. Veltman	1931–	
2000	Zhores I. Alferov	1930–	Their developing semiconductor heterostructures used in high-speed- and opto-electronics
	Herbert Kroemer	1928–	
	Jack S. Kilby	1923–	His part in the invention of the integrated circuit
2001	Eric A. Cornell	1961–	Their achievement of Bose-Einstein condensation in dilute gases of alkali atoms, and for early fundamental studies of the properties of the condensates
	Wolfgang Ketterle	1957–	
	Carl E. Wieman	1951–	
2002	Raymond Davis Jr.	1914–	Their pioneering contributions to astrophysics, in particular for the detection of cosmic neutrinos
	Masatoshi Koshiba	1926–	
	Riccardo Giacconi	1931–	His pioneering contributions to astrophysics, which have led to the discovery of cosmic x-ray sources
2003	Alexei A. Abrikosov	1928–	Their pioneering contributions to the theory of superconductors and superfluids
	Vitaly L. Ginzburg	1916–	
	Anthony J. Leggett	1938–	
2004	David Gross	1941–	Their discovery of asymptotic freedom in the theory of the strong interaction
	H. David Politzer	1949–	
	Frank Wilczek	1951–	

Answers to Selected Odd-Numbered Problems

Chapter 2

3. $\theta = \sin^{-1}\left(\dfrac{v}{c}\right)$

13. K′ travels at a speed $c/2$ in the $-x$ direction.

15. (a) 3.86×10^{-8} s   (b) $x' = 10.4$ m, $y' = 5$ m, $z' = 10$ m, $t' = 51.0$ ns

17. (a) $1 + 4.3 \times 10^{-15}$ (b) $1 + 4.7 \times 10^{-13}$
(c) $1 + 3.1 \times 10^{-12}$ (d) $1 + 3.1 \times 10^{-10}$
(e) 1.10   (f) 3.20

19. $0.14c$

21. $1.4 \times 10^{-4}c$

23. $\dfrac{4c}{5}$

25. $\Delta t = 16.7$ ns

27. $\Delta t' = 1.0 \times 10^{-5}$ s

29. 22.1 m

31. (a) $0.96c$ (b) $0.23c$

33. Each sees the other traveling at 62.1 m/s.

35. $0.60c$ and $0.88c$

37. Classical: 15 muons; relativistic: 2710 muons

39. Mary receives signals at a rate f' for t_1' and a rate f'' for t_2'. Frank receives signals at a rate f' for t_1 and a rate f'' for t_2.

45. (b) β (d) The lines are parallel.

47. Two events simultaneous in K are not simultaneous in K′.

51. 82 Hz

53. Van 2 receives signals from vans 1 and 3 at a rate
$f' = f_0\sqrt{\dfrac{1 - \beta}{1 + \beta}}$, and vans 1 and 3 receive signals from
van 2 at a rate of $f' = \dfrac{f_0}{1 + \sqrt{2}\beta}$.

59. 1.42×10^{-25} kg

61. 0.155 T

65. $0.999\,999\,984c$

67. (a) $p = 5.11$ keV/c, $K = 30$ eV, $E = 511.03$ keV
(b) $p = 51.4$ keV/c, $K = 2.6$ keV, $E = 513.6$ keV
(c) $p = 1055$ keV/c, $K = 661$ keV, $E = 1172$ keV

69. 2.25×10^{17} m

71. (a) $0.417c$ (b) $0.866c$ (c) $0.996c$

75. 2.55×10^{14} J

79. 28.3 MeV

81. $v = 0.999\,999\,561c$, $p = 1.000938$ TeV/c, $E = 1$ TeV $+$ 938 MeV

83. $v = 0.938c$, $p = 287.05$ MeV/c, $E = 306$ MeV

85. (a) $\Delta E = 17.6$ MeV   (b) approximately 0.37% of the initial rest energy

87. 2330 MeV

89. Time $= L/(\gamma v)$, number $= \dfrac{fL(1 - \beta)}{v}$

91. (a) 52   (b) 9 years; 156   (c) Frank 312, Mary 520
(d) Frank 10 years, Mary 6 years   (e) Each one agrees with (d)

93. (a) $(1 - 3.97 \times 10^{-8})c$ (b) $(1 - 8.85 \times 10^{-8})c$

95. (a) $v = 0.36c$ (b) 53.6 years old

97. (a) 7.24×10^7 m/s   (b) 0.124 Hz

99. $v = 2.47 \times 10^7$ m/s

101. (a) 228 MeV   (b) 209.23 MeV/c

Chapter 3

3. Nonrelativistically $V = 1137$ V; relativistically $V = 1141$ V

9. Lyman 91.2 nm; Balmer 364.7 nm

11. 93.6 cm between the first and second lines; 202.7 cm between the second and third lines

13. $\Delta\lambda = 29$ nm

15. (a) Paschen series: $k = 4$, $\lambda = 1875.63$ nm; $k = 5$, $\lambda = 1282.17$ nm; $k = 6$, $\lambda = 1094.12$ nm; $k = 7$, $\lambda = 1005.22$ nm; $k = 8$, $\lambda = 954.86$ nm
(b) The observed spectral lines have been Doppler redshifted.   (c) $v = 1.20 \times 10^7$ m/s

17. (a) 0.69 mm   (b) 9.89 μm   (c) 1.16 μm

19. 19.9

21. 1.447×10^{-8} m

23. 966 nm

27. 9.35 μm

33. (a) 2.06×10^{29} (b) 6.05×10^{18} (c) 2.34×10^{14}

35. $\lambda = 267.2$ nm $K = 4.64$ eV

37. 1.59×10^{15} Hz

39. $h = 4.40 \times 10^{-15}$ eV·s; $\phi = 4.1$ eV

41. 0.0413 nm

43. 0.0620 nm

45. 17.2 kV

47. 5.41 keV

49. Yes; 1.32 fm; 938 MeV

53. 2.00243 nm, a change of 0.122%

55. (a) 2.04 MeV   (b) 1.02 MeV

57. $r = 3.68$ km; 2×10^{13} nuclear arsenals

61. (a) 0.015 photons/s   (b) 6.45×10^3 photons/s

Chapter 4

5. (a) 1.69×10^{-12} m   (b) 1.48×10^{-14} m
7. 36.93
9. (a) 2170   (b) 1347
11. (a) Al: 6.04 MeV, Au: 23.7 MeV   (b) Al: 3.82 MeV, Au: 13.7 MeV
13. (a) $0.013°$ (b) The results are comparable.
15. (a) $1.75 \times 10^{-4} c$ (b) -14.4 eV
17. H: -13.6 eV; He$^+$: -54.4 eV; Li^{++}: -122.5 eV
21. $hc = 1239.8$ eV · nm; $\dfrac{e^2}{4\pi\epsilon_0} = 1.4400$ eV · nm; $mc^2 =$ 511.00 keV; $a_0 = 5.2918 \times 10^{-2}$ nm; $E_0 = 13.606$ eV
23. $n = 2$ and $n = 5$
25. (a) 13.6 eV   (b) 54.4 eV   (c) 218 eV
27. $E_{1,D} = 13.602$ eV; $E_{1,T} = 13.603$ eV
29. 2.44×10^6
31. (a) 2.84×10^{-13} m   (b) 2535 eV   (c) 0.49 nm, 1.96 nm, 4.40 nm
33. (a) $2a_0$ (b) 243 nm
35. The lines nearly match for all even n_u in the He$^+$ series. Each "matching" pair actually differs by the ratio of the reduced masses, $R_{\mathrm{He^+}}/R_{\mathrm{H}} = 1.0004$, or about 0.04%.
37. $R = 4.38889 \times 10^7$ m^{-1}
39. No, because the K$_\alpha$ lines for Pt and Au are separated by less than 10^{-12} m.
41. Helium: 122 nm for K$_\alpha$ and 103 nm for K$_\beta$; Lithium: 30.4 nm for K$_\alpha$ and 25.6 nm for K$_\beta$
43. Molybdenum: $\lambda(\mathrm{K}_\alpha) = 72.8$ pm; $\lambda(\mathrm{K}_\beta) = 60.99$ pm; $\lambda(\mathrm{K}_\gamma) = 57.82$ pm; the series limit is 54.21 pm
45. 4.33×10^{-4} eV
47. 4.13×10^{-15} eV · s
49. Magnesium: 1.00 nm for K$_\alpha$ and 31.0 nm for L$_\alpha$; Iron: 0.194 nm for K$_\alpha$ and 1.90 nm for L$_\alpha$
51. (a) 0.148   (b) 100.5 and 1.31×10^4

Chapter 5

1. $31.2°$ and $50.9°$
3. 9.47 keV; we can observe up through $n = 4$
5. 3.68×10^{-35} m; no
7. $\lambda = 6.02$ pm for the 40-keV electrons and $\lambda = 3.70$ pm for the 100-keV electrons
9. 1.73×10^{-10} m
11. (a) $hc/\sqrt{K^2 + 2Kmc^2}$ (b) $hc/\sqrt{2mc^2 K}$
13. (a) 8.27 keV   (b) 66.9 eV   (c) 0.036 eV   (d) 9.17×10^{-3} eV
15. $K = 3$ keV, $E = 514$ keV, $p = 55.4$ keV/c, $\lambda = 22.4$ pm
17. (a) 2.60×10^{-11} m   (b) 4.42×10^{-13} m
19. $d = 0.063$ nm, $\lambda = 0.122$ nm, $p = 10.2$ keV/c, $E = 511.102$ keV, $K = 102$ eV
21. 0.457 nm, 0.412 nm, 0.301 nm
23. 1.67 s
25. (a) 0.571 Hz   (b) 3.0 cm
27. (a) $\Psi = 0.006 \sin(6.5x - 275t)\cos(0.5x + 25t)$ (b) $v_{\mathrm{ph}} = 42.3$ m/s, $u_{\mathrm{gr}} = 50$ m/s   (c) 2π m   (d) 2π
29. v_{ph} is independent of the wavelength

31. $\Psi(x, 0) = -(2A_0/x)\sin(\Delta k\, x/2)\cos(k_0 x)$; width $\Delta x = \pi/\Delta k$; $\Delta k\, \Delta x = \pi$
33. The intensity is higher by a factor of 4 for the double slit.
35. $K = 2.67$ eV
37. 51.1 MeV, 1.30 MeV
39. $\Delta L = \hbar/(4\pi)$
41. 2.1×10^5 Hz
43. (a) 3.29×10^{-3} eV   (b) 0.185 nm
45. (a) 1.51 eV   (b) 2.24×10^{-6} eV; yes
47. 3.15×10^{-33} J
55.
$$\frac{E_2}{E_1} = \left[\frac{1 + h^2 c^2/(L^2 E_0^2)}{1 + h^2 c^2/(4L^2 E_0^2)}\right]^{1/2}$$
$$\frac{E_3}{E_1} = \left[\frac{1 + 9h^2 c^2/(4L^2 E_0^2)}{1 + h^2 c^2/(4L^2 E_0^2)}\right]^{1/2}$$
$$\frac{E_4}{E_1} = \left[\frac{1 + 4h^2 c^2/(L^2 E_0^2)}{1 + h^2 c^2/(4L^2 E_0^2)}\right]^{1/2}$$
59. $\Delta t = 4.0 \times 10^{-24}$ s; $\Delta E = 82$ MeV; $m = 82$ MeV/c^2, within a factor of 2 of the actual mass
61. $\hbar \le 0.081$ J · s
63. (b) The packet is centered about $x = 0$ but extends to $\pm\infty$ and repeats for every unit along the x axis.

Chapter 6

1. The function is not localized; it does not vanish at $\pm\infty$.
5. $A = 2\alpha^{-3/2}$
7. (a) The wave function does not satisfy condition 3.
(b) The wave function is not physically possible.
(c) Alter the function near $|x| = 0$ so that its derivative is continuous.
11. $A = \sqrt{2/\pi}$; probability $= 0.091$
13. $1/L$ (independent of n), in agreement with the classical result
15. (a) $E_1 = 9.40 \times 10^{-8}$ eV; $E_2 = 1.88 \times 10^{-7}$ eV; $E_3 = 2.82 \times 10^{-7}$ eV   (b) $n = 134$
17. 0.1955; 0.6090; 0.1955
19. 11.3 GeV
21. 10.5 eV, 18.1 eV, 22.6 eV, 7.5 eV, 12.0 eV, 4.52 eV
23. (a) λ is longer for the finite well.   (b) Larger λ implies lower E.   (c) When $E > V_0$ there are no bound states.
25. $Ce^{ikL} + De^{-ikL} = Be^{-\alpha L}$ $\dfrac{C}{D} = \dfrac{ik - \alpha}{ik + \alpha} e^{-2ikL}$
27. In general we have
$$\psi(x) = A \sin\left(\frac{n_1\pi x}{L}\right)\sin\left(\frac{n_2\pi y}{L}\right)\sin\left(\frac{n_3\pi z}{L}\right)$$
For $\psi_2(x)$ we can have $(n_1, n_2, n_3) = (1, 1, 2)$ or $(1, 2, 1)$ or $(2, 1, 1)$.
For $\psi_3(x)$ we can have $(n_1, n_2, n_3) = (1, 2, 2)$ or $(2, 2, 1)$ or $(2, 1, 2)$.
For $\psi_4(x)$ we can have $(n_1, n_2, n_3) = (1, 1, 3)$ or $(1, 3, 1)$ or $(3, 1, 1)$.
For $\psi_5(x)$ we can have $(n_1, n_2, n_3) = (2, 2, 2)$.
29. One quantum number is associated with each boundary condition.

31. $\Delta E = \hbar\omega$ for all n

33. $E = (n + \frac{1}{2})(4.136 \times 10^{-2})$ eV; $k = 131$ N/m

35. $\langle p \rangle = 0; \langle p^2 \rangle = \frac{1}{2}m\hbar\omega$

37. (a) $E = (n + \frac{1}{2})\hbar\omega = (n + \frac{1}{2})(0.755)$ eV
(b) 1640 nm, 821 nm, 549 nm

39. (a) $p = \sqrt{2m(E - V_0)}; \lambda = h/\sqrt{2m(E - V_0)};$
$K = E - V_0$
(b) $p = \sqrt{2m(E + V_0)}; \lambda = h/\sqrt{2m(E + V_0)};$
$K = E + V_0$

43. (a) $L = 0.229$ nm or any integer multiple thereof
(b) $L = 0.114$ nm or any odd integer multiple thereof

47. Let $E_0 = \pi^2\hbar^2/(2mL^2)$; $E_1 = (13E_0)/4$; $E_2 = 4E_0$; $E_3 = (21E_0)/4$; $E_4 = (25E_0)/4$; $E_5 = 7E_0$; only E_5 is degenerate

53. (b) $\langle x \rangle = 0; \langle x^2 \rangle = 5/(2\alpha)$

55. $V(r) = Da^2(r - r_e)^2$

57. Let $E_0 = \pi^2\hbar^2/(2mL^2)$;
$E_1 = E_0(1^2 + 1^2) = 2E_0$ $n_1 = 1, n_2 = 1$
$E_2 = E_0(2^2 + 1^2) = 5E_0$ $n_1 = 2, n_2 = 1$ or vice versa
$E_3 = E_0(2^2 + 2^2) = 8E_0$ $n_1 = 2, n_2 = 2$
$E_4 = E_0(3^2 + 1^2) = 10E_0$ $n_1 = 3, n_2 = 1$ or vice versa
$E_5 = E_0(3^2 + 2^2) = 13E_0$ $n_1 = 3, n_2 = 2$ or vice versa
$E_6 = E_0(4^2 + 1^2) = 17E_0$ $n_1 = 4, n_2 = 1$ or vice versa

59. 6.51×10^{-6}

61. (a) $E_n = \dfrac{(3Anh)^{2/3}}{2^{5/3}m^{1/3}}$

Chapter 7

5. $E = -E_0/4$, as predicted by the Bohr model

9. $\ell = 5$; $m_\ell = 0, \pm1, \pm2, \pm3, \pm4, \pm5$ $\ell = 4$; $m_\ell = 0, \pm1, \pm2, \pm3, \pm4$ $\ell = 3$; $m_\ell = 0, \pm1, \pm2, \pm3$ $\ell = 2$; $m_\ell = 0, \pm1, \pm2$ $\ell = 1$; $m_\ell = 0, \pm1$ $\ell = 0$; $m_\ell = 0$

11. $\psi_{310} = R_{31}Y_{10}$
$= \dfrac{1}{81}\sqrt{\dfrac{2}{\pi}} a_0^{-3/2}\left(6 - \dfrac{r}{a_0}\right)\left(\dfrac{r}{a_0}\right)e^{-r/3a_0}\cos\theta$
$\psi_{31\pm1} = R_{31}Y_{1\pm1}$
$= \dfrac{1}{81\sqrt{\pi}} a_0^{-3/2}\left(6 - \dfrac{r}{a_0}\right)\left(\dfrac{r}{a_0}\right)e^{-r/3a_0}\sin\theta\, e^{\pm i\phi}$

13. 36

15. 365

17. $\psi_{21-1} = R_{21}Y_{1-1} = \dfrac{1}{8}\sqrt{\dfrac{1}{\pi}} a_0^{-3/2}\left(\dfrac{r}{a_0}\right)e^{-r/2a_0}\sin\theta\, e^{-i\phi}$
$\psi_{210} = R_{21}Y_{10} = \dfrac{1}{4\sqrt{2\pi}} a_0^{-3/2}\left(\dfrac{r}{a_0}\right)e^{-r/2a_0}\cos\theta$
$\psi_{32-1} = R_{32}Y_{2-1} = \dfrac{1}{81\sqrt{\pi}} a_0^{-3/2}\left(\dfrac{r}{a_0}\right)^2 e^{-r/3a_0}$
$\times \sin\theta\cos\theta\, e^{-i\phi}$

19. $0, \pm\hbar$

23. $30°$

25. Seven different states: with $B = 0$, $E = -E_0/25 = -0.544$ eV; with the field on, the levels are changed by $\Delta E = 0$ for $m_\ell = 0$, $\Delta E = \pm1.74 \times 10^{-4}$ eV for

$m_\ell = \pm1$, $\Delta E = \pm3.47 \times 10^{-4}$ eV for $m_\ell = \pm2$, and
$\Delta E = \pm5.21 \times 10^{-4}$ eV for $m_\ell = \pm3$

27. The magnet should be designed so that the product of its length squared and its vertical magnetic field gradient is 57 T·m.

29. $n = 5$, $\ell = 3$, $m_\ell = 0, \pm1, \pm2, \pm3$, and $m_s = \pm1/2$, for a total degeneracy of 14

31. $T = 0.0456$ K

35. $r = (3 \pm \sqrt{5})a_0$

37. $r = 2a_0$

39. 0.056

41. $2s$: 1.1×10^{-15} $2p$: 2.01×10^{-26}

43. 0.9999935

45. $0.24c$

49. $E_0 = 2.53$ keV

Chapter 8

1. The first two electrons are in the $1s$ subshell and have $\ell = 0$, with $m_s = \pm1/2$. The third electron (in the $2s$ subshell) has $\ell = 0$, with either $m_s = 1/2$ or $-1/2$. With four particles there are six possible interactions: the nucleus with electrons 1, 2, 3; electron 1 with electron 2; electron 1 with electron 3; or electron 2 with electron 3. In each case it is possible to have a Coulomb interaction and a magnetic moment interaction.

3. 2, 4, 5

5. K: $4s^1$, V: $4s^23d^3$, Se: $4s^23d^{10}4p^4$, Zr: $5s^24d^2$, Sm: $6s^24f^6$, U: $7s^26d^15f^3$

7. $1.36e$

9. (a) B (b) Na (c) Ar

11. (a) As (b) Ag (c) Ho

13. $4^2P_{1/2}$

15. In the $3d$ state $\ell = 2$ and $s = 1/2$, so $j = 5/2$ or $3/2$. As usual $m_\ell = 0, \pm1, \pm2$. The value of m_j ranges from $-j$ to j, so its possible values are $\pm1/2, \pm3/2$, and $\pm5/2$. As always $m_s = \pm1/2$. The two possible term notations are $3D_{5/2}$ and $3D_{3/2}$.

17. $\pm\hbar/2, \pm3\hbar/2, \pm5\hbar/2, \pm7\hbar/2$

21. 7.33×10^{-3} eV; 63.4 T

23. The $2s$ to $1s$ transition is forbidden by the $\Delta L = \pm1$ selection rule. The two lines result from the transition from the $2p$ level to the $1s$ level.

25. He, Ca, and Sr have single ground states and may have a singlet or triplet excited states; Al has only a doublet.

29. (a) 2.4×10^{-5} eV (b) 4.53×10^{-5} eV

33. (a) 2 (b) 4/3 (c) 6/5

37. 3.86×10^{-5} eV

39. (a) Sc$^-$: $1s^22s^22p^63s^23p^64s^23d^2$; Al$^-$: $1s^22s^22p^63s^23p^2$
(b) Sc$^-$: 3^3F_2; Al$^-$: 3^3P_0

Chapter 9

3. (a) $\bar{f} = f_0$ (b) Standard deviation $= \dfrac{f_0}{c}\sqrt{\dfrac{kT}{m}}$
(c) 3.66×10^{-6} for H_2 at 293 K; 2.25×10^{-5} for H at 5500 K

5. (a) $\displaystyle\int_c^\infty F(v)\,dv = 4\pi C \int_c^\infty v^2 \exp\left(-\frac{1}{2}\beta mv^2\right)\,dv$

with $T = 293$ K and $C = \left(\dfrac{\beta m}{2\pi}\right)^{3/2}$

7. (a) $\bar{v} = 2510$ m/s, $v^* = 2220$ m/s   (b) $\bar{v} = 6480$ m/s, $v^* = 5740$ m/s

9. (a) 399 m/s   (b) 424 m/s

13. (a) $v_{\mathrm{rms}} = 1902$ m/s for H_2, $v_{\mathrm{rms}} = 511$ m/s for N_2

15. (a) $\bar{E} = \frac{3}{2}kT$ (b) $\frac{1}{2}m\overline{v^2} = \bar{E} = \frac{3}{2}kT; \frac{1}{2}m\bar{v}^2 = 1.27kT$

17. 7790 K

23. (a) 5.86×10^{28} m^{-3} (b) 5.28×10^4 K   (c) 5.28×10^6 K

25. (a) 5.50 eV   (b) 1.39×10^6 m/s

27. 8.45×10^{28} m^{-3}; one conduction electron per atom

29. (a) 1.18×10^6 m/s   (b) 1.83×10^6 m/s

35. 147 MeV

37. $E - E_F = 0.032$ eV

49. $N/V \le 1.97 \times 10^{31}$ m^{-3} for the Bose-Einstein condensate; this is about one million times the number density of an ideal gas

51. (a) 0.343 K   (b) The temperature in (a) is far below the freezing point.

Chapter 10

1. (a) 1.4×10^{-3} eV   (b) 1.04×10^{-2} eV

3. 1.1×10^{-11} m

5. $E_{\mathrm{rot}} = \dfrac{n^2\hbar^2}{2I}$, which is similar to the quantum mechanical result in the limit of large n

7. 24

9. (a) 1.46×10^{-46} kg·m^2 (b) 1.13×10^{-10} m

11. (a) 34.1 eV   (b) no

15. (a) 2.4×10^{-47} kg·m^2 (b) 478 N/m

17. 5.78×10^{-20} J, infrared

19. (a) 1.00 W   (b) 1.08 μm

21. (a) 7.37×10^{-13} m   (b) 1.06×10^{-19} m

23. In a three-level system the population of the upper level must exceed the population of the ground state. This is not necessary in a four-level system.

25. 0.315 nm

27. $\alpha = 4 - \dfrac{4}{\sqrt{2}} - 2 + \dfrac{8}{\sqrt{5}} - \dfrac{4}{\sqrt{8}} + \cdots$

31. (c) 7.4×10^{-6} K^{-1}

33. 2.58×10^{10} N^{-1}

35. (b) 9.2×10^{13} N/m^2

41. $\bar{\mu} \approx \mu$

45. $0.95T_c$; $0.71T_c$; $0.32T_c$

47. ^{201}Hg: $T_c = 4.171$ K; ^{204}Hg: $T_c = 4.140$ K

49. 130 K

51. 8.74×10^{-6} m

53. (a) 307 A   (b) 307 A is more than a factor of 100 greater

55. 2.38 m/s$^2 \approx$ g/4

57. 17.4 TeV

59. 2.97×10^{-2} J/K

Chapter 11

1. (a) 236 Ω   (b) 610 Ω   (c) 5.48×10^7 Ω

3. The voltmeter reads positive since positive charges will drift to the right.

5. 1.26×10^{-6} V/K

7. 5×10^{-7} V

9. (a) 1850 nm   (b) 1130 nm

11. (a) Spring 0.839, winter 0.559, summer 0.985
   (b) Spring 0.643, winter 0.292, summer 0.891
   (c) Spring 0.500, winter 0.122, summer 0.799

13. (a) 1.99 A   (b) 288 mA   (c) 6.00 mA

15. 2080 kg/m^3

17. (a) 0.619 nm   (b) About 2.6 times higher

21. (a) 55.5 mV   (b) −40.6 mV

23. 2.70 eV

25. Si at $T = 0°$C: $F_{\mathrm{FD}} = 7.02 \times 10^{-11}$
   Si at $T = 75°$C: $F_{\mathrm{FD}} = 1.08 \times 10^{-8}$
   Ge at $T = 0°$C: $F_{\mathrm{FD}} = 6.54 \times 10^{-7}$
   Ge at $T = 75°$C: $F_{\mathrm{FD}} = 1.41 \times 10^{-5}$

27. (a) At $T = 77$ K: $\dfrac{I_f}{I_r} = -1.5 \times 10^{98}$

   At $T = 273$ K: $\dfrac{I_f}{I_r} = -4.9 \times 10^{27}$

   At $T = 340$ K: $\dfrac{I_f}{I_r} = -1.7 \times 10^{22}$

   At $T = 500$ K: $\dfrac{I_f}{I_r} = -1.3 \times 10^{15}$

29. 52.5

31. 4.2×10^{12} bits/m^2

Chapter 12

1. 54.9 MeV

3. In each case the atomic number equals the number of protons (Z) and the atomic charge is Ze.
   $^3_2$He: $Z = 2$, $N = 1$, $A = 3$, $m = 3.02$ u
   $^4_2$He: $Z = 2$, $N = 2$, $A = 4$, $m = 4.00$ u
   $^{17}_8$O: $Z = 8$, $N = 9$, $A = 17$, $m = 17.0$ u
   $^{42}_{20}$Ca: $Z = 20$, $N = 22$, $A = 42$, $m = 42.0$ u
   $^{210}_{82}$Pb: $Z = 82$, $N = 128$, $A = 210$, $m = 210.0$ u
   $^{235}_{92}$U: $Z = 92$, $N = 143$, $A = 235$, $m = 235.0$ u

5. Isotopes: ^{38}Ca through ^{52}Ca; Isobars: ^{40}S, ^{40}Cl, ^{40}Ar, ^{40}K, and ^{40}Ca; Isotones: ^{32}Mg, ^{34}Si, ^{35}P, ^{36}S, ^{37}Cl, ^{38}Ar, ^{39}K, and ^{40}Ca

7. ^{39}Ar(269 y), ^{41}Ar(1.822 h), ^{42}Ar(32.9 y), ^{43}Ar(5.37 m), ^{44}Ar(11.87 m), ^{45}Ar(21.48 s), ^{46}Ar(8.4 s);
   ^{250}No(0.25 ms), ^{251}No(0.8 s), ^{252}No(2.30 s), ^{254}No(55 s), ^{256}No(2.91 s), ^{258}No(1.2 ms), ^{259}No(58 m), ^{260}No(106 ms)

9. -1.52×10^{-3}

11. 7.25×10^{-9}; yes

13. The electrostatic force is 50 times weaker than the strong force. The gravitational force is almost 10^{38} times weaker than the strong force.

15. (b) 5.67 MeV; 4.14 MeV; 6.74 MeV

17. 7.27 MeV

19. The electron binding energy is less by a factor of 227.

23. 4.4682×10^4 MeV/c^2

25. 0.57 μg

29. 6.53×10^3 Bq

31. 120 kg

33. Yes

37. 0.09 MeV

39. β^-, β^+, and EC

43. 0.663 atm for He; 0.337 atm for Rn

47. 3.65×10^9 y for $R = 0.76$ and 9.10×10^9 y for $R = 3.1$; no

49. 6.4×10^7 y

51. 7 alpha, 4 beta

53. (a) 51.9 g of Fm and 41.1 g of Cf   (b) 0 g of Fm and between 94 and 95 g of Cf   (d) ^{240}Pu
(e) 2.34×10^7 y

61. (a) The chain that begins with ^{238}U   (c) Four days

63. $^4_2$He, $^{16}_8$O, $^{40}_{20}$Ca, $^{48}_{20}$Ca, $^{208}_{82}$Pb

Chapter 13

1. (a) $^4_2$He   (b) $^1_1$H   (c) $^{16}_8$O   (d) $^{73}_{32}$Ge   (e) $^{108}_{48}$Cd
(f) $^3_2$He

3. 0.087

5. 0.199

7. (a) ^{16}O$(n, \alpha)^{13}$C   (b) ^{16}O$(d, n)^{17}$F   (c) ^{16}O$(\gamma, p)^{15}$N
(d) ^{16}O$(\alpha, p)^{19}$F   (e) ^{16}O$(d, {}^3\text{He})^{15}$N
(f) ^{16}O$(^7\text{Li}, p)^{22}$Ne

All the products listed above are stable except the one in part (b).

9. (b) 15.9 MeV   (c) 2.34 MeV

11. (a) 2.0 MeV   (b) 2.0 MeV

15. 10.7 MeV

17. (a) 4.94×10^{17} (b) 2.06×10^9 Bq   (c) Place 1.94 kg of ^{60}Co into the reactor for one week.

19. 10.40 MeV

21. 6.98 MeV; 6.51 MeV; yes

23. No

25. $d + {}^{21}\text{Ne} \rightarrow {}^{22}\text{Ne} + p$ $d + {}^{22}\text{Ne} \rightarrow {}^{22}\text{Ne} + d$
$d + {}^{23}\text{Na} \rightarrow {}^{22}\text{Ne} + {}^3\text{He}$
$\alpha + {}^{18}\text{O} \rightarrow {}^{22}\text{Ne} + \gamma$ $\alpha + {}^{19}\text{F} \rightarrow {}^{22}\text{Ne} + p$
$\alpha + {}^{21}\text{Ne} \rightarrow {}^{22}\text{Ne} + {}^3\text{He}$ $\alpha + {}^{22}\text{Ne} \rightarrow {}^{22}\text{Ne} + {}^4\text{He}$
$d + {}^{25}\text{Mg} \rightarrow {}^{22}\text{Ne} + {}^5\text{Li}$ $d + {}^{26}\text{Mg} \rightarrow {}^{22}\text{Ne} + {}^6\text{Li}$
$d + {}^{27}\text{Al} \rightarrow {}^{22}\text{Ne} + {}^7\text{Be}$
(and so on)
$\alpha + {}^{23}\text{Ne} \rightarrow {}^{22}\text{Ne} + {}^5\text{Li}$ $\alpha + {}^{24}\text{Mg} \rightarrow {}^{22}\text{Ne} + {}^6\text{Be}$
$\alpha + {}^{27}\text{Al} \rightarrow {}^{22}\text{Ne} + {}^9\text{B}$
(and so on)

27. (a) 400 kg   (b) 1.01×10^{27} (c) 276 Bq
(d) 2.38×10^7

29. $^{236}\text{U} \rightarrow {}^{95}\text{Y} + {}^{138}\text{I} + 3n$ $^{236}\text{U} \rightarrow {}^{94}\text{Y} + {}^{140}\text{I} + 2n$
$^{236}\text{U} \rightarrow {}^{97}\text{Y} + {}^{136}\text{I} + 3n$

31. Uranium: 2.30×10^7 kWh; Coal: 8.50 kWh

33. 4.16 MeV, 7.55 MeV, 10.05 MeV, 4.97 MeV, total 26.7 MeV

35. (a) The high temperature is necessary so that the two oxygen nuclei can overcome the Coulomb barrier and be close enough for the nuclear force to be effective.
(b) 16.5 MeV, including the energy of the γ ray

37. (a) 1.5×10^{11} K   (b) 13.9 MeV, including the energy of the γ ray

39. (b) $K = \frac{3}{2}kT$ (c) 4.64×10^7 K

41. 1.69×10^{10} K

43. (a) 3.88×10^{-2} eV   (b) 2720 m/s   (c) 1.45×10^{-10} m

45. (b) 1.46 Bq   (c) 1.37×10^6 Bq

47. (a) 1.92×10^{-14} A   (b) 1.92×10^{-15} A

49. 8.95×10^{11} Bq

51. 11.6 W

53. (b) ^{192}Ir is a γ emitter; ^{32}P, ^{90}Sr, and ^{90}Y are all β^- emitters

55. (a) -13.37 MeV   (b) 33.979 u

57. (a) 8620 eV   (b) 0.607; 0.135; 0.0111

59. 6790 Bq

Chapter 14

1. 2.27×10^{23} Hz

3. The range of values is 1.3×10^{-23} s to 5.0×10^{-23} s

5. 1.24 TeV for both

7. 6.58×10^{-19} m

9. 3.57×10^{51}

11. 7.23×10^{-21} s; 1.24×10^{-20} s

13. In both (a) and (b) baryon number is not conserved.

15. 74.5 MeV

17. (a) ν_μ (b) K$^+$

21. B$^+$ $\bar{b}u$; B$^-$ $b\bar{u}$; B^0 $\bar{b}d$

23. (a) The mean lifetimes indicate weak interactions.
(b) $\Omega^- \rightarrow \Lambda^0 + K^-$, $sss \rightarrow uds + \bar{u}s$; $\Lambda^0 \rightarrow p + \pi^-$, $uds \rightarrow uud + \bar{u}d$; K$^- \rightarrow \mu^- + \bar{\nu}_\mu$, $\bar{u}s \rightarrow$ no quarks

25. 0.087

27. 2.84×10^{-7}

31. $(1 - 9.0 \times 10^{-9})c$

33. 5630 MeV

37. (a) $2mc^2$ (b) $\sqrt{2Kmc^2}$

39. (a) 1.0107   (b) 1.107   (c) 2.066

41. (a) Charge $= 0$, baryon number $= 0$, charm $= 0$, strangeness $= 0$, $L_e = +1$, spin $= 1/2$ (b) ν_e satisfy all the requirements

43. (a) 1585 MeV   (b) 336 MeV

45. (a) Allowed
(b) ν_e should be $\bar{\nu}_e$ to conserve electron lepton number
(c) Strangeness is not conserved (changes by 2 units)
(d) Strangeness is not conserved in a strong interaction
(e) Allowed

47. (a) Allowed (strangeness changes by 1 unit)
(b) Charge is not conserved
(c) Baryon number is not conserved
(d) Allowed (strangeness changes by 1 unit)

49. (a) 114.5 GeV   (b) For colliding beams the available energy is the sum of the two beam energies, or 14 TeV. This is an improvement over the fixed-target result by a factor of 122.

Chapter 15

5. 1726.5 s; $1.16 \times 10^{-5}\%$

7. 1.17×10^{-3} nm

9. 8541 Hz; $2.45 \times 10^{-13}\%$

11. 2.82 m

13. 4.19×10^{20} kg; $r_s = 6.22 \times 10^{-7}$ m; 2.11×10^{-10} solar masses

15. (a) 1.73×10^{11} kg   (b) This mass is too small to be a black hole.

17. 0.834 Hz

19. 9.61×10^{-3} Hz

21. 2.18×10^{-8} kg; 1.22×10^{28} eV

23. In both cases $t = 1.35 \times 10^{-43}$ s

Chapter 16

3. 1.57×10^{12} K; 10^{-4} s

5. 5.93×10^{9} K; 1.23×10^{12} K

7. π^{+}; 5.80×10^{10} K

9. 1.58×10^{5} K

11. 7.14×10^{17} kg/m^3, about three times as dense as a nucleon or a nucleus

13. (a) 5.00×10^{13} N/m^2 (b) 3.21×10^{33} N/m^2

15. 689 Mly

17. (a) $0.984c$ (b) 13.6 Gly

19. (a) 0.18 nucleons/m^3 (b) 4.41×10^{6} nucleons/m^3

23. 5.67×10^{-27} kg/m^3; 1.35×10^{-26} kg/m^3

27. Draw a tangent to the curve at the present time; the tangent in all cases intersects near the origin, which is $\tau = 15$ Gyr ago. The age of the universe must be less than τ.

31. 8.4×10^{-25} kg/m^3

33. Redshift = 3.79; $v = 0.92c$

37. $Q = 0.43$ MeV; maximum neutrino energy ≈ 0.43 MeV

39. (a) Deceleration for $0 < n < 1$ (b) $H = n/t$

Index